Lagrangian Interaction

An Introduction to Relativistic Symmetry
in Electrodynamics and Gravitation

Philosophy is written in that great book which ever lies before our eyes — I mean the Universe — but we cannot understand it if we do not first learn the language and grasp the symbols in which it is written. This book is written in the mathematical language, and the symbols are triangles, circles, and other geometrical figures, without whose help it is impossible to comprehend a single word of it; without which one wanders in vain through a dark labyrinth.

Galileo Galilei   1564–1642

L'expérience est la source unique de la vérité.

Henri Poincaré   1854–1912

Caminante, no hay camino, se hace camino al andar.

Antonio Machado   1875–1939

Lagrangian Interaction

An Introduction to Relativistic Symmetry
in Electrodynamics and Gravitation

Noel A. Doughty
Department of Physics, University of Canterbury,
Christchurch, New Zealand

CRC Press
Taylor & Francis Group
Boca Raton London New York

CRC Press is an imprint of the
Taylor & Francis Group, an **informa** business

First published 1990 by Westview Press

Published 2018 by CRC Press
Taylor & Francis Group
6000 Broken Sound Parkway NW, Suite 300
Boca Raton, FL 33487-2742

CRC Press is an imprint of the Taylor & Francis Group, an informa business

No claim to original U.S. Government works

This book contains information obtained from authentic and highly regarded sources. Reasonable efforts have been made to publish reliable data and information, but the author and publisher cannot assume responsibility for the validity of all materials or the consequences of their use. The authors and publishers have attempted to trace the copyright holders of all material reproduced in this publication and apologize to copyright holders if permission to publish in this form has not been obtained. If any copyright material has not been acknowledged please write and let us know so we may rectify in any future reprint.

Except as permitted under U.S. Copyright Law, no part of this book may be reprinted, reproduced, transmitted, or utilized in any form by any electronic, mechanical, or other means, now known or hereafter invented, including photocopying, microfilming, and recording, or in any information storage or retrieval system, without written permission from the publishers.

For permission to photocopy or use material electronically from this work, please access www. copyright.com (http://www.copyright.com/) or contact the Copyright Clearance Center, Inc. (CCC), 222 Rosewood Drive, Danvers, MA 01923, 978-750-8400. CCC is a not-for-profit organization that provides licenses and registration for a variety of users. For organizations that have been granted a photocopy license by the CCC, a separate system of payment has been arranged.

Trademark Notice: Product or corporate names may be trademarks or registered trademarks, and are used only for identification and explanation without intent to infringe.

Visit the Taylor & Francis Web site at
http://www.taylorandfrancis.com

and the CRC Press Web site at
http://www.crcpress.com

National Library of Australia Cataloguing in Publication
Doughty, Noel.
    Lagrangian interaction.

    Bibliography.
    Includes index.
    ISBN 0 201 41625 5.

    1. Relativity (Physics). 1. Electrodynamics. 3. Lagrange
    equations. I. Title.

530.1

Library of Congress Cataloging in Publication Data
Doughty, Noel A. (Noel Arthur)
  Lagrangian interaction.
  Includes bibliographical references.
  1. Electrodynamics. 2. Gravitation. 3. Relativity
  (Physics). 4. Symmetry (Physics). I. Title.
  QC631.D68 1989 537.6 89–18470
  ISBN 0-201-41625-5

ISBN 13: 978-0-201-41625-1 (pbk)

Cover designed by Hybert Design & Type, Maidenhead.
Typeset by the author using the TEX macro LaTEX.
TEX is a trademark of the American Mathematical Society.

To my family and friends,

For their patience and support.

Preface

The division of the subject matter of physics into a rather large number of self-contained and disjoint topics is favoured by its historical development, the research specialities of teachers and the demands of curricula. This can permit each topic to be treated to the depth required for specific vocational goals or as preparation for subsequent courses. However, the differing treatments of the foundations and goals of individual topics can also hide the features they have in common and make it difficult for a student to appreciate where each fits into the whole. This is particularly true for the differences between Newtonian and relativistic physics, for mechanics and fields, for classical and quantum properties and for the bulk of physics on the one hand and the gauge fields of fundamental particle theory on the other. The inherent chasms between such topics are often magnified by differences in notation and philosophy or in the mathematical and physical preparation required.

The mathematical and physical material which a student or teacher of physics requires in order to appreciate some of the subtleties of interaction is very extensive. It is not surprising that many discussions of the fundamental fields, particles and interactions of physics are predominantly phenomenological or essentially popularizations. If theoretical in emphasis, they are often directed at those with a very sophisticated post-graduate level of mathematical training.

This text is based on limiting the ultimate depth achieved in examining the foundations of physics and the fundamental interactions, via a study of electrodynamics and gravitation, in order to provide a considerable fraction of the different topics required in one place in a common notation and philosophy at an undergraduate level. Material conventionally taught by very different techniques, such as Newtonian and relativistic physics, is presented in a way which highlights their similarities and isolates the minute but very significant differences. The intention is to smooth the transition undergone by a typical student in going from a classical mechanical discussion, still the backbone of all introductory physics courses, to some of the topics required for advancing to a course in relativistic quantum field theory.

The main physics prerequisites for a course based on this text are the completion of standard treatments of classical mechanics including Lagrangian analysis, vector theory in three dimensions using Cartesian coordinates and a course in electricity and magnetism as far as the Lorentz equation and the integral and differential forms of Maxwell's equations. Some exposure to the use of potential theory in electrostatics, magnetostatics or Newtonian gravity will be helpful. It will be assumed the reader has completed courses in differential and integral calculus including partial differential equations. By contrast, it will not be assumed that any courses in relativity, tensor calculus, field theory or quantum mechanics have been completed although some exposure to wave mechanics as far as the Uncertainty principle will be useful.

The equations of the *Système International* (SI) will be used for all numerical data (see Appendix A.1) and almost all equations. A natural equation system suitable for the wave equations of quantum field theory will also be introduced in Chapter 6 as one of many illustrations of dimensional analysis in physics and to help make certain points about the gravitational, electromagnetic and matter fields.

The text is well-sprinkled with exercises[1]. The bulk of these are of a theoretical rather than practical nature. Although most are relatively straightforward, some may occasionally be considerably more difficult. Most will assume various levels of familiarity with the results of one or more of the previous exercises.

No attempt has been made to give a comprehensive account of each subject treated. Rather, the topics chosen are those which are needed for continuing the development of the main theme in subsequent chapters. The development of science and physics could well have been very different were it not for a few landmark events that changed the course of history. Other equally momentous events could have retarded or advanced certain aspects of scientific thought at the expense of others. For this reason, the author has deliberately tried to look at some old topics from a new viewpoint — for example, Newton's laws, Newtonian gravity, Einstein's postulates and Maxwell's equations are all treated somewhat unconventionally.

The list of *References* is representative rather than exhaustive or balanced. It is comprised of texts and articles which the author has found useful and hopes will also be useful to the reader. A number of the references are listed in a *Bibliography* at the end of each chapter. Some are quite definitely technical, especially after the opening chapters. Others are considerably less so and some are even popularizations of scientific ideas. The author hopes that a student may choose to use such references to balance improved technical competence acquired from exercises with increased qualitative understanding of the principles involved. Their reading may also counterbalance the rather unusual order in which some topics are introduced with deliberate scant regard — but no intended disrespect —

---

[1] One must learn by doing the thing; though you think you know it, you have no certainty until you try.

Trachiniae 592   Sophocles   ca. 495 – ca. 405 B.C.

for more conventional treatments.

A reasonable amount of historical and biographical material is included in the belief that it can be a considerable aid to the student as well as providing relief from the relentless introduction of new technical material. However, we urge the reader to place his trust for authenticity on either the source material or authoritative biographies by qualified historians of science. An excellent starting point that the author has found indispensable is the set of 15 volumes of the *Dictionary of scientific biography* [197], in which there are a large number of very useful articles on all well-known scientists and their pioneering discoveries.

The first and principal occurrences of important technical concepts will generally be high-lighted using italics and most of these italicized phrases will correspond to index entries. The *Reference* list is marked with the locations of citations from the body of the text but not with those from elsewhere in the list or in the chapter ending *Bibliographies*. Some citations are also indexed.

Acknowledgments

The author is deeply grateful, for the hospitality received during 1988 while on study leave from the University of Canterbury, to the following: Professor Mario Castagnino, *Departamento de Física* and the *Instituto de Astronomía y Física del Espacio, Universidad de Buenos Aires*; Professor Ron King of the *Faculty of Mathematics*, Professor David Barron of the *Department of Electronics and Computer Science* and Professor Ken Barnes of the *Department of Physics*, all of the *University of Southampton*.

I am particularly grateful to my colleagues Philip Butler, Bill Moreau, Geoff Stedman and Brian Wybourne at Canterbury, and Ed Fackerell of the University of Sydney, for valuable comments on early drafts of some of this material. The same deep gratitude goes to all my students, in particular David Wiltshire, Rod Gover, Jeremy Shaw, Graham Collins, Richard Arnold, Amanda Peet, Keith Murdoch, David Gunn, Michael Cree and Richard Easther for their stimulating questions.

I would like to express my deepest appreciation for the generosity of Donald E Knuth in making TeX widely available, for the similar approach of Leslie Lamport on LaTeX and to William Kaster (Personal TeX, Inc.), Graeme McKinstry, Rick Beatson, Trevor Brown, Sebastian Rahtz, Sue Brooks and Barbara Beeton, for their advice.

Advice and assistance from staff in numerous libraries is deeply appreciated; I am especially indebted to Madeline Juchau, Jenny Johnson and Janet Bray of the Physical Sciences Library, University of Canterbury. I am equally grateful to Joanna Marriott and Peter Pieruschka for checking the transcription and translation of a few titles from German and to Jean-Luc Le Floc'h (Paris).

The professional advice and encouragement of the editorial and production staff of Addison-Wesley in the United Kingdom, Singapore and Australia is gratefully appreciated.

The following are acknowledged with thanks for the supply of photographs: the Master and Fellows of Trinity College, Cambridge (Newton and Maxwell); Boerhaave National Museum of the History of Science and Medicine, Leiden (Lorentz); Niedersächsische Staats-und Universitätsbibliothek, Göttingen (Minkowski, Noether and Weyl); BBC Hulton Picture Library, London (Galileo); Dunsink Observatory, Dublin (Hamilton, Schrödinger); Bildarchiv preussischer Kulturesitz, Berlin (Euler); the Master and Fellows of St John's College, Cambridge (Dirac); Mansell Collection, London (Lagrange, Poincaré, Einstein).

4 November 1989

Noel Doughty
Christchurch.

Contents

Preface	vii
Figures	xvii
Biographical details	xix

I Galilean relativity and Newtonian physics — 1

1 Introduction — 3
- 1.1 Fundamental particles and interactions — 4
- 1.2 Electrodynamics — 6
- 1.3 Weak-nuclear interaction — 11
- 1.4 Strong-nuclear interaction — 15
- 1.5 Gravity — 18
- 1.6 Unifying principles — 24

2 Space and time — 27
- 2.1 Space — 28
- 2.2 Time — 29
- 2.3 Spacetime coordinates — 30
- 2.4 Homogeneity, isotropy and curvature — 32
- 2.5 Orientability of time and space — 36

3 Euclidean geometry — 40
- 3.1 Euclidean line elements — 41
- 3.2 Indexed notation and matrix multiplication — 43
- 3.3 Parity — 46
- 3.4 Charge conjugation — 52
- 3.5 Time reversal — 53
- 3.6 Rotations — 56
- 3.7 Groups — 59

4 Tensor fields in Euclidean spacetime — 62
- 4.1 Euclidean scalars and scalar fields — 62
- 4.2 Euclidean vectors and vector fields — 63

xi

	4.3	Euclidean tensors and tensor fields	65
	4.4	Symmetric and antisymmetric tensors	67
	4.5	Proper and improper tensors	69
	4.6	Permutation pseudotensor	73
	4.7	Three-dimensional tensor duality	75
	4.8	Manifest Euclidean covariance	77
5	**Spacetime kinematics and Galilean relativity**		**80**
	5.1	Introduction	80
	5.2	Measurement and simultaneity	82
	5.3	Spacetime reference frame	85
	5.4	Principle of relativity	88
	5.5	Galilean and Lorentz transformations	93
	5.6	Galilean relativity	100
	5.7	Aristotelean and Galilean gravity	104
	5.8	Fibre bundle structure of Galilean spacetime	104
6	**Newtonian mechanics**		**107**
	6.1	Newtonian dynamics and conservation	107
	6.2	Additivity of conserved quantities	113
	6.3	Interaction in Newtonian physics	114
	6.4	Potential energy of conservative systems	120
	6.5	External interaction	124
	6.6	Galilean covariance and conservation	124
	6.7	Natural equations	127
7	**Newtonian gravity**		**130**
	7.1	Introduction	130
	7.2	Gravity	131
	7.3	Newton's gravity theory	132
	7.4	Newton's gravitation constant	135
	7.5	Field equations of Newtonian gravity	138
	7.6	First-order Newtonian gravity field equations	141
	7.7	Second-order Newtonian gravity field equations	142
	7.8	Equations of motion in a gravitational field	144
	7.9	Cosmological constant	145
	7.10	Gauge freedom of Newtonian gravity	147
	7.11	Weak equivalence principle	149
	7.12	Michell and Laplace black holes	152
	7.13	Soldner's gravitational deflection of light	153
	7.14	Deviations from Newtonian gravity	157
	7.15	Newton-Cartan spacetime	160

II Lagrangian interaction and symmetry — 163

8 Euler-Lagrange equations of motion — 165
- 8.1 Introduction — 165
- 8.2 Lagrangian of a system of Newtonian particles — 166
- 8.3 Momenta and configuration space Hamiltonian — 168
- 8.4 Generalized coordinates in mechanics — 170
- 8.5 Stationary action — 174
- 8.6 Hamilton's principle of stationary action — 175
- 8.7 Functional differentiation — 178
- 8.8 Lagrangian interaction — 179
- 8.9 Newtonian gravitational Lagrangian interaction — 182
- 8.10 Constraints — 184

9 Symmetries and Noether's theorem — 189
- 9.1 Introduction — 189
- 9.2 Point transformations of configuration space — 190
- 9.3 Invariant and covariant Lagrangians — 191
- 9.4 Hamilton's modified principle of varied action — 192
- 9.5 Symmetries of Lagrangian systems — 195
- 9.6 Noether charges for mechanical systems — 197
- 9.7 Noether's theorem for mechanical systems — 200

10 Non-relativistic Lagrangian fields — 202
- 10.1 Introduction — 202
- 10.2 Mechanics to fields — 203
- 10.3 Hamilton's principle for Newtonian fields — 205
- 10.4 Poisson and Laplace equation Lagrangians — 209
- 10.5 Newtonian gravitational action — 211
- 10.6 Schrödinger Lagrangian — 214
- 10.7 Invariant and covariant field Lagrangians — 217
- 10.8 Noether's theorem for Newtonian fields — 219
- 10.9 U(1) phase invariance — Schrödinger charge — 222

11 Hamiltonian dynamics — 225
- 11.1 Introduction — 225
- 11.2 Phase space Hamiltonian — 226
- 11.3 Hamilton's equations — 228
- 11.4 Phase space Hamilton's principle — 229
- 11.5 Poisson brackets — 231
- 11.6 Non-relativistic Hamiltonian fields — 233

III Poincaré covariance and Einsteinian physics — 237

12 Special relativity — 239
- 12.1 Introduction: the electricity and the ether — 239
- 12.2 Einsteinian relativity postulates — 246
- 12.3 Lorentz transformations — 247
- 12.4 Geometry of spacetime — 250
- 12.5 Spacetime causal structure and proper time — 252
- 12.6 The dimensionality of spacetime — 257

13 Poincaré transformations — 259
- 13.1 Covariant notation — 259
- 13.2 Minkowskian spacetime isometry group — 263
- 13.3 Restricted Lorentz transformations — 266
- 13.4 Lorentz boosts and spatial rotations — 270
- 13.5 Rapidity and 'hyperbolic rotations' — 272
- 13.6 Boosts in an arbitrary direction — 273
- 13.7 Factorizing a Lorentz transformation — 274

14 Tensor fields in Minkowskian spacetime — 276
- 14.1 Scalar and vector fields — 276
- 14.2 Lorentz tensors and tensor fields — 279
- 14.3 Lorentz scalar product — 281
- 14.4 Manifest Lorentz and Poincaré covariance — 283
- 14.5 Symmetric and antisymmetric tensors — 285
- 14.6 Pseudo 4-scalars, 4-vectors and 4-tensors — 286
- 14.7 Permutation pseudotensor — 289
- 14.8 Tensor duality in Minkowskian spacetime — 291
- 14.9 Matrix generators of Lorentz transformations — 293

15 Einsteinian mechanics — 296
- 15.1 Four-velocity and four-acceleration — 296
- 15.2 Poincaré-covariant conservation laws — 299
- 15.3 Relativistic free particle equations of motion — 305
- 15.4 Relativistic Lagrangian mechanics — 307
- 15.5 Relativistic free particle Lagrangians — 308
- 15.6 Massless particles — 311
- 15.7 Einbein Lagrangian — 312
- 15.8 Relativistic Hamiltonian mechanics — 315
- 15.9 Spin of a massive relativistic system — 319
- 15.10 Helicity of a massless particle — 323
- 15.11 Spacelike surfaces and covariant 3-volume — 325
- 15.12 Gauss's theorem in Minkowskian spacetime — 327
- 15.13 Energy-momentum tensor — 329

IV Electrodynamic and gravitational fields   333

16 Relativistic Lagrangian fields   335
16.1 Field equations and Noether currents   336
16.2 Canonical energy-momentum tensor   341
16.3 Klein-Gordon field   342
16.4 Real scalar field   344
16.5 Charged scalar field. Internal symmetries   347
16.6 Free Klein-Gordon plane-wave expansion   352
16.7 Relativistic Klein-Gordon wave mechanics   353
16.8 Klein-Gordon Hamiltonian   354

17 Relativistic scalar gravity   357
17.1 Graviscalar action of Nordstrøm   357
17.2 Gravitational time dilation or red shift   362
17.3 Riemannian structure of scalar gravity   365
17.4 Graviscalar light deflection   367
17.5 Linear graviscalar perihelion shift   367
17.6 Non-linear graviscalar perihelion shift   370
17.7 Short-range gravitational interaction   374

18 Maxwell's equations   376
18.1 Introduction   376
18.2 Characterization of the electromagnetic field   382
18.3 Invariance and conservation of charge   383
18.4 The general form of the field equations   386
18.5 Covariant first-order Maxwell's equations   387
18.6 Maxwell's equations in 3-vector form   389
18.7 Second-order Maxwell's equations   393
18.8 Electromagnetic invariants   395
18.9 Electromagnetic gauge invariance   396
18.10 Boost transformation of field strengths   398
18.11 Magnetic charge and magnetic monopoles   400
18.12 Electromagnetic energy-momentum tensor   403
18.13 Maxwell and Proca Lagrangian densities   405

19 Electrodynamics   410
19.1 Lorentz law of charged particle motion   410
19.2 Current density of a classical charged particle   412
19.3 Electromagnetic coupling constant   414
19.4 Interaction: classical electrodynamic action   415
19.5 Belinfante symmetric energy-momentum tensor   419
19.6 Physical components of the Maxwell potential   421
19.7 Lorentz family of gauges   423

19.8	Maxwell polarization basis	425
19.9	Coulomb gauges	427
19.10	Hamilton and radiation gauges	427

20 Gauge principle   430

20.1	Introduction	430
20.2	Dirac field of relativistic electrons	431
20.3	Dirac equation	435
20.4	Dirac Lagrangian density	439
20.5	Plane Dirac waves	441
20.6	Dirac current	443
20.7	Chiral Dirac fields and Weyl neutrinos	444
20.8	Localization of a global symmetry	447
20.9	Spinor electrodynamics	449
20.10	CPT symmetries of Weyl and Dirac equations	454
20.11	Gauge-covariant derivative: minimal coupling	455
20.12	Scalar electrodynamics	456
20.13	Yang-Mills fields — nuclear interactions	458
20.14	Quantum field theory	463

21 Relativistic gravitation   466

21.1	Introduction	466
21.2	Rarita-Schwinger spin $\frac{3}{2}$ vector-spinor fields	466
21.3	Fierz-Pauli massless spin 2 tensor gravity	469
21.4	Experimental tests of tensor gravity	475
21.5	Einstein strong equivalence principle	480
21.6	General relativity	482
21.7	Unification and supersymmetry	492
21.8	Relativistic superstrings	496

Appendices   499

A.1	Physical constants	499
A.2	Counting components	500
A.3	Helmholtz decomposition of vector fields	501
A.4	Poincaré Lemma	502

References   504

Index   543

Figures

2.1	Two-dimensional torus	31
2.2	Parallel transport in a plane and on a sphere	35
2.3	Möbius strip and Klein bottle	37
3.1	Chiral object and its parity conjugate	47
4.1	Effect of a passive rotation on a scalar field	64
4.2	Effect of an active rotation on a scalar field	64
4.3	Describing a rotation by a pseudovector	70
5.1	Change of frame by translation and boost	94
7.1	Gravitational deflection of a light corpuscle	155
12.1	Planar causal structure of Galilean relativity	253
12.2	Conical causal structure of Minkowskian spacetime	254
13.1	Minkowski diagram of frames in relative motion	271
15.1	Normal to a particle trajectory	321
15.2	Spacelike cross-sections of spacetime	327
15.3	Closed boundary of a spacetime region	328
19.1	Spin 1 polarization basis	426
20.1	Feynman graph of quantum electrodynamics	453

Biographical details

Euclid	41
Galileo Galilei	101
Isaac Newton	108
Leonhard Euler	167
Joseph-Louis Lagrange	170
Amelia Emmy Noether	198
Erwin Schrödinger	215
William Rowan Hamilton	226
Hendrik Antoon Lorentz	248
Jules Henri Poincaré	266
Hermann Minkowski	279
James Clerk Maxwell	388
Hermann Weyl	422
Paul Adrien Maurice Dirac	434
Albert Einstein	483

Part I

Galilean relativity and Newtonian physics

CHAPTER 1

Introduction

There is no higher or lower knowledge, but one only, flowing out of experimentation[1].

Leonardo da Vinci 1452–1519

The principal goal of this textbook is the development of the foundations of the Lagrangian theory of Galilean and relativistic fields and particles at an introductory undergraduate level using primarily, as examples, classical electromagnetism and the gravitational interaction. No attempt will be made to give a standard or complete treatment of electrodynamics even at the non-quantum level, and a complete general relativistic treatment of gravity in curved spacetime will not be included. Instead, we shall concentrate on those relativistic and Lagrangian features of electromagnetism which illustrate the principles of internal and spacetime symmetries and facilitate a comparison with the other three interactions. To highlight these relativistic features we shall compare the properties of electrodynamics with those of the gravitational interaction treated both non-relativistically and with special relativity in flat spacetime.

Our introduction to the foundations of relativistic kinematics will be based on a presentation of the postulates of relativity in a form which makes no reference to the properties of light nor indeed to any particular interaction. This will permit the equations of electrodynamics and weak gravity to be developed as very simple sets of dynamical equations which follow naturally from the Relativity principle, in either Galilean or Einsteinian form.

Part I will present Galilean relativity and Newtonian physics in a way which will permit a subsequent natural transition, in Part III, to Lorentz-Poincaré relativity and Einsteinian physics. Part II will develop the Lagrangian and Hamiltonian formulations by closely analogous methods for mechanical and field systems in the familiar context of Galilean relativity. The dynamic consequences of using special relativistic kinematics for

---

[1] *The notebooks of Leonardo da Vinci* [349].

mechanical systems will be developed in Part III and extended to fields in Part IV in an examination of the fundamental equations of electrodynamics and weak gravity.

1.1 Fundamental particles and interactions

If an isolated system is divided into two or more parts, not isolated from one another, the way in which the parts interact can be used to classify physical phenomena into four categories: *electromagnetic*, *weak-nuclear*, *strong-nuclear* and *gravitational*, which are referred to as the four fundamental interactions of physics. All physical phenomena ought, in principle, to be expressible in terms of the way in which matter and energy in the form of particles and fields interact via these four interactions. In order to appreciate the distinguishing characteristics of electromagnetism and gravity, we shall describe some of the properties of each interaction in general terms in the remainder of this chapter. Some of their features are displayed in Table 1.1. Each interaction is mediated by a characteristic particle or particles called *exchange bosons*. The non-exchange fundamental particles of matter between which they act are the three families of six quarks and six leptons shown in Table 1.2 (page 12).

Although most of the concepts we shall discuss are defined, some of the terms used in the following sections are clearly far from elementary. In some cases these ideas and the terminology will be developed in later chapters from more fundamental concepts. This will not apply, however, to all the terms used, especially those required to discuss quantum properties and the nuclear interactions. Nevertheless, some initial idea of the differences between the four types will highlight the primarily classical features of the electrodynamics and weak gravity to be developed later. Depending on the reader's background, some parts of this introductory overview may be quite difficult. If that is the case, few problems should arise from skipping, or superficially reading, those parts and returning to consider this material again after some of the concepts involved have been studied quantitatively in later chapters.

The postulates of relativity impose a severe general limitation, consistent with experiment, on the mathematical forms that may be used to describe an interaction and the matter which acts as its source. All matter exhibits quantum effects at the microscopic level and three of the four interactions (all except the gravitational) have also been formulated as consistent quantum processes involving the exchange of particles. It is widely believed that the gravitational interaction must also be quantizable although the form of the quantum theory of gravity is not yet known. The requirement that an interaction be *consistently quantizable* is also a very severe restriction on the form it may take.

1.1. Fundamental particles and interactions

Table 1.1 The four fundamental physical interactions. The strengths are based on the values of dimensionless coupling constants and are given relative to unity for the strong nuclear interaction.

	Gravity	Weak-nuclear	Electro-magnetic	Strong (quark)	Strong-nuclear
Strength $\approx$	10^{-38}	10^{-13}	10^{-2}	1	1
Domain	macroscopic	β-decay	atomic & molecular	inter-quark	nuclei
Gauge charge	mass-energy	weak flavour	electric charge	colour charge	residual colour
Source particles	all particles	quarks, leptons	all charged	quarks, gluons	hadrons
Bosons	graviton	$W^{\pm}$, Z^0	photon	8 gluons	$\pi^{\pm}$, π^0
Spin	2	1	1	1	0
Mass	0	91, 81 $\frac{\text{GeV}}{c^2}$	0	0	$\approx 140\ \frac{\text{MeV}}{c^2}$
Range (m)	∞	10^{-18}	∞	10^{-15}	10^{-15}
Symmetry	ISO(1,3)	SU(2)×U(1)		SU(3)	

One of the most significant steps in the unification of different aspects of physical theory, and its reduction to a relatively small number of fundamental concepts, is the re-expression of dynamical equations of motion and field equations into forms derivable from a single characteristic function, the *Lagrangian* or the *Hamiltonian*, by means of a *Principle of stationary action*. Although such re-expression does not supply a new theory, it permits a wide range of very different physical systems, of mechanical or field character for example, to be treated uniformly. This facilitates their appearance together in an interacting system. The stationary action principle, and the Lagrangian and Hamiltonian methods they involve, have played a vital rôle in the unification programme of the last few decades.

Attempts to unify all fundamental physics in one *superinteraction*, reducing to the four separate interactions under appropriate conditions, has led to the identification of a further property that all four have in common. That feature is *gauge invariance*, crucial in the formulation of the quantum theories of the non-gravitational interactions. The properties of each interaction permit one to isolate a *Gauge principle* whose application to various collections of fundamental particles leads to the forms in which they interact electromagnetically and via the two nuclear interactions at the microscopic level. Some of the properties of the gravitational interaction are also obtained by application of the Gauge principle in a slightly different form. The Gauge principle plays a crucial rôle in the description of the Glashow-Salam-Weinberg unified theory of the electromagnetic and weak interactions, referred to as the *electroweak theory*, and in the predictions obtained from various *grand unified theories* of the electroweak and strong-nuclear interactions. The Gauge principle is therefore, not surprisingly, a central feature in the search for a quantized theory of gravity and the theories aimed at complete unification of all matter and all the four interactions in a superinteraction.

Some idea, no matter how descriptive, of the way in which all the interactions, and the matter and energy they affect, satisfy the Principles of relativity, the Principle of stationary action, Quantum principles and the Gauge principle, is therefore of value in any study of each particular interaction. The remainder of this introductory chapter will give an overview of some of the properties of the four interactions in order to better appreciate the special features of the simplest and the most complex of these interactions, the electromagnetic and the gravitational, respectively.

1.2 Electrodynamics

Electrodynamics is the interaction which determines the overall structure of atoms and molecules and the way in which they radiate and absorb light. It dominates in the determination of the behaviour of all chemical compounds and the principal properties of solids, liquids, gases and plasmas of charged particles over a wide range of temperatures and pressures. Radioactive decay of nuclei with emission of γ-radiation is governed by the electromagnetic interaction.

Electrodynamics may be identified theoretically as the relativistic interaction which has the simplest integrated source, an invariant charge, and displays left-right reflection symmetry, or *parity covariance*. Apart from those principles which are applicable to all special relativistic fields and particles, few further assumptions are necessary to construct Maxwell's equations and the Lorentz law for the motion and interaction of charged particles via the electromagnetic field. As a consequence of this characterization it can be shown that the electromagnetic field propagates at a

1.2. Electrodynamics

frame-independent ultimate speed, the upper bound on all speeds, and this in turn implies that the interaction is of long range, the details of which are expressed in the static case by Coulomb's inverse square law. A significant fraction of this text will be devoted to discussing the Relativity principles and the classical field concepts and techniques which permit a development of electrodynamics along these lines. In the process we shall develop the Lagrangian techniques for describing the interaction of non-quantum particles and fields, both non-relativistic and relativistic.

When the solutions of the dynamical equations of a system remain solutions after its variables are transformed, the system is said to possess a *symmetry* or *invariance*. We shall discover that electrodynamics exhibits, as a result of charge conservation, a *local symmetry*, namely one involving invariance under transformations containing arbitrary functions of time and space. The equations are said to exhibit *gauge invariance* and the variables, owing to their arbitrariness, are said to have *gauge freedom*. Equivalently, the parameters of the symmetry are essentially infinite in number. This occurs because the field equations do not all determine the values of the field variables at later times from those at earlier times. Some are instead *constraints* on the *initial data*, namely conditions which relate the values of the dynamical variables at one time. The equations of the four fundamental interactions all exhibit gauge invariance of one form or another, of which the electromagnetic case is the simplest and therefore the prototype.

Translations and rotations, and *boosts* from one reference frame to another moving uniformly with respect to it, make up a closed, associative set of transformations for each of which there exists an inverse transformation. All such transformations therefore comprise a *group*. A group-theoretical analysis of the ways in which the Relativity principle permits physical systems to be represented by mathematical quantities, and corresponding evolution equations, divides them into systems which either propagate at the speed of light or not. Each of those two cases is further divided into systems with integer or half-odd-integer multiples s (in units of the rationalized Planck constant $\hbar$) of intrinsic (non-orbital) angular momentum, called *spin*. Those not propagating at the speed of light have $2s+1$ orientations of their spin or polarization vector. Those with the speed of light, referred to as *null* or *massless systems*, have only two polarization modes, irrespective of their spin, if it is non-zero.

It is possible to establish Maxwell's equations uniquely starting from its characterization as a null system with spin 1. Such fields and particles will propagate at the ultimate speed in all frames with two independent transverse polarization states of vectorial character corresponding to two orientations of the one unit of intrinsic angular momentum of the *photon*, the electromagnetic quantum. We shall not use the theory of group representations in this text to examine the symmetries of electrodynamics and gravity. However, we shall make use of some of the nomenclature and basic ideas of group theory as a contribution to making group-theoretical treatments of particles and fields at least a little more accessible to undergraduate physics students.

For a third characterization of electrodynamics, one may select a single non-interacting complex field based on a parity-covariant representation (charged massive spin $\frac{1}{2}$) of the spacetime symmetries implied by the Relativity principle, appropriate to the Dirac electron-positron theory. One then notes that it has what is called an *internal global symmetry*, namely a transformation which leaves the wave equations of the field unchanged in form. An *internal symmetry* is one which acts on the field variables but does not affect the values of the external (spacetime) coordinates, comprising the arguments of the field. The *global* property means there are a finite number of symmetry parameters, like angles of rotations, that are not functions of time or position. The global symmetry is in this case simply invariance of a complex field describing charged spin $\frac{1}{2}$ particles under arbitrary constant phase changes, namely multiplication by a factor of $e^{i\xi}$ for ξ real and not dependent on time or position. It is thus a straightforward symmetry example, very much like rotation in a plane but involving the real and imaginary parts of the complex field to label the axes rather than spacetime coordinates. Such phase transformations form one of the simplest possible symmetries that a system of fields can exhibit and they play an important rôle also in the nuclear interactions.

A *phase transformation* of a field involves the multiplication of each of its components by a complex number of magnitude unity. Such a number is a 1-dimensional matrix. Furthermore, the phase factor $e^{i\xi}$ times its complex or hermitian conjugate is unity. For these two reasons the phase symmetry is said to be 1D (1-dimensional) and *unitary* (U) and is referred to as a U(1) *invariance*. From the purely transformational point of view, such an invariance is indistinguishable from a rotation by the phase angle in a plane labelled by the real and imaginary parts of the field. With no loss of generality, the phase angle ξ may be limited to the range from 0 to 2π (with 0 and 2π being identified). Because such a range is bounded and closed, the U(1) symmetry is said to be *compact*. It makes no difference in what order one multiplies a field by two successive phase factors and for this reason the U(1) invariance is *commutative* or *Abelian*. The compactness and Abelian nature of the U(1) invariance are important features of the global symmetry of the electron-positron field system (and any system of electrically charged particles) which are closely related to conservation of electric charge and to the way that system interacts electrodynamically.

The application of a technique originally due to Weyl [571], called the *Gauge principle*, to the U(1) invariance of charged, massive spin $\frac{1}{2}$ systems, will lead to a description of the way the source particles may interact. We shall describe this application of the Gauge principle in Chapter 20. It will lead to almost complete details on the nature of the mediating electromagnetic field. The form of the well-known gauge freedom of the electromagnetic potential arises naturally as a consequence of the gauge process. With the Gauge principle one finds that, in passing from a description of a free field to one involving interacting fields, the original global symmetry becomes the *local symmetry* known as gauge invariance. An important feature of the Gauge principle, as a procedure for determining the nature of the interactions between two systems, is its general applicability. It may be applied to other global symmetries to determine the nature of the other three fundamental interactions. If applied to an invariance with a sufficiently large number of parameters, the Gauge principle is potentially capable of unifying the three non-gravitational interactions in a *grand unified theory* and all four interactions in a *superunified theory*.

In its application to the 1D unitary phase symmetry of electrons and positrons (and other Dirac particles) one finds that the mediating field of the interaction which arises by *gauging* the global phase symmetry, namely by applying the Gauge principle to it to make it local, is precisely the one described by Maxwell's equations. The fact that the interaction is mediated by only one real vector field, whose quantized particle is the *photon*, is a consequence of there being only one real parameter, the phase angle ξ, in the global symmetry group. A mediating quantum arising from the gauge process is referred to as a *gauge particle*.

All particles may be characterized by their electric charge, a scalar which may be positive or negative for those interacting electromagnetically and zero for all the others. Electric charge, in contrast to mass for example, is therefore said to be *bipolar*. The total charge is given by summing the individual charges. A particle made up from two equal and opposite electric charges will be electrically neutral. Electric charge is therefore a scalar *additive* quantity. In these respects, it differs from the *spin magnitude*, $s = 0, \frac{1}{2}, 1, \frac{3}{2}, 2, \ldots$, for example.

At the microscopic level, corresponding to each particle with a given non-zero charge there always exists another distinct particle, referred to as the *antiparticle*, which is identical in mass, radioactive half-life and spin magnitude, but of opposite charge. The total value of charge is *conserved*, namely it never changes with time. Use of the Gauge principle to determine the nature of the electromagnetic interaction is closely connected to the fact that its equations may be obtained from a Lagrangian by the Principle of stationary action and that the electric charge is a conserved, additive and frame-independent quantity. All these properties are intimately related to the original U(1) invariance by a procedure, known as *Noether's theorem* (see Chapter 9), which supplies an additive, conserved quantity, in this case scalar, corresponding to any continuous symmetry (in this case U(1)) of any Lagrangian system.

An important property of non-quantum electrodynamics that sets it apart from all the other three interactions is that it obeys a *linear super-*

position principle: two sets of fields may be added or subtracted and the resulting equations will describe another physically realizable system interacting electrodynamically. This property is a result of the fact that the equations of motion and the field equations are linear in the electromagnetic fields in source-free regions or where the source is external and thus prescribed. The linearity of electrodynamics is related to the fact that the underlying global symmetry has only one parameter and is thus Abelian.

Classical electrodynamics is the study of the electromagnetic interaction of charged particles where their quantum structure is unimportant and in the approximation where photon numbers are very high. The existence of a classical limit to electrodynamics is closely related to the long range of the interaction and to the fact that the exchange quanta, photons, are of integer spin, namely spin 1. They therefore obey *Bose-Einstein statistics* which place no limit on the number of quanta in each state and are called *bosons*. This permits the construction of *coherent states*, which behave essentially classically, corresponding to large numbers of photons. Because the photon is a gauge particle and a boson, it is called a *gauge boson*.

The relations satisfied by the quantum field operators of Bose-Einstein particles have a non-quantum limit in which the variables commute with one another, an essential property of a classical system. The non-quantum structure of the basic electrodynamic equations is believed to be essentially completely understood — procedures exist, at least in principle if not in practice, to solve any conceivable classical electrodynamic problem. Electrodynamics may be quantized by the techniques of *quantum field theory* (QFT) where the interaction between charged particles is understood in terms of the *exchange* of particles (photons) representing the quantum properties of the electromagnetic field. The charged particle interaction is said to be *mediated* by the electromagnetic field and its quanta. The electromagnetic source, which classically is comprised of point particles, is first replaced by a field description, as for the mediating electromagnetic field itself. Quantization of each field, mediating and source, then gives rise to the corresponding particles, photons for the electromagnetic field and the various charged particles (electrons, positrons, protons, etc. or, at a more fundamental level, quarks) for the source fields.

Particles decaying electromagnetically do not change their identity. If we constructed a table showing which particles can decay into others by each interaction, the table for electrodynamics would consist of only diagonal entries. For this reason, the electromagnetic interaction is said to be *diagonal*. This is a property which distinguishes it from the nuclear interactions.

The rules for the quantization of a classical system, such as a collection of fields describing charged particles interacting electrodynamically, do not guarantee that the resulting equations are consistent. This is partly a result of ambiguity in the choice of the ordering of some operator products, namely those whose classical analogues are non-vanishing Poisson brackets. Maxwell's equations for the electromagnetic field, and the Dirac equation for electrons and positrons, nevertheless may be consistently quantized to give the equations of *quantum electrodynamics* (QED). These equations are also generally considered to be well-understood and are in excellent accord with experiment. Nevertheless, they suffer from an apparent defect in that some of the ostensibly measurable quantities involved *diverge* — that is, they assume values which are infinite. QED is therefore said to be a *non-finite* theory. However, once certain well-defined rescaling or renormalizing procedures are applied, the infinite values which still occur in the theory do not involve measurable quantities. QED, although non-finite, is therefore said to be *renormalizable*, which is a minimal requirement for a satisfactory quantum theory.

Microscopically, charge is discretized (or quantized) in that it occurs only in simple integer or fractional multiples of the frame-independent charge e on a proton. The charge e is therefore an invariant measure of the *interaction strength* of electromagnetism. In a relativistic quantum context, the universal constants c (the speed of light) and the rationalized Planck constant $\hbar = h/2\pi$ (where h is Planck's constant) are at our disposal. The electromagnetic interaction strength therefore has a dimensionless measure or *coupling constant* $e^2/4\pi\epsilon_0\hbar c$ (the fine structure constant) whose value ($\approx 1/137$) is very much less than unity.

Although the electromagnetic interaction is much stronger than the weak, which is very much stronger than the gravitational, the small value of the fine structure constant implies that it is nevertheless itself of inherently low strength (see Table 1.1). This inherent weakness, even more pronounced in the weak interaction and gravity, distinguishes these three from the strong-nuclear interaction. Directly related to this result is the possibility of calculating, using the techniques of perturbative QFT, the values of measurable QED quantities, such as collision cross-sections, reaction rates and magnetic moments, to arbitrary precision, at least in principle. In practice such calculations have been carried out to give agreement with measurement to as many as 10 or 11 significant figures in some cases.

Unstable particles which decay with γ emission, as a result of the electromagnetic interaction, have lifetimes as small as 10^{-20} s, comparable to those for the strong interaction but very much shorter than those which decay weakly. Typical cross-sectional areas of fundamental particles being scattered electromagnetically are $\approx 10^{-33}$ m^2.

1.3 Weak-nuclear interaction

The *weak-nuclear interaction*, in contrast to the electromagnetic, is a short-range, parity non-covariant (or non-conserving) effect.

Table 1.2 Quark and lepton families and flavours with electric charges Q in units of the proton charge e.

Q	Quark family			Q	Lepton family		
$\frac{2}{3}$	u	c	t	0	ν_e	ν_μ	ν_τ
$-\frac{1}{3}$	d	s	b	-1	e^-	μ^-	τ^-

It operates between *leptons*, originally meaning light particles but now meaning those which are not strongly interacting, and *hadrons*. The latter are the strongly interacting particles which divide into *baryons* (fermionic hadrons) and *mesons* (bosonic hadrons). All hadrons are constructed from *quarks* the two lightest of which are the principle basic constituents of nuclear matter. Baryons were so named originally because they were relatively heavy particles (at least proton mass) while mesons were originally named because of their intermediate mass (between electron and proton masses).

The six known leptons, not counting their antiparticles, exist in three *families* or *generations* corresponding to the *electron* e^-, *muon* μ^- and *tauon* τ^-. The corresponding antiparticles are denoted by e^+, μ^+ and τ^+. To each lepton ℓ their corresponds a *neutrino* ν_ℓ. They are referred to as being of electron, muon and tauon type, denoted ν_e, ν_μ and ν_τ. The tau neutrino (and the top quark) have not been observed but additional indirect evidence supporting their existence and no further families (see Section 20.13) was announced in October 1989 from CERN (Centre Européen de Recherche Nucléaire) in Geneva and SLAC (Stanford Linear Accelerator Center) in California. Each of the three families of lepton is also referred to as a *lepton flavour*.

The neutrinos, or 'little neutral ones', are so-called because they are: electrically neutral, of zero or very low mass (compared to the corresponding lepton) and very difficult to observe. Each neutrino has a definite left or right handedness and may exist and interact separately from the antineutrino of opposite handedness. Their emission and absorption due to the weak interaction violates left-right symmetry and parity conservation. Corresponding to each family their exists an additive lepton number which is conserved in weak interactions. *Weak flavour* is a generic term for the conserved lepton numbers.

The weak interaction is identified as an effect distinct from electromagnetism and the strong-nuclear interaction by the relative slowness of certain particle decays it describes which is why it is referred to as 'weak'. Reactions proceeding by the weak interaction typically have characteristic

1.3. Weak-nuclear interaction

times or decay lives $\approx 10^{-10}$ s (and even as long as 11 minutes in the case of neutron decay) compared to $\approx 10^{-23}$ s for most strong-nuclear interactions and $\approx 10^{-20}$ s for those decaying electromagnetically. It describes the β-decay type of nuclear reaction important in *radioactivity* and occurring as a crucial part of the cycle of reactions in solar and stellar energy generation. *Cosmic rays* give rise to showers of particles some of which interact and decay weakly. Some radioactive isotopes used in medicine, and industry in general, decay as a result of the weak interaction.

The weak interaction was important in the particle processes occurring in the near singular *Big Bang* origin of the universe and plays a dominant role in the cataclysmic phase of a *supernova* explosion. Its effects were therefore readily seen with the naked eye in February 1987, and in the following months, when a previously very faint star increased dramatically in brightness and was recorded as Supernova 1987a. The first such relatively nearby naked eye supernova to be recorded was the event observed in China in 1054 A.D., the filamentary debris of which is now known as the Crab Nebula at the centre of which is a neutron star. Another was the bright supernova observed by Tycho Brahe in 1572 and the most recent, until 1987, was Kepler's 'nova', observed in 1604. Because of its short range, the weak interaction acts exclusively at the microscopic and therefore quantum level, a property it shares with the strong interaction, distinguishing them both from electromagnetism and gravity.

Like all the other interactions, it exhibits gauge freedom. In this case the gauge invariance is of a type referred to as *Yang-Mills*. This indicates that the underlying global symmetry group is also compact. However, in contrast to the electromagnetic case, it is non-Abelian since some pairs of transformations of the global symmetry, relating collections of fields and particles with similar properties, do not commute. This non-commutativity of the symmetry transformations is an indication that the field does not satisfy a linear superposition principle and hence that, even if one suppresses the source term in the interaction, there is interaction between the parts of the mediating field itself, called *self interaction*.

The original theory of the weak interaction was developed by Fermi as a direct *contact* or *point interaction*; namely one which does not involve the exchange of particles comparable to the photons of electromagnetism. The Fermi theory suffered from the serious defect of not being renormalizable. In the modern theory of the weak decays and reactions the interaction is developed jointly with electromagnetic effects as the *electroweak interaction*. In addition to a U(1) phase invariance (as for electromagnetism) the global symmetry group describes a very close relationship between pairs of fields, each pair corresponding to one lepton and its corresponding neutrino or a pair of quarks. These pairing relationships are characterized by the properties of a set of 2×2 matrices which when multiplied by their complex transpose give the identity matrix. They are therefore referred to as 2D *unitary* matrices and denoted by the symbol U(2). They also have unit determinant, a property which means they are referred to as being *special* (S) unitary matrices.

The invariance is therefore described as being of SU(2) type and called *weak isospin*, where isospin arises from the name *isotopic spin* or *isobaric spin* for a similar but distinct SU(2) symmetry among hadrons. The latter is a global symmetry epitomizing the charge independence of the strong interaction between, for example, a pair of protons, a pair of neutrons or one of each. The reference to 'spin' in isospin arises because complex SU(2) matrices are also closely related to real 3×3 rotation matrices. Their multiplication of a non-relativistic spin $\frac{1}{2}$ wave function (or field), a Pauli spinor, has the effect of a rotation of the system the spinor describes. Although mathematically related the three applications of SU(2) are physically quite distinct.

The SU(2) matrices provide the simplest possible generalization of the U(1) phase invariance of electrodynamics to a symmetry containing more than one parameter, each with compact range. Each SU(2) matrix has in fact three independent real parameters. This has the consequence that the gauge particles which arise, as a result of applying the Gauge principle, and which mediate the weak interaction, are three in number. At extremely high energy ($\gg$ 1 GeV) they are almost indistinguishable from the photon in that each of them behaves as a massless spin 1 particle. Like photons, they are gauge bosons but differ in being unstable with very short half lives.

Furthermore, the weak interaction gauge particles actually observed at relatively low energy ($\ll$ 1 GeV) have non-zero rest mass acquired by a quantum field theoretic process referred to as *spontaneous symmetry breaking*. This is a term used to describe an interaction between the gauge particles and the quantum *vacuum* (or ground state) in which the gauge symmetry corresponding to masslessness at high energies is broken at low energies by an asymmetry of the vacuum rather than dynamically by a term in the field equations or Lagrangian of the theory.

The term *spontaneous* is used because of the similarity of this process, in a quantum field context, to the breaking of the rotational symmetry, about a point for example, of a sample of magnetizable material when the temperature is lowered. In such a process the system passes spontaneously at a critical temperature into a magnetized state of less symmetry, such as rotational symmetry about an axis. Such a change can occur spontaneously since it is a second-order phase transition, namely one which does not involve a change of energy. Alternatively, one says that, in a system with spontaneous symmetry breaking, the dynamical symmetry is still present but is *hidden* by the lack of symmetry of the ground state. The breaking of the gauge-invariance of the ground state, and the loss of masslessness of the mediating particles, is directly related to the short range of the weak-nuclear interaction and the fact that it is weak.

The weak interaction may be consistently quantized. The Yang-Mills gauge invariance of the weak interaction plays a crucial role both in the renormalizability of the quantized theory and in its partial unification with electromagnetism. The latter was confirmed by the discovery [21] of its three gauge particles, the charged and neutral *weakons*, or *intermediate vector bosons*, $W^{\pm}$ and Z^0, with the UA1 (underground area 1) detector of the SPS (super proton synchrotron) proton-antiproton collider at CERN in late 1982. The particles were observed with rest masses consistent with the predictions of the electroweak theory.

In a relativistic quantum context, the existence of a non-zero rest mass implies a corresponding finite, rather than infinite, range over which the interaction mediated by such a particle is effective.

Exercise 1.1 *Show (a) using dimensional analysis arguments and (b) from the Heisenberg uncertainty principle that the observed weakon rest masses ($\approx 81 \, \text{GeV}/c^2$ for $W^{\pm}$ and $91 \, \text{GeV}/c^2$ for Z^0 particles) imply that the corresponding relativistic quantum length or range should be about 10^{-18} m. (In contrast, the corresponding electromagnetic quantity is infinite due to the zero rest mass of the photon.)*

Because of its short range, the weak interaction has no directly observable attractive or repulsive effects on macroscopic objects. On the contrary, it describes the changes of identity that certain microscopic particles undergo in some nuclear decays and corresponding reactions. It is therefore said to be *non-diagonal*, a property it shares with the strong-nuclear interaction. Very few particles, among them the photon, fail to interact weakly.

1.4 Strong-nuclear interaction

The *strong-nuclear interaction* was first formulated as the short-range, parity-conserving effect operating between the parts of a nucleus of an atom, namely the protons and neutrons. These are referred to collectively as *nucleons*, now known to be formed from triplets of quarks, each of which have spin $\frac{1}{2}$ and fractional electric charge of $\pm\frac{1}{3}e$ or $\pm\frac{2}{3}e$. Because of its short range, like the weak interaction, it acts exclusively at the microscopic or quantum level. No particles are preferentially emitted or absorbed with one or other handedness as a result of strong interactions.

At the nuclear rather than quark level of structure, its properties may be described by the theory formulated by Yukawa [597, 598] in 1935 in which the interaction involves the exchange of *pions*, particles with non-zero rest mass, zero spin and integral electric charge, which are now known to consist of a quark and antiquark bound together. Theories based on Yukawa's *particle exchange* ideas are now used in the quantum description of all interactions. Such theories of primarily fermionic matter fields interacting via boson exchange are often referred to as being of Yukawa type in contrast to contact or point interactions of fermions as in the Fermi theory of β-decay.

Exercise 1.2 *Use the observed masses ($\approx 140\,\text{MeV}/c^2$) of pions to show that the range of the strong-nuclear interaction is $\approx 10^{-15}$ m.*

The existence of the strong interaction was first able to be deduced from the observed stability of nuclei despite the potentially disruptive electromagnetic repulsion caused by the presence of many protons occupying a very small fraction (in some cases $\approx 10^{-15}$) of the volume of a whole atom. The strong interaction provides the binding at the nuclear level analogous to electromagnetism at the atomic and molecular level and the gravitational interaction on a terrestrial and astronomical scale.

Like the weak interaction it also has identity changing or *non-diagonal* character leading to the instability of certain heavy particles including some nuclei. Radioactive decay by the emission of α particles is governed by the strong-nuclear interaction. So also is the *fusion* of high-energy light nuclei into heavier nuclear species as occurs naturally in the interiors of stars and in the nucleosynthesis era a few minutes after a singular or almost singular origin of the universe. The strong-nuclear interaction governs the natural *fission* of very heavy nuclei into lighter nuclei and the use of such isotopes in medicine, in the generation of electric and mechanical power and in industry in general. It also governs the explosive stages of nuclear weapons whether they operate by fission or by fusion.

The fundamental details of the strong nuclear force are now based on the interaction of *quarks*, the elementary constituents of protons, neutrons and pions and similar composite particles referred to collectively as *hadrons*, which means strongly interacting. The quarks interact by the exchange of particles called *gluons*, so-called because they provide the binding at the quark level. Hadrons are those strongly interacting composite particles which may, unlike their quark and gluon constituents, have a separate existence at large distances from other particles.

Quarks and gluons on the other hand are said to be *confined* within hadrons since they are never observed at large distances from the partners with which they combine to form a composite bound particle. Hadrons divide into *mesons*, of integer spin, and *baryons*, of half-odd-integer spin.

In order to classify the large number of hadrons systematically, and relatively simply, one requires six different types of *quark flavour*. The six quark flavours and the corresponding symbols are *up* (u), *down* (d),

1.4. Strong-nuclear interaction

charm (c), *strange* (s), *top* (t) and *bottom* (b). To each type there corresponds a conserved quantity. The quarks are grouped in three families or generations with two flavours each: $\{u,d\}$, $\{c,s\}$ and $\{t,b\}$ with electric charges $\{\frac{2}{3},-\frac{1}{3}\}$ of the proton charge e in each pair. To each quark q there is an antiquark $\bar{q}$ of opposite electric charge. Only the quarks of the first family are observed as constituents of stable matter. The second and third families differ mainly in their higher rest energies, the top quark being presumably so high in rest mass that it has not yet been observed directly. Despite the close similarity between the quarks and leptons as indicated by Table 1.2 (page 12), the former have six flavours, the latter only three.

To each of the six quarks one assigns a charge-like property called *colour*. The use of colour in high-energy physics does not imply the familiar subjective sensation where different colours are related to the different physical wavelengths in an ordinary beam of light. Instead, it is a precisely defined quantum property like electric charge that permits similar particles to be classified and related systematically. Since it has some properties analogous to ordinary colour it is therefore given the same name. The properties of colour charge in high-energy physics which are analogous to those of ordinary colour are:

- the existence of three bipolar, primary values, referred to as red, blue and green, corresponding to the existence of three types of quark (for each of the six flavours) and eight types of gluon;

- the antiparticle of each particle with colour is another coloured particle and a composite particle made up of the two together behaves as a particle without colour. The antiparticles may therefore be assigned an opposite value or polarity of colour referred to as antired, antiblue and antigreen; and,

- the existence of composite particles, made up of three particles or antiparticles, one of each colour or anticolour, that behave as particles without colour.

The three types of colour charge (each bipolar) are a generalization of the existence of only one type of electric charge (also bipolar). Colour is charge-like since it has values (red, blue and green) which are conserved, frame-independent (scalar) and additive. Examples of colourless particles, made up from a coloured particle and an antiparticle of complementary colour, are pions (for example, $\pi^+ = \{u\bar{d}\}$) and kaons (for example, $K^- = \{s\bar{u}\}$), which are comprised of quark-antiquark states. Protons (uud) and neutrons (udd) are colourless particles comprised of three coloured particles (quarks), one of each primary colour.

At the interquark and gluon level, the strong interaction, like the weak, has a gauge invariance based on a compact, non-Abelian group and thus is also of Yang-Mills type. The existence of three colours of quark means that the invariance involves three fields, one for each colour of quark.

It turns out to be a symmetry referred to as SU(3) involving the transformation of collections of three similar fields and particles by the 8-parameter group of 3×3 unitary (U) matrices with unit determinant (S). Application of the Gauge principle leads to eight corresponding mediating gauge particles, all massless and of spin 1, which are the gluons.

The exchange of pions between nucleons approximately models the overall low-energy or longer range features of the strong nuclear interaction. The pions, being the lightest hadrons, will be associated with the longer range features. As for the weak interaction, the field equations are non-linear in the mediating (gluon) fields. A feature of the strong force which distinguishes it from the other non-gravitational interactions is that the gluons which are interchanged between quarks have colour and anticolour themselves and are therefore also acted on by the strong interaction. This contrasts with the absence of electric charge on the photon. The short range of the strong interaction, and hence its essentially microscopic and therefore quantum nature, is closely related to this non-linear property.

Being essentially quantum in nature, and having a charge called colour in place of electric charge, the theory of the strong interaction is called *quantum chromodynamics* (QCD), by analogy with quantum electrodynamics. The theory is renormalizable and quantum-mechanically consistent. However, because of the strength of the interaction — its dimensionless coupling constant is of the order of or greater than unity — perturbative techniques cannot be used with much hope of success in analysing the bound states of hadrons. Alternative means of carrying out calculations, such as lattice gauge theory, are being actively pursued, permitting numerical results to be obtained where perturbative techniques fail. Strong nuclear reaction times and decay rates are $\approx 10^{-20}$ s. Photons, intermediate vector bosons ($W^{\pm}$ and Z^0), leptons and the graviton (as yet unobserved) are all particles which have no colour and therefore do not interact strongly.

1.5 Gravity

The *gravitational interaction* may be characterized as the parity-covariant interaction which is sourced by, and acts on, all forms of mass-energy, all with the same universal strength or coupling constant. The gravitational interaction is therefore said to be *universal*. In its standard form, it is long-range and attractive. Newtonian gravity is a non-relativistic non-quantum many-body theory which satisfactorily accounts for weak gravitational phenomena involving slow bodies over small space regions and time intervals. It is formulated as a theory in which the effect of changes of a source are perceived instantaneously at arbitrarily large distances. It is therefore said to be an *action at a distance* theory. The Newtonian description of gravity

1.5. Gravity

effectively gives the interaction an infinite propagation speed contradicting the principles of Einsteinian relativity which are required for an exact description of spacetime kinematics. It is therefore regarded as an approximate theory which must be replaced by Einsteinian relativity, either special or general, to obtain more precise results of broader applicability.

The *positivity of energy* or mass — in contrast to the bipolarity of electric charge — and the universality of gravity, mean that its effects accumulate with increase in the masses of the interacting parts. Despite the fact that gravity is by far the weakest of all the four interactions, this means that on a large scale (laboratory, terrestrial, astronomical or cosmological), gravitational effects dominate those of all the other interactions. Furthermore, the universality of gravity, and the positivity of its source, means that it is impossible to isolate totally any system from the effects of gravity in the way that various metallic shields can be used to effectively isolate a system from electric and magnetic fields.

Einstein's *general theory of relativity* is a theory of non-quantum gravity consistent with the Principle of causality (that causes must always precede effects) and the Principles of special relativity and therefore involves propagation at finite speeds. The space and time of general relativity are *dynamic* constituents of nature, affected by the matter whose relative separations they also describe. Unike Newtonian gravity, general relativistic gravitation is applicable to the very large-scale effects of cosmology. It is currently the accepted theory for the explanation of all non-quantum gravitational phenomena including those involving bodies at high speeds and in regions where the effects of gravity are strong. Even in situations where speeds are non-relativistic and gravity is relatively weak, such as in the solar system, but where very high precision is used, general relativity provides the most satisfactory explanation for the small observed deviations from Newtonian predictions. Examples are the deflection of light and other electromagnetic radiation passing near a massive body like the sun and the anomalous precession of the perihelion of a planet or of the periastron in a binary star system. Gravitational red-shifts are also naturally accounted for in the general theory.

In a relativistic formulation, the long-range nature of the gravitational interaction is closely related to the fact that its source, the mass-energy and momentum of matter and radiation, is conserved. As for the electromagnetic field, the propagation of gravitational effects in vacuo is always at the same velocity, the speed of light, and this is again related to its source being conserved. Its integrated source is the energy-momentum 4-vector whose first component, the mass-energy, is — unlike electric charge and the colour charge of the strong interaction — not frame-independent and does not occur in equal and opposite values. Gravitational effects therefore propagate via a field having, like electromagnetism, two independent transverse polarization components. However, for non-quantum gravity, these are of tensor rather than vector structure.

The tensor nature of gravity, as opposed to the vector character of the electrodynamic field, can be traced to the fact that the integrated source of gravity is not an invariant, like charge, but the 4-vector of energy and momentum. Correspondingly, the mediating field can be shown to have a spin 2 form meaning that its quantum description, once correctly formulated, must involve the exchange of particles, which in addition to being massless, have two units of intrinsic angular momentum. The fact that the gravitational interaction has a well-defined non-quantum limit is closely related to the masslessness, and consequent long range, of the field and to the integral value of its spin. As in the electromagnetic case, the exchange particles, *gravitons*, will obey Bose-Einstein statistics permitting the formation of coherent states with large occupation numbers, described by commuting variables in the non-quantum limit.

One of the most important properties of gravity, implied by its universality, and naturally incorporated into the theory of general relativity, is that the gravitational field itself also gravitates. The fact that the gravitational field has energy implies that it also has mass. Thus it is itself a source of gravity and is acted on itself by the gravitational field. The gravitational interaction is therefore non-linear and the effects of changes leading to an increase in the gravitational effects can be highly regenerative. A change may increase the attraction due to gravity which in turn increases the gravitational potential energy which in turn increases the gravitational effects. The consequences of such non-linear regenerative effects can be dramatic, particularly in certain astrophysical situations such as the death of stars, the structure of the nuclei of galaxies and in *cosmology*, the study of the universe as a whole.

Stars are objects in which there is an equilibrium between the thermal pressures of expansion arising from nuclear energy generation in their cores and the contraction effects of gravitation. The mass, temperature, size and luminosity of the sun make it typical of the large majority of stars, those referred to as *main sequence* type. Astrophysical studies show [280, 512] that three possible final products of stellar evolution, when nuclear fuels sustaining a star are exhausted, are a *white dwarf*, a *neutron star* and a *black hole*. A typical white dwarf is a compact star of similar mass to an ordinary main sequence star like the sun but with a radius about 100 times smaller. General relativistic effects are involved in determining the stability of such stars but are not important in calculating their overall structure. The latter is determined by an equilibrium between the gravitational attraction and the repulsion caused by the *degeneracy pressure* of fermions arising as a result of the Heisenberg uncertainty and the Pauli exclusion principles.

A neutron star is also typically of about a solar mass but is smaller than the sun by factors of about 100,000. General relativistic effects are important in determining the structure of such objects as well as their stability. There is an upper limit to the mass, approximately three solar masses, of a stable neutron star. Stars which are considerably more massive than the sun cannot lose sufficient mass during their evolution to end as white dwarfs or neutron stars. When the thermonuclear fuels providing the thermal energy and pressure supporting such objects against gravity are exhausted, the regenerative effects can create gravitational fields so strong that gravity overcomes all possible repulsive effects consistent with the Principles of causality and relativity. Massive stars which do not eject a large enough fraction of their mass during evolution therefore end their lives as objects in which all the matter has collapsed to an extraordinarily dense state, currently modelled by the infinitely dense source at the singularity of a black hole. Many models of galaxies with very active nuclei and of quasars (quasi-stellar sources) depend on the presence of a central black hole.

Relativistic gravity, even in its source-free form, has non-linear field equations as a result of the fact that gravity itself gravitates. In this respect, it resembles the strong interaction where the corresponding property is that the gauge boson particles (gluons) which mediate the interaction between hadrons (composed of coloured particles) also have colour themselves. The dynamical potentials of relativistic gravitation exhibit gauge freedom but the gravitational gauge invariance is of a different character to those of the other three interactions. The global symmetry which is gauged is the group of all spacetime translations, rotations and Lorentz boosts from one frame to another moving uniformily with respect to it. These spacetime or external transformations, collectively known as the Poincaré group, affect the space and time coordinates. They are similar to the internal Yang-Mills groups, which affect fields, in being non-Abelian and this is directly related to the non-linearity of the unsourced field equations. The differences from Yang-Mills symmetries are caused partly by the external nature of the gravitational gauge group (of Poincaré) and by the presence of transformations which have non-compact parameters. Translations are obvious examples of such transformations but those members of the Lorentz group which contain boosts are also non-compact as the bounding range $-c < v < c$ of the boost parameter, the component v of the velocity in the direction of the boost, is not closed since the end-points $\pm c$ are not included.

Therefore the underlying global symmetry that leads to gravity on gauging is not of Yang-Mills type. Gauging does not lead to a dynamic mediating field for each of the original ten parameters. Gravity, like electromagnetism, is mediated by just one real dynamical field which, if it could be consistently quantized, would have a single quantum, a massless spin 2 particle called the *graviton*. Being charge-free like the photon, it is its own antiparticle. The fact that gravity gravitates corresponds to the fact that a graviton has energy-momentum, the source of gravitation. However, the gravitational interaction has not yet been consistently quantized. Unlike the other three, the gravitational interaction — or more specifically the general-relativistic description of it — is not only non-finite but is also *non-renormalizable* according to the standard quantum field theory of particles. However, most physicists expect the quantum principles which have been so fruitful in formulating the equations of three of the interactions to apply also to a refined form of the gravitational interaction. The quantized theory being sought is referred to as *quantum gravity*. The graviton, as yet unobserved, is the exchange gauge boson of quantum gravity.

One of the most promising candidate theories for the quantization of gravity involves the application of quantum consistency criteria to *supersymmetric string* models. *Supersymmetry* is an invariance discovered in 1974 in which integer spin particles (whose Bose-Einstein statistical behaviour classes them as *bosons*) are each partnered by very similar half-odd-integer spin particles (whose *Fermi-Dirac statistics* class them as *fermions*). String theories [93] are based on quantum principles where the fundamental quanta are not particles but extended 1-dimensional systems equivalent to an infinite tower of particles of increasing spin and energy. The effective particle content of a string theory at the energies of laboratory experimentation, including those available from the most energetic particle accelerators, will be the lowest energy states. These must match the observed spectrum of elementary particles for the theory to be successful. String theories currently provide the most promising framework for eventual unification of gravity with the other interactions at extremely high energy ($\approx 10^{19}$ GeV). Some of these theories are also less plagued by the divergences of standard quantum field theory of particles and may even be finite. However, string theories have yet to predict an experimentally verifiable numerical result at currently accessible energies (≤ 100 GeV) and it is therefore too early in their development to know how successful this exciting programme might be.

Although the classical equations of gravity (general relativity) are in complete agreement with all long-range macroscopic observations, they have another unsatisfactory feature besides their non-renormalizability. They predict that measurable quantities, such as densities and temperatures, may diverge under certain conditions, that is take on infinite values, even at the classical level. Examples of such *singularities* are contained in the general relativistic models of the initial *Big Bang* — the singular or near-singular origin of the universe — and at the central feature of a black hole. That the actual universe obeys a cosmology which is currently best modeled with such a near-singular origin is implied by observations, at least back to within a few minutes of an initial Big Bang. Such information is provided by the highly isotropic *cosmic microwave radiation* which is approximately thermal with temperature ≈ 2.9 K.

Singularities in a physical theory may or may not involve the divergence of measurable quantities. A *physical singularity* is one involving the divergence of observable quantities, as opposed to a mere *coordinate singularity*, which may be removed or shifted by a change in coordinates. Exposure to the radiation from a physical singularity would imply physically unacceptable arbitrariness and lack of predictability in one's measurement. Physical singularities are normally regarded as fatal flaws in a theory. They indicate a need to refine the equations, however minutely.

The term *black hole* was coined by Wheeler [575] in 1968 to refer to the astrophysical objects [512], at that time unobserved, corresponding to unusual solutions of Einstein's field equations for general relativistic gravitation. The idea of a black hole was first suggested [265, 377], on the basis of Newtonian gravity, by Michell[2] as long ago as 1784 and a similar calculation was presented by Laplace [310] in 1795. Black hole solutions to Einstein's equations correspond to regions of spacetime in which all matter has collapsed to a central singularity causing the creation of a closed enveloping mathematical surface called the *event horizon* which constitutes a boundary between those events (outside) which are observable from a distance and those events (on or inside) which are not. Black hole singularities can therefore in a certain sense be considered exempt from the lack of predictability arising at the singularity. Although they involve a measurable quantity, the effects of it diverging are hidden from us by the event horizon. This protection from the singularity's non-predictability is called the *cosmic censorship hypothesis*.

A past singularity from which we are not protected in this way — for example the Big Bang — is referred to as a *naked singularity*. The prediction by general relativity of singularities — clothed or not — and its non-renormalizability, with all the difficulties this causes in unifying the four interactions, are arguments in favour of modification of the gravitational interaction, or of quantum mechanics, on a very small scale. This might even entail modification of the underlying notions of space and time in the microscopic domain, presumably as a result of applying quantum principles.

Exercise 1.3 *Use dimensional analysis to show that the only relativistic, gravitational (constants c, G) length ℓ associated with a given local mass density (ρ) is of order $\ell = c/\sqrt{G\rho}$. Evaluate this quantity for several values of ρ ranging from the density of the universe as a whole ($\approx 10^{-28}$ kg.m^{-3}) to the densities of extremely dense neutron stars ($\approx 10^{15}$ kg.m^{-3}). (Relativistic gravitation theory, as outlined in Chapter 21, provides a rigorous derivation of the relation between radii of curvature of spacetime and the local density of matter.)*

Exercise 1.4 *Show, by dimensional analysis, that there is no fundamen-*

---
[2] John Michell (1724–1793), English astronomer.

tal intrinsic length (independent of the details of individual particles and fields) associated with (a) non-gravitational relativistic quantum field theory (constants c, ℏ) (b) non-relativistic quantum gravity (ℏ, G) and (c) classical (non-quantum) relativistic gravitation. (d) Determine a fundamental length, independent of the properties of any particular particle or field, for (relativistic) quantum gravity and show that its numerical value is about 10^{-35} m. (e) Determine the corresponding energy (in GeV). (f) Can we expect that new developments in particle accelerator construction, leading to higher energies, will soon permit a direct probing of the properties of spacetime at the quantum level, and thus a direct measurement of quantum gravity effects?

The quantities in Exercise 1.4, parts (d) and (e), are referred to as the *Planck length* and *Planck energy* [436, pp. 479–480] in honour of Max Planck[3] who first pointed out [286, pp. 26–27] that a fundamental length can be constructed from the three constants we now associate with relativistic quantum physics and the gravitational interaction.

Classically, we take spacetime to be infinitely smooth as exemplified by the assumption that it can be coordinatized by four real parameters $\{t, x, y, z\}$. The Planck length is the dimension at which one can reasonably expect this smoothness to break down as a result of quantum effects giving perhaps some sort of foam-like or stringy structure or topology to spacetime. Gravity and spacetime are so intimately linked that quantum-gravitational phenomena would also be expected to appear at the same dimension.

1.6 Unifying principles

It is a remarkable recent development — of the last two decades — that a considerable part of the description of all of the four interactions at the fundamental level may be understood on the basis of a small number of unifying principles, four of which have been emphasized in this overview and warrant special mention.

The first of these, is the *Principle of relativity*. In its restricted special relativistic form it is a *kinematic* concept governing therefore the free motion of fields and particles, both classical and quantum. It embodies the notion that the laws of physics should be independent of the displacement in spacetime, and the rotation and uniform relative motion of observers. The second, the *Principle of stationary action*, greatly facilitates the consideration together of physical systems of very different nature, in particular classical and quantum particles and fields and the analysis of their symmetries and quantization. The third, the *Gauge principle*, is *dynamic* in

---

[3] Max Karl Ernst Ludwig Planck (1858–1947), German theoretical physicist.

1.6. Unifying principles

nature governing the manner in which the free fields or particles interact or couple. In some cases, it governs the way a system may be self-coupled in circumstances where the individual parts cannot exist free. Although basically classical, it has certain inherently quantum features. It may also be formulated as a relativity principle which demands that certain internal and external properties of a physical system be completely observer independent, in particular independent of the observer's location and relative motion, inertial or not. Extension to complete independence of the location, orientation and state of motion of the observer is the crucial ingredient of general relativity, the relativistic theory of the gravitational interaction, which is therefore a theory simultaneously embodying the Gauge coupling ideas and an unrestricted Einsteinian relativity principle. The fourth, the *Quantum principle*, is the demand that all physical systems be *consistently quantizable*. The expectation that this principle applies to all of physics is based on the enormous phenomenological success of those theories based on it and the possibility of contradictory results if one part of physical theory (gravity) is classical while the rest is quantized.

The partially unified description which can now, in the late 1980s, be given for the interactions of physics requires, of course, a number of other principles. It also makes direct appeal to a vast array of experimental data and observational results for much of which there is as yet no fundamental explanation. Nevertheless, these four principles have provided enormous contributions to a common understanding of all four interactions. Shared features may be explained in similar ways and variations can be attributed to fundamental differences in the way our description of each interaction embodies the general principles.

What is even more surprising is the extent to which a small number of principles of the above type go a long way toward determining the basic properties of the various forms that matter may take and the ways it may interact. Some physicists and applied mathematicians, not for the first time, are asking whether it may even be possible to discover eventually a not too large number of qualitative principles, similar to those given above, and to show that they suffice to determine uniquely, essentially by combinatorics, a large fraction of the quantitative properties of the universe, including the nature of the background spacetime, the matter for which it provides the means for measuring separations between events, and the way it interacts. Such a set of principles, a 'theory of everything' would drastically reduce the gap between our best mathematical model of the universe around us and physical reality itself. Although such a development is possible, history should teach us to be wary of nature revealing further layers of complexity beyond each breakthrough in the unification process.

The cornerstone of physics is measurement. Fundamental physics has progressed at an incredible pace in recent decades as a result of the interplay of experimental and theoretical ideas. The greatest crisis facing the unification programme appears to be the ever-increasing cost of probing nature experimentally at higher energies and smaller distances. Combined with the knowledge that complete unification requires energies so high, namely the Planck energy of $\approx 10^{19}$ GeV, this cost makes one wonder whether direct observations of sufficiently high energy may ever be possible. Unless indirect observations supply sufficient clues, fundamental physics appears to face the extreme danger of seeking a better model of nature without the possibility of testing the predictions of competing theories.

We shall not be attempting to discuss the fine details of the unification programme in this text. Our contact with such topics will be secondary. The primary goal will be to develop the theories of Galilean and of special relativity in Euclidean and Minkowskian spacetimes at an introductory level, using the classical electrodynamic interaction and weak gravity as examples, to describe properties that have parallels in all the interactions. We hope that the material chosen will provide some insights into the principles underlying general relativistic gravitation as well. Although the emphasis will be on the non-quantum character of the source and mediating particles and fields, we shall allow many of the choices made at the classical level to be influenced by the knowledge that one wishes eventually to quantize consistently the systems being studied.

Ways of investigating Nature and knowing all that exists, every mystery ... every secret[4].

A'h-mosé the Scribe   ca. 1650 B.C.

Bibliography:

Davies P C W 1987 *Superforce* [92].
Davies P C W and Brown J R (eds.) 1988 *Superstrings* [93].
Kippenhahn R 1983 *One hundred billion suns* [280].
Quinn H R et al., 1989 *Teachers' resource book on fundamental particles*... [466].
Trefil J S 1980 *From atoms to quarks* [545].
Trefil J S 1983 *The moment of creation* [546].

---

[4] Title of the Rhind Papyrus (on Egyptian mathematics), quoted in Newman [394].

CHAPTER 2

Space and time

Nothing exists except atoms and empty space; everything else is opinion[1].

Democritus of Abdera   ca. 460 – ca. 370 B.C.

Before discussing the Relativity principle (see Chapter 5), the universally accepted and more intuitive geometrical properties attributed to space and time separately will be examined. In particular, the nature and origins of the notions of space and time merit some attention along with their homogeneity, orientability, natural topology, isotropy and flatness.

The atomistic philosophy originated with Leucippus[2], the probable founder of the School of Abdera in Thrace in the fifth century B.C. It claims that both empty space, and the matter comprised of atoms (from *atomos*, physically indivisible) which filled it, were real. The changing world was understood in terms of the isolation of groups of atoms, in contrast to the widely accepted teachings of the Eleatic School of Parmenides[3] which held that everything that existed had always done so and could never change. Democritus, a student of Leucippus, was the the most famous of the atomists.

From the dynamical behaviour of matter and energy in small regions at very low densities or at very large distances from dense concentrations, in particular from conservation laws of momentum and angular momentum, it is still found to be useful to postulate the existence of a background *space*.

The mathematical properties of the space represent physical properties, such as relative separations of material and non-material objects. An inert space constitutes our macroscopic non-quantum notion of the non-gravitational *vacuum* at a given instant of time.

---

[1] References [104, 281] and Mackay [353].
[2] Leucippus (fl. fifth century B.C.), Greek philosopher.
[3] Parmenides of Elea (ca. 515 B.C. – ca. 450 B.C.), Greek natural philosopher.

2.1 Space

A fundamental constituent of any material non-quantum physical system is the *classical particle*, namely an object with non-zero mass which is large enough not to require a quantum treatment yet so small compared with all other characteristic lengths of the system to permit it to be represented at each instant of time as a point in space. In an advanced treatment, all particles arise from the quantization of a function of time and space called a *field* which we shall introduce in Chapter 3. The rôle of the background space is to provide a means for describing the relative locations of physical particles, the parts of a material *body* (a collection of classical particles) and other non-material objects such as fields, that make up our models of the universe.

Space permits us to order quantitatively the relative locations of particles, and parts of fields and bodies, among which we may include certain usually fictitious collections of particles referred to as *reference frames*. There is no difficulty in deciding that physical space has three dimensions corresponding, for example, to the notions of up and down, left and right and forward and backward. This description also shows us that physical space permits us to identify two *orientations*, such as left to right and right to left, along each direction. Mathematicians, correspondingly, have well-defined and rigorous procedures for describing the dimensions, ordering properties and *orientability* of mathematical spaces. To describe physical phenomena quantitatively we must provide an appropriate mathematical model of these and other properties of nature. We must in fact choose a 3-dimensional orientable mathematical space providing spatial ordering in each direction.

In the real world, the presence of matter anywhere and its interaction via gravity implies that the background space, even at the non-quantum level, is not inert but part of the dynamics, affected by the surrounding matter. (The same applies to time.) Strictly, one should therefore introduce the concepts of space and matter (and time) in a self-consistent way, each participating in the overall dynamics. For reasons of simplicity, initially we are choosing to ignore the effect of matter on the background space. In the context of a general relativistic treatment of gravitation, the vacuum can be shown [591] to be the lowest energy state of the dynamical system. We shall discuss the matter content initially on the assumption that its density is so low that the background space is essentially indistinguishable from the vacuum state. Equivalently, we may restrict consideration to sufficiently small regions to make it unnecessary to allow dynamically for the full effects of gravity provided we choose a suitable frame of reference and treat gravity as an external prescribed interaction of matter.

Even when general relativistic gravitational effects are not significant, it is only in a classical context that we can consider the vacuum to be inert.

A quantum field theoretic analysis shows that the lowest energy state, the vacuum, is in fact very active on a microscopic scale ($\approx 10^{-17}$ m) being comprised of a wide variety of particles which are not directly observable. However, such *virtual particles* give rise, indirectly, to observable phenomena. We may initially ignore this property of the vacuum by limiting attention to low energies and therefore to large spatial intervals ($\gg 10^{-17}$ m).

2.2 Time

Time is defined so that motion looks simple[4].

John Archibald Wheeler   1911–

Motion of bodies relative to one another, or change in relative position, leads us to the notion of *time*. Time provides a 1-dimensional (1D) ordering and orientation of the states of a body similar to, but independent of, the ordering and orientation in any given spatial direction. The orientations of physical time are the two directions of *past* and *future*.

Time has often been likened, not just in poetry but also in scientific treatments, to a forever flowing river. While this imagery captures some of the essence of time as an ordering parameter, the idea is not popular in modern physical theory. Physical time will be treated here as a 1D space. Material objects will be regarded as moving through both space and time. We shall use the term *event* to refer to a single point in space at a given time. A classical particle may therefore be represented by a locus of events $z(t)$, in space x and time t. It will be assumed that a non-quantum particle traces out a continuous and smooth sequence of events $z(t)$ in space and time. To the poetic idea of flowing time, will correspond the notion of an object never ceasing to pass through a series of events of ever-increasing time. By contrast, it need not move spatially relative to some other object.

The motions of particles may be used to define standards of time. A means must be provided for relating the measure of time provided by a standard particle clock to the notion of time and secondary clock standards at other locations. We are not concerned with this question of *distant simultaneity* in this chapter.

Although space and time are clearly distinguishable concepts, we shall see that they are in many ways so similar and interrelated that we shall collectively refer to them as *spacetime* irrespective of whether the context is that of Galilean or Einsteinian relativity.

---

[4]Misner C W, Thorne K S and Wheeler J A *Gravitation* [384, p. 23]. Note also Poincaré: *Le temps doit être défini de telle façon que les équations de la mécanique soient aussi simples que possible* [440, p. 6].

2.3 Spacetime coordinates

The quantum-mechanical *reduced Compton wavelength* $\lambda = \hbar c/E$ corresponding to the most energetic particles produced to date (1989) in laboratory accelerators, namely those with energies E equal to about 100 GeV, are $\approx 10^{-18}$ m with corresponding light-time intervals $\approx 10^{-25}$ s. The notion of particle length depends on that of coordinates and it therefore appears meaningful to allow spacetime to be continuously and smoothly coordinatizable to distances which are very small compared with any macroscopic dimensions.

Consequently, we assume that the one dimension of time and the three dimensions of space are each a real *manifold*, namely spaces which may be smoothly labelled or coordinatized by real numbers, t, x, y or z, which we call *spacetime coordinates*. This continuity and smoothness assumption for particle paths in spacetime may in fact be partly responsible for the divergences that arise in QFT and the difficulties that arise in the quantization of the gravitational interaction. Equally, the emphasis placed until recently on those fields which on quantization lead directly to particles, rather than higher dimensional objects, may be the cause of such difficulties. However, these possibilities will not be investigated here. There are an infinite number of ways to label all points in spacetime with a set of four coordinates $\{t, \mathbf{x}\} = \{t, x, y, z\}$ denoting the times and locations measured relative to a given standard clock and relative to some length standard. The standard clock, the length standard and a distant simultaneity procedure used together will comprise a *spacetime reference frame*. Nevertheless, we shall consider the location in spacetime itself, an *event*, to be a physical concept, independent of any one coordinatization.

A mathematical space which is coordinatizable may have certain properties which are independent of continuous and smooth changes in its shape. The effects of such changes may be examined by, for example, making smooth and continuous but otherwise arbitrary changes to its scale, independently in each direction at each location. Such properties are said to be *topological* and exist even for spaces for which no notion of the length between points exists. The *dimension* of a space is a topological invariant. So also is orientability. The real line $\mathbf{R} = \{x\}$ ($-\infty < x < \infty$) and the circle $\mathbf{S}^1 = \{\theta\}$ ($0 \leq \theta \leq 2\pi$, with 0 and 2π identified) represent the two distinct continuous and smooth 1-dimensional topologies. They differ in their *connectivity* properties, namely in the way the points are connected to one another.

The connectivity, continuity and differentiability properties of the real line $\mathbf{R}$ and its products $\mathbf{R}^n = \mathbf{R} \times \mathbf{R}^{n-1}$ ($n \geq 2$) with itself are referred to as the *natural* or *trivial topology*. Since any path in each $\mathbf{R}^n$ can be deformed continuously to a point, they are said to be *simply-connected* and are assigned to *genus* 0. The 2D surfaces of a sphere and an ellipsoid

2.3. Spacetime coordinates

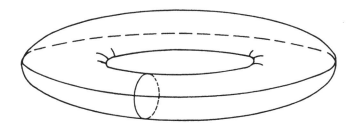

Figure 2.1 Two-dimensional torus.

are said to have the same smooth continuous topology S^2 but they differ in connectivity from the surface of a 2D *torus*, as shown in Figure 2.1, which is said to have the topology $\mathbb{T}^2 = S^1 \times S^1$ of the product of two circles. The sphere, having no holes or handles, is also simply-connected and therefore assigned genus 0, while the torus is said to be genus 1. One may extend the idea of a surface with the topology of a torus to n-dimensions by requiring it to have the same topology as the product $\mathbb{T}^n = S^1 \times \mathbb{T}^{n-1}$ of n circles.

In modern physics, the topology of spacetime and of the extended objects, such as fields, located within it, are not taken for granted but form a part of the structure of the universe to be investigated and experimentally established. No quantum breakdown in topology on small scales is expected (see Exercise 1.4) until spacetime is probed to separations down to near the Planck length, $(G\hbar/c^3)^{1/2}$, of about 10^{-35} m.

We choose to assume at this point that time and each space dimension all have the natural topology of the 1D real line and that space and spacetime therefore also have the trivial topology of the products $\mathbf{R}^3 = \mathbf{R} \times \mathbf{R} \times \mathbf{R}$ and $\mathbf{R} \times \mathbf{R}^3$. This may seem, intuitively, to be an obvious property to be required of any space which is to model the background spacetime but in fact is a simplifying assumption extending local conditions to infinity, namely *globally*, in a way that is not valid when gravity is taken fully into account.

In fact, astrophysical observations of cosmological significance are not yet sufficiently numerous or precise to permit the determination of whether the universe as a whole is even spatially infinite in extent or not and whether or not it will continue to expand indefinitely. Even if the Universe has the trivial topology overall, this still does not preclude it from having unusual topological properties in a specific locality, an example being the distortion caused by the presence of a black hole. In fact, observations are never carried out over infinite ranges — any open interval which is small compared to the local characteristic lengths (radii of curvature) arising as a result of gravity would do equally well for most observations. Since the magnitude of the radii of curvature in the vicinity of the solar system (see Exercise 1.3) are of the order of light years this need not be a significant limitation for many purposes. We shall thus, for simplicity, assume here that the range of each of the four coordinates $\{t, \mathbf{x}\}$ of a *spacetime frame* is the entire real line **R**.

2.4 Homogeneity, isotropy and curvature

Some hypotheses are dangerous — first and foremost those which are tacit and unconscious. ... Here again, there is a service that mathematical physics may render us[5].

Henri Poincaré   1854–1912

The existence of coordinates labelling the points of a manifold, including those whose range is the whole real line, does not itself imply the ability to measure distances nor that there must exist coordinates which are rectangular Cartesian. The real line **R** is not identical, for example, to the 1D Euclidean space, $\mathbf{E}^1$. However, observations show that spacetime coordinates can be more than just convenient labels for events. In certain reference frames, to be introduced later, some combinations of spatial and temporal *coordinate intervals* (differences in coordinates between particles or parts of bodies and fields) may be measured with frame-independent results and therefore correspond to invariant properties of the pairs of events they join, independent of the coordinates used to locate the events. The intervals between events in spacetime are in some sense metric or *geometric*, namely measurable.

Mathematical spaces with *non-Euclidean geometries* were discovered by Gauss[6] [188], Lobatchevsky[7] and Bólyai[8] between 1799 and 1832 (see also Weinberg [566, p. 5] and Stewart [521]), and subsequently developed by Riemann[9], among others. Previously, scientists had never doubted that physical space and time were each *Euclidean*, not just locally but globally as well. In other words, Euclid's fifth postulate, that parallel lines never meet, was assumed to be applicable to the mathematics of physical space. In fact, it was considered so obvious a property of any space that many of the greatest mathematicians and philosophers consumed a great deal

---

[5] *Science and hypothesis* [445].
[6] Carl Friedrich Gauss (1777–1855), German astronomer and mathematical physicist.
[7] Nicolai Ivanovitch Lobatchevsky (1792–1856) Russian mathematician.
[8] János Bólyai (1802–1860), Hungarian mathematician.
[9] Bernhard Riemann (1826–1866), German mathematician.

2.4. Homogeneity, isotropy and curvature

of energy in attempts, in vain, to prove that it was a logical consequence of the first four of Euclid's postulates. However, as for the topology of spacetime, its geometry is now considered a part of the structure to be determined from experiment. The determination of such properties is an advanced topic that may be approached by first limiting the scope of one's theory to those systems able to be described by Euclidean concepts.

Consequently, in the next chapter, on Euclidean geometry, we shall describe the mathematical properties of the 1D and 3D Euclidean spaces in a way which highlights their successful use in the modelling of physical phenomena and prepares the material required in subsequent chapters. Such spaces are ingredients in the mathematical models which represent certain spatial and temporal properties of material particles, bodies and fields, relative to the special class of inertial reference frames which we shall introduce in Chapters 5 and 6. We shall find in fact that the spacetime models of Galileo and Newton, based on such Euclidean spaces, are adequate for describing the behaviour of the material bodies of conventional Newtonian mechanics. The lower the speeds and the lower the densities of matter (relative to characteristic lengths and the strength of gravity), the more precisely does the model represent real physical systems, at least locally. The justification for this assumption, and its deficiencies, can then be examined after constructing a dynamics that is based on it and comparing the consequences implied by the Euclidean properties with those of actual physical phenomena.

We shall anticipate the discussion of Euclidean geometry by giving here a largely qualitative description of the physical properties it models, with no attempt at mathematical rigour. Mathematically, n-dimensional Euclidean spaces $\mathbf{E}^n$ ($n \geq 1$) are intrinsically *flat* and orientable *metric* spaces with the trivial topology. They admit a coordinatization in which the metric or distance formula is homogeneous and isotropic. Their metric property implies, first, that given any two points, there is a well-defined *distance* between them and, second, given any three points then two of them subtend a well-defined *angle* at the third. In rectangular Cartesian coordinates, the distance is given by the Pythagorean formula.

The *homogeneity* implies that the distance is dependent only on the separations between the points and not on the coordinates of the points themselves. The *isotropy* means that the distance and angle depend only on the magnitude of the differences between the coordinates of the vertices of a triangle and not on the direction of the separation between any two vertices relative to any other direction. The *intrinsic flatness*, a crucial ingredient of *euclidicity*, means that the interior angles of finite triangles, each of whose sides is a *geodesic* (the shortest path between points) always add to twice the sum of the angles between the two arms of an orthonormal basis or, equivalently, that parallel lines never meet. The orientability will be discussed in the next section.

At some stage it will be necessary to interpret the physical relevance of each of these properties and as we gain a deeper understanding of that interpretation we shall at the same time be justifying the arbitrary choice of Euclidean space and appreciating its deficiencies or limitations. At this point we shall expand the meaning of some of the above mathematical properties and give some advanced indication of the physical significance that will emerge in the subsequent dynamics based on them. To carry this out we shall require some idea of what is meant by a closed or isolated physical system, to be discussed in detail in Chapter 6. By an *isolated system* we shall mean one for which we may consider the *internal interactions*, namely the effects giving rise to changes in relative velocities of its parts, to dominate over those between it and the rest of the universe. We presume that it is sufficiently large that we may ignore the effects on the isolated system of an observer.

The distance formula of the Euclidean space provides the quantitative representation of the measurement of time intervals along a particular particle path in the 1D case or of spatial intervals in the 3D case. The homogeneity of the 1D Euclidean space representing time will correspond to the observation that a quantity identifiable as the *energy* of an isolated system is always conserved. Similarly, the homogeneity in the 3D case will embody the observation that the linear *momentum* of an isolated system is always conserved. The angle measuring property of the metric of Euclidean space permits the construction of *orthonormal bases* which correspond to the selection of sets of *rectangular Cartesian* coordinates to label preferred reference frames. The isotropy of the angle measure embodies the existence of a quantity identifiable as the *angular momentum* which, in an isolated system, is always conserved. These results may also be interpreted as embodying the fact that all dynamical measurements on the internal properties of isolated systems permit us to conclude that absolute position, time and angular orientation are unobservable.

Intrinsic flatness is rarely mentioned in introductory treatments of the Euclidean and pseudo-Euclidean spaces of Newtonian and special relativistic physics. Nevertheless, it is a crucial property that not all spaces share and we shall give a non-technical description here of its implications. We are not concerned with the *extrinsic curvature* or flatness of a space as a result of the way in which it might be embedded in a space of higher dimensions, such as for example a 1D curve or the 2D surface of a cylinder embedded in ordinary 3D Euclidean space.

We are instead concerned with the *intrinsic curvature* of a space as a result of properties it has independently of any other higher dimensional space of which it may or may not be a subspace, namely in which it is *embedded*. In fact, although a 1D path in 3D Euclidean space may clearly be curved, it cannot have intrinsic curvature of any form. A metric space with intrinsic flatness is one in which the angles of finite geodesic triangles

2.4. Homogeneity, isotropy and curvature

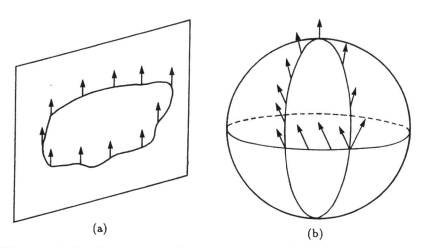

Figure 2.2 Parallel transport of a vector around a closed curve (a) in a plane $\mathbf{E}^2$ and (b) on a the surface of a sphere $\mathbf{S}^2$.

always sum to 180° and such a definition is only meaningful in spaces of at least two dimensions. The Euclidean spaces $\mathbf{E}^n$ ($n \geq 2$) are all of this intrinsically flat type, essentially by definition. This is also the case for the surfaces of cylinders which, although they apparently have curvature, are in fact intrinsically flat, according to the above definition. Flatness of a space is the property which permits us to translate copies of an object, such as a directed interval $d\mathbf{x}$, a vector $\mathbf{V}$ or an orthonormal triad $\{\mathbf{e}_x, \mathbf{e}_y, \mathbf{e}_z\}$, holding them always parallel to themselves, throughout the space, as shown in Figure 2.2(a) for $\mathbf{E}^2$, and find that the copies are still parallel to the original after passing around any closed path in the space.

Not all spaces are intrinsically flat. On intrinsically curved spaces, like the 2D surface $\mathbf{S}^2$ of a sphere, such *parallel transport* around a closed curve does not leave a basis parallel to itself as illustrated in Figure 2.2(b). We normally envisage the surface of a sphere as a 2D subspace of ordinary Euclidean space but some of its properties are quite independent of such an embedding. The surface of a sphere is an example of a 2-dimensional space of *constant curvature* and of simply-connected but non-trivial topology in contrast to the Euclidean plane which is also of constant curvature (of value zero) but with the natural topology. By convention, $\mathbf{S}^2$ is said to have *positive curvature*.

Curvature will not be a topic to be studied in detail in this text. The main reason for introducing the concept is for better appreciation of some of the properties of Euclidean space which are often taken for granted. Apart from an overall scale factor, there are only three 2D spaces of constant curvature, the third of which, the space of Gauss, Lobatchevsky and Bólyai, with constant *negative curvature*, cannot easily be visualized since it cannot be embedded completely within 3D Euclidean space. The embedding of any small part of it in ordinary space will however appear similar to a saddle. An excellent diagram is available in Thurston and Weeks [539]. There are also precisely three 3D spaces of constant curvature [384, 566], of which the zero curvature case is ordinary Euclidean space, the positive curvature case being referred to as the 3-sphere, S^3.

A space is said to be *locally-Euclidean* if any small region can be put into 1-1 correspondence with a small part of a Euclidean space. The 2D surface of a sphere is, for example, locally planar. Restricting consideration of representations of physical space to those which are modelled by 3D Euclidean space is equivalent to confining our attention to the *local phenomena* (in particular, non-gravitational) of a more general *non-Euclidean* or *Riemannian* space which is locally-Euclidean. The representation of physical spacetime by 4D spaces whose 3D spatial parts are locally-Euclidean is sufficiently general to encompass the present state of knowledge of all the interactions of physics based on low energy observations ($\leq 100\,\text{GeV}$). One therefore loses no generality by limiting consideration to a space which is globally Euclidean; one may simply re-interpret the results obtained as being valid not globally but just locally. This will be necessary not only to describe gravitation but also some non-gravitational phenomena such as the quantum interference effects discussed by Ehrenberg and Siday [126] in 1949 and Aharonov and Bohm [6] in 1959. These arise from quantum analyses of an electromagnetic field confined to non-simply connected regions of space.

2.5 Orientability of time and space

The *orientability* of a space is a topological invariant. For a 1D Euclidean space **E** it is simply the ability to select one or other of the directions along the space, namely to choose a single basis vector, and know that continuous transformation, namely translation, within the space away from and back to the original location, cannot reverse its direction. There are no non-orientable 1D spaces. For one dimension, the only continuous transformations preserving the distance are translations.

Suppose that, in a Euclidean space of dimension 2 or higher, we select an orthonormal basis and translate a copy of it throughout the space around any closed curve, no significance being attached to whether or not parallel transport is used. The orientability of such a space refers to the possibility of carrying out this process with the guarantee that the transported basis can afterwards be rigidly rotated so that every transported basis vector matches the direction of its original. In testing for orientability, we may rotate the basis copy to achieve the correspondence but we are not free to

2.5. Orientability of time and space

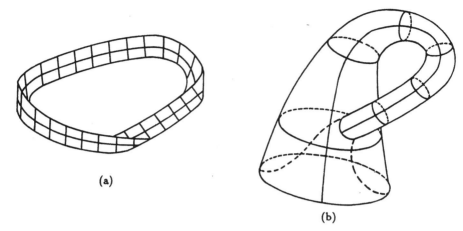

Figure 2.3 (a) Möbius strip and (b) Klein bottle.

do so in testing for flatness.

A space may be orientable but not flat, the surface of a sphere being one example, and a space may be flat but non-orientable. That orientability is a non-trivial assumption is seen by noting that it is not a property of all spaces. Consider for example, the movement and intercomparison of copies of a 2D basis $\{e_1, e_2\}$ around the surface of either the Möbius strip or a Klein bottle, each shown in Figure 2.3. The strip discovered by Möbius[10] in 1858 has zero intrinsic curvature and was the first example of a non-orientable surface.

The orientability of 1D Euclidean space is manifest when applied to time in our ability to distinguish the past from the future. Time is, however, not only orientable; we observe that macroscopic systems evolve in only one direction, toward the future. This property is referred to as *time's arrow* [86], a term first used by Eddington[11] [124, p. 68] in 1928. A vast literature exists on the nature of time and in particular on its orientability and arrow. The same is true for some of the consequences of the orientability of space although there exists no parallel in space to the arrow of time. Some elementary treatments of these and related topics are listed at the end of the chapter for further reading.

Most discussions mention a variety of different phenomena which permit one to distinguish between the past and the future and to identify the direction of time's arrow. Among these are the *Second law of thermody-*

---

[10] Augustus Möbius (1790–1868), German astronomer and mathematician. Like Gauss and Hamilton, Möbius was employed as an astronomer, but made his most significant contributions in mathematics.

[11] Arthur Stanley Eddington (1882–1944), English astronomer and relativist.

namics, the expansion of the universe, the decay of the long-lived neutral kaon, K_L^0, propagation of light via retarded signals and the subjective notion of time of the human mind. We make no attempt to relate these different aspects of time orientability. We shall simply apply the property to any time coordinate we use to label quantitatively and order a sequence of events related to the evolution of a non-quantum physical system. Such systems always move in the one temporal direction, from the past to the future. The vast majority of the microscopic laws of physics do not alter if one reverses the direction of all velocities. Microscopic systems governed by such laws cannot therefore be used to identify a direction to time. However, some microscopic phenomena, such as the decay of the neutral kaon, can be used for this purpose. Furthermore, the behaviour of large systems also gives rise to irreversibility and the Second law of thermodynamics, which states that the entropy or disorder of an isolated system never decreases. All such large systems always evolve toward a more probable state and this evolution permits the determination of time's arrow.

The orientability of 3D space means that all ordered basis triads (for example e_x, e_y and e_z in that order) can be divided into two *handedness* types, referred to as *right-handed* and *left-handed*. Reversal of a single axis, or all three axes together, changes the handedness of a basis. An alternative term for handedness coined by Lord Kelvin[12], and widely used in chemistry and high-energy physics, is *chirality*. In certain fundamental physics applications, chirality is assigned positive and negative unit values corresponding to right and left-handed, respectively.

... time, dark time, secret time, forever flowing like a river[13]...

              Thomas Clayton Wolfe 1900–1938

The world and time had both one beginning. The world was made, not in time, but simultaneously with time[14].

              Saint Augustine of Hippo 354–430

Bibliography:

Aharoni J 1972 *Lectures in mechanics* [5].
Flood R and Lockwood M 1986 *The nature of time* [166].
Gardner M 1964 *The ambidextrous universe* [185].
Hawking S 1988 *A brief history of time* [226].

---

[12] William Thomson, Baron Kelvin of Largs (1824–1907), Scots physicist.
[13] *Of time and the river*, 1935.
[14] *On the beginning of time* [23].

Landsberg P T (ed.) 1982 *The enigma of time* [307].
Layzer D 1975 *Arrow of time* [314].
Morris R 1984 *Time's arrows* [389].
Schutz B F 1980 *Geometrical methods of mathematical physics* [499].
Skylar L 1974 *Space, time and spacetime* [513].

CHAPTER 3

Euclidean geometry

Lagrange, in one of the later years of his life, imagined that he had overcome the difficulty (of the parallel axiom). He went so far as to write a paper, which he took with him to the Institute, and began to read it. But in the first paragraph something struck him which he had not observed: he muttered: 'Il faut que j'y songe encore', and put the paper in his pocket[1].

Augustin de Morgan   1808–1871

In his *Elements* (Στοιχεῖ), Euclid presented a logical, ordered compilation of the mathematical knowledge of the Greek civilization of the 4th century before Christ. Bulmer-Thomas [54] describes the work as 'having exercised an influence on the human mind greater than that of any other work except for the Bible'. Although the logic of the ancient Greeks does not always meet the very stringent demands of modern mathematics, the *Elements* is nevertheless considered to be not only a definitive treatise on what we now know as Euclidean geometry, synonymous with geometry itself until the early 19th century, but the epitome of the logical and concise presentation of axioms, postulates, theorems and propositions. 'Majestically he proceeds by rigorous steps from one proved proposition to another, using them like stepping-stones until the final goal is reached' [54].

The genius of Euclid, known even to his contemporaries as 'the Geometer', is no better demonstrated than in his fifth postulate on parallels. So frequently thought to be provable from the other four, it has since the 19th century been accepted as an unprovable characteristic of one particular type of geometry, *Euclidean*. Geometrical methods of carrying out mathematical calculations dominated the early analysis of observations, as in the work of Brahe[2] and Kepler[3] on planetary motions. The same is true of much of the mathematical work on mechanics and gravitation of Newton in the *Principia* [395]. Newton still favoured geometrical methods

---

[1] *Budget of Paradoxes* 1872 (London), cited in Mackay [353].
[2] Tycho Brahe (1546–1601), Danish astronomer.
[3] Johannes Kepler (1571–1630), German astronomer and physicist.

> Euclid ($O\ \Sigma\tau o\iota\chi\epsilon\iota\omega\tau\acute{\eta}s$)
>
> fl. Alexandria   ca. 330–265 B.C.
>
> ... there is no royal road to geometry.
> [Reply to Ptolemy Soter, ruler of Egypt, who once asked him if there were a shorter way to the study of geometry than the *Elements*.]
>
> Give him three obols, since he must needs make a gain out of what he learns.

even after having solved many problems using the new non-geometrical methods of infinitesimal calculus, differential and integral, which he expressly invented for the purpose and initially mistrusted because of its lack of rigorous foundation at that time.

Modern mathematics, and mathematical physics, is number-based rather than geometrical. Nevertheless, the geometric spirit of Euclid pervades every facet of modern theoretical physics. Geometry is intimately connected to the theory of *groups* (see Section 3.7) as embodied in the 19th century *Erlangen programme* [521] initiated by Klein[4], one task of which was to determine the group corresponding to each geometry and to set up the corresponding invariants. The geometric spirit is central to Minkowski's approach to the space and time of special relativity. Its most profound manifestation in physics results from Einstein's discovery that one interaction, gravitation, can be interpreted as a purely geometric phenomenon, albeit non-Euclidean. This extends to the other three interactions via the highly geometric notions of gauge invariance.

3.1 Euclidean line elements

Many physical properties of material particles and bodies, and of certain fields and wave functions, may be represented quantitatively in terms of the coordinates of a spacetime formed as a product $\mathbf{E}^1 \times \mathbf{E}^3$ of a 1D space with ordinary 3D space, both Euclidean. We shall refer to the product $\mathbf{E}^1 \times \mathbf{E}^3$ as *Euclidean* or *Aristotelean spacetime* [431, 433]. We shall examine here some of the primarily mathematical properties of such a space without being overly concerned with rigour. (We shall later introduce the spacetime

---

[4]Christian Felix Klein (1849–1925), German mathematician.

of special relativity and denote it by $\mathbf{E}^{1,3}$. However, it should be noted that the expression Euclidean spacetime is also used in Einsteinian relativity for a postive-definite mathematical space $\mathbf{E}^4$, the real cross-section of complex Minkowski spacetime, $\mathbf{E}^{1,3}(\mathbf{C})$. Unlike the one being introduced here, $\mathbf{E}^4$ is 4-dimensionally geometric.)

The trivial topology, continuity, smoothness, flatness, orientability and (local) homogeneity of space and time and its isotropy are the properties which make Euclidean spacetime useful in the representation of the spacetime of non-gravitational Newtonian physics.

These *euclidicity* properties arise as a result of using a time coordinate which is a point in 1D Euclidean space $\mathbf{E}^1$ and positions (at each instant of time) which are points in the 3D product $\mathbf{E}^3 = \mathbf{E}^1 \times \mathbf{E}^1 \times \mathbf{E}^1$. Each of these two spaces in the product space $\mathbf{E}^1 \times \mathbf{E}^3$ is separately a *metric* or *geometric* space. We ignore the fact that a 4D geometry can be defined on $\mathbf{E}^1 \times \mathbf{E}^3$. Such a property is not a part of Galilean or Newtonian physics but separate 1D and 3D metrics are local properties of all physical theory, including Minkowskian and Einsteinian. Instead we concentrate on the fact that the first of these spaces is the real Euclidean line endowed with the trivial, positive-definite, distance $|dt|$. The second is the 3D real Euclidean number space endowed (at each instant of time) with a *scalar product* of two displacements, $\mathbf{x}$ to $\mathbf{x} + d\mathbf{x}_1$ and $\mathbf{x}$ to $\mathbf{x} + d\mathbf{x}_2$, given in an *orthonormal basis* (of *cartesian coordinates*) by,

$$d\mathbf{x}_1 \cdot d\mathbf{x}_2 = dx_1 dx_2 + dy_1 dy_2 + dz_1 dz_2 \ . \tag{3.1}$$

Although we consider only small intervals here, in flat spacetime these formulae also apply to arbitrarily large intervals $|\Delta t|$ and $\Delta \mathbf{x}_1 \cdot \Delta \mathbf{x}_2$. From the 3D scalar product we may extract the corresponding trivial positive *distance ds* where,

$$ds^2 \equiv (ds)^2 = dx^2 + dy^2 + dz^2 \equiv d\mathbf{x}^2 \ , \tag{3.2}$$

and also determine the angle θ between $d\mathbf{x}_1$ and $d\mathbf{x}_2$ from

$$\cos \theta = \frac{d\mathbf{x}_1 \cdot d\mathbf{x}_2}{\sqrt{(d\mathbf{x}_1 \cdot d\mathbf{x}_1)(d\mathbf{x}_2 \cdot d\mathbf{x}_2)}} \ . \tag{3.3}$$

The forms of the two scalar products $|dt|$ and $d\mathbf{x}_1 \cdot d\mathbf{x}_2$ also represent the observed fact that time and length intervals, and angles, are unaffected by a reversal of $d\mathbf{x}$ to $-d\mathbf{x}$ and are thus unaffected by a change in handedness of the frame used to measure them. The squared distance ds^2 is called the *line element* of $\mathbf{E}^3$ and is clearly also *positive-definite*; namely, it is non-negative and it is zero if and only if the interval $d\mathbf{x}$ is zero. The temporal separation $|dt|$ is well-defined for any pair of events in $\mathbf{E}^1 \times \mathbf{E}^3$ irrespective of their spatial separation $d\mathbf{x}$ and, conversely, the spatial distance ds is well-defined irrespective of the value of dt.

The Aristotelean world view endows Euclidean spacetime with a distinguished family of curves in $\mathbf{E}^1 \times \mathbf{E}^3$, those which represent the loci of particles at rest, namely those with constant position x for all t. The state of rest in the *Aristotelean world view* is privileged. Time, space and motion are all absolute in Euclidean spacetime, all velocities being relative to those objects, like the earth at the centre of the universe, which are at rest. The questions of distant simultaneity and spatial coincidence at different times will be discussed further for Galilean, Newtonian, Minkowskian and Einsteinian spacetimes in Chapters 5, 7, 12 and 21.

3.2 Indexed notation and matrix multiplication

The standard vector notation $d\mathbf{x}$ for the coordinate intervals of Euclidean space is not easily adapted to the discussion of the tensor quantities needed in many non-relativistic physics applications. Nor is it easily generalized to describe the vector and tensor quantities of Einsteinian relativity. For these reasons we introduce here an explicitly indexed notation for the coordinates and vector components of 3D Euclidean space in which we denote the three coordinates x, y and z by x^k ($k = 1, 2, 3$),

$$\mathbf{x} = \{x^k\} = \{x^1, x^2, x^3\} = \{x, y, z\} . \tag{3.4}$$

We reserve all lower-case Latin letters from near the middle of the alphabet, $i, j, k, \ldots, p, q, r$ to denote such Euclidean *spatial indices*, all of which have the range $1, 2, 3$. We shall always label coordinates with raised indices.

The displacement $d\mathbf{x}$, rather than the coordinates $\mathbf{x}$ themselves, will be our prototypical *3-vector*. We introduce here the *Einstein summation convention* to abbreviate any sum over the x, y and z components of any vector such as $d\mathbf{x} = \{dx^k\}$ using a pair of repeated lower case Latin indices one of which we shall write raised and the other lowered,

$$dx^k dx_k \equiv \sum_{k=1}^{3} dx^k dx_k . \tag{3.5}$$

To each set of vector components which have a raised index (but not the coordinates) we shall define a corresponding set with lowered indices. For this purpose we use the 3D *Kronecker delta*

$$\delta_{kl} = 1 \text{ for } k = l \quad \text{or} \quad \delta_{kl} = 0 \text{ for } k \neq l , \tag{3.6}$$

according to, for example,

$$dx_k = \delta_{kl} dx^l . \tag{3.7}$$

By creating Kronecker delta symbols with the same values but with raised and mixed indices, δ^{kl} and $\delta^k{}_l$ (or δ^k_l), related according to

$$\delta^{km} \delta_{ml} = \delta^k{}_l , \tag{3.8}$$

we can obtain raised components in terms of lowered according to

$$dx^k = \delta^{kl}\,dx_l\,. \tag{3.9}$$

The actual values of dx_k are clearly identical to those of dx^k.

An index which is not summed over is called a *free index*. We shall choose to write all free indices at the same level in all terms of (almost) every indexed equation. Since our definition later of vector quantities does not include the coordinates x^k themselves, we shall not normally lower the indices on coordinates. In the present Euclidean context, we could equally well use a summation convention in which both indices are lowered or both raised. However, the transition to the notation for Einsteinian relativity will be simpler if we use raised and lowered indices in the summation convention.

The above conventions imply that the 3D scalar product and line element are given by the formulae,

$$d\mathbf{x}_1 \cdot d\mathbf{x}_2 = (dx_1)^k (dx_2)_k \quad \text{and} \quad ds^2 = dx^k dx_k\,. \tag{3.10}$$

A product such as $dx^k dx^l$ is a two-index array with a total of nine possible entries. An arbitrary linear combination of all nine,

$$g_{11}\,dx^1 dx^1 + g_{12}\,dx^1 dx^2 + g_{21}\,dx^2 dx^1 + \ldots + g_{33}\,dx^3 dx^3\,, \tag{3.11}$$

can, with the above summation convention, be compactly written as,

$$g_{1k}\,dx^1 dx^k + g_{2k}\,dx^2 dx^k + g_{3k}\,dx^3 dx^k = g_{kl}\,dx^k dx^l\,. \tag{3.12}$$

Since $g_{12}\,dx^1 dx^2 + g_{21}\,dx^2 dx^1$ can be rewritten as $\tfrac{1}{2}(g_{12} + g_{21})dx^1 dx^2 + \tfrac{1}{2}(g_{21} + g_{12})dx^2 dx^1$ we may always rearrange the terms in these sums, without altering the value of the sum, and define a new set of coefficients $\bar{g}_{kl} = \tfrac{1}{2}(g_{kl} + g_{lk})$ which satisfy $\bar{g}_{kl} = \bar{g}_{lk}$. We shall assume this has been done and drop the distinguishing bar; thus we always assume g_{kl} to be symmetric in k and l, $g_{kl} = g_{lk}$, in a sum over the symmetric $dx^k dx^l$. (See also Exercise 4.6.)

We may use Equation 3.7 to write the 3D scalar product and line element in the form of the right side of Equation 3.12, namely,

$$d\mathbf{x}_1 \cdot d\mathbf{x}_2 = \delta_{kl}\,dx_1^k dx_2^l \quad \text{and} \quad ds^2 = \delta_{kl}\,dx^k dx^l\,, \tag{3.13}$$

the second of which is an example of Equation 3.12 in which g_{kl} has the specific values of δ_{kl}.

As a result of Equation 3.13, we say that the Kronecker delta plays the role of a *metric* for $\mathbf{E}^3$ since it can be used to invariantly determine the lengths of segments and the angles between them from the components

3.2. Indexed notation and matrix multiplication

dx^k of the small intervals dx in any particular frame. Raised and lowered components permit us to express the invariant line element in several equivalent forms as,

$$ds^2 = \delta_{kl}\,dx^k dx^l = dx^k dx_k = \delta^{kl}\,dx_k dx_l \;. \tag{3.14}$$

We may clearly change any free index in an equation to any other unused index provided we change it in all terms. Similarly, we may alter a pair of *dummy indices* (those which are summed over) in any term of an equation to any other index not used in that term. More than two identical indices in one term is ambiguous and never used. To avoid this it will sometimes be necessary, during index gymnastics, either to re-label a pair of dummy indices or to re-label one or more free indices in every term of an equation.

We shall make a few comments here on the way in which matrix multiplication indices are manipulated when we wish to maintain a summation convention involving raised and lowered indices. To express the identity transformation $\mathbf{x} = \mathbb{1}\mathbf{x}$ in indexed form we are forced, by the one up one down rule of our summation convention, to write it as $x^k = \delta^k{}_l x^l$ in which the 3D *Kronecker delta* $\delta^k{}_l$ is an indexed version of the 3×3 identity matrix $\mathbb{1}$. The identity matrix $\mathbb{1} = \{\delta^k{}_l\}$ is a special trivial example of a transformation matrix. This means that the components of any transformation matrix, O say, should also be written as $O = \{O^k{}_l\}$ with its indices one up and one down. Irrespective of the level, we shall use the first index to label the rows of a matrix and the second for the columns.

The multiplication of two transformation matrices $R = \{R^k{}_l\}$ and $O = \{O^k{}_l\}$ will therefore be written

$$(RO)^k{}_l = R^k{}_m O^m{}_l \;. \tag{3.15}$$

The transpose O^T of a transformation matrix $O = \{O^k{}_l\}$ will have components $(O^\mathrm{T})^k{}_l = O_l{}^k$.

The levels of the matrix multiplication indices may need to be adjusted for other types of matrices. Suppose we multiply a transformation matrix $O^k{}_l$ by a matrix $g = \{g_{kl}\}$ whose natural form has both indices lowered (such as a metric matrix). Consistent indexing, with the summed matrix multiplication indices one up and one down, shows that we must write it in the form,

$$(gO)_{kl} = g_{km} O^m{}_l \;. \tag{3.16}$$

Similarly, the transpose O^T of O left multiplied into g will be indexed as

$$(O^\mathrm{T} g)_{kl} = (O^\mathrm{T})_k{}^m g_{ml} = O^m{}_k g_{ml} \;, \tag{3.17}$$

where the fact that the two multiplication indices m are not adjacent in the last product indicates the transpose of O is involved.

Exercise 3.1 *(a) Show that each of the metrics $d\mathbf{x}_1 \cdot d\mathbf{x}_2$ and $|dt|$ are invariant with respect to the following coordinate transformations:*

$$t \rightarrow \bar{t} = -t \tag{3.18}$$
$$t \rightarrow \bar{t} = t + t_0 \tag{3.19}$$
$$\mathbf{x} \rightarrow \bar{\mathbf{x}} = \mathbf{x} + \mathbf{a} \tag{3.20}$$
$$\mathbf{x} \rightarrow \bar{\mathbf{x}} = O\mathbf{x}, \quad O^{\mathrm{T}}O = \mathbb{1}, \tag{3.21}$$

comprising a time reversal, translation in time and space and an orthogonal transformation with real 3×3 matrix $O = \{O^{\bar{k}}{}_k\}$ with transpose O^{T}, t_0 and each component of $\mathbf{a}$ being constants ranging over the whole real line.
(b) Show that all components of O are bounded according to $|O^{\bar{k}}{}_k| \leq 1$.

It may also be shown that the transformations of Exercise 3.1 are the only ones which preserve the value of both of the two scalar products. We shall show how to establish a similar result in Section 13.2 and the same procedure could be used here. These transformations may be referred to as the Euclidean *isometry* transformations, namely those which preserve the metrics in the separate factors of the $\mathbf{E}^1 \times \mathbf{E}^3$ spacetime.

3.3 Parity

Spatial orientability permits us to define a self-inverse operation **P** in spacetime called *parity reversal*, namely a reflection of each of the three spatial directions with respect to a given point.

Since **P** does not depend on any continuous parameters, it is referred to as a *discrete* transformation relating the coordinates of the first system to those of the second. Since parity reversal is *self-inverse*, namely satisfying $\mathbf{P}^2 = \mathbb{1}$, its eigenvalues are ± 1. Owing to the fact that the physical states of quantum systems are undetermined to the extent of an arbitrary phase factor, phase changes may also appear in the time-reversal transformation rules of quantum variables. One cannot actually carry out an active parity transformation to reverse the chirality of a macroscopic object without breaking it down into a parity-invariant set of pieces and reconstructing. At the quantum level, where one is dealing with probability distributions and particles arising by quantizing fields, transitions involving a discrete change between states of opposite chirality need not destroy the system undergoing the transition.

Parity may be considered as an abstract operation **P** having a specific representation, called a *realization*, on each type of physical variable. A partial realization of the concept of parity will firstly be given from an *active* (and non-quantum) point of view using a single frame and two different

3.3. Parity

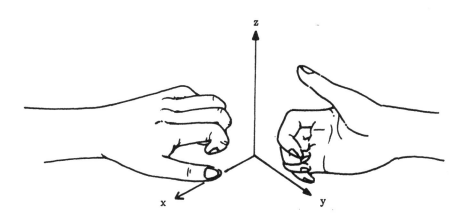

Figure 3.1 Chiral object and its parity conjugate.

material objects. Suppose we have a closed surface in 3D space, the interior of which is an extended (non-planar) object of uniform density whose only important properties are the relative locations of the particles of which it is comprised. As an explicit example, we may consider a 'left hand' as shown on the left in Figure 3.1. We may describe it by giving the coordinates of the points on its surface.

Suppose we also have a separate *reflected* object, identical to the original, except that each part at **x** is comprised of the part at $-\mathbf{x}$ in the original object. The object related in this way to our example in Figure 3.1 will be a 'right hand' as shown on the right in the figure. Reflection may take place with respect to any point but for convenience we make use of homogeneity to relabel coordinates and consider reflection in the origin, $\mathbf{x} = \mathbf{0}$.

The new object may be called the *parity transform* or *parity conjugate* of the original, with respect to the origin. The parity transformation is the operation which links the original object to its conjugate. We are not concerned with the fact that the transformation cannot physically convert one object to the other without destroying it. If the new object may be superimposed on the original by translation and rotation only (which is not the case for the example of a hand), then the original object is essentially indistinguishable from the transform. It may therefore be considered to be *parity invariant*, also referred to as a state of even parity with parity eigenvalue 1. We say it has well-defined parity. We may also say that the original object is not handed or is *non-chiral*, a term which we shall also use for states of well-defined parity of eigenvalue -1, which are not realizable with the present example. Any composite construct made up of a material object and its parity transform will be non-chiral, or of even parity. A sphere and a cube are each non-chiral. Although the hand in our example is clearly chiral, a left and a right hand (conjugate to the left) together comprise a non-chiral object of even parity.

The operation we have defined on macroscopic objects is single-valued with parity eigenstates of only even parity. It is therefore only a partial realization of the parity operation. In order to illustrate the fact that parity is in general a two-valued operation, we may consider realizing its action not on material points but on functions $f(\mathbf{x})$ of the three rectangular cartesian coordinates of Euclidean space. We then define the operation of parity on the functions $f(\mathbf{x})$ by a transformation to the new functions $g(\mathbf{x})$ according to,

$$f(\mathbf{x}) \rightarrow g(\mathbf{x}) = \mathbf{P}f(\mathbf{x}) = f(-\mathbf{x}) \ . \tag{3.22}$$

The functions $f(\pm\mathbf{x})$ are conjugate to one another, analogous to the conjugacy of a handed object and its parity transform, but allow negative function values and hence states of odd parity as well. Eigenstates of well-defined parity will be functions which are even in $\mathbf{x}$, namely $f(-\mathbf{x}) = f(\mathbf{x})$, examples being $\mathbf{x}^2$ and $\cos(\mathbf{k}\cdot\mathbf{x})$, or odd, namely $f(-\mathbf{x}) = -f(\mathbf{x})$, such as $\mathbf{x}$ or $\sin(\mathbf{k}\cdot\mathbf{x})$. Most functions, such as $e^{\kappa\cdot\mathbf{x}}$ and $e^{i\mathbf{k}\cdot\mathbf{x}}$, will be neither even nor odd. However, such functions can always be decomposed into two parts using the operators $\mathbf{P}_\pm = \frac{1}{2}(\mathbb{1} \pm \mathbf{P})$ and, since $\mathbf{P}\mathbf{P}_\pm = \pm\mathbf{P}_\pm$, those two parts are even and odd. Functions of $\mathbf{x}$ are therefore of even, odd or *mixed parity* and the same property applies to all complete realizations of the parity operation.

Exercise 3.2 *Verify that* $\mathbf{P}\mathbf{P}_\pm = \pm\mathbf{P}_\pm$. *Show that the operators* $\mathbf{P}_\pm$ *are idempotent,* $\mathbf{P}_\pm^2 = \mathbf{P}_\pm$, *and orthogonal* $\mathbf{P}_\pm\mathbf{P}_\mp = 0$.

Whenever a set of mutually orthogonal operators sum to give the identity (as do $\mathbf{P}_\pm$) and are each *idempotent*, they are referred to as a set of *projection operators*. The decomposition of the identity $\mathbb{1}$ into a sum of projection operators is referred to as a *completeness relation*. The operation P_x defined by $P_x\mathbf{V} = \mathbf{e}_x(\mathbf{e}_x\cdot\mathbf{V})$, and similarly for P_y and P_z, extract the projection of a vector $\mathbf{V}$ along each axis, and are well-known examples of projection operators, the completeness relation being $\sum_k \mathbf{e}^k\mathbf{e}_k\cdot = \mathbb{1}$. The operators $\mathbf{P}_\pm$ are also such a pair of operators projecting from a general state or object of possibly mixed parity to those of well-defined parity, ± 1.

We may use the term *chiral* to describe any physical variable (wave function, field or particle state) which is not an eigenstate of parity. Every chiral variable is therefore transformed by $\mathbf{P}$ into neither the original variable nor its negative. The field variables describing either of two circularly polarized states of light are chiral (see Equation 14.44). If one is the parity conjugate of the other, their sum and difference are non-chiral, one being of even and the other of odd parity. Other parity conjugate chiral variables are the fields (see Section 20.7) describing a neutrino and its antineutrino. For every new class of chiral variable, we either choose to assign the property of left- and right-handedness (or unit negative and positive chirality) arbitrarily to the variable and its conjugate or, if there is an obvious natural relationship with the bases of frames or other chiral objects (such as human hands) we label them accordingly. From a parity eigenstate, such as $\cos(\mathbf{k}\cdot\mathbf{x})$, we may construct chiral states, in the form $\epsilon e^{i\mathbf{k}\cdot\mathbf{x}}$ for example, where ϵ is a polarization vector, by linear superposition with adjusted phases.

There is no physical impediment, in principle, to the existence of a parity conjugate of any macroscopic object, although some such objects do appear predominantly in one or other chiral form. Many examples of chiral systems are also found in chemical compounds, both inorganic and organic. However, the existence of such chirality does not itself imply that nature is chiral. The chirality of such composite particles may arise as an accident if the fundamental constituents are non-chiral, which is the case for all the constituents of chemical compounds [195]. Neutrinos, for example, are not constituents of stable matter in the same way as protons, neutrons and electrons. At the high-energy particle level, it is an observed fact that *nature is chiral*. This was first attributed to the existence of neutrinos, which are intrinsically of one handedness or the other and do not appear only in non-chiral combinations. It is now considered to be a consequence of the form of the weak-nuclear interaction.

Since it does not involve the time coordinate or evolution, the parity operation, unlike time reversal, may be used to classify the spatial distribution of interacting fields and particle states as chiral or non-chiral (of even or odd parity). Conservation of the product of the contributions $(-)^l$ to parity where l is the orbital angular momentum quantum number of a state was formalized by Wigner [581] in 1927 as a generalization of *Laporte's rule* [312] forbidding atomic transitions between certain states. Particle states, composite or fundamental, may be eigenstates of parity and are then classified using the parity eigenvalues, ± 1. Some particle states are not eigenstates of parity.

We may re-express the operation of parity reversal *passively* using a single physical system and two sets of coordinates, one obtained by reflection of the other with respect to the origin. The coordinates may be those used in the equations describing the spatial distribution of the system at a given time. Such reflection in 3D Euclidean space reverses the direction of each unit vector of the basis and thus takes a left-handed basis triad into a right-handed one and vice-versa. In terms of coordinates we have $\mathbf{x} \to \tilde{\mathbf{x}} = -\mathbf{x}$ and (and $t \to \tilde{t} = t$). On quantities other than spacetime coordinates, such as vectors, tensors, spinors, wave-functions, fields or particle states, the passive operation of parity will take other corresponding forms, which must be determined in each case.

In common with time-reversal, parity may also be used to classify the evolution or interaction of a system as being *parity-conserving* or not. A parity-conserving interaction is one in which there is no preferential creation or destruction of subsystems of one chirality or the other, as occurs, for example, in transitions between states of opposite parity. A non-chiral system will then remain non-chiral after interaction and a chiral system will remain chiral. It is an observed fact that electrodynamics, gravity and strong-nuclear reactions, are parity-conserving interactions while the weak-nuclear interaction is not. Its lack of parity conservation results from the creation or destruction of chiral particles, namely neutrinos, without accompanying particles differing only in chirality.

The parity operator acts directly on particle wave functions (or state vectors) which, for multi-particle systems, are products of wave functions. The combined parity of a multi-particle system is therefore a product of the parity eigenvalues of the individual particles. This multiplicative property of parity is shared by other discrete symmetries such as charge conjugation, time reversal and exchange of identical particles. It contrasts with the additive nature of the conserved quantities corresponding to continuous symmetries, such as energy or electric and other gauge charges. The additive nature of the latter arises from their origin as eigenvalues of hermitian generators of the unitary operators which act on wave functions and states.

If the evolution of a physical system conserves parity, the reflection of the coordinates of space with respect to the origin must be accompanied by a corresponding variation, or *covariance*, of the physical quantities describing the system in such a way that the the evolution equations retain their form, or display *form invariance*, in the reflected coordinates. A parity-conserving interaction may therefore also be described as being *parity-covariant*. For parity-conserving interactions, the active and passive descriptions of the parity transformation will be equivalent. With non-conservation of parity, the passive transformation may lead to equations which do not describe the evolution of a physical system. We shall use the gravitational and electrodynamic cases to demonstrate that in the construction of a theory to describe a given interaction, the restriction to parity covariance severely limits the form of the equations.

In particle physics, two ways in which the states of a system may fail to be parity eigenstates have a common origin, but differ somewhat in the terminology used for each case. For fields and quanta of integer spin, the properties of being either *selfdual* or *antiselfdual* (see Section 14.8), an example being the circularly polarized states of light, imply a lack of definite parity. These properties are paralleled for fields and quanta of half-odd-integer spin by a specific property which endows handedness and which is referred to by the term *chirality* (see Section 20.7) which we have also used for handedness in general. No integer-spin particles are observed to exist in selfdual or antiselfdual states without corresponding antiselfdual and selfdual states. However, chiral fundamental particles of spin $\frac{1}{2}$ exist separately as neutrinos and antineutrinos.

The discovery of *parity violation* in 1957 is described in many textbooks and introductory articles. Among the latter are Morrison [390], Wigner [584], Gardner [185], Maglic [355, pp. 93–162] and Overseth [414]. Early in 1956, it was known that one even parity unstable particle, the spinless θ^+ particle, or K_θ^+ meson (of mass $\approx 500\,\text{Mev}/c^2$) decayed by the weak interaction into an even parity state of two pions, $K_\theta^+ \to \pi^+ + \pi^0$, each of odd parity. The τ^+ particle, or K_τ^+ meson, decayed into an odd parity state of three pions, $K_\tau^+ \to \pi^+ + \pi^+ + \pi^-$. The two K^+ mesons had identical mass, electric charge and half life ($\approx 1.2 \times 10^{-8}$ s) (and presumably the same spin) and were therefore apparently indistinguishable. Either they were identical, in which case the the K^+ meson did not conserve parity in its decay to three pions, or they were actually distinct particles despite the fact that they had the same values for all parameters normally used to decide particle identity. This enigma [391, pp. 5, 12, 364] was referred to as the $\theta - \tau$ puzzle by Yang [594, p. 398] and Lee [315, p. 10].

Over a period of 30 years all electromagnetic, strong-nuclear and gravitational interactions had been observed to be consistent with the conservation of parity. The parity violation alternative for the weak interaction was at that time almost universally considered to be an unacceptable explanation since it would imply that nature is chiral at the fundamental level. Conservation of the energy, momentum, angular momentum and centre-of-mass motion of an isolated system was known to be directly linked to the homogeneity and isotropy of empty spacetime (as we shall see in Chapters 6 and 15) and it was presumed that the universal observance of such conservation laws would extend to similar results based on other spacetime transformations such as parity and time reversal.

Later in 1956, Lee and Yang [317] confronted the theory of the weak interaction with the observed data to discover that there was no conflict between the experimental results and the violation of parity. They claimed there was one positive K^+ meson, or *kaon*, violating parity in the decay to three pions. Since this constituted an *ad hoc* solution to the $\theta - \tau$ puzzle, they recognized the need for further evidence in other reactions and suggested other weak interaction experiments, one involving cobalt-60 nuclei, to decide the issue. The publication of their paper in October of that year did not arouse widespread immediate interest [355, pp. 104, 132]. However, Chien-Shiug Wu devised an experiment involving the radioactive β decay of very low temperature cobalt-60 atoms in a strong magnetic field. If parity was not conserved, the rate of emission of electrons would be different parallel and antiparallel to the field direction. The results of the experiment published by Wu and her collaborators [592], and similar independent work by Garwin et al. [187] and Friedman and Telegdi [176], all received in January 1957, demonstrated that left-right symmetry is not an exact symmetry of nature. Initial reactions to these results included speculation on the possibility of the asymmetry being a property of space itself.

A host of experiments [171, p. 212] in 1957 confirmed the parity nonconservation. The results indicated maximal violation of parity in weak interactions for beta, muon and *hyperon* decay, the latter being unstable hadrons whose decay products include nucleons.

It was later pointed out by Grodzins, [33, p. 88] and [213], that evidence of parity violation had in fact been detected and reported by Cox et al. [82] in 1928 in the decay of a radioactive isotope of radium but not recognized as such. A similar result was obtained by Chase [64] in 1930. Despite Wigner's analysis [581] in 1927, parity conservation (which was not then referred to as such) was a new concept at that time and no reason existed to suspect it was not universally valid. The theoretical possibility of nonconservation of parity was raised prior to the work of Lee and Yang. It was first discussed by Purcell and Ramsey [463] in 1950. A detailed analysis was available in the classic paper on *intrinsic parity* by Wick, Wightman and Wigner [578] in 1952.

The overthrow of parity conservation was a dramatic and surprising development and is one of the strongest arguments in favour of relentlessly questioning and examining the bases of even the most cherished beliefs which make up the foundations of physics. An extensive elementary discussion of handedness in nature can be found in Martin Gardner's text *The ambidextrous universe* [185].

3.4 Charge conjugation

In contrast to the properties of particle mass, the existence of electric charges of two types, positive and negative, allows one to define the internal (non spacetime) self-inverse operation $\mathbb{C}$ of *charge conjugation*. From a system with charges q_i ($i = 1, 2, \ldots$), this operation gives another with identical mechanical properties such as mass, energy, and linear and angular momentum, but with opposite charges $-q_i$.

The charge conjugation of a quantum particle arose in 1928 from the form of the Dirac equation [106], the negative energy electron states of which (then thought to be protons) were interpreted by Dirac using the *hole theory* [107] in 1929. In 1931, Weyl [574] showed that if the new positive electron particle existed, the *positron*, it must be of opposite charge and the same mass, now known as the *antiparticle* of the original electron. This first antiparticle, the positron, was discovered by Anderson [14] in 1932 and confirmed in large numbers soon after by Blackett and Occhialini [41]. Dirac had shown [108] in 1930 that his theory allowed for the mutual annihilation of electrons and positrons, a phenomenon detected in 1933 by Thibeau [532] and Joliot [267]. The modern form of the charge conjugation operation [262, 410] was set out by Kramers [288] in 1937. Eigenstates of the charge conjugation operation have well-defined *charge parity* given by the eigenvalues ±1.

We shall see in Chapter 19 that a classical system of charges interacting electromagnetically can be set up so that it is charge conjugation invariant. The same is clearly true for gravitation and applies also to the quantum character of the electromagnetic and strong interactions. For the Fermi theory of β-decay in use in 1957, the maximal violation of parity meant that, from the same experiments, charge conjugation symmetry must also be violated if the system was to have invariance under time reversal (see next section). That this was necessarily so was shown by Lee et al. [319] in 1957. Some of the shock caused by the discovery of parity nonconservation in weak interactions was softened by the immediate observation that $K^{\pm}$ and β-decay data did not imply violation of the combined internal and spatial operation, CP, of charge conjugation and parity. This also meant that asymmetry of space itself could be avoided [304].

The decay of negative muons to electrons and neutrinos, namely $\mu^- \to e^- + \nu_\mu + \bar{\nu}_e$ (where $\bar{\nu}$ denotes an antineutrino) proceeds by the weak interaction. The electrons emerge left-handed as a result of maximal violation of parity. The charge conjugate decay is $\mu^+ \to e^+ + \bar{\nu}_\mu + \nu_e$. Charge conjugation does not affect the mechanical property of handedness (which for electrons is related to the relative orientation of momentum and spin). Consequently, charge conjugation invariance would in this case imply that the positrons also emerge left-handed. By demonstrating that they were all right-handed, Macq et al. [354] in 1958 and Culligan et al. [83] in 1959 showed that there was maximal violation of charge conjugation invariance [391]. Similar effects are observed in nuclear β-decay and in the decay of charged pions, $\pi^- \to \mu^- + \bar{\nu}_\mu$ and $\pi^+ \to \mu^+ + \nu_\mu$.

3.5 Time reversal

The orientability of time makes it meaningful to formulate the other main discrete spacetime transformation, somewhat misleadingly called the *time reversal* of the evolution of a physical system, which we shall denote by T. Time reversal clearly cannot actually involve the reversal of physical time but we shall see shortly why it is so-named.

Let us anticipate the setting up of dynamical equations and suppose that we have a non-quantum physical system whose time evolution is described by its trajectory of coordinates and velocities (or momenta) given at each time. Terms such as *trajectory*, *path*, *worldline* and *history* are used, in reference to a particle, to mean not just the curve or locus it traces in space but the specification also of which locations on the curve are reached at each time.

To define the transformation $\mathbb{T}$ actively, we use only one coordinate frame and consider two related physical systems. The transformation which, from each trajectory of a given physical system, constructs for another system, a trajectory with the same locus of points (one point being common to the two trajectories at a given instant) but with reversed velocities at each point along the path, is the *active time-reversal* operation. The instant when they are at the same point is the moment of time-reversal. We may time reverse at any instant, but by convention we make use of the temporal homogeneity to adjust the zero of time to carry out the reversal at $t = 0$. The time reversal operation $\mathbb{T}$ is *self-inverse*, namely it satisfies $\mathbb{T}^2 = 1$ and the time-reversal operator will therefore have eigenvalues of ± 1. Time reversal is a transformation which is not able to be obtained from the identity by continuous processes.

In the above active description, the systems move forward in time on both sets of trajectories and time reversal may be interpreted as *velocity reversal*. The new trajectories may or may not be included in the physical trajectories of the original system. If they are, the original system is said to be *time-reversal invariant*. The orientability of time and time's arrow, should not be confused with the lack of invariance of certain physical laws under the operation of time-reversal invariance.

The classical laws of all fundamental non-quantum systems, and many which are quantum-mechanical, are time-reversal invariant. It should be clearly understood that time-reversal invariance is a property of the evolution or interaction or change with time of certain physical systems. In contrast to parity, it is not an operation which can be used to assign an odd or even time-reversal eigenvalue to certain objects or particle states since the latter are concepts which apply at a given instant of time.

We may describe the time-reversal transformation *passively* by considering a single set of trajectories from the point of view of two different coordinate systems, related by a reflection of the time coordinate, which we also denote by $\mathbb{T}$, namely $t \rightarrow \bar{t} = -t$ (and $\mathbf{x} \rightarrow \bar{\mathbf{x}} = \mathbf{x}$). Given a particle trajectory described by coordinates $\mathbf{z}(t)$ and velocities $\dot{\mathbf{z}}(t)$ which are functions of time t in the first frame, the time-reversed trajectory will be obtained by the same coordinate values $\mathbb{T}\mathbf{z}(t) = \mathbf{z}(-t)$ traced in the opposite time direction with reversed velocity values $\mathbb{T}\dot{\mathbf{z}}(t) = -\dot{\mathbf{z}}(-t)$. In this passive description the time-reversed trajectory is traced out forwardly in the new time $\bar{t}$ or toward the 'coordinate past' in the t-system. This latter result is the origin of the term *time reversal*. Since coordinates are just mathematical labels in a passive transformation, this does not imply that any physical system evolves towards the past.

If a system is time-reversal invariant, then the new trajectory of the passive description will be included in the physical trajectories of the original system, and we may use this fact as an alternative way to define time-reversal invariance. For systems which are time-reversal invariant, the active and passive descriptions will be equivalent. If the system is not time-reversal invariant, the new trajectories will not be physical trajectories of the original system. The active operation of time-reversal then leads to a new system while a passive transformation may lead to dynamical trajectories which are not physical.

The quantities used in the equations describing a time-reversal invariant system must transform under T in a way which corresponds to the variation in the coordinates. The quantities must in fact *covary* in such a way as to leave the form of the equations unchanged. We may refer to such a system as having *time-reversal covariance*. Each physical quantity will transform in a specific way under time reversal. The details of each transformation will be determined from the corresponding behaviour of the quantities in terms of which it is defined, or by convention if it is fundamental. The components of simple classical variables are either unchanged or reverse sign.

Anticipating again the setting up of dynamical equations, we note that the kinetic energy of an isolated Newtonian system is a quadratic function of the momenta or velocities and the potential energy is time-independent and depends only on the magnitude of the spatial separation between the parts. Since these two quantities characterize a Newtonian system, its dynamics at the non-quantum level will be parity and time-reversal covariant. Classical electrodynamics and all gravitational phenomena are included in this category.

The time-reversal operation is therefore clearly of greatest importance in quantum systems for which it was first formulated by Wigner [582] in 1932 and given an equivalent often used form by Schwinger [410, 505] in 1951. Fundamental phenomena governed by quantum electrodynamics and strong-nuclear (and thus essentially quantum) interactions are also time-reversal invariant. Furthermore, under very general conditions (see Chapter 16), it can be shown [262, 429, 523] that quantum field theoretic processes must be symmetric under the combined operation, CPT, of all three discrete transformations together. In keeping with the very general conditions of the CPT theorem, no violations of CPT symmetry have been observed.

Time reversal invariance does not, however, apply to the weak-nuclear interactions (all of which are also essentially quantum), the first example having been provided by the decay modes of the neutral kaon, K^0, for which time reversal violations were discovered by Christenson, et al. [66] in 1964. More precisely, the experiments of Christenson et al. demonstrate violation of CP symmetry [1, 391] from which violation of time reversal invariance in weak interactions is deduced from the assumption of CPT invariance. A review of particle symmetries with a clear analysis of C, P and 𝕋 is available in Wick [579].

The violation of time-reversal invariance in fundamental phenomena occurs only on a minute scale, involving unstable particles apparently of little importance for the stability of ordinary matter. However, it may be that understanding why some very rare reactions violate time-reversal invariance could provide a vital clue for the programme of unification of all the interactions.

Many irreversible macroscopic processes violate time reversal but these phenomena may be described in terms of time-reversal invariant microscopic processes.

3.6 Rotations

With regard to the daily rotation, why should we not admit that the appearance is in the heavens and the reality in the earth[5]?

Nicolas Copernicus   1473–1543

The parity transformation **P** in $\mathbf{E}^3$ is represented or realized by the 3×3 matrix $P = -\mathbb{1}$ and is therefore also an orthogonal matrix, $P^T P = \mathbb{1}$, which commutes, $OP = PO$, with all other orthogonal matrices and has negative determinant, $\det P = -1$. A *rotation* may be defined mathematically as the homogeneous transformation of the coordinates $\mathbf{x} \to \bar{\mathbf{x}}$ in $\mathbf{E}^3$ which preserves the scalar product $d\mathbf{x}_1 \cdot d\mathbf{x}_2$ (and therefore distances and angles) and also the handedness of a frame or object (and therefore excludes a reflection of the axes). A rotation is therefore another example of an operation represented by an orthogonal transformation.

Exercise 3.3 *Show that orthogonal transformation matrices O satisfy $\det O = \pm 1$. Let the orthogonal matrices with positive determinant be denoted by R. Show that those orthogonal matrices O with $\det = -1$ may be expressed uniquely as the product (in either order) of some orthogonal matrix R with $\det R = 1$ and the parity matrix, $P = -\mathbb{1}$.*

All orthogonal matrices are therefore of just two types: those denoted R which do not include a parity reversal and which therefore mathematically represent a rotation and those which include a rotation and a parity reversal in a product, $PR = RP$. The set $\{O\}$ of all 3×3 orthogonal matrices is therefore obtained from the set $\{R, RP\}$ in which R is allowed to range through all the rotation matrices satisfying $R^T R = \mathbb{1}$ and $\det R = 1$. The rotation matrices R satisfying these conditions represent transformations which are able to be obtained from the identity transformation by

---

[5] *De revolutionibus*, I, p. 8, cited by Rosen [484].

3.6. Rotations

continuously varying the elements of the matrix. Equivalently, the parameter space of the rotation matrices is *connected*, although it happens not to be simply-connected.

Exercise 3.4 (a) Show that reflection in the plane $z = 0$ is equivalent to reflection in the origin followed by a rotation of angle π about the z-axis. (b) Use the condition $R^T R = \mathbb{1}$ to show that if an infinitesimal rotation matrix δR is re-expressed in the form $\mathbb{1} - \delta\omega$, then $\delta\omega$ is antisymmetric, namely that $\delta\omega_{kl} = \delta_{km}\delta\omega^m{}_l$ satisfies $\delta\omega_{kl} = -\delta\omega_{lk}$. How many independent parameters are there in $\delta\omega$? (Appendix A.2 gives combinatoric formulae showing how to count the number of components in indexed quantities with various symmetries.)

The 7-parameter Euclidean isometry transformations of Exercise 3.1 are thus reflection and translation in time, and reflection, translation and rotation in space:

$$t \to \bar{t} = -t \qquad (3.23)$$
$$\mathbf{x} \to \bar{\mathbf{x}} = -\mathbf{x} \qquad (3.24)$$
$$t \to \bar{t} = t + t_0 \qquad (3.25)$$
$$\mathbf{x} \to \bar{\mathbf{x}} = \mathbf{x} + \mathbf{a} \qquad (3.26)$$
$$\mathbf{x} \to \bar{\mathbf{x}} = R\mathbf{x}, \qquad R^T R = 1, \qquad \det R = 1 . \qquad (3.27)$$

Although some interactions fail to be time-reversal and parity-reversal invariant, the conservation laws of energy, momentum and angular momentum are universal and the translation and rotation transformation laws, Equations 3.25 to 3.27, are, correspondingly, exact invariance transformations of all local physics. Since no frames in relative motion are involved here, these principles are valid in both non-relativistic and relativistic physics and form an implicit part of the Principle of relativity (see Section 5.4) in either Galilean or Einsteinian form. We shall therefore first examine further the mathematical properties of quantities which have relatively simple rotation and translation laws.

Corresponding to each spatial coordinate x^k we define a unit basis vector $\mathbf{e}_k$ parallel to a line of varying x^k and, equivalently, basis vectors $\mathbf{e}^k$ normal to the planes of constant x^k. With the use of rectangular cartesian coordinates in Euclidean space, the two sets of bases are identical and we may represent this in a way which respects our indexing convention by:

$$\mathbf{e}_k = \delta_{kl}\, \mathbf{e}^l \quad \text{and} \quad \mathbf{e}^k = \delta^{kl}\, \mathbf{e}_l . \qquad (3.28)$$

The orthonormality of the coordinates and bases takes the form,

$$\mathbf{e}_k \cdot \mathbf{e}_l = \delta_{kl} \quad \text{and} \quad \mathbf{e}^k \cdot \mathbf{e}^l = \delta^{kl} . \qquad (3.29)$$

Any displacement $d\mathbf{x}$ may be expressed in terms of the orthonormal bases of unit vectors $\{\mathbf{e}_k\} = \{\mathbf{e}^k\} = \{\mathbf{e}_x, \mathbf{e}_y, \mathbf{e}_z\}$ as

$$d\mathbf{x} = dx^k \mathbf{e}_k = dx_k \mathbf{e}^k = dx\,\mathbf{e}_x + dy\,\mathbf{e}_y + dz\,\mathbf{e}_z \; . \tag{3.30}$$

If $R = \{R^{\bar{k}}{}_k\}$ denotes the matrix for a rotation from coordinates x^k to $x^{\bar{k}}$ then $R^{-1} = \{(R^{-1})^k{}_{\bar{k}}\} = \{(R^T)^k{}_{\bar{k}}\} = \{R_{\bar{k}}{}^k\}$ is the matrix for the inverse rotation from $x^{\bar{k}}$ to x^k. Nevertheless, we shall generally, somewhat perversely, abbreviate $(R^{-1})^k{}_{\bar{k}}$ not as $R_{\bar{k}}{}^k$ but as $R^k{}_{\bar{k}}$, the bar on the lower index indicating that it refers to the components of R^{-1}, not those of R. This convention will permit a close analogy between our discussions of Euclidean and Minkowskian tensors. From Equation 3.27 and the constancy of R, the components of $d\mathbf{x}$ transform under a rotation according to

$$dx^{\bar{k}} = R^{\bar{k}}{}_k\, dx^k \;, \qquad R^{\bar{k}}{}_k = \frac{\partial x^{\bar{k}}}{\partial x^k} \;, \tag{3.31}$$

for which the inverse transformation is

$$dx^k = R^k{}_{\bar{k}}\, dx^{\bar{k}} \;, \qquad R^k{}_{\bar{k}} = \frac{\partial x^k}{\partial x^{\bar{k}}} \;. \tag{3.32}$$

These equations also show that a raised index in a denominator must be considered as a lowered index overall.

Exercise 3.5 *Apply the definition of Equation 3.7 to show that dx_k transform according to*

$$dx_{\bar{k}} = R_{\bar{k}}{}^k\, dx_k = (R^{-1})^k{}_{\bar{k}}\, dx_k \;, \tag{3.33}$$

which we rewrite as

$$dx_{\bar{k}} = R^k{}_{\bar{k}}\, dx_k \; . \tag{3.34}$$

The conditions expressing the reciprocity of the partial derivatives in each direction, namely the chain rules

$$\frac{\partial x^{\bar{k}}}{\partial x^k}\frac{\partial x^k}{\partial x^{\bar{l}}} = \delta^{\bar{k}}{}_{\bar{l}} \quad \text{and} \quad \frac{\partial x^k}{\partial x^{\bar{k}}}\frac{\partial x^{\bar{k}}}{\partial x^l} = \delta^k{}_l \;, \tag{3.35}$$

here become

$$R^{\bar{k}}{}_k R^k{}_{\bar{l}} = \delta^{\bar{k}}{}_{\bar{l}} \quad \text{and} \quad R^k{}_{\bar{k}} R^{\bar{k}}{}_l = \delta^k{}_l \; . \tag{3.36}$$

The spatial separation between two nearby events $\mathbf{x}$ and $\mathbf{x} + d\mathbf{x}$ (at the same time) is a concept which is independent of the spatial basis,

$$d\mathbf{x} = dx^{\bar{k}}\,\mathbf{e}_{\bar{k}} = dx^k\,\mathbf{e}_k = \cdots \; . \tag{3.37}$$

Consequently, a rotational change in the coordinates affecting the values of dx^k and dx_k, must be accompanied by an inverse transformation,

$$\mathbf{e}_{\bar{k}} = R^k{}_{\bar{k}}\,\mathbf{e}_k \quad \text{or} \quad \mathbf{e}^{\bar{k}} = R^{\bar{k}}{}_k\,\mathbf{e}^k \;, \tag{3.38}$$

of the basis.

3.7 Groups

The various transformations linking two Euclidean reference frames $\{t, \mathbf{x}\}$ and $\{\bar{t}, \bar{\mathbf{x}}\}$ for $\mathbf{E}^1 \times \mathbf{E}^3$ illustrate a number of mathematical properties of great importance in modern physics. Consider, for example, the pair of transformations comprised of the identity transformation $\mathbb{1}$ and the parity transformation, $\mathbf{P}$. By definition of $\mathbf{P}$, we have $\mathbf{P}^2 = \mathbb{1}$ and $\mathbf{P}^{-1} = \mathbf{P}$. Similarly, from the definition of the identity, $\mathbb{1}\mathbf{P} = \mathbf{P} = \mathbf{P}\mathbb{1}$ and $\mathbb{1}^{-1} = \mathbb{1}$ and we have $\{\mathbb{1}^2, \mathbb{1}\mathbf{P}, \mathbf{P}\mathbb{1}, \mathbf{P}^2\} = \{\mathbb{1}, \mathbf{P}\}$. Consequently, the pair $\{\mathbb{1}, \mathbf{P}\}$, are a set of transformations whose repeated action, one after the other, gives one of the original transformations — the set is said to be *closed*. One of the set is the *identity* transformation which combines with the other to leave it the same. For each member of the set, there exists another member, called its *inverse*, which combines with it to give the identity. The combination of two transformations is obtained by performing one followed by the other and such composition is always *associative*. Whenever a set of objects, here transformations, has a law of combination or *composition* defined which satisfies the above four criteria, it is said to be a *group*.

Let us summarize the above definition more generally and therefore abstractly. A set of objects $G = \{g_i\}$ $(i = 1, 2, \ldots,)$ for which a *composition law* is defined forming a new object $g_2 g_1$, called the *product* of g_1 followed by g_2, from any two members g_1 and g_2, is a *group* if it satisfies the following properties:

- **Closure:**
  if g_1 and g_2 are two members then the product $g_2 g_1$ is a member.
- **Identity:**
  there exists a right identity member, denoted $\mathbb{1}$, satisfying $g\mathbb{1} = g$.
- **Inverse:**
  there exists a unique right inverse g^{-1} to g satisfying $gg^{-1} = \mathbb{1}$.
- **Associativity:**
  any three members satisfy: $g_3(g_2 g_1) = (g_3 g_2) g_1$.

The above properties imply that $\mathbb{1}$ is also a left identity, $\mathbb{1}g = g$, that g^{-1} is a left inverse, $g^{-1}g = \mathbb{1}$, and that the identity is unique.

Exercise 3.6 *Show that each of the following sets of transformations is a group: (a) the four discrete spacetime operations $\{\mathbb{1}, \mathbf{P}, \mathbf{T}, \mathbf{PT}\}$ where $\mathbf{P}$ and $\mathbf{T}$ are the parity and time-reversal operations, (b) all translations $T^3 = \{T_\mathbf{a}\}$ in 3D Euclidean space $\mathbf{E}^3$ where $T_\mathbf{a}$ is $\mathbf{x} \to \mathbf{x} + \mathbf{a}$ in which $-\infty < a^k < \infty$, and (c) all rotation matrices in $\mathbf{E}^3$, namely all 3×3 real matrices R satisfying $R^T R = \mathbb{1}$ and $\det R = 1$.*

The abstract 2-member group in which the non-identity member is self-inverse is referred as the *cyclic group of order 2* denoted $\mathbf{Z}_2$ since its structure is identical to the two integers $\{0,1\}$ with respect to a composition law consisting of addition modulo 2. The pairs of transformations $\{1,P\}$, $\{1,T\}$ and $\{1,C\}$, where C is the charge conjugation operation, are three further examples, each of which is called a *realization*, of the abstract group Z_2. Since P and T commute, and since each of the four transformations in $\{1,P,T,PT\}$ can be expressed as a product of a member of $\{1,P\}$ with a member of $\{1,T\}$, they are said to form a *direct product group*, which we write $\{1,P\}\otimes\{1,T\}$, a realization of the abstract group $Z_2\otimes Z_2$.

Groups were first used extensively to solve mathematical problems by the brilliant young French revolutionary, Galois[6], who applied them to the symmetries of polynomial equations to determine whether the equations were soluble in terms of radicals. He showed, in particular, that the solutions of the general quintic equation could not be written in that form. Galois's first papers on the topic were written at the age of 17.

The widespread use of groups in physics, particularly those comprised of transformations, is a direct consequence of the common occurrence of physical situations or states which are *equivalent*. In the present context, two sets of coordinates $\{t,\mathbf{x}\}$ and $\{\bar{t},\bar{\mathbf{x}}\}$ are equivalent labels for describing events in the homogeneous and isotropic Euclidean spacetime $\mathbf{E}^1\times\mathbf{E}^3$. The notion of equivalence is used in a precise sense in mathematics and in the mathematical representation of physical phenomena. A set of objects $\{I_i\}$ ($i=1,2,\ldots$) (in our example, spacetime reference frames) are said to satisfy an *equivalence relation* denoted $I_i \equiv I_j$ (all i,j) if and only if the relation is *reflexive* ($I_i \equiv I_i$ for each i), *symmetric* ($I_i \equiv I_j$ implies $I_j \equiv I_i$) and *transitive* ($I_i \equiv I_j$ and $I_j \equiv I_k$ implies $I_i \equiv I_k$).

Exercise 3.7 *Show that the transitivity, reflexivity and symmetry of the concept of an equivalence relation imply that transformations linking the equivalent objects (here coordinate systems) have the necessary properties to constitute a group.*

Detest it as much as lewd intercourse; it can deprive you of all your leisure, your health, your rest, and the whole happiness of your life[7].

Farkas Bólyai   1775-1856

Bibliography:

Adair R K 1988 *A flaw in a universal mirror* [1].
Davies P C W 1977 *Space and time in the modern universe* [87].

---

[6]Evariste Galois (1811–1832), French mathematician who died at the age of 20 from wounds received in a duel.
[7]From a letter to his son János (1802–1860), Hungarian mathematician, warning him to give up his attempts to prove the Euclidean fifth postulate on parallels [353].

Davies P C W 1983 *God and the new physics* [90].
Frauenfelder H et al., 1975 *Nuclear and particle physics* [172].
Gardner M 1964 *The ambidextrous universe* [185].
Gibson W M et al., 1976 *Symmetry principles in ... particle physics* [195].
Hestenes D 1986 *New foundation for classical mechanics* [247].
Morrison P 1957 *The overthrow of parity* [390].
Overseth O E 1969 *Experiments in time reversal* [414].
Poincaré J H 1982 *Science and hypothesis* [445].
Reichenbach H 1958 *Philosophy of space and time* [470].
Sachs R G 1987 *Physics of time reversal* [489].
Sakurai J J 1964 *Invariance principles and elementary particles* [490].
Wick G C 1958 *Invariance principles of nuclear physics* [579].
Wigner E P 1965 *Violations of symmetries in physics* [584].

CHAPTER 4

Tensor fields in Euclidean spacetime

Dear Sir   April 23, 1953

Development of Western Science is based on two great achievements; the invention of the formal logical system (in Euclidean geometry) by the Greek philosophers, and the discovery of the possibility to find out causal relationship by systematic experiment (Renaissance). In my opinion one has not to be astonished that the Chinese sages have not made these steps. The astonishing thing is that these discoveries were made at all[1].

Sincerely yours,   Albert Einstein

In this chapter, the geometrical properties of the Euclidean spacetime that lies at the foundation of Galileo-Newtonian physics and gravity theory will be developed further for use in subsequent chapters. In particular, scalar, vector and tensor fields will be treated in a way which makes their generalization to the corresponding Minkowskian tensors a relatively straightforward process.

4.1 Euclidean scalars and scalar fields

A *Euclidean scalar* or *3-scalar* is a single quantity ϕ, possibly a function of time t, which remains unchanged in value,

$$\phi \to \bar{\phi} = \phi \,, \tag{4.1}$$

under rotations and spatial translations. Examples are the 3-volume element $d^3x = |dx\, dy\, dz|_{dt=0}$, the mass m, electric charge q and energy E of a particle, the Hamiltonian H, Lagrangian L or action S of an isolated mechanical or field system, a temporal interval dt and the scalar product $d\mathbf{x}_1 \cdot d\mathbf{x}_2$ of two space intervals. Time translations and boosts from one frame to another moving uniformly with respect to it are not included in

---

[1]Letter to J E Switzer [460] cited in Mackay [353].

the definition of a Euclidean scalar, nor in the other Euclidean concepts to be introduced in this chapter. The reason is that the majority of the Newtonian physical quantities do not remain invariant with time and of those which do, some like the energy of an isolated system do not remain invariant under boosts. On the contrary, most evolve in ways which depend on dynamical considerations rather than the purely kinematical properties with which we are concerned at this stage. Those special quantities which do remain unchanged in time, however, are of considerable importance in the dynamics of the systems they apply to.

A *Euclidean scalar field* is a function $\phi = \phi(t, \mathbf{x})$ of time and position which, under spatial translations $\mathbf{x} \to \bar{\mathbf{x}} = \mathbf{x} + \mathbf{a}$ or a rotation $\mathbf{x} \to \bar{\mathbf{x}} = R\mathbf{x}$, acting simultaneously on $\mathbf{x}$ and on the functional form of the field, is left unchanged in numerical value,

$$\phi(t, \mathbf{x}) \to \bar{\phi}(t, \bar{\mathbf{x}}) = \phi(t, \mathbf{x}) , \tag{4.2}$$

namely

$$\bar{\phi}(t, \mathbf{x}) = \phi(t, \mathbf{x} - \mathbf{a}) \quad \text{and} \quad \bar{\phi}(t, \mathbf{x}) = \phi(t, R^{-1}\mathbf{x}) . \tag{4.3}$$

The spatial arguments are important in these definitions. Figure 4.1 which may, for example, be interpreted as a cross-section of the equal temperature surfaces near a hot object in some medium, illustrates the way the change in the argument of a scalar field leaves the values unaltered at each location when the basis is altered by a passive rotation. A slight change in the labelling of the diagram to make it refer to only one frame, as shown in Figure 4.2, permits it to be re-interpreted as the active rotation of an object. Examples of 3-scalar fields are those functions of time and position which describe temperature and density distributions $T(t, \mathbf{x})$ and $\rho(t, \mathbf{x})$, probability distributions $|\psi(t, \mathbf{x})|^2$ of spinless particles, the electromagnetic scalar potential $\phi(t, \mathbf{x})$ and the 3-scalar products of vector and tensor fields. The *3-divergence* of a vector and the *Laplacian*,

$$\nabla \cdot \mathbf{V} = \{\partial_k V^k\} \quad \text{and} \quad \nabla^2 = \partial_k \partial^k = \partial_x^2 + \partial_y^2 + \partial_z^2 , \tag{4.4}$$

are a 3-scalar field and a 3-scalar operator, respectively, where $\partial_k \equiv \partial/\partial x^k$.

4.2 Euclidean vectors and vector fields

Suppose a set of three quantities $\mathbf{V} = \{V^k\} \equiv \{V_k\}$ remain unchanged in value under spatial translations and transform under rotations in the same way as the components of the spatial interval $d\mathbf{x}$, namely

$$\mathbf{V} \to \bar{\mathbf{V}} = R\mathbf{V} \quad \text{or} \quad V^k \to V^{\bar{k}} = R^{\bar{k}}{}_k V^k . \tag{4.5}$$

Then $\mathbf{V}$ is said to be a *Euclidean vector* or *3-vector*. Despite the fact that the vector $\mathbf{V}$ is more than just a set of three quantities $\{V^k\}$, its

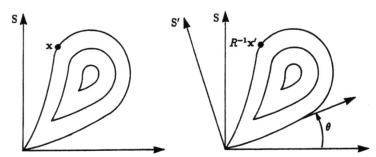

Figure 4.1 Effect of a passive rotation on a scalar field such as a temperature distribution.

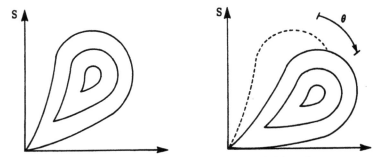

Figure 4.2 Effect of an active rotation on a scalar field such as a temperature distribution.

components in one particular basis e_k, often not specified, may be used to define it. The vector **V** will remain unchanged under rotations due to simultaneous changes in its components V^k and in the basis e_k and may invariantly describe quantities having magnitude and direction. Examples of 3-vectors are the 3-velocity $\mathbf{u} = \dot{\mathbf{z}} = d\mathbf{z}/dt$, 3-acceleration $\mathbf{a} = \dot{\mathbf{u}} = d\mathbf{u}/dt$ and 3-momentum **p** of a particle of mass m and trajectory $\mathbf{z}(t)$. We may define equivalent lowered index components $V_k = \delta_{kl} V^l$ with the same values as V^k which may be retrieved from V_k using $V^k = \delta^{kl} V_l$.

Exercise 4.1 *Explain why the coordinates x^k are not the components of a vector according to the above definition.*

The invariant 3-vector **V** may be expanded in any frame in the forms,

$$\mathbf{V} = V^{\bar{k}} \mathbf{e}_{\bar{k}} = V^k \mathbf{e}_k = \ldots = V_{\bar{k}} \mathbf{e}^{\bar{k}} = V_k \mathbf{e}^k = \ldots . \qquad (4.6)$$

A set of three functions $\mathbf{V}(t, \mathbf{x}) = \{V^k(t, \mathbf{x})\} = \{V_k(t, \mathbf{x})\}$ of time and position, which remain unchanged in value under spatial translations, and transform under a rotation according to,

$$V^k(t, \mathbf{x}) \to V^{\bar{k}}(t, \bar{\mathbf{x}}) = R^{\bar{k}}{}_k V^k(t, \mathbf{x}) , \qquad (4.7)$$

namely,

$$\mathbf{V}(t, \mathbf{x}) \to \bar{\mathbf{V}}(t, \mathbf{x}) = R\mathbf{V}(t, R^{-1}\mathbf{x}) \quad \text{or} \quad V^{\bar{k}}(t, \mathbf{x}) = R^{\bar{k}}{}_k V^k(t, R^{-1}\mathbf{x}) , \qquad (4.8)$$

is said to be a *Euclidean vector field* or *3-vector field*. Note that although we may write $\{V^k\} = \{V_k\}$ to mean the invariant vector $\mathbf{V}$, we shall respect our indexing convention by not equating an individual V^k with V_k despite the fact that the two sets of components have identical values. Examples of 3-vector fields are the electric and magnetic fields, $\mathbf{E}(t, \mathbf{x})$ and $\mathbf{B}(t, \mathbf{x})$, the electromagnetic vector potential $\mathbf{A}(t, \mathbf{x})$ and the velocity field $\mathbf{u}(t, \mathbf{x})$ of a fluid. In what follows we shall not always explicitly distinguish between fields and the corresponding quantities which are not functions of position nor shall we always explicitly include the arguments of a field quantity. Another example of a Euclidean vector field is the *gradient* $\nabla \phi = \{\partial_k \phi\}$ of a scalar field $\phi(t, \mathbf{x})$.

Exercise 4.2 (a) *If ϕ is a scalar field, show that the components $\partial_k \phi$ of its gradient $\nabla \phi$ transform in the same way as the components V_k of a vector field. (b) Show that the 3-divergence $\nabla \cdot \mathbf{V}$ of a vector field $\mathbf{V}$ is a scalar field. (c) Show that the Laplacian ∇^2 acts on any field quantity as a Euclidean scalar operator.*

The trajectory coordinates $\mathbf{z}(t)$ of a particle do not comprise a vector since they are not invariant under spatial translations. We ought strictly to use a difference of coordinates $\mathbf{z} - \mathbf{d}$, where $\mathbf{d}$ are the coordinates of some arbitrary fixed point, whenever we require a vector description of a particle trajectory. We shall not often use such a clumsy notation but the reader should remember to at least make a mental substitution where necessary to avoid inconsistency.

4.3 Euclidean tensors and tensor fields

If $\mathbf{U}$ and $\mathbf{V}$ are each 3-vectors or 3-vector fields, then the set of nine quantities $U^k V^l$ ($k, l = 1, 2, 3$) rotate according to

$$U^k V^l \to U^{\bar{k}} V^{\bar{l}} = R^{\bar{k}}{}_k R^{\bar{l}}{}_l U^k V^l . \qquad (4.9)$$

We may consider the array $\{R^{\bar{k}}{}_k R^{\bar{l}}{}_l\}$ with rows labelled by the set of indices $\bar{k}\bar{l}$ (with range $\bar{1}\bar{1}, \bar{1}\bar{2}, \cdots, \bar{3}\bar{3}$), and columns labelled by kl (with range $11, 12, \cdots, 33$), to be a 9×9 matrix abbreviated $R \otimes R$ and called the *direct product matrix* of R with itself. Correspondingly, $U^k V^l$ are said to be the components of the *direct product* of the vectors **U** and **V**, written $\mathbf{U} \otimes \mathbf{V} = \{U^k \otimes V^l\}$. Consider now the set of nine quantities $\{T^{kl}\}$ labelled by two Euclidean indices kl, each of which behaves under rotations as a 3-vector index, namely

$$T^{\bar{k}\bar{l}}(t,\bar{\mathbf{x}}) = R^{\bar{k}}{}_k R^{\bar{l}}{}_l T^{kl}(t,\mathbf{x}) \quad \text{or} \quad T^{\bar{k}\bar{l}}(t,\mathbf{x}) = R^{\bar{k}}{}_k R^{\bar{l}}{}_l T^{kl}(t, R^{-1}\mathbf{x}) \ . \quad (4.10)$$

If the values remain unchanged under spatial translations, then $\{T^{kl}\}$ are said to be the components of a *Euclidean tensor field* of *rank* 2. We may form three other equivalent forms of the tensor components,

$$T^k{}_l = \delta_{lm} T^{km}, \quad T_k{}^l = \delta_{km} T^{ml}, \quad T_{kl} = \delta_{km} \delta_{ln} T^{mn} \ , \quad (4.11)$$

and hence end up with four equivalent ways of expanding the one invariant or geometric tensor **T** in terms of components and basis vectors,

$$\mathbf{T} = T^{kl} \mathbf{e}_k \otimes \mathbf{e}_l = T^k{}_l \mathbf{e}_k \otimes \mathbf{e}^l = T_k{}^l \mathbf{e}^k \otimes \mathbf{e}_l = T_{kl} \mathbf{e}^k \otimes \mathbf{e}^l \ . \quad (4.12)$$

The transformation laws of lowered or mixed components follow immediately. For example,

$$T_{\bar{k}\bar{l}} = R^k{}_{\bar{k}} R^l{}_{\bar{l}} T_{kl} \ . \quad (4.13)$$

The orthogonality conditions $R^T R = \mathbb{1} = RR^T$ on a rotation matrix may be systematically indexed, for example,

$$\delta_{\bar{k}\bar{l}} = (R\mathbb{1}R^T)_{\bar{k}\bar{l}} = R_{\bar{k}}{}^k \delta_{kl} (R^T)^l{}_{\bar{l}} \ , \quad (4.14)$$

to give two relations

$$\delta_{\bar{k}\bar{l}} = R^k{}_{\bar{k}} \delta_{kl} R^l{}_{\bar{l}} = R_{\bar{k}}{}^k \delta_{kl} R_{\bar{l}}{}^l \ , \quad (4.15)$$

and the equivalence of the two forms may also be seen by noting that $R^T = R^{-1}$ and our convention that $(R^{-1})^k{}_{\bar{k}}$ is abbreviated to $R^k{}_{\bar{k}}$. Thus, for example,

$$\delta_{\bar{k}\bar{l}} = R^k{}_{\bar{k}} R^l{}_{\bar{l}} \delta_{kl} \ , \quad (4.16)$$

showing why the Kronecker delta is called a *numerical* tensor, also referred to as an *isotropic tensor*. Homogeneity of $\mathbf{E}^3$ is embodied in the constancy of δ_{kl} and its isotropy in the equality of δ_{11}, δ_{22} and δ_{33}. Indeed, the homogeneity and isotropy are the origins of the existence of the trivial tensor δ_{kl}.

Tensors and tensor fields $T^{kl\cdots}{}_{m\cdots}(t, \mathbf{x})$ of arbitrary rank $r \geq 0$, are defined using one rotation matrix for each of the r indices. Scalars and vectors are considered to be tensors of rank 0 and rank 1 and the term tensor will therefore sometimes include scalars and vectors. Any linear combination, with constant coefficients, of the components of a set of tensors of the same rank (choosing of course the components with the same index levels) is a component of another tensor of the same rank.

Exercise 4.3 *Let δ_{ijkl} be defined to have the value 1 if $i = j = k = l$ and 0 otherwise. Examine a specific component under a rotation with a specific angle to show that δ_{ijkl} is not a numerical tensor.*

Exercise 4.4 *(a) Show that the components $\partial_j T^{kl\cdots}{}_{m\cdots}(t, \mathbf{x})$ of the gradient $\nabla \mathbf{T}$ of a tensor $\mathbf{T}$ of rank r transform as the components of a tensor of rank $r + 1$. (b) What are the tensorial properties of the time derivative, $\dot{\mathbf{T}} = \partial_t \mathbf{T}$, of a tensor $\mathbf{T}$ of rank r?*

Let $\mathbf{S}$ and $\mathbf{T}$ be tensors of rank s and t respectively. Then we refer to the set of quantities,

$$\mathbf{S} \otimes \mathbf{T} = \{ S^{kl\cdots}{}_{m\cdots} T^{p\cdots}{}_{qr\cdots} \}, \qquad (4.17)$$

which transform as a tensor of rank $(s + t)$, as the *tensor product* of $\mathbf{S}$ and $\mathbf{T}$.

Suppose we have two spatially-indexed quantities $a = \{a^{\cdots k \cdots}\}$ and $b = \{b^{\cdots l \cdots}\}$. Then the process of letting two such indices be identical, with one up and one down, and thereby forming the implied sum, or sum of products,

$$a^{\cdots k \cdots}{}_k = \delta^k{}_l a^{\cdots l \cdots}{}_k \quad \text{or} \quad a^{\cdots k \cdots} b^{\cdots}{}_k = \delta^k{}_l a^{\cdots l \cdots} b^{\cdots}{}_k \qquad (4.18)$$

is referred to as the *contraction* of one index on the other. Each contraction reduces a tensor of rank $r \geq 2$ to a tensor of rank $r - 2$.

4.4 Symmetric and antisymmetric tensors

A second-rank indexed quantity S^{kl}, tensor or otherwise, is *symmetric* if $S^{kl} = S^{lk}$. Similarly, A^{kl} is *antisymmetric* if $A^{kl} = -A^{lk}$.

Exercise 4.5 *Show that if S^{kl} is symmetric and A^{kl} is antisymmetric, then*

$$S_{kl} A^{kl} \equiv 0. \qquad (4.19)$$

The *symmetric part* $T^{(kl)}$ and *antisymmetric part* $T^{[kl]}$ of a second-rank indexed quantity T^{kl} are defined by,

$$T^{(kl)} = \tfrac{1}{2}(T^{kl} + T^{lk}) \quad \text{and} \quad T^{[kl]} = \tfrac{1}{2}(T^{kl} - T^{lk}), \qquad (4.20)$$

while the *trace* of a second-rank tensor is given by the scalar resulting from the contraction $\delta_{kl} T^{kl} = T^k{}_k$ of its two indices either with the Kronecker delta or with one another. It follows from Equation 4.20 that,

$$T^{kl} = T^{(kl)} + T^{[kl]}. \qquad (4.21)$$

Consequently, $T^{kl} = T^{(kl)}$ and $T^{[kl]} = 0$ are each conditions for T^{kl} to be symmetric while $T^{kl} = T^{[kl]}$ or $T^{(kl)} = 0$ imply that T^{kl} is antisymmetric. The numerical tensor $\frac{1}{3}T^m{}_m\delta^{kl}$ is called the *trace part* of the tensor T^{kl}.

Exercise 4.6 *Show that if A^{kl} is antisymmetric, then $A^{kl}T_{kl} = A^{kl}T_{[kl]}$ while if S^{kl} is symmetric then $S^{kl}T_{kl} = S^{kl}T_{(kl)}$.*

Exercise 4.7 *Show that the symmetric, antisymmetric and trace parts of the tensor transform $T^{\bar{k}\bar{l}}$ of T^{kl} are the tensor transforms of the symmetric, antisymmetric and trace parts of the original tensor T^{kl}.*

Exercise 4.8 *Construct a traceless tensor from T^{kl} and the universal tensor δ_{kl} and show that it is an invariant part of T^{kl}.*

The symmetric, antisymmetric, trace, and traceless parts may be constructed on any pair of indices of a tensor of any rank 2 or higher. The fact that each of these 4 quantities is an *invariant part* of the original tensor is one of the reasons why they are so important.

Exercise 4.9 *Show that the operations of taking the the trace, the traceless symmetric part and the antisymmetric part of a second rank tensor, thereby dividing it into three invariant parts,*

$$T_{kl} = \tfrac{1}{3}T^m{}_m\delta_{kl} + T_{[kl]} + (T_{(kl)} - \tfrac{1}{3}T^m{}_m\delta_{kl}) , \qquad (4.22)$$

form a set of projection operations.

Because the three parts of T_{kl} in Equation 4.22 have no non-trivial invariant subparts, each part is said to be *irreducible*. They are also referred to as the spin 0, spin 1 and spin 2 parts of T_{kl}, having 1, 3 and 5 independent components, respectively, namely $2s + 1$, the *spin multiplicity* for spin s.

The *completely symmetric part* $T_{(klm)}$ and the *completely antisymmetric part* $T_{[klm]}$ of the rank 3 indexed quantity T_{klm} are defined by:

$$T_{(klm)} = \frac{1}{3!}(T_{klm} + T_{lmk} + T_{mkl} + T_{lkm} + T_{kml} + T_{mlk}) , \qquad (4.23)$$

and,

$$T_{[klm]} = \frac{1}{3!}(T_{klm} + T_{lmk} + T_{mkl} - T_{lkm} - T_{kml} - T_{mlk}) . \qquad (4.24)$$

These second- and third-rank concepts extend in an obvious way to give the completely symmetric and completely antisymmetric parts on any p indices of an object of rank r ($\geq p$) in N dimensions by normalizing with a factor of $1/p!$ the sum of all $p!$ permutations or the sum of all $p!/2$ even permutations minus the sum of all $p!/2$ odd permutations. The *completeness* of the symmetry of these objects refers to the fact that they are symmetric or antisymmetric on every pair of the p indices involved.

Exercise 4.10 *Use the rules given in Appendix A.2 to determine the number of independent components in $T_{(klm)}$ and $T_{[klm]}$ and hence show that these two parts alone cannot suffice for a complete decomposition of T_{klm} into two terms as was possible for T^{kl} in Equation 4.21. Does this result depend in any way on the dimensions N of the space (or spacetime) involved, or equivalently, on the range of the indices?*

Although we have chosen to use a coordinate system to define tensors, it is not necessary to do so. Coordinate-independent methods with tensors [384, 499] provide very powerful tools in the analysis of many physical systems.

4.5 Proper and improper tensors

If we leave a 3-vector such as $d\mathbf{x}$ unchanged in space and reverse the 3 spatial axes, the components of the 3-vector in the new frame will be the negative of those in the original frame. Any such vector, behaving in the same way as $d\mathbf{x}$ itself when the spatial axes are reversed is called a *proper* or *polar* 3-vector. Many other quantities having a magnitude and an axis or direction can be represented by 3-component vectors behaving like $d\mathbf{x}$ under translations and rotations. However, not all of these behave in the same way as $d\mathbf{x}$ under reversal of all three spatial axes.

Consider, for example, the description of a rotation itself. A physical rotation may be defined by specifying a directed axis $\hat{\boldsymbol{\theta}}$ and an angle θ of rotation about that axis according to the right-hand rule as shown in Figure 4.3. These quantities may be combined to describe the rotation by using a vector $\boldsymbol{\theta} = \theta\hat{\boldsymbol{\theta}}$ and $\boldsymbol{\theta}$ will behave in the same way as the vector $d\mathbf{x}$ if the basis is rotated.

Exercise 4.11 *Give a simple counter example to show that the rotation described by the vector $\boldsymbol{\theta} = \boldsymbol{\theta}_1 + \boldsymbol{\theta}_2$ is not in general identical to the result of combining the two rotations described by $\boldsymbol{\theta}_1$ and $\boldsymbol{\theta}_2$.*

Exercise 4.12 *Rotations were defined mathematically in Chapter 3 in terms of certain 3×3 matrices R. To characterize a physical rotation by a 3-vector $\boldsymbol{\theta}$ and expect the two concepts to match implies that there is some way to extract from R a vector $\boldsymbol{\theta}$ giving the axis $\hat{\boldsymbol{\theta}}$ and angle θ of rotation. Devise a means of determining $\boldsymbol{\theta}$ from a given matrix R. (For small angles $\Delta\boldsymbol{\theta}$, the answer is contained in Exercise 4.18. Not all the material has been explicity set out here for answering this exercise for finite angles. No subsequent material depends on it.)*

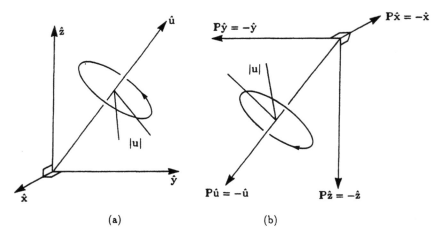

Figure 4.3 Describing a rotation by a pseudovector, (a) before reflection and (b) after reflection.

Reversal of the axes will change the right-hand rule to a left-hand one in the characterization of a physical rotation. For the vector θ to continue to describe the same rotation, it must be reversed by the parity operation. Its components will therefore, unlike those of the vector $d\mathbf{x}$, remain unaltered by a parity reversal. Such a vector is called an *axial vector* or *pseudovector*.

Let us suppose that under translations and rotations $\phi(t,\mathbf{x})$ and $\tilde{\phi}(t,\mathbf{x})$ are scalar fields, $V^k(t,\mathbf{x})$ and $\tilde{V}^k(t,\mathbf{x})$ are 4-vector fields while $T^{kl\cdots}{}_{m\cdots}(t,\mathbf{x})$ and $\tilde{T}^{kl\cdots}{}_{m\cdots}(t,\mathbf{x})$ are tensor fields. Suppose also that $\phi(t,\mathbf{x})$, $V^k(t,\mathbf{x})$ and $T^{kl\cdots}{}_{m\cdots}(t,\mathbf{x})$ transform under parity $\mathbf{P}$ according to,

$$\bar{\phi}(t,\mathbf{x}) = \mathbf{P}\,\phi(t,\mathbf{x}) = \phi(t, P^{-1}\mathbf{x})$$
$$V^{\bar{k}}(t,\mathbf{x}) = \mathbf{P}\, V^k(t,\mathbf{x}) = -V^k(t, P^{-1}\mathbf{x})$$
$$T^{\bar{k}\bar{l}\cdots}{}_{\bar{m}\cdots}(t,\mathbf{x}) = \mathbf{P}\, T^{kl\cdots}{}_{m\cdots}(t,\mathbf{x}) = (-1)^{\text{rank}}\, T^{kl\cdots}{}_{m\cdots}(t, P^{-1}\mathbf{x}) \quad (4.25)$$

so that there is an overall sign change for each index. Then $\phi(t,\mathbf{x})$, $V^k(t,\mathbf{x})$ and $T^{kl\cdots}{}_{m\cdots}(t,\mathbf{x})$ are said to be a *proper 3-scalar*, a *proper 3-vector* (or *polar 3-vector*) and a *proper 3-tensor*, respectively. (Although $P^{-1}=P$ we retain the superscript in the argument, $P^{-1}\mathbf{x}$, to ease the comparison with other transformations which are not self-inverse.) Without the spacetime arguments, these expressions will provide definitions of non-field proper 3-scalars, 3-vectors and 3-tensors.

Exercise 4.13 *Show that we may collect together the rotation and reflection properties of all these proper quantities in one formula,*

$$T^{\bar{k}\bar{l}\cdots}{}_{\bar{m}\cdots}(t,\mathbf{x}) = O^{\bar{k}}{}_k O^{\bar{l}}{}_l \cdots O^m{}_{\bar{m}} \cdots T^{kl\cdots}{}_{m\cdots}(t, O^{-1}\mathbf{x})\,, \quad (4.26)$$

4.5. Proper and improper tensors

using transformation by an orthogonal matrix O.

Examples of quantities which are designated as proper 3-scalars are the 3-volume element $d^3x = |dx\,dy\,dz|$, the mass m and electric charge q of a particle, and the Lagrangian L, Hamiltonian H or action S of a parity-covariant mechanical system. Examples of proper 3-scalar fields are temperature and density distributions, $T(t, \mathbf{x})$ and $\rho(t, \mathbf{x})$, and the Lagrangian and Hamiltonian densities of a parity-covariant field system. Examples of proper 3-vectors are the displacement $d\mathbf{x}$, the gradient operator ∇ and the 3-momentum $\mathbf{p}$ of a system, mechanical or otherwise. The electromagnetic 3-vector potential $\mathbf{A}(t, \mathbf{x})$ is an example of a proper 3-vector field. In the Hamilton gauge (see Chapter 18), the electromagnetic field is described entirely in terms of a proper 3-vector implying that an electromagnetic wave and its quantum, the photon, are of odd parity. The product $dx^k dx^l$ is an example of a proper tensor. A non-trivial example of a proper tensor field is the *stress tensor* $\mathbf{T} = \{T^{kl}(t, \mathbf{x})\}$ of a fluid. This is the tensor whose kl component is the k-th component of the *stress*, or force per unit area, on the area normal to the l-direction, the diagonal components being the pressure in each direction. The Kronecker delta δ_{kl} is a trivial example of a proper 3-tensor since it is defined to be invariant and is a tensor of rank two. Consequently, the 3D line element ds^2 is a proper scalar.

Suppose that $\tilde{\phi}(t,\mathbf{x})$, $\tilde{V}^k(t,\mathbf{x})$ and $\tilde{T}^{kl\cdots}{}_{m\cdots}(t,\mathbf{x})$, although transforming under rotations in the same way as $\phi(t,\mathbf{x})$, $V^k(t,\mathbf{x})$ and $T^{kl\cdots}{}_{m\cdots}(t,\mathbf{x})$, transform under parity according to,

$$\begin{align}
\mathbf{P}\,\tilde{\phi}(t,\mathbf{x}) &= -\tilde{\phi}(t, P^{-1}\mathbf{x}) \\
\mathbf{P}\,\tilde{V}^k(t,\mathbf{x}) &= \tilde{V}^k(t, P^{-1}\mathbf{x}) \\
\mathbf{P}\,\tilde{T}^{kl\cdots}{}_{m\cdots}(t,\mathbf{x}) &= -(-1)^{\text{rank}}\,\tilde{T}^{kl\cdots}{}_{m\cdots}(t, P^{-1}\mathbf{x})\,.
\end{align} \qquad (4.27)$$

Then $\tilde{\phi}(t,\mathbf{x})$, $\tilde{V}^k(t,\mathbf{x})$ and $\tilde{T}^{kl\cdots}{}_{m\cdots}(t,\mathbf{x})$ are said be a *pseudo-3-scalar*, *pseudo-3-vector* (or *axial 3-vector*) and a *pseudo-3-tensor* respectively. The term *improper* is also used to refer to such quantities. These last three transformation laws differ from those for proper quantities only in the presence of an extra overall sign under parity P or improper orthogonal transformations, $O = PR$. Proper scalars and pseudovectors have even parity. Pseudoscalars and polar vectors have odd parity, examples being the spin 0 and 1 fields of pions and photons. All four are non-chiral quantities.

Exercise 4.14 *Show that we may combine the transformation laws of these pseudo quantities under rotations and parity reversal, as*

$$\tilde{T}^{\bar{k}\bar{l}\cdots}{}_{\bar{m}\cdots}(t,\mathbf{x}) = (\det O)\, O^{\bar{k}}{}_k O^{\bar{l}}{}_l \cdots O^m{}_{\bar{m}} \cdots \tilde{T}^{kl\cdots}{}_{m\cdots}(t, O^{-1}\mathbf{x})\,. \qquad (4.28)$$

where O is an orthogonal matrix.

The presence of the determinant factor in Equation 4.28 but not in 4.26, gives the sign differences (under parity reversal) that distinguish proper and improper quantities from one another. The pseudo-3-scalar is defined so that it transforms in the same way as the product of a proper and a pseudo vector. (We shall not always include a tilde (˜) on improper tensors.)

We shall see later that one example of a pseudo-3-scalar field is the electromagnetic scalar product $\mathbf{E}\cdot\mathbf{B}$. If magnetic charge were discovered, in the form of magnetic monopoles for example, then it would be a pseudo-3-scalar. The large-scale structure of the strong-nuclear interaction is described by a pseudoscalar field the quantization of which leads to pions which therefore have odd parity. An example of a pseudo-3-vector is the orbital angular momentum $\mathbf{l} = \mathbf{z}\times\mathbf{p}$, about the origin, of a particle with trajectory $\mathbf{z}(t)$ and momentum $\mathbf{p}$. (The translation invariance of $\mathbf{l}$ may be made evident by expanding the definition to $\mathbf{l} = (\mathbf{z}-\mathbf{d})\times\mathbf{p}$, the angular momentum about $\mathbf{d}$.) Since the 3-vector *spin* $\mathbf{s}$ of a system of particles is the orbital angular momentum of the particles about their centre of mass, it will also be a pseudo-3-vector. By analogy, for a fundamental massive particle, we assume the same property for the *intrinsic spin* angular momentum, namely any angular momentum it has in the frame in which it is at rest. Although such angular momentum is regarded as non-orbital, in a classical context it would be presumed to originate in the composite nature of the particle at sizes smaller than all other characteristic lengths associated with its dynamics. Quantum particles may, however, have intrinsic spin even in the absence of any known internal structure.

The tensor product and the scalar product of two proper tensors or two pseudotensors (each of arbitrary rank) are each proper tensors while the product of a proper and a pseudotensor is another pseudotensor. The electric intensity $\mathbf{E}$ and the magnetic induction $\mathbf{B}$ are examples of a polar 3-vector and a pseudo-3-vector field, respectively. (An explanation for this difference will be given in Chapter 18.) This means that $\mathbf{E}^2$ and $\mathbf{B}^2$ are both proper 3-scalars while $\mathbf{E}\cdot\mathbf{B}$ is a pseudo-3-scalar. However, it should be noted that in 3D the presence of a cross product, which has no direct analogue as a vector in higher or lower dimensions, effectively introduces an extra pseudo factor as we shall see later in this chapter. Thus, in the equation for the orbital angular momentum, $\mathbf{l} = (\mathbf{z}-\mathbf{d})\times\mathbf{p}$, about $\mathbf{d}$ in which the vector $\mathbf{z}-\mathbf{d}$ and the 3-momentum $\mathbf{p}$ are both polar, the angular momentum is nevertheless axial. Similarly, in the expression of the Lorentz law of charged particle motion (see the second part of Equation 19.2), the axial properties of $\mathbf{B}$ and of the cross-product ($\times$) cancel to make the overall term polar as for the other two terms in the equation.

A theory all of whose dynamical equations contain terms of the same rank and index level, which are either all proper tensors, or all pseudotensors, will retain its form not only under continuous Euclidean transformations (translations and rotations) but also under parity transformations. It will therefore be a parity-covariant or parity-conserving theory. Examples are the theories of the electromagnetic, gravitational and strong-nuclear interactions. The presence of terms of opposite type (proper and pseudo) added to one another in the same dynamical equation in the theory of the weak interaction indicates that it is not parity-covariant.

4.6 Permutation pseudotensor

The fact that the Kronecker delta is a numerical Euclidean tensor is a direct consequence of the existence of the orthogonality condition on a rotation matrix. That condition in its unindexed form, $R^T R = 1 = R R^T$, is clearly frame-independent which leads to the invariance of δ_{kl}. Since the rotation matrices satisfy precisely one other such condition, namely $\det R = 1$, we can expect at most one other numerical Euclidean tensor.

Consider a 2-dimensional matrix $M = \{M^A{}_B\}$ where A and B each have range $\{1,2\}$. The determinant of M, namely $M^1{}_1 M^2{}_2 - M^1{}_2 M^2{}_1$, can be written using the summation convention on A and B as,

$$\det M = \epsilon^{AB} M^1{}_A M^2{}_B \,, \tag{4.29}$$

where the 2D Levi-Cività *permutation symbol*, ϵ^{AB}, is antisymmetric in A and B with values,

$$\begin{aligned} \epsilon^{AB} &= +1 \quad \text{for } AB = 12 \\ &= -1 \quad \text{for } AB = 21 \\ &= 0 \quad \text{for } AB = 11 \text{ or } 22 \,. \end{aligned} \tag{4.30}$$

Thus $\epsilon^{12} \det M = \epsilon^{AB} M^1{}_A M^2{}_B$ and since this is true for any pair of indices in place of $\{12\}$, it follows that (with C and D each also having range 1,2),

$$(\det M) \epsilon^{CD} = \epsilon^{AB} M^C{}_A M^D{}_B \,. \tag{4.31}$$

In the same way, a 3×3 matrix $M^k{}_l$ with Euclidean indices $k, l = 1, 2, 3$ has determinant given by,

$$\det M = \epsilon^{klm} M^1{}_k M^2{}_l M^3{}_m \,, \tag{4.32}$$

where the *permutation symbol* in 3D is defined, numerically and invariantly (with respect to all basis changes whether rotations, translations or reflections) by,

$$\begin{aligned} \epsilon^{klm} &= +1 \quad \text{for } klm \text{ an even permutation of 123} \\ &= -1 \quad \text{for } klm \text{ an odd permutation of 123} \\ &= 0 \quad \text{for } klm \text{ any other combination.} \end{aligned} \tag{4.33}$$

One transposition of an adjacent pair of indices in a permutation symbol changes the sign (in any dimension, here 3), namely $\epsilon^{klm} = -\epsilon^{lkm}$ and $\epsilon^{klm} = -\epsilon^{kml}$. Cyclic permutation of the indices is equivalent in 3D to two such transpositions and therefore $\epsilon^{klm} = \epsilon^{lmk}$. In view of Equation 4.32, the presence of $\det O$ in Equation 4.28 is thus an immediate indication that the permutation symbol may be a pseudo quantity relevant to parity reversal. Since by definition $\epsilon^{\bar{1}\bar{2}\bar{3}} = \epsilon^{123} = 1$, application of Equation 4.32 to an orthogonal matrix, for which $(\det O)^2 = 1$, leads to

$$\epsilon^{\bar{1}\bar{2}\bar{3}} = (\det O)\, O^{\bar{1}}{}_k O^{\bar{2}}{}_l O^{\bar{3}}{}_m\, \epsilon^{klm} . \qquad (4.34)$$

Equation 4.34 applies for all components in the new (barred) frame giving,

$$\epsilon^{\bar{k}\bar{l}\bar{m}} = (\det O)\, O^{\bar{k}}{}_k O^{\bar{l}}{}_l O^{\bar{m}}{}_m \epsilon^{klm} . \qquad (4.35)$$

Comparison with Equation 4.28 shows that ϵ^{klm} is an improper numerical tensor, called therefore the 3D *permutation pseudotensor*. Its existence and properties are clearly a direct consequence of the orientability of each of the three dimensions of space in the same way that the metric tensor δ_{kl} is a result of its homogeneity and isotropy.

From the metric and permutation tensors one may consider building up tensor products such as $\delta^{kl}\delta^{pq}$ or $\delta^{pq}\epsilon^{klm}$ ($\pm$ other permutations of the indices) as potential numerical tensors. Some of the most important of these are the 3D *generalized Kronecker deltas* of rank 2 and 3 defined by:

$$\delta^{kl}_{pq} = \begin{vmatrix} \delta^k_p & \delta^l_p \\ \delta^k_q & \delta^l_q \end{vmatrix} \equiv \delta^k_p \delta^l_q - \delta^k_q \delta^l_p , \qquad (4.36)$$

and,

$$\begin{aligned}\delta^{klm}_{pqr} &= \begin{vmatrix} \delta^k_p & \delta^l_p & \delta^m_p \\ \delta^k_q & \delta^l_q & \delta^m_q \\ \delta^k_r & \delta^l_r & \delta^m_r \end{vmatrix} \\ &\equiv \delta^k_p \delta^l_q \delta^m_r + \delta^k_q \delta^l_r \delta^m_p + \delta^k_r \delta^l_p \delta^m_q - \delta^k_q \delta^l_p \delta^m_r - \delta^k_p \delta^l_r \delta^m_q - \delta^k_r \delta^l_q \delta^m_p ,\end{aligned} \qquad (4.37)$$

in terms of which the permutation tensors satisfy the following identities:

$$\epsilon^{klm} \epsilon_{pqr} = \delta^{klm}_{pqr} \qquad (4.38)$$

$$\epsilon^{klm} \epsilon_{pqm} = \delta^{kl}_{pq} \qquad (4.39)$$

$$\epsilon^{klm} \epsilon_{plm} = 2\delta^k_p \qquad (4.40)$$

$$\epsilon^{klm} \epsilon_{klm} = 3! . \qquad (4.41)$$

These primarily manipulative identities are often valuable in simplifying certain expressions by effectively transferring a permutation tensor factor from one side of an equation to another. They can therefore be considered as a means of 'inverting' the permutation tensor.

The last three of the above identities all follow from the first by contraction. To prove the first, we note that Equation 4.37 shows that the right side of Equation 4.38 is completely antisymmetric in klm and completely antisymmetric in pqr. From Equation 4.37, the right side can take on only the values ± 1 or 0. The left side of the identity also has the same properties from the definition of the permutation tensor. It suffices therefore to check that the equality is true for one particular combination of index values, such as $klm = 123 = pqr$, and this is verified immediately.

Exercise 4.15 *Show that the permutation pseudotensor satisfies the following identity,*

$$\epsilon^{klm}\epsilon^{pmr} + \epsilon^{lpm}\epsilon^{kmr} + \epsilon^{pkm}\epsilon^{lmr} = 0 , \qquad (4.42)$$

in which the first of the two m indices in each term could be lowered if one wished to strictly obey the one up one down convention for summed indices.

The above identity plays an important rôle in studies of the rotation and restricted Euclidean groups using the theory of a *Lie algebra* [30]. Stedman [520] contains diagram techniques for generating and proving identities involving Kronecker and permutation tensors.

4.7 Three-dimensional tensor duality

The existence of the permutation tensor means that any completely antisymmetric tensor (of rank less than or equal to the dimension, which here is 3) will have an alternative equivalent form called its *tensor dual*. In this context, a scalar and a vector are considered to be degenerate examples of antisymmetric tensors. This dual is obtained, up to a constant factor decided by convention, by contracting all its indices with as many as are necessary of those on the permutation tensor.

Exercise 4.16 *Let $F^{kl} = -F^{lk}$ be an antisymmetric 3-tensor. Use permutation tensor identities and the form of the generalized Kronecker deltas to show that each of the following two relations,*

$$F^k = \tfrac{1}{2}\epsilon^{klm}F_{lm} \quad \Leftrightarrow \quad F^{kl} = \epsilon^{klm}F_m , \qquad (4.43)$$

can be determined from the other. Alternatively, show they are equivalent by writing out both equations explicitly in terms of individual components.

F^{kl} and the 3-vector $\mathbf{F} = \{F^k\}$ are said to be *dual* to one another in $\mathbf{E}^3$. They are alternative equivalent ways of describing the same physical information. If one is a proper quantity the other will be improper. A factor of $\frac{1}{2}$ is included in double sums over the product of two antisymmetric tensors, as in the first of the above equations, to allow for the fact that the actual number of non-zero components in F^{kl} is double the number which are independent. One example of dual quantities, from the Newtonian theory of fluids, is the vorticity pseudo-3-vector and the corresponding antisymmetric vorticity 3-tensor.

Exercise 4.17 *If s^{kl} and ω^{kl} are the antisymmetric duals of the 3-vectors $\mathbf{s}$ and $\boldsymbol{\theta}$ (which may characterize spin and rotation, for example), show that,*

$$\tfrac{1}{2}\omega_{kl}s^{kl} = \boldsymbol{\theta}\cdot\mathbf{s} , \qquad (4.44)$$

from which it also follows that,

$$\tfrac{1}{2}\omega_{kl}\omega^{kl} = \theta^2 \quad \text{and} \quad \tfrac{1}{2}s_{kl}s^{kl} = \mathbf{s}^2 . \qquad (4.45)$$

The components of the *cross product* $\mathbf{W} = \mathbf{U}\times\mathbf{V}$ of two 3-vectors $\mathbf{U}$ and $\mathbf{V}$ are defined by,

$$W^k = (\mathbf{U}\times\mathbf{V})^k = \epsilon^{klm}U_lV_m , \qquad (4.46)$$

which shows, for example, that if $\mathbf{U}$ and $\mathbf{V}$ are both polar or both axial vectors then $\mathbf{W}$ is an axial vector. If one is polar and the other is axial, then the cross product is polar. The components of the 3D *curl* of a 3-vector $\mathbf{V}$, written $\boldsymbol{\nabla}\times\mathbf{V}$, curl $\mathbf{V}$ or rot $\mathbf{V}$ are defined as,

$$(\boldsymbol{\nabla}\times\mathbf{V})^k = \epsilon^{klm}\partial_l V_m . \qquad (4.47)$$

The cross product and the curl, as vector operations, are peculiar to 3D Euclidean space $\mathbf{E}^3$. Each is dual to a second-rank antisymmetric tensor, namely $2(U_lV_m - U_mV_l)$ and $2(\partial_lV_m - \partial_mV_l)$, respectively. The tensor alternatives always exist in a Euclidean space $\mathbf{E}^n$ of arbitrary dimension $n \geq 2$ (or $\mathbf{E}^{1,n-1}$ if the metric is not positive-definite but Lorentzian) and have $\tfrac{1}{2}n(n-1)$ independent components. Their duals can therefore be equivalent to n-dimensional vectors only if $n = 3$. This is one of the main reasons why non-indexed vector analysis is not useful in special relativity (where $n = 4$) to the same extent as in non-relativistic work (where $n = 3$).

Exercise 4.18 *Let $\delta\boldsymbol{\theta}$ be the pseudo-3-vector dual of $\delta\omega$ where $\delta R = \mathbf{1} - \delta\omega$ is an infinitesimal rotation (see Exercise 3.4). Show that the effect of the small rotation on a vector $\mathbf{V}$ is given by $\delta\mathbf{V} = (\delta R)\mathbf{V} = \mathbf{V} + (\delta\boldsymbol{\theta}\times\mathbf{V})$ and draw a diagram to illustrate this result.*

Euclidean spacetime $\mathbf{E}^1 \times \mathbf{E}^3$ has no other geometrical properties comparable to the homogeneity, isotropy and orientability which give rise to δ_{kl} and ϵ^{klm}. Consequently, the only universal tensors available for use in the construction of scalar, vector and tensor equations in 3D are δ_{kl} and ϵ^{klm} and their combinations. We shall constantly make use of this fact.

Exercise 4.19 *Imagine that you have established communication with intelligent beings in another part of the Galaxy and supplied details to them on how to construct a basis triad. Your correspondents report that the triads duly fabricated are of just two types which they have called + and −. Can you devise a means of determining, by the exchange of coded signals, which of the sets + or − corresponds to our left-handed or right-handed? Neither correspondent may travel to the other nor send photographs nor look at the same chiral object as the other by using a telescope to examine star groupings and distances, for example.*

The above problem was implicit [185] in the studies of Kant[2] on handedness and appears in many philosophical and scientific analyses since.

4.8 Manifest Euclidean covariance

The two fundamental properties of tensor and pseudotensor transformation laws, evident from Equations 4.26 and 4.28, is that they are linear and homogeneous. There are no terms quadratic or higher in the tensor components and no terms in which the tensors are absent. Each component of a Euclidean tensor (proper or improper) of arbitrary rank in a new frame is a homogeneous, linear combination, $\pm O^{\bar{k}}{}_k O^{\bar{l}}{}_l \cdots O^m{}_{\bar{m}} \cdots T^{kl\cdots}{}_{m\cdots}$, of the components $T^{kl\cdots}{}_{m\cdots}$ of the tensor in the original frame. This means that if all the components are zero in one frame then all components will be zero in any frame. Herein lies the great power of vector and tensor analysis. To ensure that equations satisfy the appropriate Principle of relativity or covariance (see Chapter 5), whether it be rotational or orthogonal (or Lorentz), it is necessary and sufficient to write these equations so that every term in them is a tensor of the same rank. (If a distinction in level is useful, we require the levels to match also.) For example, in the 3D orthogonal case, the equation might be of the form,

$$A^{kl\cdots}{}_{m\cdots} + B^{kl\cdots}{}_{m\cdots} + \cdots = C^{kl\cdots}{}_{m\cdots}, \tag{4.48}$$

in the first frame. Such an equation can be written as $T^{kl\cdots}{}_{m\cdots} = 0$ where (suppressing indices) $T = A + B + \cdots - C$ is also a tensor. Thus, after an orthogonal transformation, the equation becomes $T^{\bar{k}\bar{l}\cdots}{}_{\bar{m}\cdots} = 0$, where

---

[2]Immanuel Kant (1724–1804), German philosopher, especially of science.

$\bar{T} = \bar{A} + \bar{B} + \cdots - \bar{C}$, and this equation is just,

$$A^{k\bar{l}\cdots}{}_{\bar{m}\cdots} + B^{k\bar{l}\cdots}{}_{\bar{m}\cdots} + \cdots = C^{k\bar{l}\cdots}{}_{\bar{m}\cdots}, \tag{4.49}$$

which has the same form as the original equation. A 3-tensor expression guarantees the *form invariance* of physical equations, under rotational or orthogonal transformations, as required by the Principle of relativity. From the active view-point, a solution of the equation is mapped to another solution by transformations which leave the equation form-invariant. The use of algebraic, as opposed to differential, 3-scalar, 3-vector and 3-tensor equations can be referred to as *rotational covariance*. Sometimes the term *manifest covariance* is used. Tensors are useful not just in 3D Euclidean space but whenever transformations are linear and homogeneous and therefore able to be described in terms of matrices.

The continuous Euclidean transformations of Equations 3.25 to 3.27, with translations included, are not homogeneous in the coordinates. They are not expressible solely as a 3×3 matrix transformation on **x** (although they can be written in terms of 4×4 matrix equations [30, 516]). The coordinate transformation law with displacements included does not lead to 3-tensor laws.

The Principle of relativity demands we also include translations in the transformations leaving the equations of a system form-invariant. When applying the time translation part of the Relativity principle to time-dependent functions, such as the trajectory $z(t)$ or momentum $\mathbf{p}(t)$ of a particle, we may consider the values $\mathbf{p}(t)$ and $\mathbf{p}(t + \Delta t)$ of the variable at nearby events $\mathbf{z}(t)$ and $\mathbf{z}(t + \Delta t)$ along the trajectory. In the limit this involves us with time differentials dt and derivatives $d\mathbf{p}/dt$ with respect to time of the dynamical variables. Whereas rotation-covariant and parity-covariant equations are essentially algebraic, *translational* and therefore *Euclidean covariance* clearly introduces ordinary differential equations with time as the independent variable, the *equations of motion* of the system.

To include translations in both space and time when dealing with a tensor field, $\psi(t, \mathbf{x})$ of any rank (with indices suppressed), we consider the values $\psi(t, \mathbf{x})$ and $\psi(t + \Delta t, \mathbf{x} + \Delta \mathbf{x})$ of the field at different nearby events in spacetime. In the limit, this involves us with coordinate differentials dt and $d\mathbf{x}$ and hence also with derivatives ∂_t, ∂_x, ∂_y and ∂_z, of the field $\psi(t, \mathbf{x})$ with respect to time and space. Euclidean covariance on fields thus clearly introduces partial differential equations, the 'equations of motion' of the field or *field equations*. Partial differential equations were first used to solve physical problems by d'Alembert[3] in 1747.

The differences and derivatives that appear in the equations of motion or field equations transform tensorially and the field equations are thus partial differential equations having Euclidean covariance. In order to

---

[3] Jean le Rond d'Alembert (1717–1783), French philosopher and mathematician.

4.8. Manifest Euclidean covariance

maintain rotational covariance the spatial derivatives ∂_x, ∂_y and ∂_z in the field equation must appear together in their Euclidean covariant combination $\nabla = \{\partial_k\} = \{\partial^k\}$.

Exercise 4.20 *Show that the only Euclidean covariant spatial derivative operations of first or second order acting on a scalar ϕ or vector $\mathbf{V}$ are: $\nabla\phi$, $\nabla^2\phi$, $\nabla\cdot\mathbf{V}$, $\nabla^2\mathbf{V}$, $\nabla\times\mathbf{V}$ and $\nabla\times(\nabla\times\mathbf{V})$.*

Vector notation in its modern form, including the gradient ∇, the divergence $\nabla\cdot\mathbf{V}$ or div $\mathbf{V}$ and the curl of a vector, $\nabla\times\mathbf{V}$, were introduced by Heaviside[4] in a paper entitled *Electromagnetic induction and its propagation* in which Maxwell's equations were re-expressed in their now familiar 3-vector form. The paper is comprised of a series of 47 sections which appeared one by one from 1885 to 1887 in *The Electrician*, reprinted in reference [229].

Bibliography:

Borisenko A I et al., 1979 *Vector and tensor analysis with applications* [43].
Chorlton F 1976 *Vector and tensor methods* [65].
Lovelock D et al. 1975 *Tensors, differential forms and variational principles* [344].

---

[4] Oliver Heaviside (1850–1925), English physicist and electrical engineer.

CHAPTER 5

Spacetime kinematics and Galilean relativity

Absolute space, in its own nature, without relation to anything external, remains always similar and immovable. Relative space is some movable dimension or measure of the absolute spaces; which our senses determine by its position to bodies; and which is commonly taken for immovable space ... .

Absolute, true, and mathematical time, of itself, and from its own nature, flows equably without relation to anything external, and by another name is called duration; relative ... time, is some sensible and external (whether accurate or unequable) measure of duration by means of motion, which is commonly used instead of true time[1].

Isaac Newton   1642–1727

5.1 Introduction

Until the end of the 15th Century, the concepts of time, space and motion were dominated by the geometry of Euclid, the philosophy, physics and doctrine of Aristotle[2] and the astronomy of Ptolemy[3]. According to Aristotle, every body in the universe had its natural place and natural motion [20, Bk. iii, ch. ii, verse 301-a-22]. All sublunar bodies [387] were composed of one of the four elements of earth, fire, water or air. The celestial bodies were comprised of a fifth element, the *ether* — transparent, rigid, all-pervasive and changeless, not found in the sublunar sphere of the earth. The only simple motions were straight and circular, the perpetual perfectly circular motion being the mode of the heavenly bodies around the earth at the centre of the universe. By contrast, earth and water were ponderable or heavy elements which 'fell' rectilinearly toward their natural place at the centre of the earth. Fire and air, being non-heavenly also

---

[1] *Scholium* of the *Principia* [395, p. 6].
[2] Aristotle (384–322 B.C.), Greek philosopher whose writings cover the topics we now identify as mathematics, physics, astronomy, meteorology, psychology and biology.
[3] Claudius Ptolemaeus (100–170), Greek astronomer and mathematician.

5.1. Introduction

moved in straight lines but, being 'light', moved away from the centre. The universe was understood to be finite in size, having a radius equal to the distance from the earth at the centre to the heavenly sphere of fixed stars on which moved the sun, moon and planets.

The earliest heliocentric system dates from the time of Heraclides[4] in the 4th century B.C. Aristarchus[5] also proposed a heliocentric system in the following century, referred to by Archimedes[6] [18]. In it he assumes the earth moves in an orbit whose radius compared to that of the fixed stars is the same as the ratio of the centre of a sphere to its surface. Archimedes assumes by 'radius of the centre' that Aristarchus means the radius of the earth. Aristarchus's ideas were not widely known and were not part of the standard astronomy of the western world which was based on the Ptolemaic epicycles of circles on circles, the perfect heavenly motion.

In 1514, the first draft of *De Revolutionibus* [79] of Copernicus[7] appeared anonymously. Known as the *Commentariolus* [78], it was based on detailed and meticulous observations of the sun, moon and stars. In it, Copernicus very cautiously advanced his theory of the earth's motion about the sun, and provided ample evidence to support his dramatic and dangerous hypothesis, not published in full until his death. To adopt a heliocentric system, Copernicus also needed to emphasize the enormous size of the cosmos, by stating that the ratio of the earth's radius to its orbital radius is much less than the ratio of the orbit to the distance of the fixed stars. We now know that ratio to be smaller than 10^{-4} if we consider stars at about 10 light years to be essentially 'fixed'. Copernicus declined to state whether he considered the universe to be finite or not.

By adopting a heliocentric system, Copernicus had ascribed to earth the same heavenly properties of perpetual motion shared by all the celestial bodies. He had thus taken the first steps in unifying the dynamics of the earth with those of extraterrestrial objects. The daily and seasonal changes of the heavens were more easily accounted for in the heliocentric theory and the different types of planetary motions were correctly based on placing Mercury and Venus on orbits about the sun inside that of the earth. In his statements about the daily rotation and the annual orbital motion of the earth, Copernicus argued that for certain phenomena, customarily used as evidence for a stationary earth, the appearance would not alter if the earth moved and the sun was almost motionless, near the centre of the universe.

> The centre of the earth is not the centre of the universe. We revolve around the sun like any other planet. The earth's immobility [is] due to an appearance[8].

---

[4] Heraclides Ponticus (ca. 388 – ca. 315 B.C.), Greek astronomer and geometer.
[5] Aristarchus of Samos, (310–230 B.C.), Greek mathematician and astronomer.
[6] Archimedes (ca. 287–212 B.C.), Greek mathematician, physicist and inventor.
[7] Nicolas Copernicus (1473–1543), Polish priest and astronomer.
[8] *Commentiarolus* [78, pp. 58–59].

These statements contain the germ of the ideas of relativity, at least for frames which are displaced and rotated with respect to one another, to be incorporated into Galileo's principles of motion later in the 16th century along with the relativity of frames in uniform motion.

In a heliocentric system, one expects to observe a change, due to parallax, in the direction of celestial objects such as stars when viewed from different parts of the earth's orbit. No such parallax was observed at the time and this was used by supporter's of the system to conclude correctly that stars were further away than had hitherto been supposed. Although Bradley[9] [45] attempted to measure stellar parallax in 1728, the magnitude (< 0.75 arc seconds) was far too small and he discovered instead the phenomenon of *stellar aberration* of magnitude 20.5 arc seconds. This is the arrival of light at the telescope at apparently slightly different angles at different times of the year due to the change in direction of the earth's motion and therefore of the telescope. The detection of *stellar parallax* did not take place until 300 years after the *Commentiarolus*. The first published results were the measurement of 0.3 arc seconds for 61 Cygni by Bessel in 1837. Although the British astronomer, Thomas Henderson, had carried out observations in South Africa a few years earlier, he did not immediately reduce and publish them.

The earth in the Copernican system could no longer serve as the attracting point for all heavy bodies, a rôle it played in the Aristotelian-Ptolemaic universe in which it happened to be at the centre. Rosen [78, 484] discusses how Copernicus 'put forward a revised conception of gravity, according to which heavy objects everywhere tended toward their own centre'.

> For my part, I think that gravity is nothing but a certain natural striving with which parts have been endowed ... so that by assembling in the form of a sphere they may join together in their unity and wholeness[10].

This concept is a forerunner of the many-body, multi-centre, universal law of gravitation of Newton to appear nearly a century and a half later.

The development of Galilean and Newtonian mechanical and gravitational ideas were only possible after the establishment of a quantitative analysis of kinematics by Galileo. Galilean kinematics will be considered in the remainder of this chapter alongside the analogous analysis carried out by Einstein 300 years later.

5.2 Measurement and simultaneity

Some of the mathematical properties of Euclidean or Aristotelean spacetime have been examined in previous chapters. These features provide the

---

[9] James Bradley (1693–1762), English atronomer.
[10] *Commentiarolus* [78, 484].

5.2. Measurement and simultaneity

capability of representing the measurability of time and space intervals. They also permit the natural incorporation of the exact conservation laws of energy, momentum and angular momentum applicable to all isolated systems and the time-reversal invariance and parity conservation properties of classical systems and most quantum interactions. The appropriate background spacetime can be described as a non-metric product, $\mathbf{E}^1 \times \mathbf{E}^3$, of a metric Euclidean time coordinate with a metric 3D Euclidean space.

All *measurement* is based on observations of the simultaneity of two events. For example: 'a collision occurred *when* the needle of the ammeter was opposite calibration point 3.0'. The length ℓ of an object is measured by noting that '*when* one end is coincident with the zero mark of a standard length, the other is opposite the mark ℓ on the standard device'. The genius of Einstein [128] was to note that such observations are understood to be at the same time in some frame and are thus meaningless without some means to synchronize two clocks at different locations. This is also true in Newtonian physics. The synchronization procedure used to establish the concept of *distant simultaneity* will affect all measurements and must be consistent with observation. The need for such a procedure was recognized by Poincaré [440] as early as 1898, and again in 1904, in applying light signals [446] to analyse the contraction hypothesis used in the Lorentz electron theory and the resulting 'apparent' constancy of the speed of light.

We could arbitrarily hypothesize that simultaneity is *absolute* in the sense of frame independence and of being unaffected by the presence of nearby matter. Such a hypothesis was implicit in pre-Newtonian physics and explicitly set out as a fundamental postulate by Newton in his Principia. We could then set out the details of Newtonian dynamics in conventional form and observe that the system obeys the Galilean relativity principle.

Alternatively, we could use the standard Einsteinian procedure, as embodied in his second postulate on the constancy (frame independence) of the speed of light, to proceed directly to the kinematics and dynamics of special relativity. Either procedure is satisfactory in its own domain of validity, with special relativity having a broader applicability. It includes Newtonian physics as an approximation outside of which the latter is in disagreement with observations and experiment.

Rather than take either of the above routes we prefer instead to base our analysis primarily on the Principle of causality and only the first postulate, the Principle of relativity, and use them to limit drastically the ways in which we may specify the character of distant simultaneity. We shall see then that there are only two ways in which these Principles may be applied. Only one of those, Einsteinian relativity, leads to a spacetime which is geometric and that case is in full accord with local observations. The other, Galilean relativity, does not provide a spacetime with a 4D metric geometry and is only approximately in agreement with observations under certain restricted conditions of validity.

In the pre-Galilean spacetime of Aristotle, motion was understood to take place in an *absolute space*, fixed to the earth at the centre of the universe, and was measured with respect to an *absolute time*. Absoluteness implied there was one privileged spacetime reference frame, that of observers (with clocks) at rest on the earth, with respect to which all space and time displacements, rotations and velocities were to be measured. Absoluteness implied that motion or the presence of matter could not affect spacetime measurements. It was therefore implicitly understood that two observers, no matter where they were located relative to one another and irrespective of their relative velocity, would always agree on whether two events were simultaneous or not, and on the value of the time interval between those events if they were not simultaneous. Simultaneity was understood to be absolute.

Although Newton postulated the existence of absolute space and absolute time, unaffected by the presence of matter, he also makes it clear that they are unobservable[11], and that measurements refer to relative time and relative space. To the author, Newton appears to be making a distinction very similar to the one we now make between coordinates, as unobservable event labels, and measurable intervals between events. Newton's use of his absolute space and time seems to be limited to the explicit assumption that space and time are unaffected by matter and, implicitly, that simultaneity is absolute. If the laws of physics are such that they permit one to determine the velocity v of one frame relative to another, from measurements made entirely within the one frame, then we may conclude that an absolute frame (the one with $v = 0$) exists. But Newton's laws do not permit this. The absoluteness of simultaneity is the essential difference between Newtonian and Minkowskian spacetime. Their independence of the matter content is a feature they have in common. Galilean relativity and Newtonian physics survive the non-existence of absolute space — but they do not survive the relativity of simultaneity.

Newton's absolute time and space were not accepted without criticism. In his own day, there was a lengthy debate with Leibniz[12] in which Newton's viewpoint was defended by Clarke[13]. To Leibniz, the universe was an infinitude of simple, pointlike, immaterial entities called *monads* — also endowed with spirituality and consciousness. Space was simply the relations between different monads and time was the relation between successive states of the one monad. Neither had an existence of its own; Leibniz denied the possibility of empty space and time which Newton permitted. Despite the soundness of some aspects of Leibniz's spacetime philosophy,

---

[11]*Principia* [395, p. 10].
[12]Gottfried Wilhelm Leibniz (1646–1716), German mathematician and natural philosopher.
[13]Samuel Clarke (1675–1729), English philosopher of science and mathematician.

the Newtonian kinematic view triumphed because of the incredible success of his dynamics of interaction based on it.

Newton's absolute space and time began to be criticized heavily in the decades leading up to the presentation of the special relativity theory [128] in 1905. The penetrating analyses of Mach[14] [352], in which distant matter affects the local inertial properties, were particularly influential. However, some of the criticism of that period and since could well be based on a lack of appreciation of the significance of Newton's distinction between absolute and relative concepts. Instead of directing our criticism at the unobservable Newtonian concepts (which are also numerous in modern physical theories), we should concentrate on the measurable kinematic feature that has not survived in Newtonian physics, namely the absoluteness of simultaneity, and on a dynamic feature requiring modification, that space and time are unaffected by the presence of matter.

5.3 Spacetime reference frame

There shall be standard measures of wine, beer, and corn — the London quarter — throughout the whole of our kingdom, and a standard width of dyed, russet and halberject cloth — two ells within the selvedges; and there shall be standard weights also[15].

Magna Carta 1215

Electromagnetic phenomena have provided such precise measurements of time and length over long periods of time that the SI unit of time, the *second*, is defined [274] as the duration of 9 192 631 770 periods of the radiation corresponding to the transition between the two hyperfine levels of the ground state of the caesium-133 atom. The General Conference of Weights and Measures recommended in 1975, and confirmed in 1983, that the *metre* be defined as the length [274] of the path travelled by light in vacuum during a time interval of 1/299 792 458 of a second. The standard of length is thus defined exactly in terms of the time standard in such a way that the speed of light in vacuum has the invariant exact value [274] of $c = 2.99792458$ m s^{-1}. In essence, what used to be a speed of light measurement has now become a comparison of the primary time (and length) standard with a secondary (length) standard.

To appreciate why radiation is used in the definition of both time and length standards requires an understanding of the need for a synchronization procedure relating the times on an array of clocks distributed throughout space, as emphasized by Poincaré [447] in 1904 and clarified by Einstein [128] in 1905. A set of clocks, which are calibrated to display

---

[14] Ernst Mach (1838–1916), Austrian physicist, physiologist and philospher.
[15] Chapter House, Salisbury Cathedral.

identical times when coincident, may be distributed throughout space and brought to rest relative to a master standard clock, and thus relative to one another, provided one has a notion of what it means to send a signal of constant velocity relative to the master clock. The velocity must be the same from master clock to any other clock and also in the reverse direction. Constancy of round-trip signal times between all pairs of clocks can then be used as a criterion for determining when all the clocks are at rest with respect to one another. The master clock can then send a signal to all the others stating the time of emission. On receipt of the signal, all other clocks may set their time to the emission time plus one half of the round-trip time to the master clock. Any event at one clock is then said to be simultaneous with a distant event at another clock provided both clocks read the same time for the respective events.

It is standard practice for the synchronization to be carried out by using electromagnetic radiation in vacuo. This is the case in *Gedanken* or thought experiments of the type used by Poincaré [447] in examining the contraction hypothesis. It was used by Einstein [128] in 1905 when examining the nature of the measurement process and in forming the second postulate embodying his conclusions. It is also the case in practical measurements of metrology. This is a consequence of the high precision and stability of optical and electromagnetic measurements and the close relation at the turn of the century between the non-Newtonian character of the equations of electromagnetism and the development of relativity.

However, use of electromagnetic signals not only conveniently implements the necessary synchronization process but it also partly characterizes the nature of radiation and therefore of electrodynamics. In order to stimulate the reader into separately considering the synchronization process and the characterization of light in Einstein's second postulate, an alternative equivalent procedure will be presumed here in which the synchronization is understood to be carried out by constant velocity signals (in each frame) of any form except electromagnetic radiation. We will consider the *Gedanken* experiment of Einstein to be replaced by an equivalent synchronization procedure and notion of distant simultaneity without any immediate consequential implications for the nature of electromagnetic radiation.

Elementary discussions of the process of synchronization and the possibility of using non-electromagnetic signals including those with much lower (and therefore frame-dependent) speeds are available in Aharoni [4] and Brehme [49]. Constant velocity signals are not only a means to synchronization and distant simultaneity but also provide one with a notion of length based on measurement of time intervals.

We now know that gravity affects the rates of clocks and this phenomenon will be examined in Chapters 17 and 21. The extent to which one may neglect such essentially general relativistic gravitational effects on clocks depends on the separation of the clocks compared to the radii of

5.3. Spacetime reference frame

curvature of spacetime. By considering a sufficiently small region one may reduce such effects to arbitrarily low values.

Exercise 5.1 *Show that on earth, and throughout the solar system, the radius of curvature of Exercise 1.3 has a value of about one light year and is therefore very much larger than the size of the solar system itself.*

A set of identical synchronized standard clocks at fixed relative locations will be referred to as a *local spacetime frame* with coordinates $\{t, \mathbf{x}\}$. It need not be inertial. The *velocity* of a particle at a certain location in a given spacetime reference frame is the rate of change $\mathbf{u} = d\mathbf{z}/dt = \dot{\mathbf{z}}$ of its spatial displacement with respect to the time as measured by the synchronized clocks at rest in the frame and located at each successive event of the particle trajectory. The *acceleration* of a particle is given by the corresponding rate of change $\mathbf{a} = d\mathbf{u}/dt = \dot{\mathbf{u}} = \ddot{\mathbf{z}}$ of the velocity. These are the two main kinematic quantities relating particles besides the space and time intervals $d\mathbf{z}$ and dt themselves.

The existence of spacetime reference frames embodying the notion of distant simultaneity does not make any statement about the equality or otherwise of the results of measuring the time interval between two events in two such reference frames, one of which is moving with respect to the other. There exist an infinite number of such independent spacetime reference frames, those which are rotated, translated or moving with respect to one another (including accelerated motion, either linear or angular). We have no reason to presume that the time intervals on two such frames will be identical.

The fundamental rôle of time is not only to provide a measure of the separation between the events along the path of any particle, or along the history of a field, but also to place them in order. The different coordinate times of separate spacetime frames, including those in relative motion, must be consistent with this frame-independent ordering requirement. The ordering requirement on any measure of time is a fundamental property of far-reaching significance, entirely analogous to a similar demand we place on spatial coordinates. However, we are interested primarily in interaction or evolution, not just relative location or spatial geometry. Consequently, the temporal ordering of events assumes a rôle which dominates the spatial ordering or distribution, which applies principally to the specification of initial data.

We may express this ordering requirement on the temporal evolution of mechanical systems at the macroscopic (non-quantum) level in a way which does not presuppose either Galilean or Minkowskian relativity:

- the **Principle of causality** — two events on the path of a particle will occur in the same order in all spacetime reference frames.

Thus, if one event $\{t, z\}$ occurs at time t in some frame and is capable of evolving to another event $\{t + \Delta t, z + \Delta z\}$ at $t + \Delta t$ in that frame (where $\Delta t \geq 0$), then the time interval between the two events, $\{\bar{t}, \bar{z}\}$ and $\{\bar{t} + \Delta \bar{t}, \bar{z} + \Delta \bar{z}\}$, in any other frame satisfies $\Delta \bar{t} \geq 0$.

Intuition may lead us to believe that if any event occurs before any other in some frame then a particle, or part of a field system, is capable of evolving from the 'first' to the other or to cause it. This intuitive conclusion is approximately correct if one observes only relatively slow phenomena. In the case of vanishingly small velocities, it will seem that events at the same time ($\Delta t = 0$) can also cause one another. We shall see that both these intuitive conclusions are false. The first is not in fact true of all event pairs. In fact, with some event pairs whose spatial separation is very large compared to the separation in time, the 'earlier' of the pair of events, as measured in some frame, may be incapable of evolving to (or causing) the latter event no matter what physical mechanisms are considered. This will be related to the fact that being earlier in one frame need not imply earlier in all frames. Events capable of being *causally connected*, namely that one and the same non-quantum particle may be present at each event, are precisely those for which time ordering is an invariant concept. We shall return to this principle and give it a quantitative form in terms of coordinates and geometry after establishing the transformations relating spacetime coordinates in two separate frames.

5.4 Principle of relativity

The two postulates of Einsteinian relativity, in the customary form based on the original statement by Einstein[16], are the Principle of relativity and the invariance or frame-independence of the speed of light. These fundamental principles severely restrict the forms that may be taken by any laws of physics consistent with them, such laws being now supported by a vast amount of experimental data. Both postulates involve ideas recognized earlier, at least in part, by Poincaré, the first as early [438] as 1895 and the second [440] from 1898. However, both were considered to be

---

[16] Albert Einstein (1879–1955), German theoretical physicist and naturalized United States citizen (1940), is generally credited with the definitive statements of special relativity as a result of his epoch-making 1905 paper on the *Electrodynamics of moving bodies* [128], in which many of the *ad hoc* assumptions used by his immediate predecessors were placed on a simple conceptual basis in the form of his two postulates. The paper makes no reference to the scientific literature, in particular to the earlier work of Lorentz [340]and Poincaré [446, 447] in 1904. Balanced accounts of the relative contributions of all three scientists are contained in Holton [254] and Goldberg [201]. A more controversial, and largely rejected, analysis of Einstein's contribution is available in Whittaker [576, p. 40] who states that ... *Einstein published a paper which set forth the relativity theory of Poincaré and Lorentz with some amplifications, and which attracted much attention*.

5.4. Principle of relativity

experimental consequences of a fundamental contraction hypothesis of the electron theory of Lorentz.

Poincaré [438] (*Oeuvres* [451], vol. 9, p. 412), in discussing a paper by Larmor on the electromagnetic theory of light, states that, ... *it is impossible to detect the absolute velocity of matter, or better the velocity of matter relative to the ether; all that we can detect is the velocity of the matter relative to other matter*[17]. He referred to this same idea as the *Principle of relative motion*[18], originating from Galileo's principle of the same name, in 1899 in discussions of the electron theory of Lorentz and the (Newtonian) principle of reaction (*Oeuvres* [451], vol. 9, pp. 482 and 488), claiming that the latter must also apply to electromagnetic phenomena. In September 1904 and June 1905, he referred to the same idea, in the context of the electrodynamics of moving bodies [447, 448] as the *Postulat de relativité*, translated into English [447] as the Principle of relativity. His references to the constancy or apparent constancy of the speed of light [440, p. 10] first appeared in 1898 and again [442] in 1900.

Einstein's analysis of relativistic effects was closely related to those of electrodynamics. Analyses of relativity unconnected to those of electrodynamic phenomena, except for the use of light in the second postulate, appeared within a few years of the presentation of Einstein's paper. The first paper on relativity presented at an American conference, by Lewis and Tolman [323] in 1908 (commented on in detail by Goldberg [203]) established the dependence of mass on velocity. This was done by applying Einstein's postulates to the collision of two masses subject to the conservation of momentum, without reference to the mass or momentum of light.

In 1910 and 1911 (see, for example, von Ignatowsky [259], Frank and Rothe [169] and Weichert [564]), the structure of relativistic physics was explored from postulates weaker than those originally given by Einstein. In the first of these papers, von Ignatowski uses an upper bound on the velocities $|\dot{z}| \leq c^2/v$ where v is the speed of the frame relative to the ether, the principal difficulty with the results being caused by a failure to distinguish between wave and group velocities. More recent articles renewing interest in the postulates are numerous. A selection from the literature of the last two decades are available in references [22, 350, 360, 374, 501, 528, 531]. The topic is also discussed in almost every text on relativity. Some of these analyses have also shown that one can determine the form of the transformation relating frames in uniform relative motion without reference to light phenomena or electrodynamics. Furthermore, the restrictions which the postulates impose on the transformation laws

---

[17] ... il est impossible de rendre manifeste le mouvement absolu de la matière, ou mieux le mouvement relatif de la matière pondérable relatif à l'éther; tout ce qu'on peut mettre en évidence, c'est le mouvement de la matière pondérable par rapport à la matière pondérable. (My translation.)

[18] *Le principe du mouvement relatif.*

relating the coordinates of reference frames are essentially kinematic in nature and it ought to be possible to express their essence without reference to any dynamical theory based on them. In fact, the postulates suffice to determine the transformations relating reference frames in relative motion without reference even to inertial reference frames which are essentially dynamical in nature.

In order to avoid characterizing electrodynamics as part of the topic of spacetime kinematics, we shall postpone the introduction of the second postulate to the end of the chapter and discuss separately the consequences of Einstein's first postulate [128], namely:

- **the Principle of relativity** — the local laws of physics have the same form in frames of reference that are related by a uniform relative velocity (and when translated in time and translated or rotated in space).

A statement applicable to all the laws of physics can never be proved, only disproved, and must therefore be a fundamental postulated principle until shown to be invalid by conflict with experiment.

Some confusion occurs in discussions of the Principle of relativity on account of it sometimes being used, erroneously, to denote the combination of both of Einstein's postulates or implying that it is unique to special relativity. The concept of a Principle of relativity did not originate with Einstein. Galileo was aware that the laws of kinematics of a particle were unchanged by a frame change involving a uniform relative velocity and this concept, known as *Galileo's principle of relative motion*, was an essential, implicit part of Newton's laws of mechanics. (The same is of course true for translations and rotations.) Maxwell referred to 'the doctrine of the relativity of all physical phenomena' in 1877 in his text entitled *Matter and motion* [369].

Some aspects of the Relativity principle were also implicit in the 1904 version of Lorentz's *electron theory* [202, 250, 252, 371, 372, 495] of the dynamics of electrical charges and the ether, the medium in which light propagated according to Maxwell's equations. The various versions of that theory, [332] to [341], from 1892 to 1904, established an apparent (non-reciprocal) equivalence between the ether frame and a frame moving relative to it. Later versions of the theory included a length contraction, first suggested by FitzGerald[19] [53, 163] in 1889. It was also proposed independently by Lorentz [332] in 1892 and developed in his 1895 text [334], the relevant part of which is reproduced in Lorentz et al. [343]. Electrical matter was presumed to contract, in the direction of its motion relative to the ether, as a dynamical result of forces originating in the ether.

The modern interpretation of the contraction makes no reference to forces and is thus kinematic in nature, attributed directly to the applica-

---

[19] George Francis FitzGerald (1851–1901), Irish physicist.

5.4. Principle of relativity

tion of a Relativity principle and the invariance of the speed of light, or some equivalent second postulate. We have already mentioned the use of the Principle by Poincaré in his analyses of the Lorentz theory. However, none of the pre-Einsteinian theories completely abandoned the existence of the ether and the possibility of experimentally confirming or denying its existence. Poincaré appears to have believed that an appropriate modification of the ether theory may have permitted the reinterpretation of the observations as providing experimental support for the Principle of relativity.

The most important work of Lorentz and Poincaré, on the dynamics of the electron, immediately preceded Einstein's presentation [128] of special relativity as a kinematic rather than a dynamic phenomenon. Compared to the work of Lorentz and Poincaré, Einstein's was independent of any of the details of intermolecular electromagnetic interactions or of the existence of the ether. The demand that one principle apply to all of physics, namely the mechanics of material particles and of electromagnetic fields — the same demand was made of Galileo-Newtonian mechanics — was not a new feature of Einstein's presentation, compared to Galilean relativity, Instead, one new feature was the second postulate and its implications for the relativity of simultaneity. A second new feature was the retention of the Principle of relativity and the inclusion of the second postulate as unverifiable, conventional, and general principles rooted in a deep understanding of the nature of measurement.

Retention of the Relativity principle as a kinematic concept, despite the apparent contradiction with the second postulate, dispenses with the concept of an absolute frame and hence with the ether. If the speed of light was interpreted as being relative to an ether associated with an absolute frame, then an observer moving relative to the ether would get different values for the speed of light. Maxwell's equations would then be frame-dependent contradicting the Relativity principle. Conversely, if the laws of physics are not the same in frames in relative motion of velocity $\mathbf{v}$, but depend on $\mathbf{v}$, then we may use those laws to determine the value of $\mathbf{v}$ and obtain an absolute frame as the one relative to which the velocity in those laws is $\mathbf{v} = 0$. Consequently, the experimental evidence for or against the existence of the ether were of prime importance to Einsteinian relativity.

At the turn of the century, the most important of these experiments were those which had been carried out by Michelson[20] [378] in 1881 and Michelson and Morley [379] in 1887. They used the motion of the earth about the sun, and hence presumably in different directions in the ether, to demonstrate, with disbelief and disappointment in their experiment, that the electromagnetic wave nature of light could not in fact be based on motion relative to an absolute space. The influence of the results of

---

[20] Albert Abraham Michelson (1852–1931), American physicist and metrologist.

Michelson and Morley on Einstein's analysis of space and time is difficult to assess and not direct. However, their influence on the work of Lorentz, FitzGerald and Poincaré was direct and incontrovertible. Although the results of the experiment were of crucial importance at the time, the validity of Einsteinian principles now rests on a huge body of other experimental data [175, 224, 585, 587, 588] where the effects are very much easier to detect. One recent experiment used to demonstrate independence of any supposed speed v_{lab} of the laboratory relative to the ether, reported by Isaak [260] in 1970, is based on using the Mossbäuer effect to look for Doppler variations in photons of energy 14.4 keV from decay of ^{57}Fe nuclei. The results show that $v_{lab} < 0.05\,\mathrm{m\,s^{-1}}$, which is so small compared with the orbital velocity of the earth ($\approx 30,000\,\mathrm{m.s^{-1}}$), that it effectively implies no motion relative to the ether or, alternatively, absence of any ether at all.

An object which is not symmetric about a given axis will not appear the same to two observers, one of which is rotated relative to the other about that axis. Analogously, the Principle of relativity does not demand that a given object or phenomenon will appear identically to different observers. Rather, it requires the relations or laws governing the distribution in space or the interactions of parts of that object leading to evolution, to be expressed in the same form using the reference frame of either observer, and this requirement is corroborated by experiment.

Since the coordinate labels of events are varied when changing frames, the physical quantities describing phenomena taking place at those events must also vary and in precisely such a way that the physical equations involving them do not vary in form. We say that the physical quantities must *covary* and for this reason the Principle of relativity is also a kinematic *Principle of covariance*.

The Principle of relativity has been quoted in terms of *passive* transformations between two reference frames or sets of observers. We may also use an *active* formulation in which there is only one reference frame and the transformation converts a given physical system into another system. Since the Relativity principle is satisfied, the new system will also be physically allowable and the two descriptions, active and passive, are equivalent. For some purposes, a passive description is more convenient.- This is particularly so if spacetime reflection operations are involved since the latter are not physically realizable as operations on material objects or observers.

The content of the homogeneity of time and the homogeneity and isotropy of space is included in the Relativity principle (either explicitly or implicitly) as form invariance of physical laws with respect to translations and rotations. The extra ingredient is the requirement, originating with Galileo [120, 181, 182], that invariance must also apply when uniform relative motion is involved. The Principle of relativity implies the equivalence of frames which immediately implies that the transformations taking one spacetime frame to another must form a group.

The Relativity principle restricts, but does not completely determine, the way in which systems interact. The details of interaction require other principles as a guide to their formulation. The Principle of relativity applies equally to both the so-called non-relativistic physics (*Galilean relativity*) and to relativistic physics (*Einsteinian* or *special relativity*). To be in full accord with the kinematics of physical systems, it must be complemented by the second postulate of special relativity or its equivalent.

The history of the years leading up to Einstein's theory will be discussed further in Chapter 12 specifically devoted to Einsteinian relativity.

5.5 Galilean and Lorentz transformations

You ask what if the velocity be greater than that of light? I have often asked myself that but got no satisfactory answer. The most obvious thing to ask in reply is 'Is it possible?'[21].

George Francis FitzGerald   1851–1901

We use the term *boost* to refer to the transformation relating two frames one of which moves with respect to the other at a uniform velocity **v**. There are only two ways, consistent with the Principles of causality and relativity, by which the coordinates of one frame may be related to those of another, boosted with respect to it. We shall give an elementary demonstration of this result here following the procedures used by Lee and Kalotas [316] in 1975 and Lévy-Leblond [321] in 1976.

The Principle of relativity will permit us to establish the existence of a universal positive constant c (possibly infinite) which acts as an upper bound on all speeds and is the sole constant in the transformation linking the coordinates of the two frames.

Consider a transformation from one spacetime frame S with coordinates $\{t, \mathbf{x}\}$ to another $\bar{S}$ with coordinates $\{\bar{t}, \bar{\mathbf{x}}\}$. We seek the most general form of the transformation such that the first postulate, the Principle of relativity, is observed. Let the transformations be,

$$t \rightarrow \bar{t} = \bar{t}(t, \mathbf{x}) , \qquad (5.1)$$
$$\mathbf{x} \rightarrow \bar{\mathbf{x}} = \bar{\mathbf{x}}(t, \mathbf{x}) , \qquad (5.2)$$

in which, for equivalence, $\bar{t}(t, \mathbf{x})$ and $\bar{\mathbf{x}}(t, \mathbf{x})$ must be invertible functions. We first make use of the spatial isotropy and homogeneity of $\mathbf{E}^3$ to restrict consideration, without loss of generality (by rotation and spatial translation of one frame), to two frames with all three spatial axes aligned and with the x-axis of each also parallel to the instantaneous direction $\hat{\mathbf{v}}$ of relative

---
[21] Letter to Oliver Heaviside, dated 4 February 1889.

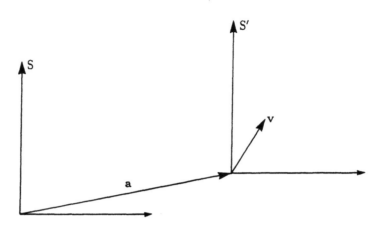

Figure 5.1 Change of frame by a translation (a) and a boost of velocity v.

velocity **v**, as shown in Figure 5.1. We appeal to homogeneity to make a translation in time for one frame to fix both sets of clocks to zero at the moment when the origins coincide. Equivalently, we consider time and space intervals, $\Delta t \to \Delta \bar{t} = \Delta \bar{t}(\Delta t, \Delta x)$ and $\Delta x \to \Delta \bar{x} = \Delta \bar{x}(\Delta t, \Delta x)$ with $\Delta \bar{y} = \Delta y$ and $\Delta \bar{z} = \Delta z$ between two events. By isotropy of space, these intervals cannot depend on Δy and Δz which must remain unchanged for a boost in the x-direction. By homogeneity of space and time, they depend linearly on Δt and Δx and not on the coordinates of the ends of the interval. These points are covered in more detail in Resnick [472], for example. Thus the intervals are given by,

$$\Delta \bar{t} = \alpha \Delta t + \beta \Delta x \qquad (5.3)$$
$$\Delta \bar{x} = \gamma \Delta x + \delta \Delta t . \qquad (5.4)$$

in which the coefficients α, β, γ and δ cannot depend on the coordinates of t and x. Homogeneity of spacetime thus suffices to show that the transformations must be linear. (Mariwalla [359] shows that this result may in fact be established without appealing to homogeneity. This follows since there are only three distinct 1-parameter groups acting in one spatial dimension and all imply linear transformation equations.)

If there are no parameters in $\Delta \bar{t}(\Delta t, \Delta x)$ and $\Delta \bar{x}(\Delta t, \Delta x)$, then the transformation will be independent of the x-component v of the velocity of relative motion and we would have $\Delta \bar{t} = \Delta t$ and $\Delta \bar{x} = \Delta x$ as in Exercise 3.1. In that case, both Δt and Δx (and, indeed, $\Delta \mathbf{x}$) would be absolute as in the Aristotelean concept of spacetime which preceded the notions introduced by Galileo at the end of the 16th century. Since we seek the most general transformation, we allow it to depend on at least the

5.5. Galilean and Lorentz transformations

relative speed v.

Suppose first that there is at least one other independent parameter w in the functions, namely $\Delta \bar{t}(\Delta t, \Delta x, v, w)$ and $\Delta \bar{x}(\Delta t, \Delta x, v, w)$. Then we can, in general, solve for these two parameters leaving the functional form (and any additional parameters) undetermined. Specific values for space and time intervals in one frame will correspond to arbitrary intervals in other frames which will not therefore be equivalent to the first, violating the Principle of relativity. We conclude that there can be at most one parameter, v, in the functions, $\Delta \bar{t}(\Delta t, \Delta x, v)$ and $\Delta \bar{x}(\Delta t, \Delta x, v)$. Since a fixed point, the origin say, in $\bar{S}$ is described by $d\bar{x}/d\bar{t} = 0$ or $\Delta \bar{x} = 0$, and by $\dot{x} = v$ in S, we have $\delta(v) = -v\gamma(v)$. The concept of being at rest in one particular frame is equivalent to being in *uniform motion* in a general frame. Frames will only be equivalent if the instantaneous speed v of relative motion is in fact constant. Replacing $\alpha(v)$ by $\gamma(v)\lambda(v)$ and $\beta(v)$ by $-\gamma(v)\mu(v)$, means the transformations are of the form,

$$\Delta \bar{t} = \gamma(v)[\lambda(v)\Delta t - \mu(v)\Delta x]$$
$$\Delta \bar{x} = \gamma(v)[\Delta x - v\Delta t] , \qquad (5.5)$$

in which $\gamma(v)$, $\lambda(v)$ and $\mu(v)$ are three functions of v to be determined. In order that motion be continuous and smooth in one frame if it is in the other, we require all three functions to be smooth and differentiable to any required order.

Although the equations describing some interactions are not parity invariant, time and length intervals are unaffected by the handedness of the frame. Consequently, parity transformed frames with coordinates $(t, -x)$ and $(\bar{t}, -\bar{x})$ must be equivalent to one another and thus related by the above equations for some speed parameter u,

$$\Delta \bar{t} = \gamma(u)[\lambda(u)\Delta t + \mu(u)\Delta x]$$
$$-\Delta \bar{x} = \gamma(u)[-\Delta x - u\Delta t] . \qquad (5.6)$$

Comparison of the two sets of equations gives,

$$\gamma(u) = \gamma(v), \quad u\gamma(u) = -v\gamma(v), \qquad (5.7)$$

and

$$\lambda(u)\gamma(u) = \lambda(v)\gamma(v), \qquad \mu(u)\gamma(u) = -\mu(v)\gamma(v) , \qquad (5.8)$$

from which we may deduce that:

$$u = -v, \quad \gamma(-v) = \gamma(v), \quad \lambda(-v) = \lambda(v) \text{ and } \mu(-v) = -\mu(v) . \quad (5.9)$$

The same results are achieved by appealing to time-reversal invariance applied to spacetime intervals.

We have made all the progress possible using form invariance with respect to translations, rotations and reflections, namely reduction to the determination of three functions of v, two of which are even and one of which is odd. We now demand that the relation linking the two frames have all the three properties of an equivalence relation, namely reflexivity ($S \equiv S$), symmetry ($S \equiv \bar{S} \Rightarrow \bar{S} \equiv S$) and transitivity ($S \equiv \bar{S}$ and $\bar{S} \equiv S' \Rightarrow S \equiv S'$). Alternatively, we demand that the transformations themselves have three of the properties of a group, namely the existence of an identity, the existence of an inverse and closure, the associativity property of the group being automatically true for transformations.

The first of the three conditions, the existence of an identity (or reflexivity of equivalence) in Equation 5.5, means that there must be a value of v for which $\mu = 0$. That value must be $v = 0$. Therefore we have $\mu(0) = 0$ and $\gamma(0) = 1 = \lambda(0)$. The second condition, existence of an inverse (or symmetry of equivalence) requires Equations 5.5 to have an inverse (of speed parameter w say) of the same form, namely given by,

$$\Delta t = \gamma(w)[\lambda(w)\Delta \bar{t} - \mu(w)\Delta \bar{x}] \quad (5.10)$$
$$\Delta x = \gamma(w)[\Delta \bar{x} - w\Delta \bar{t}] . \quad (5.11)$$

Inverting Equation 5.5 and identifying the result with this inverse gives,

$$w = -\frac{v}{\lambda(v)} \quad (5.12)$$

$$\lambda(w) = \frac{1}{\lambda(v)} \quad (5.13)$$

$$\mu(w) = -\frac{\mu(v)}{\lambda(v)} \quad (5.14)$$

$$\gamma(w) = \frac{1}{\gamma(v)}\left[1 - \frac{v\mu(v)}{\lambda(v)}\right]^{-1} , \quad (5.15)$$

(in which invertibility requires $\lambda(v) \neq 0$, $\gamma(v) \neq 0$ and $\lambda(v) \neq v\mu(v)$).

From Equations 5.12 and 5.13, we have,

$$\lambda\left(-\frac{v}{\lambda(v)}\right) = \frac{1}{\lambda(v)} \quad \Rightarrow \quad \lambda\left(\frac{v}{\lambda(v)}\right) = \frac{1}{\lambda(v)} , \quad (5.16)$$

where the second equation results from the evenness of $\lambda(v)$. Following Lévy-Leblond [321], we define $\zeta(v) = v/\lambda(v)$ which gives a functional equation, $\zeta(\zeta(v)) = v$ or $\zeta^{-1}(v) = \zeta(v)$, for $\zeta(v)$, which also satisfies $\zeta(0) = 0$ and $(d\zeta/dv)_0 = 1$. Thus $\zeta(v)$ is tangent to $\zeta(v) = v$ at $v = 0$ and the fact that $\zeta(v)$ and $\zeta^{-1}(v)$ are symmetrically related with respect to $\zeta = v$ shows that $\zeta = v$ giving $\lambda(v) = 1$ which, from Equation 5.12, implies $w = -v$. This shows that the transformation inverse to the one with parameter v

5.5. Galilean and Lorentz transformations

will have parameter $-v$. (It is also then necessary that $v\mu(v) \neq 1$ in addition to $\gamma(v) \neq 0$.) Equation 5.14 supplies nothing new but Equation 5.15 for $\gamma(w)$ now relates $\gamma(v)$ and $\mu(v)$ according to,

$$\gamma(v) = \frac{1}{\sqrt{1 - v\mu(v)}}, \qquad (5.17)$$

where we choose the positive square root since $\gamma(0) = 1$ and we must have $v\mu(v) < 1$. The original transformation has now been brought into the form,

$$\begin{aligned} \Delta \bar{t} &= \gamma(v)[\Delta t - \mu(v)\Delta x] = \frac{1}{\sqrt{1 - v\mu(v)}}[\Delta t - \mu(v)\Delta x] \\ \Delta \bar{x} &= \gamma(v)[\Delta x - v\Delta t] \quad = \frac{1}{\sqrt{1 - v\mu(v)}}[\Delta x - v\Delta t] , \end{aligned} \qquad (5.18)$$

containing only one unknown function, $\mu(v)$ or $\gamma(v)$.

We now apply the closure requirement (or transitivity of equivalence) on the set of x-direction boosts. A boost v_1 followed by another v_2 must give another with parameter $V = V(v_1, v_2)$ of the form,

$$\begin{aligned} \Delta \bar{t} &= \gamma(V)[\Delta t - \mu(V)\Delta x] & (5.19) \\ \Delta \bar{x} &= \gamma(V)[\Delta x - V\Delta t] . & (5.20) \end{aligned}$$

Exercise 5.2 *Compare Equations 5.19 and 5.20 with those obtained from Equation 5.18 for a boost with parameter $v_1 \neq 0$ followed by one with $v_2 \neq 0$ to show that $\mu(v_1)v_2 = v_1\mu(v_2)$.*

Since the result of the exercise implies that,

$$\frac{\mu(v_1)}{v_1} = \frac{\mu(v_2)}{v_2} = \alpha = \text{constant} , \qquad (5.21)$$

the ratio $\mu(v)/v$ is some real finite universal constant α (possibly zero) having the dimensions of inverse velocity squared. In terms of this constant α, we have $\mu(v) = \alpha v$. The transformations of Equation 5.18 now have the form,

$$\begin{aligned} \Delta \bar{t} &= \gamma(v)[\Delta t - \alpha v \Delta x] \\ \Delta \bar{x} &= \gamma(v)[\Delta x - v\Delta t] , \end{aligned} \qquad (5.22)$$

and the one remaining function of v is,

$$\gamma(v) = \frac{1}{\sqrt{1 - \alpha v^2}}, \qquad (5.23)$$

with boost speeds limited by $\alpha v^2 < 1$. The above calculation [321] makes no use of the second postulate.

Exercise 5.3 *Show that*

$$V = \frac{v_1 + v_2}{1 + \alpha v_1 v_2} . \tag{5.24}$$

gives the x-component of the velocity of the combined transformation of a boost v_1 followed by another v_2.

We are left with three possibilities according to whether α is negative, zero or positive:

- If $\alpha < 0$, we put $\alpha = -1/K^2$ where K is a real constant with the dimensions of a velocity. We let K be positive. The transformation becomes,

$$\Delta \bar{t} = \frac{\Delta t + v\Delta x/K^2}{\sqrt{1 + v^2/K^2}} \tag{5.25}$$

$$\Delta \bar{x} = \frac{\Delta x - v\Delta t}{\sqrt{1 + v^2/K^2}}, \tag{5.26}$$

with a velocity addition law,

$$V = \frac{v_1 + v_2}{1 - v_1 v_2/K^2} . \tag{5.27}$$

This solution for the transformation is ruled out as a boost because it does not obey the Principle of causality.

Exercise 5.4 *Show that if $\Delta t > 0$, for any pair of events, then a value of the boost parameter v exists for which Equation 5.25 implies that $\Delta \bar{t} < 0$.*

No trajectories exist along which causality is satisfied as is required for particle paths. Alternatively, given a boost $v_1 > 0$, then there exist boosts in the same direction, namely those with $v_2 > K^2/v_1$ for which the addition formula predicts that the combined boost has velocity $V < 0$, namely in the opposite direction to the two being combined.

- If $\alpha = 0$, we arrive at the *Galilean boost* transformation,

$$\begin{aligned}\Delta \bar{t} &= \Delta t \\ \Delta \bar{x} &= \Delta x - v\Delta t ,\end{aligned} \tag{5.28}$$

in which the speed v of the boost satisfies $0 \leq v < \infty$, on which there no finite upper bound, and the range of a component v^k of the velocity is $-\infty < v^k < \infty$. The *velocity addition formula* is then,

$$\mathbf{V} = \mathbf{v}_1 + \mathbf{v}_2 , \tag{5.29}$$

5.5. Galilean and Lorentz transformations

and we have $\gamma(v) = 1$ and $\mu(v) = 0$ for all v. With the Galilean transformation relating boosted frames, there is likewise no finite upper bound on the speed $u = |\dot{z}|$ of a particle. The Galilean boost transformation is in accord with the Principle of causality since it implies $\Delta \bar{t} = \Delta t$ and consequently, from $\Delta t \geq 0$ we obtain $\Delta \bar{t} \geq 0$. Any interval with non-decreasing time is a valid separation for causal effects and any path with increasing time is a valid particle path. Experiments, for example those on the variation of the energy of a system with speed, show that this second choice, the Galilean boost, is valid as an approximation to first order in the ratio of the boost or particle speed, v or u, to the speed of light provided that ratio is small, but is not experimentally verified otherwise.

- If $\alpha > 0$, we put $\alpha = 1/c^2$ (with $c > 0$) to define a new real positive finite universal constant c having the dimensions of velocity. From $\alpha v^2 < 1$, we have $0 \leq v < c$ and $1 \leq \gamma < \infty$. Any component v^k of the velocity of a boost of this type is restricted to the range $-c < v^k < c$ which, because it does not include the end-points $\pm c$, is a non-compact interval. From Equation 5.22 we obtain,

$$\begin{aligned} \Delta \bar{t} &= \gamma(\Delta t - \frac{v \Delta x}{c^2}) \\ \Delta \bar{x} &= \gamma(\Delta x - v \Delta t) \\ \Delta \bar{y} &= \Delta y \\ \Delta \bar{z} &= \Delta z \,, \end{aligned} \quad (5.30)$$

where

$$\gamma(v) = \frac{1}{\sqrt{1 - v^2/c^2}} \,, \quad (5.31)$$

and a *speed addition formula*,

$$V = \frac{v_1 + v_2}{1 + v_1 v_2/c^2} \,, \quad (5.32)$$

for two boosts in the same direction.

The above equations relating $\{\Delta t, \Delta x\}$ and $\{\Delta \bar{t}, \Delta \bar{x}\}$ were first presented by Voigt [552] in 1887 in a theoretical analysis of the Doppler effect based on the use of an elastic model of the luminiferous ether and matter to describe optical properties. Essentially equivalent results were obtained independently by Lorentz [336] in 1899. The above form was given by Lorentz [340] in 1904, prior to the presentation of special relativity by Einstein, although with a different dynamic rather than kinematic interpretation. Poincaré first referred to these transformations [446] in 1904 as *Lorentz transformations*, thus honouring the major contribution of Lorentz to electrodynamic theory and the Relativity principle. In line with modern practice, we shall refer to them as the *Lorentz boost transformations* in order to reserve the term *Lorentz transformation* for the covariant combination of a Lorentz boost and a rotation. Poincaré was also the first [448, 449] (*Oeuvres* [451, vol. 9, pp. 490 and 513]) to use group theory to show that the Lorentz transformation left Maxwell's equations unchanged in form, referring to the set of transformations as the *Lorentz group*.

Exercise 5.5 *Show that the Lorentz boost transformation is consistent with the Principle of causality for events related by $\Delta x \leq c\Delta t$ provided the boost speed satisfies $v < c$.*

There are currently no observations which conflict with the interpretation this transformation gives to the effect of boosting locally from one frame to another. The causal intervals obtained from demanding that $\Delta t \geq 0$ implies $\Delta \bar{t} \geq 0$ are those for which $|\Delta \mathbf{x}| \leq c\Delta t$ showing that causal signals have a velocity range of $-c \leq u^k \leq c$ which includes $\pm c$ as an achievable limit. The speed c is therefore an *ultimate speed*, an idea suggested privately by FitzGerald [197] in 1889, by Lorentz [338, 340] from 1900 and Poincaré [446] in 1904 as a necessary consequence of extending the Relativity principle to electromagnetic phenomena according to the electron theory of Lorentz. However, the tentative nature of Poincaré's conclusions are illustrated by his statement: *Perhaps likewise, we should construct a whole new mechanics ... where inertia increasing with velocity, the velocity of light would become an impassable limit.*

The Galilean relativity of the second of the above three boost cases permitted by the Relativity principle may be considered as the limit of this third Lorentz boost as $1/c \to 0$ for finite particle and boost speeds, or equivalently, as the $c \to \infty$ limit. Our analysis cannot provide a value for the ultimate speed nor any connection between it and the speed of propagation of any physical phenomenon. Lorentz boost transformations and Einsteinian kinematics will be discussed in Chapter 12.

5.6 Galilean relativity

For it may be that there is no body really at rest [that is, in a state of absolute rest], to which the places and motions of others may be referred[22].

Isaac Newton   1642-1727

Analysis of the Relativity and Causality principles has shown that the relative locations and time intervals of objects measured in two spacetime frames in uniform relative motion may be related in one or other of only two ways. If we restore the transformations not included during the discussion of relative uniform motion, then the first of these alternatives is

---

[22]*Principia* [395, p. 10].

5.6. Galilean relativity

Galileo Galilei

b. Pisa, Italy
15 February 1564

d. Arcetri, Italy
8 January 1642

Courtesy of BBC Hulton Picture Library, London.

I do not hold and have not held this opinion of Copernicus since the command was intimated to me that I must abandon it ...
[Cited by Gingerich [198].]

the *restricted Galilean group* of transformations, namely

$$t \to \bar{t} = t + t_0 \qquad (5.33)$$
$$\mathbf{x} \to \bar{\mathbf{x}} = R(\mathbf{x} - \mathbf{v}t + \mathbf{a}), \qquad R^T R = \mathbb{1}, \quad \det R = 1, \qquad (5.34)$$

and the pair of time and space reflections,

$$t \to \bar{t} = -t \qquad (5.35)$$
$$\mathbf{x} \to \bar{\mathbf{x}} = -\mathbf{x}, \qquad (5.36)$$

all of which taken together comprise the full *Galilean group* of isometry transformations of the spacetime of non-gravitational Newtonian physics. Since the individual transformations do not always commute with one another, we conventionally assume that the spacetime translations are followed by a boost which is followed by a rotation. We have included the reflections in the above set since Newtonian physics was traditionally applied to non-quantum systems none of which violate parity or time reversal invariance.

These Galilean transformations may be recognized as implying *absolute simultaneity*, $|\Delta \bar{t}| = |\Delta t|$, as postulated explicitly by Newton and universally accepted until the birth of Einsteinian relativity. Thus two events which are simultaneous ($\Delta t = 0$) in one frame will, in Galilean relativity, be simultaneous ($\Delta \bar{t} = 0$) in any other frame and, more generally, all frames will lead to the same result for the value of the time interval between a pair of events.

The hypothesis of absolute simultaneity, the assumption of additivity of boost velocities (according to Equation 5.29), or the absence of a finite upper bound on the speeds of boosts or particles, are all equivalent ways of supplementing the first postulate, the Principle of relativity, with a second postulate leading to Galilean relativity. We may then combine the first and second postulates to form,

- **The Hypothesis of Galilean covariance** — the local laws of physics have the same form in frames of reference that are related by a Galilean transformation.

We have called this a *hypothesis* since it is not exactly corroborated by experiment. Spacetime, $\mathbb{G} = \{\mathbb{E}^1, \mathbb{E}^3\}$, endowed with frames whose coordinates are related by Equations 5.33 to 5.36, has 1D and 3D Euclidean subspaces $\mathbb{E}^1$ and $\mathbb{E}^3$ but is no longer a simple product $\mathbb{E}^1 \times \mathbb{E}^3$. $\mathbb{G}$ may be referred to as *Galilean spacetime*. Motion in such a spacetime obeys Galilean kinematics. The Galilean relativity of non-gravitational Newtonian mechanics was well-known prior to the importance attached to the Relativity principle by Einstein. However, it was generally observed as a property of dynamics constructed in other ways rather than as a fundamental kinematic symmetry principle concerning the measurement of space and time to be used to restrict the form of any dynamical equations.

The Galilean transformations may be interpreted passively in terms of one system viewed from two reference frames corresponding to two sets of observers. Laws which remain unchanged in form with respect to these transformations may or may not describe real physical systems. In fact, it is a matter of experimental verification that they do locally describe real physical systems to first order in the ratio v/c provided it is small compared to unity. In such circumstances, an active description in terms of a single reference frame and a transformation from one physical system to another will be equivalent to the passive description.

The concept of a Euclidean scalar (see Section 4.1) may be readily extended to that of a *Galilean invariant* by requiring that it have the additional property of not changing in value in a Galilean boost. The coordinate length $|\Delta \bar{\mathbf{x}}| = |\Delta \mathbf{x}| \pm v |\Delta t|$ between two events (for **v** parallel to $\Delta \mathbf{x}$) is not a Galilean invariant. *Proper Newtonian length* is, by definition, the coordinate length in the rest frame of an object or equivalently the length between two simultaneous events at the end-points, thereby ensuring its Galilean in-

5.6. Galilean relativity

variance. Other examples of Galilean invariants are the 3-volume element $d^3x = (dxdydz)_{dt=0}$, defined using proper lengths, the time interval dt between two events, the mass m and the electric charge q of a particle. Provided it depends only on the difference between pairs of Cartesian coordinates at the same time, and not on the individual coordinates, the Lagrangian L of a mechanical system will also be Galilean-covariant, along with the action S determined from it.

Exercise 5.6 *Show that the time rate of change $\partial_t \phi$ of a Galilean invariant field $\phi(t, \mathbf{x})$ transforms under a Galilean boost according to,*

$$\partial_{\bar{t}} \bar{\phi} = \partial_t \phi + \mathbf{v} \cdot \nabla \phi , \qquad (5.37)$$

and is thus not a Galilean invariant.

Let a *Galilean 3-vector* be a Euclidean 3-vector which remains invariant under a Galilean boost. The displacement $d\mathbf{z}$ of a particle along its trajectory $\mathbf{z}(t)$ during a time interval dt and the velocity $\dot{\mathbf{z}}$ of a particle are not Galilean 3-vectors.

Exercise 5.7 *(a) Show that the change in velocity $d\dot{\mathbf{z}}$ and the acceleration $\ddot{\mathbf{z}}$ of a particle of trajectory $\mathbf{z}(t)$ are Galilean 3-vectors. (b) Show that the displacement $d\mathbf{z}$ of a particle is not a Galilean 3-vector but that the separation $\Delta \mathbf{z} = \mathbf{z}_2(t) - \mathbf{z}_1(t)$ between the spatial locations of two particles, 1 and 2, at the same time t, is a Galilean 3-vector. (c) Show that the gradient $\nabla \phi$ of a Galilean scalar field ϕ is invariant,*

$$\bar{\nabla} \bar{\phi} = \nabla \phi \qquad (5.38)$$

under Galilean boosts and is thus a Galilean 3-vector with ∇^2 being a Galilean invariant operator.

The foregoing results may be used to check the Galilean covariance of the dynamical equations of a system of non-quantum particles and fields. If we attempt to construct such a theory, the change in position $d\mathbf{z}$, along a particle path, a particle velocity $\dot{\mathbf{z}}$ or the rate of change with time $\partial_t \phi$ of a scalar field, cannot be included in dynamical equations unless each is compensated by the presence of other terms which are similarly not Galilean covariant. However, we may very easily incorporate velocity differences, accelerations, gradients of scalars and the Laplacian operator in equations describing spatial distributions or dynamics and still maintain Galilean covariance. One of the reasons why the fundamental equations of motion of Newtonian physics are second-order differential equations, rather than first-order, is that Galilean covariance may be incorporated more simply with the former.

Most of the unusual features of Galilean, as opposed to Euclidean covariance, and the Poincaré covariance of special relativity, are related to the absence of any 4D Galilean metric or scalar product in Newtonian spacetime. It is for this reason that we have referred to Galilean invariants rather than Galilean scalars.

Outside the range of validity of non-gravitational Newtonian physics, physical systems will not be described by equations remaining invariant under the Galilean boost transformations. Under such circumstances, passive and active transformations will not be equivalent, and the passive transformation loses physical meaning. The active transformation will always give a new physical system but its description need not be included in that of the original Newtonian system.

5.7 Aristotelean and Galilean gravity

The Aristotelean view of the fall of bodies was that the rate depended on the weight of the body: *Similarly, a larger quantity of gold or lead moves downwards faster than a smaller and so with all heavy bodies* [20, Bk. iv, ch. ii, verse 309-b-15].

Galileo was the first to introduce time quantitatively into the description of physical phenomena. His laws concerning accelerating and falling bodies were based on direct observations, rather than on a combination of observation and principles related to the cause of the acceleration. They were thus *empirical* laws. Although gravity provided the acceleration leading to the falling motions he studied, many of his observations and principles were primarily kinematic in nature. By using inclined planes to slow the motion of falling bodies, he was able to measure their time of fall much more precisely with the very limited precision clocks available at the time. This permitted him to show that the final velocity does not depend on the weight nor on the distance down the inclined plane, but only on the vertical distance covered and was proportional to the square of the elapsed time. He also showed that all bodies, irrespective of their weight, were accelerated by gravity at the same rate. This last result, now known as the *Principle of uniqueness of free fall* or the *Weak equivalence principle* (see Section 7.11) is the major new property of gravitation discovered by Galileo.

5.8 Fibre bundle structure of Galilean spacetime

Equation 5.28 shows that the 3D Euclidean scalar product of Equation 3.13 is not invariant under boosts. Consequently, we must not forget that, for Gaililean invariance, the intervals dx in Equation 3.13 cannot be arbitrary coordinate displacements between pairs of events, possibly at different

5.8. Fibre bundle structure of Galilean spacetime

times. Instead they must be proper Newtonian lengths, namely coordinate intervals between two events at the same time ($dt = 0$), for which Equation 5.28 gives $d\bar{\mathbf{x}} = d\mathbf{x}$ which leaves the 3D scalar product boost invariant. Without the Principles of either special or general relativity there is no 4D metric providing one with a distance between events which are not only at different times, $dt \neq 0$, but also at different spatial locations, $d\mathbf{x} \neq 0$. In contrast to Euclidean spacetime, one may specify only a unique boost-invariant distance between the two events in Newtonian physics if those events occur at the same time ($dt = 0$).

The spacetime $\mathbf{G}$ of Galilean relativity or kinematics, labelled by $\{t, \mathbf{x}\}$ and transforming according to the 10-parameter Galilei group of transformations is thus not a 4-dimensionally geometric spacetime in the same sense as ordinary Euclidean space, $\mathbf{E}^3$. The time difference between any pair of events is well-defined but it is not meaningful to speak of the spatial separation of events which are not simultaneous. Similarly, there is no notion of orthogonality between the time direction and any of the spatial directions.

Technically, the spacetime of Newtonian physics is a smoothly coordinatizable space or *manifold* of a type referred to as a *fibre bundle* [499] in which the unique 1D absolute time space $\mathbf{E}^1$ is referred to as the *base manifold*, at each point of which is attached a *fibre* consisting of an infinite number of 3D Euclidean spaces $\mathbf{E}^3$. Those spaces correspond to the frames which are related to one another either by spatial translation or have a common origin, but are rotated or in relative uniform motion. The set of all spatial frames at each time is referred to as a *bundle*. In addition, Galilean spacetime has a preferred family of straight lines, the loci of free particles and it thus has an *affine space* character compatible with its fibre bundle structure in the sense of all fibres being parallel. Galilean spacetime, with its 10-parameter isometry group is weaker in structure than the Euclidean spacetime product $\mathbf{E}^1 \times \mathbf{E}^3$ with its 7-parameter isometry group.

This completes our discussion of spacetime kinematics. We shall return to Equations 5.30 in Chapter 12 after first discussing the dynamical equations of Newtonian physics that are based on Galilean relativity and applying them to the formulation of Newtonian gravity and the Schrödinger equation.

It is bad enough to know the past; it would be intolerable to know the future[23].

William Somerset Maugham   1874–1965

There is not even a meaning to the word *experience* which would not presuppose the distinction between past and future.

Carl-Friedrich von Weizsäcker 1912–

---

[23] Mackay [353].

Bibliography:

Bunge M 1973 *Philosophy of physics* [55].
Cohen I B 1985 *Birth of a new physics* [72].
Galilei G 1610 *The sidereal messenger* [180].
Gingerich O 1982 *The Galileo affair* [198].
Goldberg S 1984 *Understanding relativity* [203].
Holton G 1960 *On the origins of the special theory of relativity* [254].
Kuhn T S 1957 *The Copernican revolution* [290].
Mook D E and Vargish T 1987 *Inside relativity* [387].
Resnick R 1968 *Introduction to special relativity* [472].
Schutz B F 1980 *Geometrical methods of mathematical physics* [499].

CHAPTER 6

Newtonian mechanics

I know not what I may appear to the world, but to myself I seem to have been only like a boy playing on the sea-shore, and diverting myself in now and then finding a smoother pebble or a prettier shell than ordinary, whilst the great ocean of truth lay all undiscovered before me[1].

Isaac Newton   1642–1727

Galilean relativity adequately describes the local kinematics of events in terms of the coordinates of frames in the case of small relative velocities $v \ll c$ and low matter densities. By low density, we mean $\rho \ll c^2/G\ell^2$ where ℓ is a characteristic size of the system (see Exercise 1.3 and Section 21.6) and G is the gravitation constant. Some of the features of Galilean relativity and the Newtonian dynamics based on it will be explored in this and succeeding chapters in order to better appreciate the differences that arise in Einsteinian or special relativity to be developed in Chapter 12.

6.1 Newtonian dynamics and conservation

The local non-quantum physics of material bodies at slow speeds is based almost entirely on Newton's three laws of mechanics. The first of these laws, the *Principle of inertia*, and some aspects of the other two, were implicit in the work of Galileo [179, 182] on motion and mechanics in 1592. The complete set of dynamical laws, crucial in the development of mechanics, were first presented by Newton [395] in 1687 in his text, *Philosophiae naturalis principia mathematica*, namely the *Mathematical principles of physics (or natural philosophy)*, commonly referred to as the *Principia* [70].

It is possible to present these laws in a manner which highlights the Galilean relativity they implicitly assume. Furthermore, their essential content may be expressed in a way which eases their extension to non-mechanical systems such as wave functions or fields, both classical and

---

[1] Brewster [50, vol. 2, chap. 27].

Isaac Newton

b. Woolsthorpe, Lincs., England
25 December 1642

d. Kensington, London
20 March 1727

Engraving, Wren Library, Cambridge,
of the Kneller portrait (1689), Portsmouth Estates.
Courtesy of the Master and Fellows, Trinity College.

Whence arises all that order and beauty we see in the world?

Physics, beware of metaphysics.

quantum. Galilean and Einsteinian kinematics are the only two kinematical relationships consistent with the Causality and Relativity principles and the difference between their forms, although extremely significant, is minute. One expects therefore that it ought to be possible to formulate Galilean-covariant dynamical laws, equivalent to those of Newton, in a way which closely resembles the standard Poincaré-covariant form of Einsteinian dynamics.

Let us recapitulate the situation thus far. We presume that we have a kinematic theory involving well-defined notions of distant simultaneity, reference frame, location, motion, time, velocity and acceleration. The dynamical concepts we now require are: inertia, inertial frame, isolated body, constant velocity, mass, momentum, energy, angular momentum, conservation and the origins of acceleration in the crucially important concept of *interaction* or *force*. In short, we require many of those notions which were first formulated by Newton in his laws of mechanics. Some of these concepts will be simply definitions — a central core will contain the essential physics of Newtonian mechanics.

6.1. Newtonian dynamics and conservation

So far we have only applied the Relativity principle to the transformation of the spacetime coordinates $t, \mathbf{x}$. Kinematics based on the Relativity principle supply us with ten parameters, $\mathbf{a}$, $\boldsymbol{\theta}$, $\mathbf{v}$ and t_0, corresponding to translations ($\mathbf{a}$ and t_0), rotations ($\boldsymbol{\theta}$) and boosts ($\mathbf{v}$), which leave unchanged the scalar products $|dt|$ (for arbitrary $d\mathbf{x}$) and $d\mathbf{x}_1 \cdot d\mathbf{x}_2$ (for $dt = 0$), and the affine fibre bundle structure of Galilean spacetime. However, the principle is hypothesized to apply also to the dynamics of all the laws of physics. This means that our laws must be expressible in terms of Galilean-covariant quantities in such a way that the laws of physics are unchanged in form with respect to the 10-parameter Galilean transformations, Equations 5.33 to 5.36. This in turn means we must use Euclidean-covariant quantities only in combinations which also display boost invariance. The Galilean invariants and vectors of kinematics available to us are time intervals dt, displacements $d\mathbf{x}$ at constant time $(dt = 0)$, accelerations $\mathbf{a}$ or changes $d\dot{\mathbf{z}}$ in velocity.

Newton's laws are general principles that any interaction between material bodies must satisfy. We should not need the details of any specific interaction at this stage. To formulate the laws, we shall introduce one new primitive physical quantity, of dynamical character, the *mass*. A kinematic characterization of a non-quantum particle is one which is described by a smooth trajectory $\mathbf{z}(t)$ in Galilean spacetime. We now extend that idea by defining a particle dynamically as one which is described by a trajectory $\mathbf{z}(t)$ and a non-zero mass m. The Galileo-Newtonian mass is defined to be an internal additive property, conserved in time, whose rôle is to characterize, in a Galilean invariant the *inertia* of the material particle, namely the extent of its change in velocity in response to an interaction of a given strength, with another body. By *internal* one means that, unlike velocity and acceleration, it is not obtainable from $\mathbf{z}$ and t. By *additive*, we mean that the mass of a composite body is the sum of the masses of its parts.

The way in which the mass characterizes the inertia and isolation of a body will be defined implicitly, along with the ten external concepts of *momentum*, *angular momentum*, *centre of mass motion* and *energy*. Each is defined to be proportional to the inertial mass of the system with such dimensions that their products with $\mathbf{a}$, $\boldsymbol{\theta}$, $\mathbf{v}$ and t_0, respectively, all have the same dimensions. The definitions are designed to ensure that all ten quantities of momentum, angular momentum, centre of mass motion and energy, as well as mass itself, are conserved. This is done in a way which links the conservation to the fact that each of the ten spacetime quantities — position, direction, velocity and time — cannot be determined absolutely; only differences are significant.

The definitions must also implicitly determine the dynamical notions of *inertial frame* and *isolated body*. Failure to constitute an isolated body will lead to the concept of *interaction*. Introducing these eleven quantities (one internal and ten external) in a way which demands that they be conserved and additive will permit a partial 'derivation' of the conventional form of Newton's laws, with the advantage of incorporating the parallel between the ten conserved quantities. One justification for this rather unorthodox presentation is that the conservation laws of momentum, angular momentum, centre of mass motion, energy and mass transcend Newtonian physics and remain valid not only in special relativistic mechanics but also where fields are involved. Morever, their applicability extends to quantum physics where the classical notions of position, momentum and force, which are central to Newton's laws, have limited usefulness.

Let the path of a material system characterized by one internal variable, the mass m, be specified by the external position coordinates t and $\mathbf{z}(t)$ in some frame or, equivalently, the 3-vector relative position $\mathbf{z}-\mathbf{d}$, where $\mathbf{d}$ is any fixed point used for reference. Then one may postulate:

- the **Newtonian conservation laws of Galilean relativity** — if the motion of the system, relative to a given reference frame, is such that m and its,

    - *3-momentum* $\mathbf{p} = m\dot{\mathbf{z}}$,
    - *orbital angular momentum* $\mathbf{l} = (\mathbf{z}-\mathbf{d})\times\mathbf{p}$ about $\mathbf{d}$, and
    - *centre of mass motion* $\mathbf{k} = \mathbf{p}t - m\mathbf{z}$,

    are all additive quantities which are conserved with time, and if a Euclidean scalar additive *energy* $E = E(m, \mathbf{z}, \dot{\mathbf{z}})$ may be constructed which is conserved as a result of time translation invariance, then the frame is said to be an *inertial frame* and the system is said to behave as an *isolated Newtonian particle* of mass m.

Exercise 6.1 *(a) Confirm that momentum, angular momentum and centre of mass motion, $\mathbf{p}$, $\mathbf{l}$ and $\mathbf{k}$, are Euclidean but not Galilean vectors. (b) Show that, nevertheless, the conservation laws are Galilean invariant statements valid in any frame moving uniformly with respect to a given inertial frame.*

The forms chosen for the conserved quantities may appear to be quite arbitrary. That there is some naturalness to their structure can be seen by noting that the first two parallel the forms of coordinate translations $\Delta\mathbf{z} = \mathbf{a}$ and rotations, $\Delta\mathbf{z} = \Delta\boldsymbol{\theta}\times(\mathbf{z}-\mathbf{d})$, about $\mathbf{d}$. Re-expressing the second (about the origin $\mathbf{d} = 0$) in its dual form, $l^{kl} = z^k p^l - z^l p^k$, or writing out the three components of the cross product, shows that the third has a similar structure as a difference between two products, except that, as is appropriate for a boost, the spatial coordinates $\mathbf{z}$ have been replaced by time t and the momentum by mass.

6.1. Newtonian dynamics and conservation

We may also note that the dimensions of $\mathbf{a}\cdot\mathbf{p}$, $\boldsymbol{\theta}\cdot\mathbf{l}$ and $\mathbf{v}\cdot\mathbf{k}$ are all the same. Each of the 10 external conservation laws may correspond in a similar way to properties of the Euclidean subspaces $\mathbf{E}^1$ and $\mathbf{E}^3$ of Galilean spacetime, $\mathbf{G} = \{\mathbf{E}^1, \mathbf{E}^3\}$, if we require the product $t_0 E$ to have the same dimensions as the other three. We define the *kinetic energy* to be $T = \mathbf{p}^2/2m$, and require it to contribute additively in the construction of the total conserved energy E. The energy requires special consideration since it corresponds to time translation which almost inevitably involves interaction.

Exercise 6.2 *Show that the kinetic energy $T = \mathbf{p}^2/2m$ is, up to a conventional scale factor, the only Euclidean scalar in m, $d\mathbf{z}$ and dt with the required dimensions of energy. (We note that, like the other conserved mechanical quantities, it is proportional to the mass m.)*

The expressions chosen for the conserved quantities appear more natural from the perspective of a low-velocity approximation to Einsteinian relativity (see Chapter 15).

It may appear superfluous to include $\mathbf{l}$ and $\mathbf{k}$ in the above laws since the conservation of each follows from those of $\mathbf{p}$ and m. For example, the standard expression $\mathbf{p} = m\dot{\mathbf{z}}$ for the momentum of a Newtonian particle in terms of the mass and the velocity is very closely related to the form of the expression for the centre of mass motion and its conservation. However, the additivity of these quantities is also a crucial ingredient in any results to be deduced from the above requirements and does not follow from the properties of the momentum.

The addition of a constant term, m_0, $\mathbf{p}_0$, $\mathbf{l}_0$, $\mathbf{k}_0$ and T_0 to each of the conserved quantities does not alter the essential physical properties of additivity and conservation. Each is therefore arbitrary, by definition, to the extent of an additive finite constant. Equivalently, the zero of each is fixed by convention and it is the difference between two values which is physically significant, not its actual value. Conservation of $\mathbf{p}$ and $\mathbf{l}$ (about $\mathbf{d}$) ensures that angular momentum $\mathbf{l}_c = (\mathbf{z}-\mathbf{c})\times\mathbf{p}$ about any other reference point $\mathbf{c}$ is also conserved. In a quantum context, this additivity corresponds to the conserved quantity being the eigenvalue of an hermitian generator of a continuous unitary symmetry. This contrasts with the multiplicative nature of conserved quantities arising from discrete symmetries.

In the geometry of Euclidean spacetime, $\mathbf{E}^1 \times \mathbf{E}^3$, considered mathematically, there is no distinction between the time dimension and any one spatial dimension considered on its own. The only distinction between time and space is their difference in dimensionality. At the kinematic level, the introduction of the Galilean relativity principle augments that distinction between time and space by hypothesizing the absoluteness of simultaneity. At the dynamic level, we see the distinction between the two increasing still further with time assuming its rôle as evolution parameter, and space providing, at least for mechanical systems, the arena in which are defined the dynamical variables, such as $z(t)$, that depend on the evolution parameter. This distinction is immediately apparent in the very concept of conservation, in which time plays an ever present part. Some of these distinctions are softened by the introduction of Einsteinian relativity and of fields, but the distinction at the physical level remains real and profound.

If the conservation laws are satisfied in some frame, then that frame $\{t, \mathbf{x}\}$ is inertial, by definition. It immediately follows that any frame $\{t, \bar{\mathbf{x}}\}$, which is translated, rotated (but not rotating) or boosted with respect to the first, is also inertial. In the case of a boost, this follows since the new quantities,

$$\bar{\mathbf{p}} = \mathbf{p} - m\mathbf{v}, \quad \bar{\mathbf{l}} = (\bar{\mathbf{z}} - \bar{\mathbf{d}}) \times \bar{\mathbf{p}} \quad \text{and} \quad \bar{\mathbf{k}} = \bar{\mathbf{p}}t - m\bar{\mathbf{z}} , \qquad (6.1)$$

are conserved in the new frame with boosted coordinates $\bar{\mathbf{x}} = \mathbf{x} - \mathbf{v}t$ and centre of mass $\bar{\mathbf{z}}(t)$. It follows that there always exist inertial frames, called *rest frames* in which the velocity, 3-momentum, kinetic energy and angular momentum of the isolated particle are zero. The rest frames will be related to one another by translations of the origin and rotations of the axes. Among those rest frames will be one frame, the *centre of mass frame* of the isolated particle, for which the centre of mass motion is zero and in which the centre of mass is at the origin. We have now clearly reached:

- **Newton's first law of motion**, Galileo's *Principle of inertia* — every isolated particle continues in its state of rest or of uniform motion in a straight line in any inertial frame. Alternatively, following a standard translation of the opening section of the *Principia* [395], the first editions of which were written in Latin: *Every body continues in its state of rest or of uniform motion in a straight line except in so far as it is compelled by impressed forces to change that state.*

The answer Newton's first law gives to the query of what keeps an astronomical object in perpetual motion at almost constant velocity is not a supernatural *primum mobile* but *nothing* at all. The standard form of the first law, still the one used in almost all textbook treatments, is sometimes criticized for its logic. However, such criticisms can only be upheld if the first law, as expressed by Newton, is considered in isolation from the other two which was clearly not Newton's intention. Excellent elementary critiques of Newton's laws, in conventional form, are given by Eisenbud [141] and Weinstock [567].

The solution for the trajectory $z(t)$ of an isolated body, in an arbitrary inertial frame, is obtained trivially from the mass and momentum conservation, namely from $\ddot{z}(t) = 0$, as

$$z(t) = at + b . \qquad (6.2)$$

The constants a and b are uniquely determined in terms of either the initial position $z(0)$ and initial velocity $\dot{z}(0)$ or from the positions $z(t_1)$ and $z(t_2)$ at times t_1 and $t_2 \neq t_1$, according to,

$$z(t) = \dot{z}(0)\, t + z(0) \quad \text{or} \quad z(t) = \frac{z(t_2) - z(t_1)}{t_2 - t_1}(t - t_1) + z(t_1) \;. \qquad (6.3)$$

If Newtonian physics were exactly true, the only limits on the values of the relative locations, times and velocities are,

$$0 < |t_2 - t_1| < \infty \quad \text{and} \quad 0 < |z_2 - z_1|, |\dot{z}_2 - \dot{z}_1| < \infty \;. \qquad (6.4)$$

One reason why the centre of mass motion attracts little attention as a conserved quantity can be based on the following quantum arguments. The quantum Hamiltonian operator $\hat{H}$, corresponding to the classical energy E, is invariably included in a complete set of commuting observables for an isolated system because of its connection with evolution. So also are the momentum operator $\mathbf{P}$ and one component L_z of the angular momentum operator, $\mathbf{L}$. Once $\hat{H}$, $\mathbf{P}$ and L_z are included, we cannot also include the other angular momentum components since they do not commute among themselves. Likewise, the components of the centre of mass motion operator $\mathbf{K}$ do not commute with the components of the angular momentum (see Exercise 11.6) and they cannot be added to a standard energy and angular momentum basis. A similar classical argument would use Poisson brackets.

6.2 Additivity of conserved quantities

A non-quantum particle is an object whose size is negligible compared to other lengths which are relevant to its dynamics. Whether an object is considered a particle or a composite body will often depend only on circumstances. If we examine a collection of isolated particles at a more coarse scale, we may treat the overall system as one isolated particle. The application of the conservation laws to the parts and the whole must be consistent and the requirement of additivity of the conserved quantities helps ensure that this is so. The additivity of mass and momentum permits us to determine their values for a composite body in terms of those of its parts of mass m_i at locations z_i ($i = 1, 2, \ldots,$) according to,

$$m = \sum_i m_i \quad \text{and} \quad \mathbf{p} = \sum_i \mathbf{p}_i = \sum_i m_i \dot{z}_i \;. \qquad (6.5)$$

The additivity $\mathbf{k} = \sum_i \mathbf{k}_i$ of the centre of mass motion then implies the usual relation between the *centre of mass* of the composite body and those of its parts,

$$z = \frac{\sum_i m_i z_i}{\sum_i m_i} \;. \qquad (6.6)$$

The total angular momentum (about **d**) and kinetic energy of a composite body will be given similarly by,

$$\mathbf{j} = \sum_i \mathbf{l}_i = \sum_i (\mathbf{z}_i - \mathbf{d}) \times \mathbf{p}_i \quad \text{and} \quad T = \sum_i T_i = \sum_i \frac{\mathbf{p}_i^2}{2m_i} \,. \tag{6.7}$$

Based on Galilean relativity and spacetime, we have set up a model of Newtonian dynamics in such a way that, by definition, an isolated body moves in a straight line in an inertial frame with conserved momentum, angular momentum and centre of mass motion. It should be noted, however, that this essentially kinematic part of Newtonian physical principles is little more than very useful conventions and definitions. The essential physics lies elsewhere.

6.3 Interaction in Newtonian physics

For it's well known, that Bodies act one upon another by the Attractions of Gravity, Magnetism, and Electricity, and these Instances shew the Tenor and Course of Nature, and make it not improbable but that there may be more attractive Powers than these[2].

Isaac Newton   1642–1727

In addition to defining an inertial frame, an isolated body and its conserved mechanical quantities, the conservation laws also say a great deal about *dynamics* or *interaction*, namely about the physics of change involving the motion of particles which are not isolated. Let us now use the additivity to examine the relation between the conservation laws based on Galilean relativity and the second and third of Newton's Laws governing motion of material bodies.

Suppose we have found at least one combination of body and reference frame which our observations permit us to classify, to a certain precision, as an isolated body in an inertial frame. Having identified an inertial frame, we may use it to examine bodies which are isolated overall ($d\mathbf{p}/dt = 0$) but the parts of which are not always isolated from one another. Let us consider division into two subsystems 1 and 2. At some times the separate parts may have non-zero rates of change of their individual momenta,

$$\frac{d\mathbf{p}_1}{dt} = -\frac{d\mathbf{p}_2}{dt} \neq 0 \,. \tag{6.8}$$

They may, for example, initially be isolated and evolve freely, approach one another and then show evidence of not being isolated, with compensating variation of their momenta to ensure that the entire system is isolated overall. Then we say that the two subsystems are *mutually interacting* or that

---

[2] *Opticks* [396, p. 376].

6.3. Interaction in Newtonian physics

the whole system is *self-interacting*. For the interaction of only two constant mass parts, from Equation 6.8 we obtain $m_1/m_2 = |\ddot{z}_2|/|\ddot{z}_1|$, which may, at least in principle, be used as the basis for measuring the masses of objects relative to a standard mass from measurements of accelerations. This relation, along with the appearance of mass as an overall proportionality constant in each conserved mechanical quantity, illustrates the fact that mass is the primitive dynamic quantity inseparable from the notion of interaction, as propounded by Mach [352].
We now postulate:

- the **Hypothesis of Newtonian causality** — all material change depends on a material cause external to it and the effect of the cause takes place instantaneously.

This basic tenet of Newtonian physics is consistent with the Principle of causality (since $\Delta t = 0$ implies $\Delta \bar{t} = 0$). Radioactive decay provides one example of a spontaneous phenomenon showing that quantum processes are not subject to Newtonian causality. Consistent with this hypothesis and Galilean relativity, we also postulate:

- the **Newtonian hypothesis of superposition of interaction** — the rate of change of momentum $d\mathbf{p}_i/dt$ of any part i of an interacting system at time t due to a part j is defined to be the value it would have if all other parts were instantaneously removed at time t from influencing i and depends only on the displacement $\mathbf{z}_j(t) - \mathbf{z}_i(t)$. The net rate of change of momentum $d\mathbf{p}_i/dt$ of any part i of an interacting system is the sum of the contributions due to its interactions with each other part.

This principle was referred to by Mach as the *law of parallelograms*. Both hypotheses imply that Newtonian mechanical interaction involves instantaneous *action at a distance*, a concept which does not survive the transition from Galilean to Einsteinian relativity. The dynamics may be considered to be *spatially non-local*. If we define the *force* of one member i on the other j, in each mutually interacting pair, to be the values,

$$\mathbf{f}_{ij} \equiv \mathbf{f}_{i\,\text{on}\,j} = \left.\frac{d\mathbf{p}_j}{dt}\right|_{\text{due to }i} \quad \text{and} \quad \mathbf{f}_{ji} \equiv \mathbf{f}_{j\,\text{on}\,i} = \left.\frac{d\mathbf{p}_i}{dt}\right|_{\text{due to }j}, \quad (6.9)$$

of the rates of change of momentum in a system composed of just the two parts i and j, then the superposition law permits us to re-express the definitions of Equation 6.9 as,

- **Newton's second law of motion** — the rate of change of momentum of a body is given by the net force acting on it,

$$\frac{d\mathbf{p}_i}{dt} = \sum_{j \neq i} \mathbf{f}_{ji}, \quad (6.10)$$

or more closely following Newton's words: *Change in motion is proportional to the impressed force and takes place in the direction of the straight line in which the force acts.*

This law is thus primarily a statement of the superposition principle along with a definition of force to supplement the first law, much as the first law re-expresses the definition of an inertial frame. However, Equations 6.8 and 6.9 together also contain another very significant result, namely:

- **Newton's third law of motion** (the *Principle of reaction*) **part I** — the interaction forces between two Newtonian bodies are equal and opposite,
$$\mathbf{f}_{ij} = -\mathbf{f}_{ji} , \qquad (6.11)$$
or translating Newton more literally: *To every action there is an equal and opposite reaction.*

This result, a re-expression of momentum conservation, follows despite the presence of many interactions, one for each pair $\{i,j\}$, since these pairs may be considered singly and then superimposed.

Exercise 6.3 *Show that the total orbital angular momentum of a composite body about an arbitrary point* $\mathbf{d}$, *comprised of the sum,* $\mathbf{j} = \sum_i \mathbf{l}_i$, *of the orbital angular momenta* $\mathbf{l}_i$ *of the constituent particles about* $\mathbf{d}$, *may be decomposed into two parts according to,*
$$\mathbf{j} = \mathbf{l} + \mathbf{s} , \qquad (6.12)$$
where $\mathbf{l}$ *and* $\mathbf{s}$, *given by,*
$$\mathbf{l} = (\mathbf{z}-\mathbf{d})\times\mathbf{p} \quad \text{and} \quad \mathbf{s} = \sum_i (\mathbf{z}_i-\mathbf{z})\times\mathbf{p}_i , \qquad (6.13)$$
are, respectively, the orbital angular momentum, of the total mass considered to be moving at the location $\mathbf{z}$ *of the centre of mass with the velocity* $\dot{\mathbf{z}}$ *of the centre of mass, and the sum of the orbital angular momentum of the parts about the centre of mass.*

The quantity $\mathbf{s}$ is referred to as the *spin angular momentum* of the composite system. The projection $h = \mathbf{j}\cdot\hat{\mathbf{p}} = \mathbf{s}\cdot\hat{\mathbf{p}}$ of the (spin) angular momentum of a system in the direction of its momentum is called the *helicity*.

Exercise 6.4 *Show (a) that the spin of a system is invariant under translations and boosts and is thus a Galilean vector (b) that the orbital angular momentum* $\mathbf{l}_c$ *about* $\mathbf{c}$ *is given in terms of* $\mathbf{l}_d$, *about* $\mathbf{d}$, *by* $\mathbf{l}_c = \mathbf{l} - (\mathbf{c}-\mathbf{d})\times\mathbf{p}$ *and (c) that, although spin is not conserved, helicity is conserved but is not a Galilean invariant.*

6.3. Interaction in Newtonian physics

The orbital, spin and total angular momentum of a system may alternatively be represented by the proper antisymmetric tensors,

$$l^{kl} = \epsilon^{klm} l_m, \quad s^{kl} = \epsilon^{klm} s_m \quad \text{and} \quad j^{kl} = \epsilon^{klm} j_m = l^{kl} + s^{kl}, \quad (6.14)$$

dual to the pseudo-3-vectors l, s and j, respectively.

The consequences of requiring that **j** be conserved may be easily examined in the centre of mass frame where **l** = **0**.

Exercise 6.5 *Use the centre of mass frame to show that conservation of total orbital angular momentum d**j**/dt = **0** implies that $(\mathbf{z}_i - \mathbf{z}_j) \times \mathbf{f}_{ij} = \mathbf{0}$.*

The result of the above exercise provides us with a re-expression of angular momentum conservation, as:

- **Newton's third law of motion, part II** — the interaction forces between any two parts of a Newtonian body are along the line joining the instantaneous locations of the parts; or following Newton: *The mutual actions of any two bodies are always equal and oppositely directed along the same straight line.*

One of the most significant features of Newton's laws is their general applicability, believed to be universal until the late 19th century. This is partly because they are expressed in terms of general properties of a body, its position and mass, not in terms of its detailed structure such as size, shape or composition.

At the non-fundamental level, the concept of *force* is extraordinarily powerful since it can summarize in a macroscopic quantity the net effects of an enormous amount of information at the microscopic level which generally cannot be examined in detail. For example, the contact forces of pressure and reaction when a person pushes against a solid wall are difficult to conceive in terms of rates of change of momentum without examining microscopic phenomena, which is also usually quite unnecessary in analysing static systems.

Furthermore, interest may primarily be in one part of an isolated system in the approximation where the details and dynamics of the remainder are not relevant, except in so far as they affect the first part. This is the case where a low mass system rapidly changes its velocity under the action of a much more massive system whose state remains relatively unchanged. The action of the larger system on the smaller may then be given or prescribed, often said to be an *external interaction*, the reaction of the system of interest on the external agent being neglected. Such would be the case for the motion of a pendulum or projectile in the gravitational field of the earth. The reaction on the earth is of little interest.

Another example is the motion of a few electrically charged particles in an electromagnetic field due itself to the huge numbers of other charges inside a magnet or electromagnet. The effect of the few particles of interest on those in the electromagnet and hence the changes in the main electromagnetic field are of little interest. (We note, however, that this electromagnetic example would take us outside the realm of validity of Newtonian physics.)

At the fundamental level, one is often very much more interested in the construction of *closed systems*, namely those in which all interacting parts are included in the system which is therefore isolated as a whole. In such cases, the rates of change of momenta of all parts are important and the concept of force, especially as an external prescribed agent, may be less useful. One may deal just as easily with the rates of change of the momenta themselves and emphasize the importance of all parts by concentrating on *interaction* rather than force. This point of view arises naturally from the conservation laws we have used above and is the approach we shall emphasize.

The examination of the details of interaction are greatly simplified by noting that the motion of a given non-quantum particle at a certain time t may be examined as a *local* effect, determined predominantly by its interactions with the nearest particles whose locations at times immediately to the past of t are the most relevant. This fact is related to the decrease in the strength of all interactions with increase in the separation between the interacting parts. Notwithstanding these remarks, it should not be forgotten that there may exist intrinsically *global* aspects of the dynamics of some systems, whether non-quantum or quantum.

When interaction takes place, the momentum $\mathbf{p}_i(t)$ of any interacting part is no longer constant in time. The interaction is said to be the *cause* of the acceleration. The Euclidean 3-vector nature of such momenta is confined to its rotational and space translation properties. The covariance of a Newtonian system under time translations applies to the whole system and not to its parts. The rate of change of momentum $d\mathbf{p}_i/dt$ of an interacting part is, however, still boost invariant. Since $d\mathbf{p}_i/dt$ is non-zero, the trajectory in an inertial frame of the i-th part of an interacting system will not be a straight line.

The essence of the Newtonian dynamical problem is to characterize the distinct types of interaction which occur in nature by determining the nature of the families of trajectories in each case. The characterization of the infinite number of possible trajectories into families is systematized by dividing the problem into two parts. The first gives expression to the temporally local nature of physical theory, by the formulation of a *differential equation of motion* that all trajectories of a given family must satisfy, the general solutions to which will contain arbitrary constants. The second is the specification of *boundary conditions* to determine the arbitrary constants appropriate to a particular trajectory.

A *temporally local theory* of mechanics is one in which the evolution of

6.3. Interaction in Newtonian physics

the variables describing the trajectory is governed by an equation of motion which is an ordinary differential equation in the trajectory variables, $z_i(t)$, and a finite number of its time derivatives, $\dot{z}_i(t)$, $\ddot{z}_i(t)$, ..., all evaluated at the same time t. Such an equation is determined, in the Newtonian mechanical case, from knowledge of the value of $d\mathbf{p}_i/dt$, or $m_i\ddot{z}_i(t)$, in terms of the locations of all the other particles ($j \neq i$), relative to the one of interest, at the same time t.

In Newtonian physics, one is almost invariably concerned with ordinary differential equations of second order arising by specifying the nature of the interaction using an explicit expression for the forces $\mathbf{f}_{ij}$ of the form,

$$\mathbf{f}_{ij} = \mathbf{f}_{ij}(t) = \mathbf{f}\big(z_i(t) - z_j(t)\big) \, , \tag{6.15}$$

where the dependence on the coordinates of the separate trajectories must be a difference, at the same time t, in order to preserve the Galilean covariance. For this form, the interacting equations of motion will automatically be time-reversal covariant. A *generalized Newtonian system* is one in which dependence on the relative velocities $\dot{z}_i(t) - \dot{z}_j(t)$ is also permitted. Although not strictly in accord with the conventional formulation of Newtonian mechanics, and not necessarily time-reversal covariant, it constitutes a natural generalization, especially in the context of a Lagrangian treatment, as we shall show in Chapter 8. The specification of an interaction via such explicit expressions for the force on a part of the system leads to a temporally local theory with a second-order ordinary differential equation for $z_i(t)$.

Exercise 6.6 *Show that the total kinetic energy $T = \sum_i \mathbf{p}_i^2/2m_i$ of a composite isolated body can be decomposed into a sum,*

$$T = \frac{\mathbf{p}^2}{2m} + \sum_i \frac{\bar{\mathbf{p}}_i^2}{2m_i} \, , \tag{6.16}$$

of the kinetic energy $\mathbf{p}^2/2m$ of the whole mass, at the centre of mass, moving with the centre of mass momentum $\mathbf{p}$ and the kinetic energy $\sum_i \bar{\mathbf{p}}_i^2/2m_i$ of the parts relative to the centre of mass (located at z), where the momenta relative to z are $\bar{\mathbf{p}}_i = \mathbf{p}_i - m_i \dot{z}$.

The observation of slowly moving systems from a particular frame permits us to classify the frame as inertial according to whether or not the system obeys Newton's laws in that frame. Since any other uniformly boosted frame is also inertial, uniform motion is relative, not absolute, in Newtonian physics, a property which it has in common with special relativity. However, the acceleration of a frame can be determined from within the frame by Newton's laws and is therefore an *absolute* concept in Newtonian physics, a property which also carries over to special but not general relativity.

> ... it comes about that we are inclined to ascribe to force a certain reality of its own. As in this case, so in other cases, the word *force* is just a name for some quantities, occurring in our mathematical formulae[3].
>
> Hendrik Antoon Lorentz   1853–1928

6.4 Potential energy of conservative systems

The rate of change $\dot{T} = \sum_i \dot{z}_i \cdot \dot{p}_i$ of the total kinetic energy of a composite system of particles need not be zero and T is thus not, in general, conserved. For a Newtonian system, we insist that we must be able to construct a conserved quantity, the *total mechanical energy* E of the system corresponding to the time translation invariance of Galilean relativity.

In a temporally local Newtonian theory of mechanics, in which any time dependence in $dp_i(t)/dt$ arises from its dependence on the trajectory variables, we may ask how this function $\dot{p}_i(\ldots z_i(t) - z_j(t)\ldots)$ varies with z_i. Let us concentrate on a part comprised of just one particle i interacting with one other j. We abbreviate $z_i - z_j$ by z and generally suppress the subscript i, for the moment. We are interested in variations of $\dot{p}(z)$ with z that may be formed in a covariant way by, for example, combining $\dot{p}$ itself with the gradient ∇_z with respect to the trajectory coordinates z. Helmholtz's theorem[4], a discussion of which is given in Appendix A.3, shows that any vector function such as $\dot{p}(z)$ may be decomposed into a longitudinal or conservative part which is curl-free and a transverse or solenoidal part which is divergence-free.

Consequently, we are interested in how such combinations as $\nabla_z \cdot \dot{p}$ and $\nabla_z \times \dot{p}$ might be related to scalars and vectors formed from the quantities m and z that dynamically characterize the trajectory of a Newtonian particle. We should expect both these quantities to be well-defined in terms of the dynamical variables. However, the second of these, the curl of $\dot{p}$, is a pseudovector and cannot be equated to a polar vector such as z or $g(|z|)z$, for example, without violating parity covariance, such violation being totally foreign to Newtonian physics and indeed to all non-quantum dynamics. Neither must it violate time reversal covariance.

One very important class of Newtonian dynamical systems, in which the curl of $\dot{p}$ is determined in accordance with parity and time reversal covariance, is comprised of those interactions for which the curl vanishes, $\nabla_z \times \dot{p}(z) = 0$. Systems in which all interactions are of this form are said to be *conservative*. *Central interactions*, those with $\dot{p} \propto z/|z|^p$, are clearly of this type, in which $g(|z|) = 1/|z|^3$ corresponds to the inverse square laws

---

[3] *Inaugural address*, 1878, Leiden [599].
[4] Hermann Ludwig Ferdinand von Helmholtz (1821–1894), German theoretical physicist and physiologist.

6.4. Potential energy of conservative systems

of Coulomb and Newton for electrostatics and gravistatics, respectively. In the simply-connected Euclidean space $\mathbf{E}^3$, the vanishing curl of a 3-vector function is a necessary and sufficient condition (see Appendix A.4) for the existence of a 3-scalar function, $V_1(\mathbf{z}(t))$, of position $\mathbf{z}$ in terms of which the vector function is given as a gradient according to,

$$\mathbf{f} = \dot{\mathbf{p}} = -\mathbf{\nabla}_{\mathbf{z}} V_1(\mathbf{z}(t)) \ . \tag{6.17}$$

The subscript on V_1 indicates that a single pair of particles is involved. Since $\dot{\mathbf{p}}$ and $\mathbf{\nabla}_{\mathbf{z}}$ are Galilean vectors, $V_1(\mathbf{z})$ must be Galilean invariant and thus a function $V_1(|\mathbf{z}|)$ of the magnitude of $\mathbf{z}$. Consequently, Galilean covariance shows that any conservative interactions will be *central*. The invariant V_1 is clearly undetermined to the extent of an arbitrary constant (independent of $\mathbf{z}$ but possibly time-dependent) and is called the *interaction potential energy* of the pair of particles. We may assume that all pairs of particles are interacting according to the same fundamental potential.

Restoring the suppressed index i (with no summation implied over i) gives us *Lagrange's formula* [296],

$$\dot{\mathbf{p}}_i = -\mathbf{\nabla}_i V_1(|\mathbf{z}_i(t) - \mathbf{z}_j(t)|) \ , \tag{6.18}$$

where $\mathbf{\nabla}_i = \partial/\partial \mathbf{z}_i$. Superimposing the contributions to $\dot{\mathbf{p}}_i$ from all particles j gives,

$$\dot{\mathbf{p}}_i = -\mathbf{\nabla}_i \sum_{j \neq i} V_1(|\mathbf{z}_i - \mathbf{z}_j|) \ . \tag{6.19}$$

This result was first obtained in 1777 by Lagrange [296] who used the idea of a potential function with calling it such. Laplace [308] showed in 1782, for gravitation, that the function $V(|\mathbf{z}|)$ satisfied the equation, $\nabla^2 V = 0$, now known as *Laplace's equation*, in regions free of matter. In 1812, Poisson [453] pointed out the usefulness of such a potential function in electrical applications and also showed [454] in 1813 that, in the presence of matter of density ρ, the gravitational potential satisfied the equation, $\nabla^2 V = 4\pi G \rho$. If we deal only with closed (isolated) systems, then the only contributions to the potential energy will be from internal interactions.

We note that $\mathbf{\nabla}_i V_1(|\mathbf{z}_i - \mathbf{z}_j|) = -\mathbf{\nabla}_j V_1(|\mathbf{z}_j - \mathbf{z}_i|)$. We may now integrate the relations in Equation 6.19 along the evolutionary trajectories of each particle from an initial configuration $\{\mathbf{z}_i(t_{\text{init}})\}$ to another $\{\mathbf{z}_i(t_{\text{final}})\}$ according to,

$$\begin{aligned}
\sum_i \int_{t_{\text{init}}}^{t_{\text{final}}} d\mathbf{z}_i \cdot \frac{d\mathbf{p}_i}{dt} &= -\sum_{i,j \neq i} \int_{t_{\text{init}}}^{t_{\text{final}}} d\mathbf{z}_i \cdot \mathbf{\nabla}_i V_1(|\mathbf{z}_i - \mathbf{z}_j|) \\
&= -\sum_{i,j<i} \int_{t_{\text{init}}}^{t_{\text{final}}} d(\mathbf{z}_i - \mathbf{z}_j) \cdot \mathbf{\nabla}_i V_1 \\
&= -\sum_{i,j<i} [V_1]_{\text{init}}^{\text{final}} \ ,
\end{aligned} \tag{6.20}$$

to obtain,

$$T_{\text{final}} - T_{\text{init}} = \sum_i \int_{t_{\text{init}}}^{t_{\text{final}}} d\left(\frac{\mathbf{p}_i^2}{2m_i}\right) = -V(|\mathbf{z}_i-\mathbf{z}_j|)_{\text{final}} - V(|\mathbf{z}_i-\mathbf{z}_j|)_{\text{init}},$$
(6.21)

where,

$$V = \sum_{i,j<i} V_1(|\mathbf{z}_i - \mathbf{z}_j|). \tag{6.22}$$

The *potential energy* $V = V(\ldots, \mathbf{z}_i, \ldots, \mathbf{z}_j, \ldots)$ of the system, like the kinetic energy, is not conserved. However, the above calculation, resulting in Equation 6.21, shows that, for conservative interaction, we may construct a total *mechanical energy*, kinetic plus potential,

$$E = T + V = \sum_i T_i + \sum_{i,j<i} V_1(|\mathbf{z}_i - \mathbf{z}_j|) = \text{constant}, \tag{6.23}$$

which is conserved, thereby satisfying a fundamental requirement of a Newtonian system. This result was first expressed by Lagrange [297] in 1780 as a consequence of Newton's laws and the inverse-square law of gravitation.

The total energy will be additive only in the sense that inclusion of an extra particle will add a term to the kinetic energy and add a contribution to the potential for its interaction with each of the other particles. It is not obvious how one can expect a purely mechanical system interacting in a non-local way (instantaneous action at a distance) to have a total mechanical energy which is additive in the same simple way as mass, momentum, angular momentum, centre of mass motion and kinetic energy. This is related to the impossibilty of locating the potential energy of interaction between two Newtonian particles. Additivity of the energy E, corresponding to the additivity of m, $\mathbf{p}$, $\mathbf{l}$ and $\mathbf{k}$, will result when the interaction is expressed locally, in space and time, with the aid of a dynamic field. For the gravitational field, this can only be done in a relativistic context as will be shown in Chapters 17 and 21. The Schrödinger equation can be considered an example of a classical field equation of Galilean relativity which is local in space and time with a simple well-defined additive and conserved local energy.

In terms of the total potential energy V, Lagrange's formula expressing Newton's second law for the rate of change of momentum of a part of a closed system interacting 'conservatively', becomes,

$$\frac{d\mathbf{p}_i}{dt} = -\boldsymbol{\nabla}_i V. \tag{6.24}$$

The fundamental nature of energy conservation [142, 291] was recognized by several scientists in the mid-nineteenth century, among them

6.4. Potential energy of conservative systems

Carnot[5], Mayer[6], Rankine[7] and Helmholtz. In 1847, Helmholtz ([232] and [269, pp. 156-162]) extended *Carnot's principle* of the 'impossibility of unlimited moving force' (now kinetic energy), applied in the context of heat, to a mathematical 'Principle of conservation of living force' or *vis viva* (kinetic energy). Although arising from mechanics, and thus encompassing gravitation, the new Principle was to be applicable to all branches of physics, in particular heat and electrodynamics. Helmholtz originally used the phrase 'Erhaltung der Kraft' translated as conservation of either force or energy. He later used the term 'Energie'. The conservation implied the inclusion of 'tension forces' (potential energy) and Rankine's phrase *conservation of energy* was adopted as being the most suitable.

In 1850 Clausius[8] [68] applied these ideas to a separation of the *First law of thermodynamics* on the equivalence of mechanical work and heat from the *Second law of thermodynamics* on non-decreasing entropy.

The *mechanistic world view*, developed extensively by Laplace, based on Newtonian mechanics and the determinism implied by the equations of motion, retained adherents until late in the nineteenth century. It had as its goal the explanation of all physical phenomena, from specified initial conditions, according to the solution of ordinary differential equations describing the motion of Newtonian particles subject to central interactions like gravity and Coulomb's law for charged particles. Newtonian mechanics was thus *atomistic* in character, dealing with bodies considered to be collections of concrete material particles.

One can now identify a number of limitations of the Newtonian world picture. No mechanism is provided for Newtonian interaction and, with instantaneous action at a distance, or infinite speed of propagation, it is difficult to see how one could ever be conceived. In addition to particles, fields are useful tools for modelling nature. Furthermore, determinism is extensively softened when quantum effects are introduced into interaction. In addition, Newtonian concepts must be replaced locally by Lorentz relativity, with the loss of absolute simultaneity, and globally by general relativity, with the introduction of curved spacetime. By contrast, the Kantian dynamical philosophy advocates the construction of a theoretical mechanics from generalizations of the evidence of our senses and confronts its new predictions with further sensory evidence. The primordial features are the consequences of interactions rather than the existence of concrete particles. Current physical theory as exemplified by relativistic quantum field theory embodies concepts from both world views, and others, achieving a certain degree of reconciliation by the notions of field-particle (or wave-particle) duality.

---

[5] Nicolas Leonard Sadi Carnot (1796–1832), French physicist.
[6] Julius Robert Mayer (1814–1878), German physician and physicist.
[7] William John Macquorn Rankine (1820–1872), British engineer and physicist.
[8] Rudolph Julius Emmanuel Clausius (1822–1880), German physicist.

6.5 External interaction

Suppose we choose to separate a closed Newtonian system into a part of large mass (the particles of which are labelled by $i = N+1, N+2, \ldots, n$) whose dynamics are not important, except for their effect on a low mass part ($i = 1, 2, \ldots, N$) of primary interest whose effect on the first part is so small that it may be neglected. Then we may split the total interaction potential into two parts, the *internal potential energy*, $V_{\text{int}}(z_i)$ ($i = 1, 2, \ldots, N$) comprised of only those terms involving mutual interaction of the parts of primary interest, and therefore dependent on only the first N dynamical variables, and the *external potential energy*, $V_{\text{ext}}(z_i)$ ($i = 1, 2, \ldots, n$), relative to the small part, which depends on all the dynamical variables. Its dependence on the first N is relevant to the dynamics of the small part but the only aspect of its dependence on the remaining variables z_i ($i = N+1, \ldots, n$) is the way in which it varies with time. We may therefore choose to specify or prescribe the external potential energy in the form,

$$V_{\text{ext}} = V_{\text{ext}}(z_i, t) \quad (i = 1, 2, \ldots, N) \tag{6.25}$$

where the functional dependence on z_i and t contains the effect on the system of interest of the external or prescribed interaction. A potential energy which depends explicitly on time in this way is an indication that we are not dealing with a closed system. Since Newton's third law, of reciprocity of action and reaction, is not automatically taken into account by such a procedure, we should not be too surprised if it does not appear to obey Galilean covariance or the conservation laws appropriate to a closed system. No attempt will be made here to analyse the dynamics of non-conservative systems or those for which dissipative effects are important.

6.6 Galilean covariance and conservation

Our presentation of the laws of Newtonian physics reverses the rôle of equations of motion and conservation laws. Our justification for this is the universal character of the conservation properties and their deep connection with the symmetries of isolated systems under transformations from one inertial frame to another, namely by translations, rotations and boosts. To get some idea of this profound relationship, let us see how the conservation of energy is intimately linked to the form invariance of the equations of motion of an isolated Newtonian system with respect to translations in time. Suppose that we consider a small time translation $t \to t + \Delta t$ and form the corresponding small change in the total energy,

$$\Delta E = \frac{dE}{dt} \Delta t$$

6.6. Galilean covariance and conservation

$$\begin{aligned}
&= \frac{d}{dt}\left[\sum_i\left(\frac{\mathbf{p}_i^2}{2m_i}\right) + V\right]\Delta t \\
&= \left[\sum_i \dot{\mathbf{z}}_i \cdot \frac{d\mathbf{p}_i}{dt} + \frac{\partial V}{\partial t} + \sum_i \frac{\partial V}{\partial \mathbf{z}_i}\cdot \dot{\mathbf{z}}_i\right]\Delta t \\
&= \frac{\partial V}{\partial t}\Delta t \\
&= [\Delta V]_{\text{time translation}} \, .
\end{aligned} \qquad (6.26)$$

Consequently, if the equations are unchanged after displacement in time, namely if the potential energy function does not depend explicitly on t, except via its dependence on the trajectory variables $\mathbf{z}_i(t)$, then the energy will be conserved.

The above result shows the intimate connection between the choice of a spacetime, $\mathbb{G} = \{\mathbf{E}^1, \mathbf{E}^3\}$, in which the time coordinate, $t \in \mathbf{E}^1$, is homogeneous, the invariance of the equations of motion of an isolated system under translation in time and the existence of a conserved quantity known as energy. The link between the invariances of dynamical equations and the conserved quantities of the system is of enormous importance.

The above case is an example of a very general principle, referred to as *Noether's theorem*, which states that a conserved quantity exists for each continuous symmetry (one depending on continuous parameters such as $\mathbf{a}$, θ, $\mathbf{v}$ and t_0) of a dynamical system. These results are usually established in the context of a Lagrangian discussion but must of course be independent of one's particular formulation of dynamical principles. We shall explore these relationships further in a Lagrangian context in Chapter 9. However, we give another illustration here, using a similar calculation, of the conservation of the centre of mass motion corresponding to Galilean boost invariance. Let $\mathbf{v}$ be the velocity of the boost. We form the product of $\mathbf{v}$ with the rate of change with time of the centre of mass motion to obtain,

$$\begin{aligned}
\mathbf{v}\cdot\frac{d\mathbf{k}}{dt} &= \mathbf{v}\cdot\sum_i \frac{d}{dt}(\mathbf{p}_i t - m_i \mathbf{z}_i) \\
&= \mathbf{v}\cdot\sum_i\left(\frac{d\mathbf{p}_i}{dt}\right)t \\
&= -\mathbf{v}\cdot\sum_i\left(\frac{\partial V}{\partial \mathbf{z}_i}\right)t \\
&= -\sum_i \frac{\partial V}{\partial \mathbf{z}_i}(\Delta \mathbf{z}_i)_{\text{boost}} \\
&= [\Delta V]_{\text{boost}} \, .
\end{aligned} \qquad (6.27)$$

Suppose that the potential is invariant with respect to boosts, as will be the case, for example, if it depends only on differences in position at the same time or differences in velocities. Then, since the boost parameters **v** are arbitrary, the centre of mass motion will be conserved. This is, of course, precisely why we include the conservation of the centre of mass motion as part of the foundations of the laws of dynamics.

Exercise 6.7 *Form the quantities* $\mathbf{a}\cdot\dot{\mathbf{p}}$ *and* $\Delta\boldsymbol{\theta}\cdot\dot{\mathbf{l}}$, *where* $\mathbf{a}$ *and* $\Delta\boldsymbol{\theta}$ *are the parameters of a spatial translation and a rotation* $\Delta\mathbf{z} = \Delta\boldsymbol{\theta}\times\mathbf{z}$, *to show that conservation of 3-momentum and angular momentum correspond to invariance with respect to translations and rotations, respectively, namely spatial homogeneity and isotropy.*

In contrast to the universal applicability of the conservation laws, Newton's laws depend on Galilean relativity and are only approximately true. Similarly, the concept of force, which is central in their formulation, is not applicable to all physical problems, this being particularly so in a quantum context or where all parts of a system are described best by fields. In both cases the conservation laws continue to be crucially important.

Some of the relations between conservation laws and Newton's equations of motion are clarified by an analysis based on the 10-parameter Galilei group and its *central extension* [30] to the 11-parameter *Schrödinger group* by inclusion of the mass operator. One such examination, in a Hamiltonian context, is given in Appendix G of Ludwig and Falter [346].

The *stress tensor* $t^{kl}(t,\mathbf{x})$ of a Newtonian mechanical system comprising either discrete particles or a fluid continuum is defined so that the kl component is the flux of k-momentum, $p^k = m\dot{z}^k$, across a surface with normal in the l-direction. The *flux* of any quantity is the product of its density and the velocity. Since the mass density at $\mathbf{x}$ of a system of particles m_i located at $\mathbf{z}_i$ is $\sum_i m_i \delta^3(\mathbf{x} - \mathbf{z}_i)$, the k-momentum density is $\sum_i m_i \dot{z}^k_{(i)} \delta^3(\mathbf{x} - \mathbf{z}_i)$ and the stress tensor is given by,

$$t^{kl}(t,\mathbf{x}) = \sum_i m_i \delta^3(\mathbf{x} - \mathbf{z}_i(t)) \dot{z}^k_{(i)}(t) \dot{z}^l_{(i)}(t) \,, \tag{6.28}$$

and a similar expression applies to the energy flux. The idea that mechanical energy could be localized and flow was a major innovation suggested by Lodge[9] [325] in 1885 as a result of similar properties of electromagnetic radiation established by Poynting[10] [459] in 1884. Such considerations opened the way for closely analogous treatments of matter and energy and thus of particles and fields.

---

[9] Oliver Joseph Lodge (1851–1940), British physicist.
[10] John Henry Poynting (1852–1914), British physicist.

6.7 Natural equations

All numerical results will be quoted in the SI (Système International) except for a few commonly accepted units such as electron volts (eV). A summary of the SI units and their history is available in pp. 1–13 of Kaye and Laby [274] which we recommend. A table of physical and astronomical constants in SI units, based on the 1986 revision of Cohen and Taylor [74] and on Allen [8], is provided in Appendix A.1.

Certain equations are customarily expressed in forms where some common physical constants do not appear. We therefore now introduce a *natural equation system* in which there is only one dimensionally independent quantity. Such an equation or *unit system* still permits dimensional analysis to be carried out on fundamental quantities but will also considerably simplify many theoretical equations. We select the one base quantity to be length. We choose the natural equation system which is most convenient for a unified comparitive study of all four interactions of physics.

The transformations used to link SI equations to natural equations are chosen so that four universal constants of fundamental physics are, in natural units, dimensionless and have the numerical value of 1. The seven base quantities of the SI equations are length, time, mass, electric current, thermodynamic temperature, amount of substance and luminous intensity. We shall primarily be concerned with the first four of these but we shall take into account the temperature for reasons of completeness. The extension to the remaining two SI dimensions is straightforward and largely independent of the other five.

In addition to the usual constants that appear in all *practical unit systems* — the speed of light c, the rationalized Planck constant $\hbar = h/2\pi$, the gravitation constant G, Boltzmann's constant k, the charge e on a proton, particle rest masses $m_e, m_p, \ldots$, — another that also appears in SI equations is the so-called *vacuum permeability*, μ_0. A dependent constant is the *vacuum permittivity*, $\epsilon_0 = 1/\mu_0 c^2$.

The four constants $c, \hbar, k$ and μ_0 are dimensionally independent in the SI system. Each is a *universal constant* not fundamentally connected with any particular one of the four interactions (as are G and e, for example). Nor are they connected to any particular particle (as are the electron and proton masses m_e and m_p). The first two are the fundamental constants c and $\hbar$ of all relativistic quantum physics. Boltzmann's constant k provides the fundamental link between practical mechanical and thermophysical units while the vacuum permeability μ_0 is a conventional constant which ensures that electrical units are of a convenient practical size for the electrical power industry, electronics and communications. One could regard $c, \hbar$ and k in precisely the same way.

Table 6.1 Transformations from five of the SI base quantities to corresponding natural quantum field theoretic units showing the natural dimensions of each quantity as a power of length L or mass M.

Quantity	Defining relation	Natural dimension	Alternative dimension
length	$x = x_{SI}$	L	M^{-1}
time	$t = c\, t_{SI}$	L	M^{-1}
mass	$m = \frac{c}{\hbar} m_{SI}$	L^{-1}	M
temperature	$T = \frac{k}{\hbar c} T_{SI}$	L^{-1}	M
electric current	$i = (\frac{\mu_0 c}{\hbar})^{1/2} i_{SI}$	L^{-1}	M

We shall choose natural quantum field theoretic equations by deciding that these four constants shall be dimensionless and have value 1 in natural units. Let x_{SI}, t_{SI}, m_{SI}, T_{SI} and i_{SI} denote the five SI base quantities of interest with units of metre (m), second (s), kilogram (kg), kelvin (K) and ampere (A). We now determine those four combinations of powers of c, $\hbar$, k and μ_0 which, when multiplied with t_{SI}, m_{SI}, T_{SI} and i_{SI}, give quantities with dimensions which are some power of length [L] or mass [M]. The results, along with $x = x_{SI}$, are used as the new natural variables x, t, m, T and i and the transformations relating them to SI quantities are given in Table 6.1 along with their natural dimensions.

Any other quantity is transformed between the SI and natural units by products of powers of the above factors, as determined using its SI dimension. If we take

$$v \le c, \quad E = \hbar\omega, \quad E_{\text{kin}} = \tfrac{3}{2}kT \quad \text{and} \quad \nabla\cdot\mathbf{E} = \rho/\mu_0 c^2 , \qquad (6.29)$$

to be four typical well-known SI equations involving the 4 constants $c, \hbar, k$ and μ_0 then applying the above transformations gives the following natural equations:

$$v \le 1, \quad E = \omega, \quad E_{\text{kin}} = \tfrac{3}{2}T \quad \text{and} \quad \nabla\cdot\mathbf{E} = \rho , \qquad (6.30)$$

which imply that the values of these constants in this natural system are,

$$c_{\text{nat}} = \hbar_{\text{nat}} = k_{\text{nat}} = (\mu_0)_{\text{nat}} = 1 . \qquad (6.31)$$

This illustrates the fact that all SI equations may be converted to natural equations by simply assigning the value unity to those four constants (and consequently to ϵ_0 as well) namely:

$$c = \hbar = k = \mu_0 = 1 \ (= \epsilon_0) \quad \text{(natural)} . \qquad (6.32)$$

6.7. Natural equations

A few special relativistic equations which will be encountered later are,

$$x^0 = ct, \quad \gamma = 1/\sqrt{1 - v^2/c^2} \quad ds^2 = -c^2 dt^2 + d\mathbf{x}^2 \quad \text{and} \quad P_\mu = \frac{\hbar}{i}\partial_\mu , \quad (6.33)$$

The above rule shows that, in natural units, they have the form,

$$x^0 = t, \quad \gamma = 1/\sqrt{1 - v^2} \quad ds^2 = -dt^2 + d\mathbf{x}^2 \quad \text{and} \quad P_\mu = \frac{1}{i}\partial_\mu . \quad (6.34)$$

To restore an equation given in the natural system to its form in the SI, it is first necessary to know, or to specify, the SI dimension of every quantity in the natural equation. One then simply applies the above set of transformations of Table 6.1 in reverse.

Exercise 6.8 *The action S of any system has dimensions of angular momentum while the Lagrangian of a mechanical system has the dimensions of energy. Show that, in natural units, angular momentum, the action S and electric charge q are dimensionless and that the energy and Lagrangian of a mechanical system have dimensions L^{-1} of inverse length.*

By including one additional universal constant, an equation system may be constructed in which all quantities are dimensionless. One such system in which the fifth quantity is the gravitational constant G was supplied by Planck [286, 436] in 1899. The numerical values of the Planck quantities in the SI are tabulated in Appendix A.1.

...hypotheses *no fingo*

...I feign no hypotheses, for whatever is not deduced from the phenomena is to be called a hypothesis, and hypotheses, whether metaphysical or physical, whether of occult qualities or mechanical have no place in experimental philosophy. In this philosophy particular propositions are inferred from the phenomena and afterwards rendered general by induction[11].

Isaac Newton   1642–1727

Bibliography:

Aharoni J 1972 *Lectures in mechanics* [5].
Eisenbud L 1958 *On the classical laws of motion* [141].
Hestenes D 1986 *New foundation for classical mechanics* [247].
Longair M S 1983 *Theoretical concepts in physics* [326].
Marion J B et al., 1988 *Classical dynamics of particles and systems* [358].
Page L 1952 *Introduction to theoretical physics* [415].
Weinstock R 1961 *Laws of classical motion* [567].

---

[11] *Scholium* to the *Principia* [50, 395].

CHAPTER 7

Newtonian gravity

We admit that the present state of the world only depends on the immediate past ... instead of studying directly the whole succession of phenomena, we may confine ourselves to writing down its *differential equation*: for the laws of Kepler we substitute the law of Newton[1].

Henri Poincaré   1854–1912

7.1 Introduction

Gravity was the first physical interaction to be isolated as fundamental. Its terrestrial effects are immediately noticed from infancy. The ancients considered celestial objects to be composed of an element, the *ether*, of different non-material character to the four terrestrial elements — earth, water, air and fire — to account for their apparently quite different circular motion compared to the primarily rectilinear motion of falling objects.

Over the ages, observations of astronomical phemonena gradually led to the belief in the operation of some natural interaction between celestial objects. The establishment of a self-consistent, unified theory of celestial and terrestrial gravitation by Newton was also the first major step in the reduction of the phenomena of nature to a small number of fundamental interactions.

Newton's theory not only described the gravitational attraction toward the earth felt by all terrestrial objects but also accounted for the orbital motions of heavenly bodies. The modern process and philosophy of reduction and unification of the phenomena of nature by scientific methods, free of subjective or at least supernatural elements, was therefore advanced by a gigantic step with the publication of Newton's work.

---

[1] *Science and hypothesis* [445].

7.2 Gravity

Gravity may be described as the inverse-square attractive interaction acting between any pair of small masses with the same universal strength proportional to the product of the masses. However, in the most precise classical formulation of gravity, Einstein's general theory of relativity, the interaction between two masses does not obey an inverse-square law exactly [113, 114]. The above description over-specifies the essential properties of gravitation by including some (for example, the inverse-square nature and the attractiveness) which may be considered determinable, in their appropriate domain of validity, from more fundamental properties, using general principles. It is also unnecessarily restricted to a description of gravity in a Newtonian context in three spatial dimensions.

A characterization will be used here which may be applied in either a Galilean or an Einsteinian context and which leaves one with natural generalizations to explore in spacetimes with other properties, such as different dimensionality, for example. The following more general characterization is therefore chosen:

- **Gravity** is the parity and time-reversal covariant interaction which acts with the same strength between any and all forms of mass.

In the context of Newtonian gravity, matter or mass is understood, in the language of Einsteinian relativity, to be non-zero rest mass. Kinetic energy, or field energy with zero rest mass, propagating at a frame-independent finite ultimate speed, cannot be included in the source of Newtonian gravity. In Einsteinian relativity on the other hand, by matter we mean mass and its equivalent energies in any form, whether of zero rest mass or not.

The importance of relativistic principles will be illustrated by examining the extent to which this characterization permits us to determine the nature of the interaction, by applying Galilean relativity. In Part IV, we shall examine the simplest and most obvious ways to modify the theory to incorporate Einsteinian relativity.

Since all matter has mass, its choice as the source gives gravitation a *universality* unique among the four interactions. That universality may be used as a general property to severely restrict the forms of the equations which describe it and plays a particularly important role in relativistic gravitation as a result of mass-energy equivalence.

Since the characterization of gravity contains no fundamental length constant, we eliminate the possibility of a short-range contribution as would be the case if the potential included an exponential factor e^{-r/r_G}, for example, in which r_G would be regarded as the finite range. This condition in fact leads to a Newtonian potential of *long-range*, the effects extending to infinity. One says that the *range* of the gravitational interaction is infinite. The long-range nature of the gravitational interaction is a property it shares only with electromagnetism and also effectively restricts the equations which describe it. We shall see that for a simple gravity theory to be of finite range, and otherwise consistent with our characterization given above, the range would have to appear as a fundamental constant in the equations of motion, separate from the strength of the interaction.

The *parity covariance* of non-quantum theories is generally taken for granted. We have explicitly characterized gravity with this property, described in general in Chapter 2, and shall show that it may also be used to restrict the forms of gravitational equations of motion or field equations.

The *attractive* nature of gravity is a property which may be deduced in the context of a relativistic field theory if one demands that the classical theory be, in principle, quantizable and therefore at least of *positive energy*. Many observed features of the universe, in particular the stability of most astrophysical systems, are consequences of the attractiveness of gravity. This property permits an equilibrium between gravity and the repulsive effects that may arise from thermal pressures as occurs in ordinary stars. It also counteracts the quantum-mechanical *degeneracy pressure*, arising in white dwarf and neutron stars, from application of the *Pauli exclusion principle*[2] that applies to the fermionic quantum particles comprising the bulk of the matter in the universe, namely protons, neutrons and electrons. The long-term implosive instability of other constituents of the universe, such as incipient stellar or galactic black holes, when the repulsive pressures no longer suffice for equilibrium at the end of their evolutionary lives, are also related to the fact that gravity is always attractive rather than sometimes repulsive as in the case of the electromagnetic interaction.

That gravity was the first of the four interactions to be identified is a consequence of its universality and the fact that mass, its source, unlike the bipolar charges which interact electromagnetically, is always positive. Despite the incredible intrinsic weakness of gravity in comparison with the other three interactions, the positivity of the source means that its effects grow in magnitude as the source increases in mass. It consequently dominates on a large scale, from laboratory-sized samples of mass and higher.

7.3 Newton's gravity theory

And thus Nature will be very comfortable to her self and very simple, performing all the great Motions of the heavenly Bodies by the Attraction of Gravity which interceded those Bodies, and almost all the small ones of their Particles by some other attractive and repelling Powers which intercede the Particles[3].

Isaac Newton   1642–1727

---

[2] Wolfgang Pauli (1900–1958), Swiss physicist.
[3] *Opticks* [396, p. 397].

7.3. Newton's gravity theory

Newton constructed his theory of gravity in accordance with his laws of mechanics by techniques very different from those used in the twentieth century to compare and attempt to unify the interactions of physics. His theory continues to be widely used to this date as an approximate description valid in almost all ordinary circumstances subject to certain limitations such as slow velocities of objects and low density of matter. This continued phenomenological success despite its inability to explain some gravitational effects, many but not all of which are very small, is a tribute to his outstanding genius and that of his immediate predecessors in the development of physics and astronomy. In 1543, the publication took place of Nicolas Copernicus's revolutionary work [79] supporting the claims of Heraclides and Aristarchus that the motion of celestial bodies was simpler in terms of orbits about the sun. Shortly after, Tycho Brahe [47] accumulated the precise observations of cometary and planetary positions [121], published from 1585 to 1598, which permitted Johannes Kepler to establish his empirical laws of elliptic planetary motion [276] in the following two decades.

Newton first worked on an inverse square law for the gravitational interaction in 1661. He deduced the centripetal acceleration v^2/r for circular motion of radius r at constant speed v for application in his mechanical laws. Combined with $v \propto r/P$, where P is the orbital period, and Kepler's 3rd law, $P \propto r^{3/2}$, this showed that that the centripetal acceleration is proportional to $1/r^2$. Consequently, an interaction which causes this acceleration must be inverse square. Expressed in modern form using vector notation, Newton's law of universal gravitation specifies the rate of change $d\mathbf{p}_i/dt$ of the momentum $\mathbf{p}_i(t)$ at the time t of the i-th particle of a system of particles labelled by $i = 1, 2, \ldots$, in terms of the displacement, $\mathbf{z}_i(t) - \mathbf{z}_j(t)$, of the particle relative to all the other particles ($j \neq i$). The rate of change $d\mathbf{p}_i/dt$ is given as the parity-covariant and time-reversal invariant sum of an inverse square, attractive interaction, between each pair $\{i,j\}$, proportional to the product, $m_i m_j$, of the masses with the same proportionality constant G in all terms, namely,

$$\frac{d\mathbf{p}_i(t)}{dt} = -G \sum_{j \neq i} m_i m_j \frac{\mathbf{z}_i(t) - \mathbf{z}_j(t)}{|\mathbf{z}_i(t) - \mathbf{z}_j(t)|^3} , \qquad (7.1)$$

with no implied sum over i. The velocities do not appear. Using $\mathbf{p}_i = m_i \dot{\mathbf{z}}$ and cancelling the mass m_i, gives,

$$\ddot{\mathbf{z}}_i(t) = -G \sum_{j \neq i} m_j \frac{\mathbf{z}_i(t) - \mathbf{z}_j(t)}{|\mathbf{z}_i(t) - \mathbf{z}_j(t)|^3} . \qquad (7.2)$$

From Equations 6.18 and 6.24, the gravitational potential energy of one particle i due to all the others is,

$$V(t, \mathbf{x}_i) = -G m_i \sum_{j \neq i} \frac{m_j}{|\mathbf{z}_i(t) - \mathbf{z}_j(t)|} . \qquad (7.3)$$

After an exchange of letters with Hooke[4] in 1679, Newton worked on the determination of the orbit of a particle subject to an inverse square law during the period from about 1680 to 1684, inventing infinitesimal methods (differential and integral calculus) for that purpose. The first published version of the law appeared in 1684 in a manuscript entitled *De motu* (*On motion*) which later became the first part of the *Principia* [395], appearing in 1687. The definitive results, showing that bound orbits were elliptical, were published in the *Principia*. With his law of gravitation, Newton had provided a simple explanation for why the moon was not 'left behind' by its parent body, the earth, if the latter moved at high speed around the sun, as required by the heliocentric cosmology.

This work is the first really dramatic theoretical example of the use of one simple law to describe a number of empirical laws (those of Kepler), based on a huge body of precise astronomical data (principally those of Tycho Brahe). It illustrates a very important facet in the process of quantifying physical measurements, namely their description not only in mathematical terms but as simply as possible, with a minimum number of fundamental postulates, and in a way which is as general and unified as possible. Such generalizations always contain greater predictive power than the data from which it is induced. For example, in this case it permits the analysis of perturbations to the two-body motion due to additional bodies. One of the most important features of theoretical physics is the search for such simple, general and unified analyses and the deduction of all their physical consequences, most of which will not have been apparent in the observations which gave rise to the general principles. The Relativity principle, the Principle of stationary action, the Gauge principle and the Quantum postulates are further important examples of such general unifying laws.

If we include all the bodies which significantly interact gravitationally, and if no other interactions take place, Equation 7.1 describes a closed system of particles behaving overall as an isolated system and which therefore must exhibit Galilean covariance. We may in fact immediately observe that Newton's law of universal gravitation is Galilean covariant since $d\mathbf{p}_i$ and $\mathbf{z}_i(t) - \mathbf{z}_j(t)$ are Galilean 3-vectors and $|\mathbf{z}_i(t) - \mathbf{z}_j(t)|^3$, dt, m_i and G are Galilean invariants.

This Galilean covariance, comprising Euclidean covariance and boost invariance, is a very important property of Newton's theory of gravitation. The interaction terms are not just functions of the locations $\mathbf{z}_i(t_i)$ and $\mathbf{z}_j(t_j)$ of the particles, i at time t_i and j at time t_j, but involve only differences of the locations at the same time $t_i = t_j$ which must also be the time appearing in the change of velocity $\ddot{\mathbf{z}}_i(t)$ of the i-th particle. Thus

---

[4]Robert Hooke (1635–1703), English experimental physicist, Curator and Secretary of the Royal Society, London.

the implementation of the Galilean covariance is intimately linked to the fact that Newton's theory is an *action at a distance* theory; particle j may be at arbitrarily large distance and still gravitationally affect the motion of particle i, instantaneously. This property of Newton's law of gravitation implies that the interaction cannot be considered as an effect which propagates from one particle to another at some constant speed, since whatever is being propagated would have to do so at an infinite speed.

Before we criticize Newton for the notion of action at a distance, we should perhaps consider not only the continued success of the theory but also one of Newton's own views on a related topic contained in a letter (1692–93) to Richard Bentley, quoted in the *Principia* [395, p. 643]:

> That one body may act upon another at a distance through a vacuum, without the mediation of anything else, by and through which their action and force may be conveyed from one to another, is to me so great an absurdity, that I believe no man, who has in philosophical matters a competent faculty of thinking, can ever fall into it.

7.4 Newton's gravitation constant

No value for the gravitational constant G was available to Newton who concluded that it would be too small to be measured. The numerical strength of its effects was first noted in 1774 by Maskelyne[5] who observed the small deviation of 11.7 arc seconds of a plumb-line from the vertical caused by the gravitational attraction of Mt. Schiehallion in Scotland. Maskelyne's measurement was part of a determination of the average density of the earth but is now considered to be the most precise demonstration at that time of the universality of gravitation.

Cavendish[6] used the interaction of two laboratory masses in Michell's torsion balance to make a measurement in 1797 of the strength of gravity [60], referred to as *weighing the earth*, for the purpose of determining the value of the earth's density which he found to be $\rho_\oplus = 5.48\,\rho_{\text{water}}$. The value of $G = 6.754 \times 10^{-11}\,\text{kg}^{-1}\,\text{m}^3\,\text{s}^{-2}$ implied by his results is reasonably close to recent laboratory measurement, $G = (6.6726 \pm .0005) \times 10^{-11}$, reported by Luther and Towler [347, 519] in 1982 and the 1986 recommended value [74] of $G = (6.67259 \pm .00085) \times 10^{-11}$.

The gravitational constant remains the most uncertain of all the fundamental constants of physics, due largely to the extreme relative weakness of gravity which means that large (planetary sized) rather than atomic

---

[5] Nevil Maskelyne (1732–1811), English astronomer, appointed Astronomer royal in 1765, was deeply involved in the improvement of longitude determinations at sea and in the proposal for and production of the *Nautical Almanac* [197].

[6] Henry Cavendish (1731–1810), English natural philospher (physicist) and chemist.

masses would have to be used to achieve precision comparable to other physical constants. However, the masses of such large bodies are not able to be determined independently of G and laboratory sized masses must be used. Cavendish's experiment provided a crucially important result for astrophysical calculations since it then permitted the law of gravitation to be used to determine the masses of astronomical bodies, referred to as *weighing the planets and stars*, from their orbital radii and periods. His scientific work is commemorated in the naming of the Cavendish Laboratory (Department of Physics) of the University of Cambridge, site of many significant experimental and theoretical achievements in physics which include some of the work of Maxwell, Lord Rayleigh[7], J. J. Thomson[8] and Rutherford[9] to name just a few.

Despite the fact that gravitational effects appear to dominate at laboratory scales and above, gravity is a very weak effect, both intrinsically and by comparison with the other interactions. To see this, we must note that smallness is meaningless except by comparison with other quantities of the same dimension. This is because the units chosen for dimensional quantities are arbitrary and the values of such quantities in terms of the units are equally arbitrary. We therefore ask ourselves how we can construct a fundamental dimensionless quantity from G. We clearly cannot do so solely in a context of Newtonian gravitation since no other fundamental constants are contained within the theory. However, we may anticipate the eventual need to replace the approximate relativity of Galileo and Newton by Einsteinian relativity with the appearance of a fundamental constant c, the ultimate speed, whose value is given by the speed of light. We may also note that c is intimately connected with the relativistic theory of gravity and is therefore appropriate for use along with G.

It is easy to show that no dimensionless quantity can be constructed from just G and c. We must therefore look for a further quantity and the most fundamental, unconnected (like the proton charge e, for example) with any particular interaction or (like the the pion mass m_π, for example) with any particular particle, is the Planck constant $\hbar$ of quantum physics.

Exercise 7.1 *Show that the three mechanical constants G, c and $\hbar$ are dimensionally independent.*

We are still unable to form a dimensionless combination. The above three constants are all either associated with gravitation or are universal constants. In choosing a fourth quantity we seek to maintain these quali-

---

[7] John William Strutt, Third Baron Rayleigh (1842–1919), English experimental and theoretical physicist.
[8] Joseph John Thomson (1856–1946), English physicist.
[9] Ernest Rutherford, Lord Rutherford of Nelson, New Zealand (1871–1937), experimental physicist.

7.4. Newton's gravitation constant

ties. There are no other constants with the universality of c and $\hbar$. Since the source of gravitation is mass we seek a fundamental mass. Rather than choosing an arbitrary mass, we note that the smallest mass is that of the electron with the proton mass being not many powers of 10 larger. We conventionally select the proton mass m_p and find that a dimensionless combination proportional to G itself is the so-called *gravitational fine structure constant*,

$$\alpha_G = \frac{Gm_p^2}{\hbar c} = 5.90 \times 10^{-39} . \tag{7.4}$$

We see from the very small value of this constant that, as claimed, the gravitational interaction is indeed extraordinarily weak (see Table 1.1).

Since gravity and electromagnetism are both inverse square interactions we may compare their strengths by taking the ratio of the magnitudes of the interactions at any separation. To compare the strength of gravity with the nuclear interactions we must do so at distances comparable or smaller than the nuclear ranges where these interactions are significant. The ratio of the gravitational to the Coulomb interaction between two protons is given by,

$$\frac{\alpha_G}{\alpha_e} = \frac{Gm_p^2}{e^2/4\pi\epsilon_0} = 8.09 \times 10^{-37} , \tag{7.5}$$

where,

$$\alpha_e = \left(\frac{e^2}{4\pi\epsilon_0 \hbar c}\right)_{SI} = \left(\frac{e^2}{\hbar c}\right)_{Gaussian} = \left(\frac{e^2}{4\pi}\right)_{natural} , \tag{7.6}$$

with value $\alpha_e = 7.297\,350\,6 \times 10^{-3} = 1/137.036\,04$ (see Table 1.1), is the *electromagnetic fine structure constant*. (Gaussian electromagnetic units are not rationalized. To obtain them from SI units one must at least put $\epsilon_0 = 1/4\pi$.) The electromagnetic interaction is intrinsically weak since $\alpha_e \ll 1$ but gravity is very much weaker by a factor of about 10^{36}. The comparable forces between other elementary particles will not differ greatly from the value obtained for protons. For the weak nuclear constant, see Equation 20.67.

The fact that μ_0, $\hbar$ and c are of unit value in the natural units we have chosen means that, in those units, the electric charge is dimensionless. This will also be true of any unit system having only one base dimension chosen from mechanical physical quantities, namely some product of powers of mass, length and time. The fact that the coupling strength of the electromagnetic field, the constant e, is dimensionless in natural units, is closely related to the renormalizability of quantum electrodynamics. The fact that G_{nat} is not dimensionless is closely related to the non-renormalizability of the gravitational interaction.

Exercise 7.2 *(a) Use the transformation equations between SI units and natural units to show that the natural dimensions of G are [length^2]. (b) Show that this result is consistent with Equation 7.4.*

7.5 Field equations of Newtonian gravity

Newton's description of gravitation was concrete and *mechanical* in that it involved material bodies which have a finite number of degrees of freedom.

One feature common to a modern treatment of all the interactions and to their source material is their description in terms of a *local field theory*, where the local aspect applies to both spatial and temporal characteristics. The concept of a *field* was not used in Newton's time. The gravitational interaction itself was not considered to be a dynamical component of the system of interacting particles requiring its own evolution equations and capable of self-sustaining propagation independent of its sources. Fields were first introduced into physics over a century later by Faraday[10], who used the ideas of 'contiguous action' and 'lines of force' to describe spatial locality in connection with electromagnetism. The propagation of fields free of their source was first formulated by Maxwell several decades later.

We may examine an interaction such as gravity from the point of view of its mediation by a separate gravitational field with its own field equations involving derivatives with respect to spatial coordinates to describe its local distribution or field character. (Field equations may also involve time derivatives to describe the field's evolution.) An interaction mediated by a local field theory is one in which the source of the interaction may be expressed locally, as a function of time and position, and the production of the field at a given event $\{t, \mathbf{x}\}$ is determined from the source and field properties on a suitable boundary by equations involving the field ψ and a finite number of its derivatives, $\partial_t \psi$, $\partial_t^2 \psi$, ..., $\partial_k \psi$, $\partial_k \partial_l$, ..., $\partial_t \partial_k$, ..., at the events concerned. The parts of the source and field which have most influence on determining the field at a given event are then those immediately adjacent in time and space. Because of the Principle of causality, only those parts before the event concerned may influence it. The nature of the boundary will depend on the nature of the interaction and whether an evolving system or only the spatial distribution of a stationary system is being considered.

For the interaction to be entirely field-theoretic, the source material will be described by one field whose field equation contains another field describing the interaction. The latter may be referred to as the *interaction field* or *mediating field* whose quanta are referred to as *exchange particles*.

---

[10]Michael Faraday (1791–1867), English chemical physicist.

7.5. Field equations of Newtonian gravity

The equation for the mediating field will in turn contain terms involving the source fields. The electromagnetic and gravitational interactions may also be formulated in terms of mechanical sources interacting via a mediating field and we shall initially examine such a partially field-theoretic description.

Fields are thus of two forms, those like gravity and electromagnetism which are generated by a source (for example mass or electric charge), and those which are not but represent the sources themselves, such as the non-relativistic Schrödinger wave function or the Dirac field describing an electron in relativistic quantum field theory. The field equations of a sourced, or mediating field, can be recognized by the presence in them of a term, the *source* term, which does not contain the field itself.

By the introduction of a mediating field (for gravitation, for example) we can divide the interaction into a part involving the dynamics of each interacting mass and a part which should be temporally and spatially local, involving the production of a field by each mass. Newtonian gravity, however, when examined in this way does not in fact involve any time derivatives at all because of its action at a distance nature. It is not therefore a dynamical component of the whole system. Its evolution arises solely from the evolution of its sources. It is a field theory in which the field equations only determine the distribution in space and not the distribution in time. Nevertheless, it will be instructive to examine Newtonian gravity mediated by a Galilean field.

A field, whether mediating like gravity or a source such as the Dirac field of an electron, is usually considered to be an abstract quantity or set of quantities $\psi = \psi(t, \mathbf{x}) = \{\psi_j(t, \mathbf{x})\}$ ($j = 1, 2, \ldots d$) not directly observable but from which observable quantities may be constructed. For example, in a Galilean context, it may be a 3-scalar field, $\Phi = \Phi(t, \mathbf{x})$ ($d = 1$) or a 3-vector field, $\mathbf{f} = \{f^k(t, \mathbf{x})\}$ ($d = 3$) from either of which one may construct the acceleration of a particle. Being a function of position $\mathbf{x}$, it is a *non-mechanical* quantity, namely one with an infinite number of degrees of freedom given by the three-fold infinity of its values at all points $\mathbf{x}$ throughout space.

Fields not only permit the interaction to be divided into two separate more tractable operations. They constitute a generalization of mechanics that permits one to describe the source and the interaction itself in the same way, either as non-quantum fields, or as source and mediating particles obtained from the fields by quantization. They also generally permit the energy V of interaction to be localized thus giving the total energy, $E = T + V$, an additivity property more closely resembling those of the other nine quantities, $\mathbf{p}$, $\mathbf{l}$ and $\mathbf{k}$, which are conserved as a result of space-time invariances.

We seek a description of gravitation in which the interaction is described by a local field whose *integrated source* is the invariant mass. We demand that the equations of the theory be unchanged in form in changing the Euclidean basis e_k by translations, rotations and Galilean boosts. We shall also not permit our equations to violate parity covariance. The simplest way to implement Galilean covariance is to make use only of Euclidean 3-scalars, 3-vectors and 3-tensors, and their field counterparts, to demand that every term of each equation be a tensor of the same rank and to require the equation to be also boost invariant.

For inclusion in a field equation we require a *local source* distributed throughout space. That source will be the proper Galilean invariant mass density $\rho(t, \mathbf{x}) = dm/d^3x$, in terms of which we write the mass as the integral,

$$m = \int d^3x \, \rho(t, \mathbf{x}) \, . \tag{7.7}$$

Although the total mass cannot depend on t, the way in which it is distributed will in general be time-dependent. We make no attempt at this stage to determine its dynamics as a fundamental field but assume that its evolution is determined by the Newtonian mechanical equations for its constituent masses.

To construct an equation for a field quantity having $\rho(t, \mathbf{x})$ as its source, we need a proper Galilean invariant combination of the field and its derivatives, to equate to the local source. In constructing this invariant, the spatial derivatives, ∂_x, ∂_y and ∂_z, may not appear separately. To preserve Euclidean covariance, they must appear together as the gradient $\boldsymbol{\nabla} = \{\partial_k\}$. Invariant terms may be constructed from vectors by contraction with the universal Galilean numerical tensors, δ_{kl} and ϵ_{klm}, although the second of these may appear only in such a way that each term in the non-source part of the field equation is a proper rather than an improper invariant and the sum is invariant. Consequently, we seek a field equation of the form,

$$f\big(\delta_{kl}, \epsilon_{klm}, \partial_t, \partial_k, \psi^{kl\cdots}(t, \mathbf{x})\big) = \rho(t, \mathbf{x}) \, , \tag{7.8}$$

in which f is some real combination transforming as a proper invariant.

We limit attention to equations of lowest possible degree (linear if possible) and then the best compromise between lowest differential order and lowest rank of the field. If we consider only first- and second-order equations involving a scalar or vector field then we have the following possibilities:

- a proper vector field $\mathbf{f} = \mathbf{f}(t, \mathbf{x})$ satisfying a first-order equation of the form,
$$c_1 \boldsymbol{\nabla} \cdot \mathbf{f} + c_2 = \rho \, , \tag{7.9}$$

- a proper scalar field $\Phi = \Phi(t, \mathbf{x})$ satisfying a second-order equation of the form,
$$a_1 \boldsymbol{\nabla}^2 \Phi + a_2 \Phi + a_3 = \rho \, , \tag{7.10}$$

where c_1 and a_1 may be taken to be constant (independent of t and $\mathbf{x}$) by absorbing any variation in the field factor. The parameters c_2 and a_3 and the fields Φ and $\mathbf{f}$ must have a time dependence reflecting that of $\rho(t,\mathbf{x})$. Otherwise, we restrict attention to constant coefficient equations with no dependence on $\mathbf{x}$ in c_2, a_2 and a_3. The 3-divergence of $\mathbf{f}$ and the Laplacian operator in rectangular Cartesian coordinates are discussed in Exercise 4.2. We cannot include $\nabla \times \mathbf{f}$, $\nabla^2 \mathbf{f}$, $\nabla \times (\nabla \times \mathbf{f})$, $\dot{\mathbf{f}}$ or $\ddot{\mathbf{f}}$ in Equation 7.9 since they are not scalars. We do not include terms such as $a_4 \dot{\Phi}$ and $a_5 \ddot{\Phi}$ in Equation 7.10 since they are not Galilean invariant; the effect of a boost on $\dot{\Phi}$ introduces a term in $\mathbf{v} \cdot \nabla \Phi$ where $\mathbf{v}$ is the velocity of the boost.

7.6 First-order Newtonian gravity field equations

We first examine Equation 7.9. We are free to select the units of $\mathbf{f}(t,\mathbf{x})$ to which we conventionally assign the dimensions of [velocity^2.length^{-1}]. By so doing, we force the constant c_1 to have the dimensions of the reciprocal of the gravitational constant, G. We conventionally rewrite this constant as $c_1 = -1/4\pi G$ and recast the field equation in the form,

$$-\nabla \cdot \mathbf{f}(t,\mathbf{x}) + \Lambda(t) = 4\pi G \rho(t,\mathbf{x}) , \qquad (7.11)$$

where $\Lambda = \Lambda(t)$ is a function of t with the dimensions of [time^{-2}]. The field equations of a system must completely determine the field given its properties on a suitable boundary. In order to have a complete determination of $\mathbf{f}(t,\mathbf{x})$ by first-order equations in terms of its values on a boundary, the values of all Galilean-covariant first-order spatial derivatives linear in the field must be well-defined, in terms of the source. One such combination, $\nabla \cdot \mathbf{f}$, already appears on the left of the field equations. However, there is one other such combination independent of $\nabla \cdot \mathbf{f}$, namely $\nabla \times \mathbf{f}$, the pseudo-3-vector curl of $\mathbf{f}$, the value of which must be well-defined in terms of the source ρ. It is not possible to construct a pseudo-3-vector from ρ and ∇ and the only proper 3-vector of order no higher than 1 is $\nabla \rho$. We cannot include this proper 3-vector and $\nabla \times \mathbf{f}$ in the same equation without violating parity-covariance. Consequently, the only possible value for the latter quantity is zero and the combined field equations for $\mathbf{f}$ take the form:

$$-\nabla \cdot \mathbf{f} + \Lambda = 4\pi G \rho \quad \text{and} \quad \nabla \times \mathbf{f} = 0 . \qquad (7.12)$$

A 3-vector field whose divergence and curl are both specified on a suitable boundary [287] is completely determined (see also Appendix A.3) and these field equations are therefore a self-consistent expression of the effects of the source $\rho(t,\mathbf{x})$ and the parameter $\Lambda(t)$. We shall return to these equations after examining the second-order alternative.

7.7 Second-order Newtonian gravity field equations

We assign to Φ, in Equation 7.10, the dimensions of [velocity^2] in order to permit the constant a_1 to be rewritten as $1/4\pi G$ where the constant G has the dimensions of Newton's gravitational constant as before. We recast the field equation in the form,

$$(\nabla^2 + \frac{1}{\lambda^2})\Phi + \Lambda = 4\pi G\rho \qquad (7.13)$$

where λ is a real length and Λ is a function of t with the dimensions of [time^{-2}]. We use the same symbol Λ as in the first-order case in anticipation of identifying the two. Standard Newtonian gravity has no finite fundamental length or time constant and corresponds in fact to the choice of $1/\lambda = 0$ and $\Lambda = 0$. The field $\Phi = \Phi(t, \mathbf{x})$ is referred to as the *gravitational potential*. Poisson[11] first mentioned the use of a scalar function to describe the interior of attracting masses in a brief memoir in 1813, the importance of which for electrostatic theory was also recognized by him at the time. The standard Newtonian gravitational field equation is thus given by,

$$\nabla^2 \Phi = 4\pi G\rho , \qquad (7.14)$$

known as *Poisson's equation*. This equation is left unaltered by addition to Φ of a function $\alpha(t)$ with arbitrary dependence on time t, namely under $\Phi(t, \mathbf{x}) \rightarrow \Phi(t, \mathbf{x}) + \alpha(t)$. However, two solutions of Laplace's equation subject to the same boundary conditions differ by at most a constant (independent of $\mathbf{x}$). Since $\Phi(t, \mathbf{x})$ is not completely determined in terms of the source by the field equations, and since measurement must always lead to a unique finite real value of some physical variable, the field $\Phi(t, \mathbf{x})$ cannot be an observable. Even if $\Lambda \neq 0$, the (negative) gradient of Φ,

$$\mathbf{f} = -\nabla\Phi , \qquad (7.15)$$

satisfies the field equation, $-\nabla \cdot \mathbf{f} + \Lambda = 4\pi G\rho$ when $1/\lambda = 0$, and satisfies the subsidiary condition $\nabla \times \mathbf{f} = 0$ (arising from the identity $\nabla \times \nabla \equiv 0$), which determine it completely in terms of the source ρ.

The field $\mathbf{f} = \mathbf{f}(t, \mathbf{x})$ may be regarded as the *field strength* of Newtonian gravity generalized to include $\Lambda \neq 0$. Provided $1/\lambda = 0$, its field equations are identical to the first-order equations for $\mathbf{f}$ obtained separately, including the possibility of the term with $\Lambda \neq 0$. Conversely, given the first-order equations, the curl-free condition implies (see Appendix A.4) that there exists a scalar field Φ, undetermined to the extent of a function of time, such that $\mathbf{f} = -\nabla\Phi$. Consequently, one set of field equations may be derived from the other. The two sets of equations are therefore physically equivalent. Seeking a first- or second-order Galilean-covariant linear

---

[11] Siméon-Denis Poisson (1781–1840), French mathematical physicist.

7.7. Second-order Newtonian gravity field equations

field equation (with constant coefficients) and ruling out the term requiring a fundamental length, leads naturally to the field equations of Newtonian gravity apart from the extra term in Λ.

Since the transformation $\Phi(t, \mathbf{x}) \to \Phi(t, \mathbf{x}) + \alpha(t)$ leaves the field equations of the gravitational potential unchanged in form, it is referred to as a *symmetry* or *invariance* of the system described in terms of a potential. The spacetime coordinates $\{t, \mathbf{x}\}$ and the field Φ are often referred to as *external* and *internal* coordinates of the field system they describe. Since the transformation affects only the field Φ and not its arguments, the invariance involved is referred to as an *internal symmetry* of the system, in this case a translation invariance in the 1D space labelled by the values of $\Phi(\mathbf{x})$. This should be contrasted with the Galilean transformations of Equation 5.33 which, acting on $\{t, \mathbf{x}\}$ in G, embody *external* or *spacetime symmetries* of an isolated system.

In the absence of matter ($\rho = 0$ everywhere), the standard Newtonian gravity field equation reduces to Laplace's equation, $\nabla^2 \Phi = 0$, first deduced by Laplace[12] [308] in 1782 and used soon after for the purpose of treating the phenomenon of the rings of Saturn as a gravitational problem [309, p. 278].

The link between the mechanical and field descriptions is illustrated by noting that the gravitational potential is the potential energy per unit mass $\Phi = V/m$. Using Equation 7.3 with $\nabla^2(1/|\mathbf{x}|) = -4\pi\delta(\mathbf{x})$ and $m = \delta(\mathbf{x})\rho(\mathbf{x})$ for a point particle shows that $\Phi(\mathbf{x})$ satisfies Poisson's equation.

Exercise 7.3 *Let us suppose that the source ρ (and therefore the field Φ) is spherically symmetric. Use the expression*

$$\nabla^2 \Phi = \frac{1}{r^2} \partial_r (r^2 \partial_r \Phi) , \qquad (7.16)$$

for the Laplacian operator ∇^2 in spherical polar coordinates, in the case of spherical symmetry, to show that the translationally-invariant solution of Laplace's equation for Φ is an arbitrary function of t, which since it is physically equivalent to a zero potential, represents a complete absence of any gravitational field.

Exercise 7.4 *Show that the solution to Poisson's equation in the case of a source comprised of a homogeneous distribution of matter of constant density ρ for $r \leq R$ and zero density for $r > R$ is $\Phi(r) = 2\pi G\rho(r^2/3 - R^2)$ for $r \leq R$ and $\Phi(r) = -GM/r$, where M is the total mass, for $r > R$.*

Standard Newtonian gravity clearly implies that the vacuum ($\rho = 0$) has no gravitational field.

---
[12] Pierre-Simon de Laplace (1749–1827), French mathematical physicist.

7.8 Equations of motion in a gravitational field

Gravitational theory must also make a statement about how a given object responds to the gravitational field. The trajectory $\mathbf{z}(t)$ will, at all times, be determined from the value of the rate of change $\dot{\mathbf{p}}(t)$ of the 3-momentum $\mathbf{p} = m\dot{\mathbf{z}}$ of a particle due to the action of the gravitational field. We note that $d\mathbf{p}/dt$ is not only a physical (measurable) Euclidean 3-vector but is also invariant in a boost. Any field quantity to which we equate it must share these properties. The gradient of the gravitational potential is the simplest such quantity and is a Galilean 3-vector. The simplest expression for $d\mathbf{p}/dt$ will make it proportional to the gradient $\nabla\Phi$ of the potential. This will also mean that the equation of motion, like the field equation, is linear in Φ. The dimensions of Φ and of all the other factors in these two quantities, $d\mathbf{p}/dt$ and $\nabla\Phi$, are already determined and the proportionality constant must be a Galilean invariant with the dimensions of mass. The only such quantity available is the mass m of the particle being affected by the gravitational field. It is therefore natural to examine the consequences of assuming that the effect of the field Φ on matter is of the form,

$$\frac{d\mathbf{p}}{dt} = -m\nabla\Phi = m\mathbf{f} \quad \text{or} \quad \ddot{\mathbf{z}} = -\nabla\Phi = \mathbf{f} , \qquad (7.17)$$

in which the sign is arbitrarily chosen to ensure that the interaction is attractive. (In the context of relativistic Lagrangian fields, the attractiveness follows as a consequence of the universal demand for positivity of the energy, including that of the field, and the consequent stability of the vacuum based on the quantization of a field.) The gravitational acceleration deduced in this way is in accord with Newton's mechanical theory and with experiment and is clearly independent of any internal properties of the particle such as its mass, electric charge, spin or chemical composition. This is a noteworthy result which should be contrasted with the acceleration of a particle in an electromagnetic field where the result is dependent not only on the fields but also on the mass, charge and velocity of the particle (see Equation 19.2).

Exercise 7.5 *Use the results of Exercise 7.4 to show that the gravitational acceleration of a particle of mass m at a distance r from a spherically symmetric mass M is given by Newton's universal attractive inverse square law of gravity, Equation 7.2.*

We have now reproduced the Newtonian gravitational equations describing the production of a gravitational field from a matter source and its action on a particle from a simple characterization of the gravitational interaction based on its long range, attractiveness, Galilean and parity covariance and universality. These properties sufficed to show that the lowest-rank, lowest-order linear field equations and equations of motion are precisely those of Newtonian gravity supplemented by a term in Λ which we still have to examine.

We may summarize here the equations of non-relativistic Newtonian gravitation in the form,

$$\nabla^2 \Phi = 4\pi G \rho \quad \text{and} \quad \ddot{\mathbf{z}} = -\nabla \Phi . \tag{7.18}$$

These constitute a satisfactory approximation subject to the following validity conditions on field strengths, speeds (v), pressures (p), densities (ρ) and characteristic lengths (ℓ) (not all of which are independent):

$$\frac{|\Delta \Phi|}{c^2} \ll 1, \quad \frac{v^2}{c^2} \ll 1, \quad \frac{p}{\rho c^2} \ll 1, \quad \ell \ll \frac{c}{\sqrt{\rho G}} . \tag{7.19}$$

These validity conditions follow by comparing Newtonian gravity with the relativistic gravitation theories for which it is an approximation. Subject to these conditions the above equations for Newtonian gravitation follow (see Chapter 21) from those of general relativity or from special relativistic spin 2 theories of weak gravity.

7.9 Cosmological constant

Galilean covariance and observations permit a gravitational potential equation given by,

$$\nabla^2 \Phi + \Lambda = 4\pi G \rho , \tag{7.20}$$

with the possibility of $\Lambda \neq 0$, although they do place upper limits on its magnitude if it is non-zero.

Exercise 7.6 *Let there be a small spherically-symmetric mass M centred on the origin. Show that the first integral of the generalized Newtonian gravity equation, 7.20, in the vacuum region outside M, can be written,*

$$f = -\frac{\partial \Phi}{\partial r} = -\frac{GM}{r^2} + \frac{\Lambda r}{3} , \tag{7.21}$$

or, equivalently,

$$\Phi(r) = -\frac{GM}{r} - \frac{\Lambda r^2}{6} + \text{ constant.} \tag{7.22}$$

The term $\Lambda r/3$ in the field strength $f = -\partial \Phi / \partial r$, or $-m\Lambda r^2/6$ in the gravitational potential energy $m\Phi$, may be interpreted as an 'interaction' on a test particle of mass m at r. The new effect will increase with separation

r and, unlike the gravitational term, is independent of the values of G and M. It gives rise to an effect on m which is present in a vacuum, $M = 0$.

Exercise 7.7 *Show that the sign of the term which it contributes to the energy of m, implies that our convention for Λ is such that the interaction is repulsive if $\Lambda > 0$ and attractive if $\Lambda < 0$.*

This new term will clearly be of negligible importance for very small separations. Observations of galaxies at typical intergalactic separations show no detectable deviation from the inverse square law of Newtonian gravity.

Exercise 7.8 *Suppose that observations on galaxies (with mass $\approx 10^{10} M_\odot$ where $M_\odot = 2.0 \times 10^{30}$ kg is the solar mass) at typical intergalactic separations ($\approx 1\,\text{M}\ell y$) show that the effects of the new term are less than about 1% of the gravitational term. Show that this places an upper limit, $|\Lambda| \leq 10^{-34}\,\text{s}^{-2}$, on the magnitude of Λ.*

Observations and the above exercise show that, if this term is to be important, the theory of non-quantum Newtonian physics suggests it can only be so on a cosmological scale at distances $\geq 1\,\text{M}\ell y$ (or $\geq 10^{22}$ m). For this reason, and because it is independent of M and G, $\Lambda(t)$ is referred to as the *cosmological parameter* and its value $\Lambda_0 = \Lambda(t_0)$ at the present epoch $t = t_0$, as the *cosmological constant*. The term involving Λ in the generalized gravitational equation is referred to as the *cosmological interaction*, although its appearance even in the absence of any mass M shows that it is of somewhat different character to any ordinary mutual interactions between Newtonian particles.

Although the cosmological term, since it appears in the field equation of gravity, is a gravitational effect, it should be distinguished from the conventional gravitational term for which a mass acts as the source. The cosmological interaction, if it exists, occurs even in the absence of any matter at all. The possible existence of a fundamental gravitational parameter with an upper limit of about 10^{-34} s^{-2} is one of the most extraordinary properties of physics. If such a non-zero parameter exists, it will imply that nature requires the use of a dimensionless quantity with a magnitude smaller than about 10^{-120}.

To see this, let us construct a fundamental dimensionless constant proportional to Λ. To do so we need another quantity with the dimensions of time. Since the cosmological interaction is of a gravitational nature, we expect to use G in forming such a constant. We again anticipate relativistic gravity by allowing for the inclusion of the constant speed c, but these two do not suffice to form a quantity of time dimensions. Independence from non-gravitational interactions and specific particles may be maintained by including the Planck constant $\hbar$ in the analysis and this permits us to form the *Planck time*, $(G\hbar/c^5)^{1/2} = 5.4\times10^{-44}$ s. The dimensionless combination proportional to Λ is therefore,

$$\frac{G\hbar\Lambda}{c^5} \leq 10^{-120}, \tag{7.23}$$

an approximate upper bound on the *relativistic quantum cosmological constant*. There is no hint of the existence elsewhere in the laws of nature of a fundamental dimensionless physical quantity anywhere near as small as this value. If such a small parameter exists there ought to be some fundamental reason why it is so small. If the explanation is that it is in fact exactly zero, then again there ought to be a fundamental explanation for this result.

The cosmological constant was first introduced, with a positive value, by Einstein [140] in 1917 to explain the universally recognized 'fact' that the universe was static by supplying a repulsion to cancel the universal gravitational attraction. The introduction appeared to be a very natural generalization of his general relativistic theory of gravitation formulated two years earlier. The equations of the general theory (see Chapter 21) are essentially a relativistic equivalent of Equation 7.14 in which the gravitational field, tensorial in character, is intimately connected to the structure of spacetime itself. The generalization is entirely analogous to that which takes us from Equation 7.14 to Equation 7.20, which is equally natural since it leaves the equation still satisfying Galilean covariance. When Hubble [255] used observations of the red shifts of galaxies a few years later, in 1929, to demonstrate that the universe was not in fact static but expanding, Einstein's reason for introducing it vanished. He then abandoned its use and used the phrase 'the greatest blunder of my life' [184, p. 44] to describe the introduction of the cosmological constant into his relativistic gravitational field equations.

Having escaped, the cosmological constant genie has steadfastly refused to vanish again and continues to be included in most analyses of the relativistic gravitational field. In fact, those whose prime interest centres on the relativistic quantum physics of the non-gravitational interactions consider the cosmological constant to be the most interesting and important property of gravitation [193], some even say the only interesting property. This is a consequence of its connection with the lowest energy (vacuum) state of a system and the importance of the dynamic properties of the vacuum in quantum field theory.

7.10 Gauge freedom of Newtonian gravity

The origin of the term *gauge invariance* dates from early attempts to unify the two long-range interactions, electromagnetism and gravity. In 1918,

Weyl [572] examined a unified theory based on introducing a symmetry involving a scaling,

$$t \to \alpha(t,\mathbf{x})t \quad \text{and} \quad \mathbf{x} \to \alpha(t,\mathbf{x})\mathbf{x} , \qquad (7.24)$$

of the time and length dimensions by a common arbitrary function $\alpha(t,\mathbf{x})$ of spacetime position. The word *gauge*, associated with standards of length, was used in the English translation of Weyl's *Eichinvariance* and has been carried over to all forms of invariance of equations of motion or field equations containing an arbitrary function, like $\alpha(t,\mathbf{x})$, in the transformation rather than a finite number of parameters, as for example in a rotation or translation.

Let us suppose that choosing to describe the spatial distribution or evolution of a physical system using a certain set of dependent and independent variables, leads to equations in which the highest order time derivatives are of order n. The *initial data* are the independent variables, the dependent variables and the first $n-1$ time derivatives of the dependent variables. For a system of non-quantum particles, for example, we may choose the spatial coordinates $z_i(t)$ as the dependent variables with the time t as the independent evolution or path parameter. For a field $\psi(t,\mathbf{x}) = \{\psi_j(t,\mathbf{x})\}$ ($j = 1, 2, \ldots$) the values of the components $\psi_j(t,\mathbf{x})$ at all $\mathbf{x}$ are the dependent variables and the spacetime coordinates $\{t,\mathbf{x}\}$ are the independent variables corresponding to evolution and spatial distribution. There may be some compelling reason to choose a particular set of dynamical variables such as a desire to have manifest covariance or the incorporation of conservation laws automatically as identities.

Suppose there exists a transformation, containing one or more arbitrary functions of the initial data, the effect of which changes the dependent and/or independent variables in a non-trivial way involving the arbitrary functions but leaves the distribution or evolution equations unchanged in form. Then the equations describing the dynamics of the system in the chosen variables are said to exhibit *local gauge invariance*. Each of the arbitrary functions is referred to as a *gauge function* while the chosen variables are said to be *gauge-variant* or to exhibit *local gauge freedom*. In the continuous (non-mechanical) case the variable is referred to as a *gauge field* or *gauge potential*. The gauge field ideas introduced by Weyl were extended by Cartan [58, 59] in 1901 and 1923 to what he called *generalized spaces*. The mathematical term for the type of gauge field studied by Weyl and Cartan is a *linear connection* [125, 499] on a fibre bundle.

Gauge invariance effectively contains an infinite number of parameters, those required to describe the gauge function. This is in contrast to rotational, Galilean, Lorentz, Poincaré, and global Yang-Mills invariances. The latter are parameterized either by 3 angles (in the case of a rotation), by 3 angles plus 3 boost parameters (Lorentz), by 10 rotation, boost and translation parameters (Galilean and Poincaré) and in the case of SU(2) and SU(3) invariances (Yang-Mills) by 3 or 8 internal parameters or phase angles.

Since a set of physical (measurable) quantities cannot have arbitrary values, the occurrence of gauge invariance shows that the dynamical variables having the gauge freedom cannot all be physical degrees of freedom. The number of degrees of freedom cannot be greater than the original number of dynamical variables less the number of degrees of freedom in the gauge functions. The particle variables z_i of Newtonian gravity exhibit no gauge freedom.

Exercise 7.9 *Let a transformation $z_i \to \alpha(t)z_i + \beta(t)\dot{z}_i + \gamma(t)$ leave unchanged the form of Equation 7.2, for a system of particles interacting gravitationally, and show that the form invariance implies that $\beta = 0$ and that α, and γ must be constant.*

We have already seen that the gravitational potential is arbitrary to the extent of a scalar function $\alpha(t)$. According to the above criteria, the Newtonian gravitational potential has gauge freedom. In fact, since it has only one component $\Phi(t, \mathbf{x})$, as does the gauge function $\alpha(t)$, the gravitational potential has no dynamical degrees of freedom! The gravitational field equation of Poisson, Equation 7.14, is not a dynamical equation for the determination of the evolution of the potential, but an equation governing its distribution in space at a given instant of time. From the dynamical point of view, it is said to be a *constraint* on the initial spatial distribution of the potential, which must apply at all times. The term *field strength* is generally reserved for quantities which do not have gauge freedom.

Exercise 7.10 *Show, from first principles, that Equations 7.12 for the gravitational field strength $\mathbf{f}$ have no gauge freedom, and verify that this is consistent with the relation between field strength and potential.*

The field equations of electrostatics, $\nabla \cdot \mathbf{E} = \rho/\epsilon_0$ and $\nabla \times \mathbf{E} = 0$ or $\nabla^2 \phi = -\rho/\epsilon_0$, for the electric field strength $\mathbf{E} = -\nabla \phi$ or scalar potential ϕ (see Chapter 18) are closely anagolous to those of Newtonian gravity. The equations, $\nabla \cdot \mathbf{B} = 0$ and $\nabla \times \mathbf{B} = \mu_0 \mathbf{J}$ or $\nabla^2 \mathbf{A} - \nabla(\nabla \cdot \mathbf{A}) = -\mu_0 \mathbf{J}$, for the magnetostatic field strength $\mathbf{B} = \nabla \times \mathbf{A}$ or vector potential $\mathbf{A}$ are very similar. The above arguments may also be applied to the variables involved with similar conclusions regarding gauge freedom.

7.11 Weak equivalence principle

Let us initially not assume that the two masses appearing in the gravitational equations, one for the source producing the field and one for the particle acted on by the field, are identical nor that either is identical to the inertial mass. This will permit us to reveal the tacit assumptions, and their justification, involved in setting them equal. We denote by m_I the *inertial mass* which appears in the expression,

$$\mathbf{p} = m_\text{I}\dot{\mathbf{z}}, \qquad (7.25)$$

for the 3-momentum of a particle of 3-velocity $\dot{\mathbf{z}}$.

We denote the source of the gravitational field $\Phi(t,\mathbf{x})$ by $\rho_\text{AG}(t,\mathbf{x})$ where the subscript AG indicates *active gravitational mass*, namely the mass which generates the gravitational field. The field equation for a long-range interaction will therefore be,

$$\nabla^2 \Phi = 4\pi G \rho_\text{AG}. \qquad (7.26)$$

As before, we take the simplest expression for $d\mathbf{p}/dt$ and make it proportional to the gradient $\nabla\Phi$ of the potential. The proportionality constant must be a Galilean invariant with the dimensions of mass and, apart from a sign, we shall call it the *passive gravitational mass*, namely the mass on which the gravitational field acts, and denote it by m_PG. The effect of the gravitational field on matter will then take the form,

$$\frac{d\mathbf{p}}{dt} = -m_\text{PG}\nabla\Phi, \qquad (7.27)$$

in which the sign is chosen as before to ensure that the gravitational interaction is attractive.

Combining Equations 7.25 and 7.27 now gives,

$$\ddot{\mathbf{z}} = -\frac{m_\text{PG}}{m_\text{I}}\nabla\Phi. \qquad (7.28)$$

The reciprocity of action and reaction,

$$\left(\frac{d\mathbf{p}}{dt}\right)_\text{action} = -\left(\frac{d\mathbf{p}}{dt}\right)_\text{reaction}, \qquad (7.29)$$

which itself arises from the translational part of the Euclidean invariance of the whole system, requires the active and passive gravitational masses to be equivalent and we denote the common quantity, $m_\text{PG} \equiv m_\text{AG}$, by m_G giving,

$$\ddot{\mathbf{z}} = -\frac{m_\text{G}}{m_\text{I}}\nabla\Phi. \qquad (7.30)$$

Galileo [182] first measured the accelerations of inertial masses of different magnitude and different composition using an inclined plane to slow the motion. He states that Aristotle says that *'an iron ball of one hundred pounds falling from a height of one hundred cubits reaches the*

7.11. Weak equivalence principle

ground before a one-pound ball has fallen a single cubit'. I say that they arrive at the same time.

Newton was aware of the distinction between the two ways in which mass entered the mechanical equations and describes the observed equality [395], based on experiments with pendula, in the *Principia*. Careful experiments were carried out by Eötvös [143] in 1889 to compare the gravitational accelerations due to the earth on a wide variety of substances. His experiments were repeated [144] with improved precision in 1922 because of their importance to the foundations of general relativity. These showed that m_G/m_I is a universal constant of nature to very high precision (of the order of 3 parts in at least 10^9). Experiments by Dicke, Roll and Krotkov [481] in 1964 and by Braginsky and Panov [46] have improved the precision to 1 part in 10^{12} by using the acceleration due to the sun.

By choosing the dimensions and units of m_G to coincide with those of m_I, this constant can therefore be assigned the value unity requiring the acceleration of a gravitational test particle to be determined entirely from the local value of the gravitational field strength according to the second of Equations 7.18. The original Eötvös data, other Eötvös-type experiments on the earth and direct measurements of the earth's gravitational potential have assumed major importance again since 1981 as a result of reported deviations [397, 519] from Newton's inverse square law at distances beyond laboratory dimensions, ≥ 10 m, but closer than about 10,000 km. At larger distances, data from laser ranging to the LAGEOS satellite at about 100,000 km, and to the moon, leave little room for such deviations.

Putting aside the interpretation of the small observed deviations (see Section 7.14), we may abstract the essential part of the result and state it in the general form of the *Weak equivalence principle* [131, 132, 486, 585, 586], also referred to as the *Principle of uniqueness of free fall* [384]. In order to eliminate the effects of non-gravitational interactions, let us stipulate that a *gravitational test body* be electrically neutral, non-spinning, of negligible self-gravitational energy and small compared to the characteristic lengths of the gravitational field. If it is placed at an initial event and given an initial velocity there, then we have,

- the **Weak equivalence principle** — the motion of a gravitational test body is determined entirely by the local structure of the gravitational field (independently, for example, of its internal structure, composition, mass, size or velocity).

Let a mass point be at rest in S, an inertial frame with coordinates $\{t, \mathbf{x}\}$, near a large mass, at a point where the gravitational field is Φ. The gravitational acceleration in S on the particle is $\ddot{\mathbf{z}} = -\nabla\Phi \equiv \mathbf{g}$. Let S' be a frame (attached to a rocket, for example) with coordinates $\{t, \mathbf{x}'\}$ with constant acceleration $\mathbf{a} = -\mathbf{g}$ with respect to the inertial frame S. Far from all sources of gravity, a free particle will have acceleration $\ddot{\mathbf{z}} = 0$ in

S and therefore acceleration $\ddot{\mathbf{z}}' = \ddot{\mathbf{z}} - \mathbf{a} = \mathbf{g} = -\nabla\Phi$ in S'. Local observations of the acceleration of a single object will be unable to distinguish between the above two cases of acceleration relative to an inertial frame in the absence of gravitation and acceleration, relative to an inertial frame, caused by gravitation.

Locally, an acceleration relative to an inertial frame is equivalent to gravitation. Only by considering relative accelerations, as for example in tidal forces, can one detect the existence of a gravitational field. Accelerated frames and gravitational forces appear to be intimately connected. This forms the basis of the *Strong equivalence principle* (see Section 21.5) suggested by Einstein [131] as early as 1907 and in more detail [132] in 1911. In fact, Einstein's general relativistic theory of gravity arises from an extension of the Principle of relativity from frames in uniform relative motion to those whose relative motion is arbitrary and therefore accelerated in general.

7.12 Michell and Laplace black holes

Conservation of energy shows that, according to Newtonian gravity, the *escape velocity* of an object at a distance r outside a spherically symmetric mass M is $v_{\text{esc}} = \sqrt{2GM/r}$. For an object of radius R and uniform density ρ, the escape velocity is greatest on the surface where it is $v_{\text{esc}} = R\sqrt{8\pi G\rho/3}$. Consequently, if the radius is increased at constant density, the escape speed will reach the speed of light c at a limiting radius of $R_{\text{lim}} = \sqrt{3c^2/8\pi G\rho}$. For an object like the sun, where the mean density is near that of water, the limiting radius is about 1000 times the radius of the sun. The radius of such an object is $R = 2GM/c^2$.

Light was generally treated as a Newtonian corpuscle with mass prior to about 1800. One can therefore use such a calculation to argue that Newtonian gravity suggests that something unusual might occur if the physics of stellar material permitted the formation of such an object near which the gravitational acceleration is very strong.

In 1784, Michell [194, 377], in a paper in which stellar distances were calculated correctly for the first time, concluded that the light from an object of solar density and 500 times larger than the sun would fall back towards the object due to gravity, thereby making it invisible. That paper can now be regarded as the earliest known reference to the unusual system, now referred to as a *black hole* (in spacetime), that can naïvely be expected to form if a constant density object is sufficiently large. A similar calculation was carried out in 1795 by Laplace [310] who concluded: *A star having the density of the earth and 250 times the solar size would, as a consequence of gravitational attraction, allow none of its rays to reach us. It's possible that the largest luminous bodies of the universe are therefore invisible*[13].

It is now known that Newtonian laws are invalid for rays (and particles) of light and for strong gravity. We also know that the mass of stellar material of solar density cannot greatly exceed that of the sun. A more astrophysical, but still rather naïve, approach to the same problem would consider a roughly constant mass and allow the radius R to decrease as the star exhausts its thermonuclear fuel and loses its supporting pressure. We can determine a characteristic *gravitational length* associated with any given mass M and the speed of light c, namely the *Schwarzschild length*[14],

$$r_M = 2GM/c^2 = 2.95 \left(\frac{M}{M_\odot}\right) \text{km} , \qquad (7.31)$$

in which the factor of 2 is conventional. This is the only gravitational length associated, via the speed of light, with a given mass M. If the radius decreases to the Schwarzschild length, we shall again have an escape velocity equal to the speed of light.

Exercise 7.11 *Assume that Euclidean formulae are valid and show that the density at which a Newtonian black hole would form, due to contraction at constant mass M, is given by,*

$$\rho_{bh} = \frac{3c^6}{32\pi G^3 M^2} = 1.84 \times 10^{19} \left(\frac{M_\odot}{M}\right)^2 \text{kg.m}^{-3} . \qquad (7.32)$$

This result shows that by examining objects of sufficiently high mass, the escape speed may formally be comparable with that of light for densities as low as one chooses indicating that it is not sufficient to appeal to unknown properties of matter at high densities to avoid the limit.

Since light is a relativistic phenomenon, we cannot expect to form a satisfactory description of a black hole, and determine whether such objects can form as a result of stellar evolution, without a theory of gravity which is consistent with relativity theory. The conceptual difficulties that arise with a Newtonian description of a black hole are avoided using Einstein's general relativity theory.

7.13 Soldner's gravitational deflection of light

Throughout antiquity and the middle ages, it was assumed almost universally that light propagated instantaneously. Galileo was convinced that

---

[13] Reference [310, vol. II, p. 305]. *Un astre lumineux de même densité que la terre, et dont le diamètre serait deux cents cinquantes fois plus grand que celui du soleil, ne laisserait en vertu de son attraction, parvenir aucun de ses rayons jusqu'à nous; il est possible que les plus grands corps lumineux de l'univers, soient par cela même, invisibles.*

[14] Karl Schwarzschild (1873–1916), German astronomer.

light had a finite speed and devised a not very adequate method to measure it. Römer[15] was the first to demonstrate in 1676 that Galileo was right as a result of a study of the eclipses of the satellites of Jupiter which occurred at times which were inconsistent with regular motion until the light travel time across a varying fraction of the diameter of the earth's orbit during each year was taken into account. Since other perturbations were not recognized at the time, Römer's value of $230,000\,\mathrm{km\,s^{-1}}$ is not in close agreement with modern measurements. His results were not widely accepted until confirmed by the discovery of *stellar aberration* by Bradley [45] in 1728.

The finiteness of the speed of light was thus well-established at least two centuries before the advent of special relativity. The wave theory, first suggested by Hooke (in his monograph, *Micrographia*) in 1664, considered light to be a very high speed rectilinear propagation of longitudinal vibrations of a medium, the ether, in which individual wavelets spread spherically. The wave theory was developed further by Huyghens[16] [258] from about 1677. Newton made extensive investigations of light culminating in the publication in 1704 of his monograph *Opticks* [396], and at first described light in both wave terms and as a stream of material particles referred to as *corpuscles*.

Owing to difficulties in describing rectilinear propagation with spherical wavelets, Newton increasingly favoured the corpuscular theory. Newton's work on the mechanics of material particles was so influential that the wave theory was not widely considered throughout almost a century. It was revived in 1801 by Young[17], and independently in 1814 by Fresnel[18], with the new feature of *intereference*, in which form it was able to account for the rectilinear propagation. In order to account for the recently discovered *polarization* phenomena, Young suggested that light must be a vibration of the ether transverse to the propagation direction and by about 1825 the wave explanation of light was so successful that the corpuscular theory had few remaining adherents.

We know now that light is an essentially relativistic phenomenon and that the finiteness and invariance of its speed of propagation is intimately linked to that fact. Newtonian gravity is, however, non-relativistic. Nevertheless, placing ourselves in Newton's time, or shortly thereafter, we may ignore the inconsistency in using a non-relativistic gravitational field equation and the essentially relativistic characteristic of a finite speed of light to determine the extent of the Newtonian deflection expected when a light corpuscle passes near a mass such as that of the sun. Newton himself apparently did not consider the possibility of light bending. However, his

---

[15] Ole Christensen Römer (1644–1710), Danish astronomer and statesman.
[16] Christiaan Huyghens (1629–1695), Dutch mathematical physicist and astronomer.
[17] Thomas Young (1773–1829), English physicist and physician.
[18] Augustin Jean Fresnel (1788–1827), French physicist.

7.13. Soldner's gravitational deflection of light

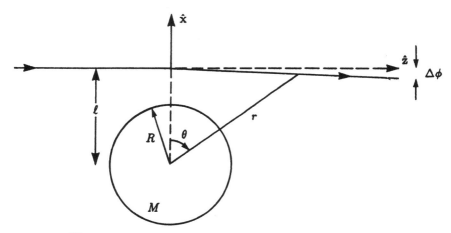

Figure 7.1 Gravitational deflection of a light corpuscle.

Query 1 in *Opticks* [396]: *Do not Bodies act upon Light at a distance, and by their action bend its Rays; and is not this action strongest at the least distance?* is, as Jaki [265] points out, typically phrased in a sufficiently general way to accomodate the idea of light deflection by gravity despite the fact that it refers in fact to reflection, refraction and diffraction.

The extent to which light of corpuscular character would be deflected by a nearby mass, if the gravitational theory used was essentially Newtonian, based on a Galilean spacetime, is very easy to calculate. Consider a spherically-symmetric mass M of radius R (for example, the sun) and a light corpuscle of mass m and speed c which approaches and passes nearby with an impact parameter $\ell \geq R$ as shown in Figure 7.1.

We may first estimate the likely magnitude of any deflection using dimensional analysis. The angular deflection will decrease with increase in the impact parameter ℓ of the ray and, being dimensionless, can be expected to be closely related to the ratio $r_M/\ell = 2GM/c^2\ell$ of the gravitational length to the impact parameter. The solar mass $M_\odot$ equals 2.0×10^{30} kg giving $r_M^\odot = 3$ km which, with $\ell \geq R_\odot$ and $R_\odot = 700{,}000$ km, shows that the deflection can be expected to be very small, namely of the order of $3/700{,}000$ radians or about a second of arc. The smallness of the deflection corresponds to the weakness, $|\Delta\Phi|/c^2 \ll 1$, of the gravitational field even near the sun where it is strongest in the solar system. Referring to Figure 7.1, we take $\mathbf{e}_z$ in the initial asymptotic direction of the ray and we take $\mathbf{e}_x$ to point from the centre of the sun to the nearest point on the extrapolation of the initial trajectory of the ray.

This means that the initial momentum of the ray will be $(0, 0, p_z)$ (with $p_z > 0$). By symmetry, and the attractiveness of gravity, the deflection will be in the negative x-direction. To first order in small quantities, the z-component p_z of momentum will be constant. We now evaluate the deflection $|\Delta\phi| = -(p_x/p_z)_{\text{final}}$ as follows:

$$\begin{aligned}
|\Delta\phi| &= -\frac{1}{p_z}\int_{-\infty}^{\infty}\frac{dp_x}{dz}dz \\
&= -\frac{1}{p_z c}\int_{-\infty}^{\infty}\frac{dp_x}{dt}dz \\
&= -\frac{1}{p_z c}\int_{\theta=-\pi/2}^{\pi/2}\left(-\frac{GMm\cos\theta}{r^2}\right)dz\,,
\end{aligned} \quad (7.33)$$

where m denotes the mass of the particle of light. Provided it is non-zero, we do not need a theory for the value of m (such as $m = E/c^2 = \hbar/c\lambda$ for a quantum of energy E and reduced wavelength λ) since it will immediately cancel with m in the $p_z = mc$ factor in the denominator. (Alternatively, one may consider this result to apply to a massless particle obtained as the limit for $m \to 0$.) Evaluation of the integral gives a deflection of,

$$|\Delta\phi| = \frac{2GM}{c^2\ell} \quad \text{(Newtonian gravity, corpuscular light)}\,. \quad (7.34)$$

This result predicts a deflection of 0.88 arc seconds, for grazing incidence on the sun, whereas the observed deflection angle is very close to twice this value [167, 586]. The essential features of this calculation, carried out in 1801 by Soldner[19] [515], who obtained a value of 0.84 arc seconds, involve finiteness of the speed of light and non-zero total mass for the light corpuscle and hence its gravitational deflection by any other mass. Soldner's work arose out of Laplace's publications in 1795 and 1799 on the influence of gravity on light and Michell's similar calculations of 1784 also led Cavendish to obtain Soldner's results, recorded in unpublished work [61, p. 437]. The same result was found independently by Einstein [132] in 1911 using the Equivalence principle in a Newtonian context, giving a value of 0.83 arc seconds, which was doubled to the correct result using general relativity in 1915 (see Chapters 17 and 21).

British expeditions to observe solar eclipses at Sobral in Brazil and on Principe Island off the west coast of Africa in 1919 confirmed the general relativistic calculation of light deflection. Although the 1919 measurements were not of high precision, they clearly confirmed the existence of an effect that was unsuspected immediately prior to Einstein's prediction and constituted a vital step in the widespread acceptance of the general relativistic theory of gravitation.

The observational confirmation in 1919 of Einstein's 1915 light bending prediction was a major scientific event. Attention was only drawn to

---

[19] Johann Georg von Soldner (1776–1833), German mathematician, astronomer and geodesist.

Soldner's work, neglected for 120 years, by Martin Näbauer, in 1921. It was again neglected until revived by Rudenko [486], Jaki [265] and Gibbons [194] in 1978. We shall discuss the reasons why this calculation gives only half the observed value in Chapter 21.

Light deflection by gravity could not be described, even with an incorrect magnitude, by the wave theory of 1801 since the necessary mass-energy equivalence was not a part of the theory. Had Soldner's work been well-known it would surely have led to experimental verification, since this would have been possible from about 1820 with the telescopy that permitted Bessel to detect stellar parallax, of well under 1 arc second, in 1837. Such an occurrence would have created a major problem for the wave theory revived in 1801 and which gained rapidly thereafter in predictive power and acceptance with a corresponding demise of the corpuscular theory.

7.14 Deviations from Newtonian gravity

Tout écart décèle une cause inconnue et peut être la source d'une découverte.

Urbain Le Verrier

Newtonian gravity calculations were used during the 19th century to analyse the non-elliptic motions of the known planets due to the perturbations of other planets. In 1846, Le Verrier[20] used the observations of known planets, especially those furthest from the sun, to predict the existence and location of a new planet, Neptune. When Johann Galle, at the Berlin Observatory, looked at that location a few weeks later to find the faint disk of a planet, confirming the existence of Neptune, Newtonian gravity theory achieved one of its greatest victories, reinforcing its apparently unassailable place at the centre of physical theory.

However, the observations for Mercury, the innermost planet and the one with the highest eccentricity ($e = 0.2056$), showed that its line of apses still advanced by 38 arc seconds per century (now 42.56 ± 0.94 [384]) after all known perturbations and the general precession, due to the observer's frame being non-inertial, had been taken into account. We now know that these *perihelion shift* observations were the first to show that Newtonian gravity must be modified under certain conditions. The most important test of general relativistic gravitation in the first few decades of its history was its description [135, 136, 137, 138, 139] of the deviation of a planetary orbit from the ellipse predicted by Newtonian gravity. This effect will be discussed in the context of relativistic theories of gravitation in Chapters 17 and 21.

Another non-Newtonian phenomenon which is now very well established by observation and measurement is the effect that gravity has on the

---

[20] Urbain Jean Joseph Le Verrier (1811–1877), French astronomer and mathematician.

rates of clocks. This is known both as *gravitational time dilation* of clocks and the *gravitational red shift* of signals. Let t be the time on a set of *coordinate clocks* fixed in an inertial reference frame $\{t, \mathbf{x}\}$ with the earth at the origin and non-rotating relative to the distant stars. We presume the times on all the coordinate clocks have been synchronized in rate, *in situ*, by the exchange of constant velocity signals with a master clock. In the context of Galilean relativity and Newtonian gravity, t will be the universal time of Newton.

We now take two identical atomic or *particle clocks*, C_l and C_u, whose rates, as measured in a frame in which they are at rest, is given in terms of atomic constants such as the proton charge e and the electron mass m_e. (We could use sets of clocks to permit them to be checked by intercomparison with others spatially coincident with them). We fix C_l at a location $\mathbf{x}$ in the frame of the earth at a height where its static gravitational potential is $\Phi(\mathbf{x})$ and we similarly place C_u at $\mathbf{x} + \Delta\mathbf{x}$ where the potential is $\Phi + \Delta\Phi$. We make no attempt to alter the rate of either set of particle clocks. If necessary, each set may be constructed on site from atomic components to ensure the they are not affected by transport through the gravitational field.

We now count periods of each particle clock from coordinate time t_i to a final time t_f, as measured on the coordinate clocks spatially coincident with each particle clock and determine the difference $\Delta P = P_u - P_l$ between the periods of the two particle clocks. Atomic clocks have been sufficiently precise during the last three decades to show that the $\Delta P \neq 0$. We call $z = -(\Delta P/P)_{\text{clock}}$ the *gravitational red shift factor* for reasons which arise from the close relationship, in relativistic theories, between gravitational dilation of the times of particle clocks and the effect of gravitation on the periods of signals, $(\Delta P/P)_{\text{signal}} = -(\Delta P/P)_{\text{clock}}$, to be discussed in Chapter 17.

The postulated equivalence of the two particle clocks, when $\Delta\mathbf{x} = 0$, means that to first-order in the ratio of the small height difference $h = \Delta r$ to the earth's radius $R_\oplus$, we have $z \propto \Delta\Phi$ where $\Delta\Phi = gh = GMh/R_\oplus^2$. In anticipation of the observational results, we write the expression for the small red shift as,

$$z = -\left(\frac{\Delta P}{P}\right)_{\text{clock}} = \left(\frac{\Delta P}{P}\right)_{\text{signal}} = \alpha \frac{\Delta\Phi}{c^2}, \qquad (7.35)$$

where the dimensionless quantity α may be used to quantify the observational results.

The first measurements of this red-shift effect by Adams [2] in 1925 and by Moore [388] in 1928 made use of the spectral lines of hydrogen in Sirius B, the white dwarf companion of the star Sirius. The results for α were highly uncertain, ranging from 0.2 to 1.2. Measurements of the white dwarf star Eriadni B by Popper [456] as recently as 1954 were still reasonably uncertain with values between 0.3 and 1.2. For this reason, despite the importance attached to it by Einstein, the classic gravity test by redshift measurements did not constitute an important factor in establishing the characteristics of the gravitational field during the first half century of relativistic gravitation theory.

In 1960, Pound and Rebka [457] used the Mössbauer effect in laboratory experiments with γ rays over heights as small as a few metres to determine α to be 1.05 ± 0.10 and these results were improved by a refinement of the same technique in 1964 by Pound and Snider [458] who obtained 1.00 ± 0.01. Muon decay measurements by Bailey et al. [24] in 1979 increased the precision to 1.000 ± 0.001 and radar ranging rocket experiments by Vessot, et al. [555] in 1980 measured α to be 1.0 to within about 4×10^{-4}. These results are interpreted to mean that gravity affects the rate at which clocks measure time and show that for small differences in the gravitational field the *gravitational time dilation* is given by,

$$z = \frac{\Delta \Phi}{c^2}. \qquad (7.36)$$

The presence of the ultimate speed c shows that these results cannot be described by a consistent theory based on Galilean relativity, such as Newtonian gravity. They constitute a deviation from the predictions of Newtonian physics.

We have already seen how the Newtonian theory, as used by Soldner in 1801 and by Cavendish, implies *light bending* with a magnitude subsequently shown to be only half the observed value. The incorrect value should not be surprising since we cannot expect a Galilean theory to describe effects involving propagation at a finite ultimate speed. Light bending is well-known as one of the three original classic tests of the general relativistic gravitation theory presented in its final form by Einstein [136, 137, 138] in 1915. Einstein had suggested observations looking for this effect [132] at times of solar eclipse as early as 1911 after using the Principle of equivalence, applied in a Newtonian manner, to argue that bending should occur. An attempt to observe the effect in Brazil in 1912 failed because of poor weather. The outbreak of World War I terminated a German expedition to the Crimea in 1914. We shall further examine these predictions and observations of non-Newtonian gravity in Part IV.

All observations on slowly moving objects, carried out over small distances, $\ell \leq 10\,\text{m}$, in the static gravitational field of the earth near its surface, are in close agreement with Newtonian gravity. The same is true for weak, static fields and slow motion with astronomical but not cosmological separations, namely $10^5\,\text{m} \leq \ell \leq 10^{20}\,\text{m}$.

Exercise 7.12 *Consider the possibility of a Newtonian gravitational field Φ_0 arising from matter but determined by Equation 7.13 with $\Lambda = 0$ and* the mass density ρ replaced by $q_0\rho$ where q_0 is a constant dimensionless quantity related to the types of particle making up the matter and therefore composition dependent. Show that, in the absence of the Λ term, but including a term with constant $\lambda = \lambda_0$, the Yukawa expression,

$$\Phi_0 = -\frac{GMq_0}{r}e^{-r/\lambda_0} . \qquad (7.37)$$

is a solution for $\Phi_0(r)$ at large distances (where $\rho = 0$) from a compact spherically-symmetric mass M.

The constant q_0 is a dimensionless measure of the strength, relative to the long-range Newtonian gravitational potential, of a possible short-range spin 0 contribution to the potential Φ. This contribution is referred to as a *Yukawa term* as a result of similarity to the potentials appearing in Yukawa's pion exchange theory [597] of the strong interaction between the constituents of nuclei.

Theoretical discussions of the possible origins of such a term have been given greater impetus during the last decade as a result of a number of experiments suggesting that the gravitational potential may deviate from the inverse square law at intermediate distances from 100 m to 10,000 km. References to the theoretical work and a review of the experimental results are available in Paik [417], Stacey et al. [519] and Nieto et al. [397]. If observations were to show that such Yukawa terms were required, our preceding examination has shown that they may be included reasonably naturally in a second-order analysis in terms of a scalar potential Φ without violating Galilean covariance.

7.15 Newton-Cartan spacetime

The existence of a Newtonian gravitational field does not affect the Galilean spacetime concepts of simultaneity, time interval between any events and lack of spatial separation for non-simultaneous events. However, it changes the family of curves traced out by free particles, namely those free of non-gravitational interactions, from the privileged family of Galilean spacetime, $\mathbb{G} = \{\mathbf{E}^1, \mathbf{E}^3\}$. If such particles are used to characterize the spacetime, then the resulting structure, which we shall denote by $\mathbb{N}$ and call *Newton-Cartan spacetime*, has no symmetries in the presence of a generic gravitational field. It retains the fibre bundle structure of $\mathbb{G}$. To each distinct gravitational field there corresponds a distinct linear connection on the fibre bundle.

Although the Galilean and Newton-Cartan fibre bundle spacetimes have no 4D metric, the 4D curvature of both is a well-defined concept. The curvature of Newton-Cartan spacetime bears the same relationship to flat Galilean spacetime that the curved spacetime of general relativity bears to the flat special relativistic spacetime of Minkowski. Analyses of Newtonian gravitation in the spirit of general relativity using the tools of curvature were carried out by Cartan [59] in 1923 and 1924 and by Friedrichs [177] in 1928. For further details on Newton-Cartan spacetime, we refer the reader to Trautmann [542, 544], Chapter 12 of Misner, Thorne and Wheeler [384] and Penrose [431, 433].

When Newton saw an apple fall, he found ...
A mode of proving that the earth turn'd round
In a most natural whirl, called gravitation,
And thus is the sole mortal who could grapple
Since Adam, with a fall or with an apple[21].

George Gordon [Lord Byron]   1788–1824

Bibliography:

Davies P C W 1980 *Search for gravitational waves* [88].
Einstein A 1911 *On the influence of gravitation on the propagation of light* [343].
Jaki S L 1978 *Johann Georg von Soldner and gravitational bending of light* [265].
Ohanian H C 1976 *Gravitation and spacetime* [404].
Penrose R 1972 *Black holes* [432].
Ruffini R and Wheeler J A 1971 *Introducing the black hole* [487].
Stacey F D et al., 1987 *Geophysics and the law of gravity* [519].
Thorne K S 1974 *The search for black holes* [537].
Thorne K S et al., 1986 *Black holes: the membrane paradigm* [538].

---

[21] *Don Juan* 10, 11.

Part II

Lagrangian interaction and symmetry

CHAPTER 8

Euler-Lagrange equations of motion

In any change which occurs in nature, the sum of the products of each body multiplied by the space it traverses and by its speed (referred to as *the action*) is always the least possible[1].

Pierre Louis Moreau de Maupertuis[2]   1698–1759

8.1 Introduction

An early appearance of a *Principle of stationary action* in physics was the statement by Hero[3] in his *Catoptrica (Optics)* [236] that the reflection of light at a plane mirror was such that it took the shortest possible path. A simple calculation shows that the shortest path is the one with equal angles of incidence and reflection. Fermat[4] reformulated the idea [153] as a *Principle of least time* in 1657 by postulating that a light ray takes the path which requires the least time even if it has to deviate from the shortest path in space: *nature operates by the simplest and most expeditious ways and means*, and in 1661 applied this optical minimum principle to media of different refractive indices [153]. *Fermat's principle* was not only capable of giving the correct law of reflection but also led to the law relating the angles of incidence and refraction at an interface to the ratio of the refractive indices of the two media, discovered experimentally in 1621 by Snell[5].

Maupertuis first stated a minimum principle in mechanics [362] in 1746. Maupertuis's principle was intended to apply not only to inanimate objects but universally, including biological phenomena. However, it was somewhat vaguely expressed, with supernatural aspects:

---

[1] *Sur les lois du mouvement* 1746 in *Oeuvres* [362, vol. 2, p. 273]. My translation.
[2] French astronomer, biologist and mathematical physicist.
[3] Hero of Alexandria (fl. A.D. 62), Greek mathematician, physicist and inventor.
[4] Pierre de Fermat (1601–1665), French linguist, lawyer and amateur mathematician.
[5] Willebrord Snell van Royen (1591–1626), Dutch mathematician.

> The laws of movement thus deduced [from this principle], being found to be precisely the same as those observed in nature, we can admire the application of it to all phenomena, in the movement of animals, in the vegetation of plants, in the revolution of the heavenly bodies: and the spectacle of the universe becomes so much the grander, so much the more beautiful, so much worthier of its Author ... .
>
> These laws, so beautiful and so simple, are perhaps the only ones which the Creator and Organizer of things has established in matter in order to effect all the phenomena of the visible world[6].

The young Italian, Lagrange, who later became one of the most distinguished mathematicians in France, was the first to place the minimum principles on a solid mathematical foundation as more general stationary laws. For this purpose, Lagrange [261, 293, 294, 301] and Euler [145, 146, 147, 596] devised, perfected and explicated from about 1744 to 1764 a new mathematical technique now referred to as the *calculus of variations*, a name coined by Euler.

8.2 Lagrangian of a system of Newtonian particles

Let us use Cartesian coordinates to consider the simple case of a closed conservative interacting system comprised of unconstrained Newtonian particles ($i = 1, 2 \ldots$) with total kinetic energy $T = T(\dot{z}_i)$ and internal potential energy of interaction given by the Galilean invariant, $V = V(z_i(t))$, the details of which we leave unspecified. Using Lagrange's formula [296] of Equation 6.24, the equations of motion may be re-written in the form,

$$\frac{d}{dt}\left(\frac{\partial T}{\partial \dot{z}_i}\right) = -\frac{\partial V}{\partial z_i}, \qquad (8.1)$$

in which the partial derivative with respect to a particular z_i assumes all other variables, z_j ($j \neq i$), are held constant with a similar assumption for the partial derivatives with respect to $\dot{z}_i$. The total derivative with respect to time, d/dt, is based on assuming that all the arguments of $\partial T/\partial \dot{z}_i$ have been expanded as functions of time.

The kinetic energy $T = T(\dot{z}_i)$ does not depend on z_i and the potential energy $V = V(z_i)$ is independent of $\dot{z}_i$. We may therefore rewrite the equations of motion in the form,

$$\frac{d}{dt}\left(\frac{\partial(T-V)}{\partial \dot{z}_i}\right) = \frac{\partial(T-V)}{\partial z_i}, \qquad (8.2)$$

---

[6] *Oeuvres* [362, vol. 1, pp. 44–45], cited in Glass [200].

8.2. Lagrangian of a system of Newtonian particles

Leonhard Euler

b. Basel, Switzerland
15 April 1707

d. St. Petersburg, Russia
18 September 1783

Portrait by Emanuel Handmann, Kunstsammlung, Basel.
Courtesy of the Bildarkiv preussischer Kulturbesitz, Berlin.

$e^{\pi i} = -1$.

and note that the difference $T - V$ appears naturally on both sides of the equation. We see here one of the many advantages of the Lagrangian formulation. All the dynamical information about a closed interacting system may be encapsulated in the properties of one Galilean invariant function of the Cartesian coordinates and velocities,

$$L = L(t) = L(\mathbf{z}_i, \dot{\mathbf{z}}_i) = T(\dot{\mathbf{z}}_i) - V(\mathbf{z}_i), \qquad (8.3)$$

called the *Lagrangian*, in terms of which the equations of motion take the very simple and elegant form,

$$\frac{d}{dt}\left(\frac{\partial L}{\partial \dot{\mathbf{z}}_i}\right) = \frac{\partial L}{\partial \mathbf{z}_i}. \qquad (8.4)$$

Lagrange [298, p. 300] (see also [297, p. 24] and [299, 300]) was, in 1788, the first to express the dynamics of mechanical systems using the single invariant function now known by his name. The associated equations of motion are sometimes referred to as *Lagrange's equations*. In recognition of the important contributions of Euler [145], from as early as 1744, to the calculus of variations and its application to mechanics, they are also referred to as the *Euler-Lagrange equations*.

The treatise of Lagrange [298] is a primarily theoretical work based on analysis and algebra. Indeed, at first sight the Lagrangian method appears only to provide an elegant mathematical framework within which one may use the known equations of motion of a system to determine a corresponding Lagrangian from which one may easily obtain the equations of motion again. The fundamental problem of solving the dynamical equations to determine the trajectory does not seem to have been advanced by construction of the Lagrangian. We therefore need to justify the heavy emphasis placed on the Lagrangian procedures in physics ever since their inception with continuing importance in the last few decades and this we shall do throughout the remainder of the text.

The fact that the operations of $\partial/\partial \dot{z}_i$ and $\partial/\partial z_i$ generate vectors from invariants and that d/dt preserves invariance, guarantees that the equations of motion will be Galilean covariant if the Lagrangian is constructed to be a Galilean invariant. This is a second very important feature of the Lagrangian technique.

8.3 Momenta and configuration space Hamiltonian

The momentum $\mathbf{p}_i = m_i \dot{\mathbf{z}}_i = \partial T/\partial \dot{\mathbf{z}}_i$ of the i-th particle is given in terms of the Lagrangian function by a simple expression,

$$\mathbf{p}_i = \frac{\partial L}{\partial \dot{\mathbf{z}}_i}, \qquad (8.5)$$

linear in the Lagrangian. The same is therefore also true of the angular momentum and the centre of mass motion each of which is linear in $\mathbf{p}_i$. Although the kinetic energy is a quadratic in the momentum, it and the total energy may also be obtained from the value of an expression linear in the Lagrangian, by noting that energy is obtained from the evaluation, using the physical path $[\mathbf{z}_i(t)]_{\text{physical}}$, of the function $T+V = 2T-L$ of the coordinates $\mathbf{z}_i$ and velocities $\dot{\mathbf{z}}_i$. To see this, we simply note that,

$$2T = \sum_i \dot{\mathbf{z}}_i \cdot \mathbf{p}_i = \sum_i \dot{\mathbf{z}}_i \cdot \frac{\partial L}{\partial \dot{\mathbf{z}}_i}. \qquad (8.6)$$

We now define the *Hamiltonian* function, in a Lagrangian context, to be that function of the coordinates and velocities, namely,

$$H = H(\mathbf{z}_i, \dot{\mathbf{z}}_i) = \sum_i \dot{\mathbf{z}}_i \cdot \frac{\partial L}{\partial \dot{\mathbf{z}}_i} - L, \qquad (8.7)$$

which gives the *energy* when evaluated with the actual physical trajectory values,

$$E = H([\mathbf{z}_i(t)]_{\text{physical}}, [\dot{\mathbf{z}}_i(t)]_{\text{physical}}). \qquad (8.8)$$

8.3. Momenta and configuration space Hamiltonian

We see therefore that all ten external quantities which are conserved as a result of Galilean spacetime symmetries may be obtained from expressions linear in the one Galilean covariant Lagrangian function of the system. This is another of the very important properties of the Lagrangian which extends to other types of symmetries. It greatly facilitates the determination of the conserved quantities of the system.

The link between the conserved quantities and the spacetime symmetries of a closed system is also particularly transparent when using a Lagrangian. Consider, for example, symmetry of a system with respect to translation in space, a small translation being $\Delta \mathbf{x} = \Delta \mathbf{a}$. We have,

$$\begin{aligned} \Delta \mathbf{a} \cdot \dot{\mathbf{p}} &= \sum_i \Delta \mathbf{z}_i \cdot \frac{d}{dt}\left(\frac{\partial L}{\partial \dot{\mathbf{z}}_i}\right) \\ &= \sum_i \Delta \mathbf{z}_i \cdot \frac{\partial L}{\partial \mathbf{z}_i} \\ &= (\Delta L)_{\text{space translation}} \, . \end{aligned} \qquad (8.9)$$

Since $\Delta \mathbf{a}$ is arbitrary, the vanishing of the right side implies the vanishing of $\dot{\mathbf{p}}$ and the constancy of $\sum_i \mathbf{p}_i$. Thus momentum conservation is seen to be related to the symmetry of a system under spatial translation, as characterized by a translationally invariant Lagrangian function.

Consider next the symmetry of an isolated system with respect to translations in time. This does not mean that there is no evolution — rather, the expression of that evolution must, for an isolated system, have the same form in frames related by $t \to t + \Delta t$. In the Lagrangian formulation, the comparable requirement, since T and V do not then depend on t for an isolated system, is $\partial L/\partial t = 0$. Now,

$$\begin{aligned} -\Delta t \frac{d}{dt}\left(\sum_i \dot{\mathbf{z}}_i \cdot \frac{\partial L}{\partial \dot{\mathbf{z}}_i} - L\right) &= \Delta t \left\{\frac{dL}{dt} - \sum_i \dot{\mathbf{z}}_i \cdot \frac{d}{dt}\left(\frac{\partial L}{\partial \dot{\mathbf{z}}_i}\right) - \sum_i \ddot{\mathbf{z}}_i \cdot \frac{\partial L}{\partial \dot{\mathbf{z}}_i}\right\} \\ &= \Delta t \left\{\frac{dL}{dt} - \sum_i \dot{\mathbf{z}}_i \cdot \frac{\partial L}{\partial \mathbf{z}_i} - \sum_i \ddot{\mathbf{z}}_i \cdot \frac{\partial L}{\partial \dot{\mathbf{z}}_i}\right\} \\ &= \Delta t \frac{\partial L}{\partial t} \\ &= (\Delta L)_{\text{time translation}} \end{aligned} \qquad (8.10)$$

which shows that the Hamiltonian function, and thus the energy, is conserved if we have time translation symmetry.

Exercise 8.1 *Use the Lagrangian properties to establish the relation between rotational and boost symmetries and the conservation of angular momentum and the centre of mass motion.*

Joseph-Louis Lagrange

b. Turin, Italy
25 January 1736

d. Paris, France
10 April 1813

From a bust in the Library, Institut Français.
Courtesy of the Mansell Collection, London.

Je me propose ici de généraliser ce même principe, et d'en faire voir l'usage pour en resoudre avec facilité toutes les questions de Dynamique [295, p. 365].

On ne trouvera point de Figures dans cet ouvrage.
[*Mécanique analytique* [298], preface, p. I.]

8.4 Generalized coordinates in mechanics

It is not always convenient to analyse a Newtonian mechanical system using Cartesian coordinates. The geometry of a particular problem may mean that certain non-Cartesian coordinates are preferable. An example would be the analysis of the motion of a pendulum using the angle it makes with the vertical. The existence of restricting relations, called *constraints*, among the coordinates and either the velocities or momenta of a system, as for example in the motion of a particle confined to some surface or curve, is another situation where the use of non-Cartesian coordinates may make a solution of the dynamics easier. Such relations are one type of constraint to which dynamical variables may be subject.

The Lagrangian formulation is particularly suited to the use of non-Euclidean and/or constrained dynamical variables. The reader should note

8.4. Generalized coordinates in mechanics

that the non-Euclidean variables of interest in the present context are not alternative coordinates for the locations of events $\{t, \mathbf{x}\}$ in Galilean space-time, such as curvilinear or general coordinate systems, but alternative variables for the degrees of freedom z_i of a mechanical system. Of course, the two uses of non-Cartesian coordinates are related.

For simplicity, we shall first examine the case without constraints. Consider, therefore an isolated Newtonian mechanical system, with a finite number of degrees of freedom, such as a system of n particles of mass m_i and $3n$ coordinates z_i ($i = 1, 2 \ldots, n$). We can choose any $3n$ independent invertible functions,

$$q = q(t) = \{q_j(t)\} = \{q_j(z_i(t))\} \qquad (j = 1, 2, \ldots, 3n) , \qquad (8.11)$$

of the original $3n$ dynamical coordinates z_i, whose inverses we may write,

$$z_i = z_i(q_j(t)) . \qquad (8.12)$$

Such dynamical variables $q = \{q_j(t)\}$ of a mechanical system are referred to as *generalized coordinates*. The *generalized velocities* are defined to be,

$$\dot{q} = \dot{q}(t) = \{\dot{q}_j(t)\} = \left\{\frac{dq_j(t)}{dt}\right\} . \qquad (8.13)$$

Any actual set of numerical values of $\{q_j(t)\}$ describes an allowable *configuration* of the system at the time t and the real space each of whose points is labelled by a set of possible values of $q(t)$ is called the *configuration space* of the system. The range of each generalized coordinate is determined by the nature of the physical system. If the generalized coordinates chosen happen to be Cartesian, the configuration space will be isomorphic to the $3n$-dimensional space comprised of n products $\mathbf{E}^3 \times \cdots \times \mathbf{E}^3$ of 3D Euclidean space. Provided the generalized coordinates are defined in terms of differences $|z_i(t) - z_j(t)|$ in the locations of two particles at the same time t, it will not be difficult to maintain Galilean covariance of a system described with non-Cartesian coordinates.

Differentiating Equation 8.12 with respect to time, and using the summation convention on j, gives,

$$\dot{z}_i = \frac{dz_i}{dt} = \frac{\partial z_i}{\partial q_j}\dot{q}_j . \qquad (8.14)$$

Thus $\dot{z}_i$ depends on $\dot{q}_j$ and, via $\partial z_i / \partial q_j$, on q_j.

Exercise 8.2 *Show that,*

$$\frac{\partial \dot{z}_i}{\partial \dot{q}_j} = \frac{\partial z_i}{\partial q_j} \quad \text{and} \quad \frac{d}{dt}\frac{\partial z_i}{\partial q_j} = \frac{\partial \dot{z}_i}{\partial q_j} . \qquad (8.15)$$

using the fact that the total time derivative d/dt is defined to be,

$$\frac{d}{dt} = \frac{\partial}{\partial t} + \dot{q}_j \frac{\partial}{\partial q_j} + \ddot{q}_j \frac{\partial}{\partial \dot{q}_j} , \qquad (8.16)$$

when acting on a function of t, q_j and $\dot{q}_j$.

In terms of the generalized coordinates, the kinetic energy is then given by,

$$T = T(q_j, \dot{q}_j) = \tfrac{1}{2} \sum_i m_i \frac{\partial z_i}{\partial q_j} \cdot \frac{\partial z_i}{\partial q_k} \dot{q}_j \dot{q}_k = \tfrac{1}{2} \dot{q}_j \frac{\partial T}{\partial \dot{q}_j} . \qquad (8.17)$$

which now includes a dependence on q_j (in $\partial z_i/\partial q_j$ and $\partial z_i/\partial q_k$) as well as on $\dot{q}_j$, the last equality following either direct — or using the fact that T is a homogeneous quadratic in $\dot{q}$ and applying Euler's theorem.

Using Equations 8.15 in Equation 8.4 to make a change of variable from the Cartesian variables z_i to the generalized coordinates q_j shows that the Newtonian dynamical equations for the accelerations $\ddot{z}(t)$ have precisely the same expression in generalized coordinates, namely the *Euler-Lagrange equations* of motion,

$$\frac{d}{dt}\left(\frac{\partial L}{\partial \dot{q}_j}\right) = \frac{\partial L}{\partial q_j} . \qquad (8.18)$$

Now, however, the Lagrangian is given by,

$$L = L(t) = L(q, \dot{q}) = T(q, \dot{q}) - V(q) , \qquad (8.19)$$

and the *configuration space Hamiltonian function* by,

$$H = H(q_j, \dot{q}_j) = \dot{q}_j \frac{\partial L}{\partial \dot{q}_j} - L , \qquad (8.20)$$

both having quadratic dependence on $\dot{q}_j$.

If we define *generalized momenta* π^j, conjugate to each generalized coordinate q_j, analogously to Equation 8.5, then the Euler-Lagrange equations are equivalent to the pair of equations,

$$\pi^j = \frac{\partial L}{\partial \dot{q}_j} \quad \text{and} \quad \dot{\pi}^j = \frac{\partial L}{\partial q_j} . \qquad (8.21)$$

Although these equations appear to be first order in t, in the Lagrangian formulation the first of these equations is simply a definition of π^j and the true equations of motion are the second set, which are effectively second-order.

In the *Lagrangian formulation* of mechanics, it is always understood that $\partial L/\partial q_j$ is carried out with all other q_k ($k \neq j$) and all $\dot{q}_j$ kept constant. Similarly, $\partial L/\partial \dot{q}_j$ is performed with all other $\dot{q}_k$ ($k \neq j$) and all

8.4. Generalized coordinates in mechanics

q_j constant. Thus $\{q_j, \dot{q}_j\}$ are assumed initially to be mathematically independent variables, despite the fact that knowledge of the physical path $q_j(t)$ traced by a system implies knowledge of the velocities $\dot{q}_j(t)$.

Of the factors involving T and V that appear formally in the Euler-Lagrange equations, namely $\partial V/\partial q_j$, $\partial T/\partial \dot{q}_j$, $\partial T/\partial q_j$ and $\partial V/\partial \dot{q}_j$, only the last of these contributes nothing in the Lagrangian formulation of Newton's equations for conservative interaction. The contribution of the third is a consequence solely of the introduction of non-Cartesian dynamical variables. The form of the Euler-Lagrange equations shows, however, that the Lagrangian formulation admits a natural generalization in which the fourth factor also contributes, namely when the potential is velocity dependent and the Lagrangian has the form,

$$L = L(t) = L(q, \dot{q}) = T(q, \dot{q}) - V(q, \dot{q}) \;. \tag{8.22}$$

This is the most general form of the Lagrangian of an isolated mechanical system whose equations of motion are given in terms of the Lagrangian by the Euler-Lagrange equations. The motion of a charged particle in an external electromagnetic field is an example of a dynamical system in which the external interaction is 'non-conservative', in the sense of a non-vanishing curl of the force field. The interaction is in fact described by Euler-Lagrange equations with a potential which depends linearly on the velocities. Despite such systems being referred to as 'non-conservative', there is of course no violation of the conservation laws when the source of the external interaction is taken into account which cannot be done for the electromagnetic field in the context of a purely mechanical description.

We shall return to a re-appraisal of some of the concepts implied by the existence of the generalized momenta in Chapter 11 on Hamiltonian dynamics. It should be noted that the momenta are generalized for two reasons. First, the coordinates are no longer necessarily Cartesian and, secondly, we are allowing the potential that appears in the Lagrangian to include a dependence on the generalized velocities. We shall limit consideration to Lagrangians whose dependence on $\dot{q}$ is no higher than quadratic thereby limiting the Euler-Lagrange equations to no higher than second-order.

We shall refer to the expression,

$$L^j = L^j(q, \dot{q}) = \frac{\partial L}{\partial q_j} - \frac{d}{dt}\left(\frac{\partial L}{\partial \dot{q}_j}\right) , \tag{8.23}$$

as the *Euler derivative* [526] of the Lagrangian with respect to the variable q_j, although it is also referred to as the *Lagrangian derivative* [249]. Its vanishing gives the Euler-Lagrange equations.

8.5 Stationary action

The term *action* used by Maupertuis [362] in 1746 is employed in a variety of specific ways in discussing the dynamics of a mechanical system. All of these use a physical quantity — with angular momentum dimensions — comprised of a sum or integral over the constituents of the system. The action thus formed is then used in a global principle of stationarity to establish the local equations governing the evolution of the system or its distribution in space. One form of action is the integral, over the distance ds along some arbitrary curve $\mathcal{C} = \{s(t)\}$,

$$A = A[\mathcal{C}] = A[s(t), t_i, t_f] = \int_{s(t_i)}^{s(t_f)} ds\, m\, v(s) \quad \left(\equiv \int_{t_i}^{t_f} dt\, p\, \dot{z}\right) \quad (8.24)$$

of the momentum mv of a particle of mass m. This is equivalent to taking the integral with respect to time of the momentum times velocity. Euler [145] used such an expression in 1744 to describe the motion of a particle in a central force in terms of the minimization of the integral with respect to variations in the path $s(t)$ taken by the particle from initial point to final point. The actual path followed is the one which makes the expression for $A[s(t)]$ stationary with respect to small changes in the path. Except in unusual circumstances, the path followed is the one which minimizes the action. A Principle of least action equivalent to this is established in Goldstein [206, pp. 228–235, edn. 1] using the action,

$$A = A[\mathcal{C}] = A[q(t), t_i, t_f] = \int_{t_i}^{t_f} dt \sum_i p_i \dot{q}_i, \quad (8.25)$$

in terms of the generalized coordinates of a multi-particle system and we examine a related expression in Chapter 11.

However, another use of the term *action* is for the integral of the Lagrangian function with respect to time over a compact interval from an initial point $\{q_j(t_i)\}$ to a final point $\{q_j(t_f)\}$ along an arbitrary curve $\mathcal{C}$ in configuration space. We may parameterize $\mathcal{C}$ by its generalized coordinates $q(t)$, and the action of a mechanical system is thus defined to be,

$$S = S[\mathcal{C}] = S[q(t), t_i, t_f] = \int_{t_i}^{t_f} dt\, L(q_j, \dot{q}_j, t) . \quad (8.26)$$

This is the meaning we shall attach to the expression *action of a mechanical system* in this text along with analogous meanings in the context of non-mechanical systems such as fields (see Chapters 10 and 16) and in the Hamiltonian formulation (see Chapter 11).

An important feature of the Lagrangian formulation of the Principle of stationary action is the fact that the curve $\mathcal{C} = \{q(t)\}$ along which the integral is calculated is undetermined and arbitrary, with possibly fixed arbitrary end-points $q(t_i)$ and $q(t_f)$, before the stationarity procedure has been carried out. The important physical property of the Lagrangian in a variational approach is the way in which the function $L(q, \dot{q}, t)$ depends on its arguments. For this reason the path $q(t)$ that appears in the Lagrangian prior to applying the stationary principle is referred to as a *virtual path* or *history* that the system may conceivably follow between the fixed endpoints. The determination of the actual trajectory the system follows, namely the physical path, $[q(t)]_{\text{physical}}$, along with specification of each point q reached at a given time, is the goal of the variational procedure.

In the variational approach, the essence of the dynamics is contained not in the values of the Lagrangian but in the form of the function $L(q, \dot{q}, t)$. For each time t along each virtual path $q(t)$ between t_i and t_f, the form of the Lagrangian supplies a number $L(t)$. The Lagrangian of a mechanical system, where the number of degrees of freedom is finite or denumerably infinite, is therefore a *function*. The action also produces a number from its argument, the virtual path $q(t)$, which is itself a function of t. The action therefore determines a number from one function $q(t)$ via another $L(q, \dot{q}, t)$. A quantity which determines a number from a function or set of functions is referred to as a *functional*, and the action $S[q(t)]$ of a mechanical system is therefore often referred to as the *action functional*.

8.6 Hamilton's principle of stationary action

The solutions of second-order differential equations of motion contain two arbitrary constants which may be determined by specifying the initial values $q(t_i)$ and $\dot{q}(t_i)$ of the generalized coordinates and velocities. From those initial values, the local properties embodied in the differential equations of motion, serve to evolve the system forward to a subsequent time t_f. However, two values $q(t_i)$ and $q(t_f)$ of the generalized coordinates alone, the second of which is reachable from the first, also suffice to determine the path followed. Hamilton's principle is such a global procedure for the determination of the physical path equivalent to the local procedure supplied by the Euler-Lagrange differential equations of motion.

To determine the effect of putting the variation of the action to zero, one uses the calculus of variations invented by Euler and Lagrange. Ordinary differential calculus of functions involves the variation of a dependent variable to give the variation of the function. The calculus of variations is a procedure which essentially extends the ordinary differential calculus to one in which whole functions are varied to give the consequent variation in a dependent function or functional. We shall require only the small first-order change in the action functional $S[q(t)]$ due to a small variation in the virtual path function $q(t)$.

Variations in the virtual path are of two types, those involving functional variation of the independent variable t and those in which the dependent variable q is functionally varied. We concentrate first on variations only of this latter form and distinguish them from the more general variations, to be considered later, in which t is also varied, by the use of the variational symbol δ_0 rather than δ. Thus we denote such limited variations in $q(t)$ by $\delta_0 q(t)$ and the consequential variations in $\dot{q}(t)$, $L(q,\dot{q},t)$ and $S[q]$ by $\delta_0 \dot{q}(t)$, $\delta_0 L$ and $\delta_0 S$, which are defined by:

$$\delta_0 \dot{q}(t) = \frac{d}{dt}(\delta_0 q(t)) \tag{8.27}$$

$$\delta_0 L\big(q(t),\dot{q}(t),t\big) = L\big(q(t)+\delta_0 q(t),\dot{q}(t)+\delta_0 \dot{q}(t),t\big) - L\big(q(t),\dot{q}(t),t\big)$$

$$= \frac{\partial L}{\partial q_j}\delta_0 q_j(t) + \frac{\partial L}{\partial \dot{q}_j}\delta_0 \dot{q}_j(t) \tag{8.28}$$

$$\delta_0 S[q(t)] = \int_{t_i}^{t_f} dt\, \delta_0 L\big(q(t),\dot{q}(t),t\big) . \tag{8.29}$$

We may regard $q = \{q_j\}$ as a column vector labelled by j and $\partial L/\partial q$ and $\partial L/\partial \dot{q}$ as row vectors. The time arguments have been included throughout these definitions to emphasize the distinction between ordinary variation and functional variation. This distinction should be constantly borne in mind although we shall not continue always to include such arguments.

Since there is no functional variation of the time t, the region of integration is the same both before and after variation of the path. The mathematical significance of the variational procedure is that $q(t)$ has effectively been varied to a function $q(t) + \delta_0 q(t)$ with $\delta_0 q(t) = \epsilon\, \alpha(t)$ where $\alpha(t)$ is arbitrary, except for vanishing at the end-points, and ϵ is a small constant in powers of which each function may be expanded. We shall only be interested in expansion to first non-vanishing order in ϵ.

Combining Equations 8.28 and 8.29 shows that the variation in the action is:

$$\delta_0 S[q(t)] = \int_{t_i}^{t_f} dt\, \left\{\frac{\partial L}{\partial q_j}\delta_0 q_j(t) + \frac{\partial L}{\partial \dot{q}_j}\delta_0 \dot{q}_j(t)\right\} . \tag{8.30}$$

Using Equation 8.27 in 8.30, and integrating the second term by parts, gives:

$$\delta_0 S[q(t)] = \int_{t_i}^{t_f} dt\, \left[\frac{d}{dt}\left\{\frac{\partial L}{\partial \dot{q}_j}\delta_0 q_j(t)\right\} + \left\{\frac{\partial L}{\partial q_j} - \frac{d}{dt}\left(\frac{\partial L}{\partial \dot{q}_j}\right)\right\}\delta_0 q_j(t)\right]$$

$$= [\pi^j \delta_0 q_j(t)]_{t_i}^{t_f} + \int_{t_i}^{t_f} dt\, L^j\, \delta_0 q_j(t) . \tag{8.31}$$

Since it is evaluated on the boundary of the path, namely the value at t_f less the value at t_i, the first term is called an *end-point contribution* or, in

8.6. Hamilton's principle of stationary action

a field context, a *surface term*. This completes the purely mathematical part of the calculation using functional variation.

For the physical interpretation of the results we use a variation $\delta_0 q_j$ of $q_j(t)$ which is required to vanish at the end-points,

$$\delta_0 q_j(t_i) = 0 = \delta_0 q_j(t_f) . \tag{8.32}$$

There is, however, no similar restriction on the variation of the velocities. With such variations the surface term must vanish. Since the variations $\delta_0 q_j(t)$ are arbitrary for $t_i < t < t_f$, the vanishing of the variation of the action to first order in the path variation requires the *Euler-Lagrange equations* (8.18) to be satisfied. We have therefore:

- **Hamilton's principle of stationary action** — the Euler-Lagrange equations of motion of a Lagrangian system are necessary conditions for the vanishing, $\delta_0 S[q_j(t)] = 0$, of the first-order functional variation of the action with respect to variations $\delta_0 q_j(t)$ of the dynamical variables $q_j(t)$, held constant at the end-points of a compact time interval, there being no variation of the independent variable t.

An important characteristic of the Lagrangian procedure of Hamilton's principle is that variations $\delta_0 \dot{q}_j(t)$ of the velocities are not independent of the variations $\delta_0 q_j(t)$ of the coordinates. Included in the variations of the trajectory are those which trace the same locus in configuration space at different speeds at each point. The action need only be stationary — it is generally an extremum, but may not be. If it is an extremum, it is almost always a minimum, but in Chapter 15 we shall encounter a well-known case where the stationary action is in fact a maximum.

Given a Lagrangian, we can therefore determine the equations of motion which correspond to Hamilton's principle. Conversely, given an equation of motion we can seek a Lagrangian from which it can be derived via the Euler-Lagrange equations. Not all Lagrangian functions lead to a consistent set of Euler-Lagrange equations.

Exercise 8.3 *Show that the Euler-Lagrange equations for a system with one coordinate q and Lagrangian $L = L(q, \dot{q}) = \dot{q} + q$ are mathematically inconsistent.*

The Euler-Lagrange equations are not a sufficient condition for a stationary action. We shall henceforth assume that we are dealing only with Lagrangian functions whose Euler-Lagrange equations are at least mathematically consistent. We shall see later that not all equations of motion involving $q(t)$ and $\dot{q}(t)$ will be derivable from a Lagrangian. This point is examined further by Santilli [494]. The Lagrangian philosophy of mechanics and fields selects from all possible equations of motion a subset that apparently occupies a privileged place in physical theory.

For those equations that are expressible in Euler-Lagrange form, the Lagrangian is not unique (see Currie and Saletan [84]). The homogeneity of Equation 8.18 in L shows that re-scaling of L by any constant (independent of t, q_j and $\dot{q}_j$) will not alter the Euler-Lagrange equations. However, two Lagrangians may differ by a total time derivative of a function of t and q, referred to as a *divergence*, as a result of the form of the corresponding term in a Lagrangian field theory, and still lead to the same Euler-Lagrange equations.

Exercise 8.4 *Show that the two Lagrangians L and $L + d\Lambda/dt$, where $\Lambda = \Lambda(t,q)$ is an arbitrary scalar function of the time t and q (but not $\dot{q}$) lead to the same Euler-Lagrange equations since the divergence $d\Lambda/dt$ only contributes a surface term which vanishes due to the boundary conditions.*

8.7 Functional differentiation

The process of obtaining the variation of a functional $S[q(t)]$ with respect to the function $q(t)$ may be formalized mathematically using the notion of *functional derivative*. We do not require an extensive treatment of the topic in this text but it will be useful to state a few of the results to establish the connection with more complete treatments. A functional $F[q] = F[q(t)]$ of $q(t)$ is continuous if,

$$\lim_{\epsilon \to 0} F[q + \epsilon \alpha] = F[q] \qquad (8.33)$$

for all functions $\alpha(t)$ from a specified set of functions. $F[q]$ is said to be differentiable if,

$$\left.\frac{dF[q + \epsilon \alpha]}{d\epsilon}\right|_{\epsilon=0}, \qquad (8.34)$$

exists. If this expression is a linear functional of $\alpha(t)$ in the form of an integral over the independent variable t, which may be written as,

$$\left.\frac{dF[q + \epsilon \alpha]}{d\epsilon}\right|_{\epsilon=0} = \int dt\, \alpha(t) \frac{\delta F}{\delta q(t)}, \qquad (8.35)$$

then $\delta F/\delta q(t)$ is called the *functional derivative* of $F[q]$ with respect to the function $q(t)$. Denoting $\epsilon \alpha(t)$ by $\delta q(t)$, we can make a Taylor's expansion to first order in ϵ to obtain,

$$\delta F = \int dt\, \delta q(t) \frac{\delta F}{\delta q(t)}. \qquad (8.36)$$

The functional derivative generalizes readily to any number of independent variables and may then be used to write Hamilton's principle of stationary action in the form,

$$\frac{\delta_0 S}{\delta_0 q_j(t)} = L^j = \frac{\partial L}{\partial q_j} - \frac{d}{dt}\left(\frac{\partial \mathcal{L}}{\partial \dot{q}_j}\right) = 0 , \qquad (8.37)$$

in which it is seen that the functional derivative of the action with respect to the generalized coordinate $q(t)$ is the Euler derivative whose vanishing gives the Euler-Lagrange equations.

8.8 Lagrangian interaction

Let us consider the division of an isolated system with generalized coordinates $q(t) = \{q_1(t), q_2(t)\}$ into two subsystems with coordinates $q_1(t)$ and $q_2(t)$. (Indices which label the individual coordinates comprising q, q_1 and q_2 are suppressed.) Since kinetic energy is postulated and observed to be an additive quantity we may split the total kinetic energy T into two parts,

$$T(q, \dot{q}) = T_1(q_1, \dot{q}_1) + T_2(q_2, \dot{q}_2) . \qquad (8.38)$$

Since the system as a whole is isolated, the total potential energy $V(q)$ contains no explicit time dependence. If it is possible to divide the potential energy $V = V(q) = V(q_1, q_2)$ additively into two parts,

$$V(q, \dot{q}) = V_1(q_1, \dot{q}_1) + V_2(q_2, \dot{q}_2) , \qquad (8.39)$$

then the Lagrangian similarly divides into two parts, each with no explicit time dependence,

$$L(q, \dot{q}) = L_1(q_1, \dot{q}_1) + L_2(q_2, \dot{q}_2) . \qquad (8.40)$$

There will then be a consequent division of the Euler-Lagrange equations into two disjoint sets of dynamical equations,

$$\frac{d}{dt}\left(\frac{\partial L_1}{\partial \dot{q}_1}\right) = \frac{\partial L_1}{\partial q_1} \quad \text{and} \quad \frac{d}{dt}\left(\frac{\partial L_2}{\partial \dot{q}_2}\right) = \frac{\partial L_2}{\partial q_2} , \qquad (8.41)$$

corresponding to the two subsystems, each of which will clearly be isolated from one another. For there to be interaction of the two parts, it is necessary and sufficient that the potential energy contain a part which depends on the separate coordinates and velocities in such a way that it is not divisible additively into two independent parts. The Euler-Lagrange equations of each part will incorporate the dynamical variables of the other part via the non-trivial dependence of the potential term on both sets. For this reason the interaction of two systems is also referred to as their *coupling*.

Let us suppose that the total potential energy may be divided additively into parts V_1 and V_2 involving only the coordinates and velocities of subsystems 1 and 2, respectively, and a remaining part V_{intn} involving the coordinates and/or velocities of both parts,

$$V = V_1 + V_2 + V_{\text{intn}} \ . \tag{8.42}$$

The division of the isolated system into two interacting parts will then permit the division of its total Lagrangian into three parts,

$$L = L_{\text{free, 1}} + L_{\text{free, 2}} + L_{\text{intn, 1 and 2}} \ , \tag{8.43}$$

where,

$$\begin{aligned} L_{\text{free, 1}} &= T_1(q_1, \dot{q}_1) - V_1(q_1, \dot{q}_1) \\ L_{\text{free, 2}} &= T_2(q_2, \dot{q}_2) - V_2(q_2, \dot{q}_2) \\ L_{\text{intn, 1 and 2}} &= -V_{\text{intn}}(q_1, q_2, \dot{q}_1, \dot{q}_2) \ , \end{aligned} \tag{8.44}$$

corresponding to contributions to the total Lagrangian from the separate kinetic and internal potential energies of each subsystem, referred to as the *free* contributions to the whole system, and the interacting part involving the non-separable part of the potential energy of interaction.

A very common situation is to know the dynamical equations and Lagrangians for two separate systems each of which is free of interaction with the other and wish to determine ways in which the two may interact subject to certain very general symmetry conditions. This corresponds to a knowledge of the separate Lagrangians $L_{\text{free, 1}}$ and $L_{\text{free, 2}}$ and the desire to construct the Galilean invariant term $L_{\text{intn, 1 and 2}} = -V_{\text{intn}}$ in the Lagrangian corresponding to their interaction. We shall provide numerous illustrations of this procedure starting with the trivial example in the next section of the Newtonian gravitational interaction.

Let us first summarize some of the properties of the Lagrangians and actions we have been manipulating that we shall see have general applicability not only in classical mechanics subject to the Galilean relativity principle but also in quantum mechanics, Einsteinian relativity, both special and general, and with continuous systems such as fields.

The **action** $S[q]$ of an isolated system:

- is a real functional of a compact virtual path, parameterized by generalized coordinates q, in configuration space,
- is generally assigned the dimensions of angular momentum,
- is required to be invariant or covariant with respect to the symmetries of the system,
- may be divided into a sum of free parts plus mutual interactions, each of which shares the above properties, and

8.8. Lagrangian interaction

- is expressed as the integral of the Lagrangian (or Lagrangian density), with respect to the independent variable(s), over a compact arbitrary region between two 'points' of the configuration space of the system.

The **Lagrangian** $L = L(q, \dot{q})$ or **Lagrangian density** $\mathcal{L} = \mathcal{L}(\psi, \dot{\psi})$ of an isolated system:

- is an hermitian function or functional of the generalized coordinates and velocities,

- has the dimensions of action divided by the dimensions of the volume element of the space of independent variables,

- is a scalar density in the space of the independent variables, with respect to the symmetry transformations,

- is a scalar with respect to all transformations which do not alter the volume element of the space of independent variables, and

- may be divided into a sum of free parts plus mutually interacting parts, each of which shares the above properties.

Some of the items listed above for the action and the Lagrangian encapsulate properties which have emerged from the way we have constructed them for mechanical systems. However, we have endeavoured to express those properties in ways which not only apply to Newtonian mechanical systems but to a wide range of physical phenomena, mechanical and field, relativistic or not and classical or quantum. The notion of a *covariant Lagrangian* or action will be introduced in the next chapter. Certain of these properties, such as the hermiticity of the Lagrangian and the action, anticipate the application of Hamilton's principle to systems where it is convenient to use dynamical variables q or ψ which are not real. Examples are quantum systems described by Schrödinger wave functions and the complex fields whose quantization leads to particles such as electrons, protons, quarks, pions or neutrinos, which are charged, either electrically or otherwise.

The hermiticity requirement on the Lagrangians of such wave functions and fields then ensures that observable quantities expressible in terms of the Lagrangian, such as energy and momentum, are, nevertheless, real. *Path-integral quantization* [160, 488] is based on the existence of a non-quantum Lagrangian and action obeying a stationary action principle. The hermiticity of the Lagrangian means that the appearance of the real action in the standard phase factor $e^{iS/\hbar}$ of the path integral ensures that the stationary path is the strongest contributor to the probability of transition between initial and final states. Non-stationary paths will also contribute to the probability of a quantum process but with decreasing magnitude the more they deviate from the stationary path due to the increasingly rapid fluctuations in $e^{i(S-S_{\text{stat}})/\hbar}$, leading to destructive interference, unless ($|S - S_{\text{stat}}| \ll \hbar$).

For reasons which follow from relativistic considerations (see Chapter 15) the stationary and non-stationary paths are often referred to as being *on-shell* and *off-shell*, respectively.

8.9 Newtonian gravitational Lagrangian interaction

It is straightforward to write down the internal potential energy of gravitational interaction and the kinetic energy of an isolated system of particles and thereby determine a suitable Lagrangian. However, Lagrangians provide one of the principal means currently being used in research to deduce the structure of new forms of interaction between particle, field and string systems. Many of these are adding to a deeper understanding of all four interactions, to progress in their unification and to the quantization of gravity. In order to gain some idea at an elementary level of how these techniques can be used, let us presume that we accept all the principal Lagrangian properties as they apply to any mechanical system and attempt to construct a theory of an interaction involving Newtonian particles, and having the full Galilean symmetry (including with respect to reflections) without actually knowing the details of the interaction we are trying to construct.

A Newtonian particle is characterized dynamically by its mass m and its trajectory variable $z(t)$ whose evolution is parameterized by the absolute time t of Galilean relativity. A particle, by definition, has no internal structure and therefore has a free Lagrangian comprised solely of the scalar kinetic contribution which must be constructed from the mass m and the square $\dot{z}^2$ of the velocities since the coordinates themselves are not Galilean vectors. The dimensions of the Lagrangian must be those of action divided by time, namely energy, and the only combination with these dimensions is a dimensionless multiple of $m\dot{z}^2$. We fix the scale by convention and choose the *free Newtonian particle Lagrangian* to be,

$$L_{\text{part}} = \tfrac{1}{2} m \dot{z}^2 . \tag{8.45}$$

Consider two such particles, $i = 1, 2$, which are interacting. Their total Lagrangian will have the form,

$$L = L_{\text{part, 1}} + L_{\text{part, 2}} + L_{\text{intn}} = \tfrac{1}{2} m_1 \dot{z}_1^2 + \tfrac{1}{2} m_2 \dot{z}_2^2 - V_{\text{intn}} , \tag{8.46}$$

in which our sole remaining task is to determine the nature of the last term, the Galilean invariant interaction potential energy,

$$V_{\text{intn}} = V_{\text{intn}}(|z_1 - z_2|, |\dot{z}_1 - \dot{z}_2|) . \tag{8.47}$$

We have now reached the end of the straightforward and largely deductive application of the Lagrangian technique in this instance. To proceed further we must appeal to the detailed observational knowledge that may be

8.9. Newtonian gravitational Lagrangian interaction

available or theoretically try any conceivable way of combining the variables of the system, starting with the simplest ways, and comparing the predictions of each with experiment. Whatever method is used will involve a high component of inductive reasoning typical of research in physics. In the original formulation of gravitation (using equations of motion directly rather than the equivalent Lagrangian technique), Newton had the empirical laws of Kepler as his main guide.

Let us suppose that our observations suggest an interaction whose strength increases with the size or mass of the participating 'particles'. We choose first one of the simplest ways to incorporate this observation by taking the strength of the interaction to be proportional to the mass of a participating particle, $V \propto m$. (There is no absolute uniqueness in this process — in a Galilean context, we could, for example, have chosen to make the interaction strength dependent on the volume of the gravitating body, although this would make little sense in the case of a particle.) If we have no reason to presume the system is asymmetrical with respect to the interchange of the pair of particles, we choose an interaction which is proportional to the product of the masses of the interacting pair, $V \propto m_1 m_2$.

Suppose we now observe that the interaction decreases with the increasing separation of the two parts and does not appear to depend on the relative velocities of the participating masses. Many simple forms of dependence on velocities can be ruled out by appealing to time-reversal invariance. To be conservative, it must be central. We incorporate these results in a very simple way by demanding that the interaction term depend only on position and that it be proportional to some negative power of the separation between the masses, using the locations of each at the same time to preserve Galilean invariance,

$$V \propto \frac{m_1 m_2}{|z_1(t) - z_2(t)|^p} \ . \tag{8.48}$$

By choosing the magnitude of the separation between the two masses we have also preserved the symmetry under interchange of the two particles.

If there is to be no further dependence on the locations of the particles, the only remaining factor is a constant, one of whose rôles must be to ensure that the term has the right dimensions. Let us assume this factor is a universal constant G_p (namely one which is independent of t, m_i, z_i and $\dot{z}_i$) called the *coupling constant* of the interaction. It must have the dimensions $[G_p] = [\text{kg}^{-1}\,\text{m}^{2+p}\,\text{s}^{-2}]$. We may assume that it is positive and write the interaction energy, without loss of generality, in the form,

$$V = V(z_1, z_2) = \pm G_p \frac{m_1 m_2}{|z_1(t) - z_2(t)|^p} \ , \tag{8.49}$$

in which the plus and minus signs give (for $p > 0$) repulsive and attractive interactions, respectively. The attractive interaction with potential linear

($p = 1$) in the inverse separation, generalized to an arbitrary number of particles, gives the mechanical Lagrangian,

$$L_{\text{Newt grav}} = \tfrac{1}{2} \sum_i m_i \dot{\mathbf{z}}_i^2 - G \sum_{j \neq i} m_i m_j \frac{\mathbf{z}_i(t) - \mathbf{z}_j(t)}{|\mathbf{z}_i(t) - \mathbf{z}_j(t)|^3}, \qquad (8.50)$$

of the Newtonian theory of gravitation, except of course for the unknown numerical value of the constant $G = G_1$ having units of $\text{kg}^{-1}\,\text{m}^3\,\text{s}^{-2}$, to be determined from experiment.

The above simple but somewhat artificial example involves a great deal of hindsight and a certain amount of luck but illustrates the value of generalization, the power of methods based on maximum symmetry and simplicity and the need to make full use of any information available. This is particularly the case for the use of the symmetries of a system (such as Galilean covariance) in limiting as far as possible the wide range of options that the form of an interaction may take.

8.10 Constraints

The Lagrangian formulation in generalized coordinates is well-suited to situations where some of the convenient coordinates of a system are not independent of others. Such systems are said to contain *constraints*.

The generalized Newton's equations of motion, for a closed unconstrained system of n particles ($i = 1, 2, \ldots, n$) of masses m_i, trajectories $\mathbf{z}_i(t)$ and potential energy of interaction $V = V(\mathbf{z}_i, \dot{\mathbf{z}}_i)$, are $3n$ differential equations of the form,

$$m_i \frac{d^2 \mathbf{z}_i(t)}{dt^2} = -\frac{\partial V(\mathbf{z}_i, \dot{\mathbf{z}}_i)}{\partial \mathbf{z}_i}. \qquad (8.51)$$

If at least one of the independent equations of a dynamical set contains a second time derivative of a dynamical variable, the equations are said to be second-order. Not only is the above system of equations of second-order as a whole but every equation has one term involving a second derivative and each dynamical variable $\mathbf{z}_i$ is involved in one of those second derivatives. Each equation may, at least in principle, be integrated for the trajectories $\mathbf{z}_i(t)$ given the initial conditions $\mathbf{z}_i(0)$ and $\dot{\mathbf{z}}_i(0)$. Every equation in the system participates in the evolution forward in time of the *initial data* comprised of values of the variables and their first derivatives and is therefore regarded as a true dynamical equation. Similarly, every variable has a unique value at each time, determined from the initial data, and is therefore said to be a true dynamical variable. This is all obvious, normal and characteristic of unconstrained or *deterministic* Newtonian mechanical systems. Each of the equations of the unconstrained set describes one independent dynamical degree of freedom of the system.

8.10. Constraints

The mathematical expression of the independence of the $3n$ coordinates in the above equations is the non-vanishing of the determinant of the $3n \times 3n$ matrix of coefficients of the highest derivatives, here of second-order. That matrix, for the unconstrained system in Cartesian coordinates, is the identity. A *constrained system* is then one for which the determinant of the matrix of coefficients of highest derivatives vanishes, examples being the gauge invariant equation of motion (15.25) of a relativistic particle, the second-order Maxwell's equations (18.22) and the gravitino and graviton field equations, 21.4 and 21.34. The number of constraints will be related to the rank of that matrix.

Since we have up to this point constructed the generalized Euler-Lagrange equations from those in Cartesian coordinates in terms of an equal number of generalized coordinates obtained by invertible functions, the same characteristics will apply to the Lagrangian formulation. However, Euler-Lagrange equations are not confined solely to Newtonian mechanical systems where all coordinates originate from Cartesian ones. Furthermore, in other contexts, such as relativistic equations of motion and field equations, the Euler-Lagrange equations may not be second-order; for example, the wave equations of the fields whose quantization leads to particles of half-odd-integer spin are first-order differential equations some forms of which are derivable from Lagrangians. In such cases, initial data will be comprised of only the values of the dynamical variables and the criteria for characterizing an equation as a true dynamical equation or not must be modified accordingly to the case where the highest derivatives are first-order.

One way for a Newtonian mechanical system of n particles to be governed by equations of motion in the $3n$ Cartesian coordinates, such that not all of them are dynamically independent, is by the existence of $3n - N$ real relations ($1 \leq N < 3n$) among the coordinates of the form,

$$\phi_\ell(t, z_i) = 0 \quad \text{for} \quad \ell = 1, 2, \ldots, 3n - N \ . \tag{8.52}$$

Such coordinate relationships are referred to as time-dependent *holonomic constraints*. An example of a system with time-independent holonomic constraints is a body constructed from a set of masses some of which are maintained at constant separations, $|z_i - z_j|$.

Holonomically constrained systems are extremely widespread. Their analysis within the Lagrangian framework is treated in a wide selection of classical dynamics texts, such as Fetter and Walecka [158], Goldstein [206], Hestenes [247], Kibble [279], Landau and Lifschitz [305] and Marion [358], to name just a few. Other types of constrained systems are of more immediate interest in the material we wish to examine. For this reason we shall assume that many of the ways in which a system may be constrained, including such holonomic constraints, have been taken care of by a suitable choice of a reduced number of independent generalized coordinates, namely N functions,

$$q = q(t) = \{q_j(t)\} = \{q_j(z_i(t))\} \qquad (j = 1, 2, \ldots, N) \qquad (8.53)$$

of the original $3n$ coordinates z_i, to obtain a description of the system which is free of the $3n - N$ holonomic constraints.

For a Lagrangian system, whether obtained from a Newtonian description in Cartesian coordinates, or in any other way, the Euler-Lagrange equations do not express the generalized accelerations $\ddot{q}_j$, if they are not Cartesian, directly as functions of t and the variables q_j and $\dot{q}_j$. The latter are the variables whose values at a given instant are the apparent *initial data* to be evolved forward by the equations of motion.

Exercise 8.5 *Show that Equation 8.18 has the general form,*

$$\frac{\partial^2 L}{\partial \dot{q}_j \partial \dot{q}_k} \ddot{q}_k = \frac{\partial L}{\partial q_j} - \frac{\partial^2 L}{\partial \dot{q}_j \partial q_k} \dot{q}_k - \frac{\partial^2 L}{\partial t \partial \dot{q}_j}, \qquad (8.54)$$

where the right side is some function of $\{t, q, \dot{q}\}$.

To be able to obtain all $\ddot{q}_k$ in terms of t, q and $\dot{q}$ it is necessary that the Hessian matrix of coefficients of $\ddot{q}_k$ have non-vanishing determinant,

$$\det\left(\frac{\partial^2 L}{\partial \dot{q}_j \partial \dot{q}_k}\right) \neq 0 \quad \text{(non-singular second-order Lagrangian)}. \qquad (8.55)$$

An obvious example is the unconstrained case of $N = 3n$ in Equations 8.52 and 8.53 for which Equation 8.54 reduces, in Cartesian coordinates, to 8.51. Whether or not Equation 8.55 holds in general will depend on the form of the Lagrangian, as a function of $q(t)$ and $\dot{q}(t)$, which in turn will depend on the nature of the physical system and, equally importantly, on the generalized coordinates chosen to describe it.

For a *non-singular Lagrangian system*, one may, in principle, solve for the $\ddot{q}_j(t)$ and integrate each to determine the trajectory from the initial conditions. Each of the coordinates q_j will then be referred to as a *true dynamical variable* and each equation containing a $\ddot{q}_k$ term with non-zero coefficient will be a *true dynamical equation*. Their role is to evolve the initial data forward in time.

If the choice of a certain set of generalized coordinates q_j to describe a mechanical system is such that it leads to a *singular Lagrangian*, then this implies that not all the equations using those variables are true second-order equations of motion. If r is the Hessian rank ($0 \leq r < N$), then for certain forms of singular Lagrangian, there may be $N - r$ combinations of

8.10. Constraints

the coordinates and equations which effectively do not contain a second-order term but are relations of the form,

$$\phi_c(t, q_j, \dot{q}_j) = 0 \qquad (c = 1, 2, \ldots, N - r) , \tag{8.56}$$

between the coordinates and the velocities. Such relations, if they are not holonomic constraints, but involve the velocities in an essential non-integrable way, may be referred to as *Lagrange constraints* on the initial data $q_j(t)$ and $\dot{q}_j(t)$ rather than equations of motion. In that case $N - r$ of the equations will be Lagrange constraints and the remaining r will be true dynamical equations. The solution of the equations of motion will provide definite functions of t in terms of the initial data for r of the generalized coordinates. The remaining $N - r$ accelerations will appear in those r solutions as arbitrary functions of t. The variables then exhibit *local gauge freedom* (see Section 7.10) and the system has *local gauge invariance*. Such systems are of considerable importance in the unification program of high-energy particle physics where they occur in relativistic systems which may be mechanical, such as particles and strings, or continuous systems, such as collections of fields, whose quantization may lead to particles or collections of particles. For other forms of singular Lagrangian, it may not be possible to form constraint equations. Instead, the singular nature may be manifest as identities relating the equations of motion. This also leads to gauge freedom.

Many of the statements made above about initial data, dynamical equations and constraints are valid only for a second-order Lagrangian system as occurs in classical mechanics. For an order-n Lagrangian system, initial data are comprised of the values of the generalized coordinates and their first $n - 1$ time derivatives and the statements concerning dynamical equations and constraints must be modified accordingly. The only other case, besides second order, of interest here will be those of first order where the initial data involve only the values of the generalized coordinates at a given time. In that case, $\partial^2 L / \partial \dot{q}_j \partial \dot{q}_k = 0$ for all $\{j, k\}$ and the condition of non-singularity becomes,

$$\det\left(\frac{\partial^2 L}{\partial q_j \partial \dot{q}_k}\right) \neq 0 \quad \text{(non-singular first-order Lagrangian)}. \tag{8.57}$$

If $\partial L/\partial \dot{q}_j = 0$ for a particular value of j, the corresponding coordinate q_j is said to be non-dynamic or *non-propagating*. Lagrangians containing a non-dynamic variable are singular. A mechanical Lagrangian is said to be *non-dynamic* or *dynamic* according to whether all $\partial L/\partial \dot{q}_j$ are zero or not. The singular or non-singular nature of the Lagrangian of a system is of crucial importance in the passage from the Lagrangian formulation to the Hamiltonian description which we shall discuss, for non-singular systems, in Chapter 11. Non-dynamic Lagrangians are singular, and the standard (canonical) Hamiltonian is then formally the negative of the Lagrangian. We shall postpone further discussion of singular systems, having Lagrangian constraints, until Chapter 15.

Il n'est pas sûr que la Nature soit simple. Pouvons-nous sans dangers faire comme si elle l'était?

Henri Poincaré   1854–1912

Bibliography:

Fetter A L et al., 1980 *Theoretical mechanics of particles and continua* [158].
Goldstein H 1980 *Classical mechanics* [206].
Kibble T W B 1985 *Classical mechanics* [279].
Lanczos C 1962 *The variational principle of mechanics* [303].
Landau LD and Lifschitz E M 1976 *Mechanics* [305].
Marion J B and Thornton S T 1988 *Classical dynamics* [358].

CHAPTER 9

Symmetries and Noether's theorem

Symmetry is one idea by which man through the ages has tried to comprehend and create order, beauty and perfection.

Hermann Weyl 1885–1955

9.1 Introduction

Conserved quantities play a major rôle in the analysis of evolving dynamical systems. Their use permits one to solve some problems without detailed knowledge of the dynamics and they can greatly help in the search for complete solutions of the full dynamical equations.

The ten basic external conserved quantities of an isolated Newtonian system are intimately linked to the form invariance of physical laws with respect to the Galilean symmetry transformations. Their conservation transcends the Newtonian mechanical context, remaining valid, with minor re-expression, for classical fields, in the Einsteinian physics of special and general relativity, in quantum mechanics and in quantum field theory.

In the search for a unified theory of all the interactions of physics, a leitmotif has been the search for larger symmetries encompassing all the fundamental subsystems to be incorporated. Even when such larger symmetries do not hold exactly under all circumstances, it is often useful to include them as a property which is then broken at a later stage. This may occur *dynamically* by an explicit term in the field equations or Lagrangian or *spontaneously* as a quantum phenomenon affecting the properties of the ground state.

For this reason, it is very important to know what constitutes a symmetry of a system, whether geometrical or dynamical, and whether there are corresponding conserved quantities. For a large class of symmetries, the conserved quantities may be determined by a fairly automatic procedure which we shall examine in this chapter.

9.2 Point transformations of configuration space

A set of generalized coordinates may be chosen for a given mechanical system in an infinite number of ways. The dynamics of the system will be independent of the particular choice and this may be demonstrated explicitly using *differential geometry* [499, 526], the general theory of coordinate transformations. The most general transformations of the generalized coordinates q and velocities $\dot{q}$ into $\bar{q}$ and $\dot{\bar{q}}$ that transform the original Euler-Lagrange equations into the new Euler-Lagrange equations in the barred coordinates are natural transformations of intrinsic importance in the study of mechanical systems. They are therefore referred to as *canonical transformations* and are analysed in detail in Sudarshan and Mukunda [525].

A subset of the canonical transformations of a mechanical system are those transformations $q(t) \rightarrow \bar{q}(\bar{t})$ of the points of configuration space of the form,

$$t \rightarrow \bar{t} = \bar{t}(t) \tag{9.1}$$
$$q \rightarrow \bar{q} = \bar{q}(q(t),t) , \tag{9.2}$$

called *point transformations*. Examples in which $t \rightarrow \bar{t} = t$ or $t + t_0$ or $-t$ are the translations, rotations, boosts and reflections which make up the full set of Galilean spacetime symmetry transformations.

Under such a transformation, the number which the action supplies for a given $q(t)$ between the two points $q(t_i)$ and $q(t_f)$ provides a definition of the transformed Lagrangian function, $\bar{L}(\bar{q}, \dot{\bar{q}}, \bar{t})$, of the new variables, namely,

$$\bar{S}[\bar{q}] = \int_{\bar{t}_i}^{\bar{t}_f} d\bar{t}\, \bar{L}(\bar{q}, \dot{\bar{q}}, \bar{t}) = \int_{t_i}^{t_f} dt\, L(q, \dot{q}, t) = S[q] . \tag{9.3}$$

We note, however, that it does not supply full details on how the functional form of the new and old Lagrangians differ when both are considered as functions of the same variables. It is the product of dt and L, and not the Lagrangian itself, which remains of the same value under transformations in which the time variable is functionally altered. It follows from Equation 9.3 that the Lagrangians before and after the transformation are related according to,

$$\bar{L}(\bar{q}, \dot{\bar{q}}, \bar{t}) = \frac{\partial t}{\partial \bar{t}} L(q, \dot{q}, t) , \tag{9.4}$$

$\partial t/\partial \bar{t}$ being the inverse Jacobian of the time transformation. The Lagrangian is therefore not an invariant but a *density* with respect to such transformations of the time variable just as the mass density $\rho(\mathbf{x})$ is a density with respect to ordinary space. The Jacobian equals 1, $|\partial t/\partial \bar{t}| = 1$, for Galilean transformations and the Lagrangian is thus a Euclidean scalar and a Galilean invariant if the system is isolated. The above transformation means that the Euler derivatives in the two sets of coordinates will be related according to,

$$\bar{L}^j(\bar{q},\dot{\bar{q}},\bar{t}) = \frac{\partial t}{\partial \bar{t}} \left|\frac{\partial q}{\partial \bar{q}}\right| L^j(q,\dot{q},t) \,, \tag{9.5}$$

in which $|\partial q/\partial \bar{q}|$ is the inverse Jacobian for the transformation of the dependent variables.

Since the vanishing of the Euler derivative gives the Euler-Lagrange equations, we see indeed that those equations describe the dynamics again after a point transformation, confirming that such a transformation is a special case of the more general canonical transformations. This result is sometimes referred to as the *canonical invariance* of the Euler-Lagrange equations. Let us note, however, that this does not necessarily imply that the two sets of Euler-Lagrange equations are of identical form. The functional form of the Euler derivative L^j need not be identical to that of $\bar{L}^j$. We must require such *form invariance* for the transformation to be a symmetry.

9.3 Invariant and covariant Lagrangians

An *invariant Lagrangian* $L(q,\dot{q},t)$ with respect to a transformation $\{t, q(t)\}$ to $\{\bar{t}, \bar{q}(\bar{t})\}$, is defined to be one for which the new Lagrangian of Equation 9.4, evaluated with the old variables, is identical to the original Lagrangian,

$$\bar{L}(q,\dot{q},t) = L(q,\dot{q},t) \,. \tag{9.6}$$

The equations of motion given by $L^j(q,\dot{q}) = 0$ and $\bar{L}^j(\bar{q},\dot{\bar{q}}) = 0$ will have the same form for such a Lagrangian and solutions of one will be mapped to solutions of the other. The transformation will be a symmetry.

Exercise 9.1 (a) *Show that a Lagrangian which is invariant with respect to a continuous transformation, whose infinitesimal changes are the functions $\Delta t(t)$ and $\Delta q(t)$, satisfies*

$$\frac{\partial(\Delta t)}{\partial t} + \Delta L = 0 \,, \tag{9.7}$$

where,

$$\Delta L = \frac{dL}{dt}\Delta t = \left\{\frac{\partial L}{\partial q}\dot{q} + \frac{d}{dt}\left(\frac{\partial L}{\partial \dot{q}}\right)\ddot{q} + \frac{\partial L}{\partial t}\right\}\Delta t \,. \tag{9.8}$$

(b) *Verify directly, and using the equations of part (a), that the Lagrangian of Equation 8.50 for a system of gravitationally interacting Newtonian particles is invariant with respect to space translations but not with respect to boosts.*

A *quasi-invariant* or *covariant Lagrangian*, with respect to a given transformation, is one which is either invariant or whose transformed Lagrangian differs from the original Lagrangian (in both of which the arguments, either $\{q, \dot{q}, t\}$ or $\{\bar{q}, \dot{\bar{q}}, \bar{t}\}$, are the same) by a *divergence*, namely a total time derivative of a scalar function $\Lambda(q, t)$ of the time t and the generalized coordinates $q(t)$ (but not $\dot{q}$), namely,

$$\bar{L}(q, \dot{q}, t) = L(q, \dot{q}, t) + \frac{d\Lambda(q,t)}{dt} \; . \tag{9.9}$$

Such a Lagrangian is referred to as covariant because, as a result of Exercise 8.4, the equations of motion given by $L^j(q, \dot{q}) = 0$ and $\bar{L}^j(\bar{q}, \dot{\bar{q}}) = 0$ will have the same form. If the transformation is continuous and infinitesimal, then this equation still holds with Λ replaced by an infinitesimal function $\Delta\Lambda$. The transformation will still be a symmetry of the system.

Exercise 9.2 *(a) Show that a Lagrangian which is covariant with respect to a continuous transformation whose infinitesimal changes are $\Delta t(t)$ and $\Delta q(t)$, satisfies,*

$$\frac{\partial(\Delta t)}{\partial t} + \Delta L = -\frac{d(\Delta\Lambda)}{dt} \; , \tag{9.10}$$

for some infinitesimal function $\Delta\Lambda(q(t), t)$. (b) Show that the left-side of Equation 9.10 can, for a boost transformation of the the Lagrangian of Equation 8.50, be re-arranged as the time derivative of an infinitesimal function $\Delta\Lambda(z_i, t)$, and is therefore boost covariant.

9.4 Hamilton's modified principle of varied action

The definition of a Lagrangian symmetry shows that, in the context of a system described with generalized coordinates $q(t)$, we must make use of Equation 9.4, which is true whether the transformation is a symmetry or not, and Equation 9.9, which requires it to be a symmetry. This will be done by generalizing the variational procedure used in Hamilton's principle.

In contrast to the previous discussion of functional variation in Chapter 8, we now consider transformations which functionally change not only the dependent variable q but also the independent variable t to some function $\bar{t} = \bar{t}(t)$ of itself. We require that function to be invertible, $d\bar{t}/dt \neq 0$, and continuous and therefore demand that $d\bar{t}/dt > 0$, the choice of positive sign meaning the monotonic increase of our time parameter t from initial to final states carries over to $\bar{t}$.

The form of the Euler-Lagrange equations is determined by the form of the action functional $S[q(t)]$. Let us denote a small functional variation

9.4. Hamilton's modified principle of varied action

of $q(t)$, in which both q and t are varied, by the symbol δ to distinguish it from the variation δ_0 in which t is not varied ($\delta_0 t = 0$). Thus,

$$t \rightarrow \bar{t}(t) = t + \delta t(t) \tag{9.11}$$
$$q(t) \rightarrow \bar{q}(\bar{t}) = q(t) + \delta q(t) . \tag{9.12}$$

These variations, and those of any quantity which is a function of t only, will be related to $\delta_0 q(t)$ according to,

$$\delta q(t) = \delta_0 q(t) + \delta t \frac{d(q(t))}{dt} , \tag{9.13}$$

and, correspondingly,

$$\delta \dot{q}(t) = \delta_0 \dot{q}(t) + \delta t \frac{d(\dot{q}(t))}{dt} . \tag{9.14}$$

from which we may note that, unlike Equation 8.27 for $\delta_0 \dot{q}(t)$,

$$\delta \dot{q}(t) \neq \frac{d}{dt}(\delta q(t)) . \tag{9.15}$$

Exercise 9.3 *Show that:*

$$\delta \dot{q}(t) = \frac{d}{dt}(\delta q(t)) - \dot{q}(t) \frac{d}{dt}(\delta t(t)) . \tag{9.16}$$

The time differential dt in the integrand of the action $S[q(t)]$ of Equation 8.26 will now have a non-zero variation determined in terms of the Jacobian according to the Fundamental theorem of calculus,

$$d\bar{t} = \frac{d\bar{t}}{dt} dt = \left(1 + \frac{d\delta t(t)}{dt}\right) dt , \tag{9.17}$$

showing that

$$\delta(dt) = dt \frac{d(\delta t(t))}{dt} . \tag{9.18}$$

We may summarize the difference between the two types of variation, δ_0 and δ, acting on t, $q(t)$ and $\dot{q}(t)$, in the identity,

$$\delta = \delta_0 + \delta t \frac{d}{dt} , \tag{9.19}$$

in which form it applies also to the action of δ_0 and δ on $L(q, \dot{q}, t)$ where d/dt is given by Equation 8.16. We may now straightforwardly determine the first-order variation $\delta S[q(t)]$ in the action, induced by the variation $\delta q(t)$, for a Lagrangian density of given functional form. Allowing for a change in the domain of integration, from the definition of S in Equation 8.26, we obtain its variation,

$$\delta S[q(t)] = \int_{\bar{t}_i}^{\bar{t}_f} d\bar{t}\, L(\bar{q}, \dot{\bar{q}}, \bar{t}) - \int_{t_i}^{t_f} dt\, L(q, \dot{q}, t) , \qquad (9.20)$$

in which the first Lagrangian is indeed L, not $\bar{L}$. We now obtain,

$$\begin{aligned}
\delta S &= \int_{t_i}^{t_f} \Big[(dt + \delta\, dt)(L + \delta L) - dt\, L\Big] \\
&= \int_{t_i}^{t_f} dt\, \Big[\frac{d(\delta t)}{dt} L + \delta L\Big] \\
&= \int_{t_i}^{t_f} dt\, \Big[\frac{d(\delta t\, L)}{dt} + \delta_0 L\Big] \\
&= \int_{t_i}^{t_f} dt\, \Big[\frac{d}{dt}(\delta t\, L) + \Big(\frac{\partial L}{\partial q} - \frac{d}{dt}\frac{\partial L}{\partial \dot{q}}\Big)\delta_0 q + \frac{d}{dt}\Big(\frac{\partial L}{\partial \dot{q}}\delta_0 q\Big)\Big] \\
&= \int_{t_i}^{t_f} dt\, \Big[\frac{d}{dt}\Big\{\delta t\, L + \frac{\partial L}{\partial \dot{q}}\delta_0 q\Big\} + \Big(\frac{\partial L}{\partial q} - \frac{d}{dt}\frac{\partial L}{\partial \dot{q}}\Big)(\delta q - \dot{q}\,\delta t)\Big] \\
&= \int_{t_i}^{t_f} dt\, \Big[\frac{d}{dt}\Big\{\frac{\partial L}{\partial \dot{q}}\delta q - \Big(\frac{\partial L}{\partial \dot{q}}\dot{q} - L\Big)\delta t\Big\} + \Big(\frac{\partial L}{\partial q} - \frac{d}{dt}\frac{\partial L}{\partial \dot{q}}\Big)(\delta q - \dot{q}\,\delta t)\Big] \\
&= \int_{t_i}^{t_f} dt\, \Big[\frac{d}{dt}\{\pi^j \delta q_j - H\,\delta t\} + L^j(\delta q_j - \dot{q}_j\,\delta t)\Big] \\
&= [\pi^j \delta q_j - H\,\delta t]_{t_i}^{t_f} + \int_{t_i}^{t_f} dt\, L^j(\delta q_j - \dot{q}_j\,\delta t) . \qquad (9.21)
\end{aligned}$$

The $\delta_0 L$ term has been converted in the same manner here as in going from Equation 8.30 to 8.31. The summation index j has been suppressed on a number of paired factors of $q = \{q_j\}$ or $\dot{q} = \{\dot{q}_j\}$ in three lines of Equation 9.21. This equation is the result of the mathematical process of obtaining the first-order variation in the action in terms of the variations δq and δt of the dependent and independent variables in the virtual path $q(t)$ of which $S[q(t)]$ is the functional. The first term of the last line of Equation 9.21 is a *surface term*, although now depending not only on the end values of δq but also on those of δt.

To return to the physical interpretation of this result, we note that the form of the Lagrangian function $L(q, \dot{q}, t)$ is presumed to contain all the physics of the system and that the equations of motion it implies are obtained using Hamilton's principle applied to a variation δ_0 not affecting the independent variable (nor therefore the region of integration) and for which the dependent variables are fixed at the end-points,

$$\delta_0 t = 0 \text{ (all } t\text{)} \quad \text{and} \quad \delta_0 q(t_i) = 0 = \delta_0 q(t_f) . \qquad (9.22)$$

The surface term vanishes under these restrictions. The demand that $\delta_0 S = 0$, according to Hamilton's principle, requires the Euler-Lagrange equations as before.

In terms of the more general variation, we may generalize Hamilton's principle of stationary action to the following statement:

- **Hamilton's modified principle of varied action** — the dynamical path followed by a Lagrangian system in configuration space is the path about which general variations in dependent and independent variables lead to only end-point contributions to the variations in the action functional.

9.5 Symmetries of Lagrangian systems

A *symmetry* of a system of equations (whether static or evolving) is a transformation of its dependent and independent variables (for example, z and t for a dynamical system described by Cartesian coordinates) which maps solutions into other solutions of those same equations. The symmetries of a static Newtonian system, as in the theories of electrostatics, magnetostatics and Newtonian gravity, for example, will be related to the geometrical symmetries of the distribution in Euclidean space $\mathbf{E}^3$.

If the system is not static but evolving, an example being a wave function governed by the Schrödinger equation, then such transformations are closely related to the quantities which are *conserved* during the evolution of the system and their use can greatly simplify solving the dynamical equations of motion. They will be related to the geometrical symmetries of the Galilean spacetime $\mathbf{G} = \{\mathbf{E}^1, \mathbf{E}^3\}$. The symmetries will then, in accord with the Erlangen programme, be closely related to geometrical properties of the spaces defined by the dynamical variables. Incorporation of the Galilean relativity principle into the form of our dynamical laws guarantees that the Galilean transformations (which affect both dependent and independent variables of a mechanical system) are symmetries of an isolated Newtonian system. Such a system may, however, also have many other symmetries of interest.

If the system is described by a Lagrangian, to which the application of Hamilton's principle gives the equations of motion, we may determine conditions that the Lagrangian must satisfy in order that such transformations are symmetry operations of the Euler-Lagrange equations. Among these transformations of configuration space, we must seek all those which leave the Euler-Lagrange equations unchanged in form and thus lead to solutions which are also solutions of the untransformed equations.

We shall be able to use the more general Equation 9.21 to examine the properties of continuous symmetry transformations, namely those which like translations, rotations and boosts, depend on a number of continuous parameters (such as a, θ, v and t_0).

To make use of Equation 9.21 for this purpose we need to characterize the properties of the special form of transformation of q and t called a symmetry transformation. For continuous transformations we may obtain all the information we require from small variations which, if they are symmetries, we shall denote by the special symbols $\Delta q = \Delta q(t)$ and $\Delta t = \Delta t(t)$. Any symmetry transformation, continuous or not (small or finite) must by definition leave the form of the Euler-Lagrange equations unchanged which means first that the form of the Lagrangian will change according to Equation 9.4.

It should be understood that there is no conflict between the requirement that the action be numerically unchanged in the transformation which gives rise to the new Lagrangian $\bar{L}(\bar{q}, \dot{\bar{q}}, \bar{t})$ and the non-zero change in S represented by Equation 9.21 when it is functionally varied, before application of Hamilton's principle puts that variation to zero. The zero change in the numerical value of S with respect to a transformation of $q(t)$ is quite independent of the variation in $S[q(t)]$ owing to a functional variation in $q(t)$. For similar reasons, there is no conflict in the use of $\bar{L}(\bar{q}, \dot{\bar{q}}, \bar{t})$ in Equation 9.3 and $L(\bar{q}, \dot{\bar{q}}, \bar{t})$ in 9.20. In the latter case, it is precisely the change in L due to the change in arguments with unchanged functional form that is presumed to contain the physical information. Both results will be required in what follows.

The symmetry transformations are in fact intimately connected to the change in the functional form of the Lagrangian from $L(q, \dot{q}, t)$ to $\bar{L}(\bar{q}, \dot{\bar{q}}, \bar{t})$. To ensure that the two sets of equations are unchanged in form we must also demand that the Lagrangian is either invariant or covariant with respect to the transformation. Covariance implies that the difference between the old and new Lagrangian functions differ by at most the time derivative of a scalar function of time t and $q(t)$. For a continuous transformation this implies that the change in the Lagrangian under a small transformation $\{\Delta t(t), \Delta q(t)\}$ is given, in terms of an infinitesimal function $\Delta \Lambda$ (determined for each particular Lagrangian and each transformation using Equation 9.10), according to,

$$\bar{L}(q, \dot{q}, t) = L(q, \dot{q}, t) + \frac{d(\Delta \Lambda(q, t))}{dt}. \tag{9.23}$$

We already have an expression, Equation 9.21, for the variation of the action owing to an arbitrary functional variation in t and $q(t)$, which we may use for the case of the special type of variation, $\Delta t(t)$ and $\Delta q(t)$, that constitutes an infinitesimal continuous symmetry transformation.

We may now determine an alternative formula for the variation of the action which applies specifically to a symmetry transformation. Using first Equation 9.23 with arguments $\{\bar{q}, \dot{\bar{q}}, \bar{t}\}$ in the definition, Equation 9.20, of the variation of the action, and then substituting Equation 9.4, we have:

$$\begin{aligned}
\Delta S &= \int_{\bar{t}_i}^{\bar{t}_f} d\bar{t}\, L(\bar{q},\dot{\bar{q}},\bar{t}) - \int_{t_i}^{t_f} dt\, L(q,\dot{q},t) \\
&= \int_{\bar{t}_i}^{\bar{t}_f} d\bar{t} \left[\bar{L}(\bar{q},\dot{\bar{q}},\bar{t}) - \frac{d(\Delta\Lambda(\bar{q},\bar{t}))}{dt} \right] - \int_{t_i}^{t_f} dt\, L(q,q,t) \\
&= - \int_{\bar{t}_i}^{\bar{t}_f} d\bar{t}\, \frac{d(\Delta\Lambda(\bar{q},\bar{t}))}{dt} \\
&= - \int_{t_i}^{t_f} dt\, \frac{d(\Delta\Lambda(q,t))}{dt} \,, \quad (9.24)
\end{aligned}$$

where the last line follows by retaining only first-order terms in small quantities. Applying Equation 9.21 with the general variation $\{\delta t, \delta q(t)\}$ replaced by the specific small symmetry transformation $\{\Delta t(t), \Delta q(t)\}$ to obtain an alternative expression for ΔS and subtracting from Equation 9.24 gives a zero integral over an arbitrary range from t_i to t_f showing that the integrand must vanish. The resulting equation is,

$$\frac{d}{dt}\left\{ \pi^j \Delta q_j - H \Delta t + \Delta\Lambda \right\} + \left(\frac{\partial L}{\partial q} - \frac{d}{dt}\frac{\partial L}{\partial \dot{q}} \right)(\Delta q - \dot{q}\Delta t) = 0 \,. \quad (9.25)$$

Since we have not yet used the equations of motion, this differential equation is a direct consequence of the transformation being a symmetry. If the system is symmetric with respect to a particular transformation, then this equation will hold not only for the on-shell (physical) path $q(t)$ actually followed by the system from t_i to t_f but also along all off-shell (virtual) paths in configuration space between those two end-points. The properties of systems along paths other than the classical stationary ones are of crucial importance in quantum applications since all paths, on or off shell, contribute to the quantum probability of transition from one state to another.

Equation 9.25 does not result from demanding an invariant action, $\Delta S = 0$. Rather, it suffices for the action to be covariant, namely the integral of a covariant Lagrangian. Invariance of the action and Lagrangian is a commonly occurring special case corresponding to $\Delta\Lambda = 0$.

9.6 Noether charges for mechanical systems

In the case of a small translation in space, the variations in the dependent and independent variables are of the form,

$$\Delta t(t) = 0 \quad \text{and} \quad \Delta z_i(t) = \Delta \mathbf{a} \,. \quad (9.26)$$

Only in very special cases do the variations $\Delta t(t)$ and $\Delta q(t)$ turn out to be either constant or zero. In general, $\Delta t(t)$ are non-zero functions of t while

Amelia Emmy Noether

b. Erlangen, Germany
23 March 1882

d. Bryn Mawr, Pennsylvania
14 April 1935

Courtesy of the State University, Göttingen.

$\Delta q(t)$ are functions of $\{t, q(t)\}$. We may generalize the parameter notation from **a** to ω^a ($a = 1, 2, 3$) and trivially rewrite these relations, using a as a summation index, in a generalized form as,

$$\Delta t = \frac{\partial \Delta t(t)}{\partial \Delta \omega^a} \Delta \omega^a \quad \text{and} \quad \Delta z_i = \frac{\partial \Delta z_i(t)}{\partial \Delta \omega^a} \Delta \omega^a \quad (a = 1, 2, 3), \quad (9.27)$$

where, for translations,

$$\{\Delta \omega^a\} = \Delta \mathbf{a}, \quad \frac{\partial \Delta t(t)}{\partial \Delta \omega^a} = 0 \quad \text{and} \quad \frac{\partial \Delta z_j(t)}{\partial \Delta \omega^a} = 1. \quad (9.28)$$

The advantage in writing such a simple symmetry transformation in such an apparently complicated way may not be immediately obvious. However, some symmetry transformations are more complicated in their expression than space translation. Nevertheless, for a very wide range of continuous symmetries, a small change in either t or $q(t)$ will depend linearly on the small constant changes $\Delta \omega^a$ in the parameters.

Let us suppose that a mechanical system has an arbitrary *Lie group* symmetry, namely under transformations which depend on not just 3 but a finite number p of independent continuous parameters ω^a ($a = 1, 2, \ldots, p$) which do not depend on the time t where a is a collective index labelling all the parameters. It may, for example, be a single Euclidean index (for example, $k = 1, 2, 3$ for x, y, z) or it may collectively represent several such

9.6. Noether charges for mechanical systems

indices such as $\{\omega^a\} = \{\mathbf{a}, \boldsymbol{\theta}, \mathbf{v}, t_0\}$ for Galilean transformations. Replacing the Cartesian variables z_i of the special case of translations by the appropriate generalized coordinates q_j permits us to further recast the above equations in the form,

$$\Delta t = \frac{\partial \Delta t(t)}{\partial \Delta \omega^a} \Delta \omega^a \quad \text{and} \quad \Delta q_j = \frac{\partial \Delta q_j(t)}{\partial \Delta \omega^a} \Delta \omega^a \quad (a = 1, 2, \ldots, p) \,. \quad (9.29)$$

If we introduce these two definitions into Equation 9.25 we obtain

$$\left[\frac{dQ_a}{dt} + \left(\frac{\partial L}{\partial q_j} - \frac{d}{dt} \frac{\partial L}{\partial \dot{q}_j} \right) \left(\frac{\partial \Delta q_j(t)}{\partial \Delta \omega^a} - \dot{q}_j \frac{\partial \Delta t(t)}{\partial \Delta \omega^a} \right) \right] \Delta \omega^a = 0 \,, \quad (9.30)$$

where,

$$Q_a = \pi^j \frac{\partial \Delta q_j(t)}{\partial \Delta \omega^a} - H \frac{\partial \Delta t(t)}{\partial \Delta \omega^a} + \frac{\partial (\Delta \Lambda)}{\partial \Delta \omega^a} \,. \quad (9.31)$$

Since the parameters ω^a are all independent, we arrive at,

$$\frac{dQ_a}{dt} = -\left(\frac{\partial L}{\partial q_j} - \frac{d}{dt} \frac{\partial L}{\partial \dot{q}_j} \right) \left(\frac{\partial \Delta q_j(t)}{\partial \Delta \omega^a} - \dot{q}_j \frac{\partial \Delta t(t)}{\partial \Delta \omega^a} \right) \,. \quad (9.32)$$

Both the functions $\partial \Delta t(t)/\partial \Delta \omega^a$ and $\partial \Delta q_j(t)/\partial \Delta \omega^a$ may be determined for each symmetry transformation in the same way as their trivial values of 0 and 1 were determined above for spatial translation.

Equations 9.31 and 9.32 are results of extraordinary power and elegance. Although they have been established in the context of Newtonian mechanical systems, essentially identical results apply to any Lagrangian system including relativistic mechanical systems, Galilean and Einsteinian fields, relativistic strings and membranes and systems having supersymmetry.

Some aspects of this work were published by Hamel [218] for Newtonian mechanical systems in 1904 and by Hergoltz [235] in 1911 from the stand-point of Einsteinian relativity. They were obtained in the context of conservation theorems by Emmy Noether [399] and Klein [283] in 1918 and by Bessel-Hagen [36] in 1921 and are now generally referred to as the *Noether relations*.

Equation 9.31 shows that, to each of the p parameters ω^a of a symmetry transformation of a dynamical Lagrangian system ($\partial L/\partial \dot{q}_j \neq 0$ for some j), there corresponds an additive quantity Q_a called a *Noether charge* which is a function of q, $\dot{q}$ and t. Equation 9.32 shows that the total time derivative of the Noether charge is a certain linear combination of the Euler derivatives. This results holds for any path $q(t)$ in configuration space, stationary (satisfying the equations of motion) or not.

9.7 Noether's theorem for mechanical systems

If the path concerned is one of the family satisfying the equations of motion, then Equation 9.32 reduces to

$$\frac{dQ_a}{dt} = 0 \; , \tag{9.33}$$

showing that the on-shell Noether charge of a dynamical system is conserved with time along the stationary trajectory. Since it applies only along the non-quantum stationary path, this result is often referred to as a *weak conservation law* in contrast to those symmetries, such as local gauge invariances, where the conservation laws are satisfied identically, on or off the stationary path. We have constructed, semi-automatically, p independent *constants of motion*,

$$Q_a = \text{ constant} \; , \tag{9.34}$$

which essentially consist of integrals of p of the equations of motion, leaving a reduced number of equations of motion to integrate in seeking solutions. The reason for calling the conserved quantity Q_a a 'charge' is related to the fact that electric charge (and similarly barionic, leptonic and colour charges, for example) can be shown to arise as Noether charges corresponding to continuous symmetries of the systems they apply to. We shall see that the electric charge arises in this way from a phase invariance of the source field, usually the Dirac spinor which gives a relativistic description of electrically charged non-chiral spin $\frac{1}{2}$ particles (such as electrons, protons and quarks). It should be noted that Equation 9.33 is a set of p linear combinations of the Euler-Lagrange equations of motion. We may now state:

- **Noether's theorem** — if a mechanical Lagrangian system is covariant (namely, if the action or Lagrangian is covariant) with respect to a Lie group symmetry transformation involving a finite number p of parameters which are not dependent on the time t, then there exist p linearly independent combinations of the Euler-Lagrange equations of motion which are each derivatives with respect to the time of a quantity referred to as a Noether charge, which is therefore conserved along a stationary trajectory.

Equation 9.31 for the Noether charge may be re-scaled if necessary in each case, according to convention, in order to obtain standard expressions for the conserved quantities and this will not affect Equation 9.34.

Alternative treatments of Noether's theorem may be found in Anderson [15], Fletcher [165], Hill [249], Lord [328], Ramond [467], Srivistava [517], Sundermeyer [526] and Utiyama [548], for example.

We can illustrate Noether's theorem with the familiar case of the translation invariance of a system of Newtonian particles interacting grav-

9.7. Noether's theorem for mechanical systems

itationally, for which Equation 9.31 becomes,

$$\mathbf{p} = \sum_i \frac{\partial L}{\partial \dot{z}_i} \cdot \frac{\partial \Delta z_i(t)}{\partial \Delta \mathbf{a}} + \frac{\partial(\Delta \Lambda)}{\partial \Delta \mathbf{a}} , \qquad (9.35)$$

giving a conserved quantity,

$$\mathbf{p} = \sum_i \frac{\partial L}{\partial \dot{z}_i} = \sum_i m_i \dot{z}_i , \qquad (9.36)$$

since the Lagrangian of Equation 8.50 is invariant ($\Delta \Lambda = 0$) with respect to translations. Noether's theorem thus shows us directly that corresponding to translational invariance in space, we have conservation of linear momentum. The advantage of establishing this result using Noether's theorem is that the process is completely automatic and of very general applicability, as we shall see with numerous less trivial examples.

Exercise 9.4 *Use the covariance of the gravitational Lagrangian of Equation 8.50 with respect to boosts to show that the corresponding Noether charge conserved along a stationary trajectory is proportional to the centre of mass motion.*

We shall discuss Noether's theorem for Newtonian field systems in the next chapter and for special relativistic particles and fields in Chapters 15 and 16.

Exercise 9.5 *Show that the Noether charges of a non-dynamic invariant Lagrangian are either empty (for symmetries not involving the time) or proportional to the Hamiltonian.*

Bibliography:

Anderson J L 1967 *Principles of relativity physics* [15].
Goldstein H 1980 *Classical mechanics* [206].
Hill E L 1951 *Hamilton's principle and the conservation theorems* [249].
Sundermeyer K 1982 *Constrained dynamics* [526].

CHAPTER 10

Non-relativistic Lagrangian fields

> Perhaps our classical mechanics is the *complete* analogue of geometrical optics and, as such, false .... Therefore we have to seek an undulatory mechanics[1].
>
> Erwin Schrödinger   1887–1961

10.1 Introduction

One of the great advantages of the Lagrangian formulation and stationary action principles is the way in which they relate the many different structures which are used to describe physical nature at all levels from the fundamental to the most complex. At the fundamental level, the concept of *field* plays a pre-eminent rôle in both classical (non-quantum) field theory and in the description of quantum particles via relativistic quantum field theory.

This chapter will be devoted to the extension of the Lagrangian formulation to include the fields of Galilean relativity, non-generic examples of which are the the classical fields of Newtonian gravitation, electrostatics and magnetostatics. A more representative example is the Schrödinger wave function of an electron described non-relativistically and therefore from the single particle point of view.

Let us first use dimensional analysis to gain some idea of where we might expect a fundamental theory of Galilean fields to be applicable. By fundamental we always mean that the constants involved will be those of relativity and quantum mechanics (c and $\hbar$) on the one hand, interaction or coupling constants (like G and e) and finally particle masses (such as m_e and m_p). However, we cannot include c in a Newtonian context. A field theory, by definition, involves a continuous distribution over space and the

---

[1] *Collected papers on wave mechanics* [498].

fields must therefore satisfy equations which at least contain the gradient operator ∇ acting on the field.

Let us limit attention to first- and second-order terms only in the gradient. The field equations may or may not include a derivative ∂_t with respect to time acting on the field. Let us limit this factor also to first- or second-order. In Newtonian physics, there is no demand that ∇ and ∂_t appear to the same order in a field equation. However, if both do appear to the same order, we shall require a quantity of velocity dimensions to match the terms and no such fundamental quantity is available without Einsteinian relativity.

Exercise 10.1 (a) *Show that no velocity may be constructed from G, e and a mass m and that the only velocity obtainable from e, $\hbar$ and a mass m (such as that of the electron) leads to the speed of light c.*

Equations of first-order in ∇ and second-order in ∂_t require a quantity with the dimensions of length divided by the square of time and again there is no obvious combination except for $me^6/\epsilon_0^3\hbar^4$. For an electron or proton mass, this can be recognized as a common factor involved in the theory of interaction of radiation with atomic or nuclear matter and thus leads back to the velocity of light and non-Galilean relativity.

We are therefore left with three cases. The first is an equation which is first-order in ∂_t and second-order in ∇, an example being the Schrödinger dynamics of massive particles such as the electron. The second and third cases involve ∇ on its own in either first or second order. Each of the latter can be used to describe either electrostatics or Newtonian gravity. These then are the fundamental theories for which we can expect a Galilean-covariant field theory to be relevant, although they far from exhaust the applications of field theory to non-relativistic physical phenomena. Sudarshan and Mukunda [525], for example, give a detailed Lagrangian analysis of the dynamics of sound waves in an ideal gas.

10.2 Mechanics to fields

In discussing the dynamics of a conservative mechanical system of particles of trajectories $\mathbf{z}_i(t)$, one encounters a potential energy function $V(\mathbf{z}_i, t)$ or generalized potential energy function $V(\mathbf{z}_i, \dot{\mathbf{z}}_i, t)$. As the notation indicates, these are functions of the trajectory variables, not of the spatial coordinates defined at all points $\mathbf{x}$ of Euclidean space, $\mathbf{E}^3$. The evolution of a Newtonian system is presumed to be continuous in time. Newton's laws, being differential equations in the time, t, are local laws with respect to time that determine the state of a system at a given time in terms of the state a short time earlier.

In the interests of dividing the interaction between two Newtonian mechanical systems into separate processes which are more tractable individually, we may attempt to extend this locality in time to a similar *locality* in space (Faraday's *contiguous action*) by introducing a new type of physical quantity which is not only a function of the time coordinate t but also of space $\mathbf{x}$. We could argue that the interactions implied by the properties of the generalized mechanical potential energy $V(\mathbf{z}_i, \dot{\mathbf{z}}_i, t)$ will act at any point of space where there happens to be some particle i. It makes no difference to the actual interaction if we assume the existence of an interaction field at all points, whether a particle is present or not, generated of course by the actual particles.

The advantages of such a procedure are, first, that it will permit the interaction to be divided into generation of the field and reaction of matter to the field. Second, it permits us to generalize the scope of the very powerful concept of locality, from being purely temporal, to locality in space as well. In so doing, we shall also be able to replace the conserved total energy of Newtonian mechanics by one whose additivity is more simply evident by including the potential energy of interaction via the energy of a mediating field.

We shall let the symbol $\psi = \psi(t, \mathbf{x}) = \{\psi_j(t, \mathbf{x})\}$ ($j = 1, 2, \ldots, d$) denote an arbitrary Newtonian field with a finite number d of components $\psi_1(t, \mathbf{x}), \ldots, \psi_d(t, \mathbf{x})$. In a specific example, there may be only one component, as in the case of a Schrödinger wave function of a spinless particle, the electrostatic potential field $\phi(t, \mathbf{x})$ or the Newtonian gravitational potential field $\Phi(t, \mathbf{x})$. In others the label j may denote the Euclidean index $k = 1, 2, 3$ for x, y, z as in the case of the electric field $\mathbf{E}(t, \mathbf{x})$, the magnetic induction $\mathbf{B}(t, \mathbf{x})$, the electromagnetic 3-vector potential $\mathbf{A}(t, \mathbf{x})$ or the Newtonian gravitational field strength $\mathbf{f}(t, \mathbf{x})$. We may also use ψ as an abbreviation for the tensor field $\mathbf{T} = \{T^{kl\ldots}{}_{m\ldots}(t, \mathbf{x})\}$ where j now collectively denotes $\{k, l, m, \ldots, \}$.

Finally, one further possibility ought to be at least mentioned in a Newtonian context, namely the non-relativistic description of a spin $\frac{1}{2}$ particle by a Pauli-Schrödinger double-valued (*spinor*) wave function with two complex components $\{\psi^A(t, \mathbf{x})\}$ ($A = 1, 2$). We may choose to imagine the separate components of the field organized as a column vector:

$$\psi = \psi(t, \mathbf{x}) = \{\psi_j(t, \mathbf{x})\} = \begin{pmatrix} \psi_1(t, \mathbf{x}) \\ \psi_2(t, \mathbf{x}) \\ \vdots \\ \psi_d(t, \mathbf{x}) \end{pmatrix}. \tag{10.1}$$

The fact that a system described by $\psi(t, \mathbf{x})$ is a field means that it has a nondenumerably infinite number of mathematical degrees of freedom or generalized coordinates, namely the values of the field at all spatial points $\mathbf{x}$

at some time t. A field propagates not along a 1D trajectory but throughout a whole spacetime region — not from one event to another but from an initial configuration defined throughout $\mathbf{E}^3$ at some time t_i to some final configuration also defined throughout space at a final time t_f. In passing from a mechanical system with generalized coordinates $q_j(t)$ to a field system, the independent variables have changed from t to $\{t, \mathbf{x}\}$, and the dependent variables have changed from z_i or q_j to ψ_j with $j = 1, 2, \ldots, d$. However, the analogue of the denumerable label j of the mechanical case is not just the denumerable index j in the field case. The analogous labels in the field case are comprised of the discrete label j plus the continuous labels $\mathbf{x}$.

The mathematical degrees of freedom are not the finite number of components of ψ but the nondenumerably infinite number of values of all components $\psi_j(t, \mathbf{x})$ of ψ at all locations $\mathbf{x}$ in space. To fail to appreciate this point will make the transition from mechanics to a field system more difficult. The mathematical degrees of freedom of a field are all available as possible dynamical degrees of freedom of the physical system they describe. Which are independent physical degrees of freedom will depend on the nature of the field equations which describe the spatial distribution of the field and its evolution if it is dynamic.

In Newtonian physics, not all field equations will describe the evolution of a field as does, for example, the Schrödinger equation — some, like the equations of electrostatics, magnetostatics and Newtonian gravity, are non-dynamic and describe only the spatial distribution of the field whose dynamics is determined, instantaneously, from the equations of motion of the mechanical systems which generate the field. A field description appears to place the temporal and spatial properties of a system on a more equal footing, although the full potential of this generalization is not always achievable at the Newtonian level where temporal variation and therefore propagation of a field is sometimes in conflict with Galilean covariance.

10.3 Hamilton's principle for Newtonian fields

From the generalized coordinates $\psi = \psi(t, \mathbf{x}) = \{\psi_j(t, \mathbf{x})\}$, we define the corresponding *generalized velocities* of a field,

$$\dot{\psi} = \dot{\psi}(t, \mathbf{x}) = \frac{\partial \psi(t, \mathbf{x})}{\partial t}, \qquad (10.2)$$

which are treated as mathematically independent of the coordinates in the Lagrangian approach when partial derivatives are taken with respect to ψ_j or $\dot{\psi}_j$. Note however, that given the usual initial data, namely ψ and $\dot{\psi}$ for all $\mathbf{x}$, the spatial derivatives $\nabla \psi(t, \mathbf{x})$ are known independently of the equations governing ψ and therefore cannot be regarded as independent variables at the Lagrangian (as opposed to Lagrangian density) level.

We consider initially only those fields all of whose dynamics are obtained from a *non-singular Lagrangian*,

$$L = L[\psi, \dot\psi] = L[\psi(t, \mathbf{x}), \dot\psi(t, \mathbf{x})] \quad \text{with} \quad \det\left(\frac{\partial^2 L}{\partial \dot\psi_j \partial \dot\psi_k}\right) \neq 0, \qquad (10.3)$$

involving only first-order time derivatives of ψ to no higher than second degree, as in the case of mechanical systems. There is a continuous (non-denumerable) infinity of generalized coordinates (and velocities) on which the Lagrangian of a field system depends, from which a single value of the action is determined by an integral over time. Thus, in contrast to the mechanical case, the Lagrangian is a functional [517, 525, 526] in ψ.

As for the mechanical case, the *action functional* of the field system is defined by,

$$S[\psi(t), t_i, t_f] = \int_{t_i}^{t_f} dt\, L[\psi, \dot\psi, t] . \qquad (10.4)$$

Explicit dependence on t will occur only for non-isolated systems. We define the *generalized momentum* π^j conjugate to $\psi_j(t, \mathbf{x})$, by analogy with the mechanical case, using however not an ordinary derivative, but the appropriate functional derivative (see Equation 8.36),

$$\pi^j = \pi^j(t, \mathbf{x}) = \frac{\delta L}{\delta \dot\psi(t, \mathbf{x})} . \qquad (10.5)$$

Since the Lagrangian is a functional of the field and of a finite number (here one) of its time derivatives, the theory is said to be *temporally local*. This does not guarantee that it is also *spatially local*. However, we are interested only in *local field theories*, in space and in time. We therefore restrict attention to systems described by a Lagrangian expressible as an integral over 3-space,

$$L = L[\psi, \dot\psi] = \int_R d^3x\, \mathcal{L}(\psi, \boldsymbol{\nabla}\psi, \ldots, \dot\psi, \boldsymbol{\nabla}\dot\psi, \ldots, t, \mathbf{x}) , \qquad (10.6)$$

of a function $\mathcal{L}$ (not a functional), called the *Lagrangian density*, of the field and a finite number of its partial derivatives with respect to time and space (in which can occur only single derivatives with respect to time). In such cases, the generalized momenta will also be local functions of ψ and its partial derivatives. As defined, the Galilean Lagrangian density has the dimensions $[\hbar\, L^{-3}\, T^{-1}]$ of energy density. It is common practice in field theory to refer to the Lagrangian density as the *Lagrangian*. Explicit dependence on t and $\mathbf{x}$ will not occur for isolated systems.

If one considers only Lagrangian densities which are functions of the field, its first time derivative and purely spatial derivatives,

$$\mathcal{L} = \mathcal{L}(\psi, \dot\psi, \boldsymbol{\nabla}\psi, \ldots) , \qquad (10.7)$$

10.3. Hamilton's principle for Newtonian fields

then the generalized momenta are given by,

$$\pi^j = \pi^j(t, \mathbf{x}) = \frac{\partial \mathcal{L}}{\partial \dot{\psi}(t, \mathbf{x})} , \qquad (10.8)$$

which are now completely analogous to the mechanical equivalents even to the extent of the appearance of ordinary (non-functional) derivatives.

The Lagrangian form of Equation 10.7, restricted to functions with only first-order time derivatives in which derivatives appear to no higher than second degree, is also sufficiently general to encompass all the fundamental systems we need to discuss. Limitation to second degree in first derivatives is equivalent to limitation to Euler-Lagrange differential equations which are linear in the first and second derivatives. Generalization to *higher derivative* Lagrangians is straightforward if required.

For a field system, we therefore consider an action given by an integration over all independent variables $\{t, \mathbf{x}\}$ throughout a compact (closed and bounded) spacetime region. Thus, with $t_i \leq t \leq t_f$ and $\mathbf{x}$ within a compact 3-volume R, we take the action to be the integral of the Lagrangian density $\mathcal{L} = \mathcal{L}(\psi, \dot{\psi}, \boldsymbol{\nabla}\psi, t, \mathbf{x})$,

$$S[\psi(t), t_i, t_f] = \int_{t_i}^{t_f} dt \int_R d^3x \, \mathcal{L}(\psi, \dot{\psi}, \boldsymbol{\nabla}\psi, t, \mathbf{x}) . \qquad (10.9)$$

The physical interpretation of a Lagrangian field and its action is analogous to the mechanical case. The functional form of the Lagrangian or Lagrangian density is presumed to contain all the details of the physics of the field system via:

- **Hamilton's principle of stationary action** for fields — the Euler-Lagrange field equations of a Lagrangian field system are necessary conditions for the vanishing, $\delta_0 S[\psi(t, \mathbf{x})] = 0$, of the first-order functional variation of the action with respect to variations $\delta_0 \psi(t, \mathbf{x})$ of the field variables held constant, for all $\mathbf{x}$, at the end-points of a compact time interval, there being no variation of the independent variables, t and $\mathbf{x}$.

Given this interpretation, the actual process of determining the form of the field equations is a purely mathematical application of the calculus of variations entirely analogous to the discussion in Chapter 8. However, since a few new features arise, we shall outline the steps.

The variation of ψ_j,

$$\psi_j(t, \mathbf{x}) \to \psi_j(t, \mathbf{x}) + \delta_0 \psi_j(t, \mathbf{x}) , \qquad (10.10)$$

implies corresponding variations of $\dot{\psi}_j$ and $\boldsymbol{\nabla}\psi_j$,

$$\delta_0(\dot{\psi}_j(t, \mathbf{x})) = \partial_t(\delta_0 \psi_j(t, \mathbf{x})) \qquad (10.11)$$
$$\delta_0(\boldsymbol{\nabla}\psi_j(t, \mathbf{x})) = \boldsymbol{\nabla}(\delta_0 \psi_j(t, \mathbf{x})) , \qquad (10.12)$$

which induce variations,

$$\delta_0 \mathcal{L}(\psi_j(t,\mathbf{x}), \dot{\psi}_j(t,\mathbf{x}), t, \mathbf{x})$$
$$= \frac{\partial \mathcal{L}}{\partial \psi_j}\delta_0\psi_j + \frac{\partial \mathcal{L}}{\partial \dot{\psi}_j}\delta_0\dot{\psi}_j + \frac{\partial \mathcal{L}}{\partial \boldsymbol{\nabla}\psi_j}\cdot\delta_0\boldsymbol{\nabla}\psi_j \ , \quad (10.13)$$

in $\mathcal{L}$. Inserting this variation into the variation of the action and integrating the second and third terms by parts gives,

$$\delta_0 S = \int_{t_i}^{t_f} dt \int_R d^3x \left\{ \frac{\partial \mathcal{L}}{\partial \psi} - \partial_t\left(\frac{\partial \mathcal{L}}{\partial \dot{\psi}}\right) - \boldsymbol{\nabla}\cdot\left(\frac{\partial \mathcal{L}}{\partial \boldsymbol{\nabla}\psi}\right)\right\}\delta_0\psi(t,\mathbf{x})$$
$$+ \left[\int_R d^3x \frac{\partial \mathcal{L}}{\partial \dot{\psi}}\delta_0\psi\right]_{t_i}^{t_f} + \int_{t_i}^{t_f} dt \oint_{\partial R} d\mathbf{S}\cdot\frac{\partial \mathcal{L}}{\partial \boldsymbol{\nabla}\psi}\delta_0\psi \ , \quad (10.14)$$

where we have used Gauss's theorem to convert the initial and final 3-volume integrals of a divergence to closed integrals over the 2D surfaces ∂R bounding the spatial regions R of integration. Both these contributions involving quantities at the boundaries of the finite region of spacetime used in the action integral are referred to as *surface terms*. The index j has been suppressed in Equation 10.14.

The physical application of Hamilton's principle is now straightforward. The variation of $\psi_j(t, \mathbf{x})$ is required to be zero on the boundary,

$$\delta_0\psi_j(t_i,\mathbf{x}) = 0 = \delta_0\psi_j(t_f,\mathbf{x}) \text{ and } \delta_0\psi_j = 0 \text{ on } \partial R \ , \quad (10.15)$$

although there is no similar restriction on the variation of the velocities or gradients. The surface terms therefore contribute nothing to the variation of the action and the arbitrariness in the variations $\delta_0\psi_j(t,\mathbf{x})$ for $t_i < t < t_f$ establish the Euler-Lagrange equations,

$$\frac{\delta_0 S}{\delta_0 \psi_j} = \mathcal{L}^j = \frac{\partial \mathcal{L}}{\partial \psi_j} - \partial_t\left(\frac{\partial \mathcal{L}}{\partial \dot{\psi}_j}\right) - \boldsymbol{\nabla}\cdot\left(\frac{\partial \mathcal{L}}{\partial \boldsymbol{\nabla}\psi_j}\right) = 0 \ , \quad (10.16)$$

as the necessary conditions for the stationarity of the action. Since there are a continuous infinity of equations, one for each spatial location $\mathbf{x}$, they are referred to as the *field equations* of the system, governing not only its evolution but also its spatial distribution. The left side of these equations, arising as the expression of the functional derivative of the action, is the Euler derivative, denoted $\mathcal{L}^j$, of the field Lagrangian, in which $\partial\mathcal{L}/\partial\psi_j$, $\partial\mathcal{L}/\partial\dot{\psi}_j$ and $\partial\mathcal{L}/\partial\boldsymbol{\nabla}\psi_j$ are normal (not functional) partial derivatives of the function $\mathcal{L}$ with respect to one or other of its arguments.

A second-order *non-singular Lagrangian density* is one for which,

$$\det\left(\frac{\partial^2 \mathcal{L}}{\partial \dot{\psi}_j \partial \dot{\psi}_k}\right) \neq 0 \ . \quad (10.17)$$

If we regard $\psi = \{\psi_j\}$ as a column vector labelled by j then, since the action must be a scalar, we consider $\partial\mathcal{L}/\partial\psi$, $\partial\mathcal{L}/\partial\dot\psi$ and $\partial\mathcal{L}/\partial\nabla\psi$ to be row vectors. We have chosen to carry out the variational procedure using the action expressed in terms of the Lagrangian density which is appropriate for a local theory. We could, however, proceed from Equation 10.4 to which application of Hamilton's principle gives,

$$\frac{\delta_0 S}{\delta_0 \psi_j} = \frac{\delta_0 L}{\delta_0 \psi_j} - \frac{d}{dt}\frac{\delta_0 L}{\delta_0 \dot\psi_j} = 0 , \qquad (10.18)$$

very much as for a mechanical system except for the replacement of ordinary derivatives by the functional derivatives appropriate to a field system. This alternative form of the Euler-Lagrange field equations reduces to Equation 10.16 in the case of a local Lagrangian expressible in terms of a Lagrangian density.

Any set of fields ψ governed by a Lagrangian containing no term involving the generalized velocities ($\partial\mathcal{L}/\partial\dot\psi_j = 0$ all j) is said to be a *non-dynamic* or *non-propagating field*. The corresponding Lagrangian density is singular. The use of such a field to describe an interaction is equivalent to using an action at a distance mechanical description.

The *configuration space Hamiltonian functional* is defined in terms of the Lagrangian L by,

$$H = H[\psi, \dot\psi, t] = \int_R d^3x \sum_j \pi^j(t, \mathbf{x})\dot\psi_j(t, \mathbf{x}) - L[\psi, \dot\psi, t] , \qquad (10.19)$$

namely, as a functional in the fields and velocities (or momenta) in which it can be seen that the summation over the denumerable set of coordinates of a mechanical system must be replaced by an integral over the continuous set of spatial coordinates, plus a sum over the component label j (often suppressed) if it also exists. For a local theory, we replace the Lagrangian by the integral over the the Lagrangian density and hence also express the Hamiltonian functional as the integral,

$$H = H[\psi, \dot\psi, t] = \int_R d^3x\, \mathcal{H}(\psi, \pi, t, \mathbf{x}) = \int d^3x \left(\sum_j \pi^j \dot\psi_j - \mathcal{L}\right) , \qquad (10.20)$$

over an ordinary function $\mathcal{H}(\psi, \pi)$ of the coordinates and velocities (or momenta), called the *Hamiltonian density*, given by,

$$\mathcal{H} = \mathcal{H}(t, \mathbf{x}) = \sum_j \pi^j(t, \mathbf{x})\dot\psi_j(\psi, \pi) - \mathcal{L} = \frac{\partial \mathcal{L}}{\partial \dot\psi_j}\dot\psi_j - \mathcal{L} . \qquad (10.21)$$

10.4 Poisson and Laplace equation Lagrangians

We may now re-examine the equations of the gravitational field (or of electrostatics) in terms of Lagrangian fields.

Exercise 10.2 *Show that the Euler-Lagrange field equation of the field $\Phi = \Phi(t, \mathbf{x})$, governed by the Lagrangian density,*

$$\mathcal{L} = -\tfrac{1}{2}\nabla\Phi\cdot\nabla\Phi \, , \tag{10.22}$$

is Laplace's equation, $\nabla^2\Phi = 0$.

The field Φ of the above Lagrangian is clearly not only non-interacting but also non-propagating and therefore singular. The sign of the Lagrangian $\mathcal{L}$ leading to Laplace's equation is chosen conventionally in order to agree with the static limit of the standard form for related dynamic Lagrangians. One such Lagrangian is the one chosen for the Schrödinger equation later in this chapter. Others are the dynamic Lagrangians of the relativistic gravitational field, whether special or general relativistic, where the sign of the term corresponding to the present Lagrangian is determined by demanding energy positivity.

The same argument cannot be used here since energy, being the conserved quantity corresponding to time translation, cannot be defined for a static system for which the time evolution is not specified. A field satisfying Laplace's equation is in some respects analogous to a free Newtonian particle at rest.

One way the above Lagrangian can be altered to describe a field which is no longer *free* is to add a *self interaction* potential term $V_{\text{self}} = V_{\text{self}}(\Phi)$, which is a function of Φ itself, to the free Lagrangian to obtain,

$$\mathcal{L} = -\tfrac{1}{2}\nabla\Phi\cdot\nabla\Phi - V_{\text{self}}(\Phi) \, . \tag{10.23}$$

Let us consider such potential terms linear and quadratic in Φ (the latter with constant coefficient) by choosing a Lagrangian of the form,

$$\mathcal{L} = -\tfrac{1}{2}\nabla\Phi\cdot\nabla\Phi + a(t,\mathbf{x})\Phi + \frac{\Phi^2}{2\lambda^2} \, . \tag{10.24}$$

in which we have written the coefficient of the quadratic term as an inverse square since it must have the dimensions of inverse length squared. The dimension of a cannot be fixed unless we assign dimensions to Φ. In order that Galilean covariance be preserved, $a(t, \mathbf{x})$ must be an invariant, possibly constant. The coefficient a may have a constant part $\Lambda(t)$ (independent of $\mathbf{x}$). We therefore re-write it as,

$$a(t, \mathbf{x}) = \Lambda(t) - j(t, \mathbf{x}) \, . \tag{10.25}$$

Exercise 10.3 *Show that,*

$$(\nabla^2 + \frac{1}{\lambda^2})\Phi + \Lambda(t) = j(t, \mathbf{x}) \, , \tag{10.26}$$

is the Euler-Lagrange field equation of Equation 10.24.

The term in the Lagrangian potential which is linear in Φ gives rise to terms, $\Lambda(t)$ and $j(t,\mathbf{x})$, in the field equations which do not involve the field itself — a term such as j is referred to as an *external source*, since the details of its distribution in space are specified rather than being included in the analysis. The term quadratic in Φ can be introduced only if one also introduces a fundamental quantity λ having the dimensions of length. The linear and quadratic terms in the Lagrangian self-interaction leave the field equations linear in Φ.

In order to apply this result to the standard gravitational case we first put $1/\lambda = 0$. We may now choose to fix the dimensions of Φ to be those of the gravitational field, namely velocity squared, and thus fix Λ and j to have the dimensions of inverse time squared, which are also those of $G\rho$ where G is the gravitational constant and ρ is a mass density. The entire Lagrangian density is then rescaled by a constant factor to ensure that it has the conventional dimensions for the Lagrangian of a Galilean field, namely action divided by volume times time. This can be achieved by rewriting j as $4\pi G\rho$ and dividing throughout by $4\pi G$, in which the 4π is conventional.

We therefore arrive at the Lagrangian density of the Newtonian gravitational field generated by the external mass density ρ, and with cosmological constant Λ,

$$\mathcal{L} = -\frac{1}{8\pi G}\nabla\Phi\cdot\nabla\Phi + \left(\frac{1}{4\pi G}\Lambda(t) - \rho(t,\mathbf{x})\right)\Phi \ . \tag{10.27}$$

Before leaving these almost trivial examples of Lagrangian fields we may note that potential terms of higher degree than quadratic, such as a cubic Φ^3, quartic Φ^4, etc., will clearly contribute quadratic and cubic terms, etc., to the field equations rendering them non-linear. However, there is no natural way to include dynamic or non-linear gravitational field effects in the context of Newtonian gravity.

A similar Lagrangian may be formed for the electrostatic potential ϕ by replacing $\Phi/\sqrt{4\pi G}$ by $\phi/\sqrt{\mu_0 c}$. The free field term for the magnetostatic potential $\mathbf{A}$ will similarly be $-(\nabla A^k\cdot\nabla A_k - \nabla A^k\cdot\partial_k\mathbf{A})/2\mu_0 c$.

10.5 Newtonian gravitational action

We now have the necessary machinery to give a complete Lagrangian discussion of a closed system of Newtonian particles, the gravitational interaction of which is described by a local field. A Newtonian particle and the gravitational field are very different constructs with quite different Lagrangians, the first a quantity of energy dimensions involving scalars and vectors such as m and $\dot{z}$ and the second having dimensions of energy over volume and involving scalar and vector fields and operators, such as $\Phi(\mathbf{x})$ and $\boldsymbol{\nabla}\Phi$. Nevertheless, the actions for both systems are Euclidean invariants of the same dimensions and we shall see that the easiest way to combine two such different systems is by forming the total action of the free and interaction parts.

As first contribution, we take the free action of the particles (p),

$$S_p[\mathbf{z}_i(t)] = \tfrac{1}{2} \int_{t_i}^{t_f} dt \sum_i m_i \dot{\mathbf{z}}_i^2 \ . \tag{10.28}$$

As second contribution we take the free, non-propagating action of the gravitational field $\Phi(t,\mathbf{x})$ to be,

$$S_f[\Phi(t,\mathbf{x})] = -\frac{1}{8\pi G} \int_{t_i}^{t_f} dt \int_R d^3x \, \boldsymbol{\nabla}\Phi \cdot \boldsymbol{\nabla}\Phi \ . \tag{10.29}$$

As in the similar mechanical analysis of the gravitational interaction, we may see how far we can go in guessing a simple Lagrangian interaction consistent with the requirements of Galilean covariance between Newtonian particles mediated by the simplest possible field, namely a field with the non-propagating scalar character of Φ. The interaction term must be a Euclidean invariant with coupled dependence on Φ and $\mathbf{z}_i$. The simplest contribution from the field will be a single factor of Φ. Since this has dimensions of velocity squared, a mass will bring it to the required dimensions of energy for integration over time to form an action. Consider therefore a contribution to the total action of the form,

$$S_{\text{intn}}[\mathbf{z}_i(t), \Phi(t, \mathbf{z}_i)] = a \int_{t_i}^{t_f} dt \sum_i m_i \Phi(t, \mathbf{z}_i) \ . \tag{10.30}$$

in which a must be a dimensionless constant. We shall anticipate the determination of the Euler-Lagrange equations and put $a = -\tfrac{1}{2}$, the value which in fact gives the gravitational equations in standard form.

The interaction between particles and field (and hence between the particles themselves) will be achieved by the presence of trajectory variables $\mathbf{z}_i$ in the spatial argument of the field factor $\Phi(t, \mathbf{z}_i)$ in the interaction term of the action. The above form of the interacting term in the action is suitable for analysing the motion of particle i. However, to determine the field equations of $\Phi(t, \mathbf{x})$ we need the action expressed as an integral over space as well, which may easily be done using a property,

$$\int d^3x \, \delta^3(\mathbf{x} - \mathbf{z}) f(\mathbf{x}) = f(\mathbf{z}) \ , \tag{10.31}$$

of the 3D Dirac delta function $\delta^3(\mathbf{x} - \mathbf{z})$ to obtain,

$$S_{\text{intn}}[\mathbf{z}_i, \Phi] = -\tfrac{1}{2} \int_{t_i}^{t_f} dt \int_R d^3x \sum_i m_i \Phi(t, \mathbf{x}) \delta^3(\mathbf{x} - \mathbf{z}_i) \ , \tag{10.32}$$

10.5. Newtonian gravitational action

in which the field now appears, as required, with the argument **x**. The total action is therefore,

$$S[\mathbf{z}_i, \Phi] = -\frac{1}{8\pi G} \int_{t_i}^{t_f} dt \int_R d^3x \, \boldsymbol{\nabla}\Phi \cdot \boldsymbol{\nabla}\Phi + \tfrac{1}{2} \int_{t_i}^{t_f} dt \sum_i m_i \dot{\mathbf{z}}_i^2$$
$$-\tfrac{1}{2}\int_{t_i}^{t_f} dt \sum_i m_i \Phi(t, \mathbf{z}_i) + \frac{1}{4\pi G}\int_{t_i}^{t_f} dt \int_R d^3x \, \Lambda(t)\Phi(t,\mathbf{x}) \quad (10.33)$$

in which we have included a cosmological term as well, for completeness. Whenever convenient, we may substitute the right side of Equation 10.32 for the third term.

Exercise 10.4 (a) *Show that the Euler-Lagrange equations of motion for $\mathbf{z}_i$ determined from the Lagrangian of Equation 10.33 are,*

$$\ddot{\mathbf{z}}_i = -\boldsymbol{\nabla}_i \Phi = -\frac{\partial \Phi(t, \mathbf{z}_i)}{\partial \mathbf{z}_i} \equiv -[\boldsymbol{\nabla}\Phi(t, \mathbf{x})]_{\mathbf{x}=\mathbf{z}_i} \, . \quad (10.34)$$

(b) *Show that the Euler-Lagrange field equation for $\Phi(t, \mathbf{x})$ determined from the Lagrangian density of Equation 10.33 is Poisson's equation,*

$$\nabla^2 \Phi(t, \mathbf{x}) = 4\pi G \sum_i m_i \delta^3(\mathbf{x} - \mathbf{z}_i) \ \left(= 4\pi G \rho(t, \mathbf{x})\right) \, . \quad (10.35)$$

One of the striking features of the Lagrangian treatment of gravitation mediated by a field is the extent to which the correct result can be obtained by choosing the simplest field, a non-propagating scalar, and the simplest interaction term consisting of no more than the scalar field itself evaluated on the trajectory. Simplicity in this case leads to the inverse square law in a more direct and convincing manner than is possible without the use of a mediating field. An alteration of the sign of the third term alters the sign of the field generated by a given mass. Since a compensating sign change also appears in the equation of motion of a particle, the interaction will still be attractive.

Nevertheless, in the non-dynamic context under consideration, the sign of the free field term is arbitrary and no general field theoretic principles, such as energy positivity of free particle and field, are available to determine the interaction to be attractive. It is only in a relativistic context that one may appeal to positivity of the energy to conclude that the interaction must be attractive. However, by dividing the interaction into the effect of matter in producing the field and action of the field on the matter, the local nature of the field equations select the long-range inverse square law as a very natural form of interaction in 3D. This is also evident by direct use of Gauss's theorem.

The potential equations of electrostatics and magnetostatics,

$$\nabla^2 \phi = -\rho/\epsilon_0 \quad \text{and} \quad \nabla^2 \mathbf{A} - \nabla(\nabla \cdot \mathbf{A}) = -\mu_0 \mathbf{J} , \qquad (10.36)$$

may be handled with Lagrangians in almost exactly the same way as Poisson's equation for the gravitational potential and we need not examine them in detail here. Particles interacting via static ϕ or $\mathbf{A}$ fields may be described by Lagrangians,

$$L = \tfrac{1}{2} m \dot{z}^2 - q\phi \quad \text{and} \quad L = \tfrac{1}{2} m \dot{z}^2 + q\dot{z} \cdot \mathbf{A} . \qquad (10.37)$$

Energy will not be conserved for such systems unless the contributions from the fields are included which is possible only with dynamic fields ϕ and $\mathbf{A}$. Such fields require a fully relativistic treatment (see Chapters 18 and 19) and, for consistency, the particles require a similar analysis to be given in Chapter 15.

Entia non sunt multiplicanda praeter necessitatem.
It is vain to do with more what can be done with less[2].

Occam's razor

10.6 Schrödinger Lagrangian

The probability *wave function* describing a non-relativistic spinless particle of mass m is a single complex function $\psi = \psi(t, \mathbf{x})$ of t and $\mathbf{x}$ which satisfies the *time-dependent Schrödinger equation* [375, 498],

$$\left(-\frac{\hbar^2}{2m} \nabla^2 + V(t, \mathbf{x}) \right) \psi(t, \mathbf{x}) = i\hbar \, \dot{\psi}(t, \mathbf{x}) , \qquad (10.38)$$

in which $V(t, \mathbf{x})$ is a real potential. All terms of the equation contain the wave function ψ as a factor — there is no source term for ψ. Rather, ψ itself acts as a source for other fields describing interactions rather than as a mediating field.

The Schrödinger equation may be considered to arise from the representation theory of the Galilei and Schrödinger groups. It may be obtained more directly from the Galilean covariant Equation 6.23 using an *ad hoc* method of construction, or *quantization ansatz*. The real classical momentum and energy variables $\mathbf{p}$ and E are replaced, in the expression $H(\mathbf{x}, \mathbf{p})_{\text{physical}} = E$, by hermitian differential operators,

$$\mathbf{P} = \frac{\hbar}{i} \nabla \quad \text{and} \quad \hat{E} = i\hbar \, \partial_t , \qquad (10.39)$$

---

[2] Attributed to William of Occam, or Ockham, probably Oakham in Surrey (ca. 1300– ca. 1349), Oxford scholar in the Order of Fransiscan Friars. Occam's razor is widely used in scientific analysis with an interpretation akin to: *One should always choose the simpler of two otherwise equivalent competing descriptions of physical phenomena.*

10.6. Schrödinger Lagrangian

Erwin Schrödinger

b. Vienna, Austria
12 August 1887

d. Alpach, Austria
4 January 1961

Courtesy of Dunsink Observatory, Dublin.

acting on ψ to form the equation $H\psi = \hat{E}\psi$ from which energy eigenstates and eigenvalues may be obtained. The real variables $\mathbf{x}$ and $V(t,\mathbf{x})$ are replaced by hermitian operators $\mathbf{X}$ and $V(t,\mathbf{X})$. In the position representation, these give rise to multiplication of ψ by the values of $\mathbf{x}$ and $V(t,\mathbf{x})$. The essence of the quantization process is contained — first, in the eigenvalue equation which leads to discrete values of the energy, and — second, in the presence in Equations 10.39 of the constant $\hbar$ relating the group theoretical translation generators $\frac{1}{i}\nabla$ and $i\partial_t$ of the Galilean symmetry of $\psi(t,\mathbf{x})$ to conserved mechanical quantities $\mathbf{p}$ and E.

We may treat the Schrödinger wave function as a Euclidean scalar field, although it is not a Galilean invariant. Nevertheless, the Schrödinger equation of a closed system, whether a single free particle $(V=0)$ or a closed system of particles which are mutually interacting, will be Galilean covariant. The same equation with 2-component wave functions, referred to as the *Pauli-Schrödinger equation* [75], may be used to non-relativistically describe the wave mechanics of spin $\frac{1}{2}$ particles. Allowing the wave function and potential to be a function of a number of spatial coordinates $\{\mathbf{x}_1, \mathbf{x}_2, \ldots, \}$, on which there is appropriate symmetrization, provides a description of the wave-mechanical interaction of a number of non-relativistic particles.

It is not difficult to construct a Lagrangian, $\mathcal{L} = \mathcal{L}(\psi, \dot{\psi}, \nabla\psi, t, \mathbf{x})$, whose Euler-Lagrange equation is the Schrödinger equation. Since the wave function is complex, we may consider variations with respect to the real and imaginary parts of ψ. Alternatively, we may consider independent variations with respect to ψ and its hermitian conjugate $\psi^\dagger$ (which for one component is simply the complex conjugate, ψ^*), knowing that the two separate Euler-Lagrange equations will be the conjugates of one another, since the Lagrangian must be hermitian. We shall use the second alternative. Since the equation is linear, each term will be quadratic in ψ and its derivatives. It must be first degree in $\dot{\psi}$ and second degree in $\nabla\psi$.

We may therefore consider a Lagrangian density of the form,

$$\mathcal{L} = \mathcal{L}(\psi, \dot{\psi}, \nabla\psi, t, \mathbf{x}) = a\nabla\psi^\dagger \cdot \nabla\psi + b_1 \psi^\dagger \dot{\psi} + b_2 \dot{\psi}^\dagger \psi + c\psi^\dagger V \psi \ . \quad (10.40)$$

Exercise 10.5 (a) *Show that $\nabla\psi^\dagger \cdot \nabla\psi$ and $\psi^\dagger V\psi$ are hermitian and that, provided b_1 is pure imaginary and $b_2 = -b_1$, then $b_1\psi^\dagger\dot{\psi} + b_2\dot{\psi}^\dagger\psi$ is also hermitian. (b) Evaluate a, b_1 and c by requiring the Euler-Lagrange equations of the Lagrangian in Equation 10.40 to lead to the Schrödinger equation and its conjugate, and hence show that, up to a constant multiple,*

$$\mathcal{L} = -\frac{\hbar^2}{2m}\nabla\psi^\dagger \cdot \nabla\psi + \tfrac{1}{2}i\hbar\psi^\dagger \overleftrightarrow{\partial_t}\psi - \psi^\dagger V \psi \ , \quad (10.41)$$

where $\psi^\dagger \overleftrightarrow{\partial_t}\psi = \psi^\dagger\dot{\psi} - \dot{\psi}^\dagger\psi$. The choice of signs may be shown to correspond to positive energy (see Exercise 10.11 (c)) provided $V \geq 0$. (c) Show that

$$\frac{\partial\mathcal{L}}{\partial\psi} = -\tfrac{1}{2}i\hbar\dot{\psi}^\dagger - \psi^\dagger V, \quad \frac{\partial\mathcal{L}}{\partial\dot{\psi}} = \tfrac{1}{2}i\hbar\psi^\dagger \text{ and } \frac{\partial\mathcal{L}}{\partial\nabla\psi} = -\frac{\hbar^2}{2m}\nabla\psi^\dagger \ . \quad (10.42)$$

The conjugate relations follow directly from these. (d) Show that the Lagrangian has standard dimensions provided the wave function has dimensions $[\psi] = L^{-3/2}$ where L is length.

The last term of the Schrödinger Lagrangian is the interaction term. The first two terms are the wave function or field equivalent, of the kinetic energy of a mechanical system. We find that the free terms for the wave function Lagrangian depend not only on the time derivatives of ψ but also on their gradient. The existence of the time derivative or *dynamic term* guarantees that the wave function will be propagating rather than static.

Exercise 10.6 (a) *Show that the Schrödinger Lagrangian is non-singular. (b) Show that parity $\mathbf{P}$ is an invariance operation on the Schrödinger equation, where $\mathbf{P}\psi(t,\mathbf{x}) = \psi(t,-\mathbf{x})$ and that the same is true of time reversal $\mathbf{T}$ provided $\mathbf{T}\psi(t,\mathbf{x}) = \psi^*(-t,\mathbf{x})$ and $V(-t,\mathbf{x}) = V(t,\mathbf{x})$. Charge conjugation is not relevant to the non-relativistic Schrödinger equation.*

In natural units, where $\hbar = 1$ and the dimensions of the Lagrangian density are $L^{-3}T^{-1}$, the only factors in the dynamic term of the Lagrangian are the wave function and its time derivative. This is a feature that is used to define a natural or *canonical dimension* for a wave function or field. For wave functions or fields whose equations are first-order in time derivatives, the canonical dimensions will be $L^{-3/2}$. Relativistic wave equations for particles of half-odd-integer spin also fall into this category.

The Galilean covariance properties do not, of course, suffice to establish the wave-mechanical nature of non-relativistic particles for use as a first approximation to the dynamics of an electron in the hydrogen atom. For that one needs a physical interpretation of the wave function in terms of particles which was supplied in 1927 by the probability interpretation of Born, Heisenberg and Bohr, which as a result of the brilliant defence of those ideas by Bohr over a long period, became known as the *Copenhagen interpretation* [392] of quantum mechanics.

10.7 Invariant and covariant field Lagrangians

In extending the examination of the symmetries of a mechanical system to those of a Galilean field, we shall encounter a certain degree of complexity in the equations. These have two origins. One is the introduction of the field concept itself involving distribution over space as well as evolution in time. The other related factor is a consequence of the fact that in Newtonian physics, the evolution parameter t and the distribution variables $\mathbf{x}$ do not appear symmetrically in the Galilean transformation equations. This means it could be misleading to attempt to collect them together in one variable as we shall see is natural in the case of Einsteinian relativity.

Despite the apparent complexity of the general symmetry formulae for Galilean fields, we shall discuss the case in sufficient detail to permit a reader to derive all the equations on the grounds that understanding the field concepts in a familiar non-relativistic context will facilitate the separate introduction of relativistic effects later. The corresponding discussion for relativistic fields, to be given in Chapter 16, will by comparison be very much more compact, partly for notational reasons and partly owing to the intrinsically closer relationship between time and space.

We now consider point transformations of a field (or wave function) system involving changes,

$$\{t, \mathbf{x}\} \rightarrow \{\bar{t}, \bar{\mathbf{x}}\} = \{\bar{t}(t), \bar{\mathbf{x}}(t, \mathbf{x})\} \qquad (10.43)$$
$$\psi(t, \mathbf{x}) \rightarrow \bar{\psi}(\bar{t}, \bar{\mathbf{x}}) = \bar{\psi}(\psi(t, \mathbf{x}), t, \mathbf{x}) , \qquad (10.44)$$

of the dependent and independent variables. Use of the numerical invariance of the action, $\bar{S}[\bar{\psi}] = S[\psi]$, shows that the new Lagrangian density, as a function of the new variables, is given in terms of the old by,

$$\bar{\mathcal{L}}(\bar{\psi},\ldots) = \frac{\partial t}{\partial \bar{t}} \left|\frac{\partial \mathbf{x}}{\partial \bar{\mathbf{x}}}\right| \mathcal{L}(\psi,\ldots) , \qquad (10.45)$$

while the *Euler derivatives* with respect to a component ψ_j of ψ are related according to,

$$\bar{\mathcal{L}}^j(\bar{\psi},\ldots) = \frac{\partial t}{\partial \bar{t}} \left|\frac{\partial \mathbf{x}}{\partial \bar{\mathbf{x}}}\right| \left|\frac{\partial \psi}{\partial \bar{\psi}}\right| \mathcal{L}^j(\psi,\ldots) . \qquad (10.46)$$

An *invariant Lagrangian density* with respect to such a transformation is, as before, one for which,

$$\bar{\mathcal{L}}(\psi,\ldots) = \mathcal{L}(\psi,\ldots) . \qquad (10.47)$$

Exercise 10.7 (a) Show that

$$\partial_t(\Delta t) + \boldsymbol{\nabla} \cdot (\Delta \mathbf{x}) + \Delta \mathcal{L} = 0 , \qquad (10.48)$$

where,

$$\begin{aligned}
\Delta \mathcal{L} &= \frac{d\mathcal{L}}{dt}\Delta t + \frac{d\mathcal{L}}{d\mathbf{x}}\cdot \Delta \mathbf{x} \\
&= \left\{\frac{\partial \mathcal{L}}{\partial \psi}\dot{\psi} + \frac{d}{dt}\left(\frac{\partial \mathcal{L}}{\partial \dot{\psi}}\right)\ddot{\psi} + \partial_t \mathcal{L}\right\}\Delta t \\
&\quad + \left\{\frac{\partial \mathcal{L}}{\partial \psi}\boldsymbol{\nabla}\psi + \frac{d}{dt}\left(\frac{\partial \mathcal{L}}{\partial \dot{\psi}}\right)\boldsymbol{\nabla}\dot{\psi} + \boldsymbol{\nabla}\mathcal{L}\right\}\cdot \Delta \mathbf{x} ,
\end{aligned} \qquad (10.49)$$

is the condition for a Newtonian field Lagrangian density to be invariant with respect to a continuous transformation, whose infinitesimal changes are Δt, $\Delta \mathbf{x}$ and $\Delta \psi$. (b) Verify directly, or using the equations of part (a), that the Schrödinger Lagrangian of Equation 10.41 for a closed system of interacting wave-mechanical particles is invariant with respect to space translations but not with respect to boosts.

A *covariant Lagrangian density* is one which is either invariant or whose transformed Lagrangian density differs from the original Lagrangian (in both of which the arguments are the same) by a time derivative of a scalar $\Lambda^0(\psi,t,\mathbf{x})$ plus the 3-divergence of a vector $\boldsymbol{\Lambda}(\psi,t,\mathbf{x})$, each functions of the independent variables $\{t,\mathbf{x}\}$ and the generalized coordinates $\psi(t,\mathbf{x})$ (but not $\dot{\psi}$ or $\boldsymbol{\nabla}\psi$), namely,

$$\bar{\mathcal{L}}(\psi,\dot{\psi},\boldsymbol{\nabla}\psi,t,\mathbf{x}) = \mathcal{L}(\psi,\dot{\psi},\boldsymbol{\nabla}\psi,t,\mathbf{x}) + \frac{d\Lambda^0}{dt} + \boldsymbol{\nabla}\cdot\boldsymbol{\Lambda} . \qquad (10.50)$$

If the continuous transformation is infinitesimal, then this equation still holds with $\{\Lambda^0,\boldsymbol{\Lambda}\}$ replaced by infinitesmal functions $\{\Delta\Lambda^0,\Delta\boldsymbol{\Lambda}\}$.

Exercise 10.8 (a) Show that a Lagrangian density which is covariant with respect to a continuous transformation whose infinitesimal changes are $\{\Delta t(t), \Delta \mathbf{x}(t, \mathbf{x})\}$ and $\Delta \psi(t, \mathbf{x})$, satisfies,

$$\partial_t(\Delta t) + \nabla \cdot (\Delta \mathbf{x}) + \Delta \mathcal{L} = -\frac{d(\Delta \Lambda^0)}{dt} - \nabla \cdot (\Delta \Lambda) , \quad (10.51)$$

for some infinitesimal functions $\Delta \Lambda^0$ and $\Delta \Lambda$ of ψ, t and $\mathbf{x}$. (b) Show that the left-side of Equation 10.51 can, for a boost transformation of the Lagrangian of Equation 10.41, be re-expressed in such a form that we can conclude that the Lagrangian is boost covariant.

10.8 Noether's theorem for Newtonian fields

In order to establish Noether's theorem for Newtonian fields, we require first the variation of the action under the arbitrary general variations,

$$\{t, \mathbf{x}\} \rightarrow \{t + \delta t(t), \mathbf{x} + \delta \mathbf{x}(\mathbf{x})\} \quad (10.52)$$
$$\psi(t, \mathbf{x}) \rightarrow \psi(t, \mathbf{x}) + \delta \psi(\psi(t, \mathbf{x}), t, \mathbf{x}) . \quad (10.53)$$

The integration intervals vary by,

$$\delta(dt) = dt \frac{d(\delta t)}{dt} \quad \text{and} \quad \delta(d^3 x) = d^3 x \, \nabla \cdot (\delta \mathbf{x}) = d^3 x \frac{d}{d\mathbf{x}} \cdot (\delta \mathbf{x}) . \quad (10.54)$$

We deduce,

$$\delta \psi = \delta_0 \psi + \delta t \, \dot\psi + \delta \mathbf{x} \cdot \nabla \psi \quad (10.55)$$
$$\delta \dot\psi = \delta_0 \dot\psi + \delta t \, \ddot\psi + \delta \mathbf{x} \cdot \nabla \dot\psi \quad (10.56)$$
$$\delta \nabla \psi = \delta_0 \nabla \psi + \delta t \, \nabla \dot\psi + \delta \mathbf{x} \cdot \nabla \nabla \psi . \quad (10.57)$$

From,

$$\delta S = \int d\bar t \, d^3 \bar x \, \mathcal{L}(\bar\psi, \dot{\bar\psi}, \bar\nabla \bar\psi, \bar t, \bar{\mathbf{x}}) - \int dt \, d^3 x \, \mathcal{L}(\psi, \dot\psi, \nabla\psi, t, \mathbf{x}) , \quad (10.58)$$

we obtain,

$$\begin{aligned}
\delta S &= \int \Big[\{ dt \, d^3 x + (\delta \, dt) d^3 x + (\delta \, d^3 x) dt \} (\mathcal{L} + \delta \mathcal{L}) - \mathcal{L} \Big] \\
&= \int dt \, d^3 x \left[\left\{ \frac{d(\delta t)}{dt} + \frac{d}{d\mathbf{x}} \cdot (\delta \mathbf{x}) \right\} \mathcal{L} + \delta t \frac{d\mathcal{L}}{dt} + \delta \mathbf{x} \cdot \frac{d\mathcal{L}}{d\mathbf{x}} + \delta_0 \mathcal{L} \right] \\
&= \int dt \, d^3 x \left[\frac{d}{dt} \left\{ \delta t \, \mathcal{L} + \frac{\partial \mathcal{L}}{\partial \dot\psi} \delta_0 \psi \right\} + \frac{d}{dx^k} \left\{ \delta x^k \, \mathcal{L} + \frac{\partial \mathcal{L}}{\partial [\partial_k \psi]} \delta_0 \psi \right\} \right. \\
&\quad \left. + \left(\frac{\partial \mathcal{L}}{\partial \psi} - \frac{d}{dt} \frac{\partial \mathcal{L}}{\partial \dot\psi} - \frac{d}{d\mathbf{x}} \cdot \frac{\partial \mathcal{L}}{\partial \nabla \psi} \right) \left(\delta \psi - \delta t \, \dot\psi - \delta \mathbf{x} \cdot \nabla \psi \right) \right] \quad (10.59)
\end{aligned}$$

after the usual integrations by parts and switching between δ_0 and δ.

We may apply Equation 10.59 to obtain ΔS for the special changes Δt, $\Delta \mathbf{x}$ and $\Delta \psi$ that comprise a small continuous symmetry operation. For the latter, we have an alternative expression for ΔS in terms of the derivatives of $\Delta \Lambda^0$ and $\Delta \boldsymbol{\Lambda}$ in the most general case of a covariant, but not necessarily invariant, Lagrangian density and action. Subtracting the two expressions to obtain a vanishing integral over an arbitrary region, we equate the integrand to zero. The result may be expressed as,

$$\frac{d\rho_a}{dt} + \frac{d}{d\mathbf{x}} \cdot \mathbf{j}_a + \left(\frac{\partial \mathcal{L}}{\partial \psi} - \frac{d}{dt} \frac{\partial \mathcal{L}}{\partial \dot{\psi}} - \frac{d}{d\mathbf{x}} \cdot \frac{\partial \mathcal{L}}{\partial [\nabla \psi]} \right) \times$$
$$\times \left(\delta\psi - \delta t\, \dot{\psi} - \delta\mathbf{x} \cdot \nabla \psi \right) = 0 \ . \quad (10.60)$$

The *Noether density* ρ_a and *Noether 3-current* $\mathbf{j}_a$, corresponding to one of the p independent parameters ω^a, are given by,

$$\rho_a = \mathcal{H} \frac{\partial \Delta t}{\partial \Delta \omega^a} + \boldsymbol{\pi}^j \nabla \psi_j \cdot \frac{\partial \Delta \mathbf{x}}{\partial \Delta \omega^a} - \pi^j \frac{\partial \Delta \psi_j}{\partial \Delta \omega^a} - \frac{\partial (\Delta \Lambda^0)}{\partial \Delta \omega^a} \ , \quad (10.61)$$

and,

$$j_a^k = (\pi^j)^k \dot{\psi}_j \frac{\partial \Delta t}{\partial \Delta \omega^a} - t^k{}_l \frac{\partial \Delta x^l}{\partial \Delta \omega^a} - (\pi^j)^k \frac{\partial \Delta \psi_j}{\partial \Delta \omega^a} - \frac{\partial (\Delta \Lambda^k)}{\partial \Delta \omega^a} \ , \quad (10.62)$$

where

$$\mathcal{H} = \frac{\partial \mathcal{L}}{\partial \dot{\psi}_j} \dot{\psi}_j - \mathcal{L} \quad \text{and} \quad \pi^j = \frac{\partial \mathcal{L}}{\partial \dot{\psi}_j} \ , \quad (10.63)$$

are the Hamiltonian density and the canonical momenta. Similarly,

$$\boldsymbol{\pi}^j = \{(\pi^j)^k\} = \frac{\partial \mathcal{L}}{\partial [\nabla \psi_j]} \quad \text{and} \quad t^{kl} = -\frac{\partial \mathcal{L}}{\partial [\partial_k \psi_j]} \partial^l \psi_j + \delta^{kl} \mathcal{L} \ , \quad (10.64)$$

are spatial analogues of the canonical momenta and the *stress tensor* of the field system. The latter may be compared with Equation 6.28 for the mechanical stress tensor.

On the stationary path, Equation 10.60 becomes a *continuity equation*,

$$\frac{d\rho_a}{dt} + \frac{d}{d\mathbf{x}} \cdot \mathbf{j}_a = 0 \ , \quad (10.65)$$

relating the Noether density and current, which when integrated over a 3-volume surrounding the entire system, presumed to be finite in extent, leads to,

$$\frac{dQ_a}{dt} = -\int_R d^3x \, \mathrm{div} \cdot \mathbf{j}_a = -\oint_{\partial R} d\mathbf{S} \cdot \mathbf{j}_a \ , \quad (10.66)$$

10.8. Noether's theorem for Newtonian fields

showing that the *Noether charge* Q_a of the field system is conserved,

$$Q_a = \int_R d^3x \, \rho_a(t, \mathbf{x}) = \text{constant} , \qquad (10.67)$$

in the usual case where all fields fall off sufficiently rapidly at large spatial distances.

Some subtle changes have occurred in the expressions for the Noether quantities in the passage from a mechanical to a field system. The term containing $\pi^j \Delta q_j$ in the mechanical Noether charge of Equation 9.31 is the one which, under spatial translation, led to the conservation of momentum. That term has become proportional to the integral of $-\pi^j \Delta \psi_j$ in the Noether charge of the field system. Since the dependent variable ψ is in general unaffected by purely spacetime transformations, we see that this term contributes as a result of internal (non-spacetime) symmetry transformations. The fact that a field is also distributed over space has meant the appearance of a new term, proportional to $\pi^j \nabla \psi_j \cdot \Delta \mathbf{x}$, which takes over the rôle of ensuring that the momentum of the field system is conserved as a result of spatial translation invariance.

Consider the case of a Lagrangian system whose action is invariant ($\Lambda^0 = 0$ and $\Lambda = 0$) with respect to time translation. In that case, the corresponding conserved Noether density, current and charge are the Hamiltonian density $\mathcal{H}$, the energy flux $\pi^j \dot\psi_j$ and the Hamiltonian H of the system. For the case of a Lagrangian which is invariant under spatial translations, the Noether density, current and charge are $\pi^j \nabla \psi_j$, the k-momentum flux in the l-direction (the stress tensor) and the conserved generalized momentum of the system, which if rescaled by a constant factor in order to re-express it in terms of the hermitian operator $\frac{1}{i} \nabla$, is

$$\mathbf{p} = \frac{1}{i} \int d^3x \, \frac{\partial \mathcal{L}}{\partial \dot\psi_j} \nabla \psi_j = \frac{1}{i} \int d^3x \, \pi^j \nabla \psi_j . \qquad (10.68)$$

The third terms of Equation 10.61 and 10.62 will contribute internal charges as we shall see shortly with examples while the fourth term takes into account Lagrangians which are covariant but not invariant for a particular symmetry.

Exercise 10.9 (a) *Show that the orbital angular momentum Noether charge of a single component field ψ, corresponding to rotational symmetry, apart from a normalization factor, is given by,*

$$\mathbf{l} = \frac{1}{i} \int d^3x \, \frac{\partial \mathcal{L}}{\partial \dot\psi} \mathbf{x} \times \nabla \psi = \frac{1}{i} \int d^3x \, \pi \mathbf{x} \times \nabla \psi . \qquad (10.69)$$

with dual form,

$$l^{kl} = \frac{1}{i} \int d^3x \, \pi (x^k \partial^l - x^l \partial^k) \psi . \qquad (10.70)$$

and *(b) derive the corresponding result for boosts.*

10.9 U(1) phase invariance — Schrödinger charge

We can illustrate the above material on the Noether densities, currents and charges using the Schrödinger Lagrangian, which is invariant with respect to a global change of phase of the wave function,

$$\psi(t,\mathbf{x}) \to e^{i\alpha}\psi(t,\mathbf{x}) , \qquad (10.71)$$

where the term *global* applied to this symmetry implies that the phase α does not depend on t or $\mathbf{x}$.

The transformation involves only the field ψ and the invariance of the Lagrangian, action and wave equation is therefore an *internal symmetry*. One phase transformation followed by another in either order is a third phase transformation and it is immediately clear that all such changes form a 1-parameter Abelian group. Each member of the group is obtainable from the identity ($\alpha = 0$) by continuous variation of the parameter in the range $0 \leq \alpha \leq 2\pi$, a property of a Lie group. Since the phase factor $U = e^{i\alpha}$ is a 1×1 *unitary* matrix, $U^\dagger U = 1$, the global phase transformations are referred to as the *1-dimensional unitary group* and denoted U(1). This is one of only 2 one-parameter groups, the other being the translation group in one dimension. Since the range of α is compact, U(1) is said to be a *compact group*. The transformation may be split into its real and imaginary parts,

$$\mathrm{Re}\,\psi \to \mathrm{Re}\,\psi\,\cos\alpha - \mathrm{Im}\,\psi\,\sin\alpha \qquad (10.72)$$
$$\mathrm{Im}\,\psi \to \mathrm{Re}\,\psi\,\cos\alpha + \mathrm{Im}\,\psi\,\sin\alpha , \qquad (10.73)$$

which form illustrates that it is isomorphic to a transformation by a 2×2 real matrix O acting on the 2-dimensional column vectors,

$$\psi = \begin{pmatrix} \mathrm{Re}\,\psi(t,\mathbf{x}) \\ \mathrm{Im}\,\psi(t,\mathbf{x}) \end{pmatrix} . \qquad (10.74)$$

Since the matrices O are all orthogonal, $O^\mathrm{T} O = \mathbb{1}$ and special (S), namely of unit determinant, such transformations are referred to as *special orthogonal*, and being 2-dimensional, the matrix group is denoted SO(2). Since there is a one-to-one correspondence between their members, the two groups U(1) and SO(2) are said to be *isomorphic*, written U(1) ≈ SO(2).

From Equation 10.71, a small U(1) transformation of ψ will have the form,

$$\Delta\psi = i\Delta\alpha\,\psi \quad \text{and} \quad \Delta\psi^\dagger = -i\Delta\alpha\,\psi^\dagger . \qquad (10.75)$$

10.9. U(1) phase invariance — Schrödinger charge

The Lagrangian is invariant ($\Delta\Lambda^0 = 0$ and $\Delta\Lambda = 0$) and the Noether density for a purely internal transformation ($\Delta t = 0$ and $\Delta \mathbf{x} = 0$) of a complex variable is obtained from Equation 10.61 (rescaled by a factor of $\hbar$) as,

$$\rho = -\hbar^{-1}\left(\frac{\partial \mathcal{L}}{\partial \dot{\psi}}\frac{\partial \Delta \psi}{\partial \Delta \omega^a} + \frac{\partial \Delta \psi^\dagger}{\partial \Delta \omega^a}\frac{\partial \mathcal{L}}{\partial \dot{\psi}^\dagger}\right), \qquad (10.76)$$

from which it follows that,

$$\rho = \psi^\dagger \psi, \qquad (10.77)$$

the *Schrödinger probability density*. Noether's theorem states that it must combine with the 3-current density in a continuity equation, and its integral over space must be conserved, as consequences of the internal global U(1) phase invariance of the Schrödinger equation.

Exercise 10.10 *Show that the rescaled U(1) Noether 3-current of the Schrödinger equation is given by the Schrödinger probability current,*

$$\mathbf{j} = \tfrac{\hbar}{2mi}\psi^\dagger \overset{\leftrightarrow}{\nabla} \psi, \qquad (10.78)$$

and verify that $\dot{\rho} + \nabla \cdot \mathbf{j} = 0$.

Exercise 10.11 *(a) Use Equation 10.48 to show that the Schrödinger Lagrangian is space or time translation invariant for one particle only if the potential does not depend explicitly on either t or $\mathbf{x}$ and that it is invariant for a number of particles provided V is independent of t and depends on $\mathbf{x}_i$ only via the differences, $\mathbf{x}_i - \mathbf{x}_j$. (b) Show that the Noether charge corresponding to a space translation invariant Schrödinger Lagrangian is given by the conserved momentum,*

$$\langle \mathbf{p} \rangle = \tfrac{\hbar}{i}\int d^3x\, \psi^\dagger \nabla \psi. \qquad (10.79)$$

(c) Determine the 'stress tensor' t^{kl} of probability fluxes in the direction $\mathbf{e}_k$ across an area normal to the direction $\mathbf{e}_l$. (d) Show that the Hamiltonian for the Schrödinger Lagrangian, namely the time translation Noether charge, is given by,

$$H = \int d^3x\, \mathcal{H} = -\tfrac{\hbar^2}{2m}\int d^3x\, \nabla\psi^\dagger \cdot \nabla \psi + \int d^3x\, \psi^\dagger V \psi. \qquad (10.80)$$

Interpret the first term of the Hamiltonian and show that the sign of the Lagrangian has been chosen so that energy positivity is satisfied in the free case ($V = 0$).

Exercise 10.12 *Examine the effects of form invariance of the Schrödinger equation, 10.38, for a transformation $\psi(t,\mathbf{x}) \to \alpha(t,\mathbf{x})\psi(t,\mathbf{x}) + \beta(t,\mathbf{x})$, to show that the equation has no local gauge freedom in $\psi(t,\mathbf{x})$.*

Exercise 10.13 *Show that Equations 7.12 for the Newtonian gravitational field strength* **f** *are not derivable from a Lagrangian density in which* **f** *is the Lagrangian field to be varied in accordance with Hamilton's principle.*

Bibliography:

Aitchison I J R and Hey A J G 1982 *Gauge theories in particle physics* [7].
Barut A O 1964 *Electrodynamics and the classical theory of fields* [29].
Landau L D and Lifschitz E M 1980 *The classical theory of fields* [306].
Ramond P 1981 *Field theory: a modern primer* [467].
Soper D E 1975 *Classical field theory* [516].

CHAPTER 11

Hamiltonian dynamics

What I have just said throws some light at the same time on the rôle of general principles, such as those of the principle of least action or the conservation of energy .... They represent the quintessence of innumerable observations. However, from their very generality results a consequence ... namely, that they are no longer capable of verification[1].

Henri Poincaré   1854–1912

11.1 Introduction

The Lagrangian formulation of the dynamics of mechanical or field systems is ideal for the incorporation of symmetries and hence for ensuring the manifest covariance of the equations of motion or field equations. It greatly facilitates the determination of the conserved Noether quantities corresponding to each of the continuous symmetries of the system. Given two free systems, it is also very useful for the construction of mutual interactions consistent with any overall symmetry principles such as Galilean or Poincaré covariance. The existence of a classical Lagrangian obeying a stationary action principle is necessary for the application of the path-integral quantization procedure [488].

However, prior to the development of path-integral techniques, the Lagrangian formulation was not the most suitable to use in seeking the quantized form of a given classical system, which was traditionally based on *canonical quantization*, namely the use of Hamiltonian or canonical dynamics. The term *canonical* is applied in this context to properties which are natural or intrinsic as a result of being independent of the particular dynamical variables used. We shall re-interpret the configuration space Hamiltonian function that appears naturally in the Lagrangian formulation and in Hamilton's modified principle of varied action, to make a transition

---

[1] *Science and hypothesis* [445].

William Rowan Hamilton

b. Dublin, Ireland
4 August 1805

d. Dunsink Observatory, Dublin
2 September 1865

Courtesy of Dunsink Observatory, Dublin.

$\mathbf{i}^2 = \mathbf{j}^2 = \mathbf{k}^2 = \mathbf{ijk} = -1.$

to an alternative equivalent procedure which is more closely related to the commutator algebras of quantum observables. It is common practice to make use of both formulations of the dynamics of a given system, transforming from one to the other as convenient.

11.2 Phase space Hamiltonian

In Hamilton's modified action principle of the Lagrangian formulation of dynamics, two functions of the coordinates and velocities, appear naturally — the generalized momenta $\pi^j = \pi^j(q, \dot{q}, t)$ and the configuration space Hamiltonian function, $H = H(q, \dot{q}, t)$. These are the coefficients of the arbitrary general variations δq and $-\delta t$ in the end-point contributions to the variation of the action. From Equation 9.21 the latter is,

$$\delta S = \left[\pi^j \delta q_j - H \delta t\right]_{t_i}^{t_f} + \int_{t_i}^{t_f} \left(\frac{\partial L}{\partial q_j} - \frac{d}{dt}\frac{\partial L}{\partial \dot{q}_j}\right)(\delta q_j - \dot{q}_j \delta t), \qquad (11.1)$$

where,

$$H = \sum_j \pi^j \dot{q}_j - L \quad \text{and} \quad \pi^j = \frac{\partial L}{\partial \dot{q}_j}. \qquad (11.2)$$

11.2. Phase space Hamiltonian

These definitions, based on the variation of the action, are consistent with those which form the basis of Newtonian mechanics but apply to any non-singular mechanical system, Newtonian or not. Since the Hamiltonian appears as the coefficient of $-\delta t$, it is said to *generate* time translations. Correspondingly, the generalized momentum π^j is said to generate displacements in the coordinate q_j.

Given the primary importance of the temporal evolution of a system, and of its distribution in space, it is clear that the Hamiltonian function and the generalized momenta are intrinsically important dynamical variables of a system. One could therefore consider the possibility of using the definition of the generalized momentum in Equation 11.2 to determine the generalized velocities $\dot q$ in terms of the coordinates q and momenta π (and possibly t) and thereby eliminate velocities in favour of momenta. If the system is second-order (not all $\partial^2 L/\partial \dot q_j \partial \dot q_k$ are zero), the necessary and sufficient condition for the second of Equations 11.2 to be solvable uniquely for $\dot q_j$ in terms of q_j and π_j (and possibly t) is the non-vanishing of the determinant of the square matrix $\partial \pi^j/\partial \dot q_k$. But this matrix is simply the Hessian matrix $\partial^2 L/\partial \dot q_j \partial \dot q_k$, and the condition of invertibility of the second of Equations 11.2 is precisely Equation 8.55. The latter defines a non-singular Lagrangian guaranteeing that the Euler-Lagrange equations are all unconstrained dynamical equations each giving one acceleration in terms of coordinates and velocities (and possibly the time explicitly).

Consequently, for such *non-singular Lagrangians*, the process of elimination of velocities in favour of generalized momenta is straightforward. If the Lagrangian is constrained and therefore singular, the elimination cannot be performed and special procedures are required to make a transition to the Hamiltonian formalism. These special procedures are intimately linked to the phenomenon of local gauge freedom. We shall confine attention for the moment solely to the non-singular case.

If we carry out the above elimination, the Lagrangian may then be expressed entirely as a function $L = L(q, \pi, t)$ of q, π (and possibly t) and therefore so also may the Hamiltonian of the first of Equations 11.2, namely,

$$H = H(q, \pi, t) = \sum_j \pi^j \dot q_j(q, \pi, t) - L\big(q, \dot q(q, \pi, t)\big) \ . \tag{11.3}$$

All the physical information about a non-singular Lagrangian system is contained in the functional form of the Lagrangian and will therefore clearly also be contained in the functional form of the Hamiltonian.

The crucial step taken by Hamilton was to make a transition from the Lagrangian function of coordinates in configuration space (and velocities dependent on the coordinates) to a *characteristic function*, now called a Hamiltonian, of independent coordinates and momenta whose values define points in a new dynamical space called *phase space*. This was first done in his 1834 paper: *On a general method in dynamics* [222]. The doubling in the number of independent variables is precisely compensated by a change from second-order dynamical equations to first-order ones, Hamilton's equations. The *phase* of a Hamiltonian system at a given time is the pair of values $\{q, \pi\}$. The transformation from a Lagrangian with coordinates q and $\dot{q}$ to a Hamiltonian with coordinates q and π, according to Equation 11.3, is called a *Legendre transformation*[2].

11.3 Hamilton's equations

To determine Hamilton's equations, as a consequence of this re-formulation, we may compare the total differential of $H(q, \pi, t)$,

$$dH = \frac{\partial H}{\partial q_j}dq_j + \frac{\partial H}{\partial \pi^j}d\pi^j + \frac{\partial H}{\partial t}, \qquad (11.4)$$

which follows from the requirement that q and π be independent, with the same differential obtained from its relation to L in Equation 11.2, namely,

$$\begin{aligned} dH &= d\left(\frac{\partial L}{\partial \dot{q}_j}\dot{q}_j\right) - \frac{\partial L}{\partial q_j}dq_j - \frac{\partial L}{\partial \dot{q}_j}d\dot{q}_j - \frac{\partial L}{\partial t} \\ &= \dot{q}_j\, d\left(\frac{\partial L}{\partial \dot{q}_j}\right) - \frac{\partial L}{\partial q_j}dq_j - \frac{\partial L}{\partial t} \\ &= -\frac{\partial L}{\partial q_j}dq_j + \dot{q}_j\, d\pi^j - \frac{\partial L}{\partial t}. \end{aligned} \qquad (11.5)$$

Together these two forms of dH imply that,

$$-\left(\frac{\partial L}{\partial q_j} + \frac{\partial H}{\partial q_j}\right)dq_j + \left(\dot{q}_j - \frac{\partial H}{\partial \pi^j}\right)d\pi^j - \left(\frac{\partial L}{\partial t} + \frac{\partial H}{\partial t}\right)dt = 0. \qquad (11.6)$$

The presumed independence of the coordinates q and momenta π (for a non-singular Lagrangian) permits us to deduce that,

$$\dot{q}_j = \frac{\partial H}{\partial \pi^j}, \quad \dot{\pi}^j = -\frac{\partial H}{\partial q_j} \quad \text{and} \quad \frac{\partial L}{\partial t} = -\frac{\partial H}{\partial t}, \qquad (11.7)$$

where in this final step we have used the Euler-Lagrange equations to replace $\partial L/\partial q_j$ by $d(\partial L/\partial \dot{q}_j)/dt$ which is simply $\dot{\pi}^j$.

Provided that the Lagrangian is second-order in $\dot{q}$, the Hamiltonian will be second-order in π and the first two of Equations 11.7 will be first-order equations with unique solutions for $q_j(t)$ and $\pi^j(t)$, given the initial values $q_j(t_i)$ and $\pi^j(t_i)$. Those two equations are known as *Hamilton's equations* for a mechanical system. The extension of the formulation to

---

[2] Adrien-Marie Legendre (1752–1833), French mathematician.

non-mechanical systems is relatively straightforward in the non-singular case and we shall carry this out for Newtonian fields in Section 11.6 and see the results for relativistic particles and fields in Chapters 15 and 16.

Hamilton set up his technique initially to handle optical problems. The first manuscript, *On caustics* [219], was written in 1823 while he was still an undergraduate student. It led to the publication of the *Theory of systems of rays* in four parts [220, 221] written between 1828 and 1833. One result based on this work was an outstanding example in the history of science where a theoretical prediction was made of a completely unsuspected phenomenon in nature. This was Hamilton's assertion [220, vol. 17, p. 1] that a biaxial crystal should give rise to *conical refraction*, namely, that certain relative orientations of incident light beam and crystal will lead to a hollow cylindrical emergent beam.

The experiment to test this prediction was exceedingly difficult to carry out but conical refraction, both internal and external, was observed by Lloyd [324] in 1832, confirming Hamilton's method. At that time there was considerable controversy over whether light was a wave or corpuscular phenomenon. However, Hamilton's method had the decided advantage of remaining valid irrespective of the detailed nature of light. This is a consequence of the fact that, as we shall show in the next section, Hamilton's equations, like the Euler-Lagrange equations of the Lagrangian formulation, are obtainable from a global variational principle applicable to either detailed structure for light.

Since the Hamiltonian function is the generator of time translations, its selection as the characteristic function of the dynamics of a system requires the evolution variable, the time t, to play a priviledged rôle. This is natural in a Newtonian context since time is already special in Galilean relativity where simultaneity is absolute. We shall see that singling time out in this way is less natural in Einsteinian relativity. Indeed, there is then no unique evolution parameter and the identification of a Hamiltonian function can be a non-trivial process.

11.4 Phase space Hamilton's principle

Let us suppose that two points in the phase space of a system described by Hamilton's equations have phases $\{q(t_i), \pi(t_i)\}$ and $\{q(t_f), \pi(t_f)\}$. Then an arbitrary curve $\mathcal{C}$ in phase space from the initial to the final point may be parameterized in terms of the values of the phase along the curve in the form $\mathcal{C} = \mathcal{C}(q(t), \pi(t))$. Both $q(t)$ and $\pi(t)$ are specified independently of one another along the curve $\mathcal{C}$ in the Hamiltonian formulation. The choice of a curve in phase space implies a choice of $q(t)$ which automatically determines the values of $\dot{q}(t)$ along the curve, as in the Lagrangian formulation. However, in contrast to the Lagrangian formulation, the values of $\pi(t)$ along an arbitrary curve are not determined from those of $q(t)$ and $\dot{q}(t)$ in the Hamiltonian method. Only when the stationary path has been determined, and along with it the equations of motion, will the relations between $q(t)$, $\pi(t)$ and $\dot{q}(t)$ implied by the second of Equations 11.2, or its inverse giving $\dot{q}(t)$ in terms of $\pi(t)$, be satisfied.

We now consider an arbitrary variation $\delta t(t)$ in the independent variable, along with simultaneous arbitrary and independent variations $\delta q(t)$ and $\delta \pi(t)$ of the dependent variables, which imply a variation in $\dot{q}(t)$ given by Equation 9.16. From the definition, Equation 8.26, of the action and of the Hamiltonian, Equation 11.2, in terms of the Lagrangian, the action is,

$$S[C] = S[q,\pi] = \int_{t_i}^{t_f} dt \left\{ \sum_j \pi^j \dot{q}_j - H(q,\pi,t) \right\}, \qquad (11.8)$$

in which we consider $\dot{q}(t)$ to be re-expressed in terms of $\pi(t)$. We may now consider the extremization of this action as the starting point of a variational approach to dynamics totally independent of the Lagrangian formulation. Mathematically, of course, the two approaches will be closely related and we shall continue to abbreviate the integrand by $L = L(q,\pi,t)$ in the first step of the next equation, in order to easily carry over the first steps of the analysis leading to Equation 9.21. We obtain,

$$\begin{aligned}
\delta S &= \int_{t_i}^{t_f} dt \left[\frac{d(\delta t\, L)}{dt} + \delta_0 L \right] \\
&= \left[\left(\sum_j \pi^j \dot{q}_j - H \right) \delta t \right]_{t_i}^{t_f} \\
&\quad + \int_{t_i}^{t_f} dt \left[\sum_j \left\{ \pi^j \delta_0 \dot{q}_j + \dot{q}_j \delta_0 \pi^j - \frac{\partial H}{\partial q_j} \delta_0 q_j - \frac{\partial H}{\partial \pi^j} \delta_0 \pi^j \right\} \right] \\
&= \left[\sum_j \pi^j \left(\dot{q}_j \delta t + \delta_0 q \right) - H \delta t \right]_{t_i}^{t_f} \\
&\quad + \int_{t_i}^{t_f} dt \left[\sum_j \left\{ \left(\dot{q}_j - \frac{\partial H}{\partial \pi^j} \right) \delta_0 \pi^j - \left(\dot{\pi}^j + \frac{\partial H}{\partial q_j} \right) \delta_0 q_j \right\} \right] \\
&= \left[\sum_j \pi^j \delta q - H \delta t \right]_{t_i}^{t_f} \\
&\quad + \int_{t_i}^{t_f} dt \left[\sum_j \left\{ \left(\dot{q}_j - \frac{\partial H}{\partial \pi^j} \right) \delta_0 \pi^j - \left(\dot{\pi}^j + \frac{\partial H}{\partial q_j} \right) \delta_0 q_j \right\} \right]. \quad (11.9)
\end{aligned}$$

An integration by parts was used on the term involving $\delta_0 \dot{q}$ and we made use of $\delta q = \dot{q}\delta t + \delta_0 q$ in the last step.

A physical variational principle based on the above mathematical result must lead to Hamilton's equations since we have shown they are

11.5. Poisson brackets

equivalent to the Euler-Lagrange equations. It is now clear that one form of the variational principle of Hamiltonian dynamics states that the path taken by a Hamiltonian system from one state to another is obtained by demanding that arbitrary variations in the path between the two points in phase space which represent those states, including variations in the independent variable, give rise only to end-point contributions to the action of Equation 11.8. The integral term in the variation in the action must then vanish and, since the variations in q and π are presumed to be independent, we arrive at the required Hamilton's equations. This may be called the *phase space modified Hamilton's principle*. Alternatively, we may consider only variations δ_0 leaving the end-points fixed to arrive immediately at the same result.

Equation 11.9 may also be used to deduce the Noether charges corresponding to the continuous symmetries of a Hamiltonian system by proceeding analogously to the methods used in the Lagrangian formulation.

Exercise 11.1 *If H_1 and H_2 are the Hamiltonian functions of two isolated systems, determine the relation between H_1 and H_2 and the total Hamiltonian H of a system comprised of the two separate parts interacting via a Hamiltonian term, H_{intn}, demonstrating that it is consistent with Equation 8.43 and the application of the definition of the Hamiltonian to each term.*

Exercise 11.2 *Show that the generalized momentum of the Newtonian gravitational field (or of the fields of either electrostatics or magnetostatics) vanishes identically. State what is formally involved in making the transition from the Lagrangian to the Hamiltonian formulation in any one of these three theories.*

11.5 Poisson brackets

The link between Hamiltonian dynamics and quantum mechanics is provided by a quantity discovered by Poisson [452], now called a *bracket*. Jacobi[3], who also made major contibutions to theoretical mechanics, one of which involves what is now referred to as the *Hamilton-Jacobi equation* [206], declared the *Poisson brackets* to be 'the most important advance in the transformation of equations of motion since the first version of the *Mécanique analytique* [of Lagrange]' [261].

To see the way the Poisson brackets arise, one may examine the time variation of any dynamical quantity f of a Hamiltonian system. Such a quantity will, at a given time, have a particular value expressible in terms

---
[3] Carl Gustav Jacob Jacobi (1804–1851), German mathematician.

of the phase $\{q, \pi\}$ and is thus some function of coordinates, momenta and the time, namely $f = f(q, \pi, t)$.

Exercise 11.3 *Show that the rate of change of f with time is given by,*

$$\dot{f} = \frac{\partial f}{\partial t} + \{f, H\}, \tag{11.10}$$

where the Poisson bracket of two dynamical variables f and g (at the same time t) is defined by,

$$\{f, g\} = \sum_j \left(\frac{\partial f}{\partial q_j} \frac{\partial g}{\partial \pi^j} - \frac{\partial f}{\partial \pi^j} \frac{\partial g}{\partial q_j} \right). \tag{11.11}$$

Exercise 11.4 *Show that the* **fundamental Poisson brackets** *of the coordinates and momenta have the values,*

$$\{q_j, q_k\} = 0 = \{\pi^j, \pi^k\} \quad \text{and} \quad \{q_j, \pi^k\} = \delta^k{}_j . \tag{11.12}$$

Exercise 11.5 *Show that the Poisson bracket satisfies the* **Jacobi identity**,

$$\{e, \{f, g\}\} + \{f, \{g, e\}\} + \{g, \{e, f\}\} = 0 . \tag{11.13}$$

Exercise 11.6 *(a) For consistency of notation, let the Hamiltonian of a mechanical system be denoted by h, its value on the stationary path being the total energy E. Express each of the 10 conserved quantities,* **p, j, k** *and h, for a closed system of Newtonian particles, in Hamiltonian form (in terms of coordinates and momenta, and possibly t). (b) Calculate the Poisson brackets between them to obtain the following results,*

$$\{p^r, p^s\} = \{p^r, h\} = \{p^r, k^s\} = \{k^r, k^s\} = \{j^r, h\} = 0 \tag{11.14}$$
$$\{j^r, j^s\} = \epsilon^{rst} j_t \quad \{j^r, k^s\} = \epsilon^{rst} k_t \quad \{j^r, p^s\} = \epsilon^{rst} p_t \tag{11.15}$$
$$\{k^r, h\} = p^r . \tag{11.16}$$

The above *Poisson bracket algebra* is the realization, for a classical mechanical system, of a much more general result, referred to as the *Lie algebra* of the Galilei group of translations in time and space, rotations and Galilean boosts. For further details, see Ludwig and Falter [346] or Sudarshan and Mukunda [525].

Lie algebras are typically encountered for the first time by physics students in discussions of the wave mechanics of rotation, where the results are referred to as the commutator algebra of the orbital angular momentum operators, $\mathbf{J} = \mathbf{L} = \mathbf{X} \times \frac{\hbar}{i} \nabla$. In many instances, the relativistic counterparts of these results are encountered for the first time in the *Poincaré Lie algebra* of the corresponding operators of relativistic quantum field theory.

It is important for the student to be aware of the fact that such results are not intrinsically quantum-mechanical nor only applicable to field theory. The above exercise establishes the Lie algebra of the spacetime symmetries in a Galilean covariant, non-quantum and mechanical context. The essential ingredients are in fact the Hamiltonian formulation and the existence of some group of symmetry transformations, factors which are of almost universal importance in the analysis of any dynamical physical system.

The quantum commutator algebras of the fundamental conjugate operators $\mathbf{X}$ and $\mathbf{P}$ or of the 10 Galilean hermitian generators $\mathbf{P}$, $\mathbf{L}$, $\mathbf{K}$ and H of the Galilei group are of course simply related to the above classical results. The relationship is summarized by the replacement of the Poisson bracket $\{a, b\}$, having the dimensions of ab divided by those of action, by the commutator $[A, B]/i\hbar$ of the corresponding quantum operators. The first of the Poisson bracket relations in Equation 11.15 then becomes the *angular momentum Lie[4] algebra* so(3) of the *rotation group* SO(3),

$$[J^k, J^l] = i\hbar \epsilon^{klm} J_m \ . \tag{11.17}$$

The closed set of Poisson brackets (or corresponding quantum commutation rules) involving $\{p^r\}$ and $\{j^s\}$, namely

$$\{j^r, j^s\} = \epsilon^{rst} j_t, \quad \{j^r, p^s\} = \epsilon^{rst} p_t \quad \text{and} \quad \{p^r, p^s\} = 0 \ , \tag{11.18}$$

make up the iso(3) Lie algebra of the restricted *Euclidean group* ISO(3).

11.6 Non-relativistic Hamiltonian fields

The transition to a Hamiltonian formulation for a local system of Newtonian fields, $\psi = \{\psi_j\}$ ($j = 1, 2, \ldots, d$), is straightforward for a non-singular Lagrangian.

The generalized momenta π^j conjugate to each ψ_j are given in terms of the Lagrangian as the functional derivative of Equation 10.5 which, for the local fields we are considering, reduces to the ordinary derivative of the Lagrangian density given by Equation 10.8. Provided the Lagrangian is non-singular, the generalized velocities $\dot{\psi}_j$, can be eliminated algebraically in favour of the generalized momenta π^j, to provide the *phase space Hamiltonian density*.

The vanishing variation of the action functional,

$$S[\psi, \pi] = \int_{t_i}^{t_f} dt \int_R d^3x \left\{ \sum_j \pi^j \dot{\psi}_j - \mathcal{H}(\psi, \pi, t, \mathbf{x}) \right\} , \tag{11.19}$$

---

[4] Marius Sophus Lie (1842–1899), Norwegian mathematician.

leads to *Hamilton's equations* for a local field system,

$$\dot{\psi}_j = \frac{\partial \mathcal{H}}{\partial \pi^j} \quad \text{and} \quad \dot{\pi}^j = -\frac{\partial \mathcal{H}}{\partial \psi_j} , \qquad (11.20)$$

the partial time derivatives of the Hamiltonian and Lagrangian densities of an open system being related by,

$$\frac{\partial \mathcal{L}}{\partial t} = -\frac{\partial \mathcal{H}}{\partial t} . \qquad (11.21)$$

The *Poisson bracket* of two functionals $f(\psi, \pi, t)$ and $g(\psi, \pi, t)$ of the phase space variables (at the same time t) is defined by,

$$\{f, g\} = \int_R d^3x \sum_j \left(\frac{\delta f}{\delta \psi_j(t,\mathbf{x})} \frac{\delta g}{\delta \pi^j(t,\mathbf{x})} - \frac{\delta f}{\delta \pi^j(t,\mathbf{x})} \frac{\delta g}{\delta \psi_j(t,\mathbf{x})} \right) , \qquad (11.22)$$

in which we see again that the sum over the label j of mechanics (Equation 11.11) becomes an integral over space with a sum also over the components of the field labelled by j. The *fundamental Poisson brackets* relating the coordinates and momenta (at the same time) are,

$$\{\psi_j(t,\mathbf{x}), \psi_k(t,\mathbf{y})\} = 0 = \{\pi^j(t,\mathbf{x}), \pi^k(t,\mathbf{y})\} , \qquad (11.23)$$

and,

$$\{\psi_j(t,\mathbf{x}), \pi^k(t,\mathbf{y})\} = \delta^k{}_j \delta(\mathbf{x} - \mathbf{y}) , \qquad (11.24)$$

in which we see, in the appearance of the Dirac delta function, the essential difference again between a mechanical and field coordinate.

Exercise 11.7 *Assuming that the functional derivative can be shown to satisfy all the usual properties of a derivative, such as the Leibniz rule, for example, show that the equations of motion of a dynamical variable $f(\psi, \pi, t)$ are given by Equation 11.10 as in the mechanical case, now interpreted as a functional equation.*

Hamilton's equations for a field system, not necessarily spatially local, are,

$$\dot{\psi}_j(t,\mathbf{x}) = \frac{\delta H}{\delta \pi_j(t,\mathbf{x})} \quad \text{and} \quad \dot{\pi}^j(t,\mathbf{x}) = \frac{\delta H}{\delta \psi_j(t,\mathbf{x})} , \qquad (11.25)$$

which may be written as,

$$\dot{\psi}(t,\mathbf{x}) = \{\psi(t,\mathbf{x}), H\} , \quad \text{and} \quad \dot{\pi}(t,\mathbf{x}) = -\{\pi(t,\mathbf{x}), H\} . \qquad (11.26)$$

Exercise 11.8 *Form all Poisson brackets with one another of the 10 Galilean Noether charges of a Newtonian field system to demonstrate that* *they satisfy the same Poisson bracket algebra, Equations 11.14 to 11.16, as the corresponding mechanical quantities.*

The passage to the Hamiltonian formulation fails for the non-dynamic Lagrangians given in Section 10.5 for Newtonian gravity, electrostatics and magnetostatics.

In the Hamiltonian formulation, the difference between the Poisson bracket algebra of the non-quantum generators and the angular momentum algebra of quantum generators of the Galilean group is minute but highly significant. Replacement of the Poisson brackets of classical mechanics with the commutators of quantum mechanics introduces an extra factor of $i\hbar$ where $\hbar$ is the characteristic constant of quantum physics.

Another apparently minute difference accounts for the other revolutionary change in physical ideas of the early part of the 20th century, namely the consequences of implementing the Relativity principle with a finite rather than infinite upper bound on the velocities. The effect of choosing the kinematics corresponding to a bound on the velocities is the topic to be developed in the following chapters.

Bibliography:

Goldstein H 1980 *Classical mechanics* [206].
Ludwig W et al., 1988 *Symmetry in physics: group theory* ... [346].
Marion J B and Thornton S T 1988 *Classical dynamics* [358].
Sudarshan E C G and Mukunda N 1974 *Classical dynamics* [525].
Sundermeyer K 1982 *Constrained dynamics* [526].

Part III

Poincaré covariance and Einsteinian physics

CHAPTER 12

Special relativity

If we accept the Principle of relativity, ... either everything in nature is electromagnetic or that part which would be ... common to all phenomena would be only a façade, something pertaining to our measurement procedures[1].

Henri Poincaré 1854-1912

12.1 Introduction: the electricity and the ether

In order to appreciate the revolutionary nature of the relativity postulates in Einsteinian form, some of the history of the relevant physics, especially during the decades immediately prior to their publication, will be outlined.

In preparing historical material, particularly in this section and in the Introductions to Chapters 5 and 18, the excellent articles by Everitt [149] on Maxwell, McCormach [373] on Lorentz, Dieudonné [105] on Poincaré, and Balazs [26] on Einstein, in the *Dictionary of scientific biography* [197] have been invaluable. Jungnickel and McCormach [269] provide an indispensable survey of the growth of theoretical physics with especial emphasis on the development of electromagnetism, relativity, gravitation and quantum mechanics in Germany. Other very useful material on the Lorentz electron theory and the origins of relativity theory are available in Goldberg [202], Hirosige [250, 251, 252], Holton [254], McCormach [371, 372], Miller [380] and Schaffner [495].

The wave theory of optics originated in the 17th century. However, Newtonian mechanics was so dominant after about 1700 that Newton's corpuscular theory of light became the accepted norm during the 18th century. The Newtonians generally assumed all interaction was mechanical and the effects took place instantaneously in a void or vacuum. The followers of

---

[1] *Oeuvres* [451, vol. 9, p. 498] (July 1905), reprinted from reference [449]. My translation.

Descartes, the Cartesians, assumed the void was filled with vortices to transmit the mechanical interaction.

Prior to the rise in popularity of the wave theory of light around 1800, the transmission of electric, magnetic and gravitational effects was believed by proponents of that theory to be attributed to the existence of separate mechanical space-filling media referred to as the *electric, magnetic* and *gravitational ethers*. The ether was originally the transparent, all-pervasive, changeless and rigid yet non-resistive fifth element of the ancient Greeks from which the celestial (non-material) bodies were composed complementing the earth, water, fire and air of material bodies.

The medium of propagation of light was referred to as the *luminiferous ether*. The observations of 20.5 arc seconds of stellar aberration [45] in 1728 by Bradley were interpreted, using a Newtonian velocity addition formula, by assuming that the ether was not dragged along by the movement of the surrounding matter. The Newtonian result was adequate owing to the low value of the earth's orbital speed relative to the speed of light. Stellar aberration ruled out all so-called *ether-drag* theories. The development of the wave theory of light was revived independently by Young and Fresnel, in the early 19th century and assumed the luminiferous ether to be an elastic solid thus permitting light to be regarded as a mechanical vibration subject to Newton's laws, along with all other phenomena. Like the celestial element, the luminiferous ether had to be extremely rigid and extraordinarily dense to support waves with frequency as high ($\approx 5 \times 10^{14}$ Hz) as that of light. Since stars are visible, it must pervade all space. Since planetary motion is unaffected, it must be very tenuous or non-resistive. No absorption observations are attributable to it, so it must be transparent. Light penetrates matter — so must the ether.

The alternative to ether drag was the *ether wind* theory of Young, namely a free motion of matter such as the earth relative to the ether, presumed fixed in an absolute space with respect to which the velocity of any body could, in principle, be determined. Any ether-wind effect involving the motion of the earth about the sun must change with the reversal of direction of the earth's motion about the sun relative to the ether. The dimensionless magnitude of the effects must depend on the ratio $v_\oplus/c \approx 10^{-4}$ of the earth's speed $v_\oplus$ to the speed of light c, known from the measurements of Römer in the 17th century and confirmed by Bradley in 1728.

However, one major difficulty of the wave theory was its failure to provide an adequate explanation of some optical phenomena such as reflection and refraction. Since it assumed that light vibrations were longitudinal like sound, namely in the direction of propagation, it was also unable to explain the polarization phenomena discovered soon after its revival. The latter difficulty was overcome by correctly assuming that light vibrations were transverse to the direction of propagation.

12.1. Introduction: the electricity and the ether

The theory of electromagnetism at the same epoch, in the 1820s, was in an embryonic state with a large number of competing theories some of which were in conflict with one another and all giving incomplete descriptions of observed electric and magnetic properties. Each, like Newtonian gravity, was based on *action at a distance*, which is a simple direct means for implementing Galilean covariance.

In 1845 Faraday [150, §2151f] established a connection with optics by showing that the transverse polarization direction of a light beam was rotated about the axis of propagation by a strong magnetic field, a phenomenon now known as *Faraday rotation*. The earliest indications of spatial locality in electromagnetism, referred to as *contiguous action*, arose in the period from 1845 to 1850 in the work of Faraday [150], who developed the notion of *lines of force* to describe the new concepts now described by fields. Faraday gave an indication in 1851 (see Whittaker [576, p. 194]) of the possible electromagnetic character of light stating that *'it is not at all unlikely that if there be an aether, it should have other uses than simply the conveyance of radiations'*. He suggested, in particular, that it may be the medium for magnetic effects.

An unpublished conjecture concerning the electromagnetic character of light was made by Riemann [269, p. 176] about 1854 in an 'investigation of the connection between electricity, galvanism, light and gravity'. Riemann's investigations were stimulated by the works of Newton but based on a continuous medium not involving the 'absurdity' of direct action at a distance (see the quotation at the end of Section 7.3). In 1855, Weber[2] and Kohlrausch [562] determined a limiting velocity parameter appearing in Weber's electrodynamic theory [269, p. 145] to be $c_1 = 439,450 \text{ km.s}^{-1}$.

In an epoch-making development starting in 1861, Maxwell [364] built on Faraday's work by dispensing with action at a distance ideas completely. His local field theory of electromagnetism, published from 1861 to 1865, became widely accepted during the following two decades. It predicted that light, already understood as an essentially wave phenonenon, was an electromagnetic effect. Thus Maxwell's theory was not only consistent with the known properties of electricity and magnetism but provided a dramatic unification of optics and electromagnetism. In particular, his theory predicted the existence of purely transverse waves of electricity and magnetism having a speed in vacuo which was given in terms of what we now refer to as the electric permittivity ϵ_0 and the magnetic permeability μ_0. The numerical value of Maxwell's propagation speed, $1/\sqrt{\epsilon_0\mu_0}$ in modern terms, was very close to the value of the speed of light measured relative to the luminiferous ether, presumed to be the medium of the Maxwell waves.

Although Maxwell's original analyses [364] included a molecular vortex model of the 'electric fluid' of charged matter which generated his elec-

---
[2] Wilhelm Eduard Weber (1804–1891), German physicist.

tric and magnetic fields, he did not intend a detailed microscopic model to be an essential element in his theory. Having obtained his equations for the electrical fields propagating in the ether, Maxwell was prepared to stand by their validity irrespective of the details of the electricity. In his 1865 paper [366], he set out an electrodynamical wave propagation theory, based on Lagrangian and Hamiltonian techniques, which was totally independent of any detailed microscopic mechanism.

Maxwell applied his theory to explain many electro-optical effects but he did not use it to provide an analysis of reflection and refraction, which remained as unexplained as they were with Fresnel's theory. Such effects required not only an electromagnetic theory but a more complete theory of the interaction of the electromagnetic field with charged matter. At that time all electricity was referred to as composed of electrons, a name which arose from the Greek for amber, since rubbing the latter produced electrostatic charges. The electrons were generally regarded as components of a fluid and Maxwell developed his theory by building on a strong analogy between non-mechanical electromagnetic effects and purely mechanical phenomena associated with fluid dynamics. An excellent discussion of Maxwell's discoveries is available in the text by Longair [326].

Lorentz, in his doctoral thesis [329], presented at Leiden in 1875, gave the first comprehensive treatment of optics and electrodynamics, showing that the laws of reflection and refraction were a consequence of Maxwell's theory. This was the start of a quarter century of brilliant adjustments by Lorentz to the classical formulation of electrodynamics. Lorentz showed, as early as 1878, that the ether was the only ultimate dielectric having properties totally independent of the matter it was presumed to pervade and essentially identical to those of the vacuum.

Maxwell, in a letter to D. P. Todd of the U.S. Nautical Almanac Office, and in an article in the *Encyclopaedia Britannica* [370], suggested in 1879 that an earth-based measurement of the ether drift was theoretically possible with a pair of mirrors but deemed it unfeasible being of second-order, $(v_\oplus/c)^2 \approx 10^{-8}$, in a very small quantity. Michelson immediately accepted the challenge of such a difficult experiment and constructed his interferometer to measure the value of the ether drift, fully expecting to confirm that it existed. A null result was obtained thus creating a crisis in the understanding of optical and electromagnetic phenomena.

The 1846 partial drag theory of Stokes[3] was revived and became widely accepted as the means to retain agreement with aberration. It involved an ether which behaved rigidly for the propagation of light and elastically for slow speed solar system objects which could drag some of the ether with them to account for the null result of Michelson. In 1885, Lorentz [331] pointed out a fault in the Stokes theory and a factor of two

---

[3] George Gabriel Stokes (1819–1903), British mathematical physicist.

error in the Maxwell and Michelson predictions of the ether drift effect. He proposed a new theory of the ether and suggested the ether drift be remeasured. However, the much improved results of Michelson and Morley [378, 379] in 1887 still indicated a null ether-drift effect.

The existence of Maxwell's electromagnetic waves, and their propagation at finite speed c, was confirmed in 1887 by the brilliant and meticulous experiments of Hertz[4] [85, 239, 240, 241, 326] on radio waves with frequencies in the GHz range. Some qualitative effects [527] — such as sparks from metallic objects — which could later be attributed to the existence of electromagnetic waves, were noted prior to the work of Hertz. Hertz had to deal with signals of high frequency ν in order that their wavelength $\lambda = c/\nu$ be comparable to the size of his laboratory and concluded that Maxwell's theory explained all the observations including reflection, refraction, diffraction and interference [242]. According to Hertz: 'Electric rays are light rays of long wavelength' [243, p. 781]. In the early stages of his work, in 1887, Hertz had also discovered the photoelectric effect [238] and predicted that gravitation would also turn out to have a finite speed of propagation. Hertz died a few years later at the age of 37.

FitzGerald [163] gave the first really satisfactory explanation of the null ether-drift result in 1889 by hypothesizing a physical contraction of matter itself in the direction of its motion relative to the ether. This resulted from assuming that molecular forces (not known then to be electrical) were affected, as was known to be the case for electrical forces obeying Maxwell's equations, by the frame from which they were measured, owing to interactions between molecules and the ether. Thus the molecules in the arm of the interferometer lying in the direction of the earth's motion were assumed to be 'closer together' in the frame moving relative to the ether. Lorentz [332] suggested such a contraction independently [53] in 1892. This result, with a kinematic or 'theory of measurement', rather than dynamic interpretation, is now known as the *Lorentz-FitzGerald contraction*.

Lorentz had been developing, throughout the period since 1880, his *electron theory*, namely his *classical* (non-quantum, non-relativistic) theory of the matter carrying the electricity in a manner consistent with Maxwell's description of the electromagnetic fields generated by, and in turn affecting, the motion of the electricity. Although Heaviside was the first to set down the equation determining the acceleration of a charged particle in terms of its velocity and the values of the electric and magnetic fields at the location of the particle (see Equations 19.1 and 19.2), it goes by the name of Lorentz as a result of the completeness of his analysis of electrodynamics.

As early as 1892 Lorentz [332], in his first publication on the electron theory, wrote of 'charged particles'. He changed to the term 'charged ions' in 1895 and finally to 'electrons' in 1899 after their discovery as the first

---

[4] Heinrich Rudolf Hertz (1857–1894), German physicist.

microscopic carriers of electricity in the form of cathode 'rays', shown by Lord Kelvin to travel at less than the speed of light with a definite mass to charge ratio.

Lorentz was the first to state that a body carries a charge if it has an excess of one or other type of electron; he referred to an electric current as a flow of electrons; he stated that the electrons act as the source of the electrodynamic field; he made it clear that the field then acts on ordinary matter via the electrons embedded in material molecules.

Although rejecting a large part of the Continental model of electricity, Lorentz fused the Continental and Maxwellian theories of electrodynamics by adopting a particulate model of the electric fluid and making it consistent with the Maxwell theory of local electromagnetic field propagation at the speed of light. Lorentz [334] first gave the full electrodynamic equations in their modern vector form in 1895. The theory of (i) a separate non-mechanical electromagnetic field propagating in the ether and of (ii) mechanical matter based on electrons, *the ether and the electricity*, had been made available for a systematic analysis of the interaction between them. The predictive power of that theory was almost identical in many of its aspects with the classical parts of the relativistic theory which replaced it.

In 1892, Lorentz used an *ad hoc* set of transformations for the electrical field quantities, in going from the ether frame to another moving relative to it, and corresponding transformations for the spatial coordinates and a special *local time* coordinate in the 'moving frame'. The coordinate transformations were essentially the first-order approximations to Equation 5.30, namely with $\gamma = 1$, and therefore did not leave time intervals invariant as required by the absolute nature of simultaneity inherited from Galileo and Newton. However, he considered his local time in the moving frame to be a mathematical artifact rather than the time as measured on any clock.

In this way Lorentz had been able to adapt his theory to show that it predicted that no measurable ether-drift effects could occur to first order in $v_\oplus/c$. His sketchy suggestion of a contraction (which is a second-order effect) was an indication that null ether-drift effects to second-order could also be incorporated. As soon as the broadening of spectral lines by a magnetic field was reported by Zeeman, Lorentz was able to explain the results [334] with his by now very well-established *electron theory*, or electrodynamics. When cathode rays were discovered in 1899, he was able to identify them as the charged particles, electrons, of his theory.

In 1899, Lorentz [336] included his ideas on contraction as an integral part of his electron theory and the more complete 1899 transformations of space, time and electromagnetic quantities were essentially identical to those he presented [340] in 1904 prior to Einstein's classic paper [128] of 1905. They predicted, in accord with observation, that no ether-drift effects would be detected to second-order in v/c. A consequence of these

12.1. Introduction: the electricity and the ether

transformations was that all mass, not just the mass of electrically charged particles, must depend on the velocity of the mass relative to the ether.

During the last decade of the 19th century, electrodynamics began to supplant the *mechanical world view* of Newton. Many physicists who still adhered to Newton's theories, began to believe that unification of electrodynamics with mechanics, namely with gravitation, could arise as a result of the inertia of the electrodynamic field providing the origin of the mass of particles. Gravitation would thereby be described as a consequence of electrodynamics [338, 339, 580]. All matter was believed to be composed of charged particles and all its properties (hardness, elasticity, ductility) should be reducible to electrodynamic laws consistent with Newtonian mechanics describing interactions of charges with one another and the ether. The classical *electrodynamic world view* of ether and electricity was to be the new 'theory of everything' to replace the purely mechanistic view of Newtonian mechanics.

Poincaré, the French mathematical physicist, whom Bertrand Russell described, in the Preface to Poincaré's 1902 text *Science and Hypothesis* [445], as the most eminent scientist of his generation, had argued that Lorentz should not continue to counter null effects at a higher order with a modification to his theory only at that order. Instead, he should adopt a theory valid to all orders. Lorentz's classic 1904 paper on the topic [340] described a version of the electron theory for which no experiment, however accurate, would be capable of detecting the ether drift. The problem of the null ether-drift results were essentially reconciled with the theory of electrodynamics, although not in a way which accords completely with the modern interpretation of the Lorentz-FitzGerald contraction and spacetime measurement. Poincaré also elevated the correspondence between the two states of the Lorentz electrodynamic system, in the ether frame and a frame in motion relative to it, which was approximate in the theory of Lorentz, to an exact *Relativity principle*, which he suggested should apply to all physical phenomena.

The development, principally by Lorentz, of the classical electron theory from 1892 to 1904 constituted a striking change of direction in the course of dynamical theory. The dimensions of bodies contracted in the direction of motion relative to the ether; the masses of all particles, not just those which were charged, were velocity dependent; the speed of light was an upper limit to the speed of any material particle; and the energy and mass of a particle tended to infinity as its speed tended to that of light. The electron theory was compatible with the large part of the accepted physical theory of the time. It was not, however, without defects. It failed to withstand completely the quantum revolution heralded by the work of Planck [437] in 1900 and the equally revolutionary re-appraisal of spacetime kinematics introduced by Einstein [128] in 1905. It was, for example, unable to predict the observed wavelength distribution of blackbody radiation, already a successful result of the new quantum ideas of Planck, a theory which Lorentz was one of the first to support.

The Relativity principle was a problem in the electron theory of Lorentz and Poincaré, since it required transformations of space and time which were in conflict with the well-established Galilean equations of all mechanical phenomena. This problem was only resolved by the completely new interpretation given to the Relativity principle by Einstein. Lorentz considered the ether to be an essential part of his theory despite the fact that no observations confirmed its existence. Poincaré, the other key figure in the pre-Einsteinian development of the ideas of relativity, also expected the ether to survive in one form or another.

Lorentz and Poincaré had contributed enormously in setting the stage for the revolutionary quantum relativistic changes about to overtake the classical theory of physics, the ultimate expression of which was the dynamic theory of moving electrons, based on Maxwell's theory of the electromagnetic field. Many aspects of their work remain valid to this day, albeit with some re-interpretation.

12.2 Einsteinian relativity postulates

The first postulate of Einsteinian relativity, the Principle of relativity (see Chapter 5) is identical to the modern expression of Galilean relativity, applicable to Newtonian physics. Only the form of the transformations, dependent on the second postulate, differs. In Newtonian physics, the Relativity principle was an unconscious implicit assumption and not at first considered at all remarkable. We should regard the Einsteinian form on its own in the same way. The remarkable feature of the Einsteinian form is that it remains valid even when it is applied in conjunction with a second postulate which appears, from a Galilean point of view, to contradict it.

The customary form of the second postulate, based on the original statement by Einstein [128] in 1905 is the requirement of the invariance of the speed of light, its frame-independence — including with respect to transformations between frames in relative motion. Alternatively, one may adopt a light-free formulation (see Chapter 5) of the foundations of *Einsteinian relativity* by adopting as:

- **second postulate** — the speed of propagation of information or energy has a finite upper bound.

Einstein's reasons for the second postulate arose from a deep analysis of the nature of spacetime measurement, in which it was convenient for light to play a rôle as a means for synchronizing clocks and defining distant simultaneity. The postulate adopted here is essentially equivalent to the conventional form but will permit the examination of electrodynamics on a par with other relativistic phenomenon. The postulates were used by

Einstein [128] in his first special relativity paper in 1905 not only to derive the Lorentz transformations but to provide a definitive determination of the aberration formulae which during almost two centuries had received so many different explanations none of which had stood the test of time.

In 1915, Einstein referred to his 1905 relativity theory as *special relativity*. The qualification of 'special' refers here to the fact that it deals only with frames in uniform relative motion in contrast to general relativity which extends the principle to frames which may not only be accelerated with respect to one another but may be in an arbitrary state of relative motion. Whereas special relativity is a kinematic theory of the measurement of time and space over small intervals, the general theory is a dynamic theory of the gravitational interaction. Although the electron theory of Lorentz, and the relativistic ideas of Poincaré based on it, are not equivalent to Einstein's special relativity theory, some of their predictions are identical and many more are very similar. For these reasons, the terms *Lorentz*, *Poincaré* or *Lorentz-Poincaré relativity* are also widely used to describe the special theory of Einstein.

The second postulate does not apply to Galilean relativity and hence contains the essence of special relativity. Since the coordinate labels of events are varied when changing frames, the physical quantities describing phenomena taking place at those events must also vary and in precisely such a way that the laws involving them do not vary in form. We say that the physical quantities must *covary* and for this reason the Principle of relativity is also a *Principle of covariance*.

12.3 Lorentz transformations

The second postulate selects the Lorentz form, Equation 5.30, of the x-direction boost transformation. In it, c is a positive finite universal constant, the *ultimate speed*, which will be identified in Section 18.6 as the speed of light. The corresponding transformation for a boost in an arbitrary direction will be given in Section 13.6.

Equation 5.30 shows that a Lorentz boost of speed v_1 followed by a second in the same direction of speed v_2 is equivalent to another boost, again in the same direction, whose speed V is given by the special relativistic velocity addition formula of Equation 5.32, which can only be approximated by Equation 5.29 for small velocities $v \ll c$. If $v_1 = c$ then $V = c$ irrespective of the value of v_2, showing that propagation at the ultimate speed is frame-independent. The same is not true of speeds $v_1 < c$ where $V \neq v_1$ if v_2 is a boost speed.

We shall see in Chapter 15 that one natural consequence of the Lorentz transformation is that the inertial mass of a particle measured in a frame in which it is moving at speed u is greater, $m/\sqrt{1 - u^2/c^2}$, than

Hendrik Antoon Lorentz

b. Arnhem, Netherlands
18 July 1853

d. Haarlem, Netherlands
4 February 1928

Teylers Museum, Harlem.
Courtesy of the Boerhaave Museum, Leiden.

... the final aim of all research must be the deduction of the innumerable natural phenomena as necessary consequences of a few simple fundamental principles.
[Inaugural address [330, p. 26], Leiden, 25 January, 1878.]

the value m obtained in the frame in which it is at rest ($u = 0$). This result depends on the finiteness of c and is a property of Lorentz relativity which is in agreement with experiment. Consequently, data on particle masses at high speed show that c must in fact be finite. Such data can in principle provide a measurement [316] of the value of c. The measurements need not involve electromagnetism. They could, at least in principle, be carried out without basing synchronization on light signals.

No observations or experiments are known which invalidate Lorentz relativity. The second postulate embodies these observations by ruling out the Galilean case except as a low-speed approximation. Experiments which stringently test the validity of the Lorentz form of the boost transformation are discussed by Will [588] and Haughan and Will [224].

The Principle of causality limits the x-component v of the Lorentz boost velocity to $-c < v < c$. However, actual particle or wave-front speeds within a frame are bounded by $-c \leq u \leq c$ justifying the name *ultimate speed* for c. We may confirm this result by application of the Principle of

12.3. Lorentz transformations

causality to the Lorentz transformation, Equation 5.30,

$$\Delta \bar{t} = \gamma(\Delta t - v_x \Delta x/c^2) \,, \tag{12.1}$$

of the time interval Δt between a pair of events able to be causally related, where v_x is the boost speed in the positive x-direction. Since $\Delta t > 0$ implies $\Delta \bar{t} > 0$ for causal relationship, such event pairs must be related by $\Delta t > v_x \Delta x/c^2$ for all $v_x < c$. This relation will hold for all those spacetime intervals Δt and Δx which satisfy $\Delta x \leq c\Delta t$ but not those for which $\Delta x > c\Delta t$. If the axes are oriented arbitrarily relative to the boost, causally-related events satisfy $|\Delta \mathbf{x}| \leq c\Delta t$.

The existence of the finite velocity bound c means that all relativistic propagation will be of one of two types: either it takes place in each frame at some frame-dependent speed $u < c$ or it takes place at the invariant speed $u = c$ in all frames.

The Principle of relativity applies not only to uniform relative motion but also to inertial frames which are related by a translation in space or in time or by a rotation. In relativistic physics, one generally combines a rotation and a boost under the common name of *restricted Lorentz transformation*. All such transformations make up the *restricted Lorentz group*. The full Lorentz group denotes the extension which includes the spacetime reflections. The combination of a restricted or full Lorentz transformation with translations in space and time is called a restricted or full *Poincaré transformation*, all of which together make up the *restricted* or *full Poincaré group*. We may therefore re-state the Einsteinian postulates of relativistic physics in the weaker than customary (but sufficient) form of a:

- **Principle of Poincaré covariance** — the laws of physics are form-invariant under restricted Poincaré transformations.

Poincaré was the first to argue [448, 449] in 1905 (*Oeuvres* [451, vol. 9, pp. 490 and 513]) that the transformations, to which he attached the name of Lorentz, must form an invariance group of Maxwell's equations. In so doing, he was able to fix the value of one remaining undetermined overall constant in the transformations. Since a Lorentz or Poincaré transformation has, by definition, a finite parameter c, this covariance statement is presumed to incorporate the second postulate. The local laws of physics are therefore postulated to be *restricted Poincaré-covariant*. Most but not all of the local laws of physics are in fact also form-invariant under the full Poincaré group of transformations with reflections included.

As a consequence of the relativity postulates, some intuitively reasonable notions become either meaningless or have to be re-interpreted. *Instantaneous action at a distance* is inconsistent with an upper bound on speeds and must be replaced by *retarded action at a distance*. The concept of *absolute simultaneity*, for example, only applies as an approximation for nearby events when speeds are low. The idea of a *rigid body*, understood classically as one with invariant dimensions, is rendered meaningless by the finite time required for information, on an impact for example, to be transmitted from one part of it to another.

The notion of a static electric charge distribution with zero current has little meaning except in one particular frame. The concept of a purely electric or purely magnetic field is of little meaning for the same reasons and the same applies to a mass density without energy-momentum flux and stress or pressure.

In both Galilean and Lorentz relativity the spatial coordinate interval alters, namely $d\bar{\mathbf{x}} \neq d\mathbf{x}$, when one changes frames by boosting, although the proper Newtonian length ($|d\mathbf{x}|$ with $dt = 0$) is unchanged. In both cases, spatial coordinate separation is *relative* or frame-dependent. On the other hand, temporal coordinate intervals change, $d\bar{t} \neq dt$, in Lorentz relativity (*simultaneity* $dt = 0$ is relative) and are invariant, $d\bar{t} = dt$, in Galilean relativity (embodying the absoluteness of simultaneity). In this respect Lorentz relativity has a greater degree of symmetry between space and time than its Galilean counterpart. In fact, the spacetime of Lorentz relativity has the maximum possible symmetry between the concepts of (flat) space and (flat) time consistent with causality and with one being 3D and the other 1D.

12.4 Geometry of spacetime

We confine our attention to spacetime coordinate systems which do not involve a change of the time and space length scales. It follows, as an immediate consequence of the fact that boosts are described by Lorentz transformations, that the 4D spacetime of special relativity is *geometric* in contrast to Euclidean, Galilean and Newtonian spacetimes, $\mathbf{E}^1 \times \mathbf{E}^3$, $\mathbf{G} = \{\mathbf{E}^1, \mathbf{E}^3\}$ and $\mathbf{N}$, of Aristotelean and Newtonian physics, discussed in Chapters 2 to 7. That this is so can be seen from the existence of a 4D invariant *scalar product* $-c^2 dt_1 dt_2 + d\mathbf{x}_1 \cdot d\mathbf{x}_2$ involving the two 4D intervals $\{dt_1, d\mathbf{x}_1\}$ and $\{dt_2, d\mathbf{x}_2\}$ from one event $\{t, \mathbf{x}\}$ to two neighbouring events. Associated with this scalar product is a 4D *line element*, or distance formula,

$$ds^2 = -c^2 dt^2 + d\mathbf{x}^2 = -c^2 dt^2 + dx^2 + dy^2 + dz^2 \ . \qquad (12.2)$$

These expressions are invariant with respect to all the transformations taking one inertial frame to another inertial frame, namely with respect to the full Poincaré group of transformations (translations, rotations, reflections and Lorentz boosts). The geometric nature of spacetime encompasses not just distances but also angles and the speeds of boosts.

12.4. Geometry of spacetime

Exercise 12.1 Verify that $-c^2 dt_1 dt_2 + d\mathbf{x}_1 \cdot d\mathbf{x}_2$, and thus also ds^2 in Equation 12.2, is invariant with respect to the Lorentz boost of Equation 5.30, and indeed with respect to the full Poincaré group, including reflections.

Exercise 12.2 Show that, up to a constant factor, $-c^2 dt_1 dt_2 + d\mathbf{x}_1 \cdot d\mathbf{x}_2$ is the unique Poincaré-invariant homogeneous quadratic in the components of the two intervals $\{dt_1, d\mathbf{x}_1\}$ and $\{dt_2, d\mathbf{x}_2\}$.

Note that ds^2 is *not* here an abbreviation for the square of any real positive quantity ds but rather an abbreviation for the value of the above right side of Equation 12.2 which may be positive, negative or zero. The line element ds^2 is therefore said to be *indefinite*.

We have arbitrarily chosen a $\{-+++\}$ line element in Equation 12.2 with *signature* equal to 2, the number of positive less the number of negative coefficients. This is referred to as the *spacelike convention* for the signature of the line element. We could just as easily introduce ds^2 as $c^2 dt^2 - d\mathbf{x}^2$ with signature -2, called the timelike convention, with no change in the physics. The latter is more convenient for particle motion. The former is more useful if one is involved a great deal with the 3D subspace of spacetime at a given time in some frame or more generally with an arbitrary 3D spacelike surface (see Section 15.11). Such surfaces are especially important in a general relativistic context.

The spacelike convention also has the advantage that notation used for Cartesian tensors in Newtonian physics is the same as for the spatial components of relativistic tensors whereas, in the timelike convention, certain sign differences will appear. We have adopted the spacelike convention despite the fact that the timelike convention has been in widespread use in non-gravitational physics applications, in particular in fundamental studies of the electromagnetic and nuclear interactions.

We note that the invariant 4D scalar product vanishes if one interval $\{cdt_1, \mathbf{0}\}$ has zero spatial components and the other $\{0, d\mathbf{x}_2\}$ has zero time component. We therefore consider the time coordinate, as measured on clocks at fixed spatial coordinates to be 4-dimensionally orthogonal in spacetime, referred to as *Lorentz orthogonality*, to each of the spatial directions. No such space-time orthogonality exists in Euclidean, Galilean or Newtonian spacetime.

By analogy with the 3D case we consider the time axis to define a *spacetime direction* and we assign to that direction a unit vector $\mathbf{e}_t$ or $\mathbf{e}_0$. Equation 5.30 shows that the direction $\mathbf{e}_{\bar{t}}$ is not orthogonal in 4D to the directions $\mathbf{e}_k$; rather it is orthogonal to the directions $\mathbf{e}_{\bar{k}}$ in the frame in which the clocks measuring $\bar{t}$ are at rest.

Between 1905 and about 1910, there was little appreciation of the true significance of the special relativity of Einstein. There was a widespread belief that it was an alternative description of the electron theory of Lorentz, incorporating some of Poincaré's ideas on that theory and relativity. However, possibly as early as 1905 (see Hirosige [251]), and certainly no later than November 1907 (not 1908 as is widely reported) Minkowski began a re-interpretation of the postulates of special relativity which contributed enormously to a deeper appreciation of its revolutionary nature.

In an address [381] delivered in Göttingen in 1907, on the *Principle of relativity*, Minkowski initiated the geometric approach to relativity by introducing a 4D space in which time (or more specifically $x^4 = ict$) appeared as a fourth coordinate along with the three spatial coordinates. This was followed a few weeks later by the first paper [382] on the electrodynamics of a macroscopic body in which appears an appendix stressing the need for a reformulation of mechanics consistent with special relativity. His well-known lecture [383] on 4D spacetime, reproduced in reference [343], was given near the end of 1908 and published the following year.

The space and time of Einsteinian relativity constitute together a 4D geometric space referred to as *Minkowskian spacetime*. We shall denote it by $\mathbf{E}^{1,3}$ to indicate the signs in the metric. This will also distinguish it from 4D Euclidean space $\mathbf{E}^4$ and the non-metric spacetimes, $\mathbf{E}^1 \times \mathbf{E}^3$ and $\mathbb{G} = \{\mathbf{E}^1, \mathbf{E}^3\}$, of Aristotelean and Galilean physics.

We may geometrically characterize Minkowskian spacetime by noting that it is metric, flat, orientable, reflection-symmetric and covariantly homogeneous and isotropic. Minkowskian spacetime does not have a natural, non-trivial fibre bundle structure, corresponding to that of Galilean spacetime, since there is no unique invariant time coordinate to act as base manifold.

12.5 Spacetime causal structure and proper time

In Galilean relativity, the origin event ($t = t_0$ and $\mathbf{x} = \mathbf{x}_0$) may cause any simultaneous or later event (with $t \geq t_0$) and may be caused by any event with $t \leq t_0$, the equality holding in each case only for the action at a distance phenomena of Newtonian mechanics. The inequality is appropriate to the spatially local action, propagated at finite but unbounded speed, of a Galilean field such as a Schrödinger wave function. Galilean relativity gives an incorrect planar causal structure, depicted in Figure 12.1, in which the future of an event is the entire 'upper half' ($t > t_0$) of spacetime.

In special relativity, all events $\{t, \mathbf{x}\}$ that are capable of being causally related to the origin event $\{t_0, \mathbf{x}_0\}$ are given by $|\Delta \mathbf{x}| \leq c\Delta t$, namely,

$$(x - x_0)^2 + (y - y_0)^2 + (z - z_0)^2 \leq c^2(t - t_0)^2 \ . \qquad (12.3)$$

We choose now to suppress one of the spatial dimensions in this equation in order to include in a diagram the time t (or rather ct) along one of

12.5. Spacetime causal structure and proper time

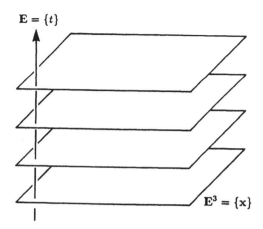

Figure 12.1 Planar causal structure of Galilean relativity

the three axes of the space of our immediate surroundings. Hence we plot only the 3D cross-section $z - z_0 = 0$ of this 4D spacetime relation, namely $(x - x_0)^2 + (y - y_0)^2 \leq c^2(t - t_0)^2$ with ct, x and y on the three orthogonal axes. We obtain the surface and interior of a double cone of apex angle $90°$, as shown in Figure 12.2. The event at the origin $\{t_0, \mathbf{x}_0\}$ is capable of causing events only inside or on the future half cone, $t > t_0$ and $|\Delta \mathbf{x}| \leq c\Delta t$, and only events in the past half cone, $t < t_0$ and $|\Delta \mathbf{x}| \leq c\Delta t$, are capable of causing an event at $\{t_0, \mathbf{x}_0\}$.

Spacetime relationships plotted in this way are called *spacetime* or *Minkowski diagrams*. Figure 12.2 illustrates a cross-section of the *causal structure* of spacetime represented by Equation 12.3. Such causal structure is said to be *conical*. However, one should always remember that Figure 12.2 is an incomplete representation of the structure fully described by Equation 12.3. If the second event $\{t, \mathbf{x}\}$ is on the surface of the 'cone', namely if $|\Delta \mathbf{x}| = c\Delta t$, then the two events can be related only by signals propagating at the ultimate speed c. The 3D spacetime surface given by,

$$(x - x_0)^2 + (y - y_0)^2 + (z - z_0)^2 - c^2(t - t_0)^2 = 0 \qquad (12.4)$$

is called the *null cone* or *light cone* of the origin event. The term null arises from the zero value of the quadratic form, and acknowledges the fact that light is not the only phenomenon whose propagation is along this surface. From Exercise 12.1, the left side of the above equation is invariant with respect to translations, rotations, boosts and reflections. The spacetime neighbourhood surrounding an event may therefore be divided frame-independently into five disjoint regions (not including the event itself). Those regions are the two half-cone surfaces, the two half-cone interiors

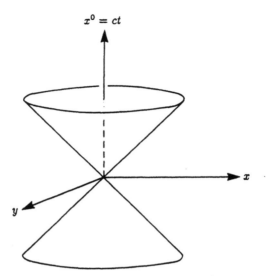

Figure 12.2 Minkowski diagram of the conical causal structure of spacetime.

and all the rest of spacetime exterior to the double cone.

Poincaré transformations are not the only ones which leave unaltered the division of spacetime near an event. Others which do so are *scale changes*, $\{t, \mathbf{x}\} \rightarrow \lambda\{t, \mathbf{x}\}$ (λ constant), which also take inertial frames to inertial frames, and more generally, *conformal transformations* [516] which preserve shape and angle but not scale. However, the latter changes will not in general change inertial frames to inertial frames. Some of the laws of physics, in particular those of purely electromagnetic nature, are invariant with respect to conformal transformations. A theory cannot be invariant under such transformations if it contains a fundamental rest mass. One way to understand this is to note that in natural units a mass has the dimensions of reciprocal length and will therefore be affected by scale changes.

If the second event is in the interior of the cone, $|\Delta \mathbf{x}| < c\Delta t$, the two events can be causally related by particles or signals with speed $u < c$. If $\{t, \mathbf{x}\}$ lies outside the cone for $\{0, 0\}$ then there are no causal ($u \leq c$) (non-tachionic) signals by which they can be related. In special relativity, instantaneous action at a distance is acausal.

The spacetime region outside the light cone, *elsewhere* relative to the apex event, is a much more extensive region than a 3D cross-section, $t = t_0$ and $\mathbf{x} \neq \mathbf{x}_0$, which in Newtonian spacetime comprises our intuitive notion of elsewhere. Every event will have its own frame-independent cone of causal structure and for some purposes spacetime may be regarded as a network of such interlocking cones [434].

When field equations, or particle equations of motion, satisfying the

12.5. Spacetime causal structure and proper time

relativity postulates are obtained they may be divided into two types, those which propagate inside the null cone at a speed $u < c$ (the actual speed will be frame-dependent) and those which propagate along the surface of the null cone at the ultimate speed in vacuo (an invariant result). These two types of equations are fundamentally different, the latter being a phenomenon with no analogue in Newtonian spacetime.

A very simple set of wave equations satisfying the Principle of relativity are Maxwell's equations and we shall see in Chapter 18 that, as a consequence of charge conservation, they fall into the category of those giving propagation at the ultimate speed. Consequently, a measurement of the speed of propagation of electromagnetic phenomena in vacuo, the *speed of light*, is in essence a measurement of the ultimate speed. Gravitational effects, described relativistically, also propagate at the ultimate speed and so also will neutrinos if it turns out that they are massless [42], as is generally assumed. The equations of electromagnetism, of gravity and possibly of neutrino propagation, are thus fundamentally relativistic.

The line element of Equation 12.2 provides a very convenient classification of event pairs (one of which may be taken at the origin) into three causal types:

- *timelike interval*: $ds^2 < 0$ giving a pair of events causally relatable with signals at frame-dependent speeds $u < c$.

- *null* or *lightlike interval*: $ds^2 = 0$ giving a pair of events causally relatable with signals at speed $u = c$.

- *spacelike interval*: $ds^2 > 0$ which implies that the pair of events cannot be causally related.

Any pair of events along the trajectory of a particle with non-zero rest mass are separated by a timelike interval. For any timelike interval we may construct a real, invariant quantity of time dimensions by the relation,

$$d\tau = \sqrt{-ds^2}/c. \qquad (12.5)$$

This quantity provides a frame-independent measure of the time interval between the two events and is called the *proper time* interval between them.

In the same way we may define an invariant *proper length* of any spacelike interval according to,

$$d\ell = \sqrt{ds^2}. \qquad (12.6)$$

Let $\{\Delta t, \Delta \mathbf{x}\}$ be the small coordinate intervals in frame S of timelike separated events. For such an interval, a frame $\bar{S}$ always exists in which the events occur at the same spatial location. This frame is called the *proper frame* of the two events. The proper frame of a small timelike interval will be the instantaneous *rest frame* of a particle for which both events are on the particle trajectory.

Exercise 12.3 *Find the velocity v of $\bar{S}$ relative to S. Show that the proper time interval is equal to the coordinate time interval in the proper frame and is the minimum of all the coordinate time intervals between the events as measured in various inertial frames.*

Exercise 12.4 *Use Equation 12.5 to establish the time-dilation formula,*

$$dt = \gamma \, d\tau = \frac{d\tau}{\sqrt{1 - v^2/c^2}} \, , \qquad (12.7)$$

giving the coordinate time interval in an arbitrary frame in terms of the proper time interval and the relative velocity v of the arbitrary and proper frames.

The kinematic phenomenon of *time dilation* leads to the well-known conundrum of the so-called *clock paradox*. This was first stated by Einstein [128] in 1905 as: 'A clock which is moved in a closed path, relative to a clock which remains at rest in an inertial frame, will record an earlier time (by $\frac{1}{2}v^2/c^2$ for small speeds v) than the stationary clock'. Discussions of this result and the *twin paradox* based on it may be found, along with numerous other 'paradoxes', in almost all relativity texts, for example French [175]. Direct experimental evidence for the clock paradox effect was obtained with terrestrial circumnavigation of atomic clocks by Hafele and Keating [217] in 1972. The time dilation formula does not imply that physical time is slowed down by movement at high speeds. It is a kinematic relation between two different measurements of the time interval between a pair of events based on a sound theory of measurement and the exclusion, corroborated by experiment, of arbitrarily large speeds of signal propagation.

Let $\{\Delta t, \Delta \mathbf{x}\}$ be the small coordinate intervals in frame S of spacelike separated events. For such an interval, a frame $\bar{S}$ always exists, the proper frame of the interval, in which the events occur at the same coordinate time.

Exercise 12.5 *Find the velocity v of $\bar{S}$ relative to S. Show that the proper length is equal to the coordinate length in the proper frame and that it is the maximum of all the coordinate lengths.*

Exercise 12.6 *Establish the Lorentz-FitzGerald contraction formula,*

$$|d\mathbf{x}| = \gamma^{-1} \, d\ell = \sqrt{1 - v^2/c^2} \, d\ell \, , \qquad (12.8)$$

relating the proper length $d\ell$ to the coordinate length $|d\mathbf{x}|$ in frame S and the relative velocity v of coordinate and proper frames.

The kinematic or *Lorentz-FitzGerald contraction* formula does not imply that an object contracts as a result of motion at high speeds. Neither are the changes in time and length intervals appearing in the Lorentz transformations simply due to distortions in the appearance that can be expected for a fast object as a result of the finite time of travel from the object to an observer or camera, granted that different parts of an object will be at different distances. The measurements referred to in the Lorentz transformations are obtained by arrays of observers, with clocks that have been previously synchronized in each frame, each event being timed by the observer spatially coincident with it. The distortions seen by one observer are more complex. A thorough analysis was given in 1959 and the results are discussed by Scott and Viner [507]. An extensive set of interesting diagrams are given in Mook and Vargish [387].

The indefiniteness of the line element may be summed up by the invariant expression,

$$ds^2 = -c^2 d\tau^2, \quad 0 \quad \text{or} \quad d\ell^2 \ . \tag{12.9}$$

Coordinates are simply labels. They do not have intrinsic (invariant) meaning. Proper intervals are invariant, however, and thus provide intervals suitable for recording intrinsic properties such as half-lives of radioactive isotopes, sizes of atoms, and characteristic wavelengths of spectral lines.

The quadratic form of each term in the line element of Equation 12.2 ensures the invariance of the metric under time-reversal and parity, the constancy of each coefficient embodies the homogeneity of space and time and the values, ± 1, of these coefficients embody not only isotropy in space but a sort of 'isotropy' linking the time axis to each of the spatial axes. We shall see that this spacetime 'isotropy' is in fact nothing but the boost invariance. For this reason, one often describes the combination of spatial rotations and boosts as *covariant* or *Lorentz rotations*.

12.6 The dimensionality of spacetime

Many physicists and mathematicians believe that the 4-dimensional nature of the observable universe may eventually be understood as a consequence of some simple physical principle. It has long been well-known [28] that 3D spaces, and 4D spacetimes, have many very special mathematical properties, some of which are unique [559]. It is also well-known that some physical properties of the universe are only possible in spacetimes with certain specific dimensions, which include the 4D spacetime of the real world. Among these are the possibility of sending coherent signals, the long-range nature of the gravitational and electromagnetic interactions, chiral symmetry (which cannot occur in odd-dimensional spacetimes) and the existence of a unique antiparticle to the electron, to name just a few properties.

During the last few years, a series of remarkable discoveries about 4-dimensional spaces, whether Euclidean $\mathbf{E}^4$ or Minkowskian $\mathbf{E}^{1,3}$ (which are topologically identical), have established their uniqueness among all the Euclidean spaces, $\mathbf{E}^n$. Specifically, it has been shown that not all 4D spaces may be classified solely by examining the effects of smooth deformations. Some claim this result, reported by Watson [559] in July 1988 and described more generally in 1987 by Stewart [521], to be one of the most important developments in mathematics this century. It could also have deep significance in physics. Indeed, part of the mathematical development has arisen from ideas first explored in gauge theories of the physical interactions.

God is subtle but he is not malicious.

Albert Einstein   1879–1955

Bibliography:

Angel R B 1980 *Relativity: the theory and its philosophy* [17].
Bergmann P G 1942 *Introduction to the theory of relativity* [32].
Bernstein J 1973 *Einstein* [34].
Clark R W 1971 *Einstein: the life and times* [67].
Cohen I B 1985 *Revolution in science* [73].
French A P 1968 *Special relativity* [175].
Goldberg S 1984 *Understanding relativity* [203].
Haughan M P and Will C M 1987 *Modern tests of special relativity* [224].
Longair M S 1983 *Theoretical concepts in physics* [326].
Lorentz H A et al., *The Principle of relativity* [343].
Møller C 1962 *The theory of relativity* [386].
Pais A 1982 *Subtle is the Lord* [418].
Pauli W 1981 *Theory of relativity* [423].
Pyenson L 1985 *The young Einstein* [465].
Rindler W 1977 *Essential relativity* [476].
Stewart I 1987 *The problems of mathematics* [521].
Watson A 1988 *Mathematics of a fake world* [559].

CHAPTER 13

Poincaré transformations

'Lorentz invariance' is a general condition for any physical theory. This was for me of particular importance because I had already previously found that Maxwell's theory did not account for the micro-structure of radiation and could therefore have no general validity[1].

Albert Einstein   1879–1955

It is possible to formulate all relativistic scalar and vector equations, and some tensor equations, entirely in terms of the time t and the 3D Euclidean coordinates x in some frame and the corresponding 3-scalars and 3-vectors. Such a formulation may be referred to as $1 + 3$ relativistic notation — it retains the advantage of similarity with the familiar 3-vector and 3-scalar notation of non-relativistic physics. However, it is also possible — and very useful for most relativistic calculations — to combine time and space together as *Minkowskian spacetime* by noting that together they have many properties that are very similar to 3D Euclidean space extended to 4D, the other principal difference being the indefiniteness of the metric.

13.1 Covariant notation

After a boost, the new time and space coordinate intervals are each a linear, homogeneous combination of both the time and space coordinate intervals in the old frame. Spacetime is not invariantly divisible into space and time separately. To help embody these features, a new time coordinate $x^0 = ct$ is defined having the dimensions of length. All events are described by spacetime coordinates $x = \{x^\mu\}$ labelled by a *Lorentz index*, here μ, for which we shall (almost) always use a lower-case Greek letter $\mu, \nu, \lambda, \ldots$, from near the middle of the alphabet, say $\{\mu\} = \{0, 1, 2, 3\} = \{0, k\}$, where $\{k\} = \{1, 2, 3\}$ labels the spatial coordinates $\{x^k\}$ as in Newtonian physics.

---

[1] Letter quoted and translated in Born *Physics and relativity* [44].

Thus we use:
$$x = \{x^\mu\} = \{x^0, x^1, x^2, x^3\} = \{x^0, x^k\} = \{ct, x, y, z\} = \{ct, \mathbf{x}\} \ . \quad (13.1)$$

We use x here without any bolding to denote the entire set of four coordinates $\{x^\mu\}$ much as $\mathbf{x}$ denotes the triple of spatial coordinates. This should cause no confusion with the very occasional use of x for one particular spatial coordinate. We shall not consider $x = \{x^\mu\}$ to be a vector in Minkowskian spacetime since coordinates no longer have this property with the relativistic introduction of gravitational effects. Furthermore, given the homogeneity of spacetime, the particular event chosen as origin of our coordinates is arbitrary — consequently, the significant vector quantities will be spacetime intervals between events. We shall take a small spacetime interval $dx = \{dx^\mu\} = \{c\,dt, dx, dy, dz\}$ as the prototypical 4-vector in spacetime since it retains this rôle locally with gravity included.

If, instead of $x^0 = ct$, an imaginary fourth coordinate $x^4 = ict$ is introduced, the resulting space would have the appearance of a 4D positive-definite Euclidean space, $\mathbf{E}^4$. This mathematical convention of an imaginary time coordinate has not been used here for several reasons. One is that hiding the indefiniteness of the metric can cause one to forget the crucial differences between indefinite and definite metric spaces. Another difficulty arises from the careful sign changes needed in time components in quantum applications where complex conjugation is of fundamental physical significance. Complex conjugation is also used in QFT as a device to efficiently manipulate the non-hermitian fields which represent charged particles and to evaluate certain path integrals on the Euclidean branch of complex Minkowskian spacetime [467]. In fact, complex conjugation is already a sufficiently widely-used device in fundamental physics for us to choose to not to use it for another all-pervasive rôle.

In order to incorporate automatically the indefiniteness of the metric, three conventions are introduced:

- We allow dx (and all spacetime vectors) to have an alternative *dual component form* equivalent, but not identical, to $\{dx^\mu\}$ in which all the indices are lowered and in which the spatial components are unchanged but the time component is of opposite sign,

$$\{dx_\mu\} \equiv \{dx_0, dx_k\} \stackrel{\text{def}}{=} \{-dx^0, dx^k\} \ . \quad (13.2)$$

We refer to $\{dx_\mu\}$ and $\{dx^\mu\}$ as the *covariant* and *contravariant* component forms respectively of the one vector dx. The change in sign corresponds to our choice of spacelike convention for the line element. (The dx_μ components are not differentials of coordinates but are equivalent to $(dx)_\mu$, namely just covariant components of dx. There is only one set of coordinates, namely x^μ. For this reason we

13.1. Covariant notation

shall also never lower the indices μ and k on coordinates, only on components of vectors.)

- We agree that the *Einstein summation convention* will apply over the range $0, 1, 2, 3$ to any pair of repeated Lorentz indices which must always be one up and one down:

$$a^\mu b_\mu \equiv \sum_{\mu=0}^{3} a^\mu b_\mu \ . \tag{13.3}$$

- All free indices (those which are not summed over) must be the same, and at the same level, in all terms of every equation.

These rules imply that:

$$\begin{aligned}(dx_1)^\mu(dx_2)_\mu &= \sum_{\mu=0}^{3}(dx_1)^\mu(dx_2)_\mu \\ &= (dx_1)^0(dx_2)_0 + (dx_1)^k(dx_2)_k \\ &= -c^2 dt_1 dt_2 + \mathbf{dx}_1 \cdot \mathbf{dx}_2 \ . \end{aligned} \tag{13.4}$$

More specifically, it implies that the 4D line element of Equation 12.2 is given by the concise formula,

$$ds^2 = dx^\mu dx_\mu \ , \tag{13.5}$$

in which the indefiniteness is embodied in the rules obeyed by the two types of Lorentz indices. In using a summation convention for Euclidean indices, $k, l, \ldots$, it made no difference whether the indices were one up and one down or both at the same level, either up or down. However, with the Lorentz indices of Minkowskian spacetime quantities, we must maintain the index levels if the convention is to incorporate the indefiniteness of the metric automatically.

The above convention means that an arbitrary linear combination of the 16 terms of the symmetric array $\{dx^\mu dx^\nu\}$ can always be written as a double sum with a symmetric coefficient $g_{\mu\nu} = g_{\nu\mu}$, in the form,

$$\begin{aligned} g_{\mu\nu} dx^\mu dx^\nu &= g_{0\nu} dx^0 dx^\nu + g_{1\nu} dx^1 dx^\nu + g_{2\nu} dx^2 dx^\nu + g_{3\nu} dx^3 dx^\nu \\ &= g_{00} dx^0 dx^0 + g_{01} dx^0 dx^1 + \cdots + g_{33} dx^3 dx^3 \ . \end{aligned} \tag{13.6}$$

By analogy with Equation 3.13 for the 3D line element, the 4D line element of Equation 12.2 can be expressed in standard metric form as a particular homogeneous quadratic in dx^μ. We introduce the symmetric *Minkowski matrix* $\eta = \{\eta_{\mu\nu}\}$ by,

$$\eta = \{\eta_{\mu\nu}\} = \text{diag}\{-1, 1, 1, 1\} \ , \tag{13.7}$$

in terms of which the Lorentz scalar product, and the line element of Equation 12.2, are expressed as,

$$(dx_1)^\mu (dx_2)_\mu = \eta_{\mu\nu}(dx_1)^\mu (dx_2)^\nu \quad \text{and} \quad ds^2 = \eta_{\mu\nu} dx^\mu dx^\nu. \tag{13.8}$$

Comparison of Equations 13.5 and 13.8 shows that

$$dx_\mu = \eta_{\mu\nu} dx^\nu, \tag{13.9}$$

indicating that $\eta_{\mu\nu}$ functions as an *index-lowering* operator between the two equivalent ways of writing the components of dx.

The level of indices is important in this notation. Both μ indices in Equation 13.9, for example, are covariant. We express the identity transformation $x = \mathbb{1}x$ in indexed form as $x^\mu = \delta^\mu{}_\nu x^\nu$ in which the 4D *Kronecker delta*,

$$\delta^\mu{}_\nu = 1 \text{ for } \mu = \nu \quad \text{and} \quad \delta^\mu{}_\nu = 0 \text{ for } \mu \neq \nu, \tag{13.10}$$

is an indexed version of the identity matrix $\mathbb{1} = \{\delta^\mu{}_\nu\}$, a special trivial example of a transformation matrix, L. This means that the components of the latter should also be written as $L = \{L^\mu{}_\nu\}$ with its indices one up and one down. Irrespective of the level, we shall use the first index to label the rows of a matrix and the second for the columns. The multiplication of two transformation matrices $\Lambda = \{\Lambda^\mu{}_\nu\}$ and $L = \{L^\mu{}_\nu\}$ will therefore be written

$$(\Lambda L)^\mu{}_\nu = \Lambda^\mu{}_\lambda L^\lambda{}_\nu. \tag{13.11}$$

The levels of the indices must be adjusted for other types of matrices. For example, the matrix multiplication of η with $L = \{L^\mu{}_\nu\}$ may be written $(\eta L)_{\mu\nu} = \eta_{\mu\lambda} L^\lambda{}_\nu$ in which the natural indexing of the right side determines the indexing on the left.

Similarly, the transpose Λ^T of Λ, left multiplied into η, will be indexed as,

$$(\Lambda^\mathrm{T} \eta)_{\mu\nu} = (\Lambda^\mathrm{T})_\mu{}^\lambda \eta_{\lambda\nu} = \Lambda^\lambda{}_\mu \eta_{\lambda\nu}. \tag{13.12}$$

We may use these results to determine a natural form for the indexed quantity corresponding to η^{-1} and thus obtain an *index-raising* operator. Indexing $\mathbb{1} = \eta^{-1}\eta$ gives,

$$\delta^\mu{}_\nu = (\eta^{-1}\eta)^\mu{}_\nu = (\eta^{-1})^{\mu\lambda} \eta_{\lambda\nu}.$$

Despite the fact that the values of η^{-1} are just $\mathrm{diag}\{-1,1,1,1\}$, the same as for η, we give them a different indexed form $(\eta^{-1})^{\mu\nu} \equiv \eta^{\mu\nu}$ consistent with the above relation which therefore becomes,

$$\eta^{\mu\lambda}\eta_{\lambda\nu} = \delta^\mu{}_\nu. \tag{13.13}$$

We may clearly change any free index in an equation to any other unused index provided we change it in all terms. Similarly, we may alter a pair of dummy (summed) indices in any term of an equation to any other index not used in that term. More than two identical indices in one term is ambiguous and never permitted. To avoid this it will sometimes be necessary either to re-label a pair of dummy indices or to re-label one or more free indices in every term of an equation.

Exercise 13.1 *Use Equations 13.9 and 13.13 to show that,*

$$dx^\mu = \eta^{\mu\nu} dx_\nu \ . \tag{13.14}$$

Explain why $\eta^\mu{}_\nu$, resulting from the raising of one index of $\eta_{\mu\nu}$, or lowering one index $\eta^{\mu\nu}$, is equal to the Kronecker delta $\delta^\mu{}_\nu$. Show that $\eta^\mu{}_\mu = 4$.

The spacetime coordinates $\{x^\mu\}$ of a set of synchronized inertial clocks will be referred to as *Minkowskian coordinates* and the basis,

$$\{\mathbf{e}_\mu\} = \{\mathbf{e}_0, \mathbf{e}_k\} = \{\mathbf{e}_0, \mathbf{e}_1, \mathbf{e}_2, \mathbf{e}_3\} \ , \tag{13.15}$$

of unit 4-vectors tangential to the lines of varying x^μ may be called a *Lorentz frame* in Minkowskian spacetime.

Exercise 13.2 *Show that the basis vectors $\mathbf{e}_\mu$ corresponding to the coordinates x^μ of $\mathbf{E}^{1,3}$ satisfy the orthonormality relation,*

$$\mathbf{e}_\mu \cdot \mathbf{e}_\nu = \eta_{\mu\nu} \ . \tag{13.16}$$

Suppose we have a Lorentz-indexed quantity $a = \{a^{\cdots\mu\cdots}{}_{\nu\cdots}\}$. Then the process of letting two such indices be identical, with one up and one down, thereby forming the implied sum,

$$a^{\cdots\mu\cdots}{}_{\mu\cdots} = \delta^\mu{}_\nu \, a^{\cdots\nu\cdots}{}_{\mu\cdots} = \eta_{\mu\nu} \, a^{\cdots\mu\cdots\nu\cdots} \ , \tag{13.17}$$

is referred to as the *contraction* of one index on the other. The two indices may also be on two separate quantities.

13.2 Minkowskian spacetime isometry group

In previous sections we introduced Minkowskian spacetime precisely so that it embodies the translation, rotation, Lorentz boost, time-reversal and parity transformations invariantly. This guarantees the observed continuity, smoothness, causal structure, homogeneity, isotropy and reflection invariance of time and space. However, it might be that the spacetime $\mathbf{E}^{1,3}$ so constructed has other symmetries leaving ds^2 invariant. It will be instructive to show that this is not the case. The demonstration will also provide us with a more compact notation for Lorentz and Poincaré transformations.

We are interested in all those coordinate transformations from one inertial frame with coordinates $\{t, x\}$ to another equivalent inertial frame with coordinates $\bar{x} = \{x^{\bar{\mu}}\} = \{c\bar{t}, \bar{x}\}$ where $\bar{x} = \bar{x}(x)$ may be any four invertible functions of the four coordinates x such that ds^2 is left invariant. We use a notation in which the individual coordinates in different frames are usually distinguished from one another by marking the index ($\bar{\mu}$) rather than the kernel (x). We follow the procedure outlined in Robertson and Noonan [480]. Using the chain rule,

$$dx^{\bar{\mu}} = \frac{\partial x^{\bar{\mu}}}{\partial x^{\mu}} dx^{\mu} , \qquad (13.18)$$

in $d\bar{s}^2 = ds^2$, namely in $\eta_{\bar{\mu}\bar{\nu}} dx^{\bar{\mu}} dx^{\bar{\nu}} = \eta_{\mu\nu} dx^{\mu} dx^{\nu}$, gives

$$\eta_{\bar{\mu}\bar{\nu}} \frac{\partial x^{\bar{\mu}}}{\partial x^{\mu}} \frac{\partial x^{\bar{\nu}}}{\partial x^{\nu}} = \eta_{\mu\nu} . \qquad (13.19)$$

We differentiate with respect to x^{π} to obtain

$$\eta_{\bar{\mu}\bar{\nu}} \left(\frac{\partial^2 x^{\bar{\mu}}}{\partial x^{\mu} \partial x^{\pi}} \frac{\partial x^{\bar{\nu}}}{\partial x^{\nu}} + \frac{\partial x^{\bar{\mu}}}{\partial x^{\mu}} \frac{\partial^2 x^{\bar{\nu}}}{\partial x^{\pi} \partial x^{\nu}} \right) = 0 . \qquad (13.20)$$

Exercise 13.3 Permute $\mu\pi\nu$ to $\pi\nu\mu$ and then to $\nu\mu\pi$ to obtain a second and third equation. Subtract the second and add the third to Equation 13.20 and make use of the symmetry of $\eta_{\bar{\mu}\bar{\nu}}$ and of $\partial_{\mu} \partial_{\nu}$ in $\mu\nu$ to show that,

$$\frac{\partial^2 x^{\bar{\mu}}}{\partial x^{\mu} \partial x^{\pi}} \eta_{\bar{\mu}\bar{\nu}} \frac{\partial x^{\bar{\nu}}}{\partial x^{\nu}} = 0 . \qquad (13.21)$$

Equation 13.21 may be considered as a set of matrix equations of the form $A_{\mu} \eta L = 0$ where,

$$L^{\bar{\nu}}{}_{\nu} = \frac{\partial x^{\bar{\nu}}}{\partial x^{\nu}} \quad \text{and} \quad (A_{\mu})_{\pi}{}^{\bar{\mu}} = \frac{\partial^2 x^{\bar{\mu}}}{\partial x^{\mu} \partial x^{\pi}} = \frac{\partial L^{\bar{\mu}}{}_{\mu}}{\partial x^{\pi}} . \qquad (13.22)$$

Equivalence of the two frames requires the transformation to be invertible so L^{-1} exists, as does η^{-1}. Hence we arrive at $A_{\mu} = 0$ whose solution is

$$x^{\bar{\mu}} = L^{\bar{\mu}}{}_{\mu} x^{\mu} + a^{\bar{\mu}} , \qquad (13.23)$$

where $L = \{L^{\bar{\mu}}{}_{\nu}\}$ and $a^{\bar{\mu}}$ are sets of real constants. So far we have shown that the transformation from x to $\bar{x}$ must be linear. The constants of integration $a^{\bar{\mu}}$ are arbitrary. However, substituting this linear transformation back into Equation 13.19 gives a condition on L, namely

$$\eta_{\bar{\mu}\bar{\nu}} L^{\bar{\mu}}{}_{\mu} L^{\bar{\nu}}{}_{\nu} = \eta_{\mu\nu} \quad \text{or} \quad L^{\mathrm{T}} \eta L = \eta . \qquad (13.24)$$

13.2. Minkowskian spacetime isometry group

This *Lorentz pseudo-orthogonality condition* is the analogue of the orthogonality condition, $R^T R = \mathbb{1}$, on a rotation matrix, as can be seen by rewriting the 3D condition as $R^T \mathbb{1} R = \mathbb{1}$ and by noting that $\mathbb{1} = \text{diag}\{1,1,1\}$ and $\{\eta_{\mu\nu}\} = \text{diag}\{-1, \mathbb{1}\}$. Equation 13.23 can thus be abbreviated as,

$$x \to \bar{x} = Lx + a \quad \text{with} \quad L^T \eta L = \eta . \tag{13.25}$$

Exercise 13.4 Show that the transformations of Equation 13.25 form a group and that those with $a = 0$ or with $L = \mathbb{1}$ each comprise subgroups.

A transformation matrix $L = \{L^{\bar{\mu}}{}_\mu\}$ satisfying Equation 13.24 also satisfies,

$$\eta_{\bar{\mu}\bar{\nu}} = L^\mu{}_{\bar{\mu}} L^\nu{}_{\bar{\nu}} \eta_{\mu\nu} , \tag{13.26}$$

obtained by indexing the equivalent relation $\eta = L^{-1,T} \eta L^{-1}$. This equation implies that $\eta_{\mu\nu}$ is a numerical Lorentz tensor (see Chapter 14) which we call the *Minkowski metric tensor* corresponding to a similar property of δ_{kl} with respect to rotations in $\mathbf{E}^3$.

The subgroup arising from the inhomogeneous terms $a^{\bar{\mu}}$ (for which $L = \{L^{\bar{\mu}}{}_\nu\} = \{\delta^{\bar{\mu}}{}_\nu\} = \mathbb{1}$) constitutes the group of spacetime translations. We shall show that the subgroup with $a^{\bar{\mu}} = 0$ and with L satisfying Equation 13.24 describes rotations, Lorentz boosts and reflections and is therefore the full Lorentz group. Consequently, Equation 13.25 describes a full *Poincaré transformation* also known, particularly in earlier work, as an *inhomogeneous Lorentz transformation*.

When it has been shown that L in Equation 13.25 contains these Lorentz transformations and nothing else, we shall have established that, for Minkowskian spacetime, the distance- and angle-preserving transformations, called the *isometry group*, is the full Poincaré group whose properties are of fundamental importance in the study of relativistic particles and fields.

Other invariances involving the coordinates of spacetime are important in physics. Some of these are: *dilatations* (scale changes); *conformal transformations* (namely area-preserving shape and scale changes) and *general coordinate transformations* of relativistic gravitation theory, namely general relativity. Each such symmetry, arising from changes in the spacetime coordinates, is called a *spacetime* or *external* symmetry in contrast to an *internal symmetry* which leaves the spacetime coordinates unchanged. Examples of the latter are those that transform a set of fields in a way that mixes the fields or field components among themselves at a particular event without affecting the value of x or multiplying a single component field by a phase factor $e^{i\alpha}$ whose phase parameter α is not dependent on the spacetime coordinates. *Supersymmetry* transformations relating bosonic and fermionic particles and fields are both external and internal.

Jules Henri Poincaré

b. Nancy, France
29 April 1854

d. Paris, France
17 July 1912

Courtesy of the Mansell Collection, London.

Science is built up of facts, as a home is built of stones; but an accumulation of facts is no more a science than a heap of stones is a house.

13.3 Restricted Lorentz transformations

The parameters $a = \{a^{\bar{\mu}}\}$ of the inhomogeneous term in Equations 13.23 and 13.25 represent dispacements in spacetime. These do not affect the coordinate intervals dx^μ which, on differentiating Equation 13.23, are transformed solely by the Lorentz matrices $L = \{L^{\bar{\mu}}{}_\mu\}$ according to,

$$dx^{\bar{\mu}} = L^{\bar{\mu}}{}_\mu dx^\mu \quad \text{where} \quad L^{\bar{\mu}}{}_\mu = \frac{\partial x^{\bar{\mu}}}{\partial x^\mu} \,. \tag{13.27}$$

Transformation matrices are not themselves tensors (see Chapter 14). Consequently, just as we never lower indices on coordinates, we shall not raise or lower their indices.

Interchanging $\bar{\mu}$ and μ in Equation 13.27, shows that the natural notation, when using barred indices, for the Lorentz matrix $(L^{-1})^\mu{}_{\bar{\mu}}$ inverse to $L^{\bar{\mu}}{}_\mu$ (taking $x^{\bar{\mu}}$ back to x^μ) is $L^\mu{}_{\bar{\mu}}$ giving,

$$dx^\mu = L^\mu{}_{\bar{\mu}} dx^{\bar{\mu}} \quad \text{where} \quad L^\mu{}_{\bar{\mu}} = \frac{\partial x^\mu}{\partial x^{\bar{\mu}}} \,. \tag{13.28}$$

13.3. Restricted Lorentz transformations

The conditions expressing the reciprocity of the partial derivatives in each direction, namely the chain rules,

$$\frac{\partial x^{\bar{\rho}}}{\partial x^{\mu}}\frac{\partial x^{\mu}}{\partial x^{\bar{\rho}}} = \delta^{\bar{\rho}}{}_{\bar{\rho}} \quad \text{and} \quad \frac{\partial x^{\mu}}{\partial x^{\bar{\rho}}}\frac{\partial x^{\bar{\rho}}}{\partial x^{\nu}} = \delta^{\mu}{}_{\nu}\,, \tag{13.29}$$

here become

$$L^{\bar{\rho}}{}_{\mu}L^{\mu}{}_{\bar{\rho}} = \delta^{\bar{\rho}}{}_{\bar{\rho}} \quad \text{and} \quad L^{\mu}{}_{\bar{\rho}}L^{\bar{\rho}}{}_{\nu} = \delta^{\mu}{}_{\nu}\,. \tag{13.30}$$

Two matrices A and B are said to be *equivalent* if there exists a non-singular matrix S for which $A = SBS^{-1}$.

Exercise 13.5 *Use the Lorentz condition to express $L^{-1,T} \equiv (L^{-1})^T$ in terms of L and η to show that $L^{-1,T}$ and L are equivalent matrices, implying that the the two are related by a similarity transformation which can be interpreted as a basis change.*

Exercise 13.6 *Use the definition of $dx_{\bar{\rho}}$ in terms of $dx^{\bar{\rho}}$, namely Equation 13.9 in the barred frame, and re-express the result in the form,*

$$dx_{\bar{\rho}} = (\eta L \eta^{-1})_{\bar{\rho}}{}^{\mu} dx_{\mu} = (L^{-1,T})_{\bar{\rho}}{}^{\mu} dx_{\mu} = (L^{-1})^{\mu}{}_{\bar{\rho}} dx_{\mu} \equiv L^{\mu}{}_{\bar{\rho}} dx_{\mu}\,. \tag{13.31}$$

When two sets of vector components transform according to a given matrix (here L) and its inverse transpose $L^{-1,T}$, then the vectors are said to transform *contragrediently* to one another.

Since dx_{μ} are referred to as the covariant components of dx, it is natural to refer to dx^{μ} as the corresponding contravariant components. The transformation laws of the two component forms of the prototypical vector $dx = \{dx^{\mu}\} = \{dx_{\mu}\}$ are thus

$$dx^{\bar{\rho}} = L^{\bar{\rho}}{}_{\mu} dx^{\mu} \quad \text{and} \quad dx_{\bar{\rho}} = L^{\mu}{}_{\bar{\rho}} dx_{\mu}\,, \tag{13.32}$$

in which the implied transposition in the second of these expressions is only noticeable by the fact that the two μ indices of the matrix multiplication are not adjacent. Let $\partial_{\mu} = \partial/\partial x^{\mu}$ and $\partial_{\bar{\rho}} = \partial/\partial x^{\bar{\rho}}$ have contravariant forms,

$$\partial^{\mu} = \eta^{\mu\nu}\partial_{\nu} \quad \text{and} \quad \partial^{\bar{\rho}} = \eta^{\bar{\rho}\bar{\nu}}\partial_{\bar{\nu}}\,. \tag{13.33}$$

Exercise 13.7 *Show that ∂_{μ} and ∂^{μ} transform according to,*

$$\partial_{\bar{\rho}} = L^{\mu}{}_{\bar{\rho}}\partial_{\mu} \quad \text{and} \quad \partial^{\bar{\rho}} = L^{\bar{\rho}}{}_{\mu}\partial^{\mu}\,, \tag{13.34}$$

for x transforming by a Lorentz transformation $dx^{\mu} \to dx^{\bar{\rho}} = L^{\bar{\rho}}{}_{\mu} dx^{\mu}$.

Matrix transposition of Equation 13.24 leaves it unaltered showing that the number of independent conditions it imposes on a Lorentz matrix in 4D (see Appendix A.2) is $4(4+1)/2$ thus leaving $16-10=6$ independent parameters in L.

We shall demonstrate that these six parameters may be identified as three rotation and three boost parameters. We shall also show that any restricted Lorentz transformation may be reduced to a single boost followed by a single rotation or vice versa.

Exercise 13.8 *(a) Use the 00 component of the indexed version of Equation 13.24 to show that*

$$(L^{\bar{0}}{}_0)^2 - L^{\bar{k}}{}_0 L^{\bar{k}}{}_0 = 1 \ . \tag{13.35}$$

(b) Use the matrix version of Equation 13.24 to show that $\det L = \pm 1$.

It follows from Equation 13.35 that,

$$(L^{\bar{0}}{}_0)^2 \geq 1 \quad \text{and} \quad 0 \leq \frac{L^{\bar{k}}{}_0 L^{\bar{k}}{}_0}{(L^{\bar{0}}{}_0)^2} < 1 \ . \tag{13.36}$$

Rearrangement of the matrix version of Equation 13.24 gives $\eta^{-1} = L\eta^{-1}L^T$ and thus $\eta^{\bar{\mu}\bar{\nu}} = L^{\bar{\mu}}{}_\mu L^{\bar{\nu}}{}_\nu \eta^{\mu\nu}$ from which it follows that we also have

$$(L^{\bar{0}}{}_0)^2 - L^{\bar{0}}{}_k L^{\bar{0}}{}_k = 1 \ , \tag{13.37}$$

and

$$0 \leq \frac{L^{\bar{0}}{}_k L^{\bar{0}}{}_k}{(L^{\bar{0}}{}_0)^2} < 1 \ . \tag{13.38}$$

The parity operation **P**, acting on dx^μ according to $dx^{\bar{\mu}} = P^{\bar{\mu}}{}_\nu dx^\nu$, will be represented by the 4×4 matrix $\{P^{\bar{\mu}}{}_\nu\}$ given by,

$$P = \{P^{\bar{\mu}}{}_\nu\} = \operatorname{diag}\{1, -\mathbb{1}\} \tag{13.39}$$

with $\det P = -1$. Time reversal **T** will be represented by

$$T = \{T^{\bar{\mu}}{}_\nu\} = \operatorname{diag}\{-1, \mathbb{1}\} \ , \tag{13.40}$$

for which $\det T = -1$ also. P and T both satisfy Equation 13.24 and are thus Lorentz, but not restricted Lorentz, matrices.

From the first part of Equation 13.36, we see that Lorentz transformations divide into two disjoint classes according to the sign of $L^{\bar{0}}{}_0$. Those having $L^{\bar{0}}{}_0 \geq 1$ are referred to as being *orthochronous* (right time or non-time-reversing) while those with $L^{\bar{0}}{}_0 \leq -1$ can be re-expressed as the product of a time-reversal matrix T with the matrix of an orthochronous transformation. If we consider only orthochronous transformations, then those with $\det L = 1$ will not include a parity reversal while those with

13.3. Restricted Lorentz transformations

det $L = -1$ will be a product of the parity operation and a completely reflection-free Lorentz matrix, which we denote by $\Lambda = \{\Lambda^\mu{}_\mu\}$.

Exercise 13.9 *Show that the set of matrices,*

$$\mathrm{SO}(1,3) = \{\Lambda \,|\, \Lambda^\mathrm{T} \eta \Lambda = \eta, \quad \Lambda^{\bar{0}}{}_0 \geq 0, \quad \det \Lambda = 1\}, \qquad (13.41)$$

form a group.

The set of all such matrices Λ comprises the *restricted Lorentz group* SO(1,3) of reflection-free Lorentz transformations, in which S denotes special (reflection-free) and O(1,3) denotes *pseudo-orthogonal* of signature $2 = -1 + 3$ in 4D. The Lorentz group was first analysed by Poincaré [448] in 1905 in demonstrating that it left Maxwell's equations and the electron theory of Lorentz unchanged in form.

Exercise 13.10 *Use the indexed version of Equation 13.24 to show that an infinitesimal Lorentz matrix $\delta\Lambda = 1 + \delta\omega$ of components $\delta\Lambda^\mu{}_\nu = \delta^\mu{}_\nu + \delta\omega^\mu{}_\nu$ with $|\delta\omega^\mu{}_\nu| \ll 1$ must satisfy $\delta\omega_{\mu\nu} = -\delta\omega_{\nu\mu}$ where $\delta\omega_{\mu\nu} = \eta_{\mu\lambda}\delta\omega^\lambda{}_\nu$.*

Exercise 13.11 *Show that the parity and time reversal matrices, P and T, do not in general commute with a restricted Lorentz matrix Λ.*

Unlike the full Lorentz group O(1,3), the restricted Lorentz group is continuous and connected (but not simply-connected). Any member Λ of the group may be reached from the identity by continuously varying the parameters from those ($\Lambda = \mathbb{1}$) of the identity.

The restricted Poincaré group of transformations, ISO(1,3), is formed by adding the translations (denoted by I for inhomogeneous) to the restricted Lorentz group and plays a central role in the postulates of relativity.

The Einsteinian form of the Relativity principle embodies the observation that all the local laws of physics are left form-invariant by the group ISO(1,3). Such transformations can be written,

$$x \to \bar{x} = \Lambda x + a \quad \text{with} \quad \Lambda^\mathrm{T} \eta \Lambda = \eta, \quad \Lambda^{\bar{0}}{}_0 \geq 1 \quad \text{and} \quad \det \Lambda = 1. \qquad (13.42)$$

The continuous and connected property of the parameter spaces of SO(1,3) and ISO(1,3) are key properties of a *Lie group*, the spacetime translation and rotation groups, T^4 and SO(3), being other examples.

Exercise 13.12 *Find $\mathrm{sgn}\Lambda^{\bar{0}}{}_0$ and $\det L$ for each of Λ, ΛP, ΛT and ΛPT.*

Exercise 13.13 *Show that (a) an orthochronous Lorentz matrix L with $\det L = -1$ can be expressed as a product ΛP for some restricted Lorentz matrix Λ, (b) a matrix L having $L^{\bar{0}}{}_0 \leq -1$ and $\det L = -1$ can be expressed* in the form ΛT, and (c) a matrix L with $L^{\bar{0}}{}_0 \leq -1$ and $\det L = 1$ can be expressed in the form ΛPT.

In general, for a given Λ, $\Lambda P \neq P\Lambda$. However, as Λ ranges over all restricted Lorentz transformations, the sets $\{\Lambda P\}$ and $\{P\Lambda\}$ will be identical. Similar comments apply to the products of Λ with T and with both P and T. The full Lorentz group of matrices may be reconstructed from $SO(1,3) = \{\Lambda\}$ and $\{P, T\}$ in the form $O(1,3) = \{L\} = \{\Lambda, \Lambda P, \Lambda T, \Lambda PT\}$. As Λ ranges over all the restricted Lorentz transformations this set of four disjoint parts will encompass the full Lorentz group of transformations.

An analysis of the way quantum states and fields transform under the parity operation may be carried out by group representation theory applied to the relativity group including reflections. An analysis for the full Lorentz group $O(1,3)$ was given by van der Waerden [550, 551] in 1929. (A discussion of the spin $\frac{1}{2}$ fields, without use of group representation theory, will be given in Chapter 20.) A detailed analysis which included the the non-orbital contributions, called *intrinsic parity*, was given by Wick, Wightman and Wigner [578] in 1952.

13.4 Lorentz boosts and spatial rotations

Equation 5.30 describes a boost of velocity v where the x-axis is aligned parallel ($v > 0$) or antiparallel ($v < 0$) to the direction of the boost. We introduce dimensionless boost parameters $\beta = v/c$ with $-1 < \beta < 1$ and $\gamma = (1 - \beta^2)^{-1/2}$ with $1 \leq \gamma < \infty$. We also replace $\{ct, \mathbf{x}\}$ by $\{x^\mu\}$ giving,

$$\begin{aligned} d\bar{x}^0 &= \gamma(dx^0 - \beta dx^1) \\ d\bar{x}^1 &= \gamma(-\beta dx^0 + dx^1) \\ d\bar{x}^2 &= dx^2 \\ d\bar{x}^3 &= dx^3, \end{aligned} \qquad (13.43)$$

namely

$$\begin{pmatrix} d\bar{x}^0 \\ d\bar{x}^1 \\ d\bar{x}^2 \\ d\bar{x}^3 \end{pmatrix} = \begin{pmatrix} \gamma & -\gamma\beta & 0 & 0 \\ -\gamma\beta & \gamma & 0 & 0 \\ 0 & 0 & 1 & 0 \\ 0 & 0 & 0 & 1 \end{pmatrix} \begin{pmatrix} dx^0 \\ dx^1 \\ dx^2 \\ dx^3 \end{pmatrix}. \qquad (13.44)$$

We shall abbreviate this equation as $d\bar{x}^\mu = B^\mu{}_\nu dx^\nu$ where the x-direction boost transformation matrix is given by,

$$\Lambda_B = (B^\mu{}_\nu) = \begin{pmatrix} \gamma & -\gamma\beta & 0 & 0 \\ -\gamma\beta & \gamma & 0 & 0 \\ 0 & 0 & 1 & 0 \\ 0 & 0 & 0 & 1 \end{pmatrix}. \qquad (13.45)$$

13.4. Lorentz boosts and spatial rotations

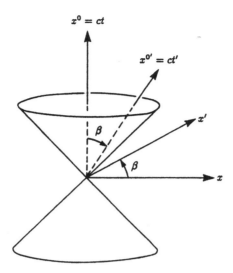

Figure 13.1 Minkowski diagram of frames in uniform relative motion.

Exercise 13.14 *Show that boosts in the x-direction are restricted Lorentz transformations by verifying that Λ_B from Equation 13.45 satisfies Equation 13.24 and satisfies the conditions for it to be reflection-free.*

The light cone of Figure 12.2 was described from the point of view of one particular frame in which the t and x directions, e_0 and e_1, correspond in the tx plane to $dx^1 = 0$ and $dx^0 = 0$, respectively. Let us now include some details of another frame boosted in the positive x or 1-direction. The new directions, $e_{\bar{0}}$ and $e_{\bar{1}}$, will correspond to $dx^{\bar{1}} = 0$ and $dx^{\bar{0}} = 0$ which, by Equation 13.43, will have Minkowski diagram slopes of $\beta < 1$ in both cases, the first relative to e_0 and the second relative to e_1. These two axes will therefore be at equal angles from the surface of the null cone and at an acute angle to one another as shown in Figure 13.1.

The direction $e_{\bar{0}}$ in any frame is clearly timelike and will therefore always lie in the interior of the future half cone, while the new spatial direction will lie outside. If we consider a third frame $\{t', \mathbf{x}'\}$ boosted along the negative x-axis of the original frame then its time and x-axis would appear in Figure 13.1 at obtuse angles to one another.

The set of events tracing out the surface of the light cone of the origin event remain completely unchanged as a whole when we alter the Minkowski diagram so that some other frame is plotted along the vertical and horizontal axes of the paper. This is because the light cone is described invariantly by $ds^2 = 0$.

A rotation, by definition, affects only spatial coordinates in some frame and hence leaves time intervals unchanged, $d\bar{x}^0 = dx^0$, in that frame. From $d\bar{x}^0 = \Lambda^{\bar{0}}{}_\mu dx^\mu$ we see that $\Lambda^{\bar{0}}{}_0 = 1$ and $\Lambda^{\bar{0}}{}_k = 0$ for a rotation. Equations 13.35 and 13.37 show that any one of these conditions, or $\Lambda^{\bar{k}}{}_0 = 0$, implies the others. We denote the remaining non-trivial part of Λ, namely $\Lambda^{\bar{k}}{}_\ell$, by $R = \{R^{\bar{k}}{}_\ell\}$. The matrix version of Equation 13.24 then shows that $R^T R = \mathbb{1}$ while $\det \Lambda = 1$ shows that $\det R = 1$. Consequently, any one of the conditions $d\bar{x}^0 = dx^0$, $\Lambda^{\bar{0}}{}_0 = 1$, $\Lambda^{\bar{0}}{}_k = 0$ or $\Lambda^{\bar{k}}{}_0 = 0$ is a necessary and sufficient condition for a restricted Lorentz transformation to be a pure rotation with 4D matrix form,

$$\Lambda_R = \begin{pmatrix} 1 & 0 \\ 0 & R \end{pmatrix} \quad \text{where} \quad R^T R = \mathbb{1} \quad \text{and} \quad \det R = 1. \tag{13.46}$$

For example, the 4D matrix given by,

$$\Lambda_R = (R^{\bar{\mu}}{}_\nu) = \begin{pmatrix} 1 & 0 & 0 & 0 \\ 0 & 1 & 0 & 0 \\ 0 & 0 & \cos\theta & -\sin\theta \\ 0 & 0 & \sin\theta & \cos\theta \end{pmatrix}, \tag{13.47}$$

describes a rotation about the x-axis or in the yz-plane.

13.5 Rapidity and 'hyperbolic rotations'

From Equation 5.32 and $\beta = v/c$, two boosts in the same direction of speed parameters β_1 and β_2 give another boost in the same direction with parameter β given by,

$$\beta = \frac{\beta_1 + \beta_2}{1 + \beta_1 \beta_2}. \tag{13.48}$$

Speed (v or β) is clearly not an *additive* parameter for Lorentz boosts. We define the *rapidity* of a boost by $\alpha = \tanh^{-1}\beta$ (with range $-\infty < \alpha < \infty$) and note that $\gamma = \cosh\alpha$ and $\gamma\beta = \sinh\alpha$. For small velocities the rapidity α and the dimensionless velocity β coincide.

Exercise 13.15 Show that, for two boosts in the same direction, the rapidity is an additive parameter.

We may re-express the boost matrix of Equation 13.45 in terms of the rapidity as

$$\Lambda_B = \begin{pmatrix} \cosh\alpha & -\sinh\alpha & 0 & 0 \\ -\sinh\alpha & \cosh\alpha & 0 & 0 \\ 0 & 0 & 1 & 0 \\ 0 & 0 & 0 & 1 \end{pmatrix} = \begin{pmatrix} \cos i\alpha & \sin i\alpha & 0 & 0 \\ \sin i\alpha & \cos i\alpha & 0 & 0 \\ 0 & 0 & 1 & 0 \\ 0 & 0 & 0 & 1 \end{pmatrix}. \tag{13.49}$$

This result, in the tx plane, should be compared with Equation 13.47 for the 4D matrix describing a rotation of angle θ in the yz-plane. This similarity between rotations and boosts should *not*, however, be overemphasized. Boosts have some characteristics which are quite different from those of rotations.

The bulk of these differences arise from the fact that the parameter range, $-1 < \beta < 1$ or $-\infty < \alpha < \infty$ is not compact as for the range $0 \leq \theta \leq 2\pi$ of a rotation parameter since the limit points ± 1 of β are not included. The boost matrix in Equation 13.49 is symmetric but the same is not true of rotation matrices. Just as we did not choose an imaginary time coordinate so also we shall not make extensive use of the mathematical device of 'imaginary angle' rotations to describe boosts.

13.6 Boosts in an arbitrary direction

So far we have chosen the x-axis in the direction of the boost. This gives the impression that a boost is a 1-parameter transformation. However, since a further two parameters are required to specify its direction in an arbitrary frame, a boost in 4D spacetime in fact requires three parameters for its specification.

It is useful to obtain the corresponding 3-parameter matrix Λ_B when the boost parameters are given by $\boldsymbol{\beta} = \{\beta_x, \beta_y, \beta_z\}$. This boost must not be considered as a sequence of three boosts in the three directions of the Cartesian axes. Rather, it is a boost of magnitude $\beta = |\boldsymbol{\beta}|$ in the direction $\hat{\boldsymbol{\beta}}$. To obtain this boost matrix from Equation 13.43 is simply a problem in spatial rotation.

Following Goldstein [206], we use the principles of rotational covariance. We require our final expression to be of the same form in all frames obtained from one another by rotation which we shall be able to ensure most easily provided we use 3-vectors and 3-scalars in a rotation-covariant way. This means that $dx^{\bar{0}}$ and $d\bar{\mathbf{x}}$ must be given respectively by a 3-scalar and a 3-vector equation constructed from similar quantities dx^0 and $d\mathbf{x}$ in the unbarred frame along with the only other 3-vector available, $\boldsymbol{\beta}$. The expression must reproduce Equation 13.43 when $\boldsymbol{\beta} = \{\beta, 0, 0\}$. This suffices to determine the coefficients in the transformation.

Exercise 13.16 *Verify that*

$$\begin{aligned}
dx^{\bar{0}} &= \gamma(dx^0 - \boldsymbol{\beta} \cdot d\mathbf{x}) \\
d\bar{\mathbf{x}} &= d\mathbf{x} + \frac{(\gamma-1)\boldsymbol{\beta}(\boldsymbol{\beta} \cdot d\mathbf{x})}{\beta^2} - \gamma\boldsymbol{\beta} dx^0 ,
\end{aligned} \qquad (13.50)$$

satisfies the above requirements for a boost in an arbitrary direction.

The boost matrix is thus,

$$\Lambda_B = \begin{pmatrix} \gamma & -\gamma\beta^T \\ -\gamma\beta & \mathbb{1} + \dfrac{(\gamma-1)\beta\beta^T}{\beta^2} \end{pmatrix}. \qquad (13.51)$$

Although it is not necessary to do so, one could check directly that this result satisfies Equation 13.24.

13.7 Factorizing a Lorentz transformation

We have now shown that boosts Λ_B and rotations Λ_R are two independent 3-parameter parts of the 6-parameter restricted Lorentz transformations Λ. We should therefore expect to be able to factorize a given Λ into a rotation Λ_{θ_1} followed by a boost Λ_{β_1} or as a boost Λ_{β_2} followed by a rotation Λ_{θ_2} in the alternative forms,

$$\Lambda = \Lambda_{\beta_1}\Lambda_{\theta_1} = \Lambda_{\theta_2}\Lambda_{\beta_2}. \qquad (13.52)$$

To establish this result we write a Lorentz matrix in partitioned form,

$$\Lambda = \begin{pmatrix} \Lambda^{\bar{0}}{}_0 & \Lambda^{\bar{0}}{}_\ell \\ \Lambda^{\bar{k}}{}_0 & \Lambda^{\bar{k}}{}_\ell \end{pmatrix}. \qquad (13.53)$$

From Equation 13.36 (restricted to a matrix Λ) it is clear that, since its magnitude is less than unity, we may take

$$\beta^k = -\Lambda^{\bar{k}}{}_0/\Lambda^{\bar{0}}{}_0, \qquad (13.54)$$

as a legitimate set of parameters for a boost. We now use Equation 13.51 to construct part of the corresponding boost matrix,

$$\Lambda_B = \begin{pmatrix} \gamma & -\gamma\beta^T \\ -\gamma\beta & \text{etc.} \end{pmatrix}, \qquad (13.55)$$

in which, from Equations 13.35 and 13.54, the value of γ will be $\Lambda^{\bar{0}}{}_0$. The details of the space-space part will not be needed. We replace β by $-\beta$ to invert this boost,

$$\Lambda_B^{-1} = \begin{pmatrix} \gamma & \gamma\beta^T \\ \gamma\beta & \text{etc.} \end{pmatrix}, \qquad (13.56)$$

and use this inverse to factor out the boost part of Λ by forming:

$$\Lambda_B^{-1}\Lambda = \begin{pmatrix} \gamma & \gamma\beta^T \\ \gamma\beta & \text{etc.} \end{pmatrix} \begin{pmatrix} \Lambda^{\bar{0}}{}_0 & \Lambda^{\bar{0}}{}_\ell \\ \Lambda^{\bar{k}}{}_0 & \Lambda^{\bar{k}}{}_\ell \end{pmatrix}. \qquad (13.57)$$

13.7. Factorizing a Lorentz transformation

By closure of the SO(1,3) group, $\Lambda_B^{-1}\Lambda$ must be another restricted Lorentz transformation.

Exercise 13.17 *By evaluating its $\bar{0}_0$ component, show that $\Lambda_B^{-1}\Lambda$ describes a pure rotation.*

We have thus shown that the original Lorentz matrix can be decomposed in at least one way into a rotation followed by a boost,

$$\Lambda = \Lambda_{\beta_1}\Lambda_{\theta_1} . \tag{13.58}$$

Exercise 13.18 *Show that we may similarly factorize the same Lorentz matrix as,*

$$\Lambda = \Lambda_{\theta_2}\Lambda_{\beta_2} , \tag{13.59}$$

by starting with $\beta_k = -\Lambda^0{}_k/\Lambda^0{}_0$, and adjusting the above steps accordingly.

In general, $\hat{\beta}_1 \neq \hat{\beta}_2$ and therefore $\theta_1 \neq \theta_2$ although $\gamma_1 = \gamma_2 = \Lambda^{\bar{0}}{}_0$ so both boosts will at least be of the same magnitude, if not in the same direction. The factors in each of these decompositions are unique since $\Lambda = \Lambda_\beta \Lambda_\theta = \Lambda_{\beta'}\Lambda_{\theta'}$ implies $\Lambda_{-\beta}\Lambda_{\beta'} = \Lambda_\theta \Lambda_{-\theta'}$. The right side of this last result is a rotation. Therefore so is the left side whose matrix will thus have its $\bar{0}_0$ component equal to 1 and will have a unit determinant.

Exercise 13.19 *Show that $(\Lambda_{-\beta}\Lambda_{\beta'})^{\bar{0}}{}_0 = 1$ if and only if $\beta' = \beta$ thus showing that the primed and unprimed factors in the alternative decompositions are identical and the decomposition is unique.*

Exercise 13.20 *Show that a boost in the x-direction and another in the y-direction do not commute and do not combine to give another boost.*

Boosts do not form a subgroup of the Lorentz group unless one restricts consideration to the boosts in one direction only.

Bibliography:

Anderson J L 1967 *Principles of relativity physics* [15].
Lord E 1976 *Tensors, relativity and cosmology* [328].
Rindler W 1977 *Essential relativity* [476].
Robertson H P and Noonan T W 1968 *Relativity and cosmology* [480].

CHAPTER 14

Tensor fields in Minkowskian spacetime

Minkowski's important contribution to the theory lies in ... introducing a formalism such that the mathematical form of the law itself guarantees its invariance under Lorentz transformations[1].

Albert Einstein   1879–1955

Many of the properties of scalars, vectors and tensors in Euclidean spacetime $\mathbf{E}^1 \times \mathbf{E}^3$, and in the Galilean spacetime $\mathbf{G} = \{\mathbf{E}^1, \mathbf{E}^3\}$ of Newtonian physics, carry over almost unchanged to the similar quantities in the Minkowskian spacetime $\mathbf{E}^{1,3}$ of Einsteinian physics.

Nevertheless, certain properties are quite different, in particular those which depend on the indefiniteness of Lorentz orthogonality. Examples are the covariant form of a spacelike 3-volume and the existence of null vectors. Others arise primarily from the appearance of an extra dimension, such as the existence of selfdual and antiselfdual tensors. Furthermore, the explicitly indexed notation widely used in many relativistic calculations is not as well-known as the 3-vector notation of Galilean relativity and Newtonian physics. Consequently, we shall develop the relativistic theory in a way which corresponds to the Newtonian material in Chapter 4, concentrating on the new features.

14.1   Scalar and vector fields

A *relativistic scalar* or *Lorentz invariant* or *4-scalar* is a single quantity ϕ which remains unchanged, $\phi \to \bar{\phi} = \phi$, when the coordinates undergo a restricted Lorentz transformation or a translation in a spacelike direction. A Lorentz scalar is also a 3-scalar but not necessarily the converse.

Examples of Lorentz scalars are the rest mass of a particle, the electric charge of an electron, the Lorentz scalar product $dx_1 \cdot dx_2$ of two spacetime intervals or the line element ds^2, the proper time interval $d\tau$ between two

---

[1] *Autobiographical notes* [497].

14.1. Scalar and vector fields

events on the trajectory of a massive particle, the Lagrangian $L = L(q, \dot{q})$ of almost all mechanical systems with coordinates q and velocities $\dot{q}$ and the action $S[q]$ or $S[\psi]$ of a mechanical or field system.

A translation in a spacelike direction will, in a particular reference frame, consist of a translation of only the spatial coordinates. However, in an arbitrary frame, it will also involve the time coordinate. Those Lorentz quantities which are also unchanged in value under all translations, timelike included, may be referred to as *Poincaré scalars* or *invariants*. The charge and rest mass of a particle and the *spacetime volume element* or *4-volume* $d^4x = c\,dt\,dx^1 dx^2 dx^3$ are examples of Poincaré scalars.

A relativistic *scalar field* $\phi = \phi(x) = \phi(ct, x, y, z)$ is a single function of the spacetime coordinates $x = \{ct, \mathbf{x}\}$ which, under a translation in a spacelike direction or under a restricted Lorentz transformation of matrix $\Lambda = \{\Lambda^\mu{}_\nu\}$, acting on x and simultaneously on the functional form of the field, remains unchanged in value,

$$x \to \bar{x} = x + a \quad \Rightarrow \quad \phi(x) \to \bar{\phi}(\bar{x}) = \phi(x) \quad \text{or} \quad \bar{\phi}(x) = \phi(x - a) \quad (14.1)$$

$$x \to \bar{x} = \Lambda x \quad \Rightarrow \quad \phi(x) \to \bar{\phi}(\bar{x}) = \phi(x) \quad \text{or} \quad \bar{\phi}(x) = \phi(\Lambda^{-1} x) . \quad (14.2)$$

Examples of 4-scalar fields are those describing temperature and density distributions, the Klein-Gordon fields whose quantization leads to spinless particles such as pions and the Lagrangian density $\mathcal{L} = \mathcal{L}(\psi, \dot{\psi}, \nabla \psi)$ of almost any field system. A theoretical example is the *Higgs field* whose interaction with the intermediate vector boson fields of the electroweak theory provide the $W^\pm$ and Z^0 particles with mass by the quantum field theoretic process of spontaneous symmetry breaking. The 4-divergence $\partial \cdot V$ of a 4-vector field V is a scalar field and the *d'Alembertian*,

$$\Box = \partial \cdot \partial = \partial_\mu \partial^\mu = -\tfrac{1}{c^2}\partial_t^2 + \nabla^2 , \quad (14.3)$$

is a 4-scalar wave operator acting on any field.

Suppose a quantity $V(x)$ has four components which may be labelled either as $V^\mu(x)$ or as $V_\mu(x)$ each being obtainable from the other by raising and lowering indices with $\eta^{\mu\nu}$ and $\eta_{\mu\nu}$ and which, under a restricted Lorentz transformation of matrix $\Lambda = \{\Lambda^\mu{}_\nu\}$, transform in the same way as dx^μ and dx_μ, namely as:

$$V^{\bar{\mu}}(x) = \Lambda^{\bar{\mu}}{}_\mu V^\mu(\Lambda^{-1}x) \quad \text{or} \quad V_{\bar{\mu}}(x) = \Lambda^\mu{}_{\bar{\mu}} V_\mu(\Lambda^{-1}x) . \quad (14.4)$$

If $V(x)$ is also unchanged in value under a translation in a spacelike direction, then $V(x)$ is called a *4-vector field* and written $V = \{V^\mu\} = \{V_\mu\}$. As for the components of the prototype 4-vector dx, V^μ and V_μ are referred to as the *contravariant* and *covariant* components of V. The definition of non-field 4-vectors is obtained in an obvious way by omission of the spacetime arguments.

Conserved four-vectors, those which also remain unchanged in timelike directions may be referred to as Poincaré 4-vectors. We shall see that an example of a Lorentz 4-vector is the 4-momentum $p^\mu = \{E/c, \mathbf{p}\}$ of a particle of energy E and relativistic 3-momentum $\mathbf{p}$. It will be conserved if the particle is isolated.

Examples of 4-vector fields are the electrodynamic 4-vector potential $A^\mu(x) = \{\phi/c, \mathbf{A}\}$ where $\phi(t, \mathbf{x})$ and $\mathbf{A}(t, \mathbf{x})$ are the scalar and 3-vector electromagnetic potentials. The *gradient* of a 4-scalar field ϕ, written $\partial \phi$ or $\nabla \phi$, is a 4-vector field.

Exercise 14.1 *If $\phi(x)$ is a 4-scalar field, show that $\partial_\mu \phi$ are the covariant components of a 4-vector field.*

Exercise 14.2 *(a) If $U = \{U^\mu\}$ and $V = \{V^\mu\}$ are both vector fields, show that $U^\mu V_\mu$ is a scalar field. (b) Let ϕ be an invariant which may be expanded in terms of a 4-vector $V = \{V^\mu\}$ in the form $\phi = U_\mu V^\mu$. Show that the coefficients U_μ must also transform as the components of a 4-vector.*

Exercise 14.3 *Show that, under a restriction of a Poincaré transformation to a Euclidean transformation, the time and space parts of a 4-vector transform as a 3-scalar and a 3-vector, respectively.*

Exercise 14.4 *Show that the spacetime volume element*

$$d^4x = dx^0 dx^1 dx^2 dx^3 = c\, dt\, dx^1 dx^2 dx^3 \, , \qquad (14.5)$$

is a Poincaré invariant and hence show that,

$$d\Sigma = \{d\Sigma_\mu\} = \{dx^1 dx^2 dx^3, dx^0 dx^2 dx^3, dx^0 dx^3 dx^1, dx^0 dx^1 dx^2\} \, , \quad (14.6)$$

is a Poincaré 4-vector.

Exercise 14.5 *Show that there are no 4-vectors with invariant numerical components, namely with the same numerical values for their components in different frames.*

Physically, we may understand the invariance of d^4x by a cancellation of a time dilation factor with a Lorentz-FitzGerald length contraction factor of reciprocal magnitude along the direction of the boost, all other parts of a Poincaré transformation also leaving the product invariant. Its Poincaré invariance also follows directly by application of the Fundamental theorem of calculus.

The *magnitude of a 4-vector* $V = \{V^\mu\}$ will be $\sqrt{-V \cdot V}$ if it is timelike, $\sqrt{V \cdot V}$ if it is spacelike and zero if it is null.

14.2 Lorentz tensors and tensor fields

Hermann Minkowski

b. Alexotas, Russia
22 June 1864

d. Göttingen, Germany
12 January 1909

Courtesy of the State University, Göttingen.

Henceforth space by itself, and time by itself, are doomed to fade away into mere shadows, and only a kind of union of the two will preserve an independent reality. [Reference [383] reprinted in *The Principle of relativity* [343], p. 75.]

14.2 Lorentz tensors and tensor fields

If U and V are two 4-vectors, or 4-vector fields, then the set of 16 quantities $\{U^\mu V^\nu\}$ Lorentz transform according to

$$U^\mu V^\nu \to U^{\bar\mu} V^{\bar\nu} = \Lambda^{\bar\mu}{}_\mu \Lambda^{\bar\nu}{}_\nu U^\mu V^\nu \;, \tag{14.7}$$

and remain unchanged in value under spacelike translations. The set of quantities $\{\Lambda^{\bar\mu}{}_\mu \Lambda^{\bar\nu}{}_\nu\}$, considered as a matrix with rows labelled by the pair $\bar\mu\bar\nu$ and columns by the pair $\mu\nu$, each ranging over $00, 01, \ldots, 33$, is the 16×16 direct product matrix $\Lambda \otimes \Lambda$ of Λ with itself. Therefore $U^\mu V^\nu$ are said to be the components of the *direct product* of the vectors U and V written $U \otimes V = \{U^\mu \otimes V^\nu\}$. Consider now a set of 16 quantities $T^{\mu\nu}(x)$ labelled by two free Lorentz indices each of which behaves as a 4-vector index in the sense that $T^{\mu\nu}$ transforms according to,

$$T^{\bar\mu\bar\nu}(\bar x) = \Lambda^{\bar\mu}{}_\mu \Lambda^{\bar\nu}{}_\nu T^{\mu\nu}(x) \quad \text{or} \quad T^{\bar\mu\bar\nu}(x) = \Lambda^{\bar\mu}{}_\mu \Lambda^{\bar\nu}{}_\nu T^{\mu\nu}(\Lambda^{-1}x) \;, \tag{14.8}$$

and remains unchanged under spacelike translations. Then $\{T^{\mu\nu}(x)\}$ are said to be the *contravariant components* of an invariant rank 2 *tensor field* $\mathbf{T}(x) = T^{\mu\nu}(x)\mathbf{e}_\mu \otimes \mathbf{e}_\nu$. Since only eight of the sixteen components of $U \otimes V$ are independent in $\mathbf{E}^{1,3}$, a tensor cannot in general be expressed as a single direct product. A *Poincaré tensor* is one which is also unchanged in value under translations in timelike directions.

A tensor whose components are unaltered by any Lorentz transformation is called a *numerical tensor*, one example being the Minkowski tensor $\eta_{\mu\nu}$, as illustrated by the inverse of the indexed form of Equation 13.24.

Exercise 14.6 *(a) Show that the only numerical tensors of second rank are constant multiples of the Minkowski tensor. (b) Show that there are no numerical tensors of rank 3 in four dimensions. (c) Show that, on restriction from Poincaré to Euclidean transformations, the time-time, space-time, time-space and space-space parts of a 4-tensor transform as a 3-scalar, two 3-vectors and a 3-tensor, respectively. (d) Show that, in $\mathbf{E}^{1,3}$, any tensor field may be expressed as the sum of two direct products made up from four independent vector fields.*

From the above definition of a second-rank tensor, via its contravariant components $T^{\mu\nu}$, we may obtain *mixed components* by leaving one index unaltered and lowering the other index as if the components were those of a 4-vector,

$$T^{\mu}{}_{\nu} = \eta_{\nu\lambda} T^{\mu\lambda} \quad \text{and} \quad T_\mu{}^\nu = \eta_{\mu\lambda} T^{\lambda\nu}, \qquad (14.9)$$

and *covariant components*,

$$T_{\mu\nu} = \eta_{\nu\lambda} T_\mu{}^\lambda = \eta_{\mu\lambda} T^\lambda{}_\nu = \eta_{\mu\lambda}\eta_{\nu\rho} T^{\lambda\rho}, \qquad (14.10)$$

by lowering the other index of either mixed form or both indices of the contravariant components. Each lowered index μ will be transformed by a factor of $\Lambda^\mu{}_{\bar\mu}$. The corresponding raising operations are similar. All these four forms are equivalent component versions of the same physical tensor $\mathbf{T} = \{T^{\mu\nu}\} = \{T^\mu{}_\nu\} = \{T_\mu{}^\nu\} = \{T_{\mu\nu}\}$ which may be reconstructed by contracting the components with products of $\mathbf{e}_\mu$ and/or $\mathbf{e}^\mu = \eta^{\mu\nu}\mathbf{e}_\nu$. The order of the $\mu\nu$ indices is generally significant in tensor components and should be preserved even in the mixed forms of the components. Placing one index immediately above another is usually ambiguous unless the tensor is symmetric in the two indices.

The *tensor algebra* properties introduced up to this point may be quite simply extended, in Lorentz reference frames, to a *tensor calculus* including partial derivatives of tensors.

Exercise 14.7 *Show that the partial derivatives $\partial_\nu V^\mu$ of the contravariant* components of a vector $V = \{V^\mu\}$ are the mixed components of a second-rank tensor, which we may write as ∂V or ∇V.

A Lorentz tensor of *rank* r is a quantity whose components have r indices each of which leads to a 4-vector factor in the transformation law of the tensor. The rule for the transformation of a rank r tensor will therefore have r factors of $\Lambda^\mu{}_\mu$ or $\Lambda^\mu{}_{\bar\mu}$, each with a different index pair. A scalar and a vector are regarded as tensors of rank 0 and 1 respectively.

Exercise 14.8 *Show that, if a tensor of rank r_1 is expanded in terms of another tensor of lower rank r_2 then the coefficients in the expansion transform as a tensor of rank $r = r_1 - r_2$.*

Partial differentiation with respect to x^μ of a tensor of rank r clearly leads to a new tensor of rank $r+1$. The contraction of a pair of Lorentz indices, on the same tensor or on two separate tensors in a product, produces a new tensor whose rank is 2 lower than the original tensor or tensor product.

Although we shall not prove it, the application of representation theory [30, 410] to the restricted Poincaré group of transformations ISO(1,3) shows that all physical quantities of local Einsteinian physics are either tensors (including scalars and vectors) or very closely related objects called *tensor-spinors* (including *spinors* and *vector-spinors*). In particular, all integer spin fundamental particles like photons, intermediate vector bosons ($W^\pm$ and Z^0), gluons, gravitons and some composite particles such as pions and kaons, are obtained by quantization of scalar, vector or tensor fields, obey *Bose-Einstein statistics* and are called *bosons*[2]. Similarly, all fundamental half-odd-integer spin particles, like electrons, quarks and neutrinos and some composite particles like protons and neutrons, arise by quantization of analogous spinor fields, obey *Fermi-Dirac statistics* and are called *fermions*[3].

The tensor fields we have defined are unchanged in value by a rotation of 2π and are therefore said to be *single-valued*. In contrast, the matrices which transform a spinor have value -1 for a rotation of 2π in the absence of interaction. Only for a rotation of 4π is a spinor restored to its original unrotated value. For this reason, spinors are *double-valued* quantities, a property which distinguishes them from tensors.

14.3 Lorentz scalar product

The *scalar product* of any two vectors in any metric space is formed by *contraction* of both together with the metric tensor of the space, or equiva-

---

[2] Satyendranath Bose (1894–1974), Indian physicist.
[3] Enrico Fermi (1901–1954), Italian physicist.

lently, by contraction of the two vectors in contragredient component forms with one another. In $\mathbf{E}^3$ the scalar product $dx_1 \cdot dx_2$ of two displacements is thus $\delta_{k\ell} dx_1^k dx_2^\ell$ and the line element is therefore $ds^2 = dx_k dx^k = \delta_{k\ell} dx^k dx^\ell$ where the Kronecker delta $\delta_{k\ell}$ plays the role of a metric tensor. Correspondingly, the Euclidean *scalar product* of any two 3-vectors $\mathbf{u} = \{u^k\}$ and $\mathbf{v} = \{v^k\}$ is $\mathbf{u} \cdot \mathbf{v} = \delta_{k\ell} u^k v^\ell$.

The *Lorentz scalar product* $U^\mu V_\mu$ of any two 4-vectors U and V is invariant. Since it may be rewitten as $U \cdot V = \eta_{\mu\nu} U^\mu V^\nu$, we see that $\eta_{\mu\nu}$ plays a similar role to that of the Kronecker delta in $\mathbf{E}^3$. The manipulation of this indefinite scalar product is of such importance in relativity that we display it in various *covariant* forms (labelled with Lorentz indices) and broken down into 1+3 notation,

$$\begin{aligned} U \cdot V &= U^\mu V_\mu = U_\mu V^\mu = \eta_{\mu\nu} U^\mu V^\nu = \eta^{\mu\nu} U_\mu V_\nu \\ &= U^0 V_0 + U^k V_k = -U^0 V^0 + \mathbf{U} \cdot \mathbf{V} \ . \end{aligned} \qquad (14.11)$$

The scalar product $V \cdot V$ of V with itself will often be denoted by V^2 despite the fact that, like ds^2, it may not be positive. The 3D gradient operator ∇ arises by differentiation with respect to Cartesian coordinates x^k namely $\nabla = \{\partial_k\}$ where $\partial_k \equiv \partial/\partial x^k$. The corresponding Minkowskian spacetime *gradient operator*, which we abbreviate by ∂, is given by,

$$\partial = \{\partial_\mu\} = \{\partial_0, \nabla\} = \left\{\tfrac{1}{c}\partial_t, \nabla\right\} \ . \qquad (14.12)$$

Since our coordinates x^μ are always written with raised indices, most vectors defined from their components, such as $dx = \{dx^\mu\}$, arise naturally in contravariant components from which we can obtain the covariant components with $\eta_{\mu\nu}$. The gradient operator is the exception to this rule as it arises naturally in covariant form. The 4-divergence $\partial \cdot V \equiv \partial_\mu V^\mu$ splits naturally according to,

$$\partial \cdot V \equiv \partial_\mu V^\mu = \partial_0 V^0 + \partial_k V^k = \partial_0 V^0 + \nabla \cdot \mathbf{V} \ . \qquad (14.13)$$

Physical measurements always lead to real numbers, the finite values of real scalars or the values of scalar fields at some specified event. These scalars may often be best understood in the context of the values of a number of other related scalars. Sets of such related scalars can arise as scalar products of vectors and tensors with one another, as for example in $\hat{u} \cdot T \cdot \hat{v} = \{\hat{u}_\mu T^{\mu\nu} \hat{v}_\nu\}$ which provides 16 scalars from a second-rank tensor when four independent directions are chosen for each unit vector $\hat{u}$ and $\hat{v}$.

Euclidean and Minkowskian spaces all have symmetric metric tensors which are used to form scalar products of vectors. Consistent with the symmetry of these metric tensors, the components of the vectors (and tensors) of such spaces commute with one another, being simply ordinary real (and possibly complex) numbers. Geometric vector spaces with antisymmetric

metric tensors also have widespread application in physics. Examples are the spaces of Pauli, Weyl and Dirac spinors which describe, respectively, non-relativistic electrons, massless neutrinos and relativistic electrons or quarks.

Exercise 14.9 *Show that the wave-mechanical operators* $H = i\hbar\partial_t$ *and* $\mathbf{P} = \frac{\hbar}{i}\nabla$ *combine to form a covariant 4-momentum operator* $P_\mu = \frac{\hbar}{i}\partial_\mu$ *in which* $P^0 = H$.

14.4 Manifest Lorentz and Poincaré covariance

As in the Euclidean case, the two fundamental properties of tensor transformation laws, linearity and homogeneity, are evident from Equation 14.8. The components, in a new frame, of a restricted Lorentz tensor of arbitrary rank are a homogeneous, linear combination, $\Lambda^{\bar{\mu}}{}_\mu \Lambda^{\bar{\nu}}{}_\nu \ldots \Lambda^{\bar{\lambda}}{}_\lambda \ldots T^{\mu\nu\cdots}{}_{\lambda\cdots}$ of the components $T^{\mu\nu\cdots}{}_{\lambda\cdots}$ of the tensor in the original frame. If all the components are zero in one frame then all components will be zero in any frame. To ensure that equations satisfy the Lorentz part of the Poincaré covariance principle, it is necessary and sufficient to write these equations so that every term in them is a tensor of the same rank and index levels. A Lorentz equation of the form,

$$A^{\mu\nu\cdots}{}_{\lambda\cdots} + B^{\mu\nu\cdots}{}_{\lambda\cdots} + \cdots = C^{\mu\nu\cdots}{}_{\lambda\cdots}, \tag{14.14}$$

will then Lorentz transform to,

$$A^{\bar{\mu}\bar{\nu}\cdots}{}_{\bar{\lambda}\cdots} + B^{\bar{\mu}\bar{\nu}\cdots}{}_{\bar{\lambda}\cdots} + \cdots = C^{\bar{\mu}\bar{\nu}\cdots}{}_{\bar{\lambda}\cdots}, \tag{14.15}$$

which has the same form as the original equation.

A 4-tensor expression guarantees the *form invariance* of physical equations under Lorentz transformations, as required by the Einsteinian form of the Principle of relativity. The use of 4-tensor equations rather than the unindexed 1 + 3 notation is often referred to as *manifest Lorentz covariance*. Tensors are useful not just in 3D Euclidean space and 4D Minkowskian spacetime but whenever transformations are linear and homogeneous. Homogeneous transformations based on matrix groups such as the unitary, U(1), U(2), U(3), ..., U(N) and special (determinant = 1) unitary groups SU(2), SU(3), ..., SU(N) or the orthogonal O(2), O(3), O(4), ..., O(N) and special orthogonal groups SO(2), SO(3), SO(4), ..., SO(N), among others, appear in the description of many physical symmetries and give rise to a wide variety of tensors.

The Poincaré transformations of Equation 13.23 are not homogeneous in the coordinates. They are not expressible solely as a 4×4 matrix transformation on x^μ. The coordinate transformation law with displacements included is thus not a 4-tensor law. To include translations when dealing with fields, as required by the Poincaré form invariance of the Principle of relativity, we proceed as in Newtonian physics to consider the values $\psi(x)$ and $\psi(x + \Delta x)$ of the fields at different nearby events in spacetime. In the limit, this involves us with coordinate differentials dt and $d\mathbf{x}$ and hence with derivatives $\partial_t \psi$ or $\nabla \psi(x)$ of fields with respect to time and space.

Whereas Lorentz-covariant equations are essentially algebraic, translational and Poincaré covariance on fields thus clearly introduces partial differential equations, the equations of motion of the field or *field equations*. These differences and derivatives that appear in the equations transform Lorentz tensorially and the field equations are thus differential equations having Lorentz covariance. However, in contrast to the Newtonian case, in order to maintain Lorentz covariance the differentials and derivatives must appear in their covariant combinations, dx^μ and ∂_μ. The form of the field equations will be partly determined by demanding that the Lorentz tensors that appear in them combine with factors of ∂_μ to give equations that are also Poincaré-covariant. We have already taken into account the necessity to preserve translational covariance in spacelike directions in our definitions of the tensorial quantities. Translation in time, however, is not so simply incorporated at the kinematic level since it is intimately connected to the evolution and therefore the dynamics or interactions of a system.

One of our tasks will be to analyse anew many of the familiar results of Newtonian physics expressed in $1 + 3$ form (using t and $\mathbf{x}$ separately with 3-scalars, 3-vectors and 3-tensors) to see how they are affected by the substitution of Lorentz-Poincaré relativity in place of Galilean relativity. We shall find that many results will have a similar expression, with subtle changes in interpretation, in the manifestly Poincaré-covariant form where the relativistic nature is no longer ambiguous or hidden. Some of these results may then be split into $1 + 3$ form to ease comparison with familiar non-relativistic results.

The indefiniteness of the geometry of Minkowskian spacetime has some surprising consequences compared with comparable results in Euclidean space. It is very important to remember that *Lorentz orthogonality* is not the same thing as ordinary spatial orthogonality. Some of the following exercises will illustrate this point.

Exercise 14.10

- *Show that* $\{A, A, 0, 0\}$ *has zero magnitude for arbitrarily large A.*
- *What is the magnitude of* $\{0, A, 0, A\}$*?*
- *What is the magnitude of* $\{q \cosh \alpha, q \sinh \alpha, 0, q\}$*?*
- *Show that* $\{\cosh \alpha, \sinh \alpha, 0, 0\}$ *is normalized, however large is α.*
- *Show that* $\{A, B, 0, 0\}$ *is normal to* $\{B, A, 0, 0\}$ *for any A or B.*

- If $T = \{T^\mu\}$ is time-like and $V \cdot T = 0$, show that V is spacelike.
- If $S = \{S^\mu\}$ is spacelike and $V \cdot S = 0$, is V timelike?

Exercise 14.11 *Show that two orthogonal null 4-vectors p^μ and w^μ must also be proportional, either parallel or antiparallel.*

The general geometrical result of Exercise 14.11 has an important application for null (massless) Poincaré fields, classical or quantum, of the type used to describe the fields of electrodynamics, linearized gravity, gravitational waves or the theory of massless neutrinos. In all those cases, p^μ and w^μ may denote, respectively, the 4-momentum and a psuedo-4-vector *helicity*, namely the projection of angular momentum in the direction of propagation (see Section 15.10).

14.5 Symmetric and antisymmetric tensors

A second-rank Lorentz-indexed quantity $S^{\mu\nu}$, tensor or otherwise, is *symmetric* if $S^{\mu\nu} = S^{\nu\mu}$. Similarly, $A^{\mu\nu}$ is *antisymmetric* if $A^{\mu\nu} = -A^{\nu\mu}$. The *symmetric part* $T^{(\mu\nu)}$ and *antisymmetric part* $T^{[\mu\nu]}$ of a second-rank indexed quantity $T^{\mu\nu}$ are defined as in Equation 4.20 while the *trace* of a second-rank tensor is given by the contraction $\eta_{\mu\nu} T^{\mu\nu} = T^\mu{}_\mu$ of its two indices. It follows that,

$$T^{\mu\nu} = T^{(\mu\nu)} + T^{[\mu\nu]} . \qquad (14.16)$$

Although the trace of a second-rank Lorentz tensor equals the sum of the diagonal components in either of the two mixed forms, it should be noted that this is not the case for its covariant or contravariant components.

Exercise 14.12 *Make a 1 + 3 split on the two indices of an antisymmetric 4-tensor $F^{\mu\nu}$ to show that, under restriction from Poincaré to Euclidean transformations, the independent parts are equivalent to a 3-vector and an antisymmetric 3-tensor.*

Exercise 14.13 *Show that the independent components of an antisymmetric Lorentz 4-tensor of rank 2 are equivalent to parts which transorm under rotations as a 3-vector and a pseudo-3-vector. Interpret the 3-vector parts of the small parameters $\delta\omega_{\mu\nu} = -\delta\omega_{\nu\mu}$ of a Lorentz transformation (see Exercise 13.10) in terms of the parameters of rotations and boosts by comparing the form of $\delta\Lambda$ with the infinitesimal forms of Equations 13.45 and 13.47.*

Exercise 14.14 *Show that the symmetric and antisymmetric parts of the tensor transform $T^{\bar\mu\bar\nu}$ of $T^{\mu\nu}$ are the tensor transforms of the symmetric and antisymmetric parts of the original tensor $T^{\mu\nu}$.*

Exercise 14.15 *Construct a traceless tensor from $T^{\mu\nu}$ and the universal tensor $\eta_{\mu\nu}$.*

The tensor $\frac{1}{4}T^\lambda{}_\lambda \eta^{\mu\nu}$ is referred to as the *trace part* of the tensor $T^{\mu\nu}$. The symmetric, antisymmetric, trace, and traceless parts may be constructed on any pair of indices of a tensor of any rank 2 or higher. The fact that they are *invariant parts* is one of the reasons why they are so important.

We also define the *completely symmetric part* $T_{(\mu\nu\lambda)}$ and the *completely antisymmetric part* $T_{[\mu\nu\lambda]}$ of the rank-3 indexed quantity $T_{\mu\nu\lambda}$ by:

$$T_{(\mu\nu\lambda)} = \frac{1}{3!}(T_{\mu\nu\lambda} + T_{\nu\lambda\mu} + T_{\lambda\mu\nu} + T_{\nu\mu\lambda} + T_{\mu\lambda\nu} + T_{\lambda\nu\mu}) , \quad (14.17)$$

$$T_{[\mu\nu\lambda]} = \frac{1}{3!}(T_{\mu\nu\lambda} + T_{\nu\lambda\mu} + T_{\lambda\mu\nu} - T_{\nu\mu\lambda} - T_{\mu\lambda\nu} - T_{\lambda\nu\mu}) . \quad (14.18)$$

These second- and third-rank concepts extend in an obvious way to give the completely symmetric and completely antisymmetric parts on any p indices of an object of rank $r \geq p$ in N dimensions.

14.6 Pseudo 4-scalars, 4-vectors and 4-tensors

So far we have dealt primarily with restricted Lorentz transformations. These do not include the matrix $P = \{P^{\bar{\mu}}{}_\nu\} = \text{diag}\{1, -1\!\!1\}$ which represents the spatial reflection (or parity) transformation **P** in $\mathbb{E}^{1,3}$.

A very important property of the electromagnetic, gravitational and strong-nuclear interactions is their parity covariance. No phenomena involving solely these three interactions shows any sign of preferential creation or destruction of handedness or chirality. Equally important is the lack of parity covariance of the weak-nuclear interaction. In order to be able to analyse and make use of these properties we shall investigate the differing ways that Lorentz scalars, vectors and tensors may transform under the parity operation.

We may extend the distinction between Euclidean tensors and pseudotensors to relativistic scalars, vectors and tensors by including the temporal components that are unaffected by a parity transformation.

Let us consider here all the orthochronous Lorentz matrices $L = \Lambda$, ΛP or $P\Lambda$. Suppose that $\phi(x)$ and $\tilde{\phi}(x)$ are scalar fields, $V^\mu(x)$ and $\tilde{V}^\mu(x)$ are 4-vector fields and $T^{\mu\nu\cdots}{}_{\lambda\cdots}(x)$ and $\tilde{T}^{\mu\nu\cdots}{}_{\lambda\cdots}(x)$ are tensor fields under proper Lorentz transformations of matrix Λ. Suppose also that all three fields without tildes transform under the orthochronous Lorentz transformations according to,

$$T^{\bar{\mu}\bar{\nu}\cdots}{}_{\bar{\lambda}\cdots}(x) = L^{\bar{\mu}}{}_\mu L^{\bar{\nu}}{}_\nu \cdots L^\lambda{}_{\bar{\lambda}} \cdots T^{\mu\nu\cdots}{}_{\lambda\cdots}(L^{-1}x) , \quad (14.19)$$

14.6. Pseudo 4-scalars, 4-vectors and 4-tensors

while those marked with tildes transform according to,

$$\tilde{T}^{\bar{\mu}\bar{\nu}\cdots}{}_{\bar{\lambda}\cdots}(\bar{x}) = (\det L) L^{\bar{\mu}}{}_{\mu} L^{\bar{\nu}}{}_{\nu} \cdots L^{\lambda}{}_{\bar{\lambda}} \cdots \tilde{T}^{\mu\nu\cdots}{}_{\lambda\cdots}(L^{-1}\bar{x}) \ . \tag{14.20}$$

Exercise 14.16 *Show that the parity transformations of ϕ and the individual components of V^{μ} and $T^{\mu\nu\cdots}{}_{\lambda\cdots}$ corresponding to Equation 14.19 are,*

$$\begin{aligned}
\mathbf{P}\phi(x) &= \phi(P^{-1}x) \\
\mathbf{P}V^{\mu}(x) &= \{V^{0}(P^{-1}x), -V^{k}(P^{-1}x)\} \\
\mathbf{P}T^{00\cdots}{}_{0\cdots}(x) &= T^{00\cdots}{}_{0\cdots}(P^{-1}x) \\
\mathbf{P}T^{k0\cdots}{}_{0\cdots}(x) &= -T^{k0\cdots}{}_{0\cdots}(P^{-1}x) \\
\mathbf{P}T^{k\ell 0\cdots}{}_{0\cdots}(x) &= T^{k\ell 0\cdots}{}_{0\cdots}(P^{-1}x) \ ,
\end{aligned} \tag{14.21}$$

so that there is an overall sign change in the components for each spatial index, while corresponding to Equation 14.20 we have,

$$\begin{aligned}
\mathbf{P}\tilde{\phi}(x) &= -\tilde{\phi}(P^{-1}x) \\
\mathbf{P}\tilde{V}^{\mu}(x) &= \tilde{V}^{\mu}(P^{-1}x) \\
\mathbf{P}\tilde{T}^{\mu\nu\cdots}{}_{\lambda\cdots}(x) &= \tilde{T}^{\mu\nu\cdots}{}_{\lambda\cdots}(P^{-1}x) \ .
\end{aligned} \tag{14.22}$$

These last transformation laws differ from those in Equation 14.21 for proper quantities only in the absence of the overall sign changes for each spatial index and its inclusion for $\tilde{\phi}$. The quantities $\phi(x)$, $V^{\mu}(x)$ and $T^{\mu\nu\cdots}{}_{\lambda\cdots}(x)$ are said to be a *proper scalar*, a *proper vector* (or *polar vector*) and a *proper tensor*, respectively, while $\tilde{\phi}(x)$, $\tilde{V}^{\mu}(x)$ and $\tilde{T}^{\mu\nu\cdots}{}_{\lambda\cdots}(x)$ are said be a *pseudoscalar*, *pseudovector* (or *axial vector*) and a *pseudotensor*, respectively. The pseudoscalar is defined in such a way that it transforms identically to the contraction of a proper and a pseudovector.

As in 3D, so also in 4D, the tensor product and the scalar product of two proper tensors or two pseudotensors (each of arbitrary rank) are each proper tensors while the product of a proper and a pseudotensor is another pseudotensor. Proper and pseudo scalars are of even and odd parity, respectively, and similar remarks hold for higher rank tensors if one considers only the physical components. For example, the tranverse 3-vector part $\mathbf{A}^{\mathrm{T}}$ of the Maxwell potential A^{μ}, a proper 4-vector, is of odd parity, a property which on quantization is shared by the photon. By contrast, gravitational waves and gravitons have even parity since the physical components can be carried by the transverse traceless (^{TT}) space-space part h_{kl}^{TT} of a proper symmetric 4-tensor, $h_{\mu\nu}$.

Without the spacetime arguments, the above expressions will provide definitions of non-field proper scalars, vectors and tensors. Examples of quantities which are designated as proper 4-scalars are the rest mass of a particle, electric charge and the Lagrangian or action of parity-covariant mechanical systems. Examples of proper 4-scalar fields are temperature and density distributions and the Lagrangian density of parity-covariant field systems. Examples of proper 4-vectors are dx^μ, ∂_μ and the 4-momentum p^μ of a system, mechanical or otherwise, while the electromagnetic potential $A^\mu(x)$ is an example of a proper 4-vector field. Nontrivial examples of proper 4-tensor fields include the electromagnetic field tensor $F^{\mu\nu}(x)$, the energy-momentum tensor $T^{\mu\nu}(x)$, and the linearized Ricci and Riemann curvature tensors, $R^{\mu\nu}_{\text{lin}}(x)$ and $R^{\mu\nu\lambda\rho}_{\text{lin}}(x)$ of weak gravitation. The Minkowski tensor $\eta_{\mu\nu}$ is a trivial example of a proper tensor since the line element of Equation 12.2, and thus of Equation 13.8, is unchanged in a parity transformation.

We shall see later that one example of a Lorentz pseudoscalar field is the electromagnetic scalar product $\mathbf{E}\cdot\mathbf{B}$ despite the fact that it does not appear to be Lorentz-invariant. If magnetic charge were discovered, in the form of magnetic monopoles for example, then it would be not just a pseudo-3-scalar but a pseudo-4-scalar. In the Yukawa theory of the strong-nuclear interaction, the pion field is a pseudo-4-scalar and thus has odd parity. An example of a pseudo-4-vector is the Pauli-Lubański spin 4-vector (see Section 15.10) while the 4D permutation symbol, to be introduced in the next section, and the duals (see Section 14.8) of the electromagnetic field tensor and of the linearized Riemann curvature tensors are examples of pseudotensors.

A theory all of whose equations contain terms which are either all proper tensors or all pseudotensors, of the same rank and index level, will retain its form under parity transformations and will therefore be a parity-covariant or parity-conserving theory. The presence of terms of opposite type (proper and pseudo) added to one another in the same equation in the theory of the weak interaction indicates that it is not parity-covariant.

We can carry out a similar classification of Lorentz tensors with respect to their transformation properties under time-reversal $\mathbf{T}$ and define t-pseudotensors according to,

$$^t T^{\bar{\mu}\bar{\nu}\cdots}_{\bar{\lambda}\cdots}(x) = (\operatorname{sgn} L^{\bar{0}}{}_0) L^{\bar{\mu}}{}_\mu L^{\bar{\nu}}{}_\nu \cdots L^{\lambda}{}_{\bar{\lambda}} \cdots T^{\mu\nu\cdots}{}_{\lambda\cdots}(L^{-1}x) , \qquad (14.23)$$

where L is now one of the matrices, Λ, ΛT or $T\Lambda$. However, the similarities between the two types of division into proper and pseudo quantities, and the fact that $T = -P$ when acting in $\mathbf{E}^{1,3}$, make it unnecessary to give full details in a classical context. Further information is available in Watanabe [560] and Jauch and Rohrlich [266].

Exercise 14.17 *Show that d^4x in Exercise 14.4 is proper under parity and improper with respect to time-reversal.*

14.7 Permutation pseudotensor

The Minkowski tensor $\eta_{\mu\nu}$ is a numerical tensor. The requirements of a numerical tensor are very stringent and thus there are very few.

Exercise 14.18 Show that the most general proper numerical tensor of rank 4 has the form, $a\,\eta_{\mu\nu}\eta_{\lambda\rho} + b\,\eta_{\mu\lambda}\eta_{\nu\rho} + c\,\eta_{\mu\rho}\eta_{\nu\lambda}$, where a, b and c are constant.

Any Lorentz-indexed 4×4 matrix $M = \{M^\mu{}_\nu\}$ will have determinant,

$$\det M = \epsilon^{\mu\nu\lambda\rho} M^{\bar{0}}{}_\mu M^{\bar{1}}{}_\nu M^{\bar{2}}{}_\lambda M^{\bar{3}}{}_\rho \,, \tag{14.24}$$

where the *permutation symbol* is defined, numerically and invariantly (with respect to all basis changes whether rotations, boosts, translations or reflections) by,

$$\begin{aligned}
\epsilon^{\mu\nu\lambda\rho} &= +1 \quad \text{for } \mu\nu\lambda\rho \text{ an even permutation of 0123} \\
&= -1 \quad \text{for } \mu\nu\lambda\rho \text{ an odd permutation of 0123} \\
&= 0 \quad \text{for } \mu\nu\lambda\rho \text{ any other combination.}
\end{aligned} \tag{14.25}$$

Since, by definition, $\epsilon^{\bar{0}\bar{1}\bar{2}\bar{3}} = \epsilon^{0123} = 1$, application of Equation 14.24 to a full Lorentz matrix L, for which $(\det L)^2 = 1$, leads to

$$\epsilon^{\bar{0}\bar{1}\bar{2}\bar{3}} = (\det L)\, L^{\bar{0}}{}_\mu L^{\bar{1}}{}_\nu L^{\bar{2}}{}_\lambda L^{\bar{3}}{}_\rho\, \epsilon^{\mu\nu\lambda\rho} \,. \tag{14.26}$$

Equation 14.26 applies for all components in the new (barred) frame giving,

$$\epsilon^{\bar{\mu}\bar{\nu}\bar{\lambda}\bar{\rho}} = (\det L)\, L^{\bar{\mu}}{}_\mu L^{\bar{\nu}}{}_\nu L^{\bar{\lambda}}{}_\lambda L^{\bar{\rho}}{}_\rho\, \epsilon^{\mu\nu\lambda\rho} \,, \tag{14.27}$$

which shows, by comparison with Equation 14.20, that the permutation symbol is a numerical pseudotensor. Its existence and properties are therefore closely related to the parity transformation and are clearly a direct consequence of the orientability of each of the four dimensions of spacetime in the same way that the metric tensor is a result of its homogeneity and isotropy.

As for any tensor in $\mathbf{E}^{1,3}$, we define covariant components by lowering each index with $\eta_{\mu\nu}$,

$$\epsilon_{\mu\nu\lambda\rho} = \eta_{\mu\pi}\eta_{\nu\kappa}\eta_{\lambda\sigma}\eta_{\rho\tau}\epsilon^{\pi\kappa\sigma\tau} \,, \tag{14.28}$$

from which it follows that $\epsilon_{0123} = -1$. It should be noted that some texts use the opposite sign conventions for ϵ^{0123} and ϵ_{0123} while some put both equal to 1.

Exercise 14.19 *Show that there are no second-rank pseudotensors except in 2D in which case the pseudotensor is the antisymmetric ϵ^{AB} ($A, B = 1, 2$) of Equation 4.30.*

Some of the most important invariant tensor combinations of several factors of $\eta^{\mu\nu}$ (or $\delta^\mu{}_\nu$) are the *generalized Kronecka deltas* defined by:

$$\delta^{\mu\nu}_{\pi\kappa} = \begin{vmatrix} \delta^\mu_\pi & \delta^\nu_\pi \\ \delta^\mu_\kappa & \delta^\nu_\kappa \end{vmatrix} \equiv \delta^\mu_\pi \delta^\nu_\kappa - \delta^\mu_\kappa \delta^\nu_\pi , \qquad (14.29)$$

$$\delta^{\mu\nu\lambda}_{\pi\kappa\sigma} = \begin{vmatrix} \delta^\mu_\pi & \delta^\nu_\pi & \delta^\lambda_\pi \\ \delta^\mu_\kappa & \delta^\nu_\kappa & \delta^\lambda_\kappa \\ \delta^\mu_\sigma & \delta^\nu_\sigma & \delta^\lambda_\sigma \end{vmatrix} \qquad (14.30)$$

$$= \delta^\mu_\pi \delta^\nu_\kappa \delta^\lambda_\sigma + 2 \text{ even permutations of } \pi\kappa\sigma$$
$$- \delta^\mu_\kappa \delta^\nu_\pi \delta^\lambda_\sigma - 2 \text{ even permutations of } \kappa\pi\sigma \qquad (14.31)$$

and

$$\delta^{\mu\nu\lambda\rho}_{\pi\kappa\sigma\tau} = \delta^\mu_\pi \delta^\nu_\kappa \delta^\lambda_\sigma \delta^\rho_\tau + 11 \text{ even permutations of } \pi\kappa\sigma\tau$$
$$- \delta^\mu_\kappa \delta^\nu_\pi \delta^\lambda_\sigma \delta^\rho_\tau - 11 \text{ even permutations of } \kappa\pi\sigma\tau . \qquad (14.32)$$

The following are very useful identities:

$$\epsilon^{\mu\nu\lambda\rho} \epsilon_{\pi\kappa\sigma\tau} = -\delta^{\mu\nu\lambda\rho}_{\pi\kappa\sigma\tau} \qquad (14.33)$$

$$\epsilon^{\mu\nu\lambda\rho} \epsilon_{\pi\kappa\sigma\rho} = -\delta^{\mu\nu\lambda}_{\pi\kappa\sigma} \qquad (14.34)$$

$$\epsilon^{\mu\nu\lambda\rho} \epsilon_{\pi\kappa\lambda\rho} = -2 \delta^{\mu\nu}_{\pi\kappa} \qquad (14.35)$$

$$\epsilon^{\mu\nu\lambda\rho} \epsilon_{\pi\nu\lambda\rho} = -3! \delta^\mu_\pi \qquad (14.36)$$

$$\epsilon^{\mu\nu\lambda\rho} \epsilon_{\mu\nu\lambda\rho} = -4! . \qquad (14.37)$$

The last four of the above identities all follow by contraction of Equation 14.33 which may be proved by the same technique as for Equation 4.38.

Whereas a cyclic change of indices leaves the 3D permutation tensor unchanged, $\epsilon^{k\ell m} = \epsilon^{\ell m k}$, it should be noted that such a rearrangement is equivalent to three transpositions in 4D and, consequently, $\epsilon^{\mu\nu\lambda\rho} = -\epsilon^{\nu\lambda\rho\mu}$.

Exercise 14.20 *Show that the only rank 4 numerical pseudotensors are scalar multiples of the permutation tensor $\epsilon^{\mu\nu\lambda\rho}$.*

Exercise 14.21 *Show that,*

$$T^{\mu\nu\lambda\rho} = a\,\eta^{\mu\nu}\eta^{\lambda\rho} + b\,\eta^{\mu\lambda}\eta^{\nu\rho} + c\,\eta^{\mu\rho}\eta^{\nu\lambda} + d\,\epsilon^{\mu\nu\lambda\rho} , \qquad (14.38)$$

is the most general form of an invariant numerical tensor of rank 4 with respect to the full Lorentz group.

The results of the above exercise may be extended to show that the most general form of an invariant proper tensor field of rank 4 has the above form with a, b and c scalar fields and d a pseudoscalar field with a similar result for the most general rank 4 pseudoscalar field with interchange of the properties of a, b and c with d. For non-field quantities, some of the coefficients vanish thus reproducing earlier results.

14.8 Tensor duality in Minkowskian spacetime

Material in Section 4.7 on 3D duality described why the cross-product and the curl, as vectors, are both objects which are peculiar to $\mathbf{E}^3$. The corresponding operations in Minkowskian spacetime are the formation of the antisymmetric tensors, $U_\mu V_\nu - U_\nu V_\mu$ and $\partial_\mu V_\nu - \partial_\nu V_\mu$, the second of which is referred to as the *covariant curl* of V_μ.

We also remind the reader that the fact that these quantities are not simply related to vectors is one of the main reasons why non-indexed vector analysis is not useful in special relativity to the same extent as in non-relativistic work. Another important reason has been the need to incorporate the indefiniteness of the metric of Minkowskian spacetime. The related result, noted in the first part of Exercise 14.13, is the main reason why covariant indexing can be avoided in favour of 1+3 notation in discussing the electromagnetic field.

We define the *dual* of a completely antisymmetric tensor in 4D by contraction with the rank 4 permutation tensor, $\epsilon_{\mu\nu\lambda\rho}$. To every second-rank antisymmetric tensor $F^{\mu\nu}$ in 4D, there corresponds another equivalent second-rank antisymmetric tensor $\tilde{F}^{\mu\nu}$ called the *dual* of $F^{\mu\nu}$, given by,

$$\tilde{F}^{\mu\nu} = \tfrac{1}{2}\epsilon^{\mu\nu\lambda\rho} F_{\lambda\rho} . \qquad (14.39)$$

The equation $\tilde{F}_{\mu\nu} = \tfrac{1}{2}\epsilon_{\mu\nu\lambda\rho} F^{\lambda\rho}$ is simply an alternative equivalent way of writing this definition. One could also use mixed indices on the permutation tensor to relate $\tilde{F}^{\mu\nu}$ to $F^{\mu\nu}$ directly, but such mixed permutation tensor components are rarely used. The complete antisymmetry of the permutation tensor guarantees that the dual tensor is also antisymmetric and thus has the same number of algebraically-independent components as $F^{\mu\nu}$. If $F^{\mu\nu}$ is a proper tensor then $\tilde{F}^{\mu\nu}$ will be a pseudotensor and vice versa.

Exercise 14.22 *Given $F^{\mu\nu} = -F^{\nu\mu}$ and Equation 14.39, establish the equivalence of $F^{\mu\nu}$ and $\tilde{F}^{\mu\nu}$ by showing that,*

$$F^{\mu\nu} = -\tfrac{1}{2}\epsilon^{\mu\nu\lambda\rho} \tilde{F}_{\lambda\rho} . \qquad (14.40)$$

Exercise 14.23 Show that:

$$\tilde{\tilde{F}}^{\mu\nu} = -F^{\mu\nu} , \qquad (14.41)$$

$$\tilde{F}^{\mu\lambda}\tilde{F}_{\nu\lambda} = F^{\mu\lambda}F_{\nu\lambda} - \tfrac{1}{2}\delta^{\mu}{}_{\nu}F^{\pi\lambda}F_{\pi\lambda} , \qquad (14.42)$$

$$F^{\mu\lambda}\tilde{F}_{\nu\lambda} = \tfrac{1}{4}\delta^{\mu}{}_{\nu}F^{\pi\lambda}\tilde{F}_{\pi\lambda} , \qquad (14.43)$$

and that,

$$F^{\pm}_{\mu\nu} = \tfrac{1}{2}(F_{\mu\nu} \pm i\tilde{F}_{\mu\nu}) , \qquad (14.44)$$

satisfy

$$i\tilde{F}^{\pm}_{\mu\nu} = \pm F^{\pm}_{\mu\nu} . \qquad (14.45)$$

$F^{+}_{\mu\nu}$ is said to be *selfdual* and $F^{-}_{\mu\nu}$ *antiselfdual*. Since $F_{\mu\nu} = F^{-}_{\mu\nu} + F^{+}_{\mu\nu}$ and $\tilde{F}_{\mu\nu} = i(F^{-}_{\mu\nu} - F^{+}_{\mu\nu})$, $F^{\pm}_{\mu\nu}$ are referred to as the *duality parts* of $F_{\mu\nu}$ and $\tilde{F}_{\mu\nu}$. For a plane wave field $F^{\mu\nu}$ of helicity 1, they correspond to the left and right circular polarized modes giving rise to quantum states of helicity ± 1. $F^{\pm}_{\mu\nu}$ may be used to describe chiral states, constructed from the parity eigenstates $F^{\mu\nu}$ or A^{μ}, of light waves and photons.

Exercise 14.24 Show that the only independent quadratic invariants constructable from an antisymmetric tensor $F^{\mu\nu}$ (and the universal numerical tensors in $\mathbf{E}^{1,3}$) may be taken to be $\tfrac{1}{2}F_{\mu\nu}F^{\mu\nu}$ and $\tfrac{1}{2}F_{\mu\nu}\tilde{F}^{\mu\nu}$ (in which the factors of $\tfrac{1}{2}$ are purely conventional).

Exercise 14.25 Let $S^{\mu\nu}$ and $\omega^{\mu\nu}$ each be proper antisymmetric tensors. Let $\mathbf{S} = \{\tfrac{1}{2}\epsilon^{klm}S^{lm}\}$ and $\boldsymbol{\theta} = \{\tfrac{1}{2}\epsilon^{klm}\omega^{lm}\}$ be the pseudo-3-vector duals of the space-space parts of $S^{\mu\nu}$ and $\omega^{\mu\nu}$ respectively and let $\mathbf{K} = \{S^{0k}\}$ and $\boldsymbol{\alpha} = \{\omega^{0k}\}$ denote the proper 3-vectors obtained from their time-space parts S^{0k} and ω^{0k}. Show (a) that,

$$\tfrac{1}{2}\omega_{\mu\nu}S^{\mu\nu} = \boldsymbol{\theta}\cdot\mathbf{S} - \boldsymbol{\alpha}\cdot\mathbf{K} , \qquad (14.46)$$

and (b) that,

$$\tfrac{1}{2}S_{\mu\nu}S^{\mu\nu} = \mathbf{S}^2 - \mathbf{K}^2 \quad \text{and} \quad \tfrac{1}{2}S_{\mu\nu}\tilde{S}^{\mu\nu} = -2\mathbf{S}\cdot\mathbf{K} . \qquad (14.47)$$

Exercise 14.26 A completely antisymmetric rank 3 tensor $F_{\mu\nu\lambda} = F_{[\mu\nu\lambda]}$ will have a 4-vector dual which, with standard normalization, is given by $F^{\mu} = \epsilon^{\mu\nu\lambda\rho}F_{\nu\lambda\rho}/3!$. Show that F^{μ} and $F_{\mu\nu\lambda}$ are equivalent (contain the same information) by determining the expression for $F_{\mu\nu\lambda}$ in terms of F^{μ}.

The dual, $F = \epsilon^{\mu\nu\lambda\rho}F_{\mu\nu\lambda\rho}/4!$, of a completely antisymmetric rank 4 tensor $F_{\mu\nu\lambda\rho}$ in 4D is a scalar. F is a proper or pseudo scalar according to whether $F_{\mu\nu\lambda\rho}$ is a pseudo or proper tensor.

Exercise 14.27 Find the expression for $F_{\mu\nu\lambda\rho}$ in terms of its dual F.

14.9 Matrix generators of Lorentz transformations

For some purposes, it is convenient to re-arrange the expression for a small Lorentz transformation in terms of a sum over the small parameters of the transformation. We shall make use of such an expression in Section 19.5 for the spin density of a field and for the *Belinfante energy-momentum tensor* that correctly allows for the spin angular momentum of a field system.

Suppose that under a Lorentz transformation, a d-dimensional field undergoes a transformation with a $d \times d$ matrix $S(\Lambda)$,

$$\psi \to S(\Lambda)\psi , \qquad (14.48)$$

as occurs, for a rank 2 Lorentz tensor field, for example, where $S(\Lambda) = \Lambda \otimes \Lambda$. Then we define one $d \times d$ *Lorentz generator matrix*,

$$S^{\mu\nu} = i\left(\frac{\partial S(\Lambda(\omega))}{\partial \omega_{\mu\nu}}\right)_{\omega=0} , \qquad (14.49)$$

corresponding to each of the 6 independent parameters $\omega_{\mu\nu}$ in $\Lambda = \Lambda(\omega)$. An independent set of indices in $\omega_{\mu\nu}$ is the antisymmetric index pair in increasing order, namely $|\mu\nu| = 01, 02, 03, 12, 23, 13$. For manifest covariance, it is convenient to use a tensor $\omega_{\mu\nu}$ with 12 non-zero values, only 6 of which are independent. Thus, we include in $\omega_{\mu\nu}$ the other 6 dependent non-zero values corresponding to $[\mu\nu]$, namely 10, 20, 30, 21, 32, 31. The same is true for the generators which are, by Equation 14.49 and the antisymmetry of $\omega_{\mu\nu}$, themselves antisymmetric $S^{\mu\nu} = -S^{\nu\mu}$ in their labels. The latter labels should not be confused with their row and column indices (here suppressed). An infinitesimal Lorentz transformation matrix $S(\delta\Lambda)$, and its effect on ψ, will be given by,

$$S(\delta\Lambda) = 1 - \tfrac{1}{2}i\,\delta\omega_{\mu\nu}S^{\mu\nu} \quad \text{and} \quad \delta\psi = -\tfrac{1}{2}i\,\delta\omega_{\mu\nu}S^{\mu\nu}\psi . \qquad (14.50)$$

The conventional factors of $\tfrac{1}{2}$ in these expressions allow for the fact that the double sum over μ and ν contains twelve non-zero terms in general whereas only six are independent. The factor of i means that $\{S^{kl}\}$ and its 3D dual $\mathbf{S} = \{S^k\}$ are hermitian matrices interpreted in quantum applications as spin operators. A factor of $\hbar$ may also be introduced into the definition of the generator matrices $S^{\mu\nu}$ in quantum applications to give them the dimensions of spin angular momentum.

The $d \times d$ matrices $\mathbf{S} = \{S^k\} = \{\tfrac{1}{2}\epsilon^{klm}S_{\ell m}\}$ and $\mathbf{K} = \{K^k\} = \{S^{0k}\}$ in the decomposition,

$$\tfrac{1}{2}i\omega_{\mu\nu}S^{\mu\nu} = i(\boldsymbol{\theta}\cdot\mathbf{S} - \boldsymbol{\alpha}\cdot\mathbf{K}) , \qquad (14.51)$$

of $S^{\mu\nu}$ generate rotations (with parameters $\boldsymbol{\theta}$) and boosts (of rapidity $\boldsymbol{\alpha}$).

Once the six $d \times d$ matrices $S^{\mu\nu}$ have been determined for each field which is being Lorentz transformed, the matrices $S(\Lambda(\omega))$ for finite values of the additive Lorentz rotation parameters can be determine by *exponentiation* according to,

$$S(\Lambda) = e^{-\frac{1}{2}i\omega_{\mu\nu}S^{\mu\nu}} . \qquad (14.52)$$

We can now calculate these matrices $S^{\mu\nu}$ explicitly for the case of ψ being the 4D Lorentz vector field V^μ. Corresponding to the six Lorentz parameters $\omega_{\mu\nu}$ we require six Lorentz generator matrices $S^{\mu\nu}$, each of which will be, in this case, a 4×4 matrix whose rows and columns will also be labelled by Lorentz indices. To distinguish the Lorentz parameter labels $|\mu\nu|$ on $S^{\mu\nu}$ from their Lorentz row and column indices we shall temporarily use lower case Greek letters from the beginning of the alphabet for the latter. Consequently, we shall label the components of the Lorentz generators as $S^{\mu\nu} = \{S^{\mu\nu\alpha}{}_\beta\}$ acting on a 4-vector V^β. From Equation 14.4, we deduce the change in V^α to be:

$$\begin{aligned} \delta V^\alpha &= (\Lambda^\alpha{}_\beta - \delta^\alpha{}_\beta)V^\beta \\ &= \delta\omega^\alpha{}_\beta V^\beta \\ &= \delta\omega_{\mu\beta}\, \eta^{\alpha\mu}\, V^\beta \\ &= \delta\omega_{\mu\nu}\, \eta^{\alpha\mu}\, \delta^\nu{}_\beta V^\beta \\ &= \tfrac{1}{2}\delta\omega_{\mu\nu}(\eta^{\alpha\mu}\delta^\nu{}_\beta - \eta^{\alpha\nu}\delta^\mu{}_\beta)V^\beta . \end{aligned} \qquad (14.53)$$

Both $S^{\mu\nu}$ and $\omega_{\mu\nu}$ are antisymmetric in μ and ν. Consequently, in order to deduce the generator $S^{\mu\nu}$ from its definition and the above equation we have ensured, by the last step, that the coefficient of the small parameters $\delta\omega_{\mu\nu}$ are explicitly antisymmetrized on $\mu\nu$ so that we may cancel $\delta\omega_{\mu\nu}$ in the identity that results from the comparison of the two. To do this we could have used the results of Exercise 4.6 applied to Lorentz tensors.

The generators are defined by Equation 14.50 applied, in this case, to $S(\Lambda) = \Lambda$ acting on $V = \{V^\alpha\}$ namely,

$$\delta\Lambda = \Lambda - \mathbb{1} = -\tfrac{1}{2}i\,\delta\omega_{\mu\nu}S^{\mu\nu} , \qquad (14.54)$$

which, in terms of the components of V, gives

$$\delta V^\alpha = -\tfrac{1}{2}i\delta\omega_{\mu\nu}S^{\mu\nu\alpha}{}_\beta V^\beta . \qquad (14.55)$$

Comparison of Equations 14.53 and 14.55 establishes the 4-vector generators to be,

$$S^{\mu\nu\alpha}{}_\beta = i(\eta^{\alpha\mu}\delta^\nu{}_\beta - \eta^{\alpha\nu}\delta^\mu{}_\beta) . \qquad (14.56)$$

Exercise 14.28 *Show that:*

$$\tfrac{1}{2}(S_{\mu\nu}S^{\mu\nu})^\alpha{}_\beta = 3\,\delta^\alpha{}_\beta = 1(1+1)\delta^\alpha{}_\beta \quad \text{and} \quad S_{\mu\nu}\tilde{S}^{\mu\nu} = 0 , \qquad (14.57)$$

for the invariants constructed from the generators of a vector field.

The results of the above exercise can be used to demonstrate that the vector field is of mixed spin 0 and 1. One way to split a 4-vector into a spin 0 and a spin 1 part is the non-covariant decomposition $\{V^\mu\} = \{V^0, \mathbf{V}\}$ since the first component transforms as a 3-scalar (spin 0) and the spatial components as a 3-vector (spin 1) under rotations. For vector fields $V(x)$ we also have the gradient operator ∂ available to covariantly project out $\partial \cdot V$ as a 4-scalar and thus spin 0 part. The wave equation satisfied by a 4-vector field may be used to ensure that only the spin 1 part is present in applications to systems required to have definite spin or helicity.

The generators of a tensor field may be constructed from first principles or, more easily, from those of a vector by using the fact that a tensor transforms as a direct product of vectors. The Lorentz generators of the Dirac field are given in Chapter 20.

Bibliography:

Jackson J D 1975 *Classical electrodynamics* [263].
Lord E 1976 *Tensors, relativity and cosmology* [328].
Lovelock D et al. 1975 *Tensors, differential forms and variational principles* [344].
Ohanian H C 1976 *Gravitation and spacetime* [404].

CHAPTER 15

Einsteinian mechanics

What led me more or less directly to the special theory of relativity was the conviction that the electromotive force acting on a body in motion in a magnetic field was nothing else but an electric field[1].

Albert Einstein   1879–1955

In Chapter 6, the fundamentals of Newtonian interaction and dynamics were reviewed on the basis of conservation laws and Galilean relativity. Having now replaced the Galilean version of the Relativity principle by the special relativistic interpretation of Einstein, we must re-examine the fundamentals of the dynamics of particles as set out by Newton.

In order to carry this out, we shall attempt to retrace almost the same steps, altering only the form of the boost transformation and making use, where convenient, of the more elegant Lagrangian and Hamiltonian formulations since we have shown these are equivalent to the description used by Newton. One factor will be crucially important in understanding the replacement of Newton's laws and that will be the existence of the universal constant c, the ultimate speed. Another important limitation on our choices will be the necessity to retrieve the Newtonian description when the rest masses are non-zero which permits the velocities to be small in some frames.

15.1 Four-velocity and four-acceleration

The kinematic characterization of a classical (non-quantum) Einsteinian particle remains almost the same — one which can be described by a smooth locus $z(t)$ in spacetime — except that now we must require that locus to be either *timelike* ($\dot{z}^2 < c^2$) or *null* $\dot{z}^2 = c^2$. The first of these

---

[1] From a letter (1952) to the Michelson Commemorative Meeting of the Cleveland Physics Society as quoted by Shankland [509].

15.1. Four-velocity and four-acceleration

two possibilities must reduce to the Newtonian results in a frame where the speed $|\dot{z}|$ is small.

The variable t is not now an absolute time but another frame dependent coordinate similar to each of the three in x which take on the values z on the trajectory. Whereas Newtonian time intervals are Galilean invariants, in Einsteinian relativity the Lorentzian coordinate time intervals dt are frame-dependent, being the zeroth components of a 4-vector dx^μ. A variety of different variables are now available for parameterizing the evolution of a mechanical system. We may, for example, use the time coordinate in any inertial frame. Since the time coordinate $\bar{t}$ of one frame will be a linear combination of all the four coordinates of some other equally valid frame, we cannot, *a priori*, rule out the use of an evolution parameter λ which depends on all four coordinates. However, the essential properties of an evolution parameter are to order the evolutionary states of a system. Consequently, we may use any monotonically increasing $(d\bar{\lambda}/d\lambda > 0)$ function $\bar{\lambda}(\lambda)$ of any other evolution parameter λ which may or may not be the time coordinate t of an inertial frame.

The use of z and t as dependent and independent variables in the trajectory $z(t)$ of a particle means that such a description of the motion is not manifestly Poincaré-covariant. To re-express the kinematic description of the trajectory in manifestly covariant form, we combine z and t together so that their changes $\{c\,dt, dz\}$ transform as a 4-vector and allow other evolution parameters besides t. Let us, for the moment, choose the evolution parameter to be the proper time $\tau(t)$ relative to an arbitrary origin $\tau(0)$ with only differences $d\tau$ being physically significant. Since proper time is not defined for null intervals, this choice restricts us in the meantime to the non-null case for which a Newtonian limit must exist.

The *proper time interval* is given in terms of t and $z(t)$ by the time-dilation formula $dt = \gamma(\dot{z})d\tau$ of Exercise 12.4 where $\dot{z}(t) = dz/dt$. That result integrates to,

$$\tau(t) = \tau(t_i) + \int_{t_i}^{t} dt \sqrt{1 - \dot{z}^2(t)/c^2} \; . \qquad (15.1)$$

From a knowledge of $z(t)$ in a particular frame we may calculate τ and t from one another. Proper time and coordinate time each have their uses. The above formula will give the same *invariant* result for $\Delta\tau$ for a specific trajectory from one event x_i to another x_f irrespective of which coordinates are used. This is one of the advantages of proper time when dealing with a specific particle or trajectory. In contrast, the proper time is said to be *non-integrable*, since the value it gives between two events is dependent on the path $z(t)$. It is not useful for labelling the events uniquely, whereas the *coordinate time interval* between two events, although not invariant, is path-independent and therefore *integrable*.

Using Equation 15.1, we replace $z(t)$ by the corresponding values of $z(\tau)$ and $ct(\tau)$ which we combine to build $z(\tau) = \{z^\mu(\tau)\} = \{ct(\tau), \mathbf{z}(\tau)\}$ providing a manifestly covariant description of coordinate intervals $dz(\tau)$ along the trajectory. First, we note that we have replaced a description using three independent coordinates $\mathbf{z}$ by one with four coordinates z^μ. The changes in $z^\mu(\tau)$ must satisfy one *constraint* on the initial data of the trajectory. By analogy with the non-relativistic 3-velocity $\mathbf{u} = \dot{\mathbf{z}} = d\mathbf{z}/dt$, we define the 4-*velocity* of a timelike relativistic particle ($d\tau \neq 0$) by,

$$u = \dot{z} = \{\dot{z}^\mu\} = \left\{\frac{dz^\mu}{d\tau}\right\}, \tag{15.2}$$

where we use the dot notation for the derivative of a trajectory variable with respect to its argument, t, τ or λ. Since dz^μ is a proper Lorentz 4-vector and $d\tau$ a proper scalar, the 4-velocity is a proper 4-vector. We define the 4-*acceleration* by the derivative $a = \dot{u} = \{du^\mu/d\tau\}$ of the 4-velocity with respect to the proper time. From the 3-velocity $\dot{\mathbf{z}} = d\mathbf{z}/dt$, one obtains the 4-velocity using,

$$u = \{u^\mu\} = \{\gamma c, \gamma \dot{\mathbf{z}}\}, \tag{15.3}$$

and, conversely, from u one may determine $\dot{\mathbf{z}}$ using,

$$\gamma = u^0/c \quad \text{and} \quad \dot{\mathbf{z}} = \{u^k/\gamma\}. \tag{15.4}$$

Exercise 15.1 *(a) Show that the 4-velocity is timelike by establishing the constraint,*

$$u \cdot u = -c^2. \tag{15.5}$$

(b) Show that the 4-velocity and 4-acceleration are orthogonal, $u \cdot a = 0$, and state whether the 4-acceleration a is time-like, null or space-like. It is instructive to do this two ways: first, by evaluating its 'square' $a \cdot a$ in an arbitrary frame (or in any convenient frame and appealing to invariance) and, second, by simply differentiating Equation 15.5. (c) Show that, in a frame where the 3-velocity and 3-acceleration are $\mathbf{u} = \dot{\mathbf{z}}$ and $\mathbf{a} = \ddot{\mathbf{z}}$, the 4-acceleration components are given by,

$$a = \{\gamma^4 \mathbf{u} \cdot \mathbf{a}/c, \gamma^4(\mathbf{u} \cdot \mathbf{a})\mathbf{u}/c^2 + \gamma^2 \mathbf{a}\}. \tag{15.6}$$

(d) What are the components of the 4-velocity and of the 4-acceleration in the instantaneous rest frame of a particle? (e) Show that the relative speed parameter γ may be expressed covariantly in terms of the 4-velocities u_1 and u_2 of the two frames it describes by $\gamma = -u_1 \cdot u_2/c^2$. (f) Show that the relative velocity of two particles ($i = 1, 2$) of mass m_i and 4-momentum p_i is $v = c\sqrt{(p_1 \cdot p_2)^2 - m_1^2 m_2^2 c^4}/|p_1 \cdot p_2|$, valid also for massless particles.

It should be noted that the spatial parts of the 4-velocity and 4-acceleration are *not* equal to the non-relativistic 3-velocity and 3-acceleration except in the low-velocity limit where $\gamma \to 1$.

15.2 Poincaré-covariant conservation laws

The conservation postulates on which we based the discussion of Newton's laws in Chapter 6 remain valid in Einsteinian physics, in the sense that we demand the existence of ten quantities, the generators of spacetime translations, rotations and boosts, whose conservation will indicate the motion is that of an isolated system in an inertial frame. All that changes is the way those quantities transform in a boost if the boost velocity is no longer small compared to c.

Since an instantaneous proper or rest frame always exists for a timelike trajectory (see Exercise 12.3) we may characterize the inertia of such a particle by its mass in the rest frame, called its *rest mass* which is the same quantity as the Newtonian mass used previously. Since it is defined by using one particular frame, the rest mass is a Poincaré-invariant which we shall denote by m.

We may now form a Poincaré-covariant version of the Newtonian conservation laws. The conserved generators of time and space displacements are the energy E and 3-momentum $\mathbf{p}$. Since displacements $\{c\Delta t, \Delta \mathbf{x}\}$ transform as a 4-vector, we combine the corresponding generators and demand that the four quantities $\{E/c, \mathbf{p}\}$ not only be conserved in an inertial frame for an isolated system, but also transform as a 4-vector. Putting $p^0 = E/c$, covariance therefore requires the Newtonian postulate of conservation of energy E and 3-momentum $\mathbf{p} = m\dot{\mathbf{z}}$ to be replaced by conservation of 4-*momentum*,

$$p = mu \quad \text{or} \quad \{p^\mu\} = \{mu^\mu\} = \{m\gamma c, m\gamma \dot{\mathbf{z}}\} = \frac{1}{\sqrt{1 - \dot{z}^2/c^2}}\{mc, m\dot{\mathbf{z}}\} \,, \tag{15.7}$$

where the last two equalities arise by making use of the kinematic result in Equation 15.3. The 4-momentum of an isolated system is thus required to be a Poincaré vector.

Because one of the expressions in Equation 15.7 is manifestly covariant, all are covariant. The equation therefore automatically holds in any frame, in particular one where $|\dot{\mathbf{z}}|$ is not small, if it holds in the rest frame $\dot{\mathbf{z}} = 0$ or in near rest frames, $|\dot{\mathbf{z}}| \ll c$, as it must since Newton's laws are then valid. Evaluating the zeroth component of this equation at vanishingly low speeds ($\gamma \to 1$), gives the following equation relating the rest energy E_{rest} to the rest mass via the ultimate speed c,

$$E_{\text{rest}} = (p^0)_{\text{rest}} c = mc^2 \,. \tag{15.8}$$

Poincaré-covariance of the standard conservation laws, consistent with the Newtonian limit, therefore demands that the rest value of the conserved generator of time displacement, the total energy, be related to the Newtonian mass m according to the *mass-energy relation* of Einstein.

Use of Lorentz rather than Galilean relativity, namely Poincaré covariance, therefore implies that inertial rest mass m has rest energy mc^2. This result was a major theoretical prediction [129] of Einstein's re-appraisal of the nature of measurement of time and space in 1905, anticipated by a few of the writings of some of his predecessors, in particular Lorentz and FitzGerald. It provided a vital step in the unification of matter and electrodynamics, endowing matter (even at rest) with energy like the electromagnetic field and the latter with mass and thus inertia. This also linked up with the quantum properties of radiation where an energy was associated by Einstein [127] in 1905 with a quantum of a particular wavelength to explain the photoelectric effect.

The zero velocity limit of the energy E is not zero as in Newtonian physics. There is, however, no conflict with the Newtonian limit since the conservation laws are unchanged if each conserved quantity is changed by a value which is constant in any particular frame. Owing to the large value of c in practical units, the energy mc^2 obtained in this way is enormously high for ordinary objects compared with its energy as a result of other causes such as motion. However, unless that rest energy is converted to other forms, such as kinetic, it will remain a physically undetected part of the arbitrary additive constant implicit in the Newtonian energy.

In a frame where $\gamma \neq 1$, Equation 15.7 gives,

$$E = p^0 c = \gamma E_{\text{rest}} = \gamma \, mc^2 = \frac{mc^2}{\sqrt{1 - \dot{z}^2/c^2}} \,, \qquad (15.9)$$

showing that the *relativistic kinetic energy* is,

$$T = E - mc^2 = (\gamma - 1)mc^2 = \left(\frac{1}{\sqrt{1 - \dot{z}^2/c^2}} - 1\right)mc^2 \,, \qquad (15.10)$$

with its limiting value of $\tfrac{1}{2}m\dot{z}^2$ for slow speeds providing the connection with Newtonian energy.

In Newton's laws, the non-zero rate of change with time of $m\dot{z}$, namely $\dot{p} = m\ddot{z} \neq 0$, for a part of a system which is isolated overall, indicates that the part is not isolated but interacting. The mass m is a measure of its acceleration for a given strength of interaction, and therefore a measure of its *inertia*. Correspondingly, we can expect that, given an isolated relativistic system, a part with 4-momentum p having a non-zero value of $dp/d\tau$ will not be isolated but interacting, and the spatial part, $d\mathbf{p}/d\tau$, will give us an indication of the strength of the interaction. However, for a given interaction strength, the acceleration which results will no longer be inversely

15.2. Poincaré-covariant conservation laws

proportional to the rest mass m, since $\mathbf{p} = \{p^k\}$ is no longer linear in $\dot{z}$ but also depends on it via the factor of γ in Equation 15.7. The quantity which characterizes the *relativistic inertia* is no longer m but γm, which is therefore called the *relativistic mass* of the system in motion with velocity $\mathbf{u} = \dot{z}$.

The prediction that inertia, as characterized by the relativistic mass $\gamma m = m/\sqrt{1 - \dot{z}^2/c^2}$, depends on the speed $u = |\dot{z}|$, and tends to infinity as $u \to c$, is another striking theoretical result [129, 130] of Einsteinian physics. It is confirmed by a huge body of experimental data, in particular with the use of particle accelerators. That data permits one, in principle, to determine the ultimate speed c. Although more precise methods exist to determine c once it is established that electromagnetic radiation propagates at the ultimate speed, particle data on masses in motion certainly suffice to demonstrate that c is finite thus showing that Galilean relativity is only approximately true. However, in 1910, when reliable experimental results confirming the variation of relativistic mass were first available, special relativity was already becoming widely accepted on other grounds, as a result of its examination over the previous five years, particularly from 1907 to 1909.

With the above natural definition of relativistic mass, the equivalence between energy and inertial mass extends to the energy and mass of a particle in motion. From Equation 15.9, energy, the conserved quantity corresponding to time-translation invariance, is thus equivalent to the inertial property of relativistic mass γm according to,

$$\text{Energy} = (\text{mass})\, c^2 \ . \tag{15.11}$$

This close connection between energy and mass is a direct consequence of the intimate relationship between time and space in Minkowskian spacetime resulting from kinematics incorporating a finite upper bound on speeds. Since its origin is in the spacetime translation invariance of isolated systems, *mass-energy equivalence* applies to all forms of energy, such as excitation, virtual particle and binding energies, and to all forms of mass. Countless experiments, such as those involving atomic (chemical) and nuclear reaction energies, confirm this equivalence.

Exercise 15.2 *A quantum of light of wavelenth λ has energy $E = hc/\lambda$ and therefore apparently has 'mass' $h/c\lambda$. Given that light always has speed c, and that inertia is a characteristic of acceleration, can a photon therefore have inertia and mass?*

Poincaré was aware [438] (*Ouevres* [451, vol. 9, p. 392]) as early as 1895 that, if radiant energy had inertia, conflicts between electromagnetic phenomena and the Newtonian law of action and reaction would be resolved but was not prepared to adopt such a radical notion at that time. In 1900, he showed that the Lorentz electron theory of electrodynamics required the reaction law to apply not only to matter but also to radiation and linked this result to the applicability of the 'Principle of relativity' (*le principe du mouvement relatif*) [442] (*Oeuvres* [451, vol. 9, pp. 471, 472, 482 and 488]) to both matter and radiation. He also concluded that the radiant momentum and inertia that this would imply was beyond the limits of measurability of radiation pressure. In September 1904, he again stated [446, p. 17] that mass-energy equivalence was a necessary consequence of applying the Principle of relativity to electrodynamic phenomena.

By definition, relativistic energy $\gamma m c^2$ and mass γm, are not scalars but the time components of a 4-vector. The spatial part $\mathbf{p} = \{p^k\}$ of the 4-momentum may be called the *relativistic 3-momentum*,

$$\mathbf{p} = m\gamma \dot{\mathbf{z}} , \qquad (15.12)$$

which only reduces to the familiar non-relativistic 3-momentum $m\dot{\mathbf{z}}$ when velocities are so low that γ can be taken equal to 1. Using $\mathbf{p}$ for the spatial part of p^μ and also for $m\dot{\mathbf{z}}$ in non-relativistic situations should never be allowed to lead to confusion of the two.

In terms of energy E and relativistic 3-momentum $\mathbf{p} = \{p^k\}$, the 4-momentum is,

$$p = \{p^\mu\} = \{E/c, p^k\} = \{E/c, \mathbf{p}\} . \qquad (15.13)$$

In the instantaneous rest frame, which always exists for a system with non-zero rest mass, the 4-momentum has the value $p = \{p^\mu\} = \{mc, \mathbf{0}\}$. Making use of this result can often simplify calculations. The results can then be re-expressed in a form valid in any frame by appealing to covariance; we shall use this procedure on numerous occasions.

Since $p = \{p^\mu\}$ is, by construction from Equation 15.7, equal to dz^μ divided by an invariant, it will transform under a Lorentz boost, in which the x-axis is taken parallel to the boost, analogously to Equation 13.43, namely;

$$\begin{aligned} p^{\bar{0}} &= \gamma(p^0 - \beta p^1) \\ p^{\bar{1}} &= \gamma(-\beta p^0 + p^1) \\ p^{\bar{2}} &= p^2 \\ p^{\bar{3}} &= p^3 . \end{aligned} \qquad (15.14)$$

Using Equation 15.13, this becomes,

$$\begin{aligned} \bar{E} &= \gamma(E - v p_x) \\ \bar{p}_x &= \gamma(p_x - vE/c^2) \\ \bar{p}_y &= p_y \\ \bar{p}_z &= p_z . \end{aligned} \qquad (15.15)$$

15.2. Poincaré-covariant conservation laws

These formulae are useful for conversion between laboratory and centre of mass frames in high-energy collision problems and, applied to a massless particle, are exceedingly useful in deriving Doppler and aberration formulae [175, 384].

The normalization condition for the 4-velocity in Equation 15.5 shows that 4-momentum must satisfy a similar relation,

$$p^2 + m^2 c^2 = 0 \quad \text{or} \quad p^\mu p_\mu + m^2 c^2 = 0, \tag{15.16}$$

which, in non-covariant notation involving E and $\mathbf{p}$, is Dirac's well-known energy relation,

$$E^2 = (\mathbf{p}c)^2 + (mc^2)^2 = (\mathbf{p}c)^2 + E_{\text{rest}}^2. \tag{15.17}$$

The kinetic energy is therefore,

$$T = \sqrt{(\mathbf{p}c)^2 + (mc^2)^2} - mc^2. \tag{15.18}$$

The components of $p = \{p^\mu\}$ may be used to label a 4D *momentum space* ($-\infty < p^\mu < \infty$) with Minkowski metric $\eta_{\mu\nu}$. In such a space, the Dirac relation defines a 3D hyperbolic surface with a separate sheet each for $E < 0$ and $E > 0$ if $m \neq 0$. The positive energy (future-pointing) sheet ($p^0 > 0$) is referred to as the *mass shell*.

In contrast to the Newtonian formula $T = \mathbf{p}^2/2m$, Equation 15.17 has 'negative energy' solutions. Classical variables arise by continuous operations on continuous variables such as $\mathbf{z}(t)$ in mechanics or the components $\psi(t, \mathbf{x})$ of fields. The positive and negative solutions cannot be connected by continuous operations on real particles (for which either $m \neq 0$ or $|\mathbf{p}| \neq 0$ or both). Consequently, requiring the low velocity limit for $m \neq 0$ to be Newtonian shows that the negative solutions may be ignored in a non-quantum context.

Exercise 15.3 (a) *Show that the speed $u = |\dot{\mathbf{z}}|$ of a particle is given in terms of its energy and the magnitude of its relativistic 3-momentum $\mathbf{p}$ by $u = |\mathbf{p}|c^2/E$ or $u = \partial E/\partial|\mathbf{p}|$. (Both these results remain valid for massless particles.) (b) Given the 4-momentum p of a particle and the 4-velocity u of an observer, we may construct an invariant $p \cdot u$ from their product. Show that $E = -p \cdot u$ is the energy of the particle in the proper frame of the observer.*

We have so far only made use of four of the ten quantities which we require to be conserved for any isolated system. Let us examine the consequences of re-expressing the remaining six in covariant form.

Exercise 15.4 *Show that the orbital angular momentum about the origin,*

$\mathbf{l} = \mathbf{z} \times \mathbf{p}$, of a particle of rest mass $m \neq 0$ can be re-expressed as,

$$l^{kl} = z^k p^l - z^l p^k ,\qquad (15.19)$$

where l^{kl} is the proper 3-tensor, dual in 3D to the pseudo-3-vector $\mathbf{l}$, and that the centre of mass motion, $\mathbf{k} = \mathbf{p}t - m\mathbf{z}$, may be re-expressed in the form,

$$ck^k \equiv l^{0k} = z^0 p^k - z^k p^0 .\qquad (15.20)$$

Hence, show that the two together transform as the six independent components,

$$l^{\mu\nu} = z^\mu p^\nu - z^\nu p^\mu ,\qquad (15.21)$$

of a proper antisymmetric Poincaré tensor.

We shall refer to $l^{\mu\nu} = (z^\mu - d^\mu)p^\nu - (z^\nu - d^\nu)p^\mu$ as the *covariant orbital angular momentum* about the event $d = \{d^\mu\}$ of the isolated system located on the trajectory $z(\tau)$ and having 4-momentum p.

In preparation for the decomposition of the total angular momentum of a massive system with internal structure into orbital and spin parts, we replace the notation $l^{\mu\nu}$ by the standard symbol $j^{\mu\nu}$ for total angular momentum. The values of $j^{\mu\nu}$ about c are related to those about d by,

$$j_c^{\mu\nu} = j_d^{\mu\nu} - (c^\mu - d^\mu)p^\nu + (c^\nu - d^\nu)p^\mu .\qquad (15.22)$$

Corresponding to the Newtonian conservation laws of Galilean relativity, and Newton's laws of motion, we may now postulate:

the Einsteinian conservation laws of Lorentz relativity — a relativistic mechanical system is said to behave as an isolated particle located at coordinates $z^\mu(\tau)$ in an inertial frame of basis $\{e_\mu\}$ if,

- 4-momentum $p^\mu = m\dot{z}^\mu$, and
- covariant angular momentum about an arbitrary fixed reference event d^μ, namely $j^{\mu\nu} = (z^\mu - d^\mu)p^\nu - (z^\nu - d^\nu)p^\mu$,

are additive, conserved Poincaré-covariant quantities, which correspond to invariance under translations $\Delta x^\mu = a^\mu$ and covariant rotations $\Delta(z-d)^\mu = \Delta\omega^{\mu\nu}(z-d)_\nu$ about d .

(Although z^μ are only coordinates, the trajectory variable $z^\mu - d^\mu$ is a 4-vector.) For slow speeds the above demand is equivalent to the Newtonian results plus mass-energy equivalence. Appealing to Poincaré covariance to demand that the ten conserved quantities be constructed explicitly for any isolated relativistic system, whether massive or null, contains the essence of the relativistic equivalent of Newton's laws.

15.3 Relativistic free particle equations of motion

We are now in a position to proceed with the description of the relativistic motion of a particle using a manifestly Lorentz-Poincaré covariant notation in complete analogy with the similar development of the dynamics of a Newtonian particle. The eleven Newtonian conservation laws for m, E, **p**, **k** and **l** are expressed compactly in a relativistic context in terms of the conservation of the ten quantities p^μ and $j^{\mu\nu}$. In a relativistic context the mass γm is not independent of $p = \{p^\mu\}$ and E and its conservation need not be separately specified as in Newtonian physics. Total rest mass m is not conserved in special relativity.

15.3 Relativistic free particle equations of motion

Let us examine three sets of equations of motion determining the trajectory variables, $\mathbf{z}(t)$ or $z(\tau)$ or $z(\lambda)$, of a free particle, namely,

$$\ddot{\mathbf{z}}(t) = 0, \qquad (15.23)$$

$$\dot{z}^2(\tau) = -c^2 \quad \text{and} \quad \ddot{z}(\tau) = 0, \qquad (15.24)$$

$$\ddot{z}^\mu(\lambda)\dot{z}^2(\lambda) - \dot{z}^\mu(\lambda)\dot{z}(\lambda)\cdot\ddot{z}(\lambda) = 0, \qquad (15.25)$$

where the origin and physical significance of the unusual third equation will be clarified as we proceed. Let us assume at least that λ is some evolution parameter (not necessarily the coordinate or proper time) for a future-pointing ($\dot{z}^0 > 0$) timelike ($\dot{z}^2 < 0$) locus. The third equation with $\lambda = t$ or τ is certainly satisfied for any solution of either of the first two. The differences between these three equations contain some of the elements of one of the most significant developments in modern physics, the importance of gauge freedom. We shall therefore examine them in detail as a means of accumulating some of the essential elements of gauge properties in a very simple mechanical context.

Equation 15.23 is the equation of motion of a free Newtonian particle and is therefore also the equation of motion of a non-null relativistic particle. The principal difference between Equations 15.23 and 15.24 is that the second is manifestly covariant.

Exercise 15.5 *Show (a) that Equations 15.23 and 15.24 are equivalent and that imposing the proper time constraint on $\dot{z}(\lambda)$ (see Equation 15.5) reduces Equation 15.25 to Equations 15.24, (b) that the variables $\mathbf{z}(t)$ and $z^\mu(\tau)$ in Equations 15.23 and 15.24 exhibit no gauge freedom and (c) that Equation 15.25 has gauge invariance under transformations $\lambda \to \bar\lambda = f(\lambda)$ where $f(\lambda)$ is an arbitrary monotonically increasing function of λ.*

In the first of these three forms, all three equations are true dynamical equations, each containing the acceleration of one of the three true dynamical degrees of freedom. The explicit constraint in the second set shows that one of the four equations of motion $\ddot{z}(\tau) = 0$ is not a true equation of motion and one of the four variables not a true physical degree of freedom. This is all very obvious. But what can we say about the third set?

It is straightforward to show that the determinant of the matrix of coefficients of $\ddot{z}^\mu$ of Equation 15.25 vanishes. (For example, one may note that $\dot{z}^\mu$ is an eigenvector of eigenvalue zero.) Its rank shows how many of the equations are true equations of motion. The fact that the gauge freedom involves one arbitrary function also implies that only three of the variables $z^\mu(\tau)$ can be true physical degrees of freedom.

Exercise 15.6 Show (a) that the 4×4 matrix of coefficients, $\eta^{\mu\nu}\dot{z}^2 - \dot{z}^\mu \dot{z}^\nu$, of the highest derivatives $\ddot{z}^\mu$ in Equation 15.25 is of rank 3, and (b) that Equation 15.25, with $\dot{z}^2 \neq 0 \neq \dot{z}^0$, implies that three of the accelerations can be expressed in terms of the velocities and a fourth acceleration as $\ddot{\mathbf{z}} = \ddot{z}^0 \dot{\mathbf{z}}/\dot{z}^0$.

The solutions of Equation 15.25 for $\mathbf{z}(\lambda)$, for given initial data $\mathbf{z}(0)$ and $\dot{\mathbf{z}}(0)$, will therefore contain one arbitrary function, $\ddot{z}^0(\lambda)$, in agreement with the implications of the gauge freedom.

We shall see with more examples that the existence of a gauge invariance in a system of dynamical equations always implies that one or more of the equations of motion is not a true dynamical equation but a constraint on the initial data. Conversely, equations of motion that contain certain types of constraints on the initial data contain gauge invariances. The choice of an explicit condition to eliminate the gauge freedom of a system is referred to as *gauge fixing* and the condition is referred to as a *gauge condition*, which should not be confused with a constraint, although the two are very closely related.

Exercise 15.7 Make a change of independent variable in Equation 15.24 from τ to $\lambda = f(\tau)$ where $f(\tau)$ is an arbitrary function of τ, except that $\Lambda(\tau) = d\lambda/d\tau > 0$, to derive Equation 15.25.

A common response to the occurrence of equations of motion with superfluous degrees of freedom is to choose a new set of dynamical variables which contain only the physical degrees of freedom. In this case one such set of variables is $\mathbf{z}(t)$. However, their use has the disadvantage, in a relativistic context, of not being manifestly covariant, and a theory involving them would need to be examined very carefully at each stage to verify that the equations were indeed covariant. There are often very good reasons for describing a system using dynamical variables which have gauge freedom. The existence of gauge freedom in the electromagnetic field, and the corresponding Yang-Mills gauge fields of the nuclear interactions, is closely related to the renormalizability [488] of the quantized theory.

15.4 Relativistic Lagrangian mechanics

We have deliberately chosen to indicate, in a non-Lagrangian context, that the free relativistic particle equations of motion may be expressed in Poincaré-covariant, gauge-invariant form in terms of an arbitrary evolution parameter, such as a monotonic function of the proper time. However, we are always interested in forming a Lagrangian to describe any system. From the Lagrangian we may obtain, normally by straightforward techniques, a Hamiltonian which provides a link to the quantization of the system via the algebra of Poisson brackets. Equation 8.45 provides a non-singular Galilean covariant Lagrangian $\frac{1}{2}m\dot{z}^2$ for the free Newtonian particle based on the variables $z(t)$.

In order to obtain a Poincaré-covariant Lagrangian, we shall set up a relativistic version of Hamilton's principle of Section 8.5 for particles $i = 1, 2, \ldots$, in terms of covariant Lagrangian variables $z_{(i)} = \{z_{(i)}^\mu(\lambda)\}$ where λ is any evolution parameter, which may, for example, be the coordinate time t, the proper time τ of some particle or indeed any function of the independent variables such that derivatives of $z_{(i)}$ with respect to it are future-pointing, whether timelike or null.

A local relativistic Lagrangian system of particles is one whose equations of motion arise by Hamilton's principle applied to an action functional,

$$S[z(\lambda)] = \int_{\lambda_i}^{\lambda_f} d\lambda \, L(q_j, \dot{q}_j, \lambda) \,, \tag{15.26}$$

in which the function $L = L(q_j, \dot{q}_j, \lambda)$ of the generalized coordinates q_j and velocities $\dot{q}_j = dq_j/d\lambda$ is the Lagrangian of the system.

We could now apply Hamilton's principle with variations δ_0 not affecting λ, as in Chapter 8. Instead, we apply the more general variations, analogous to those of Section 9.4, in which λ also varies by $\delta\lambda(\lambda)$ and, with minor adjustments to the notation, we can then specialize to the restricted variations δ_0 appropriate to Hamilton's principle to deduce that the equations of motion are given by the Euler-Lagrange equations,

$$\frac{\delta_0 S}{\delta_0 q} = L^j = \frac{\partial L}{\partial q_j} - \frac{d}{d\lambda}\frac{\partial L}{\partial \dot{q}_j} = 0 \,. \tag{15.27}$$

With minor adjustments in notation and interpretation for the material in Section 9.6, we can also deduce that the Noether charges, corresponding to continuous symmetries of the action with respect to the parameters ω^a are,

$$Q_a = -H\frac{\partial \Delta\lambda(\lambda)}{\partial \Delta\omega^a} + \pi^j \frac{\partial \Delta q_j(\lambda)}{\partial \Delta\omega^a} + \frac{\partial(\Delta\Lambda)}{\partial \Delta\omega^a} \,, \tag{15.28}$$

where,

$$\pi^j = \frac{\partial L}{\partial \dot{q}_j} \quad \text{and} \quad H = \frac{\partial L}{\partial \dot{q}_j}\dot{q}_j - L = \pi^j \dot{q}_j - L \ . \tag{15.29}$$

These expressions for the *configuration space Hamiltonian* H and *generalized momenta* π^j are the same as Equations 11.2 in the non-relativistic case, with the proviso that the independent variable may be the coordinate time t, the proper time τ or some arbitrary time parameter λ. If the Lagrangian is non-singular and $\lambda = t$ in some frame, then H and π^j will be the *canonical Hamiltonian* and *canonical momenta* in that frame. H will be a *proper time Hamiltonian* for $\lambda = \tau$.

The small quantities $\Delta\Lambda$ are functions of λ and q_j, but not $\dot{q}_j$, that appear in the expression,

$$\frac{\partial(\Delta\lambda)}{\partial\lambda} + \Delta L = -\frac{d(\Delta\Lambda)}{d\lambda} \ , \tag{15.30}$$

for those Lagrangians which are covariant but not invariant (Section 9.3). The Noether charge is conserved, $dQ_a/d\lambda = 0$, on-shell (along the stationary trajectory). In general, a is a collective index labelling the Lie symmetry parameters and may in fact be one or more indices of Lorentz or some other type. (Examples are the four translation parameters a^μ where $a = 1, 2, \ldots, p$ becomes $\mu = 0, 1, 2, 3$ or the six Lorentz 'angles' $\omega_{\mu\nu} = -\omega_{\nu\mu}$ where $a = 1, 2, \ldots, p$ may be replaced, for example, by $|\mu\nu| = 01, 02, 03, 12, 23, 13$, namely the antisymmetric pair of indices $[\mu\nu]$ with $\mu < \nu$.)

15.5 Relativistic free particle Lagrangians

We seek a Lagrangian which is manifestly covariant and contains all the information about the massive free particle trajectory in one of the three Equations 15.23 to 15.25. If we now repeat the procedure of Section 8.9 in an attempt to construct, from $z(\lambda)$ and m, a Poincaré-covariant relativistic Lagrangian for a free particle, we immediately arrive at the Poincaré-invariant combination $m\dot{z}^2$ with the dimensions of a Lagrangian. Despite the fact that its sign and scale can been chosen so that it reduces to the non-relativistic result for small velocities, apart from an additive constant $\frac{1}{2}mc^2$ of no significance in determining the equations of motion, namely,

$$L = \tfrac{1}{2}m\dot{z}^2 \ , \tag{15.31}$$

it cannot be the free relativistic particle Lagrangian. This follows since it is non-singular whereas we know that a particle described by four coordinates must be constrained. The Euler-Lagrange equations of $L = \frac{1}{2}m\dot{z}^2$ are the unconstrained equations, $\ddot{z}^\mu(\lambda) = 0$. It does not supply the corresponding

15.5. Relativistic free particle Lagrangians

constraint of Equation 15.24. We shall later see that it also leads to an incorrect Hamiltonian.

However, in a relativistic context, we have available the constant c which permits us to construct an invariant Lagrangian of the correct dimensions and Newtonian limit other than by using a term proportional to $m\dot{z}^2$ itself. Consider, for example, a Lagrangian given by some other power p of the positive invariant quadratic $-\dot{z}^2$, namely,

$$L_p = -\tfrac{1}{2} m\, p^{-1} c^{2-2p} (-\dot{z}^2)^p \,, \tag{15.32}$$

where we have used factors of p and c to ensure it contributes positively to the energy and has the correct dimensions and non-relativistic limit for all non-zero p. We have already considered the case of $p = 1$. Of all the others, the case of $p = \tfrac{1}{2}$,

$$L = -mc\sqrt{-\dot{z}^2} \,, \tag{15.33}$$

where $\dot{z} = dz/d\lambda$, may be referred to as the *geometric Lagrangian* of the relativistic free particle of locus $z(\lambda)$ and is by far the most interesting. The action for this Lagrangian is,

$$S[z(\lambda)] = -mc \int_{\lambda_i}^{\lambda_f} d\lambda \sqrt{-\dot{z}^2} \,. \tag{15.34}$$

Exercise 15.8 *Verify that the geometric Lagrangian is (a) singular, and (b) unique among those arising from powers of $-\dot{z}^2$ in having an action which is invariant under functional changes $\lambda \to f(\lambda)$ of the evolution parameter λ (1D general coordinate invariance) implying $d\lambda \to \Lambda(\lambda) d\lambda$ and a corresponding transformation $\dot{z}(\lambda) \to \Lambda^{-1}(\lambda) \dot{z}(\lambda)$ of the trajectory variable, where $\Lambda = df(\lambda)/d\lambda$. Show (c) that provided $\dot{z}^2 \neq 0$, its Euler-Lagrange equations are those of Equation 15.25, and (d) that if we choose $\lambda = t$, and thus the constraint $\dot{z}^0 = c$, before forming the Euler-Lagrange equations, to eliminate $\dot{z}^0$ in favour of independent velocities $\dot{z}$, then it leads to the correct expressions for the momenta and Hamiltonian which, on the stationary path, combine to form the 4-momentum $m\dot{z}$ (see Equation 15.7).*

Equation 15.25 of our initial set is clearly equivalent to the other two in that it describes a free particle. In fact, the way the author first encountered the equation was by carrying out a change of independent variable in the second equation involving the proper time to an arbitrary function of the proper time as evolution parameter.

The geometric Lagrangian is so named because the action of Equation 15.34 is a constant multiple, $-mc$, of the *proper path length* (see Equation 12.5),

$$\Delta s = \int_{\lambda_i}^{\lambda_f} d\lambda (-\dot{z}^2)^{1/2} \,, \tag{15.35}$$

between an initial event x_i and a final event x_f, separated from the first by a timelike interval. The length of such a path is independent of the parameter λ.

The fact that the equations of motion of a free particle correspond to a stationary action which is proportional to the path length in Minkowskian spacetime means that the path length is also stationary. Such a path is referred to as a *geodesic*. It is very easy to establish the well-known result that the stationary path between two points $\mathbf{x}_i$ and $\mathbf{x}_f$ in 3D Euclidean space $\mathbf{E}^3$ is the straight line joining them, that the stationary path is the shortest and that it is also the path followed by a free Newtonian particle in going from one point to another. The free particle path from event to event in $\mathbf{E}^{1,3}$ is clearly also stationary.

Exercise 15.9 *Show that the stationary path of a free particle in $\mathbf{E}^{1,3}$ is of maximal, not minimal, proper length.*

Since the action of Equation 15.34 leads to Equations 15.25 for the relativistic free particle, we see therefore that the Lagrangian of a free relativistic particle exhibits an intimate connection between geometry, explicit Poincaré covariance, gauge invariance and constraints.

The method in part (d) of Exercise 15.8 is equivalent to selecting $\lambda = t$ in Equation 15.34, implying $\dot{z}^0 = dz^0/dt = c$ and $\dot{z}^2 = -c^2 + \dot{\mathbf{z}}^2$, reducing the geometric action to,

$$S[z(\lambda)] = -mc^2 \int_{t_i}^{t_f} dt \sqrt{1 - \dot{\mathbf{z}}^2/c^2} \,, \qquad (15.36)$$

for which the Euler-Lagrange equations are of course $\ddot{\mathbf{z}} = 0$. Selecting λ in this way from among the infinite number of evolution parameters is an example of gauge fixing. The selection of $\lambda = t$ as gauge condition is called the *laboratory gauge*, which is closely analogous to the Coulomb and Hamilton gauges of electrodynamics (see Section 19.10). In the process we lose the manifest covariance but not the relativistic covariance itself.

Exercise 15.10 *Show that, provided $\dot{z}^2 \neq 0$, we may alternatively put $\dot{z}^0 = c$, $\dot{z}^2 = -c^2 + \dot{\mathbf{z}}^2$ and $\ddot{z}^0 = 0$ in Equation 15.25 to obtain the equation of motion $\ddot{\mathbf{z}} = 0$ directly.*

We may also fix the gauge in a covariant way, namely by imposing the gauge condition, $\dot{z}^2 = -c^2$ which means that $\lambda = \tau$, the proper time. This *proper time gauge*, analogous to the *Lorentz gauge* of electrodynamics, is just Equation 15.24 with $\ddot{z}^\mu = 0$ consistent with Equation 15.25.

Exercise 15.11 *Use the symmetry of an isolated mechanical system with* respect to small covariant translations $\Delta z^\mu = a^\mu$ and rotations about the event $d = \{d^\mu\}$, namely $\Delta(z-d)^\mu = \Delta\omega^{\mu\nu}(z-d)_\nu$, to show that (modulo a constant factor of no physical significance), the corresponding Noether charges are the 4-momentum p^μ and the total covariant angular momentum $j_{\mu\nu}$ about d. Hint: note that only six of the twelve non-zero components of $\omega^{\mu\nu}$ or $\Delta\omega^{\mu\nu}$ are independent and it may be helpful to make use therefore of a Lorentz analogue of a result from Exercise 4.6 to explicitly antisymmetrize the coefficients of $\Delta\omega^{\mu\nu}$.

15.6 Massless particles

A zero value of the rest mass in a non-relativistic situation leads to a zero value of the 3-momentum and a zero value of the energy; in fact we have no particle at all. Although a relativistic boost speed is limited by $v < c$, causality is not violated if particles or signals carry information or energy at the ultimate speed $u = c$. For such null propagation to be consistent with the limit $u \to c$ or $\gamma \to \infty$ of non-null particles and with the finiteness of momentum and relativistic energy, their rest mass (see Equations 15.7 and 15.9) must be zero.

In contrast, however, to the non-relativistic case this does not require the momentum and energy to be zero. From the massless limit of Equation 15.16, the 4-momentum will be null, $p^2 = 0$. Because of the indefiniteness of the Minkowskian spacetime metric, the time and space parts (energy and 3-momentum) may have non-zero values and cancel one another in p^2. From Equations 15.13 and 15.17 and the positivity of energy, a massless particle therefore has energy and 4-momentum related by,

$$E = |\mathbf{p}|c \quad \text{and} \quad p = \{p^\mu\} = \{|\mathbf{p}|, \mathbf{p}\} = \frac{E}{c}\{1, \hat{\mathbf{p}}\}, \qquad (15.37)$$

and the null 4-momentum is thus entirely characterized either by its relativistic 3-momentum $\mathbf{p}$ or by its energy E and its direction $\hat{\mathbf{p}}$ of propagation.

Although 4-velocity cannot be defined for massless particles, null 4-momentum exists as the massless limit of the 4-momentum of a massive particle when the rest mass and proper time interval become vanishingly small and γ tends to infinity. We take this limit as our definition of a *massless non-quantum particle*. The equations of such a particle may be used, for example, to describe a pulse of collimated electromagnetic radiation, where the pulse size is small compared to all other characteristic lengths, such as the distances over which dispersive effects manifest themselves. It may sometimes be convenient to describe propagation properties of electromagnetic radiation as if a ray behaved as a null classical particle.

The 4-momentum p^μ and angular momentum $j^{\mu\nu}$ of null parts of an isolated system must be included in the totals to ensure they are conserved.

By considering a pulse of radiation, we are able to ignore fundamental difficulties in localizing a null particle and continue to use Equation 15.21 (or its equivalent referred to an origin d^μ) for null angular momentum.

A null particle, ray, or wave front for which $p^2 = 0$, is in motion in all frames at speed c. Consequently, no frame exists in which all the spatial components of p^μ are zero. If we choose the z-axis in the direction of propagation, the x and y components of 4-momentum will be zero and from the null condition $p^2 = 0$, the 0 and z components of 4-momentum must be of equal magnitude. We can therefore always find a frame for which $p = \{p^\mu\} = \{E/c, 0, 0, E/c\}$. Use of such a frame can often simplify calculations involving null particles. The results of the calculation may be re-expressed in a form valid in any frame by using covariance.

15.7 Einbein Lagrangian

For completeness, we should like to be able to extend the Lagrangian analysis to include the possibility of null particles for which $dz^2 = 0$. We clearly cannot do this if we use the proper time as the evolution parameter. Similarly we cannot use the geometric Lagrangian for null particles since use is made of $\dot{z}^2 \neq 0$ in the extraction of its Euler-Lagrange equations.

A configuration space or *Lagrange constraint* of the form $\phi(q, \dot{q}) = 0$ may be included explicitly into a Lagrangian by a term comprised of the function ϕ multiplied by an arbitrary function of the independent variable, called a *Lagrange multiplier*, which is treated along with q as an additional generalized coordinate. This would appear to be increasing even further the number of degrees of freedom. However, provided we do not include any kinetic term for the Lagrange multiplier, it will not be a dynamical coordinate and its Euler-Lagrange equation will be the zero value of the partial derivative of the Lagrangian with respect to the Lagrange multiplier, which is simply $\phi = 0$, thus supplying the constraint we require. In the case of the massless particle, we may choose a singular Lagrangian,

$$L = L\big(z(\lambda), e(\lambda)\big) = \tfrac{1}{2} e^{-1}(\lambda) \dot{z}^2 \;, \qquad (15.38)$$

with five independent variables, $z^\mu(\lambda)$ and $e(\lambda)$, the last of which, referred to as the *einbein*, is non-dynamic. Variation of all five and then fixing the gauge by choosing $e(\lambda)$ constant gives the correct equations of motion, $\ddot{z}^\mu = 0$, and the constraint $\dot{z}^2 = 0$. For L to have standard dimensions, the non-dynamic variable $e(\lambda)$ must have the dimensions of reciprocal mass.

Exercise 15.12 *Show that the $e^{-1}(\lambda) dz/d\lambda$ and the einbein action are both reparameterization invariant provided the einbein $e(\lambda)$ transforms under $\lambda \to \bar{\lambda} = f(\lambda)$, for arbitrary monotonic $f(\lambda)$, contragrediently to*

15.7. Einbein Lagrangian

$d\lambda \to \Lambda(\lambda)d\lambda$, namely according to,

$$e(\lambda) \to e(\bar{\lambda}) = \Lambda^{-1}e(\lambda) \,, \tag{15.39}$$

where $\Lambda(\lambda) = d\bar{\lambda}/d\lambda$.

Let us now explore the Lagrange multiplier technique for incorporating a constraint condition into a massive particle Lagrangian, the defective unconstrained Lagrangian for which is Equation 15.31. Thus, instead of using the mass m in a term proportional to $m\dot{z}^2$, we replace m by an einbein Lagrange multiplier $e^{-1}(\lambda)$ and add a similar term involving the mass, the expression for which is essentially determined from dimensions, re-parameterization invariance and the form of the massless limit. Consider therefore the singular *einbein Lagrangian* with dynamic variables $z(\lambda)$ and non-dynamic variable $e(\lambda)$, namely,

$$L = L\big(z(\lambda), e(\lambda)\big) = \tfrac{1}{2}e^{-1}(\lambda)\dot{z}^2 - \tfrac{1}{2}e(\lambda)m^2 c^2 \,. \tag{15.40}$$

Exercise 15.13 Show that,

$$\frac{d}{d\lambda}(e^{-1}(\lambda)\dot{z}^\mu) = 0 \quad \text{and} \quad \dot{z}^2 = -e^2(\lambda)m^2 c^2 \,, \tag{15.41}$$

are Euler-Lagrange equations for the einbein Lagrangian of Equation 15.40.

Putting $m = 0$ into the Euler-Lagrange equations resulting from this Lagrangian, or directly into the Lagrangian, and fixing the gauge by taking $e(\lambda)$ constant, gives the correct relativistic equations of motion and constraint of a null particle, as before, in which it must be remembered that λ cannot be the proper time and thus $\dot{z}$ is not a 4-velocity. For $m \neq 0$ we may fix the gauge by selecting a constant $e(\lambda) = m^{-1}$ (thus choosing λ to be the proper time τ) to obtain the correct equations and constraint for the massive relativistic particle.

Each Lagrange multiplier in a second-order Lagrangian leads to the elimination of two apparent degrees of freedom, the multiplier itself and one of the other variables as a result of using the constraint arising from variation of the multiplier.

Exercise 15.14 Show that, provided $m \neq 0$, substitution of the positive root for $e(\lambda)$ from the constraint implied by the Euler-Lagrange equations of the einbein Lagrangian of Equation 15.40 back into the Lagrangian gives the geometric Lagrangian for a massive free relativistic particle.

We have seen that the einbein Lagrangian of Equation 15.40 can be used for a relativistic free particle which is either timelike or null. The set $\{e_\mu\}$ of four basis vectors, corresponding to the spacetime coordinates x^μ, transform contragrediently to $dx^\mu \to \Lambda^{\bar\mu}{}_\mu dx^\mu$, namely according to,

$$e_\mu \to e_{\bar\mu} = \Lambda^\mu{}_{\bar\mu} e_\mu = (\Lambda^{-1})^\mu{}_{\bar\mu} e_\mu \; . \tag{15.42}$$

They are often referred to as a *tetrad*, or by the equivalent German word *vierbein*, literally meaning 'four legs'. This is especially the case when the coordinates are not rectangular Cartesian and the vierbein [566] is then a set $\{e_\mu(x)\}$ of four 4-vector fields transforming according to a matrix $\Lambda(x)$ which depends on x. Correspondingly, the einbein is a field in the λ-space of the independent variable of the particle. It gets its name from the fact that, although it has only one component, it is not a scalar field in the λ space but transforms contragrediently to $d\lambda$.

Although a particle has a finite number of degrees of freedom in $\mathbf{E}^{1,3} = \{x\}$, its dynamic variables z^μ and the einbein $e(\lambda)$ are effectively fields in the 1D space of an arbitrary evolution parameter, λ. They may be treated as such using the techniques of Chapters 10 and 16.

The most profound example of a geometric Lagrangian theory, extensively supported by experiment, which shows similar properties to the free particle in einbein form, is Einstein's general relativistic theory of gravitation. In that theory, the gauge invariance is with respect to changes in not just t but in all four coordinates x^μ to coordinates $x^{\bar\mu}$ which are arbitrary functions $x^{\bar\mu}(x^\mu)$ of the original coordinates with corresponding changes $\mathbf{g} \to L^{-1} \otimes L^{-1} \mathbf{g}$ of the metric tensor, $\mathbf{g} = \{g_{\mu\nu}\}$ where $L = \{L^{\bar\mu}{}_\mu\} = \partial x^{\bar\mu}/\partial x^\mu$. Supergravity and relativistic strings provide further examples of geometric Lagrangians.

Some aspects of the analogy between the geometric formulation of the relativistic free particle and general relativity (see Chapter 21) are so close that the former is sometimes referred to as '1D *general relativity*'. The similarity may be highlighted by replacing $e(\lambda)$ by $\sqrt{g(\lambda)}$ in the einbein Lagrangian to obtain an action,

$$S = S[z^\mu, g] = \tfrac{1}{2} \int d\lambda \sqrt{g(\lambda)} [g^{-1}(\lambda)\dot z^2 - m^2 c^2] \; , \tag{15.43}$$

in which $g(\lambda)$ is the metric in λ-space and $m^2 c^2$ corresponds to a cosmological term (see Chapter 21) of general relativity. Since $e(\lambda)$ is non-dynamic, there is no actual gravitation in 1D.

General relativity in 4D provides another example of the quite general result [526] that the existence of a gauge invariance implies the existence of constraints. Such constraints are always of a certain specific type. However, the converse is not in general true for any type of constraint — a system with constraints need not be gauge invariant.

15.8 Relativistic Hamiltonian mechanics

The passage from a configuration space Lagrangian and Hamiltonian to a phase space Hamiltonian formalism is rarely as simple in Einsteinian mechanics as in Newtonian systems. The first reason is that the Hamiltonian is, by definition, the generator of translations with respect to the evolution parameter but there is no uniqueness in the choice of the latter. We may obtain a time coordinate Hamiltonian in each frame and also an invariant proper time Hamiltonian. If we wished, we could also determine a Hamiltonian corresponding to some arbitrary function λ of the proper time.

The other complication is not always so easy to take care of. This is a consequence of the desirability of formulating relativistic dynamics covariantly, as far as possible. We have seen that, in the case of the free relativistic particle at least (and the same is true of its interaction) manifest covariance implies the existence of constraints and therefore singular Lagrangians, whereas the Legendre transformation which forms the passage to the Hamiltonian formalism, is normally only well-defined for non-singular Lagrangians. Let us therefore examine the case of the relativistic free particle from the Hamiltonian point of view.

The generalized momenta and Hamiltonian corresponding to the defective non-singular Lagrangian of Equation 15.31 are easily shown to be $\pi_\mu = p_\mu = m\dot{z}_\mu$, the 4-momentum of the particle, and $H = \pi^2/2m$. This is not the correct Hamiltonian as a result of the Lagrangian not incorporating the constraint, $\dot{z}^2 = -c^2$, that must accompany the covariant non gauge-invariant form of the equations of motion.

Let us now use the laboratory gauge for the geometric Lagrangian, fixing the evolution parameter to be the coordinate time,

$$L = -mc^2\sqrt{1 - \dot{z}^2/c^2} \ . \tag{15.44}$$

This means abandonment of manifest covariance and taking the independent Lagrangian variables to be **z**, which we know are physical variables. Then we obtain generalized momenta,

$$\mathbf{p} = \frac{\partial L}{\partial \dot{\mathbf{z}}} = \frac{m\dot{\mathbf{z}}}{\sqrt{1 - \dot{z}^2/c^2}} \tag{15.45}$$

and Hamiltonian,

$$H = \mathbf{p}\cdot\dot{\mathbf{z}} - L = \frac{m\dot{z}^2}{\sqrt{1 - \dot{z}^2/c^2}} - (-mc^2\sqrt{1 - \dot{z}^2/c^2}) = \frac{mc^2}{\sqrt{1 - \dot{z}^2/c^2}} \ , \tag{15.46}$$

which, although not manifestly covariant, together correctly reproduce the relativistically covariant Equation 15.7.

We may ask what occurs if we use the standard formulae of Equation 15.29 to calculate the Hamiltonian using the geometric Lagrangian but without fixing the gauge, treating all four coordinates z^μ as independent to maintain covariance.

Exercise 15.15 *Show that the canonical Hamiltonians of the geometric and einbein Lagrangians of a particle, in an arbitrary gauge λ, vanish identically.*

An identically vanishing Hamiltonian implies that the standard procedure for passing from the Lagrangian to the Hamiltonian has failed, which is not surprising since the standard procedure cannot work for the singular Lagrangian of Equation 15.33. Not all singular Lagrangians lead to a vanishing Hamiltonian although another important highly non-trivial example is the canonical Hamiltonian in an arbitrary gauge of general relativity, in metric or einbein form. In both cases, new procedures must be set up.

To be certain the above point is appreciated we shall provide one of the results of Exercise 15.15. The generalized momenta corresponding to the coordinates z^μ in the geometric Lagrangian of Equation 15.33 are,

$$p_\mu = \frac{\partial L}{\partial \dot{z}^\mu} = \frac{mc\dot{z}_\mu}{\sqrt{-\dot{z}^2}} = \frac{mc\dot{z}_\mu}{\sqrt{\dot{z}_0^2 - \dot{z}^2}}, \qquad (15.47)$$

which are the correct Noether charges (see Exercise 15.11) corresponding to spacetime translation. If we now specialize to the laboratory gauge, our results agree with those obtained above. But if we do not, then formally, we have a standard (canonical) configuration space Hamiltonian of,

$$H_{\text{can}} = p_\mu \dot{z}^\mu - L = \frac{mc\dot{z}^2}{\sqrt{-\dot{z}^2}} - (-mc\sqrt{-\dot{z}^2}) = 0 . \qquad (15.48)$$

Even where the canonical Hamiltonian of a singular Lagrangian is non-vanishing, it nevertheless always has properties which prevent its direct use in Poisson bracket relations to describe the dynamics in ways which are analogous to the unconstrained case. It does not therefore provide a clear indication of how to quantize the theory. We cannot proceed to a Hamiltonian formulation, as in the case of non-singular Lagrangians, without some modification.

The general procedures for quantizing constrained Hamiltonian systems, which we shall illustrate but not develop in detail, are known as the *Dirac theory of generalized Hamiltonian dynamics* as a result of the pioneering work of Dirac [110] in 1950. The relationship between certain forms of constrained system and gauge invariance dates from the work of Bergmann and his collaborators, originating with the paper by Anderson and Bergmann [16] in 1951. Subsequent analyses were made by Kundt [292]

15.8. Relativistic Hamiltonian mechanics

in 1966 and Shanmugadhasan [510] in 1973. We also strongly recommend the texts of Sudarshan and Mukunda [525] and Sundermeyer [526] and the articles by Hanson, Teitelboim and Regge [223], Utiyama [548] and Trautman [540, 541] for further information and extensive bibliographies on constrained systems.

In the present example, the singularity of the Lagrangian means that Equations 15.47 cannot be solved to express all four generalized velocities in terms of the generalized momenta. Only three of the momenta, which we choose to be the spatial components p, are independent. Corresponding to the configuration space Lagrange constraint $\dot{z}^2 = -c^2$ there will be a *phase space constraint*, also called a *dynamical constraint*, on the momenta.

Exercise 15.16 *Show (a) that if we attempt to solve Equation 15.47 for the velocities in terms of the momenta, we can obtain,*

$$\dot{\mathbf{z}} = \frac{\dot{z}^0 \mathbf{p}}{\sqrt{\mathbf{p}^2 + m^2 c^2}}, \qquad (15.49)$$

from which, however, we cannot eliminate $\dot{z}^0$ and (b) that the fourth relation, instead of providing a solution for $\dot{z}^0$ in terms of momenta, reduces with use of Equation 15.49, to,

$$p^0 = \sqrt{\mathbf{p}^2 + m^2 c^2}, \qquad (15.50)$$

independent of $\dot{z}^0$.

Equation 15.50 is a relation between the four momenta which we may write covariantly as,

$$p^2 = p^\mu p_\mu = -m^2 c^2. \qquad (15.51)$$

This relation, the dynamical constraint of the relativistic particle, is the *mass shell condition*, obtainable directly by contraction of Equation 15.47 with itself. Its solutions with $p^0 < 0$ may be ignored in non-quantum physics as a result of requiring physical variables to have a continuous classical domain. In a quantum context, Dirac [107] argued in 1929 that such solutions cannot be ignored if instability is to be avoided when perturbations cause transitions between states with positive and negative energy.

In general, a singular Lagrangian with a configuration space or *Lagrange constraint*, $\phi(q, \dot{q}) = 0$, will lead to some phase space constraint, $\Phi(q, p) = 0$, involving coordinates as well as momenta. The fourth velocity $\dot{z}^0$, for which it is not possible to obtain an expression in terms of the independent momenta, appears in the expression for the other three as an arbitrary function of time. As a result of the Lagrange constraint, the solutions for the velocities therefore contain local gauge freedom.

The Dirac theory (see Sundermeyer [526, pp. 46-49]) shows that one may, quite generally, retain the Poisson bracket structure of the dynamical equations, over their entire phase space, in terms of a new non-canonical Hamiltonian. The latter is given by adding, to the canonical Hamiltonian, a sum of products of arbitrary functions α^c ($c = 1, 2, \ldots$) of the independent variables times each corresponding expression $\Phi_c(q, p)$ whose vanishing gives a dynamical constraint $\Phi_c(q, p) = 0$. Thus, for use in the Poisson brackets, the *effective Hamiltonian* defined over the entire phase space is,

$$H = H_{\text{can}} + \sum_c \alpha^c \phi_c(q, p) \ . \tag{15.52}$$

The canonical Hamiltonian, H_{can}, is formed by solving the expression $p_j = p_j(q_k, \dot{q}_k)$ for those velocities which are independent in terms of the independent momenta. In the case of the relativistic particle (and for general relativity), $H_{\text{can}} = 0$. Thus the effective Hamiltonian of the relativistic particle (where there is only one constraint) is,

$$H = \tfrac{1}{2} e(\lambda)(p^\mu p_\mu + m^2 c^2) \ , \tag{15.53}$$

in which $e(\lambda)$ is an arbitrary function (the factor of $\tfrac{1}{2}$ being conventional). To confirm the validity of this procedure in the present example, we may use Hamilton's equations, or evaluate the Poisson brackets, by treating all p^μ as independent.

Exercise 15.17 (a) *Show that the Hamiltonian H of Equation 15.53 combines with the standard relativistic Poisson brackets, $\dot{z}^\mu = \{z^\mu, H\}$ and $\dot{p}^\mu = \{p^\mu, H\}$, to give,*

$$\dot{z}^\mu = \{z^\mu, \tfrac{1}{2} e(\lambda)(p^\mu p_\mu + m^2 c^2)\} = e(\lambda) p^\mu \quad \text{and} \quad \dot{p}^\mu = 0 \ , \tag{15.54}$$

leading to the reparameterization invariant 4-momentum,

$$p^\mu = e^{-1}(\lambda) \dot{z}^\mu = \text{constant} \ , \tag{15.55}$$

provided $e(\lambda)$ transforms contragrediently to $d\lambda$. (b) *Show that if we substitute the above p^μ back into the phase space Lagrangian given by the canonical relation $H = p_\mu \dot{z}^\mu - L$, namely,*

$$L = p_\mu \dot{z}^\mu - \tfrac{1}{2} e(\lambda)(p^\mu p_\mu + m^2 c^2) \ , \tag{15.56}$$

then we obtain the einbein Lagrangian of Equation 15.40. (c) *Obtain the constancy of p^μ ($\propto \dot{z}^\mu$) from Hamilton's equations,*

$$\dot{z}^\mu = \frac{\partial H}{\partial p_\mu} \quad \text{and} \quad \dot{p}_\mu = -\frac{\partial H}{\partial z^\mu} \ , \tag{15.57}$$

applied to the effective Hamiltonian.

Exercise 15.18 *(a) Show that the fundamental Poisson brackets of the trajectory and momenta variables $x = \{z^\mu\}$ and $p = \{\pi_\mu\}$ of an isolated relativistic system are,*

$$\{p_\mu, p_\nu\} = 0 = \{z^\mu, z^\nu\} \quad \text{and} \quad \{z^\mu, p_\nu\} = \delta^\mu{}_\nu . \tag{15.58}$$

(b) Establish the Poisson bracket Lie algebra,

$$\{j_{\mu\nu}, j_{\lambda\rho}\} = \eta_{\mu\lambda} j_{\nu\rho} + \eta_{\nu\rho} j_{\mu\lambda} - \eta_{\mu\rho} j_{\nu\lambda} - \eta_{\nu\lambda} j_{\mu\rho} , \tag{15.59}$$
$$\{p_\mu, j_{\lambda\rho}\} = \eta_{\mu\rho} p_\lambda - \eta_{\mu\lambda} p_\rho , \tag{15.60}$$
$$\{p_\mu, p_\nu\} = 0 . \tag{15.61}$$

of the conserved generators p^μ and $j^{\mu\nu} = z^\mu p^\nu - z^\nu p^\nu$ of Einsteinian physics.

The usual quantization *ansatz* of replacing Poisson brackets involving classical variables, z^μ, p^μ and $j^{\mu\nu}$, by commutators multiplied by $(i\hbar)^{-1}$, involving the corresponding operators Z^μ, P^μ and $J^{\mu\nu}$ ($= L^{\mu\nu}$, $S^{\mu\nu}$ or $L^{\mu\nu} + S^{\mu\nu}$), gives the fundamental quantum relations, $[X^\mu, P^\nu] = i\hbar \eta^{\mu\nu}$ and,

$$[J_{\mu\nu}, J_{\lambda\rho}] = i\hbar(\eta_{\mu\lambda} J_{\nu\rho} + \eta_{\nu\rho} J_{\mu\lambda} - \eta_{\mu\rho} J_{\nu\lambda} - \eta_{\nu\lambda} J_{\mu\rho}) , \tag{15.62}$$
$$[P_\mu, J_{\lambda\rho}] = i\hbar(\eta_{\mu\rho} P_\lambda - \eta_{\mu\lambda} P_\rho) , \tag{15.63}$$
$$[P_\mu, P_\nu] = 0 . \tag{15.64}$$

Equations 15.59 to 15.61, and 15.62 to 15.64, are each a *realization*, in relativistic mechanics, classical and quantum respectively, of the *Lie algebra* iso(1,3) of the restricted *Poincaré group* ISO(1,3). Equations 15.59 and 15.62 define the Lie algebra so(1,3) of the restricted Lorentz group SO(1,3).

15.9 Spin of a massive relativistic system

The angular momentum of a massive isolated system is the conserved quantity which arises from invariance under rotations. For non-zero mass, any covariant definition of angular momentum in an arbitrary frame reduces in the rest frame to Newtonian angular momentum definable entirely in terms of rotations.

Orbital angular momentum is defined with respect to an arbitrary reference event fixed in an inertial frame. Let us denote the event by $d = \{d^\mu\}$. The *total covariant angular momentum* about d of an isolated massive system, namely the sum $j^{\mu\nu} = \sum_i l_i^{\mu\nu}$ of the covariant orbital angular momentum of its parts labelled by $i = 1, 2, \ldots$, may be decomposed,

$$j^{\mu\nu} = l^{\mu\nu} + s^{\mu\nu} , \tag{15.65}$$

into the sum of the covariant *orbital angular momentum* about d,

$$l^{\mu\nu} = (z^\mu - d^\mu)p^\mu - (z^\nu - d^\nu)p^\mu \,, \tag{15.66}$$

of the whole system of momentum $p(t)$ moving with the centre of mass $z(t)$, and the *covariant spin angular momentum*,

$$s^{\mu\nu} = \sum_i [(z_i^\mu - z^\mu)p_i^\nu - (z_i^\nu - z^\nu)p_i^\mu] \,, \tag{15.67}$$

of the parts of the system about its own centre of mass. Although the spin angular momentum is introduced here for an extended system, taking the limit of zero size can leave a classical particle with a meaningful notion of spin. Indeed, a non-quantum particle will have spin arising from the orbital angular momentum about its centre of mass of any constituents confined to a region near the centre of mass much smaller than any other lengths characteristic of the dynamics of the system.

Fundamental particles are presumed to have no internal structure with constituent orbital contributions to provide spin angular momentum. Nevertheless, elementary particle collision data show that angular momentum can only be conserved if massive particles have non-orbital angular momentum, namely relative to their centre of mass, referred to as *intrinsic angular momentum* or *intrinsic spin*.

Since the spin of a massive particle may be considered a purely rotational property, even when it is expressed covariantly we can expect it to have only three independent components, one for each parameter in a rotation. From its definition, the *spin tensor* $s^{\mu\nu}$ is antisymmetric, $s^{\mu\nu} = -s^{\nu\mu}$, leaving a maximum of six independent components.

Exercise 15.19 *Show that the spin tensor has only three independent components by establishing the condition,*

$$s^{\mu\nu}u_\nu = 0 \,, \tag{15.68}$$

where u^μ is the 4-velocity. Hint: Since this equation is manifestly covariant, it will be true in all frames if it can be established in any frame.

In the rest frame, the components s^{k0} vanish and the spin tensor reduces to the three Newtonian components s^{kl}. Although s^{k0} may not vanish in an arbitrary frame, their values may be determined from s^{kl} — their sole purpose is to render $s^{\mu\nu}$ covariant [15, 566].

Although the total covariant angular momentum $j^{\mu\nu}$ cannot be described by a vector, the fact that covariant spin $s^{\mu\nu}$ has only three independent components, like the pseudo-3-vector s describing spin non-relativistically, suggests that the relativistic spin of a system with a given

15.9. Spin of a massive relativistic system

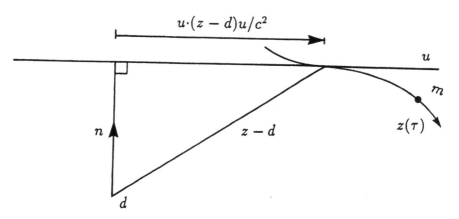

Figure 15.1 Normal from the angular momentum reference point to the instantaneous direction of motion.

4-momentum may be described covariantly by a pseudo 4-vector with only three independent components. The only 4-vectors containing $s^{\mu\nu}$ which may be constructed with the aid of the 4-velocity are proportional to the pseudo-4-vectors $\epsilon^{\mu\nu\lambda\rho}s_{\nu\lambda}u_\rho$ and $\epsilon^{\mu\nu\lambda\rho}j_{\nu\lambda}u_\rho$.

Exercise 15.20 Use Equations 15.65 and 15.66 to show that the above two pseudo-4-vectors are in fact identical.

The normalization of the pseudo-4-vector may be chosen in such a way that it reduces to $\{0,\mathbf{s}\}$ in the rest frame. The *spin pseudo-4-vector* of a massive relativistic particle is therefore defined by:

$$s^\mu = -\tfrac{1}{2c}\epsilon^{\mu\nu\lambda\rho}j_{\nu\lambda}u_\rho = -\tfrac{1}{2c}\epsilon^{\mu\nu\lambda\rho}s_{\nu\lambda}u_\rho \; . \tag{15.69}$$

Exercise 15.21 (a) Show that the components $\{s^\mu\}$ of the spin 4-vector reduce to $\{0,\mathbf{s}\}$ in the particle rest frame. (b) Show that the spin vector is orthogonal to the 4-velocity, $s \cdot u = 0$, thus establishing that it is spacelike. (c) Show, either by choosing a particular frame or by using a permutation tensor identity, that the spin tensor is retrieved from the spin 4-vector and the 4-velocity using:

$$s^{\mu\nu} = \epsilon^{\mu\nu\lambda\rho}s_\lambda u_\rho/c \; . \tag{15.70}$$

Exercise 15.22 Use a permutation identity to show that if **s** is the rest frame spin 3-vector, then $\tfrac{1}{2}s^{\mu\nu}s_{\mu\nu} = s^\mu s_\mu = \mathbf{s}^2$.

The total angular momentum $j^{\mu\nu}$ about d and the momentum p^μ suffice to determine the 4-vector displacement from d to the trajectory of a massive particle ($p^2 \neq 0$) in the spacelike direction n which is Lorentz-orthogonal to the direction p^μ of propagation, as shown in Figure 15.1. For any point d not on the trajectory, the displacement $z-d$ may be decomposed uniquely as,

$$z - d = n + \tfrac{1}{c^2} u \cdot (z-d) u \quad \text{where} \quad n \cdot u = 0 \ . \tag{15.71}$$

Exercise 15.23 *Establish Equation 15.71 by projecting it in two directions and combine it with Equation 15.65 to show that $n^\mu = j^{\mu\nu} p_\nu / p^2$.*

Exercise 15.24 *Let a particle have spin 3-vector $\mathbf{s}$ in the rest frame. Show that after a boost in the direction $\mathbf{s}$ or $-\mathbf{s}$, the spin 3-vector continues to point in the same direction.*

The *helicity* of a massive particle, in a particular frame, namely the projection $\mathbf{j} \cdot \hat{\mathbf{p}} = \mathbf{s} \cdot \hat{\mathbf{p}}$ of its (spin) angular momentum in the direction $\hat{\mathbf{p}}$ of propagation, is not a Lorentz invariant. Indeed, we can always (for $m \neq 0$) find a boost sufficiently fast in the direction $\hat{\mathbf{p}}$ that the particle moves in the direction $-\hat{\mathbf{p}}$ in the new frame. Since the direction of $\mathbf{s}$ is unchanged, the helicity in the two frames are of opposite sign.

The 4-velocity in Equation 15.69 may be replaced by the 4-momentum to obtain the alternative *'spin' 4-vector* of Pauli and Lubański [345] for a particle or field according to:

$$w^\mu = -\tfrac{1}{2} \epsilon^{\mu\nu\lambda\rho} j_{\nu\lambda} p_\rho \ , \tag{15.72}$$

and a corresponding *Pauli-Lubański 'spin' tensor* $w^{\mu\nu} = \epsilon^{\mu\nu\lambda\rho} w_\lambda p_\rho$.

Exercise 15.25 *Show that the Pauli-Lubański vector w^μ is orthogonal to the 4-momentum, $p \cdot w = 0$, and is thus spacelike if the particle is massive.*

All the results concerning the spin of a massive classical particle then follow as before by using $s^\mu = w^\mu / mc$ and $s^{\mu\nu} = w^{\mu\nu}/(mc)^2$. The Pauli-Lubański quantities have the advantage that they are also well-defined in the massless case where the ordinary spin quantities do not exist.

Exercise 15.26 *Use a permutation tensor identity to obtain the relation,*

$$w^2 = -\tfrac{1}{2} j^{\lambda\pi} j_{\lambda\pi} p^2 + j^{\pi\nu} j_{\mu\nu} p^\mu p_\pi \ , \tag{15.73}$$

for the contraction of the Pauli-Lubański vector with itself.

For non-zero mass, a rest frame exists in which $p = \{mc, \mathbf{0}\}$ giving $w^2 = \tfrac{1}{2} j^{\lambda\pi} j_{\lambda\pi} m^2 c^2 - j^{0\lambda} j_{0\lambda} m^2 c^2 = \tfrac{1}{2} j^{kl} j_{kl} m^2 c^2 = s^{kl} s_{kl} m^2 c^2 = \mathbf{s}^2 m^2 c^2$.

15.10 Helicity of a massless particle

In the massive case we could have used the spin tensor in place of the angular momentum tensor throughout this calculation, to obtain,

$$w^2 = s^2 m^2 c^2 = s^\mu s_\mu m^2 c^2 \ . \tag{15.74}$$

15.10 Helicity of a massless particle

No spin tensor $s^{\mu\nu}$ or spin 4-vector s^μ may be defined for massless particles. The first requires a reference point at the centre of mass fixed in an inertial frame and the second requires the notion of 4-velocity, neither of which exist for massless particles. The decomposition of the total angular momentum of a system into two parts according to Equation 15.65 is not possible for a null system. This is a reasonable result since spin is a purely rotational concept of a system whereas a massless particle is never at rest and its rotational properties about an arbitrary axis cannot be dissociated from its behaviour under translations. The only direction about which rotations of a null system are meaningful is the propagation direction.

However, the 4-momentum of a massless system is well-defined and although null ($p^2 = 0$), will not be zero. The momentum may be used to obtain the total angular momentum, $j^{\mu\nu} = l^{\mu\nu}$ which may be considered to be entirely orbital, using Equation 15.21, and thus may be used to form the Pauli-Lubański quantities. In the massless case, the expression must remain in terms of the total angular momentum.

Exercise 15.27 *Use Equations 15.21 and 15.73 to show that the Pauli-Lubański 4-vector of a massless particle is null.*

A massless non-quantum particle or ray is thus characterized by a null 4-momentum p and a null 'spin' pseudo-4-vector w which are Lorentz-orthogonal,

$$p^2 = p \cdot w = w^2 = 0 \ . \tag{15.75}$$

Exercise 14.11 shows that two such vectors are in fact proportional. A pair of 4-vectors which are mutually orthogonal and proportional must both be zero in a positive-definite space such as $\mathbf{E}^3$. However, this will not be the case in an indefinite metric space such as Minkowskian spacetime.

There are only two ways in which w and p may be proportional; they are either parallel $w^0 p^0 > 0$ or antiparallel $w^0 p^0 < 0$. The constant of proportionality between w and p will be a restricted Poincaré invariant. From Equation 15.72, it will be a pseudoscalar of angular momentum dimensions and, being frame-independent, is of considerable interest. We shall see that it is related to the projection of the angular momentum in the direction of the propagation, namely the *helicity*.

We may use any one of the four components of p and w to evaluate the constant of proportionality; we choose the zero component. In order to allow a sign on the helicity to distinguish between the parallel and antiparallel cases we shall define the helicity using the value of w^0 and the numerical value $|p^0|$ of the energy (which is non-negative anyway in a quantum context). Similarly, in order to define helicity as a dimensionless quantity not requiring redefinition when quantum concepts are introduced, we shall insert a factor of $\hbar$ into its definition. Consequently, we define the helicity by:

$$\lambda = \frac{w_0}{|p_0|\hbar} = -\frac{w^0}{|p^0|\hbar} \ . \tag{15.76}$$

The *helicity magnitude* $|\lambda|$ is also referred to as the *spin* of the null particle or wave.

We note that since it involves the momentum, the helicity of a massless particle, unlike the spin of a massive particle, is not simply a covariantly expressed rotation property but involves the entire Poincaré transformation, in particular the spacetime translations. We have chosen this definition to emphasize that, in contrast to the corresponding quantities $\mathbf{s}\cdot\hat{\mathbf{p}}$ for a Newtonian or massive relativistic particle (see Exercise 6.4 and Section 15.9), the helicity of a massless particle is a relativistically invariant quantity having the same value irrespective of frame. To justify its name we must relate it to the projection of the angular momentum in the direction $\hat{\mathbf{p}}$ of propagation by noting that:

$$\begin{aligned}
-w^0 = \tfrac{1}{2}\epsilon^{0\nu\lambda\rho}j_{\nu\lambda}p_\rho &= \tfrac{1}{2}\epsilon^{0klm}j_{kl}p_m \\
&= \tfrac{1}{2}\epsilon^{klm}j_{kl}p_m \\
&= j^m p_m \\
&= \mathbf{j}\cdot\mathbf{p} \\
&= \mathbf{j}\cdot\hat{\mathbf{p}}\,|\mathbf{p}| \\
&= \mathbf{j}\cdot\hat{\mathbf{p}}\,|p^0| \ ,
\end{aligned}$$

where the last line follows from the zero mass. Consequently, the helicity is given by,

$$\lambda = \frac{\mathbf{j}\cdot\hat{\mathbf{p}}}{\hbar} = \pm|\lambda| \ , \tag{15.77}$$

namely the projection of the angular momentum in the direction of propagation. The two helicity modes, $|\lambda|$ and $-|\lambda|$, of a null particle or wave of non-zero spin, correspond to left and right circular polarization. The Poincaré group irreducible representation label $j = s$ for massive spin or $j = |\lambda|$ for helicity has, in both cases, eigenvalues of $j = 0, \tfrac{1}{2}, 1, \tfrac{3}{2}, 2, \ldots$, corresponding to quanta with squared angular momentum of $j(j+1)\hbar^2$. The *spin multiplicity* in the massive case is $2s+1$ while for the massless case it is 2 for all non-zero helicities and 1 for zero helicity.

15.11 Spacelike surfaces and covariant 3-volume

The standard 3-volume element, $d^3x = |dx\,dy\,dz|$, is clearly not manifestly covariant since it has only one component but is not a Lorentz scalar. To see this, consider a boost along any axis. It will leave two factors in the volume element invariant but Lorentz contract the third. The same is true of the notion of 'all of space at a given time t', since it refers specifically to the time coordinate t of a particular frame. We shall now discuss the generalization of both these concepts to find their covariant replacements. One of the properties they must satisfy, is reduction, on choice of a suitable special case in a specific frame, to these non-covariant constructs.

Consider the rectangular Cartesian components,

$$\begin{aligned} dS_1 &= dx^2 dx^3 = dy\,dz \\ dS_2 &= dx^3 dx^1 = dz\,dx \\ dS_3 &= dx^1 dx^2 = dx\,dy \end{aligned} \qquad (15.78)$$

for the directed areal surface element $dS = \{dS_k\}$ of a 2D surface S in 3D Euclidean space $\mathbf{E}^3$. The fact that d^3x is a proper Euclidean scalar, may be used to show that dS is a proper 3-vector. Euclidean space $\mathbf{E}^3$ may be divided into two parts, by either a plane or a non-planar 2D surface on which the normals dS_1 and dS_2 at any two points satisfy $dS_1 \cdot dS_2 > 0$.

We wish similarly to divide the 4D Minkowskian spacetime into two parts, one entirely to the future and the other entirely to the past of the division, which itself must be a set of events making up all of 3D space. We may do this non-covariantly using the 3D spacelike surface $t = $ constant, in any frame. A 3D subspace of Minkowskian spacetime, called a *hypersurface*, may be defined by the constancy of some function $f(x)$ of the 4 coordinates, an example being $t = $ constant. Any 3D hypersurface whose 4-vector normal n is constant is referred to as a *hyperplane*, $t = $ constant again being an example. If a hypersurface Σ has a normal n which is everywhere timelike $n^2 < 0$, then Σ is said to be a *spacelike hypersurface*, $t = $ constant again being an example. The timelike normal n to any hypersurface, at an event where the surface is spacelike, may be *future-pointing* $n^0 > 0$ or *past-pointing* $n^0 < 0$.

The normal is thus *oriented* and may be used to distinguish the future and past sides of a spacelike surface in the same way that the normal $\mathbf{n}$ may distinguish the sides of a surface in $\mathbf{E}^3$. Furthermore, by combining the 3-volume element of a small part of a spacelike surface Σ with the direction of the normal to the surface, we may form a directed 3-volume element, $d\Sigma$, at each event on Σ. For the surface $\Sigma_{t=\text{const}}$ with constant time t, the future and past directed 3-volume elements will be $d\Sigma = dx^1 dx^2 dx^3\,\mathbf{e}_0$ or $\{\pm d^3x, \mathbf{0}\}$, with a similar result for a constant $\bar{t}$ surface in any other frame with basis $\mathbf{e}_{\bar{\mu}}$.

The covariant 3-volume analogue of the proper oriented 3-vector area $d\mathbf{S} = \{dS_k\}$ of Equation 15.78, outwardly normal to S in $\mathbf{E}^3$, for an arbitrary hypersurface Σ in $\mathbf{E}^{1,3}$, is the proper oriented 4-vector $d\Sigma = \{d\Sigma_\mu\}$ Lorentz normal to Σ, with components,

$$\begin{aligned} d\Sigma_0 &= dx^1 \, dx^2 \, dx^3 \\ d\Sigma_1 &= dx^2 \, dx^3 \, dx^0 \\ d\Sigma_2 &= dx^3 \, dx^0 \, dx^1 \\ d\Sigma_3 &= dx^0 \, dx^1 \, dx^2 \, . \end{aligned} \quad (15.79)$$

Being Lorentz orthogonal to the spacelike Σ, it is timelike ($d\Sigma \cdot d\Sigma > 0$). The proper 4-vector nature of $d\Sigma = \{d\Sigma_\mu\}$ follows from the proper scalar property (see also Exercise 14.4) of d^4x. To the orientation of $d\mathbf{S}$ to one side or the other of S in $\mathbf{E}^3$ will correspond, provided Σ is spacelike, a future-pointing ($d\Sigma_0 > 0$) or past-pointing ($d\Sigma_0 < 0$) property of the normal $d\Sigma$ to Σ. A surface Σ to the future or past of the region of interest may be considered outward oriented if $d\Sigma$ is future or past pointing, respectively. If we select $dt = 0$ in some Lorentz frame, we recover d^3x from $d\Sigma_\mu$ since then,

$$d\Sigma_\mu = \{dx^1 dx^2 dx^3, 0\} = \{\pm d^3x, 0\} \quad (t = \text{constant}) \, , \quad (15.80)$$

or,

$$d\Sigma_\mu = \{d^3x, 0\} \quad (\text{future-pointing}) \, . \quad (15.81)$$

In using such a special non-covariant 3-volume in spacetime we shall replace the label Σ by $t = \text{constant}$ and $d\Sigma_\mu$ by d^3x. The constant $\bar{t}$ surface obtained by a boost β relative to the first frame will have a signed volume element given by,

$$d\Sigma_{\bar{\mu}} = \{\gamma d^3x, \gamma \beta d^3x\} \, . \quad (15.82)$$

Constant t (or constant $\bar{t}$) surfaces in Minkowskian spacetime are examples of spacelike hyperplanes. The normals to the constant $\bar{t}$ surface will not be parallel to those for constant t if the frames are boosted relative to one another. The simplest spacelike 3-volume analogues of S in the 4D pseudo-Euclidean Minkowskian spacetime, $\mathbf{E}^{1,3}$, will be those spacelike hyperplanes Σ_i ($i = 1, 2, \ldots$) corresponding to constant time in some frame as illustrated in Figure 15.2. In Figure 15.2, the linear intersections with the paper are shown for those constant time surfaces with normal in the plane of the paper. A plurality of spaces appears in the 4D analogue of S due to the need to include not just rotated but also boosted frames. Changes in the direction of the normal to the surface will correspond to changes not only in the directions of the spatial axes of a frame (due to a rotation) but also changes in the time axis due to the boost part, Equation 13.50, of a Lorentz transformation.

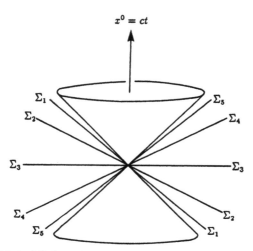

Figure 15.2 Minkowski diagram of spacelike cross-sections, Σ.

15.12 Gauss's theorem in Minkowskian spacetime

Gauss's theorem applied to a 3-vector field $\mathbf{V}(\mathbf{x})$ in $\mathbf{E}^3$ takes the form:

$$\oint_{\partial R} d\mathbf{S}\cdot\mathbf{V} = \int_R d^3x\, \boldsymbol{\nabla}\cdot\mathbf{V}\,, \tag{15.83}$$

where ∂R denotes a closed 2D boundary surface surrounding the simply-connected 3D region R. The 3-volume element d^3x is, like $d\mathbf{S}\cdot\mathbf{V}$ and $\boldsymbol{\nabla}\cdot\mathbf{V}$, a Euclidean invariant. We re-express Equation 15.83 in indexed form as,

$$\oint_{\partial R} dS_k\, V^{k\cdots} = \int_R d^3x\, \partial_k V^{k\cdots}\,, \tag{15.84}$$

in which we have allowed for the fact that nothing is altered if we add, to V^k, other (free) indices not affected by the integrations over dS_k. No part of the proofs of the 3D Gauss's theorem makes any essential use of the dimensionality of space nor of the metric signature (3 in $\mathbf{E}^3$, or 2 in the spacelike convention for the metric of $\mathbf{E}^{1,3}$). Consequently, the proof also applies to 4-vectors, which may also have other free indices not participating in the contractions with the volume element or the divergence, in Minkowskian spacetime. Consequently, Gauss's theorem in Minkowskian spacetime takes the form:

$$\oint_{\partial R} d\Sigma_\mu\, V^{\mu\cdots} = \int_R d^4x\, \partial_\mu V^{\mu\cdots}\,, \tag{15.85}$$

in which ∂R is a closed 3D boundary (it cannot be entirely spacelike) of the simply-connected 4D spacetime region R. Just as dS in 3D is a directed

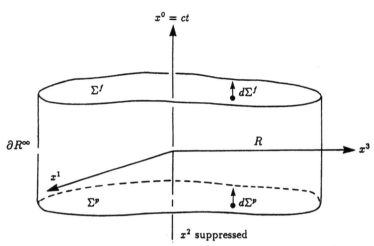

Figure 15.3 Closed boundary at infinity (∂R^∞), to the future (Σ^f) and to the past (Σ^p) of the system of interest.

normal to the 2D boundary ∂R, so also in 4D, $d\Sigma$ in Gauss's law must be a directed normal to the 3D boundary ∂R. In 4D we often choose the shape of ∂R so that it has three parts,

$$\partial R = \Sigma^f + \Sigma^p + \partial R^\infty , \qquad (15.86)$$

outside and surrounding the sources of our fields. These three parts are:

- Σ^f, a spacelike 3-surface, to the future (f) of all the events of interest, and extending to infinity in all three spatial directions. Its normal will therefore be timelike and, by convention, we take future-pointing ($d\Sigma_0 > 0$) to be the outward direction.

- Σ^p, a spacelike 3-surface to the past (p) of all of the events of interest, and also extending to spatial infinity. Past-pointing ($d\Sigma_0 < 0$) is then considered to the the outward direction.

- ∂R^∞ is a third timelike part at spatial infinity used to join Σ^f and Σ^p to obtain a closed surface.

We take ∂R^∞ far away (constant radial distance $r \to \infty$) from the source. The fields are assumed to fall off sufficiently rapidly with r that the contributions over ∂R^∞ to the integrals $\int d^3x \ldots$ will vanish (or converge appropriately) as $r \to \infty$. The three parts of a boundary are illustrated in Figure 15.3.

We now apply Gauss's theorem on ∂R to any local 4-vector source or current $J^{\mu\cdots}$ which is *conserved*, namely satisfies the local conservation law, $\partial_\mu J^{\mu\cdots} = 0$. For such a source, the integral over ∂R is zero. Since the

15.13. Energy-momentum tensor

contributions at spatial infinity vanish, we obtain:

$$\int_{\Sigma^p} d\Sigma_\mu J^{\mu\cdots} + \int_{\Sigma^f} d\Sigma_\mu J^{\mu\cdots} = 0 \ . \tag{15.87}$$

We now change the 3-volume element $d\Sigma_\mu$ in the integral on Σ^p from a past-pointing to a more conventional (inward) future-pointing $d\Sigma^p$ and obtain:

$$\int_{\Sigma^p} d\Sigma_\mu J^{\mu\cdots} = \int_{\Sigma^f} d\Sigma_\mu J^{\mu\cdots} \ , \tag{15.88}$$

There is in fact no restriction at all on the surfaces Σ^p and Σ^f which may appear here provided each is space-like and this result shows that all space-like surfaces will give the same result in the integral. We have thus shown that the integrated source, or charge,

$$Q^{\cdots} = \int_{\Sigma^p} d\Sigma \cdot J = \int_{\Sigma^p} d\Sigma_\mu J^{\mu\cdots} = \int_{t=\text{const}} d^3x J^{0\cdots} \ , \tag{15.89}$$

evaluated on any spacelike surface, is in fact independent of Σ. This is the global or integrated result corresponding to the local *continuity equation*, $\partial_\mu J^{\mu\cdots} = 0$.

15.13 Energy-momentum tensor

A single particle localized at a succession of particular events along a trajectory $z(t) = \{z^\mu(t)\}$ may also be considered to have a distribution of energy and a momentum density given as a function of spacetime position x which is singular (diverging) along the trajectory.

A particle of momentum $p^\mu(t)$ will, in a particular frame, have a spatial *density of momentum* given in terms of the 3D Dirac delta function by $p^\mu \delta^3(\mathbf{x} - \mathbf{z}(t))$ as this integrates over the 3-volume element d^3x to give the correct 4-momentum located at the appropriate spatial point on the trajectory $z(t)$. The *flux density* of its 4-momentum in the l-direction is obtained, as for any flux, by multiplying the density by the velocity dz^l/dt and, since the result is a function of $\{t, \mathbf{x}\}$ with labels μl, we denote it by:

$$T^{\mu l}(x) = p^\mu(t) \frac{dz^l}{dt} \delta^3(\mathbf{x} - \mathbf{z}(t)) \ . \tag{15.90}$$

From the way this quantity is defined, its space-space part T^{kl}, is the k-*momentum flux* across an area with normal in the l-direction, known as the *stress tensor*. Similarly, cT^{0l} is the *energy flux* in the l-direction. The labelling alone does not, however, give these quantities tensorial character which requires in addition that they transform correctly. In anticipation of a tensor result, we seek to complete both sides. Consequently, from p^μ we also form the quantity:

$$T^{\mu 0}(x) = p^\mu(t)\frac{dz^0}{dt}\delta^3(\mathbf{x} - \mathbf{z}(t)) \,, \tag{15.91}$$

which, on noting that $dz^0/dt = dx^0/dt = c$ on the trajectory, is c times the 4-*momentum density*. The zeroth component T^{00} of this quantity is therefore, using $E = p^0 c$, the *energy density* $E\,\delta^3(\mathbf{x} - \mathbf{z}(t))$. Combining these two equations gives,

$$T^{\mu\nu}(x) = p^\mu(t)\frac{dz^\nu(t)}{dt}\delta^3(\mathbf{x} - \mathbf{z}(t)) \,. \tag{15.92}$$

The 3D Dirac delta function, like d^3x, is not Lorentz-invariant. We have also not yet fixed the 4-momentum to be at the correct location along the trajectory at a given time. We therefore now multiply the right side of the above expression for $T^{\mu\nu}$ by a 1D Dirac delta function $\delta(x^0 - z^0(t))$ and integrate the result over the coordinate x^0 so that we can convert the non-invariant 3D Dirac delta into a Lorentz-invariant 4D Dirac delta function, $\delta^4(x - z(t))$. For a non-null particle we have $\int dx^0\, d/dx^0 \ldots = \int d\tau\, d/d\tau \ldots$, where τ is the proper time, and replacement of the parameter t or x^0 by the scalar parameter τ gives us the expression for the *energy-momentum tensor*, also sometimes called the stress-energy tensor, of a massive particle:

$$T^{\mu\nu}(x) = mc\int_{-\infty}^{\infty} d\tau\,\delta^4(x - z(\tau))\,\dot{z}^\mu \dot{z}^\nu \,. \tag{15.93}$$

We can now see that $T^{\mu\nu}(x)$ is a spatially and temporally proper tensor field because m, c and $d\tau\delta^4(x - z(\tau))$ are all proper scalars and $\dot{z}^\mu \dot{z}^\nu$ is a proper tensor. The energy-momentum tensor is of crucial importance in studies of the gravitational interaction.

Exercise 15.28 *Establish the relationship, under appropriate conditions, between the space-space part of Equation 15.93 and Equation 6.28.*

In order to include a null particle, we may replace t by a scalar evolution parameter λ (not the proper time τ). Noting that the 4-momentum in Equation 15.92 will be proportional to the 4-vector $\dot{z} = dz^\mu/d\lambda$, we let $p^\mu = e^{-1}(\lambda)dz^\mu/d\lambda$, where $e(\lambda)$ is an arbitrary function of λ with inverse mass dimensions, and transforming contragrediently to $d\lambda$, to obtain a reparameterization invariant energy-momentum tensor of either a massive particle (with $e(\lambda) = m^{-1}$) or a null particle as,

$$T^{\mu\nu}(x) = c\int_{-\infty}^{\infty} d\lambda\, e^{-1}(\lambda)\delta^4(x - z(\lambda))\,\dot{z}^\mu \dot{z}^\nu \,. \tag{15.94}$$

15.13. Energy-momentum tensor

The trace of the energy-momentum tensor of a null system is clearly zero.

The above results (for $m \neq 0$) hold equally if the single particle of mass m is replaced by a sum over a *swarm* of classical non-interacting point particles of mass $m_{(i)} \neq 0$ ($i = 1, 2, \ldots$) on trajectories $z_{(i)}(\lambda)$. The energy-momentum tensor of a particle system is clearly symmetric $T^{\mu\nu} = T^{(\mu\nu)}$, a property which we shall see shortly is an expression of conservation of angular momentum, and will therefore apply not only to one or more particles but also to the energy-momentum tensor of any isolated system.

From the way the energy-momentum tensor was defined, we may summarize the information it contains as follows:

$$\begin{pmatrix} T^{00} & T^{0l} \\ T^{k0} & T^{kl} \end{pmatrix} = \begin{pmatrix} \text{energy density} & (l\text{-flux of energy})/c \\ c(k\text{-momentum density}) & kl \text{ stress component} \end{pmatrix}. \tag{15.95}$$

Selecting a specific frame, Equation 15.91 shows that integration of $T^{\mu 0}(x)/c$ over all space at constant t reconstructs the 4-momentum p^μ,

$$p^\mu = \tfrac{1}{c} \int_{t=\text{const}} d^3x \, T^{\mu 0}. \tag{15.96}$$

Since the covariant 3-volume $d\Sigma_\nu$ of such a constant t surface is given by $d\Sigma = \{dx^1 dx^2 dx^3, 0\}$ we may write the reconstruction of the 4-momentum in the form,

$$p = \tfrac{1}{c} \int_\Sigma d\Sigma \cdot T \quad \text{or} \quad p^\mu = \tfrac{1}{c} \int_\Sigma d\Sigma_\nu \, T^{\mu\nu}. \tag{15.97}$$

Since this expression is covariant it remains true in any frame and for any spacelike cross-section Σ of spacetime. Since $T^{\mu\nu}$ is symmetric it does not matter on which of its indices we contract.

The fact that the 4-momentum is conserved may be expressed either non-covariantly as $dp^\mu/dt = 0$ or covariantly as $dp^\mu/d\Sigma = 0$. This means that the above integrals are independent of t or Σ. The 4D Gauss's law can now be used to establish,

$$\partial \cdot T = 0 \quad \text{or} \quad \partial_\mu T^{\mu\nu} = 0, \tag{15.98}$$

as the equation which describes 4-momentum conservation differentially. This equation is the analogue, for 4-momentum, of the continuity equation $\partial \cdot J = \partial_\mu J^\mu = 0$ for the invariant electric charge Q.

Although the isotropy of space is the key geometrical property giving rise to the conservation of angular momentum, the homogeneity is also inevitably an important factor for an extended system; rotation has little meaning for an extended system unless part of it is also translated. The energy-momentum tensor $T^{\mu\nu}(x)$ is the appropriate object with which to relativistically describe the energy-momentum densities and stress of a system distributed throughout space whether it be a swarm of particles, a fluid or a field. If we combine Equation 15.21 with 15.96 or 15.97, we obtain corresponding results for the angular momentum in the form:

$$j^{\mu\nu} = \int_\Sigma d\Sigma_\lambda\, j^{\mu\nu\ \lambda} = \int_{t=\text{const}} d^3x\, j^{\mu\nu\ 0}, \qquad (15.99)$$

where the *tensor of angular momentum density*, referred to event $d = \{d^\mu\}$, is given by:

$$j^{\mu\nu\ \lambda}(x) = \tfrac{1}{c}[(x^\mu - d^\mu)T^{\nu\lambda}(x) - (x^\nu - d^\nu)T^{\mu\lambda}(x)] = -j^{\nu\mu\ \lambda}. \qquad (15.100)$$

By the 4D Gauss's theorem, conservation of angular momentum $j^{\mu\nu}$ requires a zero divergence on the last index of the angular momentum density $\partial_\lambda j^{\mu\nu\ \lambda} = 0$. Applying this condition to Equation 15.100 gives:

$$(T^{\mu\nu} - T^{\nu\mu}) + (x^\nu - d^\nu)\partial_\lambda T^{\mu\lambda} - (x^\mu - d^\mu)\partial_\lambda T^{\nu\lambda} = 0. \qquad (15.101)$$

Thus the angular momentum of an isolated system is conserved if and only if the corresponding energy-momentum is conserved $\partial_\nu T^{\mu\nu} = 0$ and the energy-momentum tensor is symmetric $T^{\mu\nu} = T^{(\mu\nu)}$.

Extension from a single particle or a swarm of particles to a continuous distribution in the form of a fluid is straightforward. The same also applies to a continuous system consisting of one or more fields $\psi(x)$, for which explicit techniques exist for constructing the energy-momentum tensor. We shall discuss these Noether methods for relativistic fields in the following chapter.

Bibliography:

Dirac P A M 1950 *Generalised Hamiltonian dynamics* [110].
Hanson T J et al., *Constrained Hamiltonian systems* [223].
Sudarshan E C G and Mukunda N 1974 *Classical Dynamics* [525].
Sundermeyer K 1982 *Constrained dynamics* [526].

Part IV

Electrodynamic and gravitational fields

CHAPTER 16

Relativistic Lagrangian fields

Since Maxwell's time, Physical Reality has been thought of as represented by continuous fields, governed by partial differential equations, and not capable of any mechanical interpretation[1].

Albert Einstein   1879–1955

One of the many advantages of the Hamiltonian and Lagrangian descriptions, based on Hamilton's principle of stationary action, is the generality of its application. The variational analysis of relativistic Lagrangian fields may be carried out by an adjustment to the notation of the previous discussions in Parts II and III. The results are in many ways simpler than the corresponding material for Galilean fields given in Chapter 10.

Few fundamental applications make use of Galilean fields, even as an approximation to a fully relativistic treatment, as we discovered in Chapter 10. The principle examples are the non-dynamic fields of Newtonian gravity, electrostatics and magnetostatics and the spinless wave functions and spin $\frac{1}{2}$ Pauli spinors of Schrödinger theory. In contrast, there is a very rich array of relativistic Lagrangian fields. Among these are the fields associated with the names of Klein and Gordon, Dirac, Majorana, Weyl, Proca, Maxwell, Rarita and Schwinger and Fierz and Pauli. The polarization structure of such fields includes those of spins $0, \frac{1}{2}, 1, \frac{3}{2}$ and 2 with corresponding values for the intrinsic angular momentum of particles arising from them by quantization. Many of these find application in fundamental physical problems.

The *Klein-Gordon* or *scalar field* was examined by Schrödinger [498] in 1925 and independently in 1926 by Klein [284], Gordon [207] and several others. It describes spinless particles such as pions, $\pi^\pm$ and π^0, and kaons, $K^\pm$, K^0, $\bar{K}^0$. In the standard quantum model of the electroweak interaction, it is also used to describe the *Higgs field* [248] which spontaneously breaks the gauge symmetry at low energies. The Higgs boson has

---

[1] *Maxwell's influence on the development of Physical Reality* [535, p. 71].

not yet been detected. The bulk of the ordinary matter in the universe is composed of electrons, e^-, and the first family $\{u, d\}$ of quarks, q. These particles and their antiparticles, e^+ and $\bar{q}$, are described by the massive non-chiral spin $\frac{1}{2}$ *Dirac field*, the equations for which were first presented by Dirac [106] in 1928.

The three non-gravitational interactions are all described by the exchange of spin 1 particles (the photon γ, the weakons $W^{\pm}$, Z^0 and the eight gluons g) arising from the quantization of massless classical fields for which the *Maxwell field* of electromagnetism is the prototype. The *Yang-Mills fields* first introduced by Yang and Mills [595] in 1954, and which mediate the nuclear interactions by the exchange of weakons (intermediate vector bosons) and gluons, are triplets or octets of inter-related and interacting spin 1 Maxwell-like fields. Of these, the $W^{\pm}$ and Z^0 particles acquire mass by quantum interaction with the Higgs field and the gluons are confined as parts of composite particles.

Many aspects of the gravitational interaction, especially where gravity is weak, can be described by the massless spin 2 *Fierz-Pauli field* first described in Lagrangian form by Fierz and Pauli [162] in 1939. The *Rarita-Schwinger fields*, the field equations for which were discovered by Rarita and Schwinger [469] in 1941, describe the spin $\frac{3}{2}$ particles that arise in gravitational theories, such as *supergravity* [97, 98, 99, 100, 174] in which there is supersymmetry between bosons and fermions.

The *Weyl field* for describing massless 2-component complex spin $\frac{1}{2}$ particles, such as neutrinos if they propagate at the speed of light, were first presented by Weyl [573] in 1929 and rediscovered in 1957 by Lee and Yang [318], Salam [491] and Landau [304]. If neutrinos are massive, and have distinct antiparticles, they may be described by Dirac fields. If neutrino and antineutrino are indistinguishable, the *Majorana fields* first presented by Majorana[2] [356] in 1937 (see also Amaldi [9, p. 96]) may be the most appropriate. The *Proca field* described by Proca [462] in 1936, describes particles arising from a spin 1 field with mass. No observed particles correspond to Proca fields.

We shall primarily be concerned here with the Klein-Gordon, Maxwell, Dirac, and Fierz-Pauli fields, with a mention of the Weyl and Rarita-Schwinger equations, but much of the material we present may also be applied to any of the fields just mentioned.

16.1 Field equations and Noether currents

In addition to their functional dependence on the continuum of spacetime coordinates x, the fields we shall be dealing with may have a finite number of indices each having a finite number of discrete values in its range. These

---

[2] Ettore Majorana (1906–1938), Italian physicist.

16.1. Field equations and Noether currents

may arise, for example, from the behaviour of vector and tensor fields with respect to Lorentz transformations or the corresponding transformations of spinors. Specific examples are $A^\mu(x)$ for the electromagnetic 4-vector potential and $F^{\mu\nu}(x)$ for the electromagnetic field tensor. We shall continue to collectively denote all such fields by a generic symbol $\psi = \psi(x) = \{\psi_j(x)\}$ ($j = 1, 2, \ldots d$) in which ψ may have any finite number d of components $\psi_1(x), \ldots, \psi_d(x)$. In a specific example the index j will typically become one or more Lorentz, Weyl spinor or Dirac spinor indices ($\mu = 0, 1, 2, 3$, $A = 1, 2$ or $\alpha = 1, 2, 3, 4$).

As before, we may consider the separate components to be organized as a column vector:

$$\psi = \psi(x) = \{\psi_j(x)\} = \begin{pmatrix} \psi_1(x) \\ \psi_2(x) \\ \vdots \\ \psi_d(x) \end{pmatrix} . \tag{16.1}$$

The independent variables for a system of relativistic fields, corresponding to the generalized coordinates $q(\lambda)$ in a mechanical system (for example, $z(\lambda)$ for a particle) are the values of $\psi(x)$ on a spacelike surface Σ. We shall denote these variables by $\psi(\Sigma)$ or simply by $\psi(x)$, it being understood that x is restricted to some surface Σ in specifying the *initial data* to be evolved forward in time.

Since we require our Lagrangian formalism for fields to be manifestly covariant, we usually complete the generalized velocity $\partial_0 \psi(x)$ with the spatial derivatives $\nabla \psi(x)$ to form the *covariant generalized velocities* $\partial_\mu \psi(x)$. This definition gives $\partial_0 \psi$ dimensions differing by a velocity from the corresponding Galilean quantities, $\dot\psi(t, \mathbf{x})$. For the 4-vector potential A^μ, for example, the covariant velocities will be $\partial_\mu A_\nu$. Knowledge of $\psi(x)$ on Σ implies knowledge of $\nabla \psi(x)$ on Σ which is not therefore independent in the specification of initial data.

Following the procedure of Section 10.3, with suitable changes of notation, we deal from the outset only with local fields for which the action functional is therefore expressed in terms of an integral over spacetime of a Lagrangian density function of the fields and their first derivatives,

$$S = S[\psi(\Sigma), \Sigma_i, \Sigma_f] = \int_R d^4 x\, \mathcal{L}(\psi, \partial_\mu \psi) , \tag{16.2}$$

where R is a compact region of spacetime between initial and final spacelike surfaces Σ_i and Σ_f. The Lagrangian density will be *non-singular* provided,

$$\det\left(\frac{\partial^2 \mathcal{L}}{\partial_0 \psi_j\, \partial_0 \psi_k}\right) \neq 0 . \tag{16.3}$$

Classically, the system will follow a particular sequence of field configurations $\psi(\Sigma)$ from initial to final configuration. This sequence is the physical path or classical history. The Lagrangian procedure involves any path, physical or not, from $\psi(\Sigma_i)$ to $\psi(\Sigma_f)$. The convention that ψ is a column vector of components means that we may therefore regard $\partial \mathcal{L}/\partial \psi$ and the covariant *canonical momenta* $\partial \mathcal{L}/\partial(\partial_\mu \psi)$ as row vectors.

We shall consider Lagrangians only up to quadratic degree and with no essential involvement of derivatives higher than the first thus limiting the Euler-Lagrange field equations to no higher than second order. Since the action S has the dimensions $[\hbar]$ of angular momentum, the relativistic Lagrangian density has been assigned dimensions of $[\hbar\,L^{-4}]$ where L denotes the length dimension. This differs by a velocity from the dimensions of a Newtonian Lagrangian density owing the factor of c difference between $dt\,d^3x$ and d^4x in the customary expressions for the actions in terms of Lagrangians.

In order to establish the Euler-Lagrange equations and Noether's theorem, we require first the variation of the action (see Chapter 10) under the arbitrary functional variations, of independent and dependent variables,

$$x \to x + \delta x(x) \quad \text{and} \quad \psi(x) \to \psi(x) + \delta\psi(\psi(x),x) \;. \tag{16.4}$$

The integration volume element varies by,

$$\delta(d^4x) = d^4x\,\partial\!\cdot\!\delta x(x) \;. \tag{16.5}$$

We deduce,

$$\begin{aligned}\delta\psi &= \delta_0\psi + \delta x\!\cdot\!\partial\psi & (16.6)\\ \delta\partial_\mu\psi &= \delta_0\partial_\mu\psi + \delta x\!\cdot\!\partial\partial_\mu\psi \;, & (16.7)\end{aligned}$$

For the variation of the action, we obtain,

$$\begin{aligned}\delta S &= \int_{\Sigma_i}^{\Sigma_f}\{\delta(d^4x)\mathcal{L} + d^4x\,\delta\mathcal{L}\}\\ &= \int_{\Sigma_i}^{\Sigma_f} d^4x\{\partial_\mu(\delta x^\mu)\mathcal{L} + \delta x^\mu\,\partial_\mu\mathcal{L} + \delta_0\mathcal{L}\}\\ &= \int_{\Sigma_i}^{\Sigma_f} d^4x\,\partial_\mu\!\left\{\delta x^\mu\,\mathcal{L} + \frac{\partial\mathcal{L}}{\partial(\partial_\mu\psi)}\delta_0\psi\right\} + \left\{\frac{\partial\mathcal{L}}{\partial\psi} - \partial_\mu\!\left(\frac{\partial\mathcal{L}}{\partial(\partial_\mu\psi)}\right)\right\}\delta_0\psi\\ &= \int_{\Sigma_i}^{\Sigma_f} d^4x\,\bigg[\left(\delta^\mu_{\;\nu}\mathcal{L} - \frac{\partial\mathcal{L}}{\partial(\partial_\mu\psi)}\partial_\nu\psi\right)\delta x^\nu + \frac{\partial\mathcal{L}}{\partial(\partial_\mu\psi)}\delta\psi\\ &\quad + \left\{\frac{\partial\mathcal{L}}{\partial\psi} - \partial_\mu\!\left(\frac{\partial\mathcal{L}}{\partial(\partial_\mu\psi)}\right)\right\}(\delta\psi - \delta x\!\cdot\!\partial\psi)\bigg] \;,\end{aligned} \tag{16.8}$$

16.1. Field equations and Noether currents

from which, by application of Hamilton's principle (see Chapters 8 and 10), we deduce the *Euler-Lagrange equations* for a local relativistic field,

$$\partial_\mu \frac{\partial \mathcal{L}}{\partial(\partial_\mu \psi)} = \frac{\partial \mathcal{L}}{\partial \psi} \ . \tag{16.9}$$

For a small continuous Lie symmetry transformation $\{\Delta x^\mu, \Delta \psi\}$ of the system (see Chapter 9), we obtain,

$$\partial_\mu j_a^\mu + \left\{ \frac{\partial \mathcal{L}}{\partial \psi} - \partial_\mu \left(\frac{\partial \mathcal{L}}{\partial(\partial_\mu \psi)} \right) \right\} \left(\frac{\partial \Delta \psi}{\partial \Delta \omega^a} - \frac{\partial \Delta x^\mu}{\partial \Delta \omega^a} \partial_\mu \psi \right) = 0 \ , \tag{16.10}$$

where,

$$j_a^\mu = \left(\frac{\partial \mathcal{L}}{\partial(\partial_\mu \psi)} \partial_\nu \psi - \delta^\mu{}_\nu \mathcal{L} \right) \frac{\partial \Delta x^\nu}{\partial \Delta \omega^a} - \frac{\partial \mathcal{L}}{\partial(\partial_\mu \psi)} \frac{\partial \Delta \psi}{\partial \Delta \omega^a} - \frac{\partial(\Delta \Lambda^\mu)}{\partial \Delta \omega^a} \ , \tag{16.11}$$

is a *Noether current* corresponding to one of the p independent parameters ω^a. The small functions $\Delta \Lambda^\mu$ of x and ψ (but not $\partial_\mu \psi$) arise from covariant Lagrangians which are not invariant (see Section 9.3) according to,

$$\partial \cdot (\Delta x) + \Delta \mathcal{L} = -\partial \cdot (\Delta \Lambda) \ . \tag{16.12}$$

On the stationary path, referred to as *on-shell*, the Noether currents satisfy *continuity equations*,

$$\partial_\mu j_a^\mu = 0 \ . \tag{16.13}$$

As with any conserved current density, we may integrate over an entire spacelike surface to obtain a corresponding conserved *Noether charge*:

$$Q_a = \int_\Sigma d\Sigma_\mu \, j_a^\mu = \int_{t=\text{const}} d^3x \, j_a^0(t, \mathbf{x}) = \text{constant} \ , \tag{16.14}$$

in the usual case where all fields fall off sufficiently rapidly at large spatial distances.

In the Lagrangian formulation of field theory, the Relativity principle is implemented by demanding that the Lagrangian density be invariant with respect to the restricted Poincaré group ISO(1,3). One may show [262, p. 157] that any local quantum field theory described by Hamilton's principle in terms of an hermitian, restricted Poincaré-invariant Lagrangian density with commuting bosonic (integer spin) fields and anticommuting fermionic (half-odd-integer) fields is also invariant under the combined operation, **CPT**, of charge conjugation, parity and time reversal, in any order. This CPT *theorem* was established by Schwinger [506] and Pauli [429]. Invariance with respect to one of **C**, **P** and **T** implies invariance with respect to the product of the other two. Lack of invariance for one of the three implies lack of invariance for at least one of the other two.

Any field whose dynamics may be derived from such an action principle is said to be a *relativistic Lagrangian field*. Not all field equations are expressed in terms of such variables. In Section 18.13, we shall encounter an example of specific dynamical variables and corresponding field equations which correctly describe the classical dynamics of a physical system but for which a Lagrangian formulation does not exist. In that case, however, an alternative description in terms of a Lagrangian field is possible and we shall see that the Lagrangian alternative has features which make it a preferred formulation in many situations.

If a Lagrangian density $\mathcal{L}$ exists for a given system of fields then it will not be unique since $\mathcal{L} + \partial_\mu \Lambda^\mu$, where $\Lambda^\mu(x,\psi)$ are arbitrary functions of x and ψ (but not $\partial_\mu \psi$), will simply lead to an extra *surface term* in the variation of the new action and one arrives at the same Euler-Lagrange field equations. The dynamics of a classical system are entirely determined from the boundary conditions by its equations of motion, if it is mechanical, or by its field equations, if it is continuous in the spacetime coordinates. Consequently, any non-uniqueness in the Lagrangian or the action does not affect the dynamics. A quantum system with a classical analogue will have contributions to its expectation values from paths other than the stationary or classical path from initial to final state. Loosely speaking, the probabilities of their contribution will diminish with increased deviation of the path or history from the stationary one.

The *path-integral quantization* [160, 488] technique shows that the magnitudes of these quantum corrections are related to the exponential $e^{iS/\hbar}$ of the action S along the virtual path concerned. As one deviates more and more from the stationary history small variations in the path lead to more substantial variations in S giving rapidly increasing fluctuations in $e^{iS/\hbar}$ which contribute equally with all phases and therefore tend to interfere destructively. Nevertheless, the contributions off the stationary path cannot be ignored especially those nearest to that path with $|S - S_{\text{stationary}}|/\hbar \ll 1$. Consequently, the dynamics of a quantum system are determined not just by the equations of motion but by the action and Lagrangian themselves along all virtual paths. Surface terms cannot always then be ignored and the differences between classically equivalent Lagrangian densities may be significant.

A *classical symmetry* of a Lagrangian system is any transformation which leaves the equations of motion unchanged in form and this includes a transformation for which $\mathcal{L}$ is covariant but not necessarily invariant, taking $\mathcal{L}$ to $\mathcal{L} + \partial_\mu \Lambda^\mu$. However, since the action itself determines the quantum physics, a *quantum symmetry* of a Lagrangian system must leave not only the equations of motion but also the Lagrangian and the action unchanged. It may therefore turn out that a quantum system with a classical analogue does not have all the symmetries of the classical system.

Any classical symmetry which does not turn out to be a symmetry of the corresponding renormalized quantum system is referred to as an *anomaly*. There are many classical symmetries, for example the spacetime symmetries of homogeneity and isotropy, that are so fundamental that we demand they be also quantum symmetries. This demand of *anomaly freedom* is a powerful quantum principle that can severely restrict the form of a theory and thus assist in the search for uniqueness in the description of fundamental phenomena.

16.2 Canonical energy-momentum tensor

We might expect to obtain a system's energy-momentum density and total energy-momentum as the conserved Noether current and charge corresponding to a translation $x^\mu \to x^\mu + a^\mu$ in which a^μ are constants. We therefore replace the $\Delta \omega^a$ ($a = 1, 2, \ldots, p$) of Equation 16.11 by $\Delta \omega^\mu = a^\mu$ ($\mu = 0, \ldots, 3$). We consider a translational variation with no change ($\delta \psi = 0$) in the field. The derivative is then $\partial \Delta x^\mu / \partial \Delta \omega^\nu = \delta^\mu{}_\nu$. We are free to choose an overall constant factor in the expression for a Noether current in terms of the Lagrangian quantities in order that our expressions for the energy quantities have their standard forms. In this way, for an invariant Lagrangian ($\Delta \Lambda^\mu = 0$), the Noether result leads us to the conserved *canonical energy-momentum tensor* of a field,

$$\check{T}^{\mu\nu} = -c\Big(\frac{\partial \mathcal{L}}{\partial(\partial_\mu \psi)}\partial^\nu \psi - \eta^{\mu\nu}\mathcal{L}\Big) \quad \text{where} \quad \partial_\mu \check{T}^{\mu\nu} = 0 \ . \tag{16.15}$$

For classical fields, it is understood that the energy-momentum tensor $\check{T}^{\mu\nu}$, like any observable, is evaluated on-shell. Noting that $c\mathcal{L} \to \mathcal{L}_{nr}$ in the nonrelativistic (nr) limit shows that the space-space part $\check{T}^{kl}$ reduces to the stress tensor of Equation 10.64.

The conserved 4-momentum is obtained by integrating over any spacelike cross-section of spacetime according to Equation 15.97,

$$p^\mu = \tfrac{1}{c}\int_\Sigma d\Sigma_\nu \, \check{T}^{\mu\nu} = -\int_\Sigma d\Sigma_\nu \Big(\frac{\partial \mathcal{L}}{\partial(\partial_\mu \psi)}\partial^\nu \psi - \eta^{\mu\nu}\mathcal{L}\Big) \ . \tag{16.16}$$

The reality of $\mathcal{L}$ ensures the reality of $\check{T}^{\mu\nu}$ (even for complex fields ψ) and therefore the reality of the observable p^μ.

There are several difficulties with the canonical energy-momentum tensor. One of these is that it is not automatically symmetric in μ and ν, as required for angular momentum conservation. Furthermore, for gauge fields, it leads to quantities which depend directly on the gauge potentials rather than on gauge-invariant field-strength combinations. This is unsatisfactory since the components of the energy-momentum tensor are measurable quantities and cannot therefore be gauge-variant. These difficulties arise because it neglects the spin angular momentum contributions to the energy-momentum density. Rotations of a field, like those of an extended object, inevitably also give rise to translation.

To obtain a correct energy-momentum tensor that incorporates angular momentum conservation automatically (and gauge invariance if it is applicable) we must consider a variation of x^μ corresponding to a Poincaré translation combining Lorentz rotation with translation. The necessary modification will be described in Section 19.5.

The *canonical momentum density* conjugate to the field is defined by,

$$\pi = \pi^0 = \frac{\partial \mathcal{L}}{\partial(\partial_0 \psi)}, \qquad (16.17)$$

which is extended to the *covariant momentum densities* $\pi^\mu = \{\pi^0, \pi^k\}$ using,

$$\pi = \frac{\partial \mathcal{L}}{\partial(\boldsymbol{\nabla}\psi)} \quad \Rightarrow \quad \pi^\mu = \frac{\partial \mathcal{L}}{\partial(\partial_\mu \psi)}. \qquad (16.18)$$

The configuration space *Hamiltonian density* of a field can be determined from the canonical relation,

$$\mathcal{H} = c\left(\frac{\partial \mathcal{L}}{\partial(\partial_0 \psi)} \partial_0 \psi - \mathcal{L}\right), \qquad (16.19)$$

which is also the ^{00} component of the energy-momentum tensor. The *Hamiltonian* H of an entire distribution is,

$$p^0 c = H = \int d^3 x\, \mathcal{H} = c \int d^3 x \left(\frac{\partial \mathcal{L}}{\partial(\partial_0 \psi)} \partial_0 \psi - \mathcal{L}\right), \qquad (16.20)$$

illustrating the loss of manifest covariance of any procedures based on the Hamiltonian. Hamilton's equations will have the same form, Equation 11.20, as in the Galilean field case, and cannot be written in covariant form. As in the case of the relativistic particle, we can expect the naïve procedures for the transition to the Hamiltonian formalism to fail for a singular Lagrangian.

16.3 Klein-Gordon field

The simplest relativistic field is a 4-scalar $\phi = \phi(x)$. If we confine attention to first and second-order, the only non-trivial un-sourced linear Poincaré-covariant equation, with constant coefficients, that ϕ may satisfy is,

$$(\Box - a)\phi = 0 \quad \text{or} \quad (-\tfrac{1}{c^2}\partial_t^2 + \boldsymbol{\nabla}^2 - a)\phi(t, \mathbf{x}) = 0, \qquad (16.21)$$

where a is a constant with dimensions of inverse length squared.

16.3. Klein-Gordon field

Exercise 16.1 *Give a number of reasons why a first-order covariant field equation for $\phi(x)$ is of little or no physical interest.*

Equation 16.21 is referred to as the *homogeneous wave equation* for ϕ, and arises very naturally in Einsteinian relativity where the invariant velocity constant c is available. The most striking difference between this equation and many of those that appear naturally in non-quantum Newtonian physics, namely those of Newtonian gravity, electrostatics and magnetostatics, is the existence of oscillatory solutions interpretable as plane waves. From plane waves one may construct more complicated wave solutions.

In a relativistic quantum physics context, the constant a can arise from the invariant rest mass m of a massive particle according to $a = 1/\lambdabar^2$ where $\lambdabar = mc/\hbar$ is the *reduced Compton wavelength* of the particle. Equation 16.21 then becomes the relativistic equivalent of the free Schrödinger equation for a single component (spinless particle) wave function, namely the free *Klein-Gordon equation*, $(\Box - m^2 c^2/\hbar^2)\phi(x) = 0$. In natural units ($\hbar = c = 1$, see Chapter 6) this has the form,

$$(\Box - m^2)\phi(x) = 0 . \tag{16.22}$$

This equation was considered by Schrödinger [498] in 1925 during the formulation of the wave mechanics of the hydrogen atom but rejected in favour of the non-relativistic equation for reasons we now know to be related to the difference between spin 0 and spin $\frac{1}{2}$ systems. It was also published independently by Klein[3] [284] and Gordon[4] [207] and several others. As is customary in quantum field theory, equations will often be simplified throughout Part IV by using natural units. Nevertheless, practical units will also be used where convenient. The reader unfamiliar with natural units is urged to convert all equations in one system to those in the other until the conversion process is completely familiar.

With no loss of generality, let us consider the Klein-Gordon equation since, by allowing m to be an inverse length, we may retrieve the classical equivalents from it. (We may even retrieve negative values of a by allowing m to be pure imaginary although m will not then be interpretable as a mass for the field ϕ.)

The *spin* of a massive particle, whether classical or quantum, is a property arising from non-trivial characteristics of the system under rotation. Since a scalar field has only one component, it can have no such properties and is said to have zero spin, a property shared by the particles

---
[3] Oskar Klein (1894–1977), Swedish physicist.
[4] Walter Gordon (1893–1940), German physicist.

it gives rise to in quantum field theory. There are no observed fundamental particles of zero spin. The spin 0 pions for example, are composites, $\pi^+ = u\bar{d}$, $\pi^- = \bar{u}d$ and $\pi^0 = (u\bar{u} - d\bar{d})/\sqrt{2}$, of a quark and an antiquark, each of spin $\frac{1}{2}$, with antiparallel spins. The kaons are particles with the gauge charge called *strangeness* being composites, $K^+ = u\bar{s}$, $K^- = \bar{u}s$, $K^0 = d\bar{s}$ and $\bar{K}^0 = \bar{d}s$, in each of which is a strange quark s or $\bar{s}$ The quanta of a scalar field appear in the form of the *Higgs particle* [7, 248] in the *electroweak* theory of the partial unification of the electromagnetic and weak-nuclear interactions. However, no direct evidence has yet been obtained that the Higgs particles exist.

The scalar field is the prototype for many of the properties of the other fields and their corresponding quanta. Since it is free of the polarization effects of non-zero spin and of the gauge freedom which occurs for massless fields of spin ≥ 1 we shall illustrate some of the field theory of bosonic (integer spin) fields using the scalar field as an example.

16.4 Real scalar field

We seek a Lagrangian from which the Klein-Gordon equation for a real (or hermitian) scalar field $\phi(x)$ is derived using the Euler-Lagrange equations. The second-order nature of the field equation requires the Lagrangian to be quadratic in the covariant derivatives $\partial_\mu \phi$. Since the Lagrangian is a scalar in the field and its first derivatives this leads us to a kinetic contribution proportional to $\partial_\mu \phi \, \partial^\mu \phi$. The coefficient must be negative to give a positive contribution to the Hamiltonian and thus to the energy. We choose it to be a dimensionless constant by convention. This choice fixes the dimensions of ϕ to be those of inverse length, $[\phi] = \mathrm{L}^{-1}$, in natural units.

Since other integer spin fields, those comprised of vectors and tensors, are essentially sets of scalar fields transforming together with a particular polarization structure under the Lorentz group, they will all be given the same natural dimension of inverse length which we call the *canonical bosonic dimension* of quantum field theory.

If we note that the squared mass appearing in the Klein-Gordon equation has the natural dimension of inverse length squared, we see that the only scalar in ϕ by which it can be multiplied to give a term in the Lagrangian is ϕ^2. The coefficient of $m^2 \phi^2$ in this term is fixed by the requirement that the Lagrangian must lead to the Klein-Gordon equation on application of the Euler-Lagrange equations. By convention, we take the coefficient of $\partial_\mu \phi \, \partial^\mu \phi$ to be $-\frac{1}{2}$ and arrive at the following Lagrangian density for the free real scalar field,

$$\mathcal{L} = -\tfrac{1}{2} \partial_\mu \phi \, \partial^\mu \phi - \tfrac{1}{2} m^2 \phi^2 \ . \tag{16.23}$$

The quadratic nature of the terms in this Lagrangian density correspond to the linearity of the free Klein-Gordon field.

16.4. Real scalar field

To allow for *self-interaction* and an *external source* for the non-free scalar field, we may include a real potential $V(\phi(x), x)$ in the Lagrangian density,

$$\mathcal{L} = -\tfrac{1}{2}\partial_\mu \phi \, \partial^\mu \phi - \tfrac{1}{2}m^2\phi^2 - V(\phi) \,, \tag{16.24}$$

in which the sign is chosen so that positive V corresponds to positive energy. The arguments z (and t if not isolated) for a mechanical potential $V(z,t)$ are replaced here by the field $\phi(x)$ (and $x = \{t, \mathbf{x}\}$ if it is not isolated). The external nature of the source (see Chapter 6) will be reflected in the fact that, apart from the dependence on ϕ in V, the only dependence on x is explicit rather than via a dependence on other dynamical variables.

Exercise 16.2 *Show that the Euler-Lagrange equations for the Klein-Gordon Lagrangian density of Equation 16.24 are,*

$$(\Box - m^2)\phi = \frac{\partial V}{\partial \phi} \tag{16.25}$$

and that the Lagrangian density is non-singular.

Exercise 16.3 *(a) Show that Equation 16.15 for the canonical energy-momentum tensor (with $c = 1$), gives an energy-momentum tensor of,*

$$T^{\mu\nu} = \partial^\mu \phi \, \partial^\nu \phi - \tfrac{1}{2}\eta^{\mu\nu} \partial_\lambda \phi \, \partial^\lambda \phi - \tfrac{1}{2}\eta^{\mu\nu} m^2 \phi^2 - \eta^{\mu\nu} V(\phi) \,, \tag{16.26}$$

when applied to the scalar field. (b) Hence show that the energy density (and therefore the total energy) of the scalar field is positive-definite provided the potential $V(\phi)$ is positive-definite.

The canonical energy-momentum tensor of the Klein-Gordon field is itself symmetric, as required for angular momentum conservation. This is a result of it having only one component. The canonical tensors of other fields with non-zero spin will not be angular momentum conserving even if they turn out to be symmetric which is the case for the massless spin 2 Fierz-Pauli field, for example (see Chapter 21).

When the only dependence on ϕ in $V(\phi(x), x)$ is proportional to ϕ we write it as $j(x)\phi(x)$. This potential gives rise to a term j, independent of ϕ, in the field equation, resulting in the externally sourced Klein-Gordon equation,

$$(\Box - m^2)\phi(x) = j(x) \,. \tag{16.27}$$

Exercise 16.4 *Show that there is no local gauge freedom in the Klein-Gordon field ϕ satisfying Equation 16.27.*

The term $j = j(x)$ (of dimension L^{-3}) is the scalar source density. If the effect on the source j of the field ϕ can be neglected then the dynamics of the particles and fields in the source will not be of concern in a study of the ϕ field.

A potential proportional to ϕ^2 simply adds a contribution to the mass term and need not be considered separately. (A mass-squared term with a positive coefficient is of relevance to the theory of the Higgs field [168, 248] and to spontaneous breaking of the gauge invariance of the Yang-Mills fields with which it interacts.)

A potential which is cubic in ϕ such as $V(\phi) = \mu\phi^3/3!$ gives rise to the following non-linear field equation,

$$(\Box - m^2)\phi = \tfrac{1}{2}\mu\phi^2 \ . \tag{16.28}$$

Such a potential term in the Lagrangian cannot be regarded as a source term but does cause the propagation characteristics of the system to differ in essential ways from the free-field case. These effects are referred to as *self interaction* of the field ϕ and the parameter μ is called a *self-coupling constant*. The dimensions of the coupling constant are $[\mu] = L^{-1}$ which are also the dimensions of mass.

Exercise 16.5 *Show that the $\mu\phi^3$ theory is of little interest because it gives rise to an energy which is not bounded below and therefore not positive-definite.*

The absence of energy positivity violates a fundamental requirement of quantum field theory arising from a need for the lowest energy state, interpreted as the vacuum of the quantum field theory, to be stable.

If the potential is quartic, an interesting self-interacting scalar field theory arises. If we take $V(\phi) = \lambda\phi^4/4!$ we obtain a field equation,

$$(\Box - m^2)\phi = \lambda\phi^3/6 \ , \tag{16.29}$$

in which the self-coupling constant λ is dimensionless. The field equation is non-linear and the energy of the field ϕ is positive-definite for $\lambda > 0$. A common feature of the monomial potentials of even power is that their lowest energy corresponds to a zero value of the field ϕ and that zero value is a minimum. In quantum field theory, the lowest energy is defined to be the vacuum and is constructed to have no particles. Thus the vacuum *no particle state* of the ϕ^4 theory corresponds to a zero field.

Quantum field theories based on particles, in contrast to some based on strings or membranes, have a greater inherent tendency to produce quantities, such as the energy, which diverge. Measurements, on the other hand, always give rise to real, finite and single-valued quantities. If procedures exist to extract finite predictions for measurable quantities from a particle field theory, without removing all the divergences on all the intermediate quantities, then the theory is said to be *renormalizable*.

A simple criterion [262] for gaining a first indication, to be confirmed by detailed calculations, on whether an interacting field theory is likely to be renormalizable or not, involves the natural dimension of the coupling constant. If the coupling constant has the natural dimensions of a negative power of length then the theory may be referred to as *super-renormalizable*. The ϕ^3 theory falls into this category. If the coupling constant is dimensionless (in natural units), as in the case of the ϕ^4 theory, it should be renormalizable. Electrodynamics also falls into this category since its natural coupling constant is the magnitude e of the electric charge on an electron or proton which is dimensionless in natural units.

If the coupling constant has dimensions which are a positive power of L then these power counting techniques indicate that the interaction will be *non-renormalizable*. Self-interaction monomial potentials of higher than the 4th power in the field ϕ need not therefore be considered. The gravitational interaction also falls into the category of non-renormalizability since its coupling constant G (see Exercise 7.2) has dimensions of length squared in natural units. The same is true of the original Fermi theory [7] of the weak interaction (see Section 20.13).

In order to set up a theory in which some field system, rather than non-quantum particles, acts as the source of the electromagnetic field we must be able to construct, from that field system, a real, proper, single-valued, conserved current whose corresponding charge may be positive or negative. The conservation of the charge must follow automatically from the dynamics of the source field. It would also be natural if that conserved quantity arose as the Noether charge of a continuous symmetry of the source action. Noether currents are quadratic in the field quantities.

Although we can construct an hermitian current for the scalar field ϕ in the form $\frac{1}{i}\partial_\mu \phi$ it is not conserved as a result of the Klein-Gordon equation unless the mass is zero and it is not of the quadratic form required for it to be a Noether current. Furthermore, the real quadratic quantity $\frac{1}{i}\phi\partial_\mu\phi$ is also not conserved for all ϕ satisfying the Klein-Gordon equation.

The real scalar field may be used to describe uncharged particles, such as the electrically neutral pion π^0 but does not appear to have the necessary structure to function as the field corresponding to a charged source. We may suspect that a single scalar field is insufficient for the existence of the required continuous symmetry in the action and related to this could be its inability to describe both positive and negative charges. We shall therefore consider the properties of a pair of closely related real scalar fields.

16.5 Charged scalar field. Internal symmetries

The existence of spinless particle-antiparticle pairs such as the electrically charged pions $\pi^\pm$ and kaons $K^\pm$ and the electrically neutral kaons, K^0

and $\bar{K}^0$, whose principal differences are the opposite charges they carry, prompts us to consider a minimal extension from one real field to two real scalar fields ϕ_1 and ϕ_2. The charges involved in these three examples are electric charge for the $\pi^\pm$ and $K^\pm$ and one of the quark flavours, *strangeness*, for the electrically neutral kaons, K^0 and $\bar{K}^0$. Because of the observed symmetry between positive and negative charge (whether electric or strange) we wish to have these fields enter symmetrically in the theory.

Even at the free-field level, where both the fields ϕ_1 and ϕ_2 satisfy Equation 16.22, we can see the possibility of forming a real conserved current arising due to the use of the products $\frac{1}{i}\phi_1 \partial_\mu \phi_2$ and $\frac{1}{i}\phi_2 \partial_\mu \phi_1$ from which we may construct a non-trivial real combination $\frac{1}{i}(\phi_1 \partial_\mu \phi_2 - \phi_2 \partial_\mu \phi_1)$ which is also conserved provided the masses are identical, $m_1 = m_2 = m$.

In the interacting case, each will be required to satisfy a Klein-Gordon equation (with the same mass m) in which the other field may appear in the potential $V(\phi_1, \phi_2)$. The Klein-Gordon equations will be,

$$(\Box - m^2)\phi_1 = \frac{\partial V}{\partial \phi_1} \quad \text{and} \quad (\Box - m^2)\phi_2 = \frac{\partial V}{\partial \phi_2} . \quad (16.30)$$

Provided $\partial V/\partial \phi_1$ and $\partial V/\partial \phi_2$ are not functions of ϕ_1 or ϕ_2 alone then these are a pair of coupled equations describing the *mutual interaction* of ϕ_1 and ϕ_2 (or self interaction of the pair $\{\phi_1, \phi_2\}$), the details of which are dependent on the precise form of $V(\phi_1, \phi_2)$. We may derive these equations from the following Lagrangian,

$$\mathcal{L} = -\tfrac{1}{2}\partial_\mu \phi_1 \, \partial^\mu \phi_1 - \tfrac{1}{2}m^2 \phi_1^2 - \tfrac{1}{2}\partial_\mu \phi_2 \, \partial^\mu \phi_2 - \tfrac{1}{2}m^2 \phi_2^2 - V(\phi_1, \phi_2) , \quad (16.31)$$

consisting of the free Lagrangian densities of each field and the interaction term. To maintain energy positivity, and the symmetry between the two scalar fields, we must take the potential to be a real positive function $V(\phi_1, \phi_2)$, symmetric in ϕ_1 and ϕ_2, namely $V(\phi_1, \phi_2) = V(\phi_2, \phi_1)$. We now have a discrete invariance, called a Z_2 *symmetry* from the standard mathematical symbol for the finite group $\{\mathbb{1}, \mathbb{C}\}$ of order two, comprising the identity and the *charge conjugation* operation $\mathbb{C}$ interchanging the fields, $\mathbb{C} : \phi_1 \leftrightarrow \phi_2$. From an abstract group-theoretical standpoint, this 2-member set is identical to the Z_2 groups of $\{\mathbb{1}, \mathbf{P}\}$ or $\{\mathbb{1}, \mathbf{T}\}$ where $\mathbf{P}$ and $\mathbf{T}$ are the parity and time-reversal operations.

The next step is one of crucial importance for the physics of the theory which results. Instead of being satisfied with just the discrete Z_2 charge conjugation, namely charge reversal, between positive and negative charges, we consider the possibility of obtaining it as a by-product of a much stronger symmetry under continuous transformations on the fields ϕ_1 and ϕ_2, but not affecting the spacetime coordinate x. This permits the determination of a conserved current and charge by Noether's theorem and it may well be that the charge corresponds to one observed in nature.

16.5. Charged scalar field. Internal symmetries

Furthermore, suppose that we construct such a continuous symmetry with respect to some parameters ω^a which do not depend on spacetime location. It will be a global internal symmetry of the fields, not a spacetime symmetry involving transformations of $x = \{x^\mu\}$.

We need not stop at a continuous symmetry with parameters that do not depend on x. We can ask whether the parameters can be replaced by arbitrary functions $\omega^a(x)$ of the spacetime coordinates x and the field content of the system re-adjusted in such a way that it still has the extended continuous symmetry. The Relativity principle can be regarded as stating that external phenomena are observer independent — up to frame changes at constant velocity in special relativity — and for all motions in general relativity. Why not extend this observer independence to internal symmetries as well? The result of making that extension is the *Gauge principle* of interaction which will be examined in Chapter 20. The possibility of subsequently applying the Gauge principle is a second reason justifying the formation of a continuous symmetry to represent the interchange of ϕ_1 and ϕ_2.

We naturally seek the simplest possible continuous symmetry depending therefore on only one parameter α. Since we require the dynamics embodied by Equations 16.30 to be equivalent before and after the symmetry transformation, we demand that the transformations form a Lie group. The domain of the parameter must therefore be connected and the only connected, and therefore continuous, domains in 1D are the real line $\mathbf{R}$ or the circle $\mathbf{S}$. The internal continuous group on $\mathbf{R}$ is the 1D group T^1 of translations of the field values, namely $\phi \to \phi + a$ for $-\infty < a < \infty$. The internal continuous group on $\mathbf{S}$ is rotation of the the fields in a plane labelled by the values of ϕ_1 and ϕ_2, through an angle α (not dependent on x) according to,

$$\phi_1(x) \to \phi_1(x)\cos\alpha - \phi_2(x)\sin\alpha$$
$$\phi_2(x) \to \phi_1(x)\sin\alpha + \phi_2(x)\cos\alpha , \qquad (16.32)$$

where α is real and satisfies $0 \le \alpha \le 2\pi$ with 0 and 2π identified. Because the translations and rotations act on fields rather than on spatial coordinates they must be recognized as *internal* transformations of a quite different character to the *external* spacetime transformations.

Exercise 16.6 *Show that we need not consider an internal translation further since it cannot leave invariant the interacting scalar Lagrangian of Equation 16.31.*

However, the pair of kinetic terms together in Equation 16.31, and similarly the pair of mass terms together, are clearly invariant under the above rotation and the entire Lagrangian will also be invariant provided we ensure that the potential term is. The simplest way to do this and continue to respect energy positivity is to demand that the potential be a real function of $\phi_1^2+\phi_2^2$. We thus seek to obtain the conservation of charge, according to Noether's theorem, by representing the discrete symmetry between positive and negative charge by the continuous symmetry of the Lagrangian density of Equation 16.31 with respect to Equation 16.32. This has led us naturally to a system with two real fields related by the simplest Abelian compact internal Lie symmetry isomorphic to the 2D rotation group SO(2).

The internal 2-dimensional space $\{\phi_1, \phi_2\}$ in which the transformations of that symmetry act may be referred to as *charge space*. The Lagrangian that we have constructed is Poincaré-covariant. Since two scalar fields are involved, the representation of the Poincaré group ISO(1,3) is reducible; it divides into two parts, ϕ_1 and ϕ_2, each of which is separately invariant under the restricted Poincaré group. An internal symmetry must by definition commute with all external symmetries. An overall symmetry group of the Lagrangian is thus the direct product ISO(1,3)⊗SO(2) of the Poincaré group and the group of internal rotations, SO(2). Furthermore, the pair $\{\phi_1, \phi_2\}$ form an irreducible representation of the extended group.

Rather than deal with two real scalars, we choose to represent the above pair by one complex scalar field $\phi = (\phi_1 + i\phi_2)/\sqrt{2}$ and its complex conjugate $\phi^* = (\phi_1 - i\phi_2)/\sqrt{2}$ in terms of which the interacting Lagrangian of Equation 16.31 becomes,

$$\mathcal{L} = -\partial_\mu \phi^* \partial^\mu \phi - m^2 \phi^* \phi - V(\phi^* \phi) , \qquad (16.33)$$

where $V(\phi^*\phi)$ must be a real positive function of $\phi^*\phi = \frac{1}{2}(\phi_1^2 + \phi_2^2)$ and is therefore positive-definite. In terms of the new variables, the continuous symmetry of Equation 16.32 takes on the form of a U(1) *phase invariance*,

$$\phi(x) \to e^{i\alpha}\phi(x) \quad \text{and} \quad \phi^*(x) \to e^{-i\alpha}\phi^*(x) , \qquad (16.34)$$

consistent with the isomorphism between the U(1) and SO(2) symmetries. The Z_2 charge conjugation symmetry $\phi_1 \leftrightarrow \phi_2$ takes the form of the interchange $\phi \leftrightarrow \phi^*$. (We note that if ϕ is taken to be real, the resulting kinetic and mass terms will be double the correponding terms in the Lagrangian of Equation 16.24 for a real field.)

Exercise 16.7 *Verify that if ϕ^* and ϕ are regarded as independent variables, the complex conjugate Euler-Lagrange equations,*

$$(\Box - m^2)\phi = \frac{\partial V}{\partial \phi^*} \quad \text{and} \quad (\Box - m^2)\phi^* = \frac{\partial V}{\partial \phi} , \qquad (16.35)$$

correspond to their variation in Equation 16.33.

16.5. Charged scalar field. Internal symmetries

As a consequence of the hermiticity of the Lagrangian, the process of varying the two complex conjugate variables as if they were independent leads to the same result as variation of the independent real fields ϕ_1 and ϕ_2.

A field-theoretically sound interacting system of charged fields is obtained by taking the quartic self interaction with potential,

$$V = \frac{\lambda(\phi^*\phi)^2}{4!} = \frac{\lambda(\frac{1}{2}\{\phi_1^2 + \phi_2^2\})^2}{4!}, \quad (16.36)$$

with λ real. Because this ϕ^4 *theory* is renormalizable, has positive energy (for $\lambda > 0$) and a non-trivial Noether current it is widely used as a 'toy model' to illustrate quantum field theory calculations [467].

Suppose we now seek, as we did earlier for the real scalar field, a real 4-vector current that is conserved as a result of the new field equations. We now have available two real first-order 4-vector quadratics in ϕ, namely $\frac{1}{i}\phi^*\partial^\mu\phi$ and $\frac{1}{i}(\partial^\mu\phi^*)\phi$. We therefore consider:

$$j^\mu = \frac{1}{i}\{\phi^*\partial^\mu\phi + a(\partial^\mu\phi^*)\phi\}, \quad (16.37)$$

where a is an arbitrary real constant.

Exercise 16.8 *Show (a) that the current j^μ of Equation 16.37 is conserved, $\partial \cdot j = 0$, for all $V = V(\phi^*\phi)$, if and only if $a = -1$, and (b) that the Noether current corresponding to the continuous phase invariance of the complex scalar field Lagrangian is proportional to this current. (c) Verify that j^μ is real.*

The existence of a second scalar field, symmetrically and continuously related to the first, or equivalently, the replacement of a real scalar by a complex one, supplies a non-trivial real conserved *Klein-Gordon current*,

$$j = \tfrac{1}{i}\phi^*\overleftrightarrow{\partial}\phi = \tfrac{1}{i}\{\phi^*\partial\phi - (\partial\phi^*)\phi\}. \quad (16.38)$$

The Noether charge of the U(1) symmetry of complex or non-hermitian fields is appropriate for the description of any observed bipolar additive conserved quantity such as electric, leptonic, baryonic, strange, colour, charm, top and bottom charges.

Exercise 16.9 *Find the energy-momentum density of the complex scalar field and show that it is real and positive definite if V is positive.*

The demand that any Lagrangian be hermitian and Lorentz invariant ensures it transforms as a neutral scalar field. It is thus invariant under charge conjugation, a property extending to the energy-momentum tensor and 4-momentum obtained from it.

16.6 Free Klein-Gordon plane-wave expansion

In order to form an arbitrary free Klein-Gordon field, we require a basis of plane-wave solutions.

Exercise 16.10 *Verify that a plane wave $e^{\pm ik\cdot x}$ of propagation 4-vector $k^\mu = \{\omega_k, \mathbf{k}\}$ (namely of frequency $k^0 = \omega_k$ and propagation 3-vector $\mathbf{k}$) is a solution of the Klein-Gordon equation provided k satisfies $k^2 = -m^2$.*

The solutions of $k^2 = -m^2$ for $k^0 = \omega_k$ in terms of m and $\mathbf{k}$ are of both signs. Rather than including both values of k^0, for a given m and $\mathbf{k}$, we restrict k^0 to the positive value corresponding to a future-pointing propagation 4-vector which means the frequency is given by,

$$k^0 = \omega_k = \sqrt{\mathbf{k}^2 + m^2} \quad \text{where} \quad -\infty < k^j < \infty \,. \tag{16.39}$$

We must then include both exponent signs to form a basis for the solution space of free Klein-Gordon wave functions from the plane wave factors,

$$e^{\pm ik\cdot x} = e^{\pm i(-\omega_k t + \mathbf{k}\cdot\mathbf{x})} \,. \tag{16.40}$$

We need to expand an arbitrary scalar wave function in terms of plane waves. For given mass m, only three of the components $\{k^\mu\}$ are independent but we do not choose to expand $\phi(x)$ in terms of an integral $\int d^3k\, a(\mathbf{k})\ldots$ over the plane-wave functions using coefficients $a(\mathbf{k})$, since manifest covariance would be lost by the use of d^3k alone in such a procedure. Instead we first define the relativistically-invariant *bosonic momentum-space element $d\tilde{k}$* [262] according to,

$$d\tilde{k} = \frac{d^3k}{(2\pi)^3 2k^0} \,, \tag{16.41}$$

in which the $1/16\pi^3$ factor is conventional. Eventually we must also integrate over 3-volume to obtain quantities such as energy from the field $\phi(x)$ and this may be carried out covariantly using the covariant phase-space combination $d\Sigma_\mu d\tilde{k}$.

Exercise 16.11 *(a) Verify that d^3k/k^0 is a Poincaré scalar. (b) Show that,*

$$\int d\tilde{k}\, f(\mathbf{k}) = \int \frac{d^4k}{(2\pi)^4}(2\pi)\delta(k^2 - m^2)\theta(k^0) f(\mathbf{k}) \,, \tag{16.42}$$

where $\theta(t)$ is the step function:

$$\theta(t) = \left\{ \begin{array}{ll} 0 \,, & \text{for } t < 0 \\ 1 \,, & \text{for } t \geq 0 \end{array} \right\} \,, \tag{16.43}$$

and $f(\mathbf{k})$ is a smoothly-varying test function.

The automatic appearance of $\theta(k^0)$ in the 4D integral representation of the momentum-space element $d\tilde{k}$ (the range of which may include negative k^0) ensures that the propagation vector is future-pointing.

Any solution of the free Klein-Gordon equation, for a complex scalar field ϕ, may then be expanded covariantly in the form,

$$\phi(x) = \int d\tilde{k} \, [a(\mathbf{k})e^{ik\cdot x} + b(\mathbf{k})e^{-ik\cdot x}] \,. \tag{16.44}$$

The coefficients $a(\mathbf{k})$ and $b(\mathbf{k})$ in Equation 16.44 are complex numbers. If the field is real then we must take $b = a^\dagger = a^*$ where the asterisk here indicates complex conjugation, to which the hermitian conjugation ($\dagger$) reduces for fields with only one component.

16.7 Relativistic Klein-Gordon wave mechanics

Suppose we now consider the formation of Klein-Gordon quanta by treating $\phi(x)$ as a single-particle wave function. In a wave-mechanical context, where the 4-momentum operator has the form $P_\mu = \frac{\hbar}{i}\partial_\mu$, the Klein-Gordon equation is a quantum-mechanical expression, $P^2\phi = -m^2c^2\phi$, of the mass shell constraint on the magnitude of the classical 4-momentum p^μ of a non-quantum particle of mass m.

The standard *first quantization* technique imposes commutation rules (compare Equation 11.12 or 15.58),

$$[X^k, P^l] = i\hbar\delta^{kl} \,, \tag{16.45}$$

on the position and momentum operators $\mathbf{X}$ and $\mathbf{P} = \frac{\hbar}{i}\boldsymbol{\nabla}$ in the position representation. The one-particle quantum-mechanical Hamiltonian, namely $H = P^0c = i\hbar\partial_t$ acts on the plane-wave contributions to give,

$$P^0(e^{\pm ik\cdot x}) = \pm\hbar\omega_k e^{\pm ik\cdot x} \,. \tag{16.46}$$

Consequently, any term in $\phi(x)$ containing a factor of $e^{ik\cdot x} = e^{-i\omega_k t}e^{i\mathbf{k}\cdot\mathbf{x}}$ is referred to as being of *positive frequency* while those involving factors of $e^{-ik\cdot x} = e^{i\omega_k t}e^{-i\mathbf{k}\cdot\mathbf{x}}$ are said to be of *negative frequency*. In the one-particle first-quantized wave mechanics being considered here, such terms give rise to energy contributions of opposite sign. This prevents the system from having a lower bound on the energy and precludes the possibility of a stable ground state.

This is a fundamental deficiency of relativistic wave mechanics [262] which also occurs for other types of waves such as those for an electromagnetic vector potential (spin 1) or a Dirac spinor (spin $\frac{1}{2}$). These and other problems of the one-particle theory are cured by the multiparticle *second quantization* techniques of quantum field theory. The terms 'positive' and 'negative energy' are sometimes used not only in wave mechanics, but also in quantum field theory, to denote the above two types of plane-wave contribution, despite the fact that energy must be and is always positive.

Quantum theory is based on conserved transition probability and the corresponding conserved currents. In relativistic wave mechanics, a conserved probability density must be the time component of a divergence-free 4-vector probability current. For the scalar field, that conserved 4-current is given by the Klein-Gordon current of Equation 16.38.

Exercise 16.12 *Show that the positive- and negative-frequency contributions to the probability density $\rho = j^0$ of the Klein-Gordon current are of equal magnitude but of opposite sign.*

The existence of independent negative contributions (which could appear alone) in relativistic spin 0 wave mechanics prevents ρ from being interpreted as a probability density. These contributions of opposite signs can easily be traced to the presence of time derivatives in the Klein-Gordon probability density which are in turn caused by the Klein-Gordon equation being of second-order rather than first. In contrast, the Schrödinger and Dirac equations have only first-order time derivatives and this leads to a positive-definite conserved one-particle probability density which in the Schrödinger case is $\psi^*\psi$ (or $\psi^\dagger\psi$ for a multi-component wave function of non-zero spin). A relativistically covariant wave equation will have all derivatives of the same order. A non-relativistic equation will not be so constrained by either rotational covariance or by the larger Galilean covariance.

This difficulty of the Klein-Gordon theory was central to the search in the mid-1920s for equations to replace the Klein-Gordon equation and contributed to Dirac's discovery [106] in 1928 of the Dirac equation which is first-order in all derivatives. Paradoxically, the Dirac equation did not in fact solve this problem which, like the 'negative-energy' problem, was instead resolved by quantum field theory. In fact, the Dirac equation suffered from the same defects at the one-particle wave-mechanical level. Instead, the discovery of the Dirac equation provided the key to the wave equations of particles of half-odd-integer spin, a concept which was unknown prior to the discovery of intrinsic spin in quantum phenomena in the early 1920s.

16.8 Klein-Gordon Hamiltonian

The transition from a Lagrangian density to a phase space Hamiltonian density is straightforward for the non-singular Klein-Gordon field.

16.8. Klein-Gordon Hamiltonian

Exercise 16.13 Using π to denote the canonical momentum π^0, show that,

$$\mathcal{H} = \pi^*\pi + \nabla\phi^*\cdot\nabla\phi + m^2\phi^*\phi + V(\phi^*\phi) \,, \tag{16.47}$$

gives the Hamiltonian density of the complex Klein-Gordon field.

The fundamentally non-covariant nature of the Hamiltonian procedure is evident from the above expression in which the time and space parts of the covariant momenta do not appear in the covariant combination, $\pi^*_\mu \pi^\mu$.

Exercise 16.14 Determine the expressions for the energy-momentum p^μ and angular momentum $j^{\mu\nu} = l^{\mu\nu}$ of the scalar field and verify that they satisfy the Lie algebra of the Poincaré group corresponding to Equations 15.59 to 15.61 or their quantum analogues, Equations 15.62 to 15.64. (The calculation may be carried out using a real or complex ϕ which is either a classical or quantum field.)

The common action functional formalism for fields and non-quantum particles facilitates the description of the interaction of two such strikingly different systems. Furthermore, a great deal of the classical Hamiltonian and Lagrangian description of fields may be immediately re-interpreted and extended as part of a quantized field description of particles, although a great many new features must also be taken into account in that case.

An advantage of the Lagrangian formalism is the ease with which covariance is incorporated by demanding that the action and the Lagrangian be scalars. It also provides direct access to the modern path-integral techniques of quantization. In contrast, the closely related Hamiltonian method provides a relatively easy route to quantization by canonical techniques, especially at the unconstrained non-relativistic level. This can be achieved by replacing Poisson brackets of classical variables by commutators of the corresponding quantum operators. However, there is no comparably simple way [110] to make a given Hamiltonian relativistic. Covariance must be checked at each stage in the canonical quantization of relativistic fields.

Proper scalar (S) and pseudoscalar (P) fields may be constructed from an arbitrary spin 0 Klein-Gordon field ϕ by applying the parity operator $\mathbf{P}$ according to $\phi_S = (\mathbb{1} + \mathbf{P})\phi/\sqrt{2}$ and $\phi_P = (\mathbb{1} - \mathbf{P})\phi/\sqrt{2}$, respectively. The intrinsic parities of particles arising from the quantization of bosonic fields were discussed in 1938 by Kemmer [275]. Ferretti [157] suggested in 1936 that the intrinsic parity (of a pseudoscalar particle) could be determined experimentally and Panofsky et al. [420] demonstrated that the pion had an intrinsic parity of -1 in 1951. The spin $s = 0, \frac{1}{2}, 1, \ldots$, and intrinsic parity $P = \pm 1$ of a particle are often given together using the notation s^P. For example, the pion spin-parity is 0^{-1}.

Bibliography:

Barut A O 1964 *Electrodynamics and the classical theory of fields* [29].
Corson E M 1953 *Introduction to tensors, spinors and ... wave equations* [80].
Itzykson C and Zuber J-B 1980 *Quantum field theory* [262].
Landau L D and Lifschitz E M 1980 *The classical theory of fields* [306].
Ryder L H 1985 *Quantum field theory* [488].
Soper D E 1975 *Classical field theory* [516].

CHAPTER 17

Relativistic scalar gravity

> In a memoir I published four years ago, I tried to answer the question whether the propagation of light is influenced by gravitation. I return to this here, because my previous presentation of the subject does not satisfy me, and for a stronger reason, because I now see that one of the most important consequences of my former treatment is capable of being tested experimentally [132].
>
> Albert Einstein   1879–1955

    A number of features of Newtonian gravity violate the Poincaré covariance principles which it is known must replace the approximate Galilean relativity. We shall examine here the minimum changes required to correct some of these defects and then confront the predictions of the resulting relativistic scalar theory with data from the classic experimental tests of gravitation.

    Newtonian gravity theory is a classical theory expressed in terms of a 3-scalar field. In considering covariant alternatives we consider only bosonic relativistic fields with commuting components chosen therefore from among those with integer spin, namely Lorentz scalars, vectors and tensors.

17.1   Graviscalar action of Nordstrøm

We shall start by considering the simplest of the covariant fields — the 4-scalar spin 0 field $\Phi(x)$ to which, for ease of comparison with Newtonian gravity, we shall assign the same non-canonical dimensions of [velocity^2]. An alternative canonical field, with natural dimensions of inverse length, is provided by $\Phi/\sqrt{4\pi G}$. For most solar system tests, the dynamics of the matter source of the field are not important and it may be treated as external. Since Newtonian gravity is adequate for weak, static fields and slow motion, we require our field equations and equations of motion to reduce to it under those conditions.

    The universality of gravitation (see Section 7.11) requires all matter to couple to the gravitational field with the same strength or coupling constant G. In a relativistic context this must also apply to the energy. Consequently, a scalar gravitational field will have a 'Klein-Gordon form' of field equation in which the right side is proportional to the scalar external source multiplied by the universal gravitational constant G.

The gravitational potential field at a large distance r from a static mass M may be deduced using Newtonian gravity, and is of long-range inverse-square Coulomb type $\Phi \approx 1/r$. However, the regular solutions of the massive 'Klein-Gordon' equation, 16.21, with $a = 1/\lambda^2 \neq 0$, in the vacuum regions far from a static source, are of *Yukawa* type, $\Phi \approx e^{-r/\lambda}/r$ where λ is the range.

Some evidence has accumulated in support of Yukawa contributions to gravity at intermediate distances [397, 417, 519] (see Section 7.14). However, to first consider standard long-range gravity, we select a field propagating with null 4-momentum ($k^2 = 0$ or $a = 0$ in Equation 16.21), referred to as a *null* or *massless* field. We thereby also eliminate the possibility of finding a rest frame for the field component of the gravitating system whose fluctuations, *scalar gravitational waves*, would propagate at the ultimate speed c, the speed of light.

The field equation for scalar gravity must therefore be of the form,

$$\Box \Phi = G \text{ (scalar source density)} , \qquad (17.1)$$

in which constancy of G is required for universality of the interaction. From the Newtonian theory, we know that the Galilean invariant (non-zero) rest mass acts as an integrated source of gravity. The relativistic source of gravity will therefore be the mass-energy, which is not a Poincaré-invariant, but the zeroth component of a 4-vector. Covariantly expressed, the integrated relativistic source of gravity is the 4-momemtum p^μ of the source system. As the source term of a field equation, we must use the local density of matter and energy in the form of the energy-momentum tensor $T^{\mu\nu}(x)$ which is integrated over space to give p^μ according to Equation 15.97.

We cannot allow derivatives to appear in the source term since we require the covariant gravitational equations to reduce to Newtonian form under the appropriate conditions, namely for slow speeds and weak, static fields. Consequently, the only scalar source we can construct from the energy-momentum tensor is its trace $T = T^\mu{}_\mu$ as used by Nordstrøm [400] in 1913 prior to the presentation of Einstein's general theory. Nordstrøm's theory was re-examined with new insights on the relation between the special and general theories of relativity in 1959 and 1961 by Thirring [533, 534].

In the Newtonian limit of a source with small velocities $u \ll c$, the magnitudes of the components of the energy-momentum tensor are limited by,

$$\left|\frac{T^{kl}}{T^{00}}\right| \sim \frac{u}{c} \left|\frac{T^{k0}}{T^{00}}\right| \sim \frac{u^2}{c^2} , \qquad (17.2)$$

17.1. Graviscalar action of Nordstrøm

and the energy-momentum tensor effectively reduces to,

$$(T^{\mu\nu}) = \begin{pmatrix} \rho c^2 & 0 \\ 0 & 0 \end{pmatrix} = \begin{pmatrix} Mc^2\delta(\mathbf{x}) & 0 \\ 0 & 0 \end{pmatrix}. \tag{17.3}$$

Its trace T reduces to $-\rho c^2$ while the wave operator $\Box$ becomes ∇^2. For the covariant graviscalar equation to have the correct Newtonian limit, it must have the form,

$$\Box \Phi = -\frac{4\pi G}{c^2} T. \tag{17.4}$$

This equation for an externally-sourced scalar gravitational field should be compared with Equation 18.22 for the wave equation of the electromagnetic potential in terms of a prescribed current source J^μ.

We need to determine, or choose, an equation for the motion of a test particle in the gravitational field Φ. We shall do this by setting up a closed system comprising the gravitational field $\Phi(x)$ and particles with dynamical variables $z^\mu(\lambda)$ where λ is an evolution parameter for the trajectories (the individual particle labels being suppressed). The action of such a system,

$$S = \int d^4x\, \mathcal{L}_{\text{field}} + \int d\lambda\, L_{\text{part}} + \int d\lambda\, L_{\text{intn}}, \tag{17.5}$$

will be made up of three parts, the free field contribution with scalar Lagrangian $\mathcal{L}_{\text{field}} \sim -\partial_\mu \Phi\, \partial^\mu \Phi$, the free particle part with Lagrangian L_{part}, given in an arbitrary λ gauge by the einbein Lagrangian of Equation 15.40, and an interaction term L_{intn}.

All terms in the action must be Poincaré scalars. Each free part of the action must contribute positively to the energy since the interacting action may also describe free propagation in certain regions of spacetime. Dimensional arguments show that $\mathcal{L}_{\text{field}}$ will be equal to $-\partial_\mu \Phi\, \partial^\mu \Phi / Gc$ apart from a dimensionless constant which must be positive for Φ to contribute positively to the energy. Only the 1st and 3rd terms will contribute to the Euler-Lagrange equations of the field and, in order that the source in that equation is the trace T of $T^{\mu\nu}$, the interaction term must contain $T(x)\Phi(x)$ as a factor. On the trajectory we must therefore have a factor of $c \int d\lambda\, e^{-1}(\lambda) \delta^4(x - z(\lambda)) \dot{z}^2 \Phi(z)$ in the interaction Lagrangian where we have used Equation 15.94 for the energy-momentum tensor $T^{\mu\nu}$ of the particle.

For massive and massless test particles, the λ gauge may be fixed by setting the einbein $e = e(\lambda)$ to m^{-1} or 1, respectively, after the variation is complete. Only the 2nd and 3rd terms in the action will contribute to the Euler-Lagrange equations of the particles and we may use the fact that the result must reduce to the Newtonian limit to fix the overall constant in the interaction term.

We now quote the action which incorporates these criteria, namely,

$$\begin{aligned}S &= S_{\text{field}} + S_{\text{part}} + S_{\text{intn}} \\ &= -\frac{1}{8\pi G c}\int_{\Sigma_i}^{\Sigma_f} d^4x\, \partial_\mu \Phi\, \partial^\mu \Phi + \tfrac{1}{2}\int_{\lambda_i}^{\lambda_f} d\lambda\, [e^{-1}(\lambda)\dot{z}^2 - e(\lambda)m^2c^2] \\ &\quad + \tfrac{1}{c^2}\int_{\lambda_i}^{\lambda_f} d\lambda\, e^{-1}(\lambda)\dot{z}^2 \Phi(z(\lambda))\;.\end{aligned} \qquad (17.6)$$

For a canonical field, $\Phi/\sqrt{4\pi G}$, the coupling constant, $\sqrt{4\pi G}$ will appear in the numerator of the interaction term. In order to verify the field equation we rewrite the last term in the alternative form:

$$\tfrac{1}{c^2}\int_{\Sigma_i}^{\Sigma_f} d^4x \int_{\lambda_i}^{\lambda_f} d\lambda\, e^{-1}(\lambda)\delta^4(x - z(\lambda))\dot{z}^2 \Phi(x)\;.$$

Exercise 17.1 Show that the Euler-Lagrange equations of the terms in the Lagrangian depending on $z^\mu(\lambda)$ and $e(\lambda)$ in the above action, namely,

$$L_{\text{part}} + L_{\text{intn}} = \tfrac{1}{2}e^{-1}(\lambda)\left(1 + \frac{2\Phi}{c^2}\right)\dot{z}^2 - \tfrac{1}{2}e(\lambda)m^2c^2\;, \qquad (17.7)$$

are,

$$\dot{\pi}_\mu = \frac{d}{d\lambda}\left[e^{-1}(\lambda)\left(1 + \frac{2\Phi}{c^2}\right)\dot{z}_\mu\right] = e^{-1}(\lambda)(\partial_\mu \Phi)\frac{\dot{z}^2}{c^2}\;. \qquad (17.8)$$

and,

$$\left(1 + \frac{2\Phi}{c^2}\right)\dot{z}^2 = -m^2c^2 e^2(\lambda)\;. \qquad (17.9)$$

For $m \neq 0$ we select the λ gauge with $e(\lambda) = m^{-1}$ to obtain a gauge condition,

$$\left(1 + \frac{2\Phi(z(\lambda))}{c^2}\right)\dot{z}^2 = -c^2\;, \qquad (17.10)$$

and a corresponding equation of motion,

$$\dot{\pi}_\mu = \frac{d}{d\lambda}\left[\left(1 + \frac{2\Phi(z(\lambda))}{c^2}\right)\dot{z}_\mu\right] = (\partial_\mu \Phi)\frac{\dot{z}^2}{c^2}\;, \qquad (17.11)$$

which is independent of the mass of the particle and therefore satisfies the Weak equivalence principle.

In the limit of a weak field ($|\Phi|/c^2 \ll 1$), the constraint on the evolution parameter of a massive particle first reduces it to the proper time τ and then, for slow speeds ($v^2/c^2 \ll 1$), to the coordinate time t. The equation of motion then reduces correctly for static fields ($\partial_0 \Phi \ll \nabla \Phi$) to the Newtonian expression $\ddot{z} = -\nabla \Phi$ justifying our choice for the coefficient

17.1. Graviscalar action of Nordstrøm

of the interaction term. The contributions of the free field and particle to the energy-momentum tensor are available from Equations 15.94 and the canonical equation, 16.26, which for a spin 0 field is the entire energy-momentum density.

Positivity of energy of each of the two free parts, a necessary condition for any classical field theory that one may wish to quantize, has determined the gravitational interaction via a scalar field to be attractive. This feature of scalar gravity also applies to any other tensor field theory of even spin [264], in particular for spin 2 gravity as will be apparent in Chapter 21. In general relativity, the gravitational field equations themselves suffice to guarantee the positivity of energy [591] and thus also the attractiveness of gravity.

Exercise 17.2 *Noting Equation 15.94, show that the Euler-Lagrange equations for the field terms of Equation 17.6 is given as required by,*

$$\Box \Phi = -\tfrac{4\pi G}{c^2} c \int d\lambda\, e^{-1}(\lambda) \delta^4(x - z(\lambda)) \dot{z}^2 = -\tfrac{4\pi G}{c^2} T\ , \qquad (17.12)$$

and show that the interacting energy-momentum density is,

$$\begin{aligned}
T^{\mu\nu}(\Phi) &= T^{\mu\nu}_{\text{field}} + T^{\mu\nu}_{\text{part}} + T^{\mu\nu}_{\text{intn}} \\
&= \tfrac{1}{4\pi G}(\partial^\mu \Phi\, \partial^\nu \Phi - \tfrac{1}{2}\eta^{\mu\nu} \partial^\lambda \Phi\, \partial_\lambda \Phi) \\
&\quad + c \int_{-\infty}^{\infty} d\lambda \left(1 + \frac{2\Phi(z)}{c^2}\right) e^{-1}(\lambda) \delta^4(x - z(\lambda)) \dot{z}^\mu \dot{z}^\nu\ , \quad (17.13)
\end{aligned}$$

by verifying that it is conserved everywhere as a result of the interacting field equations and equation of motion.

Although we may construct a conserved energy-momentum tensor, the conservation of the source is not automatically guaranteed by a differential identity of the field as we shall see is the case for the electrodynamic interaction and for the Fierz-Pauli spin 2 and general relativistic actions of relativistic gravity. This is related to the fact that, in contrast to those theories, relativistic scalar gravity admits no gauge freedom (involving an arbitrary function) in the gravitational field Φ. The fact that Equation 17.13 is not a sum of two separate field and particle contributions is a characteristic of gravitation treated relativistically. A related property is the non-localizability of gravitational energy density and the consequent impossibility [384] of constructing a covariant expression for the total energy-momentum density of field and matter.

Exercise 17.3 *Show that* $\Box \Phi$ *and* $(1 + 2\Phi/c^2)\dot{z}^2$ *are invariant with respect to the combined internal and external transformation,*

$$\Phi(x) \ \to\ \Phi(x) + \Phi_0 + a \cdot x \qquad (17.14)$$

$$x^\mu \;\to\; x^\mu\!\left(1 - \frac{\Phi_0}{c^2}\right) - \frac{\Phi_0}{c^2} x^\lambda x^\nu (\delta^\mu{}_\lambda a_\nu + \delta^\mu{}_\nu a_\lambda - \eta_{\nu\lambda} a^\mu) \;,\quad (17.15)$$

in which Φ_0 and a are constant. Hence conclude that the interacting scalar action is covariant (quasi-invariant) with respect to this transformation.

Although this invariance is not quite a gauge freedom, since only two parameters are involved rather than an arbitrary function, it shows that the coordinates x^μ (or z^μ for the particle trajectory) do not, in the presence of scalar gravity, combine with the Minkowski tensor $\eta_{\mu\nu}$ to provide invariant spacetime intervals for interpretation as proper time and length intervals based on real clocks. The first constant in the above transformation shows that a uniform gravitational potential Φ_0, with a metric tensor $(1+2\Phi_0/c^2)\eta_{\mu\nu}$ and coordinates x^μ, is equivalent to a zero gravitational potential with coordinates obtained by a change of frame, $x \to x(1-\Phi_0/c^2)$. The transformation involving the second constant $a = \{a^\mu\}$ similarly shows that a weak constant gravitational field of acceleration $-a$ in the z direction is locally equivalent to a change, $\mathbf{x} \to \mathbf{x} - \tfrac{1}{2}\mathbf{a}t^2$, to a non-inertial frame with acceleration $\mathbf{a}$ in the z direction in the absence of a gravitational field. Relativistic scalar gravity therefore partially embodies the Strong or Einstein equivalence principle (see Chapter 21).

In the proper frame of an observer with 4-velocity u, a free particle of momentum p has conserved invariant energy $E = -p\cdot u$. Correspondingly, in such a frame, a particle interacting with a field has conserved conjugate momentum π_μ and invariant conserved energy,

$$E = -\pi_\mu u^\mu = -\pi\cdot u \;. \quad (17.16)$$

17.2 Gravitational time dilation or red shift

Let us suppose that we have two identical particle clocks which have been calibrated to read the same rate while at rest with respect to one another and in close proximity compared to the characteristic lengths over which the gravitational field changes. Let the two clocks be fixed at $\mathbf{x}$ and $\mathbf{x}+\Delta\mathbf{x}$ in the local inertial frame with Minkowskian coordinates $\{ct,\mathbf{x}\}$ in which the static source, of either a Newtonian potential or a relativistic scalar field Φ, is also fixed.

The measured effect of gravitational fields on the periods P of particle clocks (see Chapter 7) can by summarized, for small changes $\Delta\Phi$ in the Newtonian gravitational potential, by $-(\Delta P/P)_{\text{clock}} = \alpha\Delta\Phi/c^2$ with α very close to 1. The measurements are carried out by counting the periods of the particle clocks over a given coordinate time on clocks which have been previously synchronized in rate by exchanging constant velocity signals with a master clock. Since it is $(1+2\Phi/c^2)\dot{z}^2$ which is invariant and equal

17.2. Gravitational time dilation or red shift

to $-c^2$, we distinguish the Minkowskian line element and the new invariant line element, for which we use ds^2, according to,

$$ds^2_{\text{Mink}} = \eta_{\mu\nu} dx^\mu dx^\nu \quad \text{and} \quad ds^2 = (1 + \frac{2\Phi(x)}{c^2})\eta_{\mu\nu} dx^\mu dx^\nu \;, \tag{17.17}$$

In special relativity, in the absence of gravitation, the proper time interval is $d\tau_{\text{Mink}} = \sqrt{-ds^2_{\text{Mink}}}/c$. In the presence of the graviscalar field Φ, use of coordinates x^μ, implies we must define the proper time of timelike intervals according to $d\tau = \sqrt{-ds^2}/c = \sqrt{(1 + 2\Phi(z)/c^2)}d\tau_{\text{Mink}}$, which reduces to the previously used Minkowskian value at large separations from the source of Φ. For clocks at fixed spatial coordinates, so that $d\tau_{\text{Mink}} = dt$, we obtain a *gravitational time dilation* formula of,

$$d\tau = \sqrt{1 + \frac{2\Phi}{c^2}}\, dt \;. \tag{17.18}$$

If we now consider a null ray ($p^2 = 0$) by putting the mass equal to zero and fixing the λ gauge to $e(\lambda) = 1$, or use $p^\mu = e^{-1}(\lambda)\dot{z}^\mu$ in Equation 17.8, then we find that the equation of motion is,

$$\dot{\pi}_\mu = \frac{d\pi_\mu}{d\lambda} = \frac{d}{d\lambda}\Big[\Big(1 + \frac{2\Phi}{c^2}\Big)p_\mu\Big] = 0 \;, \tag{17.19}$$

in which λ is no longer the proper time, even if the field is weak. The energy of the photon, measured on the coordinate clocks, will be $E = -\pi \cdot u$ where π is the conserved photon 4-momentum and u is the 4-velocity of the coordinate clocks, calculated with the new invariant proper time τ. Since the coordinate clocks are at fixed x, we have $u = \{u^k\} = 0$ and $u^0 = c\,dt/d\tau = c/\sqrt{1 + 2\Phi/c^2}$. Consequently, the energy of the photon is,

$$E = -\eta_{\mu\nu}\pi^\mu u^\nu = -\eta_{00}\pi^0 u^0 = \pi^0 c/\sqrt{1 + 2\Phi/c^2} \;. \tag{17.20}$$

Since π_μ is conserved and $E \propto P_\gamma$ where P_γ is the period of the ray measured on the coordinate clocks, we find that the ratio of the change $(\Delta P/P)_{\text{clock}}$ from x to x + Δx, in the period of a particle clock, due to gravitational time dilation, and the similar ratio $(\Delta P/P)_{\text{signal}}$ for the change in period of the null ray, due to the scalar gravitational field, both have magnitude $|\Delta\Phi|/c^2$, in accord with observation.

The factor $z = (\Delta P/P)_{\text{signal}}$, which for small z, equals $\Delta\Phi/c^2$, is referred to as the *gravitational red shift*. (This should not be confused with the *cosmological red shift* which is also gravitational in origin but is due to the expansion of the universe.) Such an effect, attributable to the differences $\Delta\Phi$ in the gravitational field, indeed occurs and was first predicted by Einstein [132] in 1911 during the period when he was constructing the general theory of relativity. Gravitational time dilation measurements by

Vessot et al. [555] show that observed small values of the red shift factor $z = (\Delta P/P)_{\text{signal}}$ agree with the small values of the ratio $\Delta\Phi/c^2$ of the change in the Newtonian gravitational potential to the speed of light squared to within 4×10^{-4}.

Since the relativistic gravitational potential reduces to the Newtonian quantity under appropriate conditions, we see that the relativistic scalar theory is consistent with observed gravitational red shifts. With this considerable improvement in precision, time dilation provides an invaluable test of the Strong equivalence principle, independent of the Eötvös-type experiments confirming the Weak equivalence principle. However, they do not distinguish between different gravity theories if those alternatives satisfy energy conservation principles, as we invariably demand they do, at least locally in asymptotically flat spacetimes. The different field equations of distinct theories are not needed to demonstrate consistency with $z = \Delta\Phi/c^2$ in Equation 7.36. Conservation laws and the Principle of equivalence suffice to establish the local expression for the time dilation without recourse to the field equations of any one theory.

Indeed, this result was predicted by Einstein [132] in 1911 on the basis of the conservation of energy, the Weak equivalence principle and mass-energy equivalence, prior to the presentation of the full theory of general relativity between 1911 and 1915. We can summarize the argument as follows. Allow a photon of energy E_l to be emitted vertically upward from a lower level in a uniform gravitational field of strength g. As it rises, let its energy change as a result of the effect of gravitation on the propagation of light so that it arrives at a level h higher with an energy E_u. Convert the energy to a particle of inertial rest mass $m_i = E_u/c^2$ and allow that particle to free fall to the original height, gaining energy $m_G g h = m_i g h$ where we have used both mass-energy equivalence and the Weak equivalence principle ($m_G = m_i$). By conservation, the new energy value $m_i c^2 (1 + gh/c^2)$ must be identical to the original energy E_l. Since the wavelengths of photons are inversely proportional to the energies, we have $\lambda_u/\lambda_l = E_l/E_u = 1 + gh/c^2$, which shows that the red shift factor is $z = \Delta\Phi/c^2$.

To further justify the use of ds^2 rather than ds^2_{Mink} for proper spacetime intervals, we could examine real clocks based on atomic processes. Atomic phenomena, such as those arising from the structure of the hydrogen atom, are governed by quantum physics and the electromagnetic interaction. An atomic clock based on the $1s$ (ground) state of the hydrogen atom, may for our purposes be treated semi-classically [256, 533] as a particle clock whose radius and period are given by the Bohr formulae, $a_\infty = 4\pi\epsilon_0 \hbar^2/m_e e^2$ and $P_\infty = 32\pi^3 \hbar^3 \epsilon_0^2/m_e e^4$. The subscripts on r_∞ and P_∞ imply they are applicable at large distances from a localized gravitational source where proper intervals coincide with coordinate intervals. These quantities are functions of the invariant physical parameters of the electron, namely its rest mass m_e and charge magnitude e, and the fundamental constants, $\hbar$, c and ϵ_0. The atomic length and time dimensions are the basis of our definitions of the units of time and length and are by definition invariant quantities.

We may couple the electron as a classical particle with variable $z(\lambda)$ to the electromagnetic action (see Equation 19.11) and obtain the Lorentz equation for the orbital motion of the electron about the nucleus and Maxwell's equations for the vector potential A^μ. We may then deduce the Bohr results in the usual way by applying the quantization *ansatz* of one unit of orbital angular momentum $\hbar$ in the ground state. We wish to examine the differences $\Delta P/P = (P_r - P_\infty)/P_\infty$ and $\Delta a/a = (a_r - a_\infty)/a_\infty$ between these proper time and length intervals and the corresponding intervals P_r and a_r measured in a local Minkowski frame, in the presence of a gravitational field, at finite distances r from the same source. To do so, we may add a term to the electrodynamic action to describe the gravitational field and a corresponding gravitational coupling to both the electron and the electromagnetic potential.

It suffices to use a scalar gravitational field for this purpose, and the resulting electron equation of motion in the coordinate frame is the same [256, 534] as applies in the absence of the gravitational field except that where e^2 and m_e appeared before, will now appear $e^2(1 + 2\Phi/c^2)$ and $m_e(1 - 3\Phi/c^2)$. The fractional differences will be $(\Delta P/P)_{\text{clock}} = -\Delta\Phi/c^2$ and $\Delta a/a = \Delta\Phi/c^2$. In the coordinate frame, clock periods will appear to be dilated and atomic lengths will appear to be contracted as a result of the gravitational field. A light ray passing from a region of strong gravity to lesser gravity will appear, in the coordinate frame, to be reddened by the same numerical magnitude, $z = (\Delta P/P)_{\text{signal}} = \Delta\Phi/c^2$.

17.3 Riemannian structure of scalar gravity

Since real clocks and length standards correspond to the invariant proper intervals obtained from ds^2 in Equation 17.17, the physical metric corresponding to the coordinates x^μ is the symmetric tensor $\mathbf{g} = g_{\mu\nu}\mathbf{e}^\mu \otimes \mathbf{e}^\nu$ in,

$$ds^2 = g_{\mu\nu}dx^\mu dx^\nu = (1 + 2\Phi/c^2)\eta_{\mu\nu}dx^\mu dx^\nu , \qquad (17.21)$$

The invariance of ds^2 means that the components $\{g_{\mu\nu}\}$ transform as those of a general tensor (see Chapter 21). Although we can set up a local Minkowski frame with metric $\eta = \{\eta_{\mu\nu}\}$ at any event, the spacetime intervals based on it and x^μ are not invariant in the presence of a gravitational field and are therefore unobservable. Calculations may be carried out in the coordinate frame x^μ using $\eta_{\mu\nu}$ for raising and lowering and with the inclusion of the necessary factors depending on Φ but the Minkowskian spacetime will be an unobservable background.

A space with invariant intervals based on a covariant general tensor $g_{\mu\nu}$ has intrinsic curvature originating from its metric tensor and is called a *Riemann space*. Because the present example is locally pseudo-Euclidean, the spacetime of relativistic scalar gravity is a *pseudo-Riemannian* manifold. If we consider only weak gravity ($|\Delta\Phi|/c^2 \ll 1$) and localized sources, the manifold at large distances will be Minkowskian, or *asymptotically flat*, namely $g_{\mu\nu} \to \eta_{\mu\nu}$ in coordinates which are rectangular Cartesian.

We shall continue to analyse the effects of scalar gravity with the aid of the Minkowskian spacetime background. We now obtain the forms taken by the equation of motion and gauge condition on the evolution parameter when expressed in terms of non-Cartesian coordinates. With a change from Cartesian coordinates $\{x^\mu\} = \{ct, x, y, z\}$ to spherical polar coordinates $\{ct, r, \theta, \phi\}$ the Minkowskian line element,

$$ds^2_{\text{Mink}} = -c^2 dt^2 + dx^2 + dy^2 + dz^2 \equiv \eta_{\mu\nu} dx^\mu dx^\nu , \qquad (17.22)$$

may be re-expressed in the form,

$$ds^2_{\text{Mink}} = -c^2 dt^2 + dr^2 + r^2 d\theta^2 + r^2 \sin^2\theta \, d\phi^2 \equiv g^{(f)}_{\mu\nu} dx^\mu dx^\nu , \qquad (17.23)$$

where the label (f) on the diagonal metric tensor $g^{(f)}_{\mu\nu}$ emphasizes that it is a metric for (flat) Minkowskian spacetime $\mathbf{E}^{1,3}$ in non-Minkowskian coordinates. The particle plus interaction Lagrangian in Minkowskian coordinates can be written as,

$$L = L_{\text{part}} + L_{\text{intn}} = \tfrac{1}{2} e^{-1}(\lambda)\left(1 + \frac{2\Phi}{c^2}\right)\frac{ds^2_{\text{Mink}}}{d\lambda^2} - \tfrac{1}{2} e(\lambda) m^2 c^2 , \qquad (17.24)$$

showing that, in non-Minkowskian coordinates, it will be,

$$L = \tfrac{1}{2} e^{-1}(\lambda)\left(1 + \frac{2\Phi}{c^2}\right) g^{(f)}_{\lambda\pi} \dot{z}^\lambda \dot{z}^\pi - \tfrac{1}{2} e(\lambda) m^2 c^2 , \qquad (17.25)$$

in which $\dot{z} = dz/d\lambda$ and some of the metric components $g^{(f)}_{\mu\nu}(x)$ will depend on the coordinates. The corresponding Euler-Lagrange equations now follow immediately and, on fixing the gauge to $e(\lambda) = m^{-1}$, gives,

$$\frac{d}{d\lambda}\left[\left(1 + \frac{2\Phi}{c^2}\right) g^{(f)}_{\mu\lambda} \dot{z}^\lambda\right] = \partial_\mu (\Phi g^{(f)}_{\lambda\pi}) \frac{\dot{z}^\lambda \dot{z}^\pi}{c^2} , \qquad (17.26)$$

subject to a gauge condition, arising from variation of $e(\lambda)$, namely

$$\left(1 + \frac{2\Phi}{c^2}\right) g^{(f)}_{\mu\nu} \dot{z}^\mu \dot{z}^\nu = -c^2 . \qquad (17.27)$$

In terms of the tensor version,

$$g_{\mu\nu} = \left(1 + \frac{2\Phi}{c^2}\right) g^{(f)}_{\mu\nu} , \qquad (17.28)$$

of the scalar field Φ, the particle Lagrangian will be,

$$L = \tfrac{1}{2} e^{-1}(\lambda)\, g_{\mu\nu} \dot{z}^\mu \dot{z}^\nu - \tfrac{1}{2} e(\lambda) m^2 c^2 \ . \tag{17.29}$$

The Euler-Lagrange equations with gauge-fixed λ are,

$$\frac{d}{d\lambda}(g_{\mu\lambda}\dot{z}^\lambda) = \tfrac{1}{2}(\partial_\mu g_{\lambda\pi})\dot{z}^\lambda \dot{z}^\pi \ , \tag{17.30}$$

with $\dot{z}(\lambda)$ satisfying the gauge condition,

$$g_{\mu\nu}\dot{z}^\mu \dot{z}^\nu = -c^2 \quad \text{or} \quad 0 \ . \tag{17.31}$$

The new metric tensor $g_{\mu\nu}$ is no more than a certain combination of the scalar field Φ and the Minkowski metric $\eta_{\mu\nu}$, plus a possible change of variable to non-Minkowskian coordinates.

17.4 Graviscalar light deflection

In passing through a changing gravitational field $\Phi(x)$, each component p_μ of the 4-momentum of a null ray will change, according to Equation 17.19. It is not the 4-momentum of the particle alone which is conserved, but the generalized momentum π_μ which takes account of the interaction with the field. However, since all components of p^μ change in proportion, there will be no change in the direction $\hat{p}$. Scalar gravity therefore fails to account for the observed gravitational deflection of null rays. The vanishing trace of the energy-momentum tensor of any null system means that relativistic scalar gravity, with a field coupled to the trace of the matter tensor, cannot couple to a null ray.

17.5 Linear graviscalar perihelion shift

For Mercury, with an orbit of eccentricity $e = 0.2056$, the perihelion advance not accounted for by Newtonian gravity, amounts to 42.56±0.94 arc seconds per century [384]. For the Hulse and Taylor [257, 530] neutron star pulsar, PSR 1913+16, the eccentricity is $e = 0.617155$ and the periastron advance is an enormous 4.226 degrees per year. Einstein started work on a relativistic theory of gravity immediately after presenting his special relativity theory [128] in 1905 and published his first results [131] on the influence of gravity on the propagation of light in 1907. In that same year, he wrote to Habitch [508] expressing his hope of accounting for the unexplained secular changes in the perihelion movement of Mercury.

Despite its failure to account for the gravitational deflection of null rays, such as light, we shall now examine the Kepler problem of a test particle of mass m (for example, a planet) in the graviscalar field of a much larger mass M (for example, the sun). The principal purpose of this section will be to set up the notation and procedures in a Poincaré-covariant manner using the linear graviscalar equations. Parts of the analysis will then be used in a similar procedure later when the field equations are improved in attempts to correctly account for the null deflection results.

For a static source of mass M, the solution of the graviscalar field equation is identical to the Newtonian potential $\Phi = -GM/r$ and the only change in the details of the orbit must come from a covariant analysis of the equation of motion. $\Phi g^{(f)}_{\lambda\pi}$ depends only on r and θ with t and ϕ being ignorable coordinates. Since $g^{(f)}_{\mu\lambda}$ is diagonal, the equation for θ expresses $\ddot{\theta}$ as a function of $\dot{\theta}$ in which all terms are proportional to $\dot{\theta}$ or $\ddot{\theta}$. If axes are chosen so that $\dot{\theta} = 0$ initially then $\ddot{\theta}$ is also zero initially and $\theta =$ constant confirming the motion is planar. We choose our axes with $\theta = \pi/2$. The time and azimuthal equations of motion are,

$$\left(1+\frac{2\Phi}{c^2}\right)\dot{t} = \text{constant} = k, \quad \text{and} \quad \left(1+\frac{2\Phi}{c^2}\right)r^2\dot{\phi} = \text{constant} = h \ . \quad (17.32)$$

Instead of using the radial equation directly (which would complicate the analysis considerably), we use the gauge condition of Equation 17.27 which is essentially the equivalent of constancy of the Hamiltonian or energy with respect to the new proper time. In spherical polar coordinates (with $\theta = \pi/2$) this condition becomes,

$$\left(1+\frac{2\Phi}{c^2}\right)(c^2\dot{t}^2 - \dot{r}^2 - r^2\dot{\phi}^2) = c^2 \ . \quad (17.33)$$

As is customary in the Kepler problem [282], we change variables from r to $u = 1/r$. Since we seek an orbit equation $r = r(\phi)$ or $u = u(\phi)$, rather than the solution for $r = r(t)$, we eliminate $\dot{r}$ using $\dot{r} = -\dot{\phi}u^{-2}du/d\phi$ and eliminate $\dot{\phi}$ with its equation of motion. The equations we have are therefore,

$$\left(1+\frac{2\Phi}{c^2}\right)^2\dot{t}^2 = k^2 \ , \quad (17.34)$$

$$\left(1+\frac{2\Phi}{c^2}\right)^2\dot{r}^2 = h^2\left(\frac{du}{d\phi}\right)^2 \ , \quad (17.35)$$

$$\left(1+\frac{2\Phi}{c^2}\right)^2 r^2\dot{\phi}^2 = h^2 u^2 \ . \quad (17.36)$$

Substituting into the gauge condition (effectively an 'energy' equation) gives the orbit equation,

$$\left(\frac{du}{d\phi}\right)^2 + u^2 = \frac{c^2(k^2-1)}{h^2} + \frac{2GMu}{h^2} \ , \quad (17.37)$$

17.5. Linear graviscalar perihelion shift

which we differentiate to obtain,

$$\frac{d^2u}{d\phi^2} + u = \frac{GM}{h^2} . \tag{17.38}$$

The non-relativistic Kepler motion of a particle satisfies precisely this equation in which $h = r^2 d\phi/dt$ is the conserved angular momentum per unit mass. In the present context, the angular momentum per unit mass is given by the second of Equations 17.32. The solution [282] is,

$$u = \frac{1 + e \cos\phi}{a(1 - e^2)} , \tag{17.39}$$

in which the eccentricity e and semi-major axis a of the the elliptical orbit of energy E (< 0) are given by,

$$e = \left(1 + \frac{2Eh^2}{G^2 M^2 m}\right)^{\frac{1}{2}} < 1 \quad \text{and} \quad a = \frac{h^2}{GM(1 - e^2)} . \tag{17.40}$$

Consequently, although the component velocities $\dot{r}$ and $r\dot{\phi}$ of the orbit will be affected by relativistic effects, we see that the shape of the orbit $u(\phi)$ will remain an ellipse and linear scalar gravity will not account for the observed advances of the perihelia of planets or periastra of binary stars.

In order to make use of the above analysis of the orbit equation in a situation where the gravitational field is different, we re-express Equations 17.34 to 17.36 in the form,

$$g_{00}^2 \dot{t}^2 = k^2, \quad g_{rr}^2 \dot{r}^2 = h^2 \left(\frac{du}{d\phi}\right)^2 \quad \text{and} \quad g_{rr}^2 r^2 \dot{\phi}^2 = h^2 u^2 , \tag{17.41}$$

leading to an orbit equation,

$$\left(\frac{du}{d\phi}\right)^2 + u^2 = -\frac{c^2 g_{rr}(g_{00} + k^2)}{h^2 g_{00}} , \tag{17.42}$$

which is identical to Equation 17.37 on using,

$$-g_{00} = g_{rr} = \left(1 + \frac{2\Phi}{c^2}\right) = 1 - \frac{2GMu}{c^2} . \tag{17.43}$$

However, in this more general form, it may be applied to any situation where the components of the gravitational field can be re-expressed in tensor form with *isotropic coordinates*, namely those for which $g_{\theta\theta} = r^2 g_{rr}$ and $g_{\phi\phi} = r^2 g_{rr}$ if $\theta = \pi/2$ or $g_{\phi\phi} = r^2 \sin^2\theta \, g_{rr}$ otherwise. The line element in such coordinates will be,

$$g_{\mu\nu} dx^\mu dx^\nu = g_{00}(dx^0)^2 + g_{rr}(dr^2 + r^2 d\theta^2 + r^2 \sin^2\theta \, d\phi^2) , \tag{17.44}$$

which should be contrasted with the line element of Minkowskian spacetime in spherical polar coordinates, namely,

$$g^{(f)}_{\mu\nu} dx^\mu dx^\nu = -(dx^0)^2 + (dr^2 + r^2 d\theta^2 + r^2 \sin^2\theta \, d\phi^2) . \tag{17.45}$$

17.6 Non-linear graviscalar perihelion shift

The Weak equivalence principle, or the uniqueness of free fall, was discussed in Section 7.11. The Einstein (or Strong) equivalence principle links the accelerations experienced in non-inertial frames to those arising locally as a result of gravitational interaction and plays a major rôle in the formulation of general relativistic gravitation (see Section 21.5). A crucially important feature of the data which forms the basis of the Strong equivalence principle (see Section 7.11) concerns the contributions to inertial and gravitational mass from the *self-gravitational energy* of a system.

Exercise 17.4 *(a) Use a Newtonian calculation to show that a mass M of radius R will have a self-gravitational potential energy of the order of GM^2/R (the precise value depending on the details of its degree of central condensation). (b) Show that for laboratory masses this energy, as a fraction W_{grav}/Mc^2 of its rest energy, is $\approx 10^{-25}$ and that for the earth the same ratio is $\approx 5 \times 10^{-10}$.*

Nordtvedt [402] pointed out in 1968 that a difference in the contributions of W_{grav} to the inertial and gravitational masses m_{I} and m_{G} of the earth would imply, on the basis of general relativity, anomalous variations in the mean radius of the lunar orbit that could in principle be measured by laser ranging. Gravitational energy, like all energies, contributes to inertial mass according to $m_{\text{I}} = W_{\text{grav}}/c^2$. If W_{grav} did not contribute at all to m_{G}, oscillations in the lunar orbit, due to this *Nordtvedt effect* would be expected of amplitude about 10 m, decreasing to zero if it contributes equally to both types of mass. The absence of such changes in the lunar motion to a precision of about ± 0.15 m using the lunar laser ranging equipment with a corner cube reflector shows [402, 511, 589] that the mass-energy of the gravitational field contributes equally to the gravitational mass to within a few per cent. Gravity itself therefore also gravitates. The gravitational sources give rise to gravitational fields which themselves contribute to the energy of the system and thus act as a source leading to further gravitational fields.

The gravitational interaction is thus inherently non-linear as a result of the universality and mass-energy equivalence. It is this extension of the universality of the interaction to include the mass-energy of the gravitational field itself which is crucially important in the formulation of the relativistic theory of gravitation. Nordstrøm [400] applied this argument to the scalar gravity equations in 1913 to construct a gravity theory in a curved spacetime with metric tensor proportional to the Minkowski tensor and therefore said to be *conformally flat*. We shall apply the same technique here, with a localized weak source, $\Delta\Phi/c^2 \ll 1$, to examine the consequences of non-linearity in a context where the full details of space-

17.6. Non-linear graviscalar perihelion shift

time curvature may be avoided.

In the solar system, the quantity $|\Phi|/c^2$ (with the arbitrary choice of $\Phi = 0$ at large distances) has a maximum value on the surface of the sun where the numerical result is $\approx 10^{-6}$. Values of v^2/c^2 are of a similar order of magnitude but values of $p/\rho c^2$ are $\approx 10^{-12}$. The gravitational field on the surface of any ordinary *main sequence star* like the sun will have similar values of $|\Phi|/c^2$. Two other major classes of stars have masses similar to the solar mass but are much smaller, namely *white dwarf stars* and *neutron stars* [512] for which the sizes are down by factors of about 10^{-3} and 10^{-5}, respectively, compared to the sun. The dimensionless strength $|\Phi|/c^2$ of the gravitational field will in these two cases be $\approx 4 \times 10^{-3}$ and ≈ 0.1, respectively, showing that self-gravitation is still not very important in the white dwarf case but is clearly significant in neutron stars.

A fourth class of astrophysical object, such as Cygnus X-1, with masses $\approx 10\, M_\odot$, are candidate black holes, near which the gravitational field is so strong (owing to the small size of the objects) that the dimensionless strength $|\Phi|/c^2$ of the gravitational field approaches unity showing that self-gravitational effects dominate. Consequently, any attempt to describe such effects using Newtonian gravity field equations, which are linear, cannot hope to succeed.

In setting up Equation 17.4 for the graviscalar field in preparation for using it in solar system calculations, we conveniently ignored the fact that the Strong equivalence principle requires all matter and energy, including the energy-momentum tensor $T^{\mu\nu}(\Phi)$ of the gravitational field itself, to act as a source of the gravitational field. Our justification for doing so was that, for a planet, the ratio of self-gravitational energy to rest energy is $\approx |\Phi|/c^2 \approx 10^{-9}$ and thus is negligible for almost all purposes. But in relativistic gravity in the solar system one is looking for differences from Newtonian gravity predictions in a situation where the conditions validating Newtonian gravity (for example, $|\Phi|/c^2 \ll 1$) are very well satisfied. Those differences can therefore be expected to be very small and we cannot safely neglect the small effects of self-gravitational energy.

In sharp contrast to the situation for Newtonian gravity, the relativistic equations provide a natural means for determining the magnitude of the non-linear effects as a direct result of mass-energy equivalence which itself is a consequence of Poincaré covariance. As an approximation to the contribution of self-gravitational energy to the source of the scalar field, we use the trace,

$$T(\Phi) = -\tfrac{1}{4\pi G}\partial^\lambda \Phi\, \partial_\lambda \Phi\,, \qquad (17.46)$$

of the free field energy-momentum tensor $T^{\mu\nu}$. If the matter contribution to the source (the non-gravitational part) is denoted by $T(M)$, then the new non-linear field equation incorporating the universality of gravitational interaction will be,

$$\Box \Phi = -\frac{4\pi G}{c^2}(T(M) + T(\Phi)) , \qquad (17.47)$$

which gives,

$$\Box \Phi - \frac{1}{c^2}\partial^\lambda \Phi \, \partial_\lambda \Phi = -\frac{4\pi G}{c^2}T(M) . \qquad (17.48)$$

This is a non-linear equation for the gravitational field Φ. Because the term containing the highest-order derivatives (of second-order) are linear in Φ it is said to be a *quasi-linear* differential equation. Such equations play an important role in applied mathematics and theoretical physics being considerably more tractable than corresponding fully non-linear equations. Another example of quasi-linearity is supplied by Einstein's field equation (see Chapter 21) for general relativistic gravitation.

The static field due to a spherically-symmetric source of mass M will therefore satisfy,

$$\nabla^2 \Phi - \frac{1}{c^2}(\nabla \Phi)^2 = 4\pi G M \delta^3(\mathbf{x}) , \qquad (17.49)$$

or,

$$\nabla^2 \Phi - \frac{1}{c^2}(\partial^r \Phi)^2 = 4\pi G M \delta^3(\mathbf{x}) . \qquad (17.50)$$

Using $\Phi_1 = -GM/r$ as a first approximation to Φ in the small non-linear term gives,

$$\nabla^2 \Phi - \frac{G^2 M^2}{c^2 r^4} = 4\pi G M \delta^3(\mathbf{x}) . \qquad (17.51)$$

Exercise 17.5 *Solve Equation 17.51 for the small correction term Ψ in $\Phi = \Phi_1 + \Psi$ to obtain, to second-order, a final solution,*

$$\Phi = -\frac{GM}{r}\left(1 - \tfrac{1}{2}\frac{GM}{c^2 r}\right) . \qquad (17.52)$$

The correction term due to self-gravitational energy gives us a deviation from the Newtonian potential. However, this change in the field will not lead to deflection of null rays since the null condition on the 4-momentum of the ray will still cause the right side of the equation of motion 17.8 to vanish leading to Equation 17.19.

Inserting $-g_{00} = g_{rr} = (1 - 2GMu/c^2 + G^2 M^2 u^2/c^4)$ into Equation 17.42 for the orbit of a massive particle gives,

$$\frac{d^2 u}{d\phi^2} + u = \frac{GM}{h^2} - \frac{GM}{h^2}\left(\frac{GMu}{c^2}\right) . \qquad (17.53)$$

17.6. Non-linear graviscalar perihelion shift

The last term in this equation represents a very small deviation from the Newtonian equation. Let $u_1 = (1 + e\cos\phi)/a(1 - e^2)$ be the solution of the latter. We seek a perturbed solution $u = u_1 + u_2(\phi)$ where the small correction will satisfy,

$$\frac{d^2 u_2}{d\phi^2} + u_2 = -\epsilon \frac{GM}{h^2}(1 + e\cos\phi), \tag{17.54}$$

where $\epsilon = G^2 M^2/h^2 c^2 = GM/c^2 a(1-e^2)$. For Mercury, this small quantity has the value of about 1.5×10^{-8}.

The solution of this equation will comprise a complementary function plus a particular integral [19]. Following the argument in Adler, Bazin and Schiffer [3], we note that the former, a solution of the homogeneous equation, will be a small quantity added to u_1 which is an expression of the same type. The constants e and M/h^2 in the combined expression will be precisely the ones which are determined from observations of the perturbed orbit. We therefore ignore this part. The particular integral of the first term on the right side, $-\epsilon GM/h^2$, implies changes in the distance which are $\approx 1.5 \times 10^{-8} a$ which for Mercury is $\approx 1\,\mathrm{km}$. This is below observational limits on the value of the semi-major axis a of Mercury's orbit and is thus unobservable.

Exercise 17.6 *Show that the particular integral of the last term, namely the solution of,*

$$\frac{d^2 u_2}{d\phi^2} + u_2 = -\epsilon \frac{GM}{h^2} e\cos\phi, \tag{17.55}$$

is

$$u_2 = -\frac{\epsilon GM}{h^2} \tfrac{1}{2} e\phi \sin\phi. \tag{17.56}$$

Our complete solution is therefore,

$$u = \frac{GM}{h^2}\{1 + e\cos\phi - \tfrac{1}{2}\epsilon e\phi\sin\phi\}, \tag{17.57}$$

which, because $\epsilon \ll 1$, we may write as

$$u = \frac{GM}{h^2}\{1 + e\cos(\phi + \tfrac{1}{2}\epsilon\phi)\}. \tag{17.58}$$

The perihelia are defined to occur at those values of ϕ where r passes through a minimum, namely where $\phi(1 + \tfrac{1}{2}\epsilon) = 2\pi n$ for $n = 0, \pm 1, \pm 2, \ldots$, from which we find that the Newtonian value of $\Delta\phi = 2\pi$ per revolution is exceeded by $\Delta\phi_{\text{excess}} = -\pi\epsilon$. Substituting for ϵ, we determine the excess over a period of 100 years, for an orbit which has N revolutions in that period, to be

$$\Delta\phi_{\text{excess}}^{100y} = -\frac{\pi GM N}{c^2 a(1 - e^2)}. \tag{17.59}$$

Inserting the data for Mercury gives an excess of -7.19 arc seconds per century, the negative sign indicating a regression of the perihelion. The observed perihelion shift is an advance, not a regression, and of about six times the magnitude of this value.

The failure of non-linear scalar gravity to predict the observed value of the perihelion advance of planetary orbits is unrelated to the approximate nature of the process used in the above calculation. Higher-order terms in the non-linearity will only alter the above calculated value by terms which are $\approx 10^{-8}$ smaller than the observed value.

Despite the fact that scalar gravity fails to predict null deflection and predicts an incorrect value for the perihelion shift, it is nevertheless perfectly satisfactory from the field theoretic point of view. For this reason, if an alternative Poincaré field contribution (for example, tensorial) to the gravitational effects was found to account correctly for the null deflection but to predict only part of the perihelion advance, then it would not be unreasonable to consider a theory in which the gravitational field was made up of two components, one being a scalar field.

The *scalar-tensor theory* of Brans-Dicke-Jordan gravity [48, 103, 268], embodying *Mach's principle* [352, 384] that the mass of an object is related to all the other matter in the universe, was precisely of this form in a curved spacetime context. It is then a matter for confrontation of the predictions of such a theory with observation to determine the value of the parameters which quantify the ratio in which the two fields enter the theory. Current observations [585, 586, 588] essentially eliminate any significant graviscalar contribution at very large distances.

17.7 Short-range gravitational interaction

The failure of the simplest relativistic gravitational interaction leads one to consider the possibility of a description using the next simplest Poincaré field, a 4-vector potential $V^\mu(x)$.

A vector theory of gravity was considered by Lorentz [339] in 1900. The analysis may proceed by steps which are analogous [384, p. 179] to the formation of the electromagnetic interaction to be given in Section 19.4.

An important feature at that time was the demonstration by Lorentz that gravity could be made consistent with the non-Newtonian idea of finite maximum speed of propagation, namely the speed of light. However, like Newtonian gravity, it was unable to explain the anomalous perihelion observations of Mercury. A further difficulty now evident from a fully relativistic treatment is the lack of a natural way to form a covariant 4-vector local source $G^\mu(x)$ from the energy-momentum tensor $T^{\mu\nu}(x)$ consistent with the Newtonian gravity limit and thus not involving derivatives of the source fields. A third difficulty arises from the quantum requirement of

17.7. Short-range gravitational interaction

energy positivity which implies that the vector interaction will be repulsive [264].

Some theories for the quantization of gravity, and its unification with the other three interactions, provide the graviton with massive scalar and vector partners, the *graviscalar* and *graviphoton*, whose interaction potentials will be of Yukawa form. Deviations from the inverse-square law could then occur at distances intermediate between laboratory and astronomical sizes and these deviations could be of either sign if scalar and vector components were involved. Some deviations from the inverse-square of both signs have been reported recently [417, 519, 397] at scales of 100s of metres.

It can be shown quite generally [264], using the Lagrangian formulation, that positive energy interaction sourced by mass-energy will be attractive for even spin mediation and repulsive for odd spin. The inadequacy of the spin 0 and spin 1 theories alone or together for standard gravity suggest that the next simplest case should be considered and we shall examine the massless spin 2 theory in Chapter 21.

Bibliography:

Lord E 1976 *Tensors, relativity and cosmology* [328].
Nieto M M et al., 1988 *The principle of equivalence, quantum gravity* ... [397].
Ohanian H C 1976 *Gravitation and spacetime* [404].
Thirring W E 1961 *An alternative approach to ... gravitation* [534].
Wald R M 1984 *General relativity* [556].
Will C M 1981 *Theory and experiment in gravitational physics* [585].
Will C M 1984 *The confrontation between general relativity and experiment* [586].
Will C M 1986 *Was Einstein right? — Putting general relativity to the test* [587].

CHAPTER 18

Maxwell's equations

I have also a paper afloat, containing an electromagnetic theory of light, which, till I am convinced to the contrary, I hold to be great guns[1].

James Clerk Maxwell   1831–1879

18.1 Introduction

The history of electricity and magnetism is analysed in great detail by Whittaker [576].

In the 13th century, Maricourt[2] [357] demonstrated the existence of two poles in a magnet by tracing the direction of a needle laid on to a natural magnetized material and thereby published the first observation connected to one of the modern laws of electromagnetism: Gauss's law for the absence of magnetic charge.

The modern development of both topics originates with work published in 1600 by Gilbert[3] [196] who accounted for the behaviour of compasses by showing that the earth was effectively also a large magnet. Electricity and magnetism were developed as independent phenomena until about 1800. Newton [395, Bk. iii, prop. vi, cor. 5] was aware in 1687 that magnetism did not share the universality of gravitation and obtained the inverse cube law for the couple exerted by one magnet on another.

Gray[4] [208] showed in 1729 that static electricity may be transported between bodies by some substances, among them metals, which are now called conductors. Du Fay[5] [122, p. 464] discovered in 1733 that electric charge was of two types and that like charges repelled and unlike attracted.

---

[1] Letter of 5 January 1865 to his cousin Charles Cay, concerning reference [366].
[2] Pierre de Maricourt, also known as Petri Pergrinus (fl. ca. 1269), French.
[3] William Gilbert (1544–1603), English physician. Royal physician 1601.
[4] Stephen Gray (ca. 1670–1736), British physicist.
[5] Charles-François de Cisternai du Fay (1698–1739), French physicist.

18.1. Introduction

In 1746, Watson[6] [558, p. 718] put forward the theory that electrical charge may be transferred but never created or destroyed. This notion was propounded independently by Franklin[7] [69, 170] in 1747 on the basis of his own experiments. Michell [376] demonstrated in 1750 that the action of one magnet on another can be attributed to an inverse square law of force between individual poles.

In 1766, Priestley[8] [461] used the results of experiments showing absence of electrical effects inside charged hollow conducting spheres to deduce, by analogy with the laws of gravity, that the interaction followed an inverse square law. In 1785, Coulomb[9] independently invented the torsion balance to confirm [81] the inverse square law of Priestley relating the interaction between two spherically symmetric charge distributions held a fixed distance apart.

Modern equivalents of Coulomb's experiment quantify the degree to which the inverse-square nature is satisfied by noting [204] that any deviation, such as one described by $e^{-r/\lambda}/r^2$ can be entertained only if $\lambda \geq 10^{12}$ m, which is several times the radius of the earth. In quantum electrodynamics, this translates to a photon rest mass $m_\gamma \leq 10^{-50}$ kg. Coulomb verified Michell's law of force between magnetic poles and suggested it was impossible to separate the poles of a magnet without creating two more poles of opposite polarity and equal strength.

In the early 1780s, Galvani[10] [183] used the response of animal tissue to initiate the study of electrical currents produced by chemical action rather than from static electricity. Volta[11] [553, 554] showed in 1799 that *galvanism*, the mechanical response of animal tissue to contact with dissimilar metals, was not of animal origin but occurred whenever a moist substance is placed between two metals. His discovery led to the first electric cells or batteries.

In 1812, Poisson [453] formulated the concept of *macroscopic charge neutrality* as the natural state of matter and described electrification as the separation of the two types of electricity. He also extended the work of Lagrange [296] in 1777 and Laplace [308] in 1782 on the vacuum gravitational potential to regions containing a matter density (see Chapters 6 and 7) and demonstrated the utility of such ideas in determining the charge distributions on electrified bodies [453, 454].

Oersted[12] [403] noted on 21st July, 1820, after giving a lecture demonstration, that an electrical current caused a deflection of the needle of a

---

[6] William Watson (1715–1789), British physicist, physician and botanist.
[7] Benjamin Franklin (1706–1790), American statesman and scientist.
[8] Joseph Priestley (1733–1804), English chemist and theologian.
[9] Charles Augustin de Coulomb (1736–1806), French physicist.
[10] Luigi Galvani (1737–1798), Italian anatomist, physiologist and physicist.
[11] Alessandro Guiseppe Antonio Anastasio Volta (1745–1827), Italian physicist.
[12] Hans Christian Oersted (1777–1851), Danish physicist.

magnetic compass and thus initiated the process of unification of electricity and magnetism, which was to be completed, for classical effects, in the work of Maxwell. The link between electricity and magnetism had been long suspected — mariners had noted, for example, the effect of lightning on magnetic compasses. Oersted's sensational discovery was confirmed within a few months by Ampère[13] [10] and by Biot and Savart[14] [38, 39, 40] who deduced the formula now used for the strength of the magnetic effect produced by a short segment of current carrying wire. Oersted demonstrated that the action was reciprocal, with the bar magnet being affected by the current carrying conductor.

Within a week of hearing the news of Oersted's first result, Ampère had carried out significant electromagnetic experiments and had presented the results to the French Academy of Science. He expressed the results in a form applicable to the mutual interaction between two current carrying elements [11] and initiated the study of electrodynamics independently of electrostatics. He accounted for the magnetism of bar magnets and of the earth in terms of molecular electric currents. The publication of his memoir on electrodynamics [12, 13] in 1825 was one of the great milestones in the history of physics.

Ohm's law[15] [406, 407, 408, 409], relating the 'magnetic action' (current) of a galvanic (or thermoelectric) cell to the electromotive force (or temperature difference) and electrical resistance followed in 1827. Ohm also considered galvanic currents to be no different to those for conventional electricity.

In 1828, Green[16] [209] generalized the work of Lagrange, Laplace and Poisson and introduced the term *potential* for the variables which now play a vitally important rôle in the gauge description of all four of the fundamental interactions of physics. *Green's theorem* [326, 421] relating the surface and volume distributions of charge permitted the systematic and elegant determination of solutions to a wide variety of electrostatic problems.

Green's work passed unnoticed until 1846 and an equivalent result was stated without proof by Gauss in 1832. About that time, Gauss [191] (see also Jungnickel et al. [269, p. 130]) re-expressed the law of Coulomb in a more general, spatially local, form suitable for determining the electrical effects of any distribution of charges, not just point charges or those with spherical symmetry. Soon after, Gauss developed the spatially local magnetostatic relations between magnetic potential and current density. This was followed in 1835 by separate electrostatic and electrodynamic laws between charges including local forms of Coulomb's and Ampère's laws. These re-

---

[13] André-Marie Ampère (1775–1836), French mathematical and chemical physicist.
[14] Jean-Baptiste Biot (1774–1862) and Félix Savart (1791–1841), French physicists.
[15] Georg Simon Ohm (1789–1854), German experimental and theoretical physicist.
[16] George Green (1793–1841), British mathematician and natural philosopher.

18.1. Introduction

sults, including *Gauss's law* of electrostatics, remained unpublished [269, p. 130] until their appearance in his collected works [191, 192] in 1867. His reasons for not publishing were an awareness that, since they were based on instantaneous action, his electrical equations were incomplete.

Gauss [189] established fundamental experimental methods from 1832 for the measurement of absolute values of magnetic intensities in terms of mechanical units [269, p. 67] suggesting the same be done for electrical quantities. His work was supplemented by Weber and Maxwell [365], whose electrical quantities were combined with the magnetic quantities of Gauss to form the system known as *Gaussian units*.

Potential theory was applied by Gauss [190] and Weber in 1838 to the magnetism of the earth and extensively developed in 1839 as a general theory [192, vol. 5, pp. 195–242 on 199–200] applicable to all inverse-square law phenomena, in particular gravitation, electrostatics and magnetostatics. Its importance was enhanced soon after with Helmholtz's development of the principle of conservation of energy and the concept of *potential energy*.

At that stage in the development of the topic, electrodynamics was almost universally considered to be an action at a distance theory like Newtonian gravitation. From 1831, Faraday had systematically studied the possibility of greater symmetry in the interaction of electromagnetism by investigating and describing [150] the reciprocal to Oersted's discovery, *induction*, namely the production of electromotive interactions, and hence currents in conductors, by varying the degree of magnetism with time. Lenz[17] [320, p. 485] added the fact that the electromotive interactions and currents produced by changing magnetic fields always opposed the changing magnetic conditions that produced them. Between 1845 and 1850, Faraday also introduced the notion of *contiguous magnetic action* of one part of a system on its immediate neighbours. This replaced the Newtonian idea of instantaneous action at a distance by causal action via concepts now described by *fields*.

In 1845, Fechner [151] proposed a connection between Ampère's law and Faraday's law of induction in order to explain Lenz's law. Weber then introduced the idea that electric currents were moving charged particles and in 1846 proposed [561] a synthesis of electrostatics, electrodynamics and induction in which there were instantaneous electric forces between such currents and along the line joining them. Weber's theory contained a limiting velocity of electromagnetic origin whose magnitude equalled $\sqrt{2}c$.

In 1858, Riemann [473], in material published [269] posthumously in 1867, assumed equal and opposite forces but not necessarily along the line joining the current elements. He obtained Weber's result for the elec-

---

[17]Heinrich Friedrich Emil Lenz (1804–1865), German physicist (Russian Academician).

trodynamic potential ϕ from the retarded solution of the wave equation, essentially $\Box \phi = -\rho/\epsilon_0$, implying a value equal to c for Weber's electrodynamic constant. He claimed [269, pp. 179–180] that he had 'discovered the connection between electricity and optics'. In 1861, Riemann [269, p. 181], in his lectures at Göttingen [474, pp. 313–337], also showed the extent to which velocity-dependent electrical accelerations could be treated by Lagrange's theorem with conservation of energy in the same way as strictly Newtonian interactions. Weber used his own theory in 1864 to show that current oscillations in a conductor travel at $1/\sqrt{2}$ times his electrodynamic constant, namely at $310{,}740\,\text{km.s}^{-1}$. He nevertheless considered that one should not expect a deep connection between light and electricity.

Maxwell, [364] to [366], completed between 1861 and 1865 the formulation of the field equations of electromagnetism to be discussed in this chapter. In 1863, with Jenkins, he supplemented [365] the absolute magnetic units of Gauss with those for electrical quantities to form the so-called *Gaussian system* of electrodynamic equations.

In 1870, Helmholtz [233] developed a theory of electricity and also showed that Weber's theory was inconsistent with conservation of energy. In 1884, Hertz [237] developed his own expression of electrodynamics and showed that his and Helmholtz's theories both led to Maxwell's equations. He concluded that all three theories were equivalent and that the details of each were irrelevant — according to Hertz: 'Maxwell's theory is Maxwell's system of equations'. A modern expression of the essential content of Maxwell's equations, in terms of representations of the Poincaré group, can be based on the null spin 1 character of the electrodynamic field. Alternatively, they result from the localization of the U(1) phase invariance of a complex field with hermitian Lagrangian.

In 1890, Hertz [244, 245] simplified the form of Maxwell's equations, replacing all the potentials by field strengths, and deduced Ohm's law, Kirchoff's law for closed circuits, Coulomb's law and the induction law for open circuits. An even simpler form had been developed by Heaviside [229] from 1885 to 1887 using vector methods. From 1892 to 1904, Lorentz, [332] to [341], completed the description of electrodynamics by separating the electricity from the electrodynamic fields and describing the equations applicable to charged particles in motion.

In first courses on electrodynamics the electric and magnetic field 3-vectors **E** and **B** are introduced separately by describing the results of the experiments that gave rise to the law of Coulomb in the late 18th century followed during the following 100 years by the discoveries and laws of Oersted, Biot and Savart, Ampère, Gauss, Faraday, Lenz, Maxwell and Lorentz. One also implicitly incorporates the experimentally observed linear superposition property of the non-quantum electromagnetic interaction and the absence of magnetic charge.

It will be assumed that the reader is already familiar with such a pre-

18.1. Introduction

sentation of electromagnetism. This conventional procedure is a necessary initial simplification, both conceptually and analytically, and reflects the importance of experimental results in the discovery of the properties of the electromagnetic interaction. However, it can give the impression that the electric and magnetic fields are invariantly separable and that Maxwell's equations have a structure which is not simple, the origin of which is somewhat mysterious. Furthermore, this procedure makes it difficult to see that Maxwell's equations could have been discovered prior to the great bulk of experimental material that forms its foundation in the same way that Einstein set out the details of general relativity with little appeal to observations requiring a relativistic treatment of gravitation.

Familiarity with Maxwell's equations soon leads to an appreciation of the indivisibilty of the electromagnetic field, of its essentially relativistic nature and of the simplicity of its field equations. These attributes are particularly evident when Maxwell's equations are expressed in manifestly covariant form. However, if the 3-vector Maxwell's equations are simply converted into such a form, as an aside, an excellent opportunity will have been lost for acquiring, relatively easily, a deeper understanding of both electromagnetism and relativity.

The symmetry properties of wave equations have assumed considerable importance in recent decades as part of the search for a unified description of all the interactions of physics at the fundamental level. Comparison of Maxwell's equations with the evolution equations of the fields and particles of the other interactions is greatly facilitated by a description which simply displays the Poincaré covariance and gauge invariance of the electromagnetic interaction. Such a description also makes it easier to see how to generalize the equations to spacetimes of dimensions other than four, a procedure originating with Kaluza [272] in 1921 and Klein [284] in 1926. Such a process has been recently revived as a useful, indeed crucial, element in the most promising unification programmes, in particular those involving superstrings with subsequent dimensional reduction to 4D spacetime.

Einstein's formulation of relativity theory was founded on a profound analysis of spacetime measurement. It is not unreasonable therefore to imagine its discovery somewhat earlier in the history of physics, prior to the formulation of electrodynamics. In this spirit, Maxwell's equations will be presented here in a way which highlights their essentially relativistic nature, reveals the origins of their structure and the power of covariant notation. In so doing, we shall show that had special relativity been discovered some time in the 17th, 18th or early 19th century, then Maxwell's equations would almost certainly have followed immediately. Given the genius of Newton, for example, the close parallel between Newtonian and Einsteinian physics, and the opportunity provided by Soldner's prediction of light bending in 1801 (see Section 7.13), such a scenario is not completely unreasonable.

18.2 Characterization of the electromagnetic field

Once it is accepted that the foundations of relativistic physics may be developed — as in Chapters 5 and 12 — independently of the specific properties of electromagnetism, the way is open to use the quite generally applicable principles of relativity to uniquely characterize the electromagnetic interaction. Indeed, without any reference to a particular phenomenon such as electrodynamics, one may also ask what form the simplest relativistic interactions, via a field, will take.

In a Galilean context, such an approach leads to the equations of electrostatics, magnetostatics, the Schrödinger wave function and Newtonian gravity, including a possible cosmological term and short-range Yukawa contributions. We have also applied this philosophy, of examining first the simplest equations, to discuss, in an Einsteinian context, the Klein-Gordon or relativistic scalar equation. It turns out that electrodynamics may be characterized in precisely this way as the simplest possible relativistic interaction with the exception of those mediated by a scalar field.

We shall use the criterion of simplicity in a straightforward development of the covariant forms of the electromagnetic field and of Maxwell's equations and then extract the electric and magnetic field 3-vectors from the covariant field strength. In Chapter 19, we shall extend the analysis, again using the criterion of simplicity, to construct the form of the Lorentz law for the acceleration of a charged particle in an electromagnetic field. In this way we are able to exhibit from the beginning, and preserve throughout, the essentially relativistic nature and simple structure of electrodynamics.

Characterizing classical electrodynamics as a relativistic interaction implies that it has a source. To be relativistic and classical, that source and the field it gives rise to, must each be one of a 4-scalar, a 4-vector or a 4-tensor. The relativistic nature of the interaction also implies that the equations governing the variation, with time and space, of an isolated distribution of the field, namely under translations, must retain their form when the coordinates of spacetime are changed in a Poincaré transformation.

The field equations of an interaction are either parity-covariant or not. In a non-quantum context we need examine only field equations which are parity-covariant. For the electromagnetic interaction, this is a property it shares with the strong-nuclear and gravitational interactions and can be deduced from the absence, in any of the reactions it mediates, of preferential emission or absorption of electrically-charged particles of one chirality or the other. As a result of the appearance, in the laws of electrodynamics, of cross-products and right-hand rules, some students gain the impression that electromagnetic effects are inherently rotational in nature or fundamentally chiral. Superficial consideration of the results of some electrical experiments, such as those with compasses near a current carrying wire, contribute little to dispell this erroneous view. Indeed, many are those whose first encounters with electricity leave them with such impressions. Mach described his own reactions in his text [352] and Pasteur[18] is reputed to have initially concluded with astonishment that electromagnetic effects show a clear preference for one handedness. The belief that nature is not chiral is deep-seated and a careful analysis shows that this is well-founded in the electrodynamic case. Some of the origins of these initial misconceptions will be pointed out in due course.

In order to elucidate the relativistic structure of Maxwell's equations, it will not be necessary to examine their form in the presence of continuous media. It will suffice to investigate the interaction of isolated charges from which the equations in media may be constructed. For the application of Maxwell's equations to phenomena occurring in media, such as polarization and magnetization, the reader is referred to standard electrodynamic texts such as Jackson [263], for example.

18.3 Invariance and conservation of charge

Appealing to a criterion of simplicity to characterize the electrodynamic interaction, we select the local source of the mediation to be the simplest Poincaré field. After the scalar source density $j(x)$ of the Klein-Gordon field, this is a real, additive and bipolar Poincaré vector $J = \{J^\mu(x)\}$ called the *electromagnetic current density*. For maximal simplicity, we choose J^μ to be, like the covariant 3-volume $d\Sigma_\mu$, proper with respect to parity and improper with respect to time reversal. The corresponding integrated source q, the *electric charge*, will therefore be a full Poincaré scalar, defined in terms of the current density according to,

$$q(\Sigma) = \tfrac{1}{c} \int_\Sigma d\Sigma_\mu J^\mu(x) = \tfrac{1}{c} \int_{t=\text{const}} d^3x\, J^0(x) = \int_{t=\text{const}} d^3x\, \rho(t, \mathbf{x}) \,. \tag{18.1}$$

The factor of $1/c$ is to achieve conventional dimensions for J^μ and q. Charge conjugation is defined to change the sign of J^μ and q.

We have specified invariance of q with respect to the full Poincaré group of transformations, namely those which appear in the Relativity principle plus the reflections. Consequently, we are demanding that q be not only Lorentz and space-translation invariant but also unchanged with respect to spacetime reflections and time translations. In particular, the latter means that it must be conserved,

$$\frac{dq(\Sigma)}{d\Sigma} \equiv 0 \quad \text{or} \quad \partial_\mu J^\mu \equiv 0 \,. \tag{18.2}$$

---

[18] Louis Pasteur (1822–1895), French chemist and microbiologist.

The only transformation affecting the value of q is charge conjugation.

By the very nature of a source term, it must be independent of the field being sourced. We represent this feature by writing the vanishing of the divergence of the current as an identity, meaning independent of any properties of the electromagnetic field equations we seek. However, if the fundamental structure of the source is taken into account by regarding it also as a set of fields, then the conservation of the source will be related to their field equations via Noether's theorem. In fact for a closed system of charges described by Schrödinger wave functions or Klein-Gordon and Dirac fields, the existence and conservation of q will be related to global U(1) phase invariances of the source equations of Schrödinger, Klein-Gordon or Dirac.

If we do not demand that q be conserved, then our theory will contain another non-zero Lorentz invariant, $\partial_\mu J^\mu$, contradicting our assumption of maximal simplicity. It has no meaning to require the scalar local source $j(x)$ of the Klein-Gordon field to be conserved — it does not have a covariant index on which to contract the index of ∂_μ. One may therefore argue that, from the point of view of Poincaré covariance, a conserved 4-vector source is even more basic than a scalar local source.

The field has been characterized as having a 4-vector local source and the charge defined in terms of it. However, one could equally well characterize the field as having an invariant integrated source q and then expand that charge in an integral over the 4-vector covariant 3-volume $d\Sigma_\mu$ to deduce, from purely geometrical (namely, mathematical) considerations, that the local source must be a 4-vector using the quotient theorem of Exercise 14.2. We are effectively able to determine a 4-vector from a single component quantity, the charge, because the latter is an invariant.

Non-zero electric charge is not a property of all fields and their corresponding particles. This fact contrasts sharply with nature of mass-energy, the source of gravity, which is non-zero for all systems, and means that the electromagnetic interaction is not a universal interaction like gravity.

The relativistic invariance of electric charge — the fact that it has the same value in frames which are in relative motion — is a strong requirement which helps determine the ranks of the fields that can describe electromagnetism and the orders of the corresponding field equations. Charge should again be contrasted, for example, with mass-energy, which transforms not as a scalar but as the time component of the 4-vector momentum (see Equation 15.13).

Charge invariance is experimentally observed [263] to at least 1 part in 10^{21} from, for example, the neutrality of the hydrogen molecule and of the helium atom each of which is made up of (nominally) the same charges, whose average kinetic energies, in the two cases, correspond to classical motion at quite different relative speeds. An invariant of the restricted Lorentz group SO(1,3) may be split into the sum of a proper scalar

18.3. Invariance and conservation of charge

and a pseudoscalar which transform independently with respect to the full group, O(1,3). We are free to take electric charge to be the former, by convention [263].

Electric charge is carried only by non-null particles which means there will always exist a proper frame associated with a charge distribution, the local rest frame of the matter carrying the charge. In an arbitrary frame the matter will have some velocity $\dot{z}$. Equation 15.82 shows that, in a boost from the rest frame to the frame where the velocity is $\dot{z}$, the volume element $d\Sigma_\mu$ will be transformed from $\{d^3x, 0\}$ to $\{\gamma d^3x, \gamma\beta d^3x\}$. In order that the charge be invariant, the current 4-vector J^μ must transform in proportion to the 4-vector $\{\gamma, \gamma\beta\}$. From Equation 18.1, the proportionality constant will be c times the density $\hat{\rho}$ of charge in the rest frame, $e_{\hat{\rho}}$.

From the time and space parts of the 4-current density we may therefore define the *charge density* $\rho = J^0/c = \gamma\hat{\rho}$ in an arbitrary frame and the *current density*, $\mathbf{J} = \{J^k\} = \rho\dot{\mathbf{z}}$, which will transform under rotations as a 3-scalar and a 3-vector respectively. The 4-current may thus be expressed in familiar $1+3$ form as:

$$J = \{J^\mu\} = \{\rho c, \mathbf{J}\} = \{\rho c, \rho\dot{\mathbf{z}}\} = \{\gamma\hat{\rho}c, \gamma\hat{\rho}\dot{\mathbf{z}}\}, \qquad (18.3)$$

while the integrated expression of electric charge conservation reduces in a particular frame to the familiar form of $dq(t)/dt \equiv 0$.

In a $1+3$ split, the identically vanishing 4-divergence of the 4-current gives the charge continuity equation:

$$\frac{\partial\rho}{\partial t} + \mathbf{\nabla}\cdot\mathbf{J} = 0. \qquad (18.4)$$

The high precision with which certain similar atoms and molecules are known to be neutral also provides one of the experimental bases [464] for *charge quantization*, namely that the charge of each elementary particle is an integer multiple of the charge e on a proton (or $\pm\frac{1}{3}e$ and $\pm\frac{2}{3}e$ for quarks). An enormous variety of microscopic and macroscopic phenomena confirm, for example by counting the particles involved, that the charge q of an isolated system does not change with time. Conservation of electric charge is observed to be an exact law of nature — no violations of electric charge conservation have ever been detected. In this respect, electric charge is on a par with 4-momentum and angular momentum, the Noether charges corresponding to spacetime homogeneity and isotropy. Observations of the conservation of charge date from 1747 by Benjamin Franklin [170] who also introduced the convention of describing the two types of charge, its *bipolarity*, as positive and negative.

Covariance emphasizes that a source which happens to be purely a charge at rest in one frame with rest charge density $\hat{\rho}$ and zero 3-current $\{J^{\hat{\mu}}\} = \{\hat{\rho}c, 0\}$ will transform to $\{J^\mu\} = \{\rho c, \mathbf{J}\} = \{\rho c, \rho\dot{\mathbf{z}}\}$ in a frame moving at velocity $\dot{z}$ relative to the first. Although charge is the sole integrated source, charge density alone cannot be considered the sole local source of the electromagnetic field in any covariant sense.

18.4 The general form of the field equations

On the grounds of simplicity, we demand that the field equations, like those of classical mechanics, be covariant with respect to the full Poincaré group and under charge conjugation.

The electromagnetic field must be one of $\phi(x)$, $\phi^\mu(x)$, $\phi^{\mu\nu}(x)$, ..., in which the rank (and for rank two and higher the algebraic symmetry) of the field has to be decided. The proper tensor $\eta_{\mu\nu}$, and the permutation pseudotensor $\epsilon^{\mu\nu\lambda\rho}$ are universal tensors which embody all the geometrical properties of relativistic (Minkowskian) spacetime. Since all the other quantities potentially involved, $\{\eta_{\mu\nu}, \epsilon^{\mu\nu\lambda\rho}, \partial_\mu, J^\mu\}$, are real, we may assume the field is also real and that each term in the field equation is real. This is in accord with the observation that real quantities, such as accelerations of charged particles, are determined directly from the components of the electromagnetic field. It also has the consequence, in quantum field theory, that the electromagnetic quantum, the photon, is self-conjugate, namely identical to its antiparticle. Equivalently, the photon itself carries no electric charge nor any other type of charge, such as colour or lepton number.

The field equations must take the form $f(\eta_{\mu\nu}, \epsilon^{\mu\nu\lambda\rho}, \partial_\mu, \phi^{\mu\nu\cdots}) = J^\mu$ in which f is some combination of its arguments transforming overall as a real 4-vector, like J^μ. If the characterization of the source, and thus of the mediating field of the electromagnetic interaction, has been complete then all the different ways in which we may represent this characterization mathematically would be physically equivalent. The characterization of the integrated source as a real, conserved proper scalar is not sufficient to fix the form of the field equations.

We therefore appeal to Occam's razor to select those constant coefficient equations which are of the lowest possible degree (linear if possible) and then the best compromise between lowest differential order and lowest rank of the field consistent with the stated properties of the source. These criteria can in fact be satisfied non-trivially by linear field equations of which there are three low-order low-rank cases. We may reject one of these, $a\,\partial^\mu\phi = J^\mu$, involving a rank 0 scalar field ϕ for numerous reasons (see Exercise 16.1). Among these is the fact that $\Box\phi$ is not identically zero for all ϕ and the left side is thus not identically divergence-free which makes the equation inconsistent with charge conservation. We have only two cases to consider further:

• **a real second-rank spatially proper tensor field $T^{\mu\nu}$ of odd** charge parity satisfying a first-order equation of the form,

$$c_1 \, \partial_\nu T^{\mu\nu} + c_2 \, \partial^\mu T = J^\mu , \qquad (18.5)$$

and,

- a **real spatially proper 4-vector field** $A = \{A^\mu\}$, also of odd charge parity, satisfying a second-order equation of the form,

$$a_1 \, \Box \, A^\mu + a_2 \, \partial^\mu \partial \cdot A + a_3 \, A^\mu = J^\mu , \qquad (18.6)$$

where c_1, c_2, a_1, a_2 and a_3 are arbitrary real constants, $T = \eta_{\mu\nu} T^{\mu\nu}$ is the trace of $T_{\mu\nu}$, $\partial \cdot A \equiv \partial_\mu A^\mu$ is the 4-divergence of A and the *d'Alembertian* or *wave operator* is $\Box = \partial^\mu \partial_\mu = -c^{-2} \partial_t^2 + \nabla^2$, in which c is the ultimate speed. T^{0l} and A^μ must be time-reversal improper and T^{kl} proper while both $T^{\mu\nu}$ and A^μ must change sign under charge conjugation.

These alternatives may be determined as follows. The left side must have precisely one free index μ to match J^μ on the right side. To be a field equation, at least one term (on the left) must contain a partial derivative ∂. In that term we have two locations for placing this index: 1, on a field component $\phi^{\mu\cdots}$ and 2, on a derivative ∂^μ. In case 1, a ∂ in that term (there will be at least one) must have its index contracted with another index which is either A, on another index of the field tensor, giving a term such as $\partial_\nu \phi^{\mu\nu\cdots}$, or B, on another partial derivative, which leads to a term of the form $\partial^\nu \partial_\nu \phi^{\mu\cdots} \equiv \Box \phi^{\mu\cdots}$. In accordance with Occam's razor, we restrict each of these field tensors to second and first rank, respectively. Higher-rank fields would in any case only require higher-order equations. In subcase A we rename $\phi^{\mu\nu\cdots}$ as $T^{\mu\nu}$. We then note that there are no additional terms which may be included in the field equation for $T^{\mu\nu}$ besides ∂^μ acting on the trace T of $T^{\mu\nu}$. In subcase B we rename $\phi^{\mu\cdots}$ as A^μ and note that the only two additional terms which then satisfy our criteria for inclusion in the field equation are constant multiples of $\partial^\mu \partial \cdot A \equiv \partial^\mu \partial_\nu A^\nu$ and A^μ itself.

In case 2, the chosen term may either be of the form $\partial^\mu \partial \cdot A$ or $\partial^\mu T^\nu{}_\nu$ (effectively reproducing one or other of the above cases) or of the form $\partial^\mu \phi$ which we have already rejected. Any other distribution of additional indices (which must be even in number) on a partial derivative or on the field component will lead to equations two orders higher than one or other of these two possibilities and we reject these.

18.5 Covariant first-order Maxwell's equations

In the first-order case, described by Equation 18.5, the quantity $T^{\mu\nu}$ is a real second-rank tensor thus apparently allowing the field to have 16 real algebraically independent field components. However, our field equations must be such that any zero- or first-order combination of ∂_μ, $\eta_{\mu\nu}$, $\epsilon^{\mu\nu\lambda\rho}$

James Clerk Maxwell

b. Edinburgh, Scotland
13 June 1831

d. Cambridge, England
5 November 1879

Engraving by G J Stodart,
Wren Library, Cambridge,
from a photography by Fergus of Greenock.
Courtesy of the Master and Fellows, Trinity College.

We can scarcely avoid the inference that *light consists in the transverse undulations of the same medium which is the cause of electric and magnetic phenomena.* [Papers [398, vol. I, p. 500].]

and $T^{\mu\nu}$ satisfying our criteria must have a well-defined value expressible covariantly in terms of the source, J^μ.

Two spatially proper 4-vector combinations have already arisen on the left side of Equation 18.5. However, there exists one other independent non-trivial, first-order combination linear in $T^{\mu\nu}$, namely $\epsilon^{\mu\nu\lambda\rho}\partial_\nu T_{\lambda\rho}$. It must be independent, otherwise parity covariance would be broken. There are no other such terms. A 4-vector linear in $T^{\mu\nu}$ cannot be formed without the inclusion of ∂_μ since the only other objects available have even rank. With ∂_μ included, it must either be contracted on $T^{\mu\nu}$, which leads to the spatially proper 4-vector in the original equation, or it must, along with $T^{\mu\nu}$ itself, be contracted with a rank 4 object to give a 4-vector. The only one available is $\epsilon^{\mu\nu\lambda\rho}$.

If we decompose $T^{\mu\nu}$ into the sum $T^{\mu\nu} = S^{\mu\nu} + F^{\mu\nu}$ of its symmetric ($S^{\mu\nu}$) and antisymmetric ($F^{\mu\nu}$) parts, the above combination is reduced to $\epsilon^{\mu\nu\lambda\rho}\partial_\nu F_{\lambda\rho}$ using the algebraic identity of Equation 4.19 applied to

Lorentz tensors. In terms of the dual $\tilde{F}^{\mu\nu}$ of $F^{\mu\nu}$ (see Section 14.8) the above combination is proportional to the pseudo-4-vector $\partial_\nu \tilde{F}^{\mu\nu}$.

We insist that each field equation involving $F^{\mu\nu}$ be parity-covariant. This new combination can therefore be equated only to a pseudo-4-vector source. However, we appeal to our criterion of simplicity to rule out such a second source for the electromagnetic field thus requiring this new combination to vanish. Part of this result corresponds to the absence of magnetic charge as we shall see shortly. Our first-order field equations are therefore of the form $c_1(\partial_\nu S^{\mu\nu} + \partial_\nu F^{\mu\nu}) + c_2 \partial^\mu S = J^\mu$ and $\partial_\nu \tilde{F}^{\mu\nu} = 0$ where $S = S^\mu{}_\mu$. Time reversal covariance leads to the same conclusion.

However, charge conservation must be independent of any local (differential) properties of the field components $T^{\mu\nu}$; in particular, it must be independent of the details of the field equations. Application of Equation 4.19 to $\partial_\mu \partial_\nu$ and $F^{\mu\nu}$ gives an identity,

$$\partial_\mu \partial_\nu F^{\mu\nu} \equiv 0 , \qquad (18.7)$$

which despite its appearance is (like Equation 4.19) essentially algebraic and guarantees that the local form of $F^{\mu\nu}$ is not required for Equation 18.2 to be consistent with the field equation. The same is not the case for the symmetric part $S^{\mu\nu}$. Since Equation 18.2 requires $\partial_\mu \partial_\nu S^{\mu\nu} \equiv 0$, which is not in general true, $S^{\mu\nu}$ (and therefore S) itself must vanish identically.

The field components have thus been reduced to those of a spatially proper antisymmetric tensor $F^{\mu\nu}$ having six algebraically independent components satisfying first-order field equations, $c_1 \partial_\nu F^{\mu\nu} = J^\mu$ and $\partial_\nu \tilde{F}^{\mu\nu} = 0$.

We fix the dimensions and units of the electromagnetic field tensor or *field strength* $F^{\mu\nu}$, relative to those of J^μ, by arbitrarily setting $c_1^{-1} = \mu_0$ ($= 4\pi \times 10^{-7}$ N.A^{-2} in the SI system), called the *vacuum permeability*. The final first-order field equations are thus precisely *Maxwell's equations* in manifestly covariant form:

$$\partial_\nu F^{\mu\nu} = \mu_0 J^\mu \quad \text{and} \quad \partial_\nu \tilde{F}^{\mu\nu} = 0 , \qquad (18.8)$$

as we shall shortly verify. The identical vanishing of the divergence of the source current has thus constrained the left side of the sourced field equation to have an identically vanishing divergence. Equation 18.7 may therefore be called a *source constraint* on the field equations.

18.6 Maxwell's equations in 3-vector form

We may demonstrate, in a manner which will show that the electric field 3-vector and magnetic field pseudo-3-vector arise naturally and essentially uniquely, that Equations 18.8 are equivalent to Maxwell's equations in familiar 3-vector form. In the SI (Système International) equation system, the latter are:

$$\nabla \cdot \mathbf{E} = \frac{\rho}{\epsilon_0} \quad \text{(Gauss's law — electric)} \quad (18.9)$$

$$\nabla \cdot \mathbf{B} = 0 \quad \text{(Gauss's law — magnetic)} \quad (18.10)$$

$$\nabla \times \mathbf{E} = -\dot{\mathbf{B}} \quad \text{(Faraday's law)} \quad (18.11)$$

$$\nabla \times \mathbf{B} = \frac{1}{c^2}\dot{\mathbf{E}} + \mu_0 \mathbf{J} \quad \text{(Generalized Ampère's law)}, (18.12)$$

where $\epsilon_0 \equiv 1/\mu_0 c^2$ is the *vacuum permittivity*.

The independent parts of any antisymmetric 4-tensor are mathematically equivalent (see Exercise 14.13) to a 3-vector and a pseudo-3-vector. Let us therefore define a spatially proper and temporally improper 3-vector, called the *electric field strength* denoted $\mathbf{E}$ and a temporally proper spatial pseudo-3-vector called the *magnetic induction* denoted $\mathbf{B}$ by,

$$\mathbf{E} = \{E^k\} = \{cF^{0k}\} \quad \text{and} \quad \mathbf{B} = \{B^k\} = \{\tfrac{1}{2}\epsilon^{klm}F_{lm}\} . \quad (18.13)$$

(The factor of c in the first equation and $\frac{1}{2}$ in the second are purely conventional.) An identity involving two contracted 3D permutation tensors shows that the space-space part of the electromagnetic field tensor may be retrieved from the magnetic field by,

$$F^{kl} = \epsilon^{klm} B_m . \quad (18.14)$$

Divide part (a) of Equations 18.8 into $\partial_\nu F^{0\nu} = \mu_0 J^0$ and $\partial_\nu F^{l\nu} = \mu_0 J^l$ which become $\partial_k F^{0k} = \mu_0 J^0$ (since $F^{00} = 0$) and $\partial_0 F^{l0} + \partial_k F^{lk} = \mu_0 J^l$. The first of these, using Equations 18.13, is simply *Gauss's law*, Equation 18.9. Using the definition of the curl from Equation 4.47 shows that the second of these becomes Equation 18.12, incorporating *Ampère's law* and *Maxwell's displacement current*. We split the second of Equations 18.8 into $\epsilon^{0\nu\lambda\rho}\partial_\nu F_{\lambda\rho} = 0$ and $\epsilon^{k\nu\lambda\rho}\partial_\nu F_{\lambda\rho} = 0$. The properties of the permutation tensor require the last three indices in $\epsilon^{0\nu\lambda\rho}$ to be spatial and since, $\epsilon^{0klm} = \epsilon^{klm}$, the first of these equations is simply Equation 18.10, *Gauss's law* for the absence of magnetic charge, while a similar procedure applied to the last equation gives *Faraday's law*, Equation 18.11.

Exercise 18.1 *Show that,*

$$\tilde{F}^{0k} = B^k \quad \text{and} \quad \tilde{F}^{kl} = -\epsilon^{klm} E_m/c , \quad (18.15)$$

give the components of $\tilde{F}^{\mu\nu}$, the dual of $F^{\mu\nu}$.

We have shown that the conventional expression of Maxwell's equations in terms of $\mathbf{E}$ and $\mathbf{B}$ is equivalent to Equations 18.8 which, by construction, are manifestly Lorentz- and translation-covariant (indeed,

18.6. Maxwell's equations in 3-vector form

Poincaré-covariant). Consequently, as Poincaré first demonstrated [448], so also are Maxwell's Equations 18.9 to 18.12 in conventional form, despite their lack of manifest covariance. In particular, we see that Maxwell's displacement current in vacuo, $\epsilon_0 \dot{\mathbf{E}}$, cannot be eliminated without violating relativistic covariance and the Principle of relativity.

Maxwell developed his electromagnetic theory ([363] to [367]), in a series of papers from 1855 to 1868 and the equations were arranged in 3-vector form by Heaviside [229] from 1885 to 1887. For stationary fields, namely $\dot{\mathbf{E}} = 0 = \dot{\mathbf{B}}$, these equations decouple into the equations of electrostatics, $\nabla \cdot \mathbf{E} = \rho/\epsilon_0$ and $\nabla \times \mathbf{E} = 0$, and magnetostatics, $\nabla \cdot \mathbf{B} = 0$ and $\nabla \times \mathbf{B} = \mu_0 \mathbf{J}$.

The second of Equations 18.8 can be written without explicit use of the dual as $\epsilon^{\mu\nu\lambda\rho} \partial_\mu F_{\nu\lambda} = 0$ which may be written,

$$\partial_{[\mu} F_{\nu\lambda]} = 0 \; , \tag{18.16}$$

where $T_{[\mu\nu\lambda]}$ denotes the *completely antisymmetric part* (Equation 14.18) of $T_{\mu\nu\lambda}$. The rank-3 completely antisymmetric tensor in Equation 18.16 is in fact proportional to the dual of the 4-vector left-side of the second of Equations 18.8, which establishes the equivalence of Equation 18.16 and the second of Equations 18.8. Using the antisymmetry of $F^{\mu\nu}$, the six terms of the completely antisymmetric part reduce to just three,

$$\partial_{[\mu} F_{\nu\lambda]} = \frac{1}{3} (\partial_\mu F_{\nu\lambda} + \partial_\nu F_{\lambda\mu} + \partial_\lambda F_{\mu\nu}) \; , \tag{18.17}$$

and the second part of Maxwell's equations takes the alternative form,

$$\partial_\mu F_{\nu\lambda} + \partial_\nu F_{\lambda\mu} + \partial_\lambda F_{\mu\nu} = 0 \; . \tag{18.18}$$

Equation 18.18 is sometimes called a *Bianchi identity* by analogy with similar identities that occur with the Fierz-Pauli (massless spin 2) and general relativistic theories of gravity, the Yang-Mills field equations of the nuclear interactions and indeed any field equation [526] that exhibits local gauge invariance.

Equation 18.18 and the first of Equations 18.8 show that Maxwell's equations are able to be written entirely in terms of the operations of the proper quantities ∂_μ and $\eta_{\mu\nu}$ on a proper tensor $F^{\mu\nu}$. Nowhere does there appear any quantity (such as a pseudotensor) or operation (such as a 3D curl) that could permit us to conclude that there is some inherently rotational or chiral character in the electromagnetic field. The re-expression of the field equations using a pseudo 3-vector $\mathbf{B}$ is unrelated to any inherently rotational or chiral property of the electromagnetic field. Such a purely mathematical split into a 3-vector and a pseudo-3-vector can be carried out for any antisymmetric Lorentz tensor.

It was not necessary to include in the characterization of the electromagnetic interaction the requirement that its effects propagate in vacuo at the ultimate speed. Rather, this is a consequence of the other properties assumed, in particular, the simplicity and thus the conserved nature of the source.

Exercise 18.2 *Show that the first of Equations 18.8, and the alternative Equation 18.18 for the second, imply:*

$$\Box F^{\mu\nu} = \mu_0(\partial^\nu J^\mu - \partial^\mu J^\nu) \,, \tag{18.19}$$

and show that this equation splits into,

$$\Box \mathbf{E} = (\boldsymbol{\nabla}\rho + \dot{\mathbf{J}}/c^2)/\epsilon_0 \tag{18.20}$$

$$\Box \mathbf{B} = -\mu_0 \boldsymbol{\nabla} \times \mathbf{J} \,. \tag{18.21}$$

The wave operator, $\Box = -c^{-2}\partial_t^2 + \boldsymbol{\nabla}^2$, was introduced in Equation 14.3 in terms of the ultimate speed c which in turn arose in Section 5.5 as a finite universal constant of relativistic kinematics of unspecified magnitude. The result of the above exercise shows that in vacuo (where $J^\mu = 0$) the solutions for the electromagnetic field tensor $F^{\mu\nu}$ propagate as waves at the ultimate speed c. Since light is an electromagnetic phenomenon, our criteria for the electromagnetic field require the *ultimate speed* c of Lorentz relativity and the *speed of light* to be identical. The latter is therefore invariant. This means that our discussion of the foundations of relativity and of Maxwell's equations are consistent with the customary form of the relativity postulates which assume at the outset that the speed of light is invariant (and finite). Furthermore, it means that measurements of c by observing the change in inertia or energy of particles with their speed $u = |\dot{\mathbf{z}}|$ must be consistent with the value obtained from electromagnetic phenomena, which is the case.

Goldhaber and Nieto [204] concluded that experimental evidence supporting the null character of light propagation places an upper limit of about 10^{-51} kg on the rest mass of the photon. This implies that the speed of a photon is within about 1 part in 10^{13} of the ultimate speed. If the speed of light, c_{light}, was not identical to the ultimate speed c, there would be effects on the energy levels of atoms [588]. Measurements of energy levels may be interpreted as placing constraints on any deviation and the limit $|c_{\text{light}}/c - 1| \leq 10^{-20}$, was obtained in this way by Lamoreaux et al. [302] in 1986. This result may also be interpreted as a test of local Poincaré covariance.

The equality between the ultimate speed and the speed of a null phenomenon such as light can thus be accounted for by a characterization of the corresponding field (based either directly on experiment or on simplicity confirmed by experiment) whether that field be electromagnetic, gravitational or neutrino, rather than being incorporated in the relativity postulates.

For many purposes, Equations 18.8, because of their manifest covariance, have considerable advantages over the 1+3 form of Maxwell's equations. The conventional form puts emphasis on only the Euclidean subgroup of the Poincaré transformations, ignoring the boost transformations between frames in relative motion. In fact, 3-vectors are designed precisely for handling rotations and spatial translations covariantly usually as part of non-relativistic (Galilean) kinematics. Their use in a relativistic context can severely complicate some calculations for the sake of using the more familiar concepts of non-relativistic physics.

The possibility of *electromagnetic radiation*, namely the phenomenon of electromagnetic waves propagating in free space, no longer affected by the charges from which they were generated, was a totally new concept predicted by Maxwell's theory in his 1865 paper [366]. The fact that such radiation was indistinguishable from light, provided the wavelength was of the appropriate magnitude, revolutionized optics. The phenomena of reflection and refraction were soon given a fundamental explanation by Lorentz [329] in 1875 in terms of Maxwell's electromagnetic theory.

The prediction by Maxwell of radiation at speed c was first confirmed by Hertz [239, 240] in 1887 with radio waves. The subsequent discoveries of X-rays by Röntgen and γ-rays considerably extended the observed wavelength range of electromagnetic radiation.

18.7 Second-order Maxwell's equations

In order to examine the extent to which our original *ansatz* for the electromagnetic interaction leads uniquely to Maxwell's equations we need to return to examine the second-order possibility. We demand that Equation 18.6 be consistent with Equation 18.2 describing charge conservation. This implies that $a_3 = 0$ and $a_2 = -a_1$ which reduce the equation to $\Box A - \partial\, \partial \cdot A = a_1^{-1} J$. Fixing the dimensions, units and sign of A^μ in terms of those of J^μ by putting $a_1^{-1} = -\mu_0$ gives the standard Maxwell equation,

$$\Box A - \partial\, \partial \cdot A = -\mu_0 J \,, \qquad (18.22)$$

for the gauge-variant variables $A = \{A^\mu\}$ which, anticipating its connection to the field strengths, we shall call the *electromagnetic 4-vector potential*.

Exercise 18.3 *Show (a) that the electromagnetic 4-vector potential A^μ in Equation 18.22 exhibits local gauge freedom, (b) that the determinant of the matrix of terms comprised of second derivatives with respect to time in*

Equation 18.22, vanishes and (c) that the natural dimensions of the vector potential are the canonical bosonic dimension of inverse length.

We may define the electromagnetic *scalar* and *3-vector potentials* using:
$$\phi = A^0 c \quad \text{and} \quad \mathbf{A} = \{A^k\} \, . \tag{18.23}$$

A second-order differential equation may be recast as a set of first-order equations by defining appropriate new variables which involve first derivatives of the old variables, here A^μ. We choose to do this here in a manifestly covariant manner. Since Equation 18.22 can be written as $\partial_\nu(\partial^\mu A^\nu - \partial^\nu A^\mu) = \mu_0 J^\mu$ the first-order covariant variable can be given only by the *covariant curl* of A up to a constant factor of no physical significance. The curl is antisymmetric and we fix the constant by denoting that curl by $F_{\mu\nu} = F_{[\mu\nu]}$,
$$F_{\mu\nu} = \partial_\mu A_\nu - \partial_\nu A_\mu \, , \tag{18.24}$$

so that Equation 18.22, when re-expressed in terms of the new variable, becomes precisely the first of Equations 18.8. From Equation 18.24, we see that Equation 18.18 and hence its equivalent, the second of Equations 18.8, is satisfied identically.

The first-order equations are thus obtainable from Equation 18.22 and a change of variable. Conversely, given Equations 18.8 the Poincaré lemma [51, 384] (see Appendix A.4) shows that the second part acts as an *integrability condition* guaranteeing the existence of a potential A^μ in terms of which $F^{\mu\nu}$ may be expressed differentially according to Equation 18.24. We can thus re-express the first part of Equations 18.8 as Equation 18.22 (the second being satisfied identically) showing that Equations 18.8 and 18.22 are equivalent.

The definition of $\mathbf{E}$ and $\mathbf{B}$ in Equations 18.13, and ϕ and $\mathbf{A}$ in Equations 18.23, reduce Equation 18.24 to the familiar forms,
$$\mathbf{E} = -\nabla\phi - \dot{\mathbf{A}} \quad \text{and} \quad \mathbf{B} = \nabla \times \mathbf{A} \, , \tag{18.25}$$

while Equation 18.22 similarly supplies the field equations,
$$\nabla^2 \phi + \nabla \cdot \dot{\mathbf{A}} = -\rho/\epsilon_0 \, , \tag{18.26}$$

and,
$$\Box \mathbf{A} - \nabla(\nabla \cdot \mathbf{A} + \dot{\phi}/c^2) = -\mu_0 \mathbf{J} \, , \tag{18.27}$$

for the potentials. The decoupled equations of electrostatics and magnetostatics are obtained on using $\dot{\phi} = 0$ and $\dot{\mathbf{A}} = 0$.

Since Equations 18.8 have been shown to be equivalent to Equation 18.22, we see that our initial *ansatz* of a real, conserved, invariant charge, and Occam's razor, have sufficed to uniquely establish Maxwell's equations.

The first-order Maxwell's equations are often considered to be in some way more fundamental and physical than the second-order form. The latter are regarded more as a device for solving Maxwell's equations by determining the potentials, which have fewer components than the field strengths, in a de-coupled gauge (see Section 19.7) and then determining the field strengths from these. However, if we take the properties of the source q as being primordial, neither the first-order field-strength formulation nor the second-order potential form appear to be preferred in describing the dynamics of the non-quantum electromagnetic field.

We shall see shortly that the potential form must in fact be used in the Lagrangian approach to the dynamics of continuous systems. Furthermore, we shall see in Chapter 20 that it is the vector potential that enters into the equations of the interaction between charged matter when the latter is described by fields rather than by classical particles. Since some of those fields are fermionic, and therefore essentially quantum in character, the vector potential must be used in a quantum treatment of radiative interaction.

Equation 18.22 may be derived [117, 118] as the simplest representation of a field equation which describes an helicity 1 wave, namely one which is null and spin 1.

Exercise 18.4 *The characterization on which we have based the development of the electromagnetic field in 1+3 dimensions can be used to deduce analogues of Maxwell's equations in spacetimes with other than three spatial dimensions. By examining the treatment which we have given of Maxwell's equations, and any results on which that development depends, summarize the analogous development justifying the main differences, for a spacetime of 1+2 dimensions.*

18.8 Electromagnetic invariants

In any physical system it is very important to know how to construct invariants from the basic quantities involved. Invariants may be used to classify the distinct configurations a system may assume independently of the particular features it may have in any one frame. An analysis of the electromagnetic field in $1+3$ notation tells us that there are three rotation invariants: $\mathbf{E}^2$, $\mathbf{B}^2$ and $\mathbf{E}\cdot\mathbf{B}$. These need not all be Lorentz-invariant. The following exercise shows that $\mathbf{E}\cdot\mathbf{B}$ remains a relativistic invariant. Although $\mathbf{E}^2$ and $\mathbf{B}^2$ are not separately 4-scalars, the combination $\mathbf{B}^2 - \mathbf{E}^2/c^2$ is Lorentz-invariant. From the results of Exercise 14.25, the only independent quadratic invariants that can be constructed from $F^{\mu\nu}$, $\eta_{\mu\nu}$ and $\epsilon^{\mu\nu\lambda\rho}$ are

$F^{\mu\nu}F_{\mu\nu}$ and $F^{\mu\nu}\tilde{F}_{\mu\nu}$ (or any linearly independent combination of these two).

Exercise 18.5 *By comparison with Exercise 14.25, or directly, show that:*

$$\tfrac{1}{2}F^{\mu\nu}F_{\mu\nu} = \mathbf{B}^2 - \frac{\mathbf{E}^2}{c^2} \qquad (18.28)$$

$$-\tfrac{c}{4}F_{\mu\nu}\tilde{F}^{\mu\nu} = \mathbf{E}\cdot\mathbf{B} \qquad (18.29)$$

$$c\,F^{\mu\lambda}\tilde{F}_{\lambda\nu} = -\delta^{\mu}{}_{\nu}\,\mathbf{E}\cdot\mathbf{B} \qquad (18.30)$$

Exercise 18.6 *Show that the contravariant forms of the electromagnetic field tensor and its dual may be displayed as matrices, as follows,*

$$(F^{\mu\nu}) = \begin{pmatrix} 0 & +E^1/c & +E^2/c & +E^3/c \\ -E^1/c & 0 & +B^3 & -B^2 \\ -E^2/c & -B^3 & 0 & +B^1 \\ -E^3/c & +B^2 & -B^1 & 0 \end{pmatrix} \qquad (18.31)$$

$$(\tilde{F}^{\mu\nu}) = \begin{pmatrix} 0 & +B^1 & +B^2 & +B^3 \\ -B^1 & 0 & -E^3/c & +E^2/c \\ -B^2 & +E^3/c & 0 & -E^1/c \\ -B^3 & -E^2/c & +E^1/c & 0 \end{pmatrix} \qquad (18.32)$$

The equivalent mixed and covariant components will differ in signs from these in simple and regular ways. It is highly unlikely, however, that one should ever need such explicit forms for any of these tensors including those displayed above.

18.9 Electromagnetic gauge invariance

Gauge invariance originated with the attempts by Weyl [572] in 1918 to unify the electromagnetic and gravitational interactions based on a scaling, $x^{\mu} \to \alpha(x)x^{\mu}$, of the length and time dimensions by a common arbitrary function $\alpha(x)$ of spacetime position. The gauge invariance of the Newtonian gravitational field and of the relativistic particle were examined in Chapters 7 and 15.

We have noted several examples of equations of motion and field equations which do *not* display gauge invariance. Among these are the Schrödinger and Klein-Gordon equations (see Chapters 10 and 16) used to describe spin 0 particles. Others are the Weyl and Dirac equations for describing spin $\tfrac{1}{2}$ particles, to be discussed in Chapter 20.

We shall now further illustrate the ideas of gauge invariance by considering the first-order form of Maxwell's equations. In a relativistic context we need only consider transformations which are manifestly covariant.

18.9. Electromagnetic gauge invariance

If there is to be gauge invariance in Equations 18.8, the transformation must take the form $F_{\mu\nu} \to g(x)F_{\mu\nu} - f_{\mu\nu}(x)$ where the scalar and antisymmetric functions $g(x)$ and $f_{\mu\nu}$ will comprise the gauge functions. The 4-vector source of the electromagnetic field is determined independently of the propagation of the field itself and cannot therefore vary in any gauge transformation. Consequently, Equations 18.8 will be invariant if and only if $g(x) = 1$ and $\partial_\mu f^{\mu\nu} = 0 = \partial_\mu \tilde{f}^{\mu\nu}$ identically in x. Such identities require $f^{\mu\nu} = 0$ since the solution of Maxwell's equation with no source anywhere is a constant and that constant must be zero since we require the field to vanish at large distances.

Thus we see that the first-order Maxwell's equations have no gauge freedom and all six components of **E** and **B** are measurable or physical. An analysis of the dynamics of the first-order Maxwell's equations shows in fact that, once the constraints they imply on the initial data are taken into account, the number of equations precisely matches the number of components to be evolved forward in time from the initial data. We shall examine this property in more detail later in Section 19.6.

If the potentials in the second-order Maxwell's equations, 18.22, are to be gauge-variant, then the gauge transformation of A_μ must take the form $A_\mu \to A_\mu + a_\mu(x)$ with $a_\mu(x)$ an arbitrary function of spacetime position. Equation 18.22 will be invariant under this transformation if, and only if, $\partial^\nu(\partial_\mu a_\nu - \partial_\nu a_\mu) = 0$. This equation can be satisfied identically and covariantly with a non-trivial function $a_\mu(x)$ given by $a_\mu = \partial_\mu \alpha(x)$ where $\alpha(x)$, the *gauge function*, is an arbitrary scalar field.

We have shown therefore that A^μ has gauge freedom and that the second-order form of Maxwell's equations exhibits local gauge invariance. Consequently, not all the components of A^μ can be independent physical degrees of freedom of which there can be a maximum of three after the arbitrariness of the gauge function has been removed. The gauge invariance of the 4-vector electromagnetic potential thus involves a transformation of the form:

$$A_\mu \to A_\mu + \partial_\mu \alpha(x) \quad \text{or} \quad \Delta A_\mu = \partial_\mu \alpha(x) \ . \tag{18.33}$$

This gauge transformation leaves not only the first-order Maxwell's equations invariant but also the electromagnetic field tensor $F_{\mu\nu}$ itself. Since $F_{\mu\nu}$ has no gauge freedom we say that the first-order Maxwell's equations are *trivially gauge-invariant* in contrast to the wave equation, 18.22, satisfied by the electromagnetic potential. An electromagnetic potential having the form $A_\mu = \partial_\mu \alpha$ is a *pure gauge field* for which $F_{\mu\nu} = 0$ and the energy-momentum tensor (see Equation 18.49 and Exercise 19.11) correspondingly vanishes. In the stationary limit, only the spatial part of Equation 18.33, $\mathbf{A} \to \mathbf{A} + \nabla\xi$, survives as a gauge invariance of magnetostatics. The temporal part is replaced by the gauge freedom $\phi(t, \mathbf{x}) \to \phi(t, \mathbf{x}) + \alpha(t)$ of electrostatics analogous to a similar result for Newtonian gravity (see

Section 7.10)

To reduce the second-order electromagnetic potentials to a set which are physical, we must impose a restriction in order to remove the gauge freedom. The new sets of variables will be referred to as being in a particular *gauge* and the restriction is called a *gauge-fixing condition*. However, we cannot use an arbitrary restriction which just happens to give the correct number of physical degrees of freedom. Instead we must use only gauge-fixing conditions which lead to new dynamical variables which can be related to the original gauge fields by a gauge transformation. We shall examine some well-known electromagnetic gauges in the next chapter.

18.10 Boost transformation of field strengths

It is well-known that a field perceived to be purely electric **E** in one inertial frame will not be so in any other boosted with respect it but will have a non-zero value of the magnetic induction **B** in the new frame. The same is true for a field which is a purely magnetic field in one frame. Just as a boost mixes space and time coordinate intervals so also does it mix the electric and magnetic 3-vectors, **E** and **B**. To establish these results we shall use the tensor transformation law applied to the electromagnetic field tensor $F^{\mu\nu}$ choosing the x-axis aligned with the boost. We shall then apply rotational covariance to get the expression for an arbitrary boost.

$F^{\mu\nu}$ will transform according to $F^{\bar\mu\bar\nu} = \Lambda^{\bar\mu}{}_\mu \Lambda^{\bar\nu}{}_\nu F^{\mu\nu}$. This relation shows already that the different parts of the electromagnetic field are mixed in a boost. To illustrate this explicitly, make a 1+3 (space-time) split by noting that the only values of the indices $\mu\nu$ which give non-zero values of $F^{\mu\nu}$ are $0k$, $k0$ and kl with $k \neq l$. It suffices to examine only the index combinations $0k$ and kl with $k < l$. Taking first the x-component of the electric field gives,

$$\begin{aligned}
E^{\bar 1} &= c F^{\bar 0 \bar 1} \\
&= \Lambda^{\bar 0}{}_\mu \Lambda^{\bar 1}{}_\nu (c F^{\mu\nu}) \\
&= \Lambda^{\bar 0}{}_0 \Lambda^{\bar 1}{}_1 c F^{01} + \Lambda^{\bar 0}{}_1 \Lambda^{\bar 1}{}_0 c F^{10} \\
&= \gamma^2 (E^1) + (-\gamma\beta)^2(-E^1) \\
&= E^1 \, .
\end{aligned} \qquad (18.34)$$

where we have used Equation 13.45. In the same way, one finds that $E^{\bar 2} = -\gamma\beta c B^3 + \gamma E^2$ and $E^{\bar 3} = \gamma\beta c B^2 + \gamma E^3$. We use $\mathbf{v} = (v, 0, 0)$ for the relative velocity of the boost and $\mathbf{E}_\perp$ and $\mathbf{E}_\parallel$ for the projections of **E** normal and parallel to **v**. These three results may then be combined with those for the magnetic induction, to obtain the boost transformations of the electromagnetic field strengths:

$$\bar{\mathbf{E}}_\parallel = \mathbf{E}_\parallel \quad \text{and} \quad \bar{\mathbf{E}}_\perp = \gamma(\mathbf{E}_\perp + \mathbf{v}\times\mathbf{B}_\perp) \, ,$$

18.10. Boost transformation of field strengths

$$\bar{\mathbf{B}}_{\parallel} = \mathbf{B}_{\parallel} \quad \text{and} \quad \bar{\mathbf{B}}_{\perp} = \gamma(\mathbf{B}_{\perp} - \mathbf{v} \times \mathbf{E}_{\perp}) , \tag{18.35}$$

which, because they are expressed in a rotationally covariant way, remain valid for an arbitrary orientation of the axes relative to the boost.

Exercise 18.7 *Show that we may eliminate explicit reference to the projections of* **E** *and* **B** *parallel and perpendicular to the boost by writing the above equations in the form:*

$$\begin{aligned}
\bar{\mathbf{E}} &= \gamma(\mathbf{E} + \mathbf{v} \times \mathbf{B}) - \frac{\gamma - 1}{\beta^2}(\boldsymbol{\beta} \cdot \mathbf{E})\boldsymbol{\beta} , \\
\bar{\mathbf{B}} &= \gamma(\mathbf{B} - \mathbf{v} \times \mathbf{E}) - \frac{\gamma - 1}{\beta^2}(\boldsymbol{\beta} \cdot \mathbf{B})\boldsymbol{\beta} .
\end{aligned} \tag{18.36}$$

It should be clear that the apparent complexity in these results, compared to those for $F^{\mu\nu}$, is primarily due to expressing them in 1+3 notation.

The values of the two Lorentz invariants, $\mathbf{E} \cdot \mathbf{B}$ and $\mathbf{B}^2 - \mathbf{E}^2/c^2$, permit the invariant classification of electromagnetic fields into four types:

- If $\mathbf{E} \cdot \mathbf{B} \neq 0$ and $\mathbf{B}^2 - \mathbf{E}^2/c^2 \neq 0$ the field is fully general or *generic*.
- If **E** and **B** are orthogonal, $\mathbf{E} \cdot \mathbf{B} = 0$, and $\mathbf{B}^2 - \mathbf{E}^2/c^2 \neq 0$, then the field can in some frame be reduced to a *pure electric* or *pure magnetic* field, depending on the sign of the second invariant. Consequently, the field could be described as being of 'pure electric' or 'pure magnetic' type in any frame despite the fact that both 3-vectors will in general be non-zero. Conversely, if **E** or **B** is zero (but not both) in some reference frame, then the fields will be orthogonal ($\mathbf{E} \cdot \mathbf{B} = 0$) in all reference frames.
- If $\mathbf{E} \cdot \mathbf{B} \neq 0$ and $\mathbf{B}^2 - \mathbf{E}^2/c^2 = 0$ then the electric and magnetic fields are of equal magnitude in natural units (with $c = 1$) but not orthogonal.
- If $\mathbf{E} \cdot \mathbf{B} = 0$ and $\mathbf{B}^2 - \mathbf{E}^2/c^2 = 0$ but neither of **E** nor **B** is zero the field is said to be a *radiation* or *null* field. Plane electromagnetic waves fall into this category.

Exercise 18.8 *Show that* $\mathbf{E} \cdot \mathbf{B} = 0$ *and* $\mathbf{B}^2 - \mathbf{E}^2/c^2 = 0$ *for electromagnetic waves.*

Exercise 18.9 *Suppose that, in a particular frame, we have a pure electric field* **E**. *Find, in terms of that electric field, the values of the electric and magnetic fields in a frame boosted by a velocity* **v** *relative to the first and verify that in the new frame* $\bar{\mathbf{E}} \cdot \bar{\mathbf{B}} = 0$. *What is the value of* $\bar{\mathbf{B}}^2 - \bar{\mathbf{E}}^2/c^2$?

Exercise 18.10 *Suppose that an electromagnetic field has* $\mathbf{E} \cdot \mathbf{B} \neq 0$. *This result will be true in all frames and means that the field cannot be a pure* *electric or a pure magnetic field in any frame. Show that a boost in the direction of* **E**×**B** *of magnitude* $\beta = \tanh \alpha$ *for which* α *is given by,*

$$\tanh 2\alpha = \sqrt{\frac{\mu_0}{\epsilon_0}} \frac{|\mathbf{E} \times \mathbf{B}|}{\epsilon_0 \mathbf{E}^2 + \mathbf{B}^2/\mu_0} \ , \qquad (18.37)$$

will result in new **E** *and* **B** *vectors which are parallel. What does this imply about the flux of electromagnetic energy in the new frame? Note that an electromagnetic field for which* **E**·**B** $= 0$ *will not have* **E** *parallel to* **B** *in any frame as* **E**·**B** $= 0$ *is an invariant condition and such a field must be ruled out from those considered here.*

18.11 Magnetic charge and magnetic monopoles

We could ask ourselves what are the consequences of relaxing one or more of the conditions which we used to deduce Maxwell's equations. A number of obvious possibilities present themselves. One of these is to retain an invariant charge as source but not to demand that conservation be automatically incorporated in the field equation. This leads to a field which propagates at a frame-dependent speed $|\dot{z}| < c$ corresponding to particles with non-zero rest mass and spin 1. The field is given the name of Proca [462].

We could consider a complex rather than a real scalar as integrated source. Such a procedure leads to a complex electromagnetic field which on second quantization may be interpreted in terms of particles which are not self-conjugate — the quantum and its antiparticle are distinct. Unlike the photon, such particles will carry a charge of some form. We could consider the possibility of constructing a field based on an integrated source which, instead of being an invariant, is the next simplest Poincaré-covariant construct, namely a conserved 4-vector. This is precisely equivalent to the characterization of gravity as an interaction sourced by mass, and therefore energy and momentum, whose covariant form is a 4-vector p^μ. This case will be addressed in Chapter 21. Nothing new would be obtained by simply replacing a reflection invariant source in our development of Maxwell's equations by a pseudoscalar source since the classification of charge as a scalar rather than a pseudoscalar is a matter of convention, with the former being chosen because invariance is slightly simpler to handler than reversal of sign. However, another possibility is not to demand that the system be parity-covariant at all.

At this point we examine the consequences of allowing, in addition to a scalar source, one which is spatially improper and temporally proper. We return to the point where in setting up Maxwell's equations we appealed to a fundamental criterion that the only local source is a proper 4-vector current from which we deduced that the integrated source is a proper scalar.

18.11. Magnetic charge and magnetic monopoles

We now relax this criterion to admit in addition a pseudoscalar integrated source $\tilde{q}$ with corresponding pseudo-4-vector local source $\tilde{J}^\mu(x)$. We then proceed exactly as before; parity covariance still requires the first-order field equations to be Maxwell's equations except for the appearance of the new source in the second equation, $\partial_\mu \tilde{F}^{\mu\nu} = \mu_0 \tilde{J}^\nu$. We have made use of the freedom to choose the units and dimensions of $\tilde{q}$ to set them equal to those of q.

Exercise 18.11 Establish the 1+3 form, namely,

$$\nabla \cdot \mathbf{E} = \frac{\rho}{\epsilon_0} \tag{18.38}$$

$$\nabla \cdot \mathbf{B} = \mu_0 c \tilde{\rho} \tag{18.39}$$

$$\nabla \times \mathbf{E} = -\dot{\mathbf{B}} - \mu_0 c \tilde{\mathbf{J}} \tag{18.40}$$

$$\nabla \times \mathbf{B} = \frac{1}{c^2}\dot{\mathbf{E}} + \mu_0 \mathbf{J} \,, \tag{18.41}$$

of the extended Maxwell's equations, from the covariant form.

The *extended Maxwell's equations* exhibit a much greater degree of symmetry between the electromagnetic field tensor and its dual than is the case for the standard Maxwell's equations. This would be even more apparent using natural units ($\epsilon_0 = \mu_0 = c = 1$) or the variables $\mathbf{D} = \epsilon_0 \mathbf{E}$ and $\mathbf{H} = \mathbf{B}/\mu_0$ or if $\mathbf{E}/c$ was replaced by $\mathbf{E}$ with the same dimensions as $\mathbf{B}$. In particular, $\tilde{q}$ and $\tilde{J}^\mu$ have properties completely analogous to electric charge q and current J^μ but related to the magnetic induction rather than the electric field. The quantity $\tilde{q}$ is therefore referred to as *magnetic charge*.

The second Maxwell's equation, for $\tilde{F}^{\mu\nu}$, is no longer an integrability condition guaranteeing the existence of a single vector potential whose covariant curl gives the field strength. The generalization of the Poincaré lemma to this case [287] shows that two 4-vector potentials, A^μ and $\tilde{A}^\mu$, one corresponding to each type of charge, are required. In terms of these potentials the field strength and its dual are given by:

$$\begin{aligned} F_{\mu\nu} &= \partial_\mu A_\nu - \partial_\nu A_\mu - \tilde{\,}(\partial_\mu \tilde{A}_\nu - \partial_\nu \tilde{A}_\mu) \\ &= \partial_\mu A_\nu - \partial_\nu A_\mu - \tfrac{1}{2}\epsilon_{\mu\nu\lambda\rho}(\partial^\lambda \tilde{A}^\rho - \partial^\rho \tilde{A}^\lambda) \,, \end{aligned} \tag{18.42}$$

and

$$\begin{aligned} \tilde{F}_{\mu\nu} &= \tilde{\,}(\partial_\mu A_\nu - \partial_\nu A_\mu) + \partial_\mu \tilde{A}_\nu - \partial_\nu \tilde{A}_\mu \\ &= \tfrac{1}{2}\epsilon_{\mu\nu\lambda\rho}(\partial^\lambda A^\rho - \partial^\rho A^\lambda) + \partial_\mu \tilde{A}_\nu - \partial_\nu \tilde{A}_\mu \,, \end{aligned} \tag{18.43}$$

in which the tilde ($\tilde{\,}$) is used to distinguish the two types of charge and 4-vector. It is also used to denote the dual $\tilde{F}^{\mu\nu} = \tfrac{1}{2}\epsilon^{\mu\nu\lambda\rho}F_{\lambda\rho}$ of $F^{\mu\nu}$ and the duality operation on any pair of antisymmetric Lorentz indices, namely those in the covariant curls of each potential.

Exercise 18.12 *Establish the following field equations,*

$$\Box A - \partial \cdot A = -\mu_0 J \quad \text{and} \quad \Box \tilde{A} - \partial \cdot \tilde{A} = -\mu_0 \tilde{J} \,, \tag{18.44}$$

for the potentials A^μ and $\tilde{A}^\mu$.

The extended equations describe a parity-covariant system with double the number of independent degrees of freedom of the standard Maxwell equations; the modified equations nevertheless suffice to evolve the independent initial data forward in time.

A fundamental particle having magnetic charge and no higher multipole magnetic properties, such as dipole or quadrupole moment, has come to be referred to as a *magnetic monopole*. A consistent theory of point-like magnetic charge was first proposed in a quantum context by Dirac [56, 109, 178, 488]. A consequence of Dirac's theory is that the product of the charge q of any particle with the magnetic charge $\tilde{q}$ of any other particle is quantized according to $q\tilde{q} = 2\pi\epsilon_0 \hbar c n$ for integer n. The observation of a Dirac monopole anywhere in the universe with a magnetic charge which is some multiple of $2\pi\epsilon_0 \hbar c/e$ would then be directly linked to the observed quantization of free electric charges according to $q = ne$ (for integer n). Two types of Dirac monopole are possible — a north pole monopole and a south pole or *antimonopole*, which are created and annihilated in pairs. Alternative descriptions of electric charge quantization are based on non-Abelian gauge theories and the existence of quarks.

Magnetic monopoles are not a necessity in electrodynamics. However, some *grand unified theories* of the electromagnetic and nuclear interactions predict the existence of '*t Hooft-Polyakov magnetic monopoles* [455, 536] of extremely high mass, $\approx 137 m_w \approx 10^4 \, \text{GeV}/c^2$, m_w being the rest mass of the $W^\pm$ intermediate vector boson. However, they are not point-like but of topological origin. Despite extensive experimental searches there are no confirmed observations that particles exist with magnetic charge [178].

In view of the observational absence of magnetic monopoles, the electromagnetic interaction appears to be fundamentally asymmetric with respect to electric and magnetic charge and this asymmetry appears to be reflected in the asymmetric appearance of Maxwell's equations. In fact, part of the asymmetry of the equations is conventional and part is physical. We could include both types of charge in Maxwell's equations to give them an apparently more symmetrical form. The structure of the equations would then show that a certain linear combination of the two 3-vector fields involved, **E** and **B**, has vanishing 3-divergence corresponding to a vanishing of a particular linear combination of electric and magnetic charge. Rather

18.12. Electromagnetic energy-momentum tensor

than conclude that only one of the two charges (electric and magnetic) is observed, we may argue that the two charges are in the same fixed (pseudoscalar) ratio for all particles [263]. The actual form used for Maxwell's equations can then be seen as arising from the choice of a particular phase, in an invariance called *duality symmetry* [178, 384], to arbitrarily set one charge to zero so that the charge appears to be entirely electric rather than entirely magnetic or some mixture.

The transformation:

$$A^\mu \rightarrow A^\mu \cos\alpha + \tilde{A}^\mu \sin\alpha \quad (18.45)$$

$$\tilde{A}^\mu \rightarrow \tilde{A}^\mu \cos\alpha - A^\mu \sin\alpha \, , \quad (18.46)$$

with analogous rules for J^μ and $\tilde{J}^\mu$, is called a *duality rotation* of the electromagnetic field.

Exercise 18.13 *Show that the corresponding transformation for $F^{\mu\nu}$ is,*

$$F^{\mu\nu} \rightarrow F^{\mu\nu} \cos\alpha + \tilde{F}^{\mu\nu} \sin\alpha \, , \quad (18.47)$$

and that duality rotation is a symmetry of the extended equations.

The phase angle α in the above duality rotation is called the *complexion* of the electromagnetic field [385]. Any value of the complexion may be chosen to describe the physics of electromagnetism extended to include magnetic charge. In the same way the physics of the actual electromagnetic field, with a single charge, may be described equally well by choosing $J^\mu = 0$ or $\tilde{J}^\mu = 0$ (or neither zero) with any value of α. By convention we take $\tilde{J}^\mu = 0$ and a zero value for the complexion and thereby introduce a certain additional degree of apparent asymmetry into Maxwell's equations over and above the absence of magnetic charge. If we chose $\alpha = \pi/4$ with $J^\mu \neq 0 \neq \tilde{J}^\mu$ then the duality symmetry that exists between the electric and magnetic fields would be made manifest. No advantage is to be gained by using a non-zero value of the complexion.

18.12 Electromagnetic energy-momentum tensor

We shall derive the entire energy-momentum tensor $T^{\mu\nu}$ of the electromagnetic field later using Noether's theorem. In the meantime, we shall show how it can be constructed using a few plausible *ad hoc* ideas.

Energy-momentum is a measurable physical quantity and thus can have no gauge freedom. We therefore construct the energy-momentum tensor $T^{\mu\nu}$ in terms of the gauge-invariant antisymmetric electromagnetic field tensor $F^{\mu\nu}$. Equivalently, in terms of the electromagnetic potential A^μ, the only combination that may enter is its covariant curl, $\partial_\mu A_\nu - \partial_\nu A_\mu$.

$T^{\mu\nu}$ must be a proper tensor symmetric with respect to μ and ν. No symmetric tensor may be constructed from terms having only a single factor of $F^{\mu\nu}$ combined with the universal tensors $\eta_{\mu\nu}$ and $\epsilon^{\mu\nu\lambda\rho}$. Noting also that $T^{\mu\nu}$ for a particle is quadratic in the time derivatives $\dot{z}^\mu$ of the dynamical variables, we consider quadratic combinations of $F^{\mu\nu}$ which itself contains spacetime derivatives of A^μ. The indices μ and ν cannot both be on one factor of $F^{\mu\nu}$ in any term contributing to $T^{\mu\nu}$.

Consequently, we consider a term of the form $F^{\mu\lambda}F^\nu{}_\lambda$ which is symmetric in $\mu\nu$. However, we could also put neither index on a factor of $F^{\mu\nu}$ but place them instead on a factor of $\eta^{\mu\nu}$ since it is also symmetric. To maintain common dimensionality we multiply this by an invariant quadratic in $F^{\mu\nu}$, for example $F_{\lambda\pi}F^{\lambda\pi}$. We do not consider either of the other two second-rank tensors quadratic in $F^{\mu\nu}$ (see Equations 14.42 and 14.43 in Exercise 14.23) since one is improper and the other is a linear combination of the two we are already considering.

Thus we consider an electromagnetic energy-momentum tensor having the form,

$$T^{\mu\nu} = a\, F^{\mu\lambda}F^\nu{}_\lambda + b\, \eta^{\mu\nu} F_{\lambda\pi}F^{\lambda\pi} \tag{18.48}$$

where a and b are constants.

Exercise 18.14 *Show that the energy-momentum tensor of Equation 18.48 is conserved identically (for all $F^{\mu\nu}$) as a result of the free field equations (Equation 18.8 with $J^\mu = 0$) provided $b = -a/4$ and that the dimensions of a are $[a]=[\mu_0^{-1}]$.*

The overall scale of the proportionality constant may be fixed to $a = 1/\mu_0$ by coupling the electromagnetic field to any system, such as a charged particle, for which the energy details are known or by constructing $T^{\mu\nu}$ as the Noether current of spacetime translations (see Section 18.13 and Chapter 19). The electromagnetic energy-momentum tensor is therefore:

$$T^{\mu\nu} = \tfrac{1}{\mu_0}(F^{\mu\lambda}F^\nu{}_\lambda - \tfrac{1}{4}\eta^{\mu\nu} F_{\lambda\pi}F^{\lambda\pi}) . \tag{18.49}$$

with energy density:

$$\begin{aligned} T^{00} &= \tfrac{1}{\mu_0}(F^{0\lambda}F^0{}_\lambda + \tfrac{1}{4}F_{\lambda\pi}F^{\lambda\pi}) \\ &= \tfrac{1}{\mu_0}(\mathbf{E}^2/c^2 + \tfrac{1}{2}\{\mathbf{B}^2 - \mathbf{E}^2/c^2\}) \\ &= \tfrac{1}{2}(\epsilon_0\mathbf{E}^2 + \mathbf{B}^2/\mu_0) . \end{aligned} \tag{18.50}$$

We can now make a 1+3 split in the other two independent parts, T^{kl} and $T^{k0} = T^{0k}$.

Exercise 18.15 *Verify that the space-time part T^{k0} of the electromagnetic energy-momentum tensor is c times the k-momentum or $1/c$ times the* energy flux in the k-direction by showing that it reduces to the Poynting vector [459] $\sqrt{\epsilon_0/\mu_0}\mathbf{E}\times\mathbf{B}$ in 1+3 notation.

Exercise 18.16 Show that, in source regions ($J^\mu \neq 0$), the divergence of the energy-momentum tensor of the electromagnetic field alone is given by $\partial_\mu T^{\mu\nu} = J_\mu F^{\mu\nu}$, showing that conservation requires the inclusion in $T^{\mu\nu}$ of contributions from the source.

Exercise 18.17 Show that the electromagnetic stress tensor is given by:

$$T^{kl} = \frac{1}{\mu_0}\left(F^{k\lambda}F^l{}_\lambda - \tfrac{1}{4}\delta^{kl}2(\mathbf{B}^2 - \mathbf{E}^2/c^2)\right), \tag{18.51}$$

which can be reduced to:

$$T^{kl} = \tfrac{1}{2}\delta^{kl}(\epsilon_0\mathbf{E}^2 + \mathbf{B}^2/\mu_0) - (\epsilon_0 E^k E^l + B^k B^l/\mu_0), \tag{18.52}$$

with the use of a few identities involving 3D permutation tensors.

By comparing this result with the stress tensor for a fluid with anisotropic pressure we may interpret it as implying that the electromagnetic field has a (non-negative) pressure perpendicular to the local direction of the field, a negative pressure or tension parallel to the field, both of magnitude $\tfrac{1}{2}(\epsilon_0\mathbf{E}^2 + \mathbf{B}^2/\mu_0)$, and a non-zero shear stress ($T^{kl} \neq 0$ for $k \neq l$.) Faraday's analyses of electromagnetism using lines of force under tension and exerting pressure were influential in the work of Maxwell [363, 364] from 1855 to 1862 leading to his electromagnetic theory of light.

Exercise 18.18 Show that the electromagnetic energy-momentum tensor has a zero trace. This property (see Section 15.13) is shared by all systems whose propagation in vacuo is at the ultimate speed.

18.13 Maxwell and Proca Lagrangian densities

Suppose we take first the electromagnetic field tensor $F^{\mu\nu}$ satisfying the standard first-order equations in vacuo, $\partial_\mu F^{\mu\nu} = 0 = \partial_\mu \tilde{F}^{\mu\nu}$, and seek a free Lagrangian density from which these are to be derived according to the Euler-Lagrange equations, obtained by putting $\psi = \{F^{\mu\nu}\}$ in Equation 16.9,

$$\partial_\lambda \frac{\partial \mathcal{L}}{\partial(\partial_\lambda F_{\mu\nu})} = \frac{\partial \mathcal{L}}{\partial F_{\mu\nu}}. \tag{18.53}$$

Maxwell's equations in vacuo are linear and homogeneous and the Lagrangian must therefore be a homogeneous quadratic in the field variables $F_{\mu\nu}$ and their first derivatives $\partial_\lambda F_{\mu\nu}$. Since they are first-order, each term contributing towards the Lagrangian will also be first-order. A typical term will therefore be of the form $F_{\pi\rho}\partial_\lambda F^{\mu\nu}$ in which each of the five free indices must be contracted with some other index since the Lagrangian density must always be a scalar. These indices can only be contracted with one another (which effectively means contraction in pairs with factors of $\eta^{\mu\nu}$) or with factors of $\epsilon^{\mu\nu\lambda\rho}$. Inclusion of $\epsilon^{\mu\nu\lambda\rho}$ means that terms involving the dual fields $\tilde{F}^{\mu\nu}$ and their derivatives are potentially candidates. However, since both these universal tensors are of even rank it is impossible to saturate all indices and there can be no scalar Lagrangian which reproduces the first-order Maxwell's equations. The electromagnetic field tensor alone is not a Lagrangian set of fields.

We consider next the second-order form, $\Box A - \partial\partial \cdot A = 0$, of Maxwell's equations in vacuo in which the field variables are the components $A_\mu(x)$ of the 4-vector potential A. If A_μ are taken as Lagrangian generalized coordinates with corresponding covariant generalized velocities $\partial_\mu A_\nu$ then the Euler-Lagrange equations will have the form,

$$\partial_\mu \frac{\partial \mathcal{L}}{\partial(\partial_\mu A_\nu)} = \frac{\partial \mathcal{L}}{\partial A_\nu} . \tag{18.54}$$

The field equations are linear and homogeneous in A_μ and the Lagrangian must therefore be a homogeneous quadratic in the potential and its derivatives. Furthermore, since we are dealing with a second-order field equation, the first derivatives must appear quadratically in $\mathcal{L}$. Finally, since the second derivatives appear in the field equation in the combination $\partial_\mu(\partial^\mu A^\nu - \partial^\nu A^\mu)$, it suffices, in seeking the correct quadratic in the first derivatives of A^μ, to consider scalar quadratics in $F^{\mu\nu}$ (and possibly its dual via $\epsilon^{\mu\nu\lambda\rho}$) where $F^{\mu\nu}$ is considered here not as a Lagrangian field but as an abbreviation for the covariant curl of A^μ. The only such independent quadratic invariants are proportional to $F_{\mu\nu}F^{\mu\nu}$ and $F_{\mu\nu}\tilde{F}^{\mu\nu}$ since $\tilde{F}_{\mu\nu}\tilde{F}^{\mu\nu}$ is not independent of these two (see Exercise 14.23).

The Lagrangian for a free particle is a proper scalar to which we shall have to add the Lagrangian for the electromagnetic field to allow interaction to take place between charged particles. Since electrodynamics is a parity-covariant theory, this means that the Lagrangian for the free electromagnetic field must also be a proper scalar. Consequently, we take the electromagnetic Lagrangian to be proportional to the proper invariant $F_{\mu\nu}F^{\mu\nu}$.

We take the standard Lagrangian density for the free electromagnetic field to have the following SI form:

$$\begin{align} \mathcal{L} &= -\tfrac{1}{4\mu_0 c}(\partial_\mu A_\nu - \partial_\nu A_\mu)(\partial^\mu A^\nu - \partial^\nu A^\mu) \tag{18.55} \\ &= -\tfrac{1}{2\mu_0 c}(\partial_\mu A_\nu \partial^\mu A^\nu - \partial^\nu A^\mu \partial_\mu A_\nu) \tag{18.56} \\ &= -\tfrac{1}{4\mu_0 c} F_{\mu\nu} F^{\mu\nu} \quad \text{with} \quad F_{\mu\nu} = \partial_\mu A_\nu - \partial_\nu A_\mu , \tag{18.57} \end{align}$$

18.13. Maxwell and Proca Lagrangian densities

in which the constant $-1/\mu_0 c$ gives the correct dimensions and energy positivity while the factor of 4 ensures that standard forms of equations are obtained when interaction occurs, as for example with charged particles described by a Lagrangian. We shall shortly verify that this expression leads to the energy-momentum tensor of Equation 18.49, established heuristically earlier, and which we know corresponds to a positive energy density and therefore positive energy.

We can verify that this Lagrangian does lead to the unsourced second-order Maxwell's equations. From the assumed mathematical independence of the generalized coordinates and velocities, we have $\partial A_\mu / \partial A_\nu = \delta^\nu_\mu$ and $\partial_\mu A_\nu / \partial_\pi A_\kappa = \delta^\pi_\mu \delta^\kappa_\nu$. The components A^μ are of course independent of the velocities in the Lagrangian formalism. It is then straightforward to verify that $\partial \mathcal{L} / \partial (\partial_\mu A_\nu) = -(\partial^\mu A^\nu - \partial^\nu A^\mu)/\mu_0 c = -F^{\mu\nu}/\mu_0 c$. Since $\partial \mathcal{L} / \partial A_\nu = 0$, the Euler-Lagrange equations show that this Lagrangian does indeed give the correct gauge-invariant second-order Maxwell's equations $\Box A^\mu - \partial^\mu \partial \cdot A = 0$ for the gauge-variant potential A^μ.

Exercise 18.19 Show that the canonical energy-momentum tensor of the electromagnetic field A^μ is given by:

$$\check{T}^{\mu\nu} = \tfrac{1}{\mu_0}(F^{\mu\lambda}\partial^\nu A_\lambda - \tfrac{1}{4}\eta^{\mu\nu}F_{\mu\nu}F^{\mu\nu}) \, . \tag{18.58}$$

Problems with the canonical energy-momentum tensor are illustrated by the incorrect result given for the electromagnetic field in Equation 18.58 which is clearly gauge-variant and not symmetric.

The second-order equation for a non-null field A^μ with spin 1 may be formed, as in the Klein-Gordon case, by including a fundamental length λ which we presume arises quantum-mechanically as the reduced Compton wavelength of a mass. To ensure there are three independent components in A^μ, appropriate to spin 1, the initial data must be constrained by an identically vanishing divergence on A^μ. The resulting equations, in natural units,

$$(\Box - m^2)A = 0 \quad \text{and} \quad \partial \cdot A = 0 \quad (m \neq 0) \, , \tag{18.59}$$

are referred to as the second-order *Proca equations* [462] for a field whose quantization leads to massive particles of spin 1. There are no observed particles whose natural description uses the Proca field. The $W^\pm$ and Z^0 particles which combine with the photon to mediate the electroweak interaction are massive spin 1 particles but these arise from massless Yang-Mills fields, namely sets of Maxwell-like fields, which acquire mass quantum-mechanically by spontaneous symmetry breaking.

Exercise 18.20 Show (a) that the first-order Proca equations,

$$\partial_\nu F^{\mu\nu} + m^2 A^\mu = 0 \quad \text{and} \quad F_{\mu\nu} = \partial_\mu A_\nu - \partial_\nu A_\mu \, , \tag{18.60}$$

are equivalent to Equations 18.59 provided $m \neq 0$, and (b) that the second-order Proca equations show no gauge freedom. Show (c) that the Proca equations arise from a Lagrangian density (in natural units) of the form,

$$\mathcal{L} = -\tfrac{1}{4} F_{\mu\nu} F^{\mu\nu} - \tfrac{1}{2} m^2 A^2 \,, \tag{18.61}$$

in which $F^{\mu\nu}$ is an abbreviation for the covariant curl of A^μ and (d) that the second-order Proca Lagrangian (which contains the Maxwell case also when $m = 0$) is singular. (e) Show that a first-order Lagrangian for the Proca and Maxwell fields is provided (in natural units) by,

$$\mathcal{L} = \tfrac{1}{4} F_{\mu\nu} F^{\mu\nu} - \tfrac{1}{2} F^{\mu\nu} (\partial_\mu A_\nu - \partial_\nu A_\mu) - \tfrac{1}{2} m^2 A^2 \,, \tag{18.62}$$

in which the mathematically independent Lagrangian fields are the ten components $\{A_\mu, F_{\mu\nu}\}$.

Although each component of A^μ for the free Proca field satisfies the Klein-Gordon equation, the constraint $(\partial \cdot A = 0)$ means that the propagation of A^μ is not simply equivalent to the independent propagation of a set of scalar fields. The constraint does not arise from the second-order Lagrangian which is therefore incomplete. To form a non-singular Proca Lagrangian suitable for a canonical analysis and for making a transition to the Hamiltonian formalism, manifest covariance must be abandoned.

Exercise 18.21 (a) Make a 1+3 split of the first-order Proca equations (in natural units) by defining 'electric' and 'magnetic' Proca field strengths **E** and **B** with Equation 18.13 and 3-scalar and 3-vector potentials using Equation 18.23. (b) Show that only **A** and **E** are true dynamical variables. (c) Substitute the expressions for **B** and ϕ in terms of initial data on **A** and **E** into the 1+3 split of the first-order Lagrangian to obtain the following non-covariant Proca Lagrangian density,

$$\mathcal{L}_{\text{canonical}} = -\mathbf{E} \cdot \dot{\mathbf{A}} - \tfrac{1}{2} \left\{ \mathbf{E}^2 + \mathbf{B}^2 + m^2 \mathbf{A}^2 + \frac{(\boldsymbol{\nabla} \cdot \mathbf{E})^2}{m^2} \right\} \,. \tag{18.63}$$

(d) Show that the Proca Lagrangian of Equation 18.63 is non-singular. (e) Demonstrate that the Euler-Lagrange equations of the non-covariant Lagrangian are identical to the true dynamical equations of the covariant Lagrangian. (f) Show that the Hamiltonian density corresponding to the non-covariant Proca Lagrangian is positive-definite. (g) Explain why the analysis outlined in this exercise cannot all be used directly for a similar analysis of the Maxwell field.

Further details on the classical and quantized Proca fields are available in Sudarshan and Mukunda [525, p. 496] and Itzykson and Zuber [262, p. 134].

Exercise 18.22 *(a) Split the fields $\{A_\mu, F_{\mu\nu}\}$ for the first-order Maxwell Lagrangian ($m = 0$, $c = 1 = \mu_0$) into $\{\phi, \mathbf{A}, \mathbf{E}, \mathbf{B}\}$ to show that the corresponding Euler-Lagrange equations are 18.9, 18.12 and 18.25 and that the remaining free Maxwell's equations, obtained from 18.10 and 18.11 are satisfied identically. (b) State which of the four Euler-Lagrange equations are true dynamical equations. (c) Show that only one of the two non-dynamical equations provides an expression for some of the original ten variables allowing them to be eliminated in favour of the dynamical variables.*

We shall examine the Maxwell Lagrangian further in the next chapter along with the gauge invariance of Maxwell's equations.

Bibliography:

Einstein A 1931 *Maxwell's influence on ... the conception of physical reality* [535].
Jackson J D 1975 *Classical electrodynamics* [263].
Longair M S 1983 *Theoretical concepts in physics* [326].
Ohanian H 1988 *Classical electrodynamics* [405].
Page L and Adams N I, Jr. 1940 *Electrodynamics* [416].
Stratton J A 1941 *Electromagnetic theory* [522].

CHAPTER 19

Electrodynamics

Les bonnes théories sont souples[1].

Henri Poincaré 1854–1912

19.1 Lorentz law of charged particle motion

The name *electrodynamics* was introduced by Ampère [12] in 1825 to describe the interaction of current conductors. It is now used to describe the combined dynamics of the electric and magnetic fields and the electric charges which give rise to those fields and on which they act. The influence on the development of relativity of the important milestones in electrodynamics, in particular the electron theory of Lorentz, were discussed in the introduction to Chapter 12.

To complement the construction, given in Chapter 18, of Maxwell's equations for the electromagnetic field of a given source, the simplest covariant forms will be examined for the reciprocal equations describing the reaction of that field on the source in the form of non-quantum charged particles. This will complete the description of the non-quantum dynamics of a closed system of charges interacting electromagnetically.

The early sections of Chapter 15 show us how to set about seeking the explicitly covariant form for the rate at which the momentum of a charged particle (with non-zero rest mass) is changed by interaction with an electromagnetic field. The first step is to decide not to attempt to establish a formula for the non-covariant quantity $d\mathbf{p}/dt$ appearing in the familiar form of the Lorentz law. Instead, we replace the Euclidean time parameter t by an arbitrary evolution parameter λ. We then seek an equation not just involving relativistic 3-momentum $\mathbf{p}$ but instead the covariant concept of 4-momentum. This means we seek, in addition, an expression for the rate of change of charged particle energy with time.

---

[1] *Oeuvres* [442, vol. 9, p. 464], referring to the electron theory of Lorentz.

19.1. Lorentz law of charged particle motion

We are therefore interested in the way $\dot{p} = \{dp^\mu/d\lambda\}$ may depend on the electromagnetic potential $A^\mu(x)$ and on the dynamical variable $\dot{z}$ of the particle trajectory. Since momentum is a physical quantity, free of gauge freedom, we can include only the potential in the gauge-invariant combination $F^{\mu\nu} = \partial^\mu A^\nu - \partial^\nu A^\mu$.

We appeal again to a criterion of simplicity and seek parity-covariant and Poincaré-covariant particle equations of motion which are of the lowest possible degree in the dynamical variables of the field and of the trajectory. Inclusion of an undifferentiated $z(\tau)$ or $z(\lambda)$ would break the translational part of Poincaré covariance and the time-reversal symmetry so we concentrate on $\dot{z}^\nu = dz^\nu/d\lambda$. Since $\dot{p}^\mu$ is a proper 4-vector we seek a proper 4-vector combination of $\dot{z}^\nu$ and $F^{\mu\nu}$ which is linear in each. We exclude the permutation tensor or, equivalently, use of the dual $\tilde{F}^{\mu\nu}$, since this would break the parity covariance. Because of the antisymmetry of $F^{\mu\nu}$, contracting either of its indices with $\dot{z}_\nu$ gives the same result up to a physically unimportant sign.

The only covariant equation of motion, linear in the field strength $F^{\mu\nu}$ and in $\dot{z}_\nu$, is therefore $\dot{p}^\mu \propto F^{\mu\nu}\dot{z}_\nu$. The two sides will match only if a factor with the dimensions of electric charge is included in the proportionality constant on the right side. The only quantity available to achieve this is the invariant charge q of the particle. In the analysis which led to Maxwell's equation in Chapter 18, the dimensions and units of charge were fundamental and arbitrary. We now relate these, and the constant μ_0, to mechanical dimensions and units, by choosing the momentum rate of change to have the form:

$$\dot{p} = qF \cdot \dot{z} \quad \text{or} \quad \frac{dp^\mu}{d\lambda} = qF^{\mu\nu}\frac{dz_\nu}{d\lambda}, \qquad (19.1)$$

which we shall call the *Lorentz law* describing the motion of a charged particle in an electromagnetic field.

We note immediately that $\dot{p}$ remains the same if all charges are replaced by the charges of opposite sign, owing to a compensating sign change in J^μ and therefore also in A^μ and $F^{\mu\nu}$. This invariance may be referred to as classical *charge conjugation symmetry*. Furthermore, Equation 19.1 is invariant with respect to reparameterization $\lambda \to \bar{\lambda} = f(\lambda)$ of the particle's evolution variable.

To split the equations into 1+3 form, select $\lambda = \tau$ and take $\mu = 0$ giving $dp^0/d\tau = qF^{0k}\dot{z}_k = qc^{-1}E^k dz_k/d\tau$ in which we use time dilation, $dt = \gamma d\tau$, and $p^0 = E/c$, and carry out a similar reduction for the spatial components, to obtain:

$$\frac{dE}{dt} = q\dot{z}\cdot\mathbf{E} \quad \text{and} \quad \frac{d\mathbf{p}}{dt} = q(\mathbf{E} + \dot{z}\times\mathbf{B}), \qquad (19.2)$$

where $\dot{z} = dz/dt$. The manifestly covariant form, Equation 19.1, therefore contains not only the familiar Lorentz law in its spatial part but also, in its time component, the expression for the rate of change of energy of the particle. This analysis shows that, in the Lorentz law, **p** is not $m\, dz/dt$ but the relativistic 3-momentum $\gamma m\, dz/dt$ while E is the relativistic energy, γmc^2.

We have shown that the equations of classical electrodynamics, for both the field (Maxwell's equations) and the trajectory (Lorentz law) can be obtained as natural consequences of a simplicity assumption applied to a relativistic interaction. Choosing the simplest interaction leads to a Poincaré-invariant, conserved charge q acting as the source in parity-covariant equations which are linear in the electromagnetic field not only in the field equations but also in the particle equations of motion. The latter are also linear in the dynamical variables $\dot{z}$ of the particle. These results correspond to the experimentally observed *linear superposition principle* of the non-quantum electromagnetic field.

As Lorentz himself showed, his electrodynamic equations (equivalent to those reproduced here and incorporating Maxwell's equations) suffice to re-derive the experimental laws of Coulomb, Biot-Savart, Ampère, Faraday and Lenz. He also showed that they could be used to determine the nature of electromagnetic propagation in the presence of charges, the production of dipole radiation, the self-reaction of a charged particle, and provide a fundamental description of reflection, refraction, dispersion, polarization and magnetization. He was able to derive the Fresnel 'drag' coefficient, by which the speed of light is reduced in a medium of refractive index n. He showed this was not a real dragging effect but a secondary phenomenon arising from the consideration of charges in a frame in which they were moving relative to the ether. With minor but significant re-interpretation, those calculations can also be performed in a special relativistic context.

The fact that the electrodynamic laws may be built up from the postulates of relativity and a few simplicity criteria does not diminish the crucial importance of experiment in the discovery of physical law, as exemplified by the centuries of painstaking work that led to the Maxwell and Lorentz equations. Nevertheless, it illustrates, first, the power of a covariant analysis, and second, the extent to which general concepts, such as the Relativity principle, can severely constrain the laws of physics. Third, it illustrates the value of extracting all the consequences of each fundamental principle.

19.2 Current density of a classical charged particle

Maxwell's equations are expressed in terms of a source which is a continuous current density distribution $J^\mu(x)$. To complete the consideration of a closed system of classical (point) charges acted on by the electromagnetic field, and in turn acting as the source of the field, we need to determine

19.2. Current density of a classical charged particle

the current density as a function of spacetime position x corresponding to a charged point particle with a specific trajectory $z^\mu(\lambda)$. We use a method analogous to that used in constructing the energy-momentum tensor of a particle in Equation 15.93.

A charge q located at $\mathbf{z}(t)$ in a particular frame will have a charge density given non-covariantly in terms of the 3D Dirac delta function by $q\,\delta^3(\mathbf{x} - \mathbf{z}(t))$. Consequently,

$$J^0(t,\mathbf{x}) = qc\,\delta^3(\mathbf{x} - \mathbf{z}(t)) \quad \text{and} \quad J^k(t,\mathbf{x}) = q\frac{dz^k}{dt}\,\delta^3(\mathbf{x} - \mathbf{z}(t))\,. \quad (19.3)$$

We may use $c = dx^0/dt$ in the first of these expressions to make it more closely resemble the second in preparation for combining them covariantly. We then multiply each of these functions of t or z^0 by a 1D Dirac delta function $\delta(x^0 - z^0(t))$ and integrate over the trajectory parameterized by the coordinate z^0 to convert the non-invariant 3D Dirac delta into a Lorentz-invariant $\delta^4(x - z(t))$. Replacement of the parameter t or z^0 by a parameter λ gives us the 4-current of a charged particle as:

$$J^\mu(x) = qc\int_{-\infty}^{\infty} d\lambda\,\delta^4(x - z(\lambda))\,\dot{z}^\mu\,. \quad (19.4)$$

All the parts of this equation are now manifestly covariant. This equation confines the charge to only one event on the trajectory at each time. It therefore has conservation of the charge automatically built-in. We verify this differentially as follows:

$$\begin{aligned}
\partial_\mu J^\mu &= qc\int_{-\infty}^{\infty} d\lambda\,\dot{z}^\mu\frac{\partial}{\partial x^\mu}\delta^4(x - z(\lambda)) \\
&= -qc\int_{-\infty}^{\infty} d\lambda\,\dot{z}^\mu\frac{\partial}{\partial z^\mu}\delta^4(x - z(\lambda)) \\
&= -qc\int_{-\infty}^{\infty} d\lambda\,\frac{d}{d\lambda}\delta^4(x - z(\lambda)) \\
&= -qc\left[\delta^4(x - z(\lambda))\right]_{-\infty}^{\infty} = 0\,. \quad (19.5)
\end{aligned}$$

The last step follows since a real particle ($|\dot{z}| \leq c$) observed near the origin of proper time can never actually have come from or reach the infinitely distant limit points.

From the current density of Equation 19.3 we may reconstruct the charge, according to Equation 18.1, using the covariant 3-volume. These results may formally be extended by introducing a sum into the right side of Equation 19.4 if the single charge q is replaced by a *swarm* of classical point charges q_i ($i = 1, 2, \ldots$) on trajectories $z_{(i)}^\mu(\lambda_{(i)})$. However, readers are referred to Sudarshan and Mukunda [525] for a discussion of the difficulties in the interaction of relativistic multiparticle systems without using quantum field theory.

Exercise 19.1 Show that the energy-momentum tensor given by Equation 15.93 for a system of non-interacting particles of trajectories $z^\mu_{(i)}(\lambda)$, is conserved, $\partial_\mu T^{\mu\nu} = 0$, as a result of the free particle equations of motion.

Exercise 19.2 Use Equation 18.9 and Gauss's theorem to obtain **Gauss's law**, $\oint_{\partial V} \mathbf{E} \cdot d\mathbf{S} = q_V/\epsilon_0$, for the integral of the electric field $\mathbf{E}$ over the closed 2D boundary ∂V of the volume V containing total electric charge q_V and state the integral form of **Gauss's magnetic law**.

Exercise 19.3 Show that a source which is steady-state ($\dot\rho = 0$ and $\dot{\mathbf{J}} = 0$) in the frame $\{t, \mathbf{x}\}$ leads to steady-state fields ($\dot{\mathbf{E}} = 0 = \dot{\mathbf{B}}$) in that frame.

Exercise 19.4 (a) Show that, for steady-state sources, Maxwell's equations imply **Ampère's law**, $\oint_{\partial S} \mathbf{B} \cdot d\mathbf{l} = \mu_0 I_S$, for the integral of the magnetic induction $\mathbf{B}$ around the closed 1D boundary ∂S of the surface S intersected by the current I_S. (b) Establish a similar expression for **Faraday's law**.

Exercise 19.5 Use the second part of Equation 19.3 applied to a steady-state electromagnetic field to establish the **Biot-Savart law**,

$$\mathbf{B}(\mathbf{x}) = \frac{\mu_0}{4\pi} \frac{q \dot{\mathbf{z}} \times (\mathbf{x}-\mathbf{z})}{|\mathbf{x}-\mathbf{z}|^2} \quad \text{or} \quad d\mathbf{B}(\mathbf{x}) = \frac{\mu_0}{4\pi} \frac{I d\mathbf{l} \times (\mathbf{x}-\mathbf{z})}{|\mathbf{x}-\mathbf{z}|^2}, \tag{19.6}$$

for the magnetic field at $\mathbf{x}$ due to a charge q with trajectory $\mathbf{z}(t)$ or current element $I d\mathbf{l}$ at $\mathbf{z}$.

19.3 Electromagnetic coupling constant

In a closed system of charges, the interaction of the individual charges with one another and with the electromagnetic field can be described by combining Equations 18.8, 19.1 and 19.4 as,

$$\partial_\nu F^{\mu\nu} = \mu_0 qc \int d\lambda\, \delta^4(x - z(\lambda)) \dot{z}^\mu, \quad \partial_\mu \tilde F^{\mu\nu} = 0, \quad \dot p^\mu = qF^{\mu\nu}\dot z_\nu, \tag{19.7}$$

in which we have suppressed the sum over the individual charges. Equivalently, we can use similar equations in A^μ obtained by replacing the electromagnetic field tensor by the covariant curl of the 4-vector potential in the first and third of these equations, namely

$$\Box A^\mu - \partial^\mu \partial\cdot A = -\mu_0 qc \int d\lambda\, \delta^4(x - z(\lambda))\dot z^\mu, \quad \dot p^\mu = q(\partial^\mu A^\nu - \partial^\nu A^\mu)\dot z_\nu, \tag{19.8}$$

the second then being an identity.

If we put $q = 0$ in Equation 19.8 we get separate uncoupled equations for the particles and the field. These give $\ddot z^\mu = 0$ (or $\dot z^\mu =$ constant) for the particle and plane waves for the vector potential. For this reason, the charge q is referred to as a *coupling parameter*.

Since these equations are coupled only through the scalar variable q without direct involvement of the higher multipole moments, in particular the electric and magnetic dipole and quadrupole moments of the charge distribution, the charges are said to be *minimally coupled*. The fact that electric charge is quantized into simple multiples of the proton charge e permits us to use the latter as an *electromagnetic coupling constant* of significance in all electromagnetic interactions.

The numerical values of constants which are dimensionless even in equation systems, such as SI or cgs, where mass, length and time have independent dimensions, are of particular physical significance. To construct such a dimensionless coupling constant from the proton charge e requires, in cgs systems (such as Gaussian units), the introduction of two further independent dimensional constants of nature plus a third (μ_0 or $\epsilon_0 = 1/\mu_0 c^2$) in SI units. In order that the dimensionless constant based on e be primarily a characteristic of the electromagnetic interaction, those two further constants must themselves be genuinely universal having no special connection to particular particles or particular interactions between them. There are only two such universal physical constants, one being the ultimate speed c of relativity theory and the other being the quantum of action $\hbar$.

By convention, one takes the dimensionless electromagnetic coupling constant proportional to e^2, namely the fine structure constant α of Equation 7.6. Since we are dealing with a dimensionless quantity we can conclude, from the smallness of α compared to unity, that the electromagnetic interaction is of low strength. One ought therefore to be able to calculate useful quantum electrodynamic results using the methods of perturbation theory. This is confirmed by detailed calculations.

Exercise 19.6 *Suppose that the electromagnetic field is interacting with a system with a characteristic length ℓ. Determine a condition, involving ℓ and the magnitudes E and B of the electric field and the magnetic induction, which will permit the field to be treated classically.*

19.4 Interaction: classical electrodynamic action

One of the reasons for preferring a Lagrangian description of a system, rather than an equivalent description not derivable from an action principle, is the ease with which two free systems may be combined in the Lagrangian formalism to give an interacting system. A Lagrangian approach could have been used as an alternative way to construct the simple forms of the electromagnetic field equations involving a vector potential A^μ.

We shall now illustrate the formation of an interacting action using the electromagnetic coupling of a set of classical charges with dynamical variables $z^\mu(\lambda)$. Each particle will have a positive energy free term in the action of the form,

$$S_{\text{part}} = \tfrac{1}{2} \int_{\lambda_i}^{\lambda_f} d\lambda [e^{-1}(\lambda)\dot{z}^2(\lambda) - e(\lambda)\,m^2 c^2]\;, \tag{19.9}$$

where $e(\lambda)$ is the einbein of Equation 15.40.

Labels distinguishing individual particles are suppressed. This will be the first of three contributions to the interacting action. From Equations 16.2 and 18.57, the second will be the action,

$$S_{\text{field}} = -\tfrac{1}{4\mu_0 c} \int_{\Sigma_i}^{\Sigma_f} d^4 x\, F_{\mu\nu} F^{\mu\nu} = -\tfrac{1}{2\mu_0 c} \int_{\Sigma_i}^{\Sigma_f} d^4 x \left(\mathbf{B}^2 - \frac{\mathbf{E}^2}{c^2}\right)\;, \tag{19.10}$$

of the free electromagnetic field by means of which the individual charges will interact. We emphasize again that, in the second-order Lagrangian of the electromagnetic field, the generalized coordinate is $A^\mu(x)$ not $F^{\mu\nu}(x)$ which is simply an abbreviation for the covariant curl of the Lagrangian field itself.

We now require a third term to describe the interaction, which will involve $z(\lambda)$ and $A(x)$ in a product which must be a real invariant of the full Poincaré group of the correct action dimensions. We expect the parts of this term corresponding to each particle i interacting with the field to be proportional to the charge q_i of the particle which acts as a coupling parameter. Only one charge q will appear here in our condensed notation. The simplest translation-covariant variable involving $z(\lambda)$ is the particle's 4-velocity $\dot{z}(\lambda)$. The simplest interaction term satisfying these requirements is proportional to $q \int d\lambda\, \dot{z} \cdot A(z)$ which also has the correct dimensions and charge conjugation symmetry.

We must now determine the sign and magnitude of the coefficient of the interaction term and verify the magnitude already chosen for the coefficient of the free-field term. The particle term and the interaction term must combine, on variation of the path $z(\lambda)$ and the einbein $e(\lambda)$, to give the Lorentz law for the motion of the charges. Having thus fixed the interaction term we must then vary the field A^μ and verify the numerical value of the constant factor in the free electromagnetic action so that it reproduces Maxwell's equations in the standard form of our choosing.

We therefore use an interacting action of the form:

$$\begin{aligned} S &= S_{\text{field}} + S_{\text{part}} + S_{\text{intn}} \\ &= -\tfrac{1}{4\mu_0 c} \int_{\Sigma_i}^{\Sigma_f} d^4 x\, F_{\mu\nu} F^{\mu\nu} + \tfrac{1}{2} \int_{\lambda_i}^{\lambda_f} d\lambda\, [e^{-1}(\lambda)\dot{z}^2 - e(\lambda) m^2 c^2] \end{aligned}$$

19.4. Interaction: classical electrodynamic action

$$+ q \int_{\lambda_i}^{\lambda_f} d\lambda\, \dot{z} \cdot A(z(\lambda))\ , \tag{19.11}$$

in which, since the charged particle is massive, we could equally use the geometric Lagrangian (Equation 15.33) in place of the einbein in the second term. The equations of motion of the particles will arise from the Lagrangian obtained from the second and third terms in this action.

Exercise 19.7 (a) Show that the covariant generalized particle momenta of the action in Equation 19.11 are $\pi_\mu = e^{-1}(\lambda)\dot{z}_\mu + qA_\mu$ or $p_\mu + qA_\mu$ for $m \neq 0$. (b) Use the Euler-Lagrange equations for a particle with dynamical variable $z(\lambda)$ (and einbein $e(\lambda)$) to obtain the Lorentz law $\dot{p}^\mu = qF^{\mu\nu}\dot{z}_\nu$ from the above action. (c) Show that the canonical Hamiltonian of the particle vanishes, the same being true if the einbein is replaced by the geometric Lagrangian. (d) Show that $L = -mc\sqrt{-\dot{z}^2} + q\dot{z}\cdot A$ leads to

$$\pi_\mu = \frac{mc\dot{z}_\mu}{\sqrt{\dot{z}_0^2 - \dot{z}^2}} + qA_\mu\ , \tag{19.12}$$

which we may solve for three of the velocities to obtain,

$$\dot{z} = \frac{\dot{z}^0(\boldsymbol{\pi} - q\mathbf{A})}{\sqrt{(\boldsymbol{\pi} - q\mathbf{A})^2 + m^2c^2}}\ . \tag{19.13}$$

(e) Show that the fourth equation, obtainable directly from Equation 19.12, gives the mass-shell constraint,

$$(\pi^\mu - qA^\mu)(\pi_\mu - qA_\mu) + m^2c^2 = 0\ , \tag{19.14}$$

which in the proper time gauge for $m \neq 0$ is $\dot{z}^2 = -c^2$. (f) Show that the effective Hamiltonian $H_{\text{eff}} = \pi_\mu \dot{z}^\mu - L_{\text{ein+intn}}$ of the einbein particle Lagrangian,

$$L_{\text{ein+intn}} = \tfrac{1}{2}[e^{-1}(\lambda)\dot{z}^2 - e(\lambda)m^2c^2] + qA\cdot\dot{z}\ , \tag{19.15}$$

minimally coupled to the external field A_μ is given by,

$$H_{\text{eff}} = \tfrac{1}{2}e(\lambda)[m^2c^2 + (\pi_\mu - qA_\mu)(\pi^\mu - qA^\mu)]\ , \tag{19.16}$$

in accord with the Dirac prescription (see Section 15.8). (g) Show that Hamilton's equations, $\dot{z}^\mu = \partial H/\partial \pi_\mu$ and $\dot{\pi}_\mu = -\partial H/\partial z^\mu$, or the Poisson bracket relations $\dot{z} = \{z^\mu, H\}$ and $\dot{\pi}_\mu = \{\pi_\mu, H\}$, lead to $p_\mu = e^{-1}(\lambda)\dot{z}_\mu$ and the Lorentz law. (h) Show that the additive, gauge-invariant energy-momentum tensor,

$$T^{\mu\nu} = \tfrac{1}{\mu_0}(F^{\mu\lambda}F^\nu{}_\lambda - \tfrac{1}{4}\eta^{\mu\nu}F_{\lambda\pi}F^{\lambda\pi}) + \sum_i mc\int d\tau\, \delta^4(x - z(\tau))\, \dot{z}^\mu \dot{z}^\nu\ , \tag{19.17}$$

for the field and particles $z(\tau)$ (with summed particle index i suppressed) is energy and angular momentum conserving as a result of the dynamical equations coupling the charged particles via the electromagnetic field of potential A^μ and field strength $F^{\mu\nu}$.

Since they are conserved, the interacting 4-momenta π^μ will be replaced by $\frac{\hbar}{i}\partial_\mu$ on applying the quantization *ansatz*. We may choose to also denote the interacting 4-momentum by p_μ and the minimal coupling will then be expressed by the prescription, $p_\mu \to p_\mu - qA_\mu$.

In the non-relativistic limit of a massive particle, Equation 19.15 reduces to

$$L_{\text{min coupl}} = \tfrac{1}{2}m\dot{z}^2 - q\phi + q\dot{z}\cdot\mathbf{A} \qquad (19.18)$$

from which we may obtain,

$$H_{\text{min coupl}} = \frac{1}{2m}(\boldsymbol{\pi} - q\mathbf{A})^2 + q\phi . \qquad (19.19)$$

Minimal coupling is then equivalent to the prescription, $\mathbf{p} \to \mathbf{p} - q\mathbf{A}$ and $H \to H - q\phi$ where $\mathbf{p} = \frac{\hbar}{i}\nabla$ and $H = i\hbar\partial_t$.

The field equations governing A^μ will result from the first and third terms of the action. In order to apply the Euler-Lagrange equations to these two terms we need to re-express the last term as an integral over the spacetime region between the initial and final configurations. This we do by multiplying the integrand by a 4D Dirac delta function $\delta^4(x - z(\lambda))$ times d^4x and integrating over all spacetime from the initial to the final configuration. By interchanging the integration order we see immediately, using Equation 19.4, that this last term then becomes $c^{-1}\int d^4x\, J^\mu(x)A_\mu(x)$ in which the Dirac delta function has permitted us to replace $A_\mu(z)$ by $A_\mu(x)$.

Exercise 19.8 *Show that the Euler-Lagrange equations for the field A^μ, applied to the action of Equation 19.11, give the second-order sourced Maxwell's equations of Equation 18.22 or 19.8.*

We have now obtained an essentially complete Lagrangian description of a closed system of non-quantum charged particles interacting via the electromagnetic field, apart from certain questions related to the singularity of the Lagrangian and the gauge freedom of the vector potential, to be addressed in Exercise 19.12. The system we have constructed follows almost inevitably from little more than the Relativity principle, the existence of a finite upper bound to all speeds and characterization of the electromagnetic interaction as the one involving the relativistic field with the simplest integrated source.

The action given here may be extended to include a free term representing a gravitational field and factors describing its coupling to both the classical particle and to the electromagnetic field. A realistic analysis, with the capability of being extended to form a general relativistic description of gravity, would use the massless spin 2 field of Fierz-Pauli to be described in Chapter 21. A simpler exercise, using a graviscalar field, correctly describes the gravitational time dilation of atomic clocks but not the influence of gravity on light propagation. If a massive graviscalar field is used with a scalar density source that need not reduce to mass density in the Newtonian limit, the effect of such a field on gravitational time dilation may be calculated [256]. The result will in general differ from the value of $z = -(\Delta P/P)_{\text{clock}} = \Delta\Phi/c^2$ which holds for gravity theories which conserve energy and satisfy the Equivalence principle.

19.5 Belinfante symmetric energy-momentum tensor

The canonical energy-momentum tensor is the conserved current density arising from symmetry of an isolated system with respect to spacetime translations and gives the total energy-momentum when integrated over all space.

In the derivation of its form in Section 16.2 we considered only a pure translation with consequently no change ($\delta\psi = 0$) of the field. However, the canonical tensor fails to embody angular momentum conservation for fields other than the spinless Klein-Gordon field. The reason for this is that changes in coordinates of the parts of an extended system arise not only from translations but also from the effects of covariant rotations. In a small Poincaré symmetry transformation $\Delta x(x) = (\Delta L)x + \Delta a = (\mathbb{1} + \Delta\omega)x + \Delta a$, the rotations also give rise to changes in x.

A scalar field has only one component whose values remain the same under rotations and translations. But a vector or tensor field has its components intermingled in a specific way in rotations. Since the rotations do not leave ψ unchanged if it has more than one component, these changes must be taken into account in making use of Noether's theorem. The intermingling contributes not only to the structure of the angular momentum but also to the energy density itself. We may not ignore any internal contributions to j_a^μ of Equation 16.11 arising from changes $\partial\Delta\psi/\partial\Delta\omega^a$ in the field, namely $\partial\Delta\psi/\partial\Delta\omega^{\mu\nu}$ for covariant rotations, as was assumed in deriving Equation 16.15.

We consider a variation of the field action due to not just a translation but an entire Poincaré transformation. The corresponding variation of the field is given in terms of the small Lorentz parameters $\Delta\omega_{\mu\nu}$ and generators $S^{\mu\nu}$ by $\Delta\psi = -\frac{1}{2}i\Delta\omega_{\mu\nu}S^{\mu\nu}\psi$. The $d \times d$ matrices $S^{\mu\nu}$ which generate Lorentz transformations are zero for a scalar field and their form for a vector field is given in Equation 14.56. By analogy with the covariant momentum densities (see Equation 16.18), we define a *canonical spin density* by,

$$\check{S}^{\nu\lambda\,\mu} = c\frac{\partial \mathcal{L}}{\partial(\partial_\mu\psi)}iS^{\nu\lambda}\psi = -\check{S}^{\lambda\nu\,\mu}\ . \tag{19.20}$$

The variation of the action used in Noether's theorem (see Equations 16.8 and 16.11) is then given, on the stationary path, by,

$$\Delta S = -\tfrac{1}{c}\int d^4x\,\partial^\mu[\check{T}_{\mu\nu}\Delta x^\nu - \tfrac{1}{2}\check{S}_{\nu\lambda\,\mu}\partial^\lambda(\Delta x^\nu)]\ , \tag{19.21}$$

in which we have replaced $\Delta\omega^{\lambda\nu}$ by $\partial^\lambda(\Delta x^\nu)$, since we wish to transform the second term to be proportional to Δx^ν like the first. To do this, we shall integrate the second term by parts and avoid getting any contribution from a term of the form $\partial^\mu\partial^\lambda(\ldots\lambda_\mu\Delta x^\nu)$ by noting that it can be re-arranged so that there is antisymmetry in $\lambda\mu$ inside the brackets guaranteeing that this term is annihilated. Since $\check{S}_{\nu\lambda\,\mu}$ itself is not antisymmetric in $\lambda\mu$, we first construct the most general linear combination of its components having that property.

Exercise 19.9 *(a) Show that $a(\check{S}_{\nu\lambda\,\mu} + b\check{S}_{\mu\nu\,\lambda} + d\check{S}_{\nu\lambda\,\mu})$ is antisymmetric in $\lambda\mu$ provided $b = 1$. (b) Show that if, in addition to $b = 1$, we take $d = -1$ and $a = 1$, then the replacement of $\check{S}_{\nu\lambda\,\mu}$ by the* **spin density,**

$$S_{\nu\,\lambda\mu} = \check{S}_{\nu\lambda\,\mu} + \check{S}_{\mu\nu\,\lambda} - \check{S}_{\nu\lambda\,\mu} = -S_{\nu\,\mu\lambda}\ , \tag{19.22}$$

leaves the variation of the action unchanged. (c) Show that the canonical spin density may be retrieved from $S_{\nu\,\lambda\mu}$ according to $\check{S}_{\mu\nu\,\lambda} = S_{[\mu\,\nu]\lambda}$ showing that the two quantities contain the same information.

We may now integrate the second term in

$$\Delta S = -\tfrac{1}{c}\int d^4x\,\partial^\mu[\check{T}_{\mu\nu}\Delta x^\nu - \tfrac{1}{2}S_{\nu\,\lambda\mu}\partial^\lambda(\Delta x^\nu)]\ , \tag{19.23}$$

by parts and use $\partial^\mu\partial^\lambda(S_{\nu\,\lambda\mu}\Delta x^\nu) = 0$ to obtain,

$$\Delta S = -\tfrac{1}{c}\int d^4x\,\partial^\mu[(\check{T}_{\mu\nu} + \tfrac{1}{2}\partial^\lambda S_{\nu\,\lambda\mu})\Delta x^\nu]\ . \tag{19.24}$$

Following Belinfante [31], we are therefore naturally led to use the spin density $S^{\nu\,\lambda\mu}$ to define the symmetric energy-momentum tensor whose properties were also established by Rosenfeld [485], according to,

$$T^{\mu\nu} = \check{T}^{\mu\nu} + \tfrac{1}{2}\partial_\lambda S^{\nu\,\lambda\mu}\ , \tag{19.25}$$

in terms of which the variation of the action is given by,

$$\Delta S = -\tfrac{1}{c}\int d^4x\,\partial^\lambda(T_{\lambda\rho}\Delta x^\rho) = -\tfrac{1}{c}\oint d\Sigma^\lambda\,T_{\lambda\rho}\Delta x^\rho$$

$$= -\tfrac{1}{c}\oint d\Sigma^\lambda\, T_{\lambda\rho}\Delta a^\rho - \tfrac{1}{2c}\oint d\Sigma_\lambda(x^\mu T^{\lambda\nu} - x^\nu T^{\lambda\mu})\Delta\omega_{\mu\nu}\,.$$
(19.26)

The variables Δa^μ and $\Delta\omega_{\mu\nu}$ are the independent parameters of the small Poincaré transformation. Noether's theorem therefore gives us two conserved currents determinable from the Lagrangian density. These are the *Belinfante energy-momentum tensor*, as the coefficient of Δa^μ,

$$T^{\mu\nu} = c\{-\frac{\partial\mathcal{L}}{\partial(\partial_\mu\psi)}\partial^\nu\psi + \eta^{\mu\nu}\mathcal{L} + \tfrac{1}{2}\partial_\lambda(\check{S}_{\nu\lambda\mu} + \check{S}_{\mu\nu\lambda} - \check{S}_{\lambda\mu\nu})\},$$
(19.27)

and the coefficient of $\tfrac{1}{2}\Delta\omega_{\mu\nu}$, the *angular momentum density*,

$$j^{\mu\nu\,\lambda} = \tfrac{1}{c}(x^\mu T^{\lambda\nu} - x^\nu T^{\lambda\mu}) = -j^{\nu\mu\,\lambda},$$
(19.28)

their conservation being expressed by,

$$\partial_\mu T^{\mu\nu} = 0 \quad \text{and} \quad \partial_\lambda j^{\mu\nu\,\lambda} = 0\,.$$
(19.29)

Exercise 19.10 *Use the angular momentum conservation property to show that the Belinfante energy-momentum tensor $T^{\mu\nu}$ must be symmetric.*

The corresponding Noether charges, the energy-momentum and total angular momentum (orbital plus spin), are given by:

$$p^\mu = \tfrac{1}{c}\int_\Sigma d\Sigma_\nu\, T^{\mu\nu} \quad \text{and} \quad j^{\mu\nu} = \int_\Sigma d\Sigma_\lambda\, j^{\mu\nu\,\lambda} = -j^{\nu\mu}$$
(19.30)

Exercise 19.11 *Show that the spin density of the free electromagnetic field is given by,*

$$S^{\nu\,\lambda\mu} = -\tfrac{2}{\mu_0}F^{\mu\lambda}A^\nu\,,$$
(19.31)

(which we note is antisymmetric in $\lambda\mu$) and that the Belinfante formula in Equation 19.25 therefore leads to Equation 18.49.

The Belinfante-Rosenfeld result is applicable to all other non-zero spin fields such as those of Dirac, Weyl, Proca, Rarita-Schwinger and Fierz and Pauli.

19.6 Physical components of the Maxwell potential

To separate the true dynamical features from the constraint structure in Equation 18.22, and similar field equations, we make a 1+3 split giving:

$$\Box\mathbf{A} - \nabla(\partial_0 A^0 + \nabla\cdot\mathbf{A}) = -\mu_0\mathbf{J} \quad \text{and} \quad \nabla^2 A^0 - \partial^0\nabla\cdot\mathbf{A} = -\mu_0 c\rho\,,$$
(19.32)

Hermann Weyl

b. Elmshorn, Germany
9 November 1885

d. Zurich, Switzerland
8 December 1955

Courtesy of the State University, Göttingen.

from which we see that the zero component is not a true dynamical equation but a constraint on the initial data A^μ and $\dot{A}^\mu$. This constraint is easily seen to be the potential equivalent of Gauss's law, $\nabla \cdot \mathbf{E} = \rho/\epsilon_0$. There is no analogous constraint on A^μ corresponding to $\nabla \cdot \mathbf{B} = 0$ since, in potential terms, the latter is an identity, $\nabla \cdot \nabla \times \mathbf{A} \equiv 0$.

If we use the results of Appendix A.3 to decompose $\mathbf{A}$ into a transverse (divergence-free) part $\mathbf{A}^\mathrm{T}$ and a longitudinal part which, since it is curl-free, we write as the gradient of a scalar α, we have:

$$\mathbf{A} = \mathbf{A}^\mathrm{T} + \nabla \alpha \quad \text{and} \quad \nabla \cdot \mathbf{A}^\mathrm{T} = 0 \ . \tag{19.33}$$

If we now consider the free-space case ($\rho=0$), Equation 19.32 can be written as,

$$\Box \, \mathbf{A}^\mathrm{T} - \partial_0 \nabla (A^0 - \partial^0 \alpha) = 0 \quad \text{and} \quad \nabla^2 (A^0 - \partial^0 \alpha) = 0 \ . \tag{19.34}$$

The constraint is satisfied by taking A^0 to be a pure gauge field and the dynamical equation then reduces to the wave equation, $\Box \, \mathbf{A}^\mathrm{T} = 0$, for the transverse components. Although the scalar and longitudinal components need not be zero, it is clear that they effectively cancel one another and leave $\mathbf{A}^\mathrm{T}$ as the physical components of the electromagnetic field.

We may establish the above result more carefully as follows. Exercise 18.22 shows that after elimination of $\mathbf{B} = \nabla \times \mathbf{A}$ in terms of $\mathbf{A}$ in the $1 + 3$ form of the first-order free Maxwell Lagrangian of Equation 18.62

(with $m = 0$), the Euler-Lagrange equations for the true dynamical variables $\mathbf{A}$ and $\mathbf{E}$ are,

$$\dot{\mathbf{A}} = -\mathbf{E} - \nabla\phi, \quad \dot{\mathbf{E}} = \nabla\times(\nabla\times\mathbf{A}) \quad \text{and} \quad \nabla\cdot\mathbf{E} = 0 , \tag{19.35}$$

the last of which is a constraint on the initial data for $\mathbf{E}$. We now decompose $\mathbf{A}$ and $\mathbf{E}$ into their transverse and longitudinal parts,

$$\mathbf{A} = \mathbf{A}^{\mathrm{T}} + \nabla\alpha, \quad \mathbf{E} = \mathbf{E}^{\mathrm{T}} + \nabla\beta \quad \text{with} \quad \nabla\cdot\mathbf{A}^{\mathrm{T}} = 0 = \nabla\cdot\mathbf{E}^{\mathrm{T}} . \tag{19.36}$$

Only the transverse part of $\mathbf{A}$ contributes to $\mathbf{B}$ and hence, in the Hamilton gauge [346] (see Section 19.10), where $\phi = 0$, the dynamical equations are,

$$\dot{\mathbf{A}}^{\mathrm{T}} = -\mathbf{E}^{\mathrm{T}} \quad \text{and} \quad \dot{\mathbf{E}}^{\mathrm{T}} = -\nabla^2\mathbf{A}^{\mathrm{T}} . \tag{19.37}$$

Exercise 19.12 *(a) Simplify the first-order action (see Exercises 18.20, 18.21 and 18.22) for the free Maxwell field in the Hamilton gauge ($\phi = 0$), noting that integration over all space permits the use of the orthogonality (Appendix A.3) of transverse and longitudinal components, to show that all significant terms containing the dynamical variables $\mathbf{A}^{\mathrm{T}}$ and $\mathbf{E}^{\mathrm{T}}$ are given by the following non-singular Lagrangian,*

$$\mathcal{L}_{\mathrm{Max}}^{\mathrm{can}} = -\mathbf{E}^{\mathrm{T}}\cdot\dot{\mathbf{A}}^{\mathrm{T}} - \tfrac{1}{2}\{\mathbf{E}^{\mathrm{T}}\cdot\mathbf{E}^{\mathrm{T}} + \partial_k(A^{\mathrm{T}})_l\partial^k(A^{\mathrm{T}})^l\} . \tag{19.38}$$

(b) Verify that the Euler-Lagrange equations of this non-covariant but canonical Lagrangian reproduce the dynamical equations for $\mathbf{A}^{\mathrm{T}}$ and $\mathbf{E}^{\mathrm{T}}$.
(c) Determine the corresponding Hamiltonian density of the Maxwell field.

19.7 Lorentz family of gauges

The process of gauge fixing is of central importance to an analysis of the fundamental interactions and a very wide variety of techniques are available, some of which would require the development of material which would not be appropriate to our goals. We shall therefore examine only a few explicit elementary methods to set up the more well-known families of electromagnetic gauges.

A gauge transformation $A_\mu \to A_\mu + \partial_\mu\alpha$ leaves Equation 18.22, the field strength $F_{\mu\nu}$ and the energy-momentum tensor $T^{\mu\nu}$ of the electromagnetic field unchanged. Each set of A^μ constitutes an electrodynamic *gauge*, for some of which the field equations will have a simpler form. A particular gauge must be chosen to give physical meaning to the components of the potential.

A glance at Equations 18.22, 18.26 and 18.27, shows that they will be considerably simplified (by decoupling those for ϕ from those for $\mathbf{A}$) if we can impose a *gauge condition* for which the 4-divergence of the 4-vector potential vanishes everywhere at all times. A gauge condition which can be expressed in manifestly covariant form, such as the vanishing divergence of the potential, is said to lead to a *covariant gauge*. To show that this is a permissible gauge-fixing condition we may display the gauge transformation which relates the new potential to the original potential having gauge freedom. We note that the effect of the general gauge transformation $\Delta A_\mu = \partial_\mu \alpha(x)$ on the divergence is $\partial \cdot A^{\text{new}} = \partial \cdot A^{\text{old}} + \Box \alpha$ where $\partial \cdot A^{\text{old}} \neq 0$ in general and we require $\partial \cdot A^{\text{new}} = 0$. Thus α must be a solution of the inhomogeneous wave equation $\Box \alpha = -\partial \cdot A^{\text{old}} \neq 0$. This equation has non-trivial solutions [263] for $\alpha(x)$. Consequently, the equations:

$$\Box A^\mu = -\mu_0 J^\mu \quad \text{and} \quad \partial \cdot A = 0 , \qquad (19.39)$$

together comprise a permissible family of covariant electromagnetic gauges known as the *Lorentz gauges*, with retarded solutions $A^\mu(t, \mathbf{x})$ given in terms of the source $J^\mu(t, \mathbf{x})$ by the *Lienard-Wiechert potentials*,

$$A^\mu(t, \mathbf{x}) = \frac{\mu_0}{4\pi} \int d^3 x' \frac{J^\mu(t - |\mathbf{x} - \mathbf{x}'|/c, \mathbf{x})}{|\mathbf{x} - \mathbf{x}'|} . \qquad (19.40)$$

In 1+3 notation these equations are well-known as:

$$\Box \phi = -\rho/\epsilon_0, \quad \Box \mathbf{A} = -\mu_0 \mathbf{J} \quad \text{and} \quad \nabla \cdot \mathbf{A} + \dot{\phi}/c^2 = 0 . \qquad (19.41)$$

The imposition of a gauge condition has converted a constraint, namely $\nabla^2 \phi + \nabla \cdot \dot{\mathbf{A}} = -\rho/\epsilon$, into a dynamical equation, the wave equation for ϕ. This is a general feature of gauge fixing.

One of the advantages of using the Lorentz gauge is its manifest covariance. A disadvantage of gauge fixing is that the conservation of the source is no longer automatically built into the field equations and must be incorporated into the analysis 'by hand' by demanding that the field have vanishing 4-divergence. The first of Equations 19.39 does not embody the conservation of the source as an identity, the source constraint. The divergence of the left side does not vanish identically as is the case with Equation 18.22, or the first of Equations 18.8, both of which incorporate charge conservation automatically.

The above gauge invariance with a scalar gauge function tells us that there can be no more than three independent physical components among the four components A^μ. However, the Lorentz gauge condition does not remove all the gauge freedom.

Exercise 19.13 *Show that both equations in 19.39 are gauge invariant with respect to the same gauge transformation* $\Delta A_\mu = \partial_\mu \alpha$ *provided the new gauge function α satisfies the homogeneous wave equation,* $\Box \alpha = 0$.

Since the homogeneous wave equation also has non-trivial solutions, the system in Equation 19.39 is said to have *residual gauge invariance* with a scalar gauge function which demonstrates that the electromagnetic 4-vector potential can have at most two independent physical components.

To determine how many components are physical requires fixing of the residual gauge freedoms until a system is reached which has no gauge variance. Fixing the residual gauge freedom of the Lorentz gauges directly by imposing a further condition on the 4-vector potential must involve the loss of manifest covariance since there is no covariant way besides use of the 4-divergence to construct a scalar from a 4-vector and the only other first-rank object available to act on the field, namely the gradient ∂.

19.8 Maxwell polarization basis

A basis for the propagating solutions of Maxwell's equation for the vector potential in the Lorentz gauge, may be formed using an expansion in terms of plane wave factors $e^{\pm ik \cdot x}$ combined with 4-vector coefficients that depend only on the propagation 3-vector **k**.

Instead of defining only two 4-vector coefficients for the physical components of the potential, covariance may be maintained by forming four such coefficients, $\{\epsilon_\mu^{(\lambda)}(\mathbf{k})\}$ ($\lambda = 0, 1, 2, 3$). A general solution of the Lorentz gauge wave equation $\Box A_\mu = 0$ for a real potential A_μ can be written as the plane-wave expansion,

$$A_\mu(x) = \int d\tilde{k} \sum_{\lambda=0}^{3} \epsilon_\mu^{(\lambda)}(\mathbf{k}) \left[a^{(\lambda)}(\mathbf{k}) e^{ik \cdot x} + a^{(\lambda)*}(\mathbf{k}) e^{-ik \cdot x} \right], \quad (19.42)$$

where the coefficients $a^{(\lambda)*}(\mathbf{k})$ in the second term would have to be replaced by independent coefficients $b^{(\lambda)*}(\mathbf{k})$ to represent a complex potential. The set of four Lorentz 4-vectors $\{\epsilon_\mu^{(\lambda)}(\mathbf{k})\}$ ($\lambda = 0, 1, 2, 3$) form a spin 1 *polarization basis* for the space of 4-vectors. A standard 4-vector polarization basis may be defined as follows:

- $\epsilon_\mu^{(0)}(\mathbf{k}) \equiv n_\mu$ is taken to be a normalized, timelike ($n^2 = -1$), future-pointing ($n_0 > 0$) 4-vector.

- $\epsilon_\mu^{(3)}(\mathbf{k})$ is a normalized, spacelike ($\epsilon^{(3)} \cdot \epsilon^{(3)} = 1$) 4-vector orthogonal to n_μ and in the plane of k_μ and n_μ. In terms of k and n we can write,

$$\epsilon_\mu^{(3)}(\mathbf{k}) = \frac{k_\mu - (k \cdot n) n_\mu}{(k \cdot n)}. \quad (19.43)$$

- $\epsilon_\mu^{(1)}(\mathbf{k})$ and $\epsilon_\mu^{(2)}(\mathbf{k})$ are a pair of orthonormal, spacelike vectors which are orthogonal to k and n (and therefore also to $\epsilon_\mu^{(3)}(\mathbf{k})$.)

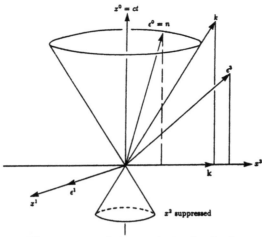

Figure 19.1 Spin 1 polarization basis

Together, the above definitions imply the *orthonormality*,

$$\epsilon^{(\lambda)}(\mathbf{k})\cdot\epsilon^{(\lambda')}(\mathbf{k}) = \eta^{(\lambda\lambda')} , \qquad (19.44)$$

and *completeness*,

$$\sum_{\lambda=0}^{3} \eta^{(\lambda\lambda)} \epsilon_\mu^{(\lambda)}(\mathbf{k})\epsilon_\nu^{(\lambda)}(\mathbf{k}) = \eta_{\mu\nu} , \qquad (19.45)$$

of $\epsilon^{(\lambda)}(\mathbf{k})$, where $\eta^{(\lambda\lambda')} = \text{diag}\{-1,1,1,1\}$. The symbol η is used because the polarization basis metric has the same value as the Minkowski metric $\eta_{\mu\nu}$. We place brackets around the indices λ and λ' to show that they are not actually tensor indices. When indices like (λ) and (λ') are repeated in an expression, no summation is implied unless explicitly indicated.

If k and n are in the $(x^0 x^3)$-plane, for example, and if we suppress the x^2 dimension, then the basis vectors relate to each other as shown in Figure 19.1. Furthermore, with $\{k^\mu\} = k(1,0,0,1)$ for null propagation in the z-direction, the basis vectors can conveniently be chosen to be,

$$\{\epsilon_\mu^0\} = \begin{bmatrix} 1 \\ 0 \\ 0 \\ 0 \end{bmatrix}, \ \{\epsilon_\mu^1\} = \begin{bmatrix} 0 \\ 1 \\ 0 \\ 0 \end{bmatrix}, \ \{\epsilon_\mu^2\} = \begin{bmatrix} 0 \\ 0 \\ 1 \\ 0 \end{bmatrix}, \ \{\epsilon_\mu^3\} = \begin{bmatrix} 0 \\ 0 \\ 0 \\ 1 \end{bmatrix}. \quad (19.46)$$

A polarization basis for rank-2 tensors (representing spin 2) can be constructed from a tensor product of vectorial (spin 1) members,

$$\epsilon_{\mu\nu}^{(\lambda_1\lambda_2)} = \epsilon_\mu^{(\lambda_1)} \otimes \epsilon_\nu^{(\lambda_2)} . \qquad (19.47)$$

19.9 Coulomb gauges

We now consider the possibilty of a gauge in which the 3-vector part **A** of the 4-vector potential A_μ is *transverse*, $\nabla \cdot \mathbf{A} = 0$. This is clearly not a covariant condition. The resulting gauge will be non-covariant. Proceeding as before, a gauge transformation $\Delta A_\mu = \partial_\mu \alpha$ will take us from a non-zero value to a zero value of this 3-divergence provided we can find a gauge function α which is a solution of Poisson's equation $\nabla^2 \alpha = -\nabla \cdot \mathbf{A}^{\text{old}} \neq 0$. This equation has non-trivial solutions which may be expressed in terms of a solution of the same equation having a singular source term on the right side. If we use the following property [263] of a 3D Dirac delta function,

$$\nabla^2 \left(\frac{1}{|\mathbf{x} - \mathbf{x}'|} \right) = 4\pi \delta(\mathbf{x} - \mathbf{x}') , \qquad (19.48)$$

a non-local solution for α is given by,

$$\alpha(t, \mathbf{x}) = -\frac{1}{4\pi} \int d^3 x' \frac{\nabla' \cdot \mathbf{A}^{\text{old}}(t, \mathbf{x}')}{|\mathbf{x} - \mathbf{x}'|} . \qquad (19.49)$$

We can thus always require the vanishing divergence of the spatial part of the 4-potential everywhere at all times and, when this is the case, we say we are in the *transverse* or *Coulomb gauge*. To appreciate this second name, we note that, in this gauge, Equation 18.22 (or Equations 18.26 and 18.27) takes the 1+3 form:

$$\nabla^2 \phi = -\rho/\epsilon_0, \quad \Box \mathbf{A} - \nabla \dot{\phi}/c^2 = -\mu_0 \mathbf{J} \quad \text{and} \quad \nabla \cdot \mathbf{A} = 0 , \qquad (19.50)$$

in the first of which **A** does not appear. This permits $\phi(t, \mathbf{x})$ to be determined at each time from the spatial distribution $\rho(t, \mathbf{x})$ of the scalar source just as if the problem was a purely electrostatic one. This result for $\phi(t, \mathbf{x})$ can be used in solving the 3-vector equation for **A**. Both the above partially-fixed Lorentz and Coulomb gauges can always be achieved, irrespective of any special conditions on the source.

Exercise 19.14 *From Equations 19.50, show that the gauge-fixed wave equation and the transversality condition have residual gauge freedom, namely that they are both gauge-invariant with respect to a harmonic gauge function α satisfying $\nabla^2 \alpha(x) = 0$.*

19.10 Hamilton and radiation gauges

The total number of physical degrees of freedom of a system cannot alter when two previously non-interacting parts begin to interact. Consequently, in order to determine the number of physically independent components of the electromagnetic field it suffices to consider the unsourced case of propagation in free space.

A gauge transformation with gauge function,

$$\alpha(t,\mathbf{x}) = -\int_0^t dt' \, \phi(t',\mathbf{x}) \,, \qquad (19.51)$$

transforms ϕ according to

$$\phi \to \phi_{\text{new}} = \phi - \partial_t \int_0^t dt' \, \phi(t',\mathbf{x}) \,, \qquad (19.52)$$

which vanishes. Consequently, whether the source is vanishing or not, we can always achieve the *Hamilton gauge* [346] in which the time component of the 4-potential vanishes. Again, this is clearly a non-covariant gauge. As before, we can have at most 3 physical components, those making up the 3-vector part $\mathbf{A}$.

Exercise 19.15 *Show that, in the Hamilton gauge (with $\phi = 0$), the equation for $\mathbf{A}$ and the gauge condition itself are gauge-invariant with respect to a gauge function which satisfies $\dot{\alpha} = 0$, implying the existence of residual gauge invariance.*

If we can satisfy $\dot{\alpha} = 0$ and also $\nabla^2 \alpha(x) = \nabla \cdot \mathbf{A}^{\text{old}}$ then we shall have a gauge which is not only transverse but which is also free of any time or scalar component. In order to satisfy the transversality, we choose the gauge function of Equation 19.49 and note that:

$$\dot{\alpha}(t,\mathbf{x}) = \tfrac{1}{4\pi} \int d^3x' \frac{\nabla' \cdot \dot{\mathbf{A}}^{\text{old}}(t,\mathbf{x}')}{|\mathbf{x}-\mathbf{x}'|} \,. \qquad (19.53)$$

Since $\nabla' \cdot \dot{\mathbf{A}}^{\text{old}}(t,\mathbf{x}') = \nabla' \cdot (-\nabla\phi' - \mathbf{E}')$ and $\phi' = 0$, this factor in the integrand reduces to $-\nabla' \cdot \mathbf{E} = \rho/\epsilon_0$. Thus we see that we can only preserve the vanishing scalar part, in imposing transversality as well, provided we have zero charge density. One therefore refers to this free-space gauge as the *radiation gauge*. If we denote the *transverse part* of $\mathbf{A}$ by $\mathbf{A}^T$, the vacuum electromagnetic equations in the radiation gauge are,

$$\Box \mathbf{A}^T = 0 \quad \nabla \cdot \mathbf{A}^T = 0 \quad \text{and} \quad \phi = 0 \qquad (19.54)$$

Exercise 19.16 *Verify that there is no residual gauge freedom in the radiation gauge.*

Since these equations have no gauge freedom, the free electromagnetic potential has two independent physical components. An electromagnetic wave, such as $\mathbf{A}^T \propto \epsilon e^{ik \cdot x}$, will have odd parity and this property will extend to the photon. Because the dynamical equation (the first of Equations 19.54) is second-order, the value and time derivative of two independent components of A^μ may be specified at an initial time. Thus a total of four independent items of initial data may be specified for the electromagnetic potential at each point of an initial spacelike surface.

The way in which the initial data is inter-related will determine the nature of the field on 'later' spacelike surfaces. In particular, it will determine any of the primarily spatial details of the evolution such as the direction of propagation of plane waves for example. For such a plane-wave one may choose to specify the four items of initial data as the magnitudes and phases of $\mathbf{A}^T$ in two independent polarizations, either linear or circular. In terms of the spin of a particle, the analogue of the two circular polarizations are the two helicity modes of the photon, one with its angular momentum parallel and the other anti-parallel to the propagation direction. Since one generally normalizes quantum wave functions so that they represent a single photon, the initial data will effectively be reduced to just two real values equivalent to the phases of each helicity mode of the photon.

Bibliography:

Barut A O 1964 *Electrodynamics and the classical theory of fields* [29].
Jackson J D 1975 *Classical electrodynamics* [263].
Misner C W et al., 1973 *Gravitation* [384].
Ohanian H C 1988 *Classical electrodynamics* [405].
Rindler W 1977 *Essential relativity* [476].

CHAPTER 20

Gauge principle

I think that there is a moral to this story, namely that it is more important to have beauty in one's equations than to have them fit experiment. If Schrödinger had been more confident of his work, he could have published it some months earlier, and he could have published a more accurate equation[1].... .

Paul Adrien Maurice Dirac   1902–1984

20.1 Introduction

In previous chapters, non-quantum particles have been used to describe the source material that generates the electromagnetic or gravitational fields and which is then acted on by those fields. We may now reduce the apparent difference between source and interaction by using a function of space and time to describe not only the mediating field but also the source material between which the interaction takes place. A spin 0 Schrödinger or spin $\frac{1}{2}$ Pauli-Schrödinger wave function may be used for this purpose where a non-relativistic treatment suffices.

The evolution of the electromagnetic field is inherently special relativistic and we can expect its mediation to be most naturally expressed and constructed in the context of a relativistic source. We shall therefore examine the use of the Klein-Gordon and Dirac fields to demonstrate that consistent coupling can be carried out in a non-quantum context to form an interacting Lagrangian which could then be used as a starting point for either path-integral or canonical quantization. Quantum particles would then arise from the many-particle techniques of quantum field theory.

Although the spin 0 scalar field is simpler than the spin $\frac{1}{2}$ Dirac field, the gauge coupling of the latter is not only more straightforward but more widely applicable to the matter of which the universe consists. We shall therefore develop sufficient Dirac theory for application of the Gauge

---

[1] *The physicist's picture of nature* [111].

principle and then return to examine a similar procedure applied to the Klein-Gordon field.

20.2 Dirac field of relativistic electrons

Because the Dirac field is non-chiral and spinorial, naming double-valued under rotations, a thorough and convincing demonstration of its properties is carried out ideally by examining the representation theory [27, 30, 410, 583] of the full Poincaré group IO(1,3) and its double covering group ISL(2,$\mathbb{C}$). The early group-theoretic analyses showing the need for Dirac fields were given by van der Waerden [550, 551] and Laporte and Uhlenbeck [313]. Such an examination is not possible here. We shall therefore describe a number of properties of the Dirac field, sufficient to write down the corresponding wave equation and Lagrangian density, admitting from the outset that several features must be accepted without detailed justification.

The *Dirac field* is a set of functions of the coordinates x which transform under translations, rotations, reflections and Lorentz boosts, in a manner appropriate to the relativistic time-reversal invariant propagation and interaction of non-null waves whose quanta are non-chiral massive spin $\frac{1}{2}$ particles. Fundamental particles with such properties are the six quarks and the massive leptons (e^-, μ^- and τ^-) and their antiparticles. Dirac particles are normally considered to have distinct antiparticles, such being the case with the quarks and non-neutrino leptons, as a result of the existence in each case of a non-zero electric charge, appearing symmetrically with equal and opposite signs for particle and antiparticle.

A self-conjugate spin $\frac{1}{2}$ particle, whose antiparticle is not distinct, is described by a *Majorana spinor field* which can be treated as a special case of the Dirac field in which all components may be taken to be real. Massless spin $\frac{1}{2}$ particles, which may include the neutrinos, are described by *Weyl spinor fields*, which may also be treated as special cases of Dirac fields (see Section 20.7).

The representation of non-chiral spin $\frac{1}{2}$ properties by a Dirac spinor ψ requires it to have four components $\psi = \{\psi^\alpha\}$ ($\alpha = 1, 2, 3, 4$),

$$\psi = \{\psi^\alpha\} = \begin{pmatrix} \psi^1 \\ \psi^2 \\ \psi^3 \\ \psi^4 \end{pmatrix}, \qquad (20.1)$$

in 4D spacetime. The equality of the dimension and the number of spin $\frac{1}{2}$ components, which is $2^{D/2}$ in D-dimensional spacetime [101], is a pecularity of four dimensions. The components are permitted to be complex to allow the representation of charge. Under Lorentz transformations, these will transform according to $\psi \to S(\Lambda)\psi$, where $S(\Lambda)$ is a 4×4 matrix determined in terms of the corresponding Lorentz matrix Λ by the representation theory for spin $\frac{1}{2}$. On the rare occasions when it is convenient to explicitly index the four components of a Dirac spinor we shall use a lower-case Greek letter from the beginning of the alphabet $(\alpha, \beta, \ldots)$. The same will be the case for labelling the rows and columns of the 4×4 transformation matrices, $S(\Lambda) = \{S(\Lambda)^\alpha{}_\beta\}$, for which an expression will be given shortly in terms of Lorentz transformation matrices, Λ.

The four components of a vector transform in an interrelated way under a restricted or full Lorentz transformation $\Lambda = \{\Lambda^\mu{}_\nu\}$ or $L = \{L^\mu{}_\nu\}$. By so doing, they may represent a field with spin 1 properties. Provided the 4-vector satisfies an appropriate partial differential wave equation, such as the Maxwell or Proca equations, it will represent the free propagation of a spin 1 system, null or non-null respectively. Analogously, a spin $\frac{1}{2}$ Dirac spinor must transform, at a particular event x, in a certain way under rotations and boosts, and satisfy a certain wave equation, the Dirac equation, to describe its translation or free propagation properties.

In order to set out these non-interacting and therefore primarily mathematical transformation properties, we shall define the four *Dirac matrices* $\gamma = \{\gamma^\mu\}$ ($\mu = 0, 1, 2, 3$) algebraically. These are any 4×4 matrices, $\gamma^\mu = \{(\gamma^\mu)^\alpha{}_\beta\}$, with rows and columns labelled by α and β ($\alpha, \beta = 1, 2, 3, 4$), that satisfy the following anticommutation ($\mu \neq \nu$) and 'self-inverse' ($\mu = \nu$) relation,

$$\gamma^\mu \gamma^\nu + \gamma^\nu \gamma^\mu = -2\, \eta^{\mu\nu}\, \mathbb{1}\,, \qquad (20.2)$$

four being the lowest dimension [262] for which a solution exists. The self-inverse property (for $\mu = \nu$) applies to γ^0 and $i\gamma^k$. This equation defines the *Dirac algebra*, an example of an anticommutation algebra first examined by Clifford[2], and known therefore as a *Clifford algebra*, another example of which is the algebra of the Pauli matrices. Many of the properties of the Dirac algebra required for analyses of spin $\frac{1}{2}$ particles were given by Pauli [424, 426, 427] in the period 1933–36. The Dirac matrices are invariant with respect to all Poincaré transformations. We shall see that their Lorentz indexing is a consequence of the transformation properties of bilinear combinations of γ^μ with two spinor factors. We define matrices with lowered indices by $\gamma_\mu = \eta_{\mu\nu} \gamma^\nu$.

An explicit numerical representation of the Dirac matrices is not necessary to describe the rotational and boost properties of Dirac spinors. However, in order to ensure the reality of observable quantities constructed from Dirac matrices and spinors, in physical applications we select from among the matrices γ^μ which satisfy the defining relation, only those which

---

[2] William Kingdon Clifford (1845–1879), English mathematician.

20.2. Dirac field of relativistic electrons

also satisfy the hermiticity conditions,

$$\gamma^{0\dagger} = \gamma^0 \quad \text{and} \quad \gamma^{k\dagger} = -\gamma^k , \tag{20.3}$$

thus also requiring γ^μ to be unitary, $\gamma^\dagger_\mu = \gamma^{-1}_\mu$.

Exercise 20.1 *Verify that, if $\mathbb{1}$ denotes the identity (here a 2×2 matrix), then*

$$\gamma^0 = \begin{pmatrix} 0 & \mathbb{1} \\ \mathbb{1} & 0 \end{pmatrix} \quad \text{and} \quad \gamma = \begin{pmatrix} 0 & -\sigma \\ \sigma & 0 \end{pmatrix} , \tag{20.4}$$

satisfy the Dirac algebra and the γ^μ hermiticity conditions, provided the matrices $\sigma = \{\sigma^k\} = \{\sigma_x, \sigma_y, \sigma_z\}$ satisfy the Pauli algebra relation,

$$\sigma^k \sigma^l + \sigma^l \sigma^k = 2\delta^{kl} \mathbb{1} , \tag{20.5}$$

and Pauli matrix hermiticity (and unitarity) conditions, $\sigma^\dagger_k = \sigma_k = \sigma^{-1}_k$.

The matrices of Equation 20.4 are referred to as the *Weyl* or *chiral representation* of the Dirac algebra. It is not necessary to use an explicit representation of the Pauli algebra. However, the standard *Pauli matrices*,

$$\sigma_x = \begin{pmatrix} 0 & 1 \\ 1 & 0 \end{pmatrix} \quad \sigma_y = \begin{pmatrix} 0 & -i \\ i & 0 \end{pmatrix} \quad \sigma_z = \begin{pmatrix} 1 & 0 \\ 0 & -1 \end{pmatrix} . \tag{20.6}$$

clearly satisfy both the above conditions.

A proper 4-vector, which may represent a spin 1 system, transforms under a full Lorentz transformation $L = \{\Lambda, P\Lambda, T\Lambda, PT\Lambda\} \in O(1,3)$ according to,

$$V^{\bar{\mu}}(x) = L^{\bar{\mu}}{}_\mu V^\mu(L^{-1}x) , \tag{20.7}$$

namely,

$$V^{\bar{\mu}}(x) = \Lambda^{\bar{\mu}}{}_\mu V^\mu(\Lambda^{-1}x) \tag{20.8}$$
$$\mathbf{P}V^\mu(x) = \{V^0(P^{-1}x), -V^k(P^{-1}x)\} \tag{20.9}$$
$$\mathbf{T}V^\mu(x) = \{-V^0(T^{-1}x), V^k(T^{-1}x)\} , \tag{20.10}$$

where **P** and **T** are the parity and time-reversal operations and $\Lambda \in SO(1,3)$ satisfies $\det \Lambda = 1$, $\Lambda^0{}_0 \geq 1$ and $\Lambda^T \eta \Lambda = \eta$. Having defined the Dirac matrices, we may now state that the corresponding transformation [262] of a spin $\frac{1}{2}$ Dirac spinor is

$$\psi \to S(\Lambda)\psi , \tag{20.11}$$

where the 4×4 Dirac spin transformation matrices $S(\Lambda)$, satisfy,

$$S(\Lambda)\gamma^\mu S^{-1}(\Lambda) = (\Lambda^{-1})^\mu{}_\nu \gamma^\nu \quad \text{and} \quad \det S(\Lambda) = 1 . \tag{20.12}$$

Paul Adrien Maurice Dirac

b. Monk Royal, England
8 August 1902

d. Tallahasse, Florida
20 October 1984

From a drawing by R. Tollast in St John's College, Cambridge.
Courtesy of the Master and Fellows.

When evaluated with a rotation Λ_R of angle 2π, the matrices have value $S(\Lambda_R(2\pi)) = -1$ which means that ψ is double-valued, or *spinorial*, over the angular domain $\{0, 2\pi\}$. The relations in Equation 20.12 are closely related to the conditions in Dirac spinor space equivalent to the Lorentz conditions, $\Lambda^T \eta \Lambda = \eta$ and $\det \Lambda = 1$. An explicit expression for $S(\Lambda)$ satisfying these conditions is given [262] by,

$$S(\Lambda) = e^{-\frac{1}{2}i\omega_{\mu\nu}S^{\mu\nu}} = e^{-\frac{1}{4}i\omega_{\mu\nu}\gamma^{\mu\nu}} . \qquad (20.13)$$

The matrices,

$$S^{\mu\nu} = \tfrac{1}{2}\gamma^{\mu\nu} \equiv \tfrac{1}{4}i(\gamma^\mu \gamma^\nu - \gamma^\nu \gamma^\mu) , \qquad (20.14)$$

are the Lorentz generators of the Dirac spinor while $\theta = \{\tfrac{1}{2}\epsilon^{klm}\omega_{lm}\}$ and $\alpha = \{\omega^{0k}\}$ are the rotation and boost parameters of the Lorentz transformation Λ. The most widespread notation for $\gamma^{\mu\nu}$, or some simple multiple of it, is $\sigma^{\mu\nu}$, which we reserve for the 2-component spinor generators.

Exercise 20.2 *Establish the following Dirac identities:*

- $\mathrm{Tr}\,(\gamma^\mu) = \mathrm{Tr}\,(\gamma^{\mu\nu}) = 0, \qquad \mathrm{Tr}\,(\gamma^\mu \gamma^\nu) = -4\,\eta^{\mu\nu}$
- $\gamma^\mu \gamma_\mu = -4, \qquad \gamma^{\mu\nu}\gamma_{\mu\nu} = 12, \qquad \gamma^\mu \gamma^\lambda \gamma_\mu = 2\gamma^\lambda$
- $\gamma^\mu \gamma^\nu = -\eta^{\mu\nu} - i\gamma^{\mu\nu}, \qquad [\gamma^\mu, \gamma^{\nu\lambda}] = 2i(\eta^{\mu\lambda}\gamma^\nu - \eta^{\mu\nu}\gamma^\lambda)$

- $(\gamma^0\gamma^\mu)^\dagger = \gamma^0\gamma^\mu, \qquad (\gamma^0\gamma^{\mu\nu})^\dagger = \gamma^0\gamma^{\mu\nu}$

20.3 Dirac equation

The above Lorentz transformational properties of a Dirac spinor comprise the part of its kinematics which is defined at an event. To prepare for a discussion of the dynamics, we need first the propagation characteristics of the non-interacting Dirac field, the quantization of which leads to free particles. Each non-interacting component propagates as a plane wave. A relativistic plane wave is at least a solution of the Klein-Gordon equation. Consequently, we require each component of the free Dirac field to satisfy $(\Box - m^2)\psi(x) = 0$.

A spin $\frac{1}{2}$ wave has two physical degrees of freedom, the direction cosines of its polarization vector, corresponding in the quantum physics of massive particles, to the spin up and spin down eigenvalues of the component of the spin angular momentum operator in a specified direction. If distinct particle and antiparticle are to be represented by such waves, the number of real dynamical degrees of freedom of a Dirac field will be four. On the other hand, the Dirac field has four complex or eight real components, the complex character permitting the Dirac particles to have charge. The wave equation it satisfies must ensure that only four solutions are linearly independent. This is analogous to Maxwell's equations ensuring that either $A^\mu = \{\phi/c, \mathbf{A}\}$ or $F^{\mu\nu} = \{\mathbf{E}, \mathbf{B}\}$ represent only two independent physical degrees of freedom, the two polarizations of a spin 1 wave or the two helicities of the photon.

The Klein-Gordon equation, $(\Box - m^2)\psi = 0$, does not interrelate the spacetime dependence of the four components of a Dirac spinor in the same way that $\Box A - \partial\partial \cdot A = 0$ (or $\Box A^\mu = 0 = \partial \cdot A$ in the Lorentz gauge) tie together the functional behaviour of the different components of the free electromagnetic potential. If the Klein-Gordon equation is the only differential condition on the Dirac spinor, the four complex components $\{\psi^\alpha\}$ will be related at each event as the parts of a spin $\frac{1}{2}$ spinor, when Lorentz rotated, but their variation from event to event is no different from that of four independent scalar plane waves. Equivalently, these properties make $\psi = \{\psi^\alpha\}$ transform correctly with respect to the full Lorentz group O(1,3) and the functions $\psi(x)$ transform appropriately with respect to the translation group T^4. However, they do not ensure that $\psi(x)$ transforms correctly under the full Poincaré group IO(1,3), considered as a whole.

Because the Dirac spinor has no Lorentz indices the Klein-Gordon equation is the unique Poincaré-covariant second-order linear equation the free field can satisfy. If the individual components are interrelated via a field equation other than the Klein-Gordon equation in such a way that the field equation implies the Klein-Gordon equation, but not the converse, then the number of real degrees of freedom will be less than the eight in the Dirac spinor. The result could in fact be as low as the number (four) required to describe the polarization properties of the waves appropiate to one spin $\frac{1}{2}$ particle and its antiparticle.

A first-order differential equation which implies a second-order equation, but not the converse, typically has a more constrained solution space than the latter. Consequently, we are led to consider the ways in which a first-order equation satisfied by $\psi(x)$ may be constructed in accordance with relativistic covariance. We naturally seek first an equation linear in the Dirac spinor. The simplest such equation is $(a^\mu \partial_\mu + b)\psi(x) = 0$ where a^μ and b are complex constants. However, if the a^μ and b are ordinary constants this equation is inconsistent with the Klein-Gordon equation which we shall see requires distinct members of a^μ to anticommute. Furthermore, the equation would not then link together the different components of $\psi(x)$.

The components of $\psi(x)$ will be interrelated if we allow them to appear, along with $\partial_\mu \psi(x)$, in a linear combination with matrix coefficients. We therefore examine the equation $(A^\mu \partial_\mu + B)\psi(x) = 0$ in which $A^\mu = \{(A^\mu)^\alpha{}_\beta\}$ and $B = \{B^\alpha{}_\beta\}$ are five 4×4 matrices, to be determined, with rows and columns labelled by spinor indices α and β. We apply the operator $-A^\nu \partial_\nu + B$ to this equation to obtain,

$$(-A^{(\mu} A^{\nu)} \partial_\mu \partial_\nu + B^2)\psi(x) = 0 \ . \qquad (20.15)$$

This equation must be equivalent to the Klein-Gordon equation. This means that $A^{(\mu} A^{\nu)} \partial_\mu \partial_\nu = -\Box = -\eta^{\mu\nu} \partial_\mu \partial_\nu$ and $B^2 = -m^2 \mathbb{1}$ where m is the mass appearing in the Klein-Gordon equation. The matrix B is therefore invertible for $m \neq 0$. By pre-multiplying $(A^\mu \partial_\mu + B)\psi(x) = 0$ by $-mB^{-1}$, we obtain $(-mB^{-1} A^\mu \partial_\mu - m)\psi(x) = 0$ and our equation may therefore be re-written in the form $(\pm i\gamma^\mu \partial_\mu - m)\psi(x) = 0$ where the matrices $\gamma^\mu \equiv \mp im B^{-1} A^\mu$ must satisfy the Dirac algebra. Indeed, this is why we defined the Dirac matrices using the Clifford condition. The lowest-dimensional matrices for which this is true are indistinguishable from the Dirac matrices γ^μ. The factor of i is inserted, by convention, in order to facilitate the construction of hermitian quantities, representing observables, from combinations of ψ, γ^μ and $\frac{1}{i}\partial_\mu$, the last of which is hermitian.

Taking the positive coefficient of γ^μ, by convention, leads to the *Dirac equation* [106] (in natural units),

$$(i\gamma^\mu \partial_\mu - m)\psi(x) = 0 \ , \qquad (20.16)$$

for the free relativistic propagation of massive non-chiral spin $\frac{1}{2}$ waves. It may be straightforwardly verified that there is no local gauge freedom of ψ in the Dirac equation.

We note the identity $(\gamma^\mu \partial_\mu)(\gamma^\nu \partial_\nu) \equiv -\Box$ and we apply the *Feynman slash* notation, defined by $\slashed{V} \equiv \gamma^\mu V_\mu$ for any 4-vector $V = \{V^\mu\}$, to

20.3. Dirac equation

the 4-momentum operator $p_\mu = \frac{1}{i}\partial_\mu$, to write the Dirac equation in the alternative form,

$$(\slashed{p} + m)\psi = 0 ,\qquad(20.17)$$

where,

$$\slashed{p} = \tfrac{1}{i}\slashed{\partial} = \tfrac{1}{i}\gamma^\mu\partial_\mu, \quad \text{and} \quad \slashed{p}^2 = -\slashed{\partial}^2 = \square . \qquad(20.18)$$

'Squaring' the Dirac equation $i\slashed{\partial}\psi = m\psi$, by operating on the left with $i\slashed{\partial}$ and multiplying on the right by m, gives the Klein-Gordon equation. These equations show that the free Dirac equation behaves in many respects as the 'square root' of the free Klein-Gordon equation. This notion may be developed more rigorously and has a parallel in a similar relationship in which the equations of supergravity [97, 98, 99, 100, 174] can be regarded as the 'square root' of general relativity.

If we perform a restricted Poincaré transformation, $x \to \Lambda x + a$ and $\psi(x) \to S(\Lambda)\psi((a, \Lambda)^{-1}x)$, to a new frame, the Dirac equation in the new coordinates may be rewritten as,

$$\left(i\gamma^\mu(\Lambda^{-1})^\nu{}_\mu \partial_\nu - m\right)S(\Lambda)\psi(x) = 0 . \qquad(20.19)$$

Exercise 20.3 *Show that the use of the first of Equations 20.12 in Equation 20.19 reduces it to the Dirac equation in the original frame thereby partially justifying the form of Equation 20.12 as being appropriate to ensure the free Dirac equation is Poincaré-covariant.*

A parity transformation $\mathbf{P}$ takes x to $\mathbf{P}x = \tilde{x} = \{x^0, -x^k\}$ and will transform $\psi(x)$ to a new Dirac field $\mathbf{P}\psi(x) = \tilde{\psi}(\tilde{x}) = \tilde{\psi}(\mathbf{P}^{-1}x)$ which we require to also satisfy the Dirac equation in the new coordinates in order to guarantee parity covariance and lack of chiral properties in ψ. Parity transformation of the original Dirac equation gives,

$$(i\gamma^\mu \tilde{\partial}_\mu - m)\tilde{\psi}(\tilde{x}) = 0 . \qquad(20.20)$$

Exercise 20.4 *Establish the identity $\gamma^0 \gamma^{\mu\dagger} \gamma^0 = \gamma^\mu$. Show that,*

$$(i\gamma^\mu \partial_\mu - m)\gamma^0 \tilde{\psi}(\tilde{x}) = 0 . \qquad(20.21)$$

by premultiplying Equation 20.20 by γ^0 and using the hermiticity of γ^0 and $i\gamma^k$.

Comparison of Equation 20.21 with $(i\gamma^\mu \partial_\mu - m)\tilde{\psi}(x) = 0$, the Dirac equation for $\tilde{\psi}(x)$, shows that,

$$\tilde{\psi}(x) = \eta_P \gamma^0 \psi(\tilde{x}) = \eta_P \gamma^0 \psi(\mathbf{P}^{-1}x) , \qquad(20.22)$$

where η_{P}, called the *intrinsic parity* [37, 578] of ψ, is a phase factor whose properties are to be determined. Space reflections are self-inverse and this fundamental property must be shared in any representation. Two parity operations in succession on the free Dirac spinor give $\eta_{\text{P}}^2 \psi(x)$ which must be physically equivalent to the original spinor $\psi(x)$. However, Dirac spinors are doubled-valued with $\pm\psi(x)$ being physically equivalent to one another. Thus $\eta_{\text{P}}^2 = \pm 1$ and the intrinsic parity may have the values $\eta_{\text{P}} = \pm 1, \pm i$, implying that $|\eta_{\text{P}}| = 1$. A single fermion cannot have a well-defined intrinsic parity. Although η_{P} may be set by convention for any given spinor, the intrinsic parities of some spinors, such as those corresponding to particle and antiparticle will be related as will also be those of particle and supersymmetric partner.

In order to form a Lagrangian, we need an invariant bilinear for any two spinors ψ and ξ, somewhat analogous to the scalar product $U \cdot V$ of two vectors. The *Pauli adjoint spinor* of ψ is defined to be $\bar{\psi} = \psi^\dagger \gamma^0$.

Exercise 20.5 *Show that the adjoint spinor $\bar{\psi}$ Lorentz transforms inversely to the Dirac spinor ξ, namely according to $\bar{\psi} \to \bar{\psi} S(\Lambda^{-1})$, thus showing that the combination $\bar{\psi}\xi$ (and therefore also $\bar{\psi}\psi$) is Lorentz-invariant. Show that $\bar{\psi}\xi$ is also a proper scalar.*

Exercise 20.6 *Hermitian conjugate the Dirac equation, multiply by appropriate factors of γ^0 and use Dirac identities, to show that the Dirac adjoint spinor field satisfies $i(\partial_\mu \bar{\psi})\gamma^\mu + m\bar{\psi} = 0$ which we re-write as,*

$$\bar{\psi}(i\gamma^\mu \overleftarrow{\partial_\mu} + m) = 0 \quad \text{or} \quad \bar{\psi}(i\overleftarrow{\slashed{\partial}} + m) = 0 \,. \tag{20.23}$$

A free Dirac equation in practical units may be constructed in the form $(i\slashed{\partial} - a)\psi = 0$ where a is a constant of inverse length dimensions which need have no connection with quantum phenomena. In a relativistic quantum context, the constant a may arise from the reduced Compton wavelength of a mass m according to $a = 1/\lambdabar = mc/\hbar$, in which case the Dirac equation in practical units is $(i\slashed{\partial} - mc/\hbar)\psi = 0$. The Dirac equation in natural units, $(i\slashed{\partial} - m)\psi = 0$, with m interpreted as a mass thus arises naturally in a relativistic quantum context, even prior to interpretation of ψ as either a wave function or quantum field operator.

Scalar, vector and tensor fields in Minkowskian spacetime were defined in Chapter 14. Their manifestly covariant use is sufficient to represent the form invariance requirement imposed by the Relativity principle for relativistic propagation of the spin 0 Klein-Gordon and spin 1 Maxwell fields, the latter in either vector or antisymmetric tensor form. A symmetric Fierz-Pauli tensor (see Chapter 21) may also describe the null spin 2 properties of weak relativistic gravitation.

We may also ask what are the necessary, as opposed to sufficient, conditions on the transformation properties of fields for them to ensure that the Relativity principle is respected with a finite ultimate speed. The answer provided by group representation theory [27, 30, 410, 583] is that physical quantities must be either tensors $\phi, \phi^\mu, \phi^{\mu\nu}, \ldots$, representing integer spin fields ($s = 0, 1, 2, \ldots$) or *tensor-spinors*, $\psi, \psi^\mu, \psi^{\mu\nu}, \ldots$, which transform as combinations of Dirac spinors (see Section 21.2) representing half-odd-integer spin fields ($s = \frac{1}{2}, \frac{3}{2}, \frac{5}{2}, \ldots$). Quantization leads to bosons and fermions, respectively.

20.4 Dirac Lagrangian density

The Dirac Lagrangian density is required to be an hermitian, single-valued proper scalar in $\psi(x)$ and $\partial_\mu \psi(x)$. The double-valued property of ψ under rotations requires an even number of factors of ψ to appear in each term of the Lagrangian. A simple proper scalar quadratic term is $\bar{\psi}\psi$ where $\bar{\psi}$ is the Dirac adjoint.

To obtain a scalar kinetic term involving $\partial_\mu \psi(x)$ requires saturation of the Lorentz index μ. This cannot be done with another derivative in the form of $\bar{\psi} \Box \psi$ or $\partial^\mu \bar{\psi} \partial_\mu \psi$ since, although both expressions are single-valued, they lead to second-order equations, not the Dirac equation. The only possibility is to make use of the Dirac matrices γ^μ to saturate the index in $\partial_\mu \psi$.

Exercise 20.7 *(a) Show that $\bar{\psi}\slashed{\partial}\psi$ is an SO(1,3) invariant and a proper scalar with respect to parity. (b) Show that $\bar{\psi}\slashed{\partial}\psi$ is neither real nor pure imaginary in general but that the combination $i(\bar{\psi}\slashed{\partial}\psi - \overline{\slashed{\partial}\psi}\psi)$ is real. We abbreviate this result as $i\bar{\psi}\overleftrightarrow{\slashed{\partial}}\psi$.*

We choose $\frac{1}{2}i\bar{\psi}\overleftrightarrow{\slashed{\partial}}\psi$ as the kinetic term in the standard Dirac Lagrangian, the sign ensuring the term contributes positive energy on quantization. Unlike the Klein-Gordon field (see Exercise 16.3), the Hamiltonian density of the Dirac field is not positive-definite prior to quantization. This choice means that the natural dimensions of a Dirac spinor are also fixed to be $[\psi]_{\text{nat}} = L^{-3/2}$ where L denotes length. All spinorial Lagrangian fields will have similar kinetic terms, differing only by having extra contracted Lorentz indices (see Section 21.2) appropriate to higher spin ($s \geq \frac{3}{2}$) and $L^{-3/2}$ therefore becomes a *canonical fermionic dimension* in natural quantum field theoretic units.

The scalar $\bar{\psi}\psi$ must be multiplied by a quantity of inverse length dimension, namely mass, to have the same dimensions as the kinetic term. The Dirac mass m will supply the extra factor and must have a coefficient of -1 to form the free Lagrangian density,

$$\mathcal{L} = \tfrac{1}{2} i \bar{\psi} \overleftrightarrow{\partial\!\!\!/} \psi - m \bar{\psi} \psi , \qquad (20.24)$$

whose Euler-Lagrange equation is the Dirac equation.

At least formally, the derivatives of the Lagrangian with respect to the field and the velocities are,

$$\frac{\partial \mathcal{L}}{\partial \psi} = -m\bar{\psi} - \tfrac{1}{2} i \bar{\psi} \overleftarrow{\partial\!\!\!/} \quad \text{and} \quad \frac{\partial \mathcal{L}}{\partial(\partial_\mu \psi)} = \tfrac{1}{2} i \bar{\psi} \gamma^\mu , \qquad (20.25)$$

with corresponding results for the adjoint fields. Since the Lagrangian density is hermitian, but the fields are complex, we may treat the Dirac spinor and its adjoint as independent Lagrangian fields. In practical units, m must be replaced by $mc/\hbar$ on the right side of Equation 20.24, which must also include an overall factor of $\hbar^{-1}$.

Exercise 20.8 *Show that the Euler-Lagrange equations for ψ and $\bar{\psi}$ obtained from Equation 20.24 give the adjoint Dirac and Dirac equations, respectively.*

The Dirac Lagrangian may be rewritten in the form,

$$\mathcal{L} = \tfrac{1}{2} \bar{\psi}(i\partial\!\!\!/ - m)\psi - \tfrac{1}{2} \bar{\psi}(i \overleftarrow{\partial\!\!\!/} + m)\psi , \qquad (20.26)$$

which shows that, evaluated on the stationary path, the Dirac Lagrangian vanishes. This fact emphasizes an important property that any Lagrangian has prior to its use in Euler-Lagrange equations. It is not a set of values for each x but a function of the arbitrary fields which appear in it, the action being a corresponding functional in the field. This contrasts, for example, with the energy-momentum tensor $T^{\mu\nu}$ of a Lagrangian field in which it is necessary to use the stationary value of the fields in the expression for $T^{\mu\nu}$ in terms of the Lagrangian to obtain the correct result.

Exercise 20.9 *(a) Show that the canonical energy-momentum tensor of the Dirac field is,*

$$\check{T}^{\mu\nu} = \tfrac{1}{2i} \bar{\psi} \gamma^\mu \overleftrightarrow{\partial^\nu} \psi , \qquad (20.27)$$

and that it is conserved, $\partial_\mu \check{T}^{\mu\nu} = 0$, if ψ is a solution of the free Dirac equation. (b) Show that the Dirac Lagrangian is non-singular and that the canonical momentum is $\pi = \tfrac{1}{2} i \psi^\dagger$. (c) Show that application of Equation 16.19 (with $c = 1$) in a form appropriate to a complex field, namely,

$$\mathcal{H} = \partial_0 \bar{\psi} \frac{\partial \mathcal{L}}{\partial(\partial_0 \bar{\psi})} + \frac{\partial \mathcal{L}}{\partial(\partial_0 \psi)} \partial_0 \psi - \mathcal{L} , \qquad (20.28)$$

leads to a Dirac field Hamiltonian density of,

$$\mathcal{H} = -\tfrac{1}{2}i\bar{\psi}\gamma\cdot\overleftrightarrow{\nabla}\psi + m\bar{\psi}\psi = -i\bar{\psi}\gamma\cdot\nabla\psi + m\bar{\psi}\psi , \qquad (20.29)$$

and compare this with the action of $i\partial_t$ on a one-particle Dirac spinor ψ. (d) Use the Dirac algebra to establish the identity, $\gamma^{\mu\lambda}\gamma_{\lambda\nu} = 2\gamma^\mu\gamma_\nu - \delta^\mu{}_\nu$, and show that evaluation of the square of the Pauli-Lubański 4-vector, using Equation 15.73 with $j^{\mu\nu}$ replaced by the Dirac spin generators $S^{\mu\nu}$, leads to $w^2 = \tfrac{3}{4}m^2c^2 = \tfrac{1}{2}(\tfrac{1}{2}+1)m^2c^2$. (This result may be used to show that the Dirac field has spin $\tfrac{1}{2}$.)

The canonical tensor $\check{T}^{\mu\nu}$ of the Dirac field leads to the correct energy-momentum p^μ but is not symmetric. Use of the Belinfante procedure to allow for the energy density of spin angular momentum, and some Dirac algebra manipulations, leads to an angular momentum conserving Dirac energy-momentum tensor given by the symmetric part $T^{\mu\nu} = \check{T}^{(\mu\nu)}$ of $\check{T}^{\mu\nu}$.

20.5 Plane Dirac waves

Since the solutions of the free Dirac equation also satisfy the Klein-Gordon equation, their dependence on the spacetime coordinates x must be of plane wave form, $e^{\pm ik\cdot x}$. The standard momentum-space measure $d\hat{k}$ for non-null (massive) fermions [262] is defined in terms of the boson measure $d\tilde{k}$ (see Equation 16.41) according to,

$$d\hat{k} = 2m d\tilde{k} = m\frac{d^3k}{(2\pi)^3 k^0} . \qquad (20.30)$$

A basis for the solution space of free Dirac spinor fields, a *spin $\tfrac{1}{2}$ polarization basis*, will be made up of the same plane-wave factors $e^{\pm ik\cdot x}$ as appear in the solutions of the free Klein-Gordon and Maxwell equations combined with spinorial factors that depend on the propagation 3-vector $\mathbf{k}$. We denote those spinorial factors by $u(\mathbf{k}) = \{u^\alpha(\mathbf{k})\}$ and $v(\mathbf{k}) = \{v^\alpha(\mathbf{k})\}$ for the positive and negative frequency cases, respectively. We therefore seek solutions of the form,

$$\psi^{(+)}(x) = u(\mathbf{k})e^{ik\cdot x} \quad \text{and} \quad \psi^{(-)}(x) = v(\mathbf{k})e^{-ik\cdot x} . \qquad (20.31)$$

These basis spinor fields will satisfy the Dirac equation provided the spinor factors satisfy,

$$(\not{p}+m)u(\mathbf{k}) = 0 \quad \text{and} \quad (\not{p}-m)v(\mathbf{k}) = 0 . \qquad (20.32)$$

Each of these is a set of four linear homogeneous equations.

Exercise 20.10 *Show that solutions exist for $u(\mathbf{k})$ and $v(\mathbf{k})$ provided that $\det(\not{p} \pm m) = 0$ implying $k^0 = \sqrt{\mathbf{k}^2 + m^2}$.*

Solutions also exist for k^0 of the same magnitude but of opposite sign but we have effectively already allowed for this alternative by including both signs of the exponent in the plane waves with $k^0 > 0$. Use of the rest frame ($\mathbf{k} = 0$) and a specific representation of γ^0 (such as Equation 20.4) shows that these equations have two linearly-independent solutions for u and two for v which we denote by $u^{(r)}(\mathbf{k})$ and $v^{(r)}(\mathbf{k})$ where $r = 1, 2$. For non-zero rest mass m, we may, in the rest frame, take these four solutions to be spinors $u^{(r)}$ and $v^{(r)}$ given by,

$$u^{(1)} = \begin{pmatrix} 1 \\ 0 \\ 0 \\ 0 \end{pmatrix} \quad u^{(2)} = \begin{pmatrix} 0 \\ 1 \\ 0 \\ 0 \end{pmatrix} \quad v^{(1)} = \begin{pmatrix} 0 \\ 0 \\ 1 \\ 0 \end{pmatrix} \quad v^{(2)} = \begin{pmatrix} 0 \\ 0 \\ 0 \\ 1 \end{pmatrix}. \quad (20.33)$$

We seek now to classify these four independent solutions using unique eigenvalues of a complete set of commuting, observable, one-particle operators, one of which we choose to take as the Hamiltonian operator. The latter, $H = i\partial_t = \gamma^0(m - i\boldsymbol{\gamma} \cdot \boldsymbol{\nabla}) = \gamma^0(\not{p} + m)$ has eigenvalue $E = \sqrt{\mathbf{k}^2 + m^2}$ for each of the $u^{(r)}$ and $-E$ for each $v^{(r)}$.

We need to remove this degeneracy by finding another hermitian operator commuting with H which will distinguish between the members in a pair with $r = 1, 2$. This distinction may be based on polarization or spin projection properties since, in the rest frame, the z-component of the Dirac 3-vector spin matrix, $\mathbf{S} = \{\frac{1}{2}\epsilon^{klm} S_{lm}\}$, namely

$$S = \frac{1}{2} \begin{pmatrix} \sigma & 0 \\ 0 & \sigma \end{pmatrix}, \quad (20.34)$$

has eigenvalues $+\frac{1}{2}$ for $u^{(1)}$ and $v^{(1)}$ and $-\frac{1}{2}$ for $u^{(2)}$ and $v^{(2)}$. However, although the total angular momentum operator $\mathbf{J}$ commutes with the Hamiltonian, the spin projection operator S_z does not as may be verified using a specific Dirac matrix representation. Nevertheless, the projection of the spin in a spacelike direction n ($n^2 = 1$) which is Lorentz-orthogonal ($k \cdot n = 0$) to the propagation 4-vector k, is conserved and the basis states are eigenstates of that spin operator with eigenvalues $\pm\frac{1}{2}$.

If the spacelike direction n is chosen so that its 3-vector part is in the same direction as the propagation 3-vector $\mathbf{k}$ in a particular frame then the projection of the spin is called *helicity* in that frame, directly analogous to the Newtonian or Einsteinian mechanical quantities with the same properties. It should be noted, however, that for non-zero mass the helicity is not a Lorentz invariant as it is for a null field or massless particle.

The Dirac equation leaves the eight real algebraic components with only four real physical degrees of freedom corresponding to both spin projections for each of the two signs of energy for one-particle states.

The most general plane-wave expansion of a non-interacting Dirac spinor field is thus,

$$\psi(x) = \int d\hat{k} \sum_{r=1}^{2} [a^{(r)}(\mathbf{k})u^{(r)}(\mathbf{k})e^{ik\cdot x} + b^{(r)*}(\mathbf{k})v^{(r)}(\mathbf{k})e^{-ik\cdot x}] . \quad (20.35)$$

20.6 Dirac current

In order to form a field source for the electromagnetic field we require a single-valued and conserved current which is proper under parity, improper under time-reversal and real. The latter property corresponds to invariance under charge conjugation. The original motivation for the abandonment of the Klein-Gordon equation and the search for a first-order alternative, leading to the discovery of the Dirac equation, was to obtain a positive-definite probability density as part of a conserved current.

It is not difficult to construct such a current from a Dirac spinor. To be single-valued it must include two factors of ψ. To be conserved as a result of the Dirac equation it cannot include the derivative ∂_μ. The only way to construct a quantity with the required Lorentz index is to include a γ^μ factor and we arrive at the *Dirac probability current* $j^\mu = \bar{\psi}\gamma^\mu\psi$ from which we construct an electric current J^μ, of particles of charge q, with:

$$J^\mu = qj^\mu = q\bar{\psi}\gamma^\mu\psi . \quad (20.36)$$

Since the time component is $\bar{\psi}\gamma^0\psi = \psi^\dagger\psi$, the probability density in a single-particle theory is positive-definite.

Exercise 20.11 Show (a) that $\bar{\psi}\gamma^\mu\psi$ is a real proper 4-vector and is conserved as a result of the Dirac equation, (b) that $J^\mu = qc\bar{\psi}\gamma^\mu\psi$ in practical units, and (c) that the Dirac Lagrangian and equation are invariant with respect to the U(1) phase invariance $\psi(x) \to e^{i\xi}\psi(x)$ where ξ is a real dimensionless parameter not dependent on the spacetime coordinates and with compact range ($0 \leq \xi \leq 2\pi$). In natural units, electric charge is dimensionless and we may conventionally rewrite the U(1) factor as $e^{iq\alpha}$ where α is a real dimensionless parameter and q is a constant which may later be interpreted as a charge and a coupling constant. (d) Show that the Noether current resulting from this continuous U(1) Lie symmetry with parameter α is proportional to the Dirac current.

In practical units, the U(1) factor will take the form $e^{iq\alpha/\hbar}$ in which α has the dimensions of [length.A^μ].

We have seen that the original difficulties of the Klein-Gordon equation that led to the discovery of the Dirac equation are overcome in the latter case since the probability density is positive-definite. However, the existence of energy eigenvalues of both signs, for the wave-mechanical Hamiltonian operator $H = i\partial_t$ applied to plane scalar waves, continues to be a problem in the single-particle interpretation of the Dirac field as a probability wave function.

These difficulties with negative energies for Klein-Gordon and Dirac fields are both solved by the methods of quantum field theory. States which would have negative energy in the one-particle theory, are then reinterpreted as antiparticles having positive energy. Paradoxically, it also turns out that quantum field theory removes the problems with the negative probabilities of the Klein-Gordon field that led to the discovery of the Dirac field.

To achieve positive energy, spinor fields are quantized according to anticommutation rather than commutation rules. The statistics of the resulting particles are Fermi-Dirac [410]. The non-quantum limit ($\hbar \to 0$) of the quantized Dirac field is not a true classical limit since classical physics conventionally involves only commuting variables. The Dirac field and other spinor fields of spin $\frac{3}{2}$, $\frac{5}{2}$, ..., are therefore in many senses essentially quantum fields, in contrast to scalar, vector and tensor fields of integer spin. The applicability of the Pauli exclusion principle limiting the occupancy of states to zero or one particle prevents one from formulating a classical description as an approximation when the number of particles is large.

In discussing the scalar field we noted the impossibility of forming a conserved Klein-Gordon Noether current unless the scalar field was complex with a Lagrangian having phase invariance. We can expect a similar limitation to appear in the Dirac case. Indeed, if the non-quantum Dirac spinor field has anticommuting (Grassmann) components, $\psi^\alpha \psi^\beta + \psi^\beta \psi^\alpha = 0$, and if it satisfies a self-conjugacy condition which makes it an effectively 'real' (Majorana) spinor, then the Dirac current and charge vanish identically.

20.7 Chiral Dirac fields and Weyl neutrinos

We shall see that the electromagnetic interaction couples a non-chiral spin $\frac{1}{2}$ Dirac field ψ via a non-chiral mediating real proper vector field A^μ of spin 1. In first-order form, the non-chiral field strengths $F^{\mu\nu}$ are also Lagrangian fields, along with A^μ.

Although chiral states may be constructed for spin 1 systems, examples being the complex selfdual and antiselfdual fields $F^\pm_{\mu\nu}$ of Exercise 14.23, the fundamental Lagrangian interaction of electrodynamics does not involve coupling via either one separately.

We may also construct chiral 'Dirac' (non-self-conjugate) fields by

20.7. Chiral Dirac fields and Weyl neutrinos

introducing the *chirality matrix* γ^5 defined by,

$$\gamma^5 = \gamma_5 = -\frac{i}{4!}\epsilon_{\mu\nu\lambda\rho}\gamma^\mu\gamma^\nu\gamma^\lambda\gamma^\rho = i\gamma^0\gamma^1\gamma^2\gamma^3 \ . \tag{20.37}$$

If we allow γ^5 to act on a non-chiral Dirac field ψ after adding it to a Dirac matrix constructed from products of γ^μ with no factors of the permutation pseudotensor, we shall be explicitly forming a field which cannot be an eigenstate of parity. In particular, we may form chiral fields by projection, from a non-chiral Dirac field, using the matrices $\frac{1}{2}(\mathbb{1}\pm\gamma^5)$.

Exercise 20.12 (a) Show that $(\gamma^5)^2 = \mathbb{1}$ and hence that $\frac{1}{2}(\mathbb{1}\pm\gamma^5)$ are projection operators. (b) Establish the following γ^5 identities:

- $\gamma^{5\dagger} = \gamma^5 \quad (\gamma^0\gamma^5)^\dagger = -\gamma^0\gamma^5 \quad (\gamma^0\gamma^5\gamma^\mu)^\dagger = \gamma^0\gamma^5\gamma^\mu$
- $i\tilde{\gamma}^{\mu\nu} = \gamma^5\gamma^{\mu\nu}$ or $\frac{1}{2}i\epsilon^{\mu\nu\lambda\rho}\gamma_{\lambda\rho} = \gamma^5\gamma^{\mu\nu}$
- $\mathrm{Tr}\,(\gamma^5) = 0 \quad \mathrm{Tr}\,(\gamma^5\gamma^\mu) = 0 \quad \mathrm{Tr}\,(\gamma^5\gamma^{\mu\nu}) = 0$
- $[\gamma^5,\gamma^{\mu\nu}] = 0 \quad \gamma^5\gamma^\mu = -\gamma^\mu\gamma^5$ or $\{\gamma^5,\gamma^\mu\} = 0 \ .$

The brackets $\{A,B\}$ denote the anticommutator $AB + BA$ of A and B.

We now form the *chiral parts* $\psi_\pm = \frac{1}{2}(\mathbb{1} \pm \gamma^5)\psi$ of the non-chiral Dirac field and justify such names by noting that these parts are parity conjugates.

Exercise 20.13 Show (a) that operation by the Dirac parity matrix on each chiral part of ψ, namely $\gamma_0\psi_\pm$, gives a multiple of the opposite chiral part and (b) that the chiral parts satisfy the γ^5 conditions, $\gamma_5\psi_\pm = \pm\psi_\pm$, showing they are eigenfields of the chirality matrix.

The γ^5 conditions show that the chiral parts each have two algebraically independent complex components or four real physical components, with half that number of physical components. The negative and positive chiral parts $\psi_\pm$ are also referred to as the *left-handed* and *right-handed* parts of the Dirac field ψ. It should be noted, however, that for a massive field, these are not Poincaré-invariant parts of the original field.

Exercise 20.14 Show that $\psi_\pm$ do not propagate independently but according to the coupled equations, $i\slashed{\partial}\psi_\pm = m\psi_\mp$, if $\psi = \psi_+ + \psi_-$ is free.

The chirality matrix cannot be used to invariantly divide a massive Dirac field into two independent left and right-handed parts. The particles which correspond to the Dirac field are not intrinsically of one chirality (handedness) or the other.

However, for spin $\frac{1}{2}$ particles satisfying the massless Dirac equation, $\slashed{\partial}\psi = 0$, the free propagation of the chiral parts are uncoupled from one another and satisfy the equations, $\slashed{\partial}\psi_\pm = 0$. Such equations describe the propagation, at speed c, of non-interacting spin $\frac{1}{2}$ massless particles of definite chirality, ± 1, or definite handedness, left or right, examples being (massless) neutrinos and antineutrinos.

Since the operation of chiral projection involves the addition of two quantities, $\mathbb{1}$ and γ^5, only one of which involves a pseudotensor quantity $\epsilon^{\mu\nu\lambda\rho}$, any wave equation involving one or other of $\psi_\pm$, not compensated by a factor or term of opposite chirality, will not be parity-covariant. Interactions based on such quantities need not conserve parity.

A wave equation for massless spin $\frac{1}{2}$ particles satisfying Poincaré-covariance was proposed by Weyl [573] in 1929 in terms of spinors with two complex components. Those spinors and their wave equation, now known as *Weyl spinors* and the *Weyl equation*, did not satisfy parity covariance and were rejected, in particular by Pauli [424, p. 226], as unsatisfactory. This was because first, they implied a zero mass, and second, there existed an essentially universal belief in the left-right symmetry of all natural phenomena. Only with the discovery of parity non-conserving weak interaction effects (see Section 3.3) in 1956 were the equations re-discovered by Lee and Yang [318], Landau [304] and Salam [491] and applied to a 2-component theory of neutrinos. It was then shown that maximum parity violation would occur if only one helicity state of the neutrino existed. The same would be true if both existed but only one couples to the charged leptons in the weak interactions

If we select the Weyl or chiral representation of the matrices of the Dirac algebra, the chirality and chiral projection matrices are,

$$\gamma^5 = \begin{pmatrix} \mathbb{1} & 0 \\ 0 & -\mathbb{1} \end{pmatrix}, \quad \tfrac{1}{2}(1-\gamma^5) = \begin{pmatrix} 0 & 0 \\ 0 & \mathbb{1} \end{pmatrix} \quad \text{and} \quad \tfrac{1}{2}(1+\gamma^5) = \begin{pmatrix} \mathbb{1} & 0 \\ 0 & 0 \end{pmatrix}, \quad (20.38)$$

showing that, if we divide a non-chiral Dirac spinor into a 2-component part ψ_R above another ψ_L, according to,

$$\psi = \begin{pmatrix} \psi_R \\ \psi_L \end{pmatrix}, \quad (20.39)$$

then the chiral parts are,

$$\psi_+ = \begin{pmatrix} \psi_R \\ 0 \end{pmatrix} \quad \text{and} \quad \psi_- = \begin{pmatrix} 0 \\ \psi_L \end{pmatrix}. \quad (20.40)$$

Exercise 20.15 *Show that, in the Weyl representation, the free massless Dirac equation divides into two 2-component equations,*

$$\dot\psi_R + \boldsymbol{\sigma}\cdot\boldsymbol{\nabla}\psi_R = 0 \quad \text{and} \quad \dot\psi_L - \boldsymbol{\sigma}\cdot\boldsymbol{\nabla}\psi_L = 0, \quad (20.41)$$

and hence show that the left and right-handed (negative and positive chirality) parts of the massless Dirac spinor correspond to negative and positive eigenvalues, respectively, of helicity $\mathbf{S} \cdot \hat{\mathbf{p}}$, each of magnitude $\frac{1}{2}$.

In contrast to the massive case, zero mass fermionic chirality eigenstates correspond to eigenstates of helicity, which in the null case is a Poincaré invariant quantity.

All covariant combinations of two spinorial factors ψ and ξ may be expressed in terms of five independent quantities. These are the scalar $\bar{\xi}\psi$, the vector $\bar{\xi}\gamma^\mu\psi$, the (antisymmetric) tensor $\bar{\xi}\gamma^{\mu\nu}\psi$, the axial tensor $\bar{\xi}\gamma^\mu\gamma^5\psi$ and the pseudoscalar $\bar{\xi}\gamma^5\psi$, their Lorentz transformation properties following by extensions of Exercises 20.5 and 20.11(a). These Dirac covariants were first discussed by Pauli [424, p. 221] in 1933 and are referred to by the letters S, V, T, A and P, respectively. They have a total of 1+4+6+4+1 = 16 independent components matching the number of elements of the 4×4 matrices γ^μ.

20.8 Localization of a global symmetry

The Maxwell and Lorentz equations of electrodynamics were developed in Chapters 18 and 19 from the Principle of relativity and the use of a source which was invariant and conserved. The ingredients of the source and the way in which the details of their dynamics determine its conservation were not then of concern. The source was taken to be prescribed or external. In Equation 19.4, we showed that a conserved source may formally be constructed from the dynamical variables $z(\lambda)$ of non-quantum charged particles using the principles of relativistic classical mechanics. A closed electrodynamic action was then set up in which a charged classical particle interacted with a mediating classical electromagnetic field.

The microscopic phenomena of electrodynamics have a quantum character. The interaction takes place via the exchange of photons and the source may be any field whose quanta are electrically charged particles. Fundamental examples are the electron e^- and its antiparticle the positron e^+ and the various quarks and antiquarks. All the observed electrically charged particles which are not known to be divisible into more fundamental constituents are non-chiral spin $\frac{1}{2}$ particles with non-zero rest mass. Consequently, we can expect such particles to be described as the quanta of a massive Dirac field subject neither to a Majorana self-conjugacy nor to a γ^5 condition (see Section 20.7) appropriate to masslessness.

To form an interacting action with Dirac source particles, we could at this point follow the technique used throughout this text of forming a Lagrangian comprised of the sum of the free Lagrangian densities of the two systems whose interaction is under consideration. We could then seek the form of an interaction term based on simplicity and subsequently compare its predictions with observation. Such a technique, although it lacks uniqueness is surprisingly powerful in selecting theories which are relevant to physical phenomena. However, any additional feature which enhances the uniqueness of the procedure, consistent with the principles of relativity, stationary action and quantizability, and leads to the interactions observed in nature, may very well contain the essence of a physical principle of widespread validity. The feature to be introduced here is the formation of an interaction by *localization* of a global symmetry of a non-interacting system.

The Lagrangian density $\mathcal{L}_D$ of the free Dirac field, given in Equation 20.24, is invariant under the phase transformations,

$$\psi(x) \rightarrow e^{iq\alpha}\psi(x) \ , \tag{20.42}$$

in which the arbitrary spacetime-independent parameter in the exponent has been split, by convention, into a constant factor q and the arbitrary parameter α. This will ensure that $|q|$ appears as a coupling constant in the interacting action and will allow the resulting field equations to appear in standard form including q explicitly. The parameter α (and therefore also q) is assumed to be dimensionless (in natural units) and of compact range $0 \leq \alpha \leq 2\pi/|q|$. This phase symmetry constitutes a *global invariance* of the free Dirac action with respect to the 1-parameter compact Lie group, U(1), sometimes referred to as *gauge invariance of the first kind*. Many of the properties one requires of an interacting gauge-invariant action, such as quantum renormalizability, are dependent on the gauge group being compact. Since U(1) is the only 1-parameter compact Lie group, the above phase invariance is the simplest example of a global gauge group to which one may apply the localization process or Gauge principle.

We have seen that the Schrödinger wave function, the complex free Klein-Gordon field and the Dirac field all exhibit an internal, global U(1) symmetry for which the conserved Noether currents are the Schrödinger, Klein-Gordon and Dirac currents. A crucial factor in the demonstration of the internal symmetry of those equations is the global nature of the internal transformation, namely the fact that the parameter α does not depend on the space and time coordinates x. We may now explore the consequences of extending the symmetry by allowing the parameter α in the U(1) symmetry of the free Dirac field to depend on x and simultaneously demanding that the system lead to equations which are again form-invariant under the extended transformation.

Motivations for this examination are, first, that it would constitute an enlargement of the scope of the Relativity principle, namely observer independence of external quantities, to quantities previously unaffected by that principle because of their internal nature. Enlargement of a symmetry principle in a general way has very frequently turned out to have fruitful results, the transition from special to general relativity being an example. Second, we have seen that the Newtonian gravitational field, the electromagnetic field and the relativistic particle all contain local gauge invariances, namely symmetries of their dynamical equations in which appear arbitrary functions of t (in the Newtonian case), $x = \{t, \mathbf{x}\}$ in the electromagnetic case, and the evolution parameter λ for the particle. These features also appear in the Fierz-Pauli (spin 2) and general relativistic treatments of gravity and in the Yang-Mills theories of the nuclear interactions.

Consequently, it seems natural to investigate the possibility of symmetries in which the parameters are *localized* to functions of x. The process of replacing each of the p parameters ω^a ($a = 1, 2, \ldots, p$) of a global Lie symmetry group, of a Lagrangian system of local fields, by spacetime dependent parameters $\omega^a(x)$ is referred to as *gauging* or *localization* of the original group or Lagrangian field system. The original Lagrangian, since it must involve gradients of the fields in order to be dynamic, will not in general be invariant with respect to the new transformation. For the system to have the larger local symmetry it must be extended by the introduction of new fields with corresponding new terms in the Lagrangian density. Let us formalize this procedure:

- **the Gauge principle** specifies a procedure for obtaining an interacting action from a free-field action which is symmetric with respect to a continuous internal global symmetry — the results of localizing or gauging the global symmetry group must be accompanied by the inclusion of additional fields, with appropriate kinetic and interaction terms in the action, in such a way that the extended Lagrangian is covariant with respect to a new extended group of local transformations.

We shall illustrate this principle by applying it to the spinor and scalar electrodynamics which, respectively, describe spin $\frac{1}{2}$ and spin 0 fields interacting electrodynamically.

20.9 Spinor electrodynamics

We gauge the phase invariance of the free Dirac Lagrangian $\mathcal{L}_D$ by replacing the parameter α by a scalar function $\alpha(x)$ of the spacetime coordinates. For a continuous group whose parameter space is connected, as is U(1), and indeed any Lie group, it suffices to consider infinitesimal transformations. If we use Δ to signify the change in a quantity as a result of the transformation, then for small values of the parameter ($\Delta\alpha \ll 1$), the infinitesimal local transformation is of the form,

$$\Delta\psi(x) = iq\,\Delta\alpha(x)\psi(x) \quad \text{and} \quad \Delta\bar{\psi}(x) = -iq\,\Delta\alpha(x)\bar{\psi}(x)\,. \qquad (20.43)$$

Exercise 20.16 *Show that the free Dirac Lagrangian is no longer invariant under the localized transformations but varies according to,*

$$\Delta \mathcal{L}_\mathrm{D} = -q\,\bar{\psi}\gamma^\mu \psi\, \partial_\mu(\Delta\alpha) = -q\, j^\mu \partial_\mu(\Delta\alpha) = -J^\mu \partial_\mu(\Delta\alpha)\ . \tag{20.44}$$

The Dirac current J^μ, being free of derivatives, remains invariant, $\Delta J^\mu = 0$, with respect to the local transformations. Gauging the global group breaks the original global symmetry of the free Lagrangian or action and the resulting non-zero variation of the Lagrangian contains a term $-q\, j_a^\mu \partial_\mu(\Delta\omega^a)$ proportional to the coupling parameter q and to the gradient of each local parameter $\Delta\omega^a(x)$ of the original group times the corresponding Noether current, j_a^μ. In the present 1-parameter example, no index a appears.

To restore the variation to zero, and to thus achieve invariance with local variations of the fields, we must cancel this term by introducing another field into the Lagrangian. The knietic term corresponding to the new field will not involve ψ and cannot cancel the unwanted term involving j^μ. This cancellation can only occur with an interaction term which involves ψ in precisely the combination $q\bar{\psi}\gamma^\mu\psi$ and a product of $\bar{\psi}\gamma^\mu\psi$ with a new field will provide the coupling. Since all the terms in a Lagrangian are scalars, the new field must have a single Lorentz index and we denote it by $A_\mu(x)$ although we must accept that it may be either a 4-vector field or the gradient $A_\mu = \partial_\mu \phi(x)$ of a scalar. The latter possibility leads to an action which on quantization is nonrenormalizable and we therefore reject it.

In order to preserve the parity covariance of the original action, we take the new field to be proper under parity, improper under time-reversal and real. We may absorb any part of the proportionality constant into the new field and we therefore conventionally take the interaction term to be $\mathcal{L}_\mathrm{I} = A_\mu J^\mu$. The local variation of this term in the extended Lagrangian is $\Delta \mathcal{L}_\mathrm{I} = J^\mu \Delta A_\mu$ since $\Delta J^\mu = 0$. The necessary cancellation of these terms occurs only if the new field transforms locally according to $\Delta A_\mu = \partial_\mu(\Delta\alpha)$ which we may recognize to be of the same form (but infinitesimal) as the gauge transformation, $\Delta A_\mu = \partial_\mu \alpha$, of the electromagnetic 4-vector. However, this has not yet been established as the necessary physical interpretation of the new field since it may also be the massive 4-vector or Proca field.

The new 4-vector field must have a free term in the interacting Lagrangian which we take in second-order form (in natural units) as,

$$\mathcal{L}_\mathrm{A} = -\tfrac{1}{4} F_{\mu\nu} F^{\mu\nu} - \tfrac{1}{2} m^2 A^2\ , \tag{20.45}$$

(where $F_{\mu\nu} \equiv \partial_\mu A_\nu - \partial_\nu A_\mu$) appropriate to A^μ being either the Maxwell

20.9. Spinor electrodynamics

or the Proca field depending on whether the mass m is zero or not. Allowing for the cancellation already discussed, the local variation of the total Lagrangian $\mathcal{L} = \mathcal{L}_{D+A+I}$ is $\Delta \mathcal{L} = -\frac{1}{2}F_{\mu\nu}\Delta F^{\mu\nu} - m^2 A^\mu \Delta A_\mu$. The local variation of A_μ shows that $\Delta F^{\mu\nu} = 0$ and, since $\Delta A_\mu \neq 0$, the mass m must be zero. The new field is thus indistinguishable from the electromagnetic potential of the Maxwell field. Gauge invariance can be shown quite generally to be intimately connected with massless Poincaré-covariant fields of spin ≥ 1.

We have made use of the Relativity principle, parity covariance, Hamilton's principle and a few of the requirements of quantum physics, in particular energy positivity and renormalizability, in applying the Gauge principle to the localization of the Dirac Lagrangian. The resulting equations require the different parts of the Dirac field to be coupled electromagnetically. Those equations, when quantized, form the basis of *quantum electrodynamics* (QED). They are obtainable from the complete interacting non-quantum 'QED *Lagrangian*' which is,

$$\mathcal{L} = \tfrac{1}{2}i\bar\psi \overleftrightarrow{\partial\!\!\!/} \psi - m\bar\psi\psi - \tfrac{1}{4}F_{\mu\nu}F^{\mu\nu} + A_\mu J^\mu , \qquad (20.46)$$

or more explicitly,

$$\mathcal{L} = \tfrac{1}{2}i\bar\psi \overleftrightarrow{\partial\!\!\!/} \psi - m\bar\psi\psi - \tfrac{1}{4}(\partial_\mu A_\nu - \partial_\nu A_\mu)(\partial^\mu A^\nu - \partial^\nu A^\mu) + q A_\mu \bar\psi \gamma^\mu \psi . \quad (20.47)$$

Exercise 20.17 *Show (a) that the Euler-Lagrange equations of the above Lagrangian, the field equations of spinor electrodynamics, are*

$$\Box A^\mu - \partial^\mu \partial \cdot A = -q\bar\psi \gamma^\mu \psi \quad (= -J^\mu) \quad \text{and} \quad (i\partial\!\!\!/ - m)\psi = -q A\!\!\!/ \psi , \quad (20.48)$$

and (b) verify that they are are invariant with respect to the finite parameter local gauge invariance, namely,

$$\psi(x) \to e^{iq\alpha(x)}\psi(x) \quad \text{and} \quad A_\mu(x) \to A_\mu(x) + \partial_\mu \alpha(x) . \qquad (20.49)$$

(c) Show that the additive, gauge-invariant energy-momentum tensor,

$$T^{\mu\nu} = F^{\mu\lambda}F^\nu{}_\lambda - \tfrac{1}{4}\eta^{\mu\nu}F_{\lambda\pi}F^{\lambda\pi} + \tfrac{1}{2i}\bar\psi \gamma^{(\mu} \overleftrightarrow{\partial^{\nu)}} \psi \qquad (20.50)$$

of the ψ fields interacting via the A^μ field is energy and angular momentum conserving as a result of the coupled dynamical equations.

The additive character of the energy-momentum tensor of closed non-gravitational systems interacting via dynamic fields should be contrasted with the description of interactions described mechanically, where a potential term appears in the Lagrangian or Hamiltonian for each pair of particles.

A local gauge invariance is sometimes also referred to as a *gauge invariance of the second kind*. Since $\bar{\psi}\gamma^\mu\psi$ is interpreted in single-particle relativistic mechanics as the probability current of a Dirac particle, the constant q arbitrarily introduced into the phase of the original global symmetry may be interpreted as the electric charge of the single Dirac particle. This interpretation carries over to second quantization. We must distinguish between the charge q whose magnitude $|q| = e$ functions as a coupling constant for the field $\psi(x)$ and the charge,

$$Q = \tfrac{1}{c}\int_\Sigma d\Sigma_\mu J^\mu(x) = \tfrac{q}{c}\int_\Sigma d\Sigma_\mu \bar{\psi}\gamma^\mu\psi \;, \tag{20.51}$$

of the field system as a whole. The equations of spinor electrodynamics decouple into those of the free electromagnetic field and a free Dirac field if $q = 0$.

Since dimensional consistency in practical units requires a factor of $\hbar$ to appear with α, either in $e^{iq\alpha/\hbar}$ with $\Delta A_\mu = \partial_\mu \alpha$ or in $\Delta A_\mu = \hbar\partial_\mu\alpha$ with $e^{iq\alpha}$, we may conclude that the gauge process is an inherently quantum phenomenon even prior to second quantization. The electromagnetic field which arises from the gauge process, and which appears in the coupled Dirac equation, is the 4-vector potential A^μ, not the electromagnetic field tensor $F^{\mu\nu} = \{\mathbf{E}, \mathbf{B}\}$. In a quantum context, the vector potential appears to be a more fundamental representation of the electromagnetic field than the field strengths $\mathbf{E}$ and $\mathbf{B}$ which dominate classically. Aharonov and Bohm [6] gave a detailed discussion of the quantum significance of the electromagnetic potentials in 1959. An earlier discussion was given by Ehrenberg et al. [126] in 1949.

We may note again that one term (J^μ) of the Maxwell field equation $(\Box A - \partial\partial \cdot A = -J)$ does not involve the Maxwell field A itself explicitly. This term is therefore a source term. There is no term not involving the Dirac field itself in the interacting Dirac equation. The form of the equations shows therefore that it is the Maxwell potential which is the mediating or interaction field between the different parts of the Dirac field. In the quantized theory, the interaction proceeds by the exchange of photons, the quanta of the mediating field.

Equations 20.48 are each non-linear owing to the coupling. Since the value of q will be a simple multiple of the proton charge e, whose square in rationalized natural units is related to the small quantity $e^2/4\pi = 1/137$, the value of q is small and a perturbative solution is appropriate. A plane wave may be used as a linear zeroth approximation to the solution for each field and its use in the terms involving the small coupling constant will provide a first approximation to the interacting fields. This in turn can be used in the interaction terms to obtain a second-order approximation at which point the non-linearity will be evident. This non-linearity will increase at each iteration. Only in the free case or when the electrodynamic

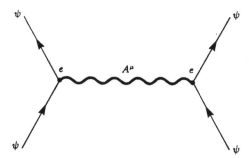

Figure 20.1 Feynman graph of quantum electrodynamics.

source is treated externally does one have approximate linearity permitting one to state that electrodynamics satisfies a *linear superposition principle*.

The analysis we have given above, although it illustrates some of the features of the equations of quantum electrodynamics, falls very much short of a quantum treatment of the electromagnetic field. The techniques for carrying out practical electrodynamic calculations by such an iterative procedure applied to the second-quantized equations obtained from the above essentially 'classical' Lagrangian can be represented by *Feynman graphs* (or diagrams) [7, 467]. We should note also that the above non-quantum equations cannot be considered truly classical because of the impossibility of a large number approximation with fermionic particles and because of the related anticommuting nature of their fields in the limit of $\hbar \to 0$.

The fundamental Feynman graph of quantum electrodynamics may be naïvely deduced from the form of the interaction term of spinor electrodynamics. It will be a 3-legged diagram with a line (representing a free particle) entering or emerging from a common origin (representing the first-order interaction) for each of the three factors, namely ψ (appearing twice) and A^μ. Two such diagrams joined together, as shown in Figure 20.1, pictorially represent one of the ways in which two Dirac fermions may interact via the exchange of virtual photons.

The Maxwell and Dirac fields A^μ and ψ are each irreducible representations of the full Poincaré group IO(1,3), the first massless and spin 1 and the second massive and spin $\frac{1}{2}$. The pair together, $\{A^\mu, \psi\}$, therefore make up a reducible representation of the Poincaré group. The local gauge group arising from the gauging of the global gauge group U(1) may be denoted U(1)^{loc} and the pair of fields may be considered, somewhat loosely, to be an irreducible representation of the group $\{\text{IO}(1,3), \text{U}(1)^{\text{loc}}\}$ made up from all Poincaré transformations and all local U_1 gauge transformations.

One of the goals of fundamental physics is to discover a symmetry group large enough to similarly contain and unify the fields whose quanta encompass all the fundamental source particles observed and those of the fields which mediate each of the four interactions between them. One would expect the set of all fundamental fields to be in some sense an irreducible representation of such a group. There is a possibility that such a unification programme will lead to a unique group whose properties make it a natural candidate for the classification of the elementary particles and the consistent quantization of all four interactions including the gravitational interaction described classically in the general theory of relativity.

20.10 CPT symmetries of Weyl and Dirac equations

Let C be a 4×4 matrix satisfying,

$$C\gamma_\mu C^{-1} = -\gamma_\mu^T \quad \text{and} \quad C^\dagger = -C = C^{-1} = -C^* = C^T . \tag{20.52}$$

Referred to as the *charge conjugation matrix*, C exists for any standard representation (Weyl, Pauli-Dirac or Majorana) of the Dirac matrices. For the Weyl representation, we may use,

$$C = i\gamma^2\gamma^0 = \begin{pmatrix} \epsilon & 0 \\ 0 & \epsilon \end{pmatrix} \tag{20.53}$$

where $\epsilon^T = -\epsilon$ and $\epsilon^{12} = 1$. The spinor $\psi^c = C\bar\psi^T$ is called the charge conjugate of ψ. Charge conjugation, $\mathbb{C} : \psi \to \psi^c$ and $A_\mu \to -A_\mu$, is an internal operation — it has no effect on the argument x of $\psi(x)$. It also has no effect on the Dirac mass m and the coupling constant q, the charge Q of the source being reversed as a consequence of the effect on ψ.

The time-reversal operation on a Dirac spinor may be defined in terms of charge conjugation, namely $\psi(x) \to \psi^t(x) = i\gamma^0\gamma^5\psi^c(-\tilde{x})$ where $\tilde{x} = \mathbf{P}x = \{t, -\mathbf{x}\}$. Its effect on the electromagnetic potential is given by $A_\mu(x) \to A_\mu^t(x) = \tilde{A}_\mu(-\tilde{x})$ while m and q remain unaltered. These discrete symmetries on classical (and anticlassical) fields have analogues [262, 410] in the actions of reflection operators on particle states in quantum mechanics and quantum field theory.

Putting $\psi_- = \psi_L$ and $\psi_+ = \psi_R$, the 2-component massless Weyl equations, 20.41, may be written in the form, $\partial_\pm \psi_\pm = 0$ where $\partial_\pm = \sigma_\pm^\mu \partial_\mu$ and $\sigma_\pm^\mu = \{\mathbb{1}, \pm\sigma^k\}$. Interaction via a gauge boson field B^μ may be similarly described by $i\partial_\pm \psi_\pm = -q_B \rlap{/}{B} \psi_\pm$ where q_B is a non-electric gauge charge of ψ.

Exercise 20.18 *Verify that parity* $\mathbf{P}$, *charge conjugation* $\mathbb{C}$ *and time-reversal* $\mathbb{T}$, *each separately leave Equation 20.47 invariant provided the components of* $\psi = \{\psi^\alpha\}$ *anticommute, namely* $\psi^\alpha\psi^\beta + \psi^\beta\psi^\alpha = 0$.

Exercise 20.19 (a) *Show that, in the chiral representation of* γ^μ, *the charge conjugation operation on 4-component Weyl spinors* $\psi_\pm$ *reduces in* the 2-component description to $\psi \to \psi^c = \pm\epsilon\psi^*$ where $\epsilon = \{\epsilon^{AB}\}$ (with $\epsilon^{12} = 1$) is the matrix $i\sigma_y$ corresponding to the 2D Levi-Cività pseudotensor of Equation 4.30. (b) Show that $\epsilon\sigma^*\epsilon^{-1} = -\sigma$ and hence show that the interacting Weyl equation is neither parity (**P**) nor charge conjugation (**C**) covariant, but is nevertheless invariant under **CP** and time reversal **T**.

The equations of spinor electrodynamics, and the similar equation for an interacting Weyl spinor, are clearly invariant under the combined operation of **CPT** in accordance with the CPT theorem.

The naturalness of γ^0 and $i\gamma^0\gamma^5$ as parity and time-reversal matrices can be seen by rewriting the first as $i\gamma^1\gamma^2\gamma^3\gamma^5$ and noting the appearance of the chirality matrix, associated with $\epsilon^{\mu\nu\lambda\rho}$, and the correspondence of the remaining gamma indices in each operation with the coordinate(s) being reflected.

20.11 Gauge-covariant derivative: minimal coupling

The gauging of the Dirac field has led to a coupling which involves only the charge explicitly and not the higher multipole moments of the electric charge distribution. For this reason the interaction arising from the gauging or localization is referred to as *minimal coupling* in contrast to the inclusion of Pauli coupling [424, p. 233] terms involving the magnetic moment directly in the Lagrangian.

Minimal coupling is adequate to account for all electromagnetic interactions. Dipole moments of particles are consequences of minimal coupling and the non-zero spin [262, 488]. The global U(1) transformation $\psi \to e^{iq\alpha}\psi$ implies that $\partial_\mu\psi \to e^{iq\alpha}\partial_\mu\psi$ since $\partial_\mu\alpha = 0$. The partial derivative $\partial_\mu\psi$ covaries with ψ with respect to the global symmetry transformation. After interactions are included by the gauge process, $\partial_\mu\psi$ no longer covaries with ψ, as a result of $\partial_\mu\alpha \neq 0$.

Exercise 20.20 *Show (a) that we can rewrite the interacting Dirac equation as,*

$$(i\gamma^\mu D_\mu - m)\psi(x) = 0 \quad \text{where} \quad D_\mu = \partial_\mu - iqA_\mu, \tag{20.54}$$

and the Maxwell field strength as,

$$F_{\mu\nu} = D_\mu A_\nu - D_\nu A_\mu. \tag{20.55}$$

(b) Show that $D_\mu\psi \to e^{iq\alpha(x)}D_\mu\psi$ under the finite transformations of Equation 20.49. (c) Show that the commutator of D_μ with D_ν is given by $[D_\mu, D_\nu] = \frac{1}{i}qF_{\mu\nu}$.

D_μ is said to be a *covariant derivative* or, more explicitly, a *gauge-covariant derivative*, with respect to the localized U(1) phase invariance. The relationship between the commutator of the gauge-covariant derivatives, acting on a spinor field $\psi(x)$, and the components of the field strength $F^{\mu\nu}$ should be compared with the Ricci identity [384] of general relativistic gravitation, quoted in Equation 21.44.

The interacting Lagrangian and the field equations of spinor electrodynamics may be each obtained from the free field equivalents by the substitution of D_μ in place of ∂_μ. In terms of the wave-mechanical 4-momentum, this substitution becomes $p_\mu \rightarrow \pi_\mu = p_\mu - qA_\mu$ which is equivalent to the well-known *minimal coupling prescription* $\mathbf{p} \rightarrow \boldsymbol{\pi} = \mathbf{p} - q\mathbf{A}$ and $H \rightarrow H_{\text{intn}} = H - q\phi$ for the 3-momentum and Hamiltonian operators $\mathbf{p}$ and $H = H(\mathbf{z},\mathbf{p})$.

20.12 Scalar electrodynamics

To give some idea of how the Gauge principle may be applied to charged particles not based on a Dirac field, we shall carry out a similar analysis with respect to the U(1) phase invariance $\phi(x) \rightarrow e^{iq\alpha}\phi(x)$ of a complex scalar field $\phi(x)$ which may also be self-interacting. Localization of the phase from α to $\alpha(x)$ leads to small changes,

$$\Delta\phi(x) = iq\,\Delta\alpha(x)\phi(x) \quad \text{and} \quad \Delta\phi^*(x) = -iq\,\Delta\alpha(x)\phi^*(x) \,. \quad (20.56)$$

in the fields.

Exercise 20.21 *Show that the complex free Klein-Gordon Lagrangian varies according to,*

$$\Delta\mathcal{L}_{\text{KG}} = -\tfrac{1}{i}q\{\phi^*\partial^\mu\phi - (\partial^\mu\phi^*)\phi\}\partial_\mu(\Delta\alpha) = -q\,j\cdot\partial(\Delta\alpha) \,. \quad (20.57)$$

Gauging the global group again breaks the original global symmetry of the free Lagrangian or action and the resulting non-zero variation of the Lagrangian contains a term proportional to the gradient of the local parameter $\alpha(x)$ of the symmetry group times the corresponding Noether current j^μ, the magnitude $|q|$ of the proportionality constant q being interpreted as a coupling constant. Due to the presence in it of derivatives, the current is not, however, invariant under the local transformation as in the Dirac case.

We must, nevertheless, at least attempt to cancel this term by introducing another field A^μ into the Lagrangian. We initially choose an interaction term of the form $\mathcal{L}_\text{I} = qA_\mu j^\mu = \tfrac{1}{i}qA_\mu(\phi^*\partial^\mu\phi - (\partial^\mu\phi^*)\phi)$.

20.12. Scalar electrodynamics

Exercise 20.22 Show that, apart from the free term for the A^μ field,

$$\Delta\mathcal{L}_{\text{KG+I}} = -q\, j^\mu \partial_\mu(\Delta\alpha) + q\, j^\mu \Delta A_\mu + 2q^2 \partial_\mu(\Delta\alpha) A^\mu \phi^* \phi, \qquad (20.58)$$

gives the local variation of the extended Lagrangian.

Even if the first two terms are cancelled by relating ΔA_μ to $\partial_\mu(\Delta\alpha)$, we now also have a term quadratic in q which can only be cancelled by an additional interaction term which must be $-q^2 A^2 \phi^* \phi$ for which the variation is $-2q^2(\Delta A_\mu) A^\mu \phi^* \phi$. This 'four-legged' contribution is referred to as a *sea-gull* term. The full interacting Lagrangian is therefore,

$$\mathcal{L} = -\partial_\mu \phi^* \partial^\mu \phi - m^2 \phi^* \phi - \tfrac{1}{4} F_{\mu\nu} F^{\mu\nu} + q A_\mu \tfrac{1}{i} q A \cdot \phi^* \overleftrightarrow{\partial} \phi - q^2 A^2 \phi^* \phi. \qquad (20.59)$$

The cancellation of all terms in $\Delta\mathcal{L}$ occurs only if the new field A^μ is a massless vector potential transforming locally according to $\Delta A_\mu = \partial_\mu(\Delta\alpha)$ and indistinguishable from the electromagnetic 4-vector.

We can see that the resulting equations require the different parts of the Klein-Gordon field to be coupled electromagnetically. Those equations, when quantized by the techniques of quantum field theory, form the basis of *scalar electrodynamics*, appropriate to an approximate description of charged pions, $\pi^\pm$, for example.

Exercise 20.23 (a) Show that the Euler-Lagrange field equations of the above Lagrangian are,

$$\Box A^\mu - \partial^\mu \partial \cdot A = -J^\mu, \quad (D^\mu D_\mu - m^2)\phi = 0 \text{ and } (D^*_\mu D^{*\mu} - m^2)\phi^* = 0, \qquad (20.60)$$

where $D_\mu = \partial_\mu - iqA_\mu$ is the gauge-covariant derivative and the total current is given by,

$$J^\mu = q\, j^\mu - 2q^2 A^\mu \phi^* \phi = q\tfrac{1}{i}(\phi^* \partial\phi - (\partial\phi^*)\phi) - 2q^2 A^\mu \phi^* \phi. \qquad (20.61)$$

(b) Verify that these equations are invariant with respect to the combined finite local gauge transformation,

$$\phi(x) \to e^{iq\alpha(x)}\phi(x) \quad \text{and} \quad A_\mu \to A_\mu + \partial_\mu \alpha(x). \qquad (20.62)$$

(c) Show that the additive, gauge-invariant energy-momentum tensor,

$$T^{\mu\nu} = F^{\mu\lambda} F^\nu{}_\lambda - \tfrac{1}{4}\eta^{\mu\nu} F_{\lambda\pi} F^{\lambda\pi} + 2\partial^\mu \phi^* \partial^\nu \phi - \eta^{\mu\nu}(\partial_\lambda \phi^* \partial^\lambda \phi + m^2 \phi^* \phi) \qquad (20.63)$$

of the ϕ fields interacting via the A^μ field is energy and angular momentum conserving as a result of the coupled dynamical equations.

The total current of the interacting system is no longer simply the coefficient of A_μ in the Lagrangian but the derivative,

$$J^\mu = \frac{\partial \mathcal{L}}{\partial A_\mu} = qj^\mu - 2q^2 A^\mu \phi^* \phi \,, \tag{20.64}$$

of the Lagrangian density with respect to A_μ, a result which also applies to the spinor electrodynamic result and remains valid in more general circumstances.

The discrete symmetry operations of $\mathbb{C}$, $\mathbf{P}$ and $\mathbb{T}$ are much simpler for scalar fields than those for spinors. If $\phi = \phi(x)$ is a proper or pseudoscalar field, then $\mathbf{P}\phi = \pm\phi$ and $\mathbb{T}\phi = \pm\phi$, and the interaction will be separately parity and time-reversal covariant, namely $\mathbf{P}\mathcal{L} = \mathcal{L}$ and $\mathbb{T}\mathcal{L} = \mathcal{L}$, provided A^μ is a 4-vector field of electromagnetic type, proper ($\mathbf{P}A^\mu = \tilde{A}^\mu$) under parity and improper ($\mathbb{T}A^\mu = \tilde{A}^\mu$) under time-reversal. The interaction will also be invariant, $\mathbb{C}\mathcal{L} = \mathcal{L}$, with respect to charge conjugation, $\mathbb{C} : \phi \leftrightarrow \phi^*$ and $\mathbb{C} : A^\mu \to -A^\mu$.

20.13 Yang-Mills fields — nuclear interactions

In 1914 Chadwick [62] observed that the electrons emitted in β-decay had a continuous energy spectrum. A discrete value was expected by energy conservation if only one particle emerged in addition to the daughter nucleus.

During the next two decades, many experiments confirmed the continuous spectrum. To avoid energy nonconservation, the existence of a new nuclear constituent, a neutral particle called the *neutron*, was hypothesized by Pauli in a letter [52, 9] in 1930. It had some of the properties of the neutron discovered in 1932 by Chadwick [63]. To make a clear distinction, Fermi suggested it be called the *neutrino* (later renamed the antineutrino). No longer considered a constituent of nuclei, its properties were outlined in more detail by Pauli [425] at the Seventh Solvay Conference in 1933. The neutrino (the 'little neutral one') was assumed to be an electrically neutral spin $\frac{1}{2}$ particle of zero or very small mass (compared to the electron mass) that must be emitted along with the electron.

Pauli's hypothesis was used by Fermi [154, 155, 156] in 1933 in the construction of the *four fermion* contact (non-exchange) theory of radioactive β-decay by the weak interaction. The latter is effectively $n \to p + e + \bar{\nu}_e$ which is equivalent to $n + e^+ \to p + \bar{\nu}_e$. It was based on a parity-covariant interaction term in the Lagrangian of the form,

$$\mathcal{L} = g\{[\bar{\psi}_p \gamma_\mu \psi_n][\bar{\psi}_e \gamma^\mu \psi_\nu] + \text{ herm. conj.}\} \,, \tag{20.65}$$

having a product of a Dirac current for the nucleons $(p + n)$ coupled with constant g to another for the leptons $(e^+ + \bar{\nu}_e)$. By comparing the results

20.13. Yang-Mills fields — nuclear interactions

predicted by his theory for the decay half lives with the experimental data, Fermi estimated the value of the coupling constant to be $g = 4 \times 10^{-63}$ J.m^3, revised later [593] to $g_F = 1.415 \times 10^{-62}$ J.m^3.

Although the weak interaction is not, like gravity, universally applicable to all particles, all weak interaction processes have been found to have the same universal strength. Wick [577] showed in 1934 that Fermi's theory could be applied to an inverse process, $n + \nu_e \rightarrow p + e$, involving an antineutrino (now called a neutrino) described by Dirac's hole theory of filled negative energy states.

In order to account for the experiments carried out in 1957, which demonstrated that parity was not conserved in weak interactions (see Section 3.3), pseudoscalar terms, involving one factor of γ^5, had to be added to the scalar terms of the Fermi theory Lagrangian in which no such factor appeared. The 1957 experiments also showed that the parity violation was maximal. This meant that the scalar and pseudoscalar terms appearing summed together in the Lagrangian had to be of equal magnitude.

If neutrinos are indeed massless, the maximal violation of parity and charge conjugation symmetry was able to be understood in terms of a new symmetry principle called *chiral invariance* introduced in 1957 by Marshak and Sudarshan [361]. A Lagrangian involving fermionic fields has chiral invariance if it remains unchanged under a continuous transformation of the spinor fields according to $\psi \rightarrow e^{i\alpha_5 \gamma_5}\psi$ where α_5 is a real parameter. This led to a new specific vector-axialvector (V-A) form for the parity non-conserving Lagrangian,

$$\mathcal{L} = \frac{G_F}{\sqrt{2}} \{[\bar{\psi}_p \gamma_\mu (1-\gamma^5)\psi_n][\bar{\psi}_e \gamma^\mu (1-\gamma^5)\psi_\nu] + \text{herm. conj.}\}, \quad (20.66)$$

in which some of the spinor fields of the Fermi form are replaced by their chiral parts coupled to the chiral massless neutrino fields. The factor of $\sqrt{2}$ is introduced so that the universal coupling constant G_F has the same value as the coupling constant g_F of the original Fermi theory. The natural dimension of G_F is an area, the same as for the gravitational constant G. Similarly to the gravitational case (see Section 7.4), we may use an electron mass m_e (appropriate to β-decay) and the universal constants c and $\hbar$ to form a dimensionless ('fine structure') weak coupling constant,

$$\alpha_w = \frac{G_F m_e^2 c}{\hbar^3} = 3.36 \times 10^{-13}, \quad (20.67)$$

confirming its low strength compared to the electromagnetic interaction (see Table 1.1).

The continuous U(1) symmetry transformations of the Schrödinger equation, the complex Klein-Gordon equation and the Dirac equation make up a compact Abelian Lie group whose gauging leads to a field theory having a local gauge invariance with one real massless gauge field A^μ, interpreted as the electromagnetic vector potential.

In 1954, Yang and Mills [595] had shown that a similar procedure could be applied to a set of matter fields related by a compact non-Abelian global Lie symmetry group in a non-interacting Lagrangian. The result was a locally gauge invariant Lagrangian containing as many massless vector (spin 1) gauge fields A_a^μ ($a = 1, 2, \ldots, p$) as there are parameters in the global group. The use of such a theory would provide a Yukawa exchange type rather than contact interaction. The $W^\pm$ intermediate bosons were introduced into the Fermi theory by Feynman and Gell-Mann [159] in 1958. (Amaldi [9, p. 95] points out that Klein [285] suggested a gauge-invariant Lagrangian for a Yukawa type theory of β-decay in 1938 which led to charged and neutral self-interacting intermediate bosons.)

Examples of such Yang-Mills groups are the *special unitary* groups in N dimensions,

$$\text{SU(N)} \equiv \{U = (U^{ab}) \mid U^\dagger U = \mathbb{1},\ \det U = 1,\ U^{ab} \in \mathbb{C}\}, \qquad (20.68)$$

with $N^2 - 1$ real parameters ($N \geq 2$) and the *special orthogonal* groups SO(N),

$$\text{SO(N)} \equiv \{O = (O^{kl}) \mid O^\text{T} O = \mathbb{1},\ \det O = 1,\ O^{kl} \in \mathbb{R}\}, \qquad (20.69)$$

with $N(N-1)/2$ parameters ($N \geq 2$). The unitary groups are natural symmetries relating complex and therefore charged fields and particles.

Some of the above groups have a number of different applications in physics that should not be confused. For example, the generating matrices of SU(2) have the same Lie algebra as those of the rotation group SO(3). There is a 2 to 1 correspondence, a homomorphism, relating a pair $\pm U(R)$ of SU(2) matrices to each 3×3 rotation matrix $R \in $ SO(3). This relation may be expressed in the form of a factor group isomorphism, SO(3) $\approx$ SU(2)/$\mathbb{Z}_2$.

Single-valued representations of SU(2) are, with respect to SO(3), either single-valued, namely Euclidean scalars, vectors and tensors, or double-valued, namely spinors, vector-spinors or tensor-spinors. The members of these representations are transformed by $(2j+1) \times (2j+1)$ matrices for $j = 0, \frac{1}{2}, 1, \frac{3}{2}, 2, \ldots$ and represent spin j particles in non-relativistic quantum mechanics. A similar result relates the Lorentz group SO(1,3) to the special linear group SL(2,$\mathbb{C}$) in the relativistic analysis of spin. The SL(2,$\mathbb{C}$) group consists of all 2×2 complex ($\mathbb{C}$) matrices (L for linear) of unit determinant (S).

A second use of SU(2) applies to the nonelectromagnetic strong-nuclear part of the interactions between hadrons such as nucleons and pions. This SU(2) symmetry leads to the strong-nuclear concept of *isotopic* or *isobaric spin*, also referred to simply as *isospin* [230]. The two nucleons $\{p, n\}$ comprise a 2D representation of SU(2) (the dimension of the representation and that of the matrices happening to coincide in this case) while the three pions ($\pi^\pm$, π^0) comprise a 3D representation. Ordinary rotations and this strong-nuclear isospin symmetry are applications of SU(2) not involving local gauge invariance.

Let the Lagrangian density of the globally symmetric collection of d Yang-Mills matter fields $\psi(x) = \{\psi_j(x)\}$ ($j = 1, 2, \ldots, d$) be denoted by $\mathcal{L}_m = \mathcal{L}_m(\psi, \partial_\mu \psi)$, then the locally gauge invariant Lagrangian density is given by,

$$\mathcal{L} = \mathcal{L}_m(\psi, D_\mu \psi) - \tfrac{1}{4} F^a_{\mu\nu} F^{\mu\nu}_a , \qquad (20.70)$$

where

$$F^{\mu\nu}_a = D^\mu A^\nu_a - D^\nu A^\mu_a . \qquad (20.71)$$

The nature of the interaction, which includes self-interaction of minimally-coupled fields A^μ_a, is thus determined by the form,

$$D_\mu = \partial_\mu - ig A^a_\mu T_a , \qquad (20.72)$$

of the gauge-covariant derivative, D_μ.

Each T_a is a $d \times d$ matrix related to the way in which the matter fields are transformed by the global gauge group, while $|g|$ is the coupling constant determining the strength of their mutual interaction via the set of p vector gauge fields A^μ_a. The T_a are the hermitian generating matrices of the particular representation of the gauge group comprising the set of matter fields, much as the spin matrices $\mathbf{S} = \{S^k\}$ generate the rotation group or $S^{\mu\nu}$ the Lorentz group in one or other of its representations, be it scalar, vector, tensor, spinor, vector-spinor or tensor-spinor. They must therefore satisfy $[T^a, T^b] = i f^{abc} T_c$ where f^{abc} are the structure constants of the Lie group.

Weak and electromagnetic interactions were long known to have many similarities such as both being vectorial in nature. All matter particles affected by the weak interaction (charged leptons and hadrons) were also affected by the electromagnetic interaction, each with its own universal coupling parameter, α_w or α_e. These are properties of an interaction governed by a gauge principle, except that for them to have a common origin with similar values for the dimensionless coupling constants, the weak bosons would have to have rest masses M_W about 100 times the proton mass m_p so that α_w evaluated with M_W was comparable to α_e.

The Glashow-Salam-Weinberg theory [7, 199, 468, 492, 493, 565] of the combined electromagnetic and weak-nuclear interaction, referred to as the *electroweak theory*, is based on the direct product group SU(2)×U(1) in which SU(2) describes a gauge symmetry quite distinct from ordinary rotations or strong-nuclear isospin. Because the group is the same, it is referred to as *weak isospin*. Since a 2D special unitary matrix has three independent real parameters, there will be a total of four real vector gauge fields, whose quanta correspond to the photon and the weakons or intermediate vector bosons, $W^{\pm}$ and Z^0. The latter were observed experimentally in a relatively small number of events with the proton-antiproton collider [21] at CERN in late 1982 with rest masses $M_W = 81$ and $M_Z = 92 \,\mathrm{MeV}/c^2$, namely about 87 and 100 m_p, thus confirming the most important predictions of the electroweak theory. Data fitted the hypothesis that the reaction $p + \bar{p} \rightarrow W^{\pm} + X$ occurred for various particles X and was followed by the decays $W^+ \rightarrow e^+ + \bar{\nu}_e$ and $W^- \rightarrow e^- + \nu_e$.

In August 1989, it was also announced that numerous Z^0 particles were observed within minutes of first operating the LEP (Large Electron-Positron) colliding ring accelerator at CERN. In October 1989, it was announced that data from the L3 and ALEPH detectors gave revised Z^0 masses (in MeV/c^2) of about 91.13 ± 0.06 and 91.17 ± 0.07. The number N_ν of neutrino species implied by the spread in rest energy in the two experiments were $N_\nu = 3.4 \pm 0.4$ and 3.3 ± 0.3, respectively. Results announced a day earlier [430] from the Stanford Linear Collider (SLC) at SLAC, also indicated three families (with $N_\nu = 2.7 \pm 0.7$). These results confirm predictions [327] based on certain grand unified theories. A fractional result could be interpreted as an indication of a new physical phenomenon such as contributions from a supersymmetric partner of the neutrino, the *neutralino*.

The Glashow-Salam-Weinberg theory of the electroweak interaction involves the coupling of (left-handed) neutrino states to the chiral parts of the Dirac fields of electrons (or muons and tauons) and quarks. Corresponding to the electromagnetic interaction term $\bar{\psi}A\!\!\!/\psi$, in the QED Lagrangian, are additional terms in which A_μ is replaced by the Yang-Mills fields A_μ^a ($a = 1, 2, 3$) of the $W^{\pm}$ and Z^0 particles with ψ replaced by either a neutrino (or antineutrino) field or a chiral part of a quark or massive lepton field.

The strong-nuclear interaction is based on the localization of an SU(3) global symmetry among collections of quark fields, the ultimate building blocks of nuclear matter. Since 3D unitary matrices have eight real parameters, localization leads to a gauge theory with eight real vector fields whose quanta are the eight gluons. The exchange of gluons between quarks gives rise to the strong-nuclear interaction at the high-energy (inter-quark) level. This SU(3) gauge symmetry with colour as the 'charge' should not be confused with the non-gauge SU(3) symmetry of the 'eight-fold-way' discovered much earlier and describing related groups of hadronic particles and excited states. The dimensionless strong coupling constant $g_s^2/4\pi$ has a value (see Table 1.1) near or greater than unity ruling out its use in a perturbation expansion.

The combination of the electroweak and strong interactions based on the product group SU(2)×U(1)×SU(3) is referred to as the *Standard model* [466] of elementary particle theory. A *grand unified theory* is one in which a simple group, such as SU(5) containing the Standard model factor group as a subgroup, provides a greater degree of unification of the electroweak and strong-nuclear interactions.

20.14 Quantum field theory

The orbital motion of a single particle acquires wave characteristics in non-relativistic quantum mechanics by the techniques of *first quantization*. The classical mechanical variables of position $\mathbf{x}$ (or $\mathbf{z}$), momentum $\mathbf{p}$, angular momentum $\mathbf{j} = \mathbf{l} + \mathbf{s} = (\mathbf{x} \times \mathbf{p}) + \mathbf{s}$ and the Hamiltonian h, satisfy either the fundamental Poisson brackets or those of the generators of the Galilean group (see Exercises 11.4 and 11.6). First quantization is carried out by imposing a commutator algebra on the corresponding quantum operators $\mathbf{X}, \mathbf{P}, \mathbf{J} = \mathbf{L} + \mathbf{S} = (\mathbf{X} \times \mathbf{P}) + \mathbf{S}$, and H.

In the position representation of the Hilbert space of quantum states, the momentum operator has the form $\mathbf{P} = \frac{\hbar}{i} \nabla$, acting on the wave functions $\psi(t, \mathbf{x}) = \langle \mathbf{x} | \psi \rangle$ formed from the state vectors $|\psi\rangle$. The resulting single particle theory is the *non-relativistic wave mechanics* of Schrödinger. This combines with the (non-wave) quantum mechanics of spin to adequately account for a great many of the properties of unbound and bound electrons in the interactions of atomic and molecular physics. For many of the dominant atomic features, a non-relativistic treatment is adequate.

One may also construct a first-quantized *relativistic quantum mechanics* by imposing a Poincaré group commutator algebra on the mechanical operators X^μ, P^μ and $J^{\mu\nu} = L^{\mu\nu} + S^{\mu\nu}$ (see Exercise 15.18). These are quantum relations corresponding to the Poisson bracket Lie algebra of the classical mechanical variables z^μ, p^μ and $j^{\mu\nu} = l^{\mu\nu} + s^{\mu\nu}$. These operators, with $P^\mu = \frac{\hbar}{i} \partial^\mu$ in the position representation, may operate on wave functions to supply relativistic equivalents of the Schrödinger or Pauli-Schrödinger equations, the form of the equation depending on the spin of the particles to be described. For spin 0 and $\frac{1}{2}$, the Klein-Gordon and Dirac equations would be appropriate. However, the relativistic wave mechanics of both these equations, and their analogues for higher spin ($s \geq 1$), contain negative probabilities and negative energies, as outlined in the discussion of the single-particle wave mechanics of the Klein-Gordon field in Chapter 16.

Furthermore, a relativistic analysis implies mass-energy equivalence which in turns implies the possibilty of *particle creation*. Even if the kinetic energies involved are insufficient to supply the rest mass energies of the real particles involved, quantum mechanics permits the temporary creation of particles of energy ΔE for periods of time $\Delta t \leq \hbar / \Delta E$ determined by the Heisenberg uncertainty principle. The energies of such *virtual particles* cannot be separately measured and in fact no measurements based on their existence can be used to provide evidence against energy conservation. Soon after creation, they annihilate again releasing the energy before it can be separately detected. Their existence can however effect measurable data, such as atomic energy levels. Since there is no limit to the number of virtual particles that can be created, relativistic quantum mechanics is an essentially *many-particle theory* and one cannot expect to formulate a relativistic theory of particles by the techniques of single particle wave mechanics.

The negative probabilities and energies predicted by relativistic quantum mechanics are avoided by using the *second quantization* techniques of *quantum field theory* which, by construction, is intrinsically of many-particle character. The reference to *second* quantization arises since it is the fields themselves which become quantum operators on a Hilbert state space and these same fields prior to quantization are solutions of wave equations which may already have some quantum properties, analogous to those of the Schrödinger wave functions. This is particularly true for matter fields satisfying the massive Klein-Gordon and Dirac equations in which the mass is accompanied by the Planck constant in practical units. Spinor fields, furthermore, cannot be given a classical interpretation.

Nevertheless, from the point of view of second quantization, the relativistic wave equations describe non-quantum fields whose components either commute or anticommute. Only in null integer spin cases, where the components commute and no factor of $\hbar$ appears in the wave equation (Maxwell's equations being a null spin 1 example) is there a meaningful non-quantum interpretation. A second example is provided by gravitational field equations for which the wave propagation has a null spin 2 character.

Quantum field theory may be based on the construction of the particle states of each non-interacting field using the appropriate non-quantum expansions of $\psi = \{\psi_j(x)\}$ in terms of a plane wave basis. These are comprised of $e^{\pm i k \cdot x}$ multiplied, in the case of non-zero spin, by polarization factors dependent only on the propagation 3-vector $\mathbf{k}$. Examples of such polarization factors are the spin 1 polarization 4-vectors $\epsilon_\mu^{(\lambda)}(\mathbf{k})$ in Equation 19.42, the spin $\frac{1}{2}$ spinors $u^{(r)}(\mathbf{k})$ and $v^{(r)}(\mathbf{k})$ of Equation 20.35 and the spin 2 tensors $\epsilon_{\mu\nu}^{(\lambda_1\lambda_2)} = \epsilon_\mu^{(\lambda_1)} \otimes \epsilon_\nu^{(\lambda_2)}$. Second quantization involves a corresponding expansion of a *quantum field operator*, also denoted by $\psi = \psi(x)$, in which the coefficients $a(\mathbf{k})$ and $b^\dagger(\mathbf{k})$ (or $a^\dagger(\mathbf{k})$ if the field is real) in the plane-wave expansion are therefore also operators which act on the Hilbert space of particle states. For non-zero spin, such operator coefficients will carry fermionic or bosonic polarization indices $(r) = (1, 2)$ or $(\lambda) = (0, 1, 2, 3)$, respectively.

The essence of second-quantization by canonical techniques is contained in the fundamental commutation or anticommutation rules which the quantum field operators must satisfy thus implying corresponding commutation or anticommutation rules for the coefficients a and b (or $a^\dagger$). The a and b become *annihilation* and *creation* operators which act on the ground state (vacuum) of the quantum fields to produce the quanta of the many-particle description.

For integer spin fields, energy positivity demands that the Poisson brackets of the classical fields be replaced by the commutators of quantum operators [262]. Such quantization leads to particles obeying *Bose-Einstein statistics* [410]. The $\hbar \to 0$ limit of the commutators is well-defined in terms of the original unquantized commuting fields and the Bose-Einstein statistics permit the construction of coherent states with large numbers of quanta with the same phase space values. The classical limit is well-defined.

Positivity of energy demands, however, that half-odd-integer spin fields be quantized by anticommutation rules [262] which give rise to particles satisfying *Fermi-Dirac statistics* [410]. The limit of the anticommutators, as $\hbar \to 0$, does not lead to the well-known commutation of classical variables but to non-quantum fields whose components anticommute with one another. Such non-quantum spinor fields are said to have *Grassman*[3] or *anticlassical* components.

The *spin-statistics theorem* [410] showing integer and half-odd-integer particles obey Bose-Einstein and Fermi-Dirac statistics, respectively, was established in a relativistic context by Fierz [161] and Pauli [428] in 1939 and 1940. An extension of the theorem to any situation where spin is well-defined was given by Sudarshan [524] in 1968.

Pure logical thinking cannot yield us any knowledge of the empirical world. All knowledge of reality starts from experience and ends in it.
Albert Einstein   1879–1955

Bibliography:

Aitchison I J R and Hey A J G 1982 *Gauge theories in particle physics* [7].
Bailin D and Love A 1986 *Introduction to gauge field theory.* [25].
de Wit B and Smith J 1986 *Field theory in particle physics* [101].
Itzykson C and Zuber J-B 1980 *Quantum field theory* [262].
Ramond P 1981 *Field theory: a modern primer* [467].
Ramond P 1983 *Gauge theories and their unification* [468].
Roman P 1969 *Introduction to quantum field theory* [482].
Ryder L H 1985 *Quantum field theory* [488].
Schweber S S 1961 *An introduction to relativistic quantum field theory* [504].
Wentzel G 1949 *Quantum theory of fields* [568].

---

[3] Hermann Günther Grassmann (1809–1877), German mathematician.

CHAPTER 21

Relativistic gravitation

The final goal of the theoretical natural sciences is to discover the ultimate invariable causes of natural phenomena[1].

Hermann L F von Helmholtz   1821–1894

21.1 Introduction

In this final chapter, the massless spin 2 tensor nature of relativistic gravity will be explored, mainly from a special relativistic point of view appropriate only to asymptotically flat, low density configurations of matter, namely those for which $\rho \ll c^2/G\ell^2$ where ℓ is a characteristic length. Some indications will be given of the links between the spin 2 theory and general relativity and a few topics of current interest, such as Rarita-Schwinger fields, supersymmetry and strings will be briefly mentioned.

21.2 Rarita-Schwinger spin $\frac{3}{2}$ vector-spinor fields

No fundamental spin $\frac{3}{2}$ or spin > 2 particles have been observed. However, there are a number of reasons why studies of the Rarita-Schwinger fields of spin $\frac{3}{2}$ are currently of some interest.

Certain properties of the spin $\frac{1}{2}$ fields that describe most of the source material in the universe are better appreciated by examining their similarities to and differences from other half-odd-integer spin fields of which those of spin $\frac{3}{2}$ are the next simplest. The notion of gauge invariance, which does not apply to spin $\frac{1}{2}$ Dirac fields, is enriched by the details of its appearance in the massless Rarita-Schwinger equations. Finally, the Rarita-Schwinger field has played a key rôle in some of the most promising attempts at unification of gravitation with the other three interactions. One of these unification frameworks is *supergravity* [97, 98, 99, 100, 174] in which the spin 2

---

[1] *Selected writings of H L F von Helmholtz* [270].

21.2. Rarita-Schwinger spin $\frac{3}{2}$ vector-spinor fields

graviton, the gauge boson of general relativity, is partnered by a spin $\frac{3}{2}$ gauge particle, the *gravitino*, in such a way that the whole theory has a boson-fermion symmetry referred to as *supersymmetry* [570]. Although supergravity alone does not successfully quantize and unify gravity with the other interactions, supersymmetry continues to be important in a more promising approach involving *relativistic strings* in dimensions higher than $1 + 3$. The extra dimensions are spatial and compact with radii comparable to the Planck length. Causality rules out the inclusion of extra time-like directions. We shall give a very brief indication here of the way the spin $\frac{1}{2}$ and 1 properties of Dirac and Maxwell-Proca fields can be combined to form the Rarita-Schwinger fields.

A natural way to construct a non-chiral entity with maximum spin $\frac{3}{2}$ is to form the direct product $V^\mu \psi = \{V^\mu \psi^\alpha\}$ of a 4-vector V^μ and a Dirac spinor $\psi = \{\psi^\alpha\}$ which Lorentz transforms according to,

$$V^\mu \psi \to \Lambda^{\bar{\mu}}{}_\mu S(\Lambda) V^\mu \psi . \qquad (21.1)$$

Any 4-vector $\psi = \{\psi_\mu\} = \{\psi_\mu{}^\alpha\}$ of four Dirac spinors (or any Dirac spinor, each of whose four components is a 4-vector), transforming according to,

$$\psi^\mu \to \Lambda^{\bar{\mu}}{}_\mu S(\Lambda) \psi^\mu , \qquad (21.2)$$

is referred to as a *vector-spinor*, in which the Dirac index α is generally suppressed. As for a 4-vector $V = \{V^\mu\}$, one may also suppress the Lorentz index on ψ_μ when indicating the whole object. (The use of ψ to denote a vector-spinor need not lead to confusion with its use for a spin $\frac{1}{2}$ Dirac spinor.) The Lorentz generators $S^{\mu\nu}$ of a vector-spinor may be constructed from the Dirac and vector generators.

Taking the *gamma trace* of ψ_μ, namely $\gamma \cdot \psi = \gamma^\mu \psi_\mu$, is an algebraic procedure for obtaining an ordinary spin $\frac{1}{2}$ Dirac spinor from the vector-spinor. For ψ^μ to have a well-defined spin of $\frac{3}{2}$, any spin $\frac{1}{2}$ part must be removed. We can remove the gamma trace part by imposing the algebraic condition, $\gamma \cdot \psi = 0$.

Exercise 21.1 *(a) Show that $\gamma \cdot \psi$ and $\gamma^\mu \gamma \cdot \psi$ transform as a spin $\frac{1}{2}$ Dirac spinor and a vector-spinor, respectively. (b) Show that, from an arbitrary vector-spinor ψ^μ, we may construct a new gamma-traceless vector-spinor using $\psi^\mu + \frac{1}{4}\gamma^\mu \gamma \cdot \psi$.*

We now seek a non-interacting field equation for $\psi^\mu(x)$, a consequence of which must be the Klein-Gordon equation establishing the field to have mass $m \neq 0$. In a field context, where ∂_μ is available, we can identify $\partial \cdot \psi$ as another spin $\frac{1}{2}$ part of ψ^μ. The field equations, and any constraints on the initial data, must remove the remaining spin $\frac{1}{2}$ parts, leaving $\psi = \{\psi_\mu\}$ with the correct number of dynamical degrees of freedom for spin $\frac{3}{2}$. We shall see that the field equation and the algebraic gamma trace constraint suffice for this purpose.

The same arguments used for the Dirac field apply here to show that a natural choice of field equation is the *massive Rarita-Schwinger equation* [469], namely the Dirac equation, $(i\slashed{\partial} - m)\psi^\mu(x) = 0$, on each of the four spinors ($\mu = 0, 1, 2, 3$) in ψ^μ, since it then follows from $\slashed{\partial}^2 = -\Box$ that $(\Box - m^2)\psi^\mu = 0$.

Exercise 21.2 *(a) Establish the identity $\gamma\slashed{\partial} = -\slashed{\partial}\gamma - 2\partial$ and show that a vector-spinor field $\psi = \{\psi^\mu\}$ satisfying the massive Rarita-Schwinger equation, and the gamma-traceless constraint, is also transverse, namely:*

$$(i\slashed{\partial} - m)\psi^\mu(x) = 0 \quad \text{and} \quad \gamma \cdot \psi = 0 \quad \Rightarrow \quad \partial \cdot \psi = 0 . \tag{21.3}$$

(b) Verify that ψ^μ is left with the correct number of real degrees of freedom, $2(\frac{3}{2})+1 = 4$, for massive spin $\frac{3}{2}$, doubled to 8 — for particle and antiparticle contributions — if it is complex.

The fundamental equation for a free massless spin $\frac{3}{2}$ potential field, analogous to the sourceless Equation 18.22 for the electromagnetic 4-vector potential, is the *massless Rarita-Schwinger equation* [469],

$$\slashed{\partial}\psi - \partial\gamma\cdot\psi = 0 , \tag{21.4}$$

which is also discussed in references [117, 118]. A second-order wave equation $\Box\psi - \partial\partial\cdot\psi = 0$ (which is not the field equation for ψ) may be deduced by operating with $\slashed{\partial}$. These equations have similarities to both the massless Dirac equation $\slashed{\partial}\psi = 0$ and Maxwell's equations (18.22) for the free 4-vector potential. Equation 21.4 for the *Rarita-Schwinger potential* is gauge-invariant under the spinorial transformation $\psi \to \psi + \partial\epsilon$ in which $\epsilon(x)$ is an arbitrary spin $\frac{1}{2}$ gauge function of Dirac type.

Exercise 21.3 *(a) Gamma trace Equation 21.4 to show that use of the gamma traceless gauge ($\gamma\cdot\psi = 0$), which leaves ψ satisfying $\slashed{\partial}\psi = 0$, is also transverse ($\partial\cdot\psi = 0$). (b) Show that the gamma-traceless gauge has residual gauge freedom with respect to a massless gauge function $\epsilon(x)$ satisfying $\slashed{\partial}\epsilon = 0$. (c) Show that the Rarita-Schwinger potential ψ has two independent physical degrees of freedom corresponding to a field with helicity components of magnitude $\pm\frac{3}{2}$.*

Equation 21.4 cannot be derived from a Lagrangian. This can be seen by noting that the Euler derivative of a Lagrangian field with a local gauge invariance must satisfy a differential identity or source condition independently of the field equation itself. In the case of the Maxwell potential, for example, where $\Box A - \partial\partial \cdot A = 0$, the left side is identically divergence-free whether A^μ satisfies the field equation or not. The same is not the case for the Rarita-Schwinger equation above.

To see why such a condition must hold for a gauge-invariant Lagrangian field, we note that for an internal variation $\delta\psi_{\mu\ldots}$ of a field $\psi_{\mu\ldots}$ of spin ≥ 1, the variation of the action is $\delta S = \int d^4x \mathcal{L}^{\mu\ldots}\delta\psi_{\mu\ldots}$ where $\mathcal{L}^{\mu\ldots}$ is the Euler derivative. If a gauge transformation $\Delta\psi_{\mu\ldots} = \partial_\mu \epsilon_{\ldots}$ with gauge function $\epsilon_{\ldots}(x)$ is an internal symmetry, we have $\int d^4x \mathcal{L}^{\mu\ldots}\Delta\psi_{\mu\ldots} = 0$ and thus $\int d^4x \mathcal{L}^{\mu\ldots}\partial_\mu \epsilon_{\ldots} = 0$. An identity of the form $\partial_\mu \mathcal{L}^{\mu\ldots} \equiv 0$ follows, the exact nature of which depends on which components of the arbitrary gauge function $\epsilon_{\ldots}(x)$ are independent.

Exercise 21.4 (a) *Verify the gauge freedom of ψ^μ in Equation 21.4 and deduce the second-order equation, $\Box \psi - \partial\partial\cdot\psi = 0$. (b) Show that, by adding a multiple of its own γ-trace to Equation 21.4, we may establish the following alternative form of the massless Rarita-Schwinger equation,*

$$\displaystyle{\not{\partial}}\psi_\mu - \partial_\mu \gamma\cdot\psi - \gamma_\mu \partial\cdot\psi - \gamma_\mu {\not{\partial}}\gamma\cdot\psi = 0 \;, \tag{21.5}$$

and verify its gauge invariance. (c) Show that the divergence of the left side of Equation 21.5 vanishes identically (namely for all ψ, stationary or not). (d) Show that Equation 21.5 implies Equation 21.4. (e) Establish the following identity involving the chirality matrix γ^5,

$$\epsilon^{\mu\nu\lambda\rho}\gamma_\rho\gamma_5 = i(\eta^{\mu\lambda}\gamma^\nu - \eta^{\mu\nu}\gamma^\lambda) - \gamma^\mu\gamma^{\nu\lambda} \;. \tag{21.6}$$

(f) Show that the Lagrangian density of the free massless Rarita-Schwinger field of the gravitino is given by,

$$\mathcal{L}_{\mathrm{RS}} = -\tfrac{1}{4}\epsilon^{\mu\nu\lambda\rho}\bar{\psi}_\mu \gamma_5 \gamma_\lambda \overset{\leftrightarrow}{\partial}_\nu \psi_\rho \;, \tag{21.7}$$

by applying the identity of Equation 21.6 to the Euler-Lagrange equation to reduce it to Equation 21.5.

Some of the classical and quantum properties of higher spin (>1) and arbitrary spin free fields and particles, both bosonic and fermionic, are reviewed and developed in Freedman and de Wit [173] and in references [76, 115, 116, 117, 118], which also cite much of the earlier work.

21.3 Fierz-Pauli massless spin 2 tensor gravity

It was demonstrated in Chapter 17 that the simplest way to construct a gravitation theory consistent with special relativity, using a scalar potential Φ, gives the correct result for gravitational time dilation but fails to predict any light deflection. In linear form it also fails to predict any shift in perihelion or periastron. If we use the fact that a graviscalar field itself contributes to the source of gravity to construct a non-linear equation, the value for the latter effect is incorrect by a factor of 6 and is retrograde rather than advancing.

Between 1905 and about 1911, Einstein explored (see [131, 132]) the possibility of constructing a theory of gravitation based on special relativity but soon abandoned that approach. Einstein's answer to this problem was to realize that the removal of a dynamical restriction of the Relativity principle he had presented [128] in 1905, namely that it dealt with privileged or special frames, those which were inertial, could provide a characterization of gravity. He had always felt that the laws of physics should be independent of the observer's frame, even for those whose relative velocity is not uniform and whether or not the relative acceleration is constant.

Einstein proposed his *general relativity theory* to incorporate these ideas and showed that it was in fact a relativistic theory of gravity. At that stage, the material resulting from the original Relativity principle became known as the *special theory of relativity* which remained valid only locally in the presence of gravitational effects. This meant that a special relativistic theory of gravity, such as that proposed by Nordstrøm [400] in 1913 using a scalar field, was essentially bypassed. We have already shown how that theory incorporates the Principle of equivalence. However, it does so only for uniformly accelerated frames. Furthermore, the dynamical equations of the field do not have gauge invariance and therefore do not automatically require or incorporate the conservation of the gravitational source.

It will be instructive to further explore the extent to which gravity and special relativity can be reconciled in order to better appreciate the essence of the contributions of general relativity. We know that the covariant source of gravity is the total energy-momentum tensor $T^{\mu\nu}$ which, for energy-momentum and angular momentum conservation, must be symmetric $T^{\mu\nu} = T^{\nu\mu}$ and divergence-free, $\partial_\mu T^{\mu\nu} = 0$. One way to understand the failure of the scalar theory concerning light deflection is based on the fact that it does not use the entire energy-momentum tensor as source but only the trace $T = T^\mu{}_\mu$. However, we know that the trace of the energy-momentum tensor of the electromagnetic field vanishes, as it must for any null wave (see Equation 15.94). An electromagnetic wave or ray cannot therefore couple to such a graviscalar field. This also suggests an obvious alternative way to attempt to construct a gravitational field consistent with special relativity, namely to use a symmetric and null tensor gravitational potential, $\bar{h}^{\mu\nu}$ ($= \bar{h}^{\nu\mu}$), linked by a wave equation to the entire energy-momentum tensor $T^{\mu\nu}$.

We follow Misner et al. [384] in using a barred variable, to reserve $h^{\mu\nu}$ for a related alternative potential. We choose a null wave in order to obtain a standard long-range potential ($\propto 1/r$) in the static limit. Goldhaber and Nieto [205] point out that data on clusters of galaxies show that they obey

21.3. Fierz-Pauli massless spin 2 tensor gravity

the $1/r$ law for the gravitational potential. This implies that any finite length scale λ of Yukawa type over long distances can be no smaller than about $1500\,\text{k}\ell\text{y} \approx 10^{19}\,\text{m}$, otherwise deviations due to the Yukawa factor $e^{-r/\lambda}$ would be detectable. In a quantized theory of gravity, this would imply that the graviton has a mass less than about $\hbar/c\lambda \approx 2\times 10^{-65}\,\text{kg} \approx 10^{-32}\,\text{eV}/c^2$.

We shall explore this spin 2 possibility by analogy with the linking of the electromagnetic potential A^μ, in the Lorentz gauge $\partial \cdot A = 0$, to the conserved current J^μ with the wave equation $\Box A^\mu = -\mu_0 J^\mu$. Let us therefore incorporate universality using the massless equations,

$$\Box \bar{h}^{\mu\nu} \propto G T^{\mu\nu} \quad \text{and} \quad \partial_\mu \bar{h}^{\mu\nu} = 0 , \tag{21.8}$$

where the second equation is a gauge condition that must apply at all x in order to ensure the equation describes spin 2. (We shall indicate in Section 21.4 how a spin 2 equation may be formed without the gauge condition.) First, we note that our re-expression of scalar gravity in Section 17.5 introduced a dimensionless tensor field $g_{\mu\nu}(x)$ (in rectangular Cartesian coordinates) obtained from $\eta_{\mu\nu}$ and Φ/c^2. We therefore conventionally choose to make $\bar{h}^{\mu\nu}$ dimensionless also, which fixes the proportionality constant in this equation to be such that it may be written,

$$\Box \bar{h}^{\mu\nu} = -\frac{16\pi G}{c^4} T^{\mu\nu} \quad \text{and} \quad \partial_\mu \bar{h}^{\mu\nu} = 0 , \tag{21.9}$$

where the dimensionless constant -16π is purely conventional.

We now choose to make a *field redefinition* of the potential to form the alternative potential,

$$h^{\mu\nu} = \bar{h}^{\mu\nu} - \tfrac{1}{2}\eta^{\mu\nu}\bar{h} , \tag{21.10}$$

where $\bar{h} = \bar{h}^\mu{}_\mu$ is the trace of $\bar{h}^{\mu\nu}$. Since $\bar{h}^{\mu\nu}$ and $h^{\mu\nu}$ may be obtained from one another by algebraic operations, they are clearly alternative expressions for the tensor field.

Exercise 21.5 (a) Show that the barring operation on any quantities $a^{\mu\nu}$ and $b^{\mu\nu}$ satisfies,

$$a^{\mu\nu}\bar{b}_{\mu\nu} = \bar{a}^{\mu\nu}b_{\mu\nu} \quad \text{and} \quad \bar{\bar{a}}^{\mu\nu} = a^{\mu\nu} , \tag{21.11}$$

and that $\bar{h}^{\mu\nu}$ may be retrieved from $h^{\mu\nu}$ according to,

$$\bar{h}^{\mu\nu} = h^{\mu\nu} - \tfrac{1}{2}\eta^{\mu\nu}h , \tag{21.12}$$

where $h = h^\mu{}_\mu = -\bar{h}$ is the trace of $h^{\mu\nu}$. (b) Show that Equations 21.9 for $\bar{h}^{\mu\nu}$ imply that $h^{\mu\nu}$ satisfy,

$$\Box h^{\mu\nu} = -\frac{16\pi G}{c^4}\bar{T}^{\mu\nu} \quad \text{and} \quad \partial_\mu h^{\mu\nu} - \tfrac{1}{2}\partial^\nu h = 0 . \tag{21.13}$$

(c) Show that $h^{\mu\nu} \to h^{\mu\nu} + \partial^\mu \xi^\nu + \partial^\nu \xi^\mu$ where the functions $\xi^\mu(x)$ are solutions of $\Box \xi^\mu = 0$, amount to a gauge freedom of $h^{\mu\nu}$ in the vacuum case ($T^{\mu\nu} = 0$), or for an external source.

Since we are using a field on which a gauge condition has already been imposed, the symmetry with gauge function $\xi^\mu(x)$ is a *residual gauge invariance*.

Equations 21.9 and 21.13 are clearly not the only equations we could impose on a tensor field having $T^{\mu\nu}$ as its source. Let us provide further justification for this choice. Equation 21.9, coupling $\bar{h}^{\mu\nu}$ to matter, implies that, in the absence of a source $T^{\mu\nu}$, the free field satisfies the massless Klein-Gordon equation, $\Box \bar{h}^{\mu\nu} = 0$, constrained by $\partial_\mu \bar{h}^{\mu\nu} = 0$. The first of these equations tells us that the plane wave solutions, $e^{\pm ik\cdot x}$, are null waves for which $k^2 = 0$, namely the frequency ω and propagation 3-vector **k** are related by $\omega = |\mathbf{k}|c$.

All observed particles have a well-defined value of spin, or helicity in the massless case, and this characteristic is shared by the corresponding wave functions or classical fields. Any field which satisfies a wave equation that implies it has a well-defined mass and a well-defined spin (or helicity if massless) will be form-invariant under the action of the Poincaré group and will therefore respect the Relativity principle. A null particle or field, having a well-defined value of helicity can have only two independent components, those for which the Pauli-Lubański 4-vector, or the corresponding eigenvalues when it acts on a field, is either parallel or anti-parallel to the direction of propagation. The four conditions in the constraint on $\bar{h}^{\mu\nu}$ reduce its number of independent components from ten to six and the existence of the residual gauge freedom with a 4-vector gauge function shows that a further four components of $h^{\mu\nu}$ are non-physical, leaving two physical components.

We could in fact evaluate the action of the Pauli-Lubański vector on $\bar{h}^{\mu\nu}$ and verify that the free field is indeed a null field of helicity components ± 2. Alternatively, one may make use of group theory to show that the equations we have selected provide a massless spin 2 irreducible representation of the Poincaré group [117, 118]. These are the reasons why we are using a field with precisely the form of $\bar{h}^{\mu\nu}$ or its equivalent $h^{\mu\nu}$.

Exercise 21.6 *Show that Equations 21.9 are form-invariant with respect to Poincaré transformations.*

The field $h^{\mu\nu}$ (or $\bar{h}^{\mu\nu}$) may be referred to as the *Fierz-Pauli field* of helicity 2 and the condition on its first derivatives is referred to as fixing the Fierz-Pauli field to the *de Donder gauge* [94]. (It is also referred to as the Lorentz, Einstein or Hilbert gauge.) This field was first examined in a

21.3. Fierz-Pauli massless spin 2 tensor gravity

Lagrangian context by Fierz and Pauli [162] in 1939.

In order to represent a gravitational field, we demand that Equations 21.13 be consistent with Newtonian gravity theory in the limit of a weak ($|h^{\mu\nu}| \ll 1$) static ($|\partial_0 h^{\mu\nu}| \ll |\nabla h^{\mu\nu}|$) field, and slow motion ($|\dot{z}| \ll c$), for which $|T^{00}| \gg |T^{k0}| \gg |T^{kl}|$. Under these conditions, we need only consider the component $T^{00} = \rho c^2$ of $T^{\mu\nu}$ leading to $T = -\rho c^2$ and $\bar{T} = \frac{1}{2}\rho c^2$. Thus $\nabla^2 h^{00} = -8\pi G\rho/c^2$ which is identical to the Newtonian gravity field equation provided $\nabla^2 h^{00} = -2\nabla^2 \Phi/c^2$. Setting the arbitrary constants of Φ and h^{00} to zero at the same location shows that,

$$h^{00} = -2\Phi/c^2 \quad \text{(Newtonian limit)}. \tag{21.14}$$

This identification of the Newtonian limit of the component h^{00} is the only restriction on the field equations of a tensor gravitational potential $h^{\mu\nu}$ satisfying Equations 21.13 in order that they be consistent with Newtonian gravity and, in particular, attractive as a result of energy positivity.

We now form an action for a non-quantum particle of trajectory $z(\lambda)$ interacting with the field $h^{\mu\nu}(x)$, namely,

$$\begin{aligned}S &= S_{\text{field}} + S_{\text{part}} + S_{\text{intn}} \\ &= -\frac{c^3}{32\pi G} \int_{\Sigma_i}^{\Sigma_f} d^4x \tfrac{1}{2} \partial^\lambda h^{\mu\nu} \partial_\lambda \bar{h}_{\mu\nu} + \tfrac{1}{2} \int_{\lambda_i}^{\lambda_f} d\lambda \, [e^{-1}(\lambda)\dot{z}^2 - e(\lambda)m^2 c^2] \\ &\quad + \tfrac{1}{2} \int_{\lambda_i}^{\lambda_f} d\lambda \, e^{-1}(\lambda) \dot{z}^\mu \dot{z}^\nu h_{\mu\nu}(z(\lambda)) \, . \end{aligned} \tag{21.15}$$

The second and third terms are based on the standard einbein expression (see Chapter 15) for the free particle in gauge-variant form. The constant factor in the third term is determined by the requirement that the particle and interaction terms be consistent with the equations of motion of a slow particle in a Newtonian gravitational field. The signs of the first two terms are fixed by energy positivity and the magnitude of the first by the requirement that the field equations of the action correspond to Equation 21.13. Renormalizing to a canonical field $h_{\mu\nu} c^{3/2}/\sqrt{G}$, of inverse length natural dimension, places $\sqrt{G}$ as a coupling constant in the numerator of the interaction term. One choice could be $\psi_{\mu\nu} = h_{\mu\nu} c^{3/2}/\sqrt{32\pi G}$.

Exercise 21.7 (a) Verify that the action of Equation 21.15 leads to,

$$\Box \bar{h}^{\mu\nu} = -\frac{16\pi G}{c^4} c \int d\lambda \, e^{-1}(\lambda) \delta^4(x - z(\lambda)) \dot{z}^\mu \dot{z}^\nu \, , \tag{21.16}$$

consistent with Equations 21.13, for Fierz-Pauli gravity, and 15.94, and show that the equations of motion of the particle are,

$$\dot{\pi}_\mu = \frac{d}{d\lambda}\left(e^{-1}(\lambda) g_{\mu\lambda} \dot{z}^\lambda\right) = \tfrac{1}{2}\left(e^{-1}(\lambda) \partial_\mu g_{\lambda\pi}\right) \dot{z}^\lambda \dot{z}^\pi \, , \tag{21.17}$$

constrained to satisfy,
$$g_{\mu\nu}\dot{z}^\mu \dot{z}^\nu = -e^2(\lambda)m^2 c^2 , \qquad (21.18)$$
where,
$$g_{\mu\nu} = \eta_{\mu\nu} + h_{\mu\nu} . \qquad (21.19)$$
(b) Show that the interacting action of Equation 21.15 is covariant, and the de Donder condition invariant, with respect to a residual gauge transformation, with a small gauge function ξ_μ satisfying $\Box \xi_\mu = 0$, combined with a small coordinate transformation, namely:
$$h_{\mu\nu} \to h_{\mu\nu} + \partial_\mu \xi_\nu(x) + \partial_\nu \xi_\mu(x) \qquad x^\mu \to x^\mu - \xi^\mu(x) , \qquad (21.20)$$
affecting both the fields and the particle variables $z^\mu(\lambda)$.

The same combined transformations (with ξ^μ unrestricted by the harmonic condition) constitute an invariance of an action coupling the variables $z^\mu(\lambda)$ to a gravitensor field not subject to the de Donder condition. Furthermore, in the absence of a source, the spin 2 pure gauge invariances (without coordinate changes) are not restricted to small ξ.

In contrast to scalar gravity, the combined field and coordinate transformation of tensor gravity contains an arbitrary function $\xi_\mu(x)$ and is therefore a local gauge invariance. Such gauge invariances involving both internal variables (here $h_{\mu\nu}$) and the external coordinates x^μ are a generalization, characteristic of relativistic gravitation, of Yang-Mills gauge transformations. A general discussion is available in the 1959 paper by Utiyama [548]. In the present context, they show that the coordinates x^μ cannot combine to give physical spacetime intervals using the metric $\eta_{\mu\nu}$ in the Minkowski line element, $ds^2_{\text{Mink}} = \eta_{\mu\nu} dx^\mu dx^\nu$. Instead, it is the combination, $g_{\mu\nu} dx^\mu dx^\nu$, with the metric $g_{\mu\nu} = \eta_{\mu\nu} + h_{\mu\nu}$, which is gauge invariant. We may still use local Minkowski coordinates, raised and lowered with $\eta_{\mu\nu}$, provided we note that, in the non-null case, proper intervals are obtained from dx^μ using $g_{\mu\nu}$. The Minkowskian spacetime is then locally valid but is not globally observable, the effective structure of the spacetime being pseudo-Riemannian with metric $g_{\mu\nu}$.

For $m \neq 0$, taking the λ gauge with constant $e(\lambda) = m^{-1}$ gives equations of motion,
$$\dot{\pi}_\mu = \frac{d}{d\lambda}(g_{\mu\lambda}\dot{z}^\lambda) = \tfrac{1}{2}(\partial_\mu g_{\lambda\pi})\dot{z}^\lambda \dot{z}^\pi \quad \text{and} \quad g_{\mu\nu}\dot{z}^\mu \dot{z}^\nu = -c^2 , \qquad (21.21)$$
in which the evolution parameter λ becomes identical to the Minkowski proper time $\sqrt{ds^2_{\text{Mink}}}/c$ at large distances from a localized source where $h_{\mu\nu} \to 0$. For a massless particle, we select any gauge with constant nonzero $e(\lambda)$, which we use to renormalize $\lambda \to e(\lambda)\lambda$ giving,
$$\dot{\pi}_\mu = \frac{d}{d\lambda}(g_{\mu\lambda}p^\lambda) = \tfrac{1}{2}(\partial_\mu g_{\lambda\pi})p^\lambda p^\pi \quad \text{and} \quad g_{\mu\nu}p^\mu p^\nu = 0 , \qquad (21.22)$$

with $p^\mu = e^{-1}\dot{z}^\mu$ being the momentum of the null particle. The conditions $p_\mu = \eta_{\mu\nu}p^\nu$ and $\eta_{\mu\nu}p^\mu p^\nu = -m^2c^2$ or 0 for a particle free of gravitational coupling have become the generalized momentum (particle plus field) given by $\pi_\mu = g_{\mu\nu}p^\nu = \eta_{\mu\nu}p^\nu + h_{\mu\nu}p^\nu$ and the relation $(\eta_{\mu\nu} + h_{\mu\nu})p^\mu p^\nu = -m^2c^2$ or 0 for the interacting particle.

The two sets of equations, 21.21 and 21.22, are those of timelike and null *geodesics*, which for massive particles are curves of extremal proper distance ds, in the spacetime of metric $g_{\mu\nu}$. The geodesic property follows in the massive case since the interacting particle action and proper distance are proportional and the Euler-Lagrange equations correspond to extremal action. The coupling of a free (geodesic) particle in Minkowskian spacetime to the gravitational field $h^{\mu\nu}$ has led to a particle which traces out a geodesic in the spacetime with metric $g_{\mu\nu} = \eta_{\mu\nu} + h_{\mu\nu}$. For this reason, it is customary to consider any particle subject only to gravitational interactions to be a free particle, also said to be in *free fall*.

The coordinate transformation invariance of massless spin 2 theory, and the associated gauge transformation of the field, shows that it incorporates invariance with respect to changes of frame with uniform relative velocity, as in special relativity without gravity, and with respect to uniform relative acceleration as in scalar gravity. However, since the transformations contain an arbitrary gauge function, there is also invariance with respect to frames in arbitrary relative motion. A gauge transformation may be used to show that, locally, any gravitational field in frame x^μ is equivalent to absence of a gravitational field is some other frame in motion relative to x^μ with either constant relative velocity, constant relative acceleration or non-constant acceleration.

21.4 Experimental tests of tensor gravity

The only difference between the equations of motion of a particle in a gravitational field described by a tensor $h_{\mu\nu}$, compared to that of a scalar field Φ, is the expression in Equation 21.19 for the tensor $g_{\mu\nu}$ compared to the corresponding Equation 17.28.

Exercise 21.8 (a) *Let M be a large spherically-symmetric static mass. Show that, in the inertial frame whose origin is at its centre of mass, the tensor field satisfying the Fierz-Pauli field equation and the de Donder gauge condition is,*

$$\{h^{\mu\nu}\} = \frac{2GM}{c^2 r} \begin{Bmatrix} 1 & 0 & 0 & 0 \\ 0 & 1 & 0 & 0 \\ 0 & 0 & 1 & 0 \\ 0 & 0 & 0 & 1 \end{Bmatrix} = \frac{r_M}{r} \begin{Bmatrix} 1 & 0 & 0 & 0 \\ 0 & 1 & 0 & 0 \\ 0 & 0 & 1 & 0 \\ 0 & 0 & 0 & 1 \end{Bmatrix} \quad (21.23)$$

where $r_M = 2GM/c^2$ is the Schwarzschild length of the mass M.

In addition to the h^{00} term, which reproduces the effect of a Newtonian potential, $h^{\mu\nu}$ has three other diagonal terms of identical magnitude.

Since ds^2_{Mink} is not gauge and coordinate invariant in the presence of the field $h_{\mu\nu}$, we obtain proper time and space intervals appropriate to real clocks from the gauge-invariant interval $d\tau = \sqrt{-ds^2}/c = \sqrt{-g_{\mu\nu}dx^\mu dx^\nu}/c$. This can be confirmed by the semi-classical coupling of a Bohr electron trajectory $z(\lambda)$ to the electromagnetic potential A^μ in the presence of the gravitational field $h_{\mu\nu}$ and determining the way the Lorentz law must be modified in the frame x^μ as a result of gravitation. The Bohr frequency, with respect to t, may be determined, showing that proper times must be calculated using $g_{\mu\nu}$.

A particle clock which is spatially fixed ($\dot{z}^k = 0$) in the frame x^μ will have proper and coordinate time intervals related by,

$$d\tau = \frac{\sqrt{-g_{00}dx^0 dx^0}}{c} = \sqrt{-g_{00}}\, dt \; . \tag{21.24}$$

The atomic proper time τ will thus coincide in rate with coordinate time t at large distances from a localized source of gravity, where $g_{00} \to \eta_{00} = -1$.

The equation of a null ray may be used, as in the scalar gravity case, to show that the *gravitational red shift* factor z of the light ray passing from position $\mathbf{x}_1$ to $\mathbf{x}_2$ is identical to the *gravitational time dilation* of particle clocks fixed at $\mathbf{x}_1$ relative to those fixed at $\mathbf{x}_2$. For a static, spherically symmetric source, the red shift factor is the ratio $\sqrt{g_{00}(r_1)/g_{00}(r_2)}$ (for $r_1 > r_2$). For small height differences h where the gravitational acceleration is $g = GM/r^2$, this reduces to the Newtonian result,

$$z = \left(\frac{\Delta P}{P}\right)_{\text{signal}} = \frac{r_M \Delta r}{2r^2} = \frac{GM \Delta r}{c^2 r^2} = \frac{(\Delta \Phi)_{\text{Newt}}}{c^2} = \frac{gh}{c^2} \; . \tag{21.25}$$

For scalar gravity, the fact that all components of $g_{\mu\nu}$ are proportional to $\eta_{\mu\nu}$ means that, in the case of a massless particle, the factor of $\eta_{\mu\nu}$ in $g_{\mu\nu}$ on the right side of the equations of motion can be contracted on the product $\dot{z}^\mu \dot{z}^\nu$ or $p^\mu p^\nu$ which causes the right side to vanish for null p^μ. This does not occur in the tensor case where $g_{\mu\nu}$ is not proportional to $\eta_{\mu\nu}$. We can therefore expect a non-zero value for *light deflection* in a tensor gravitational field. The deflection calculation is simply a covariant version of the 1801 result of Soldner [515] given in Equation 7.33 and illustrated in Figure 7.1. Since the field is weak ($|h^{\mu\nu}| \ll 1$) and the deflection small, we may put $h^{\mu\nu} = 0$ on the left of the equation of motion of the massless particle to obtain,

$$\dot{p}_\mu = \tfrac{1}{2}(\partial_\mu h_{\lambda\pi}) p^\lambda p^\pi \; , \tag{21.26}$$

in which, to first order, we may approximate p^μ on the right side by its initial free value. The normalization of p is of no significance, so we choose an initial null momentum of $p^\mu = \{1, 0, 0, 1\}$ in the frame of the mass M giving,

$$\dot{p}_\mu = \tfrac{1}{2}\partial_\mu(h_{00} + h_{zz}) , \tag{21.27}$$

since $h_{0z} = 0$. The deflection is therefore,

$$\begin{aligned}
|\Delta\phi| &= -\int_{-\infty}^{\infty} dz \frac{dp_x}{dz} \\
&= -\int_{-\infty}^{\infty} dz\, \tfrac{1}{2}\partial_x(h_{00} + h_{zz}) \\
&= -\int_{-\infty}^{\infty} dz\, \partial_x\left(\frac{2GM}{c^2 r}\right) \\
&= \frac{2GM}{c^2} \int_{-\infty}^{\infty} dz\, \frac{\cos\theta}{r^2} ,
\end{aligned} \tag{21.28}$$

leading to,

$$|\Delta\phi| = \frac{4GM}{c^2 \ell} , \tag{21.29}$$

where ℓ is the impact parameter of the null ray. Thus the deflection is 1.76 arc seconds, essentially double the value obtained by the Soldner calculation based on Newtonian gravity, and by Einstein [132] in 1911, and is in complete accord with the observations.

The first measurements of this light bending result (predicted by Einstein [135] in 1915 using general relativity) were carried out during the total solar eclipse of May 29, 1919, by British expeditions to Sobral in Northern Brazil giving a deflection of 1.98 ± 0.12 arc seconds, and to the island of Principe off the west coast of Africa where the result was 1.61 ± 0.30. These results [123] were presented to the 6 November, 1919, meeting of the Royal Society (London) by Sir Frank Dyson, Astronomer Royal. Improved precision had to await the use of long base-line radio interferometry of groups of quasars passing near sun. These measurements date from the early 1970s giving agreement with general relativity to within a few percent, and more recently, by Robertson and Carter [479] in 1984, to within a fraction of one percent.

Equation 21.29 is a significant achievement of tensor gravity in a special relativistic context. In general relativity the term h_{zz} (or h_{rr} in a spherically symmetric context) is interpreted as a contribution due to the spatial curvature of spacetime. We see therefore that a null ray or particle is equally sensitive to both contributions to the curvature, the part h_{00} which exists in any theory consistent with the Newtonian limit and the spatial curvature part which is neglected in a calculation based solely on Newtonian gravity, even if the Principle of equivalence is used. Paradoxically, relativistic scalar gravity picks up both contributions but with an incorrect sign for one so that it cancels the other to give zero deflection.

The above calculation is based on an energy-momentum tensor $T^{\mu\nu}$ originating solely in the mass M of the material source. The justification for the use of a field based on such a source is the smallness of the contribution due to the field itself. The *perihelion shift* in the orbit of a massive particle is, by definition, a deviation from Newtonian predictions. The use of a tensor field, determined linearly by Equations 21.13, can be expected, by its inclusion of the space curvature effect in h_{rr}, to lead to a non-zero perihelion shift. As in the examination of the perihelion shift for scalar gravity, $h_{\mu\nu}$ is diagonal. Incorporating the universality of gravitation, and thus the non-linear effects, in the production of the field $h^{\mu\nu}$ would mean modification of the linear solution of Equation 21.23 by higher powers of r_M/r, a small quantity in the solar system.

Since $|\dot{z}/\dot{z}^0|^2 \sim r_M/r$ for slow motion of a massive particle at distances $\sim r$ from a mass M, the right side of the equation of motion of a massive particle shows that, to the same order of precision, we must retain terms of one higher power of r_M/r in the spatial parts h_{kl} of $h_{\mu\nu}$ than are retained in h_{00}. Let us generalize our examination of the perihelion shift to any gravitational field which, in isotropic coordinates, has potentials of the form,

$$-g_{00} = 1 - \frac{r_M}{r} + \tfrac{1}{2}\beta\left(\frac{r_M}{r}\right)^2 + \cdots \quad \text{and} \quad g_{rr} = 1 + \gamma\frac{r_M}{r} + \cdots . \quad (21.30)$$

The two parameters β and γ, referred to as PPN parameters [384] (for *Parameterized Post-Newtonian approximation*), provide measures, respectively, of the non-linearity of the field equations and the degree of spatial curvature, relative to the Newtonian effects which correspond to the term proportional to $1/r$ in g_{00}.

Exercise 21.9 *Show that a relativistic gravitational field theory leading to Equation 21.30 predicts the magnitude of the light deflection to be,*

$$|\Delta\phi| = \frac{4GM}{c^2\ell}\frac{1+\gamma}{2}, \quad (21.31)$$

and (see Equations 17.42 and 17.59), a perihelion shift of,

$$\Delta\phi^{100y}_{\text{excess}} = \frac{6\pi GMN}{c^2 a(1-e^2)}\frac{2-\beta+2\gamma}{3}, \quad (21.32)$$

where N is the number of orbital revolutions of semi-major axis a and eccentricity e about the mass M in 100 years.

The observational results of light deflection and perihelion shift are consistent with PPN parameter values of $\beta = \gamma = 1$ which are also the

21.4. Experimental tests of tensor gravity

values [384, 566] obtained from general relativity. The linear Fierz-Pauli theory with $\gamma = 1$ and $\beta = 0$ gives a perihelion advance of 4/3 of the observational result. However, this failure should not be surprising given its neglect of comparable contributions from terms for which $\beta \neq 0$.

Exercise 21.10 *Show that the canonical energy-momentum tensor of the Fierz-Pauli field in the de Donder gauge is given by,*

$$\tilde{T}^{\mu\nu} = \tfrac{c^4}{32\pi G}(\partial^\mu \bar{h}^{\lambda\rho}\partial^\nu h_{\lambda\rho} - \tfrac{1}{2}\eta^{\mu\nu}\partial^\lambda h^{\pi\rho}\partial_\lambda \bar{h}_{\pi\rho}) \,. \qquad (21.33)$$

(b) Use the fact that the field $h_{\mu\nu}$ transforms in the same way as a symmetrized product $U_\mu \otimes V_\nu$, and Equation 14.56 for the 4-vector representation, to determine the corresponding spin 2 generators, $S_{\mu\nu} = \{(S_{\mu\nu})^{\alpha\beta}{}_{\gamma\delta}\}$.
(c) Determine the extra contributions required to the energy-momentum tensor of the Fierz-Pauli field, in the de Donder gauge, to allow for spin density contributions.

A non-linear field equation for the spin 2 gravitational field $h_{\mu\nu}$ may be constructed by including the energy-momentum tensor density (including the spin density) of the field itself into the source along with the contribution of the mass generating the linear field. The new contribution is second-order in $h_{\mu\nu}$ and may therefore be evaluated using the first-order solution. As shown by Thirring [534], gauge invariance may be used as a very useful guide to the second-order formulation and its solution to order $(r_M/r)^2$ (see Section 17.6 for the graviscalar field). Such a calculation leads to $\beta = 1$ identical to keeping second-order terms from the Schwarzschild solution of general relativity and demonstrates that a non-linear helicity 2 theory is consistent with perihelion advance observations.

The representation theory of the Poincaré group permits one to determine the free field equations of a null spin 2 field (see, for example, reference [117, 118]) in an arbitrary gauge, unrestricted by the de Donder gauge condition. One form of the equation is,

$$-\Box \bar{h}_{\mu\nu} + 2\partial^\lambda \partial_{(\mu}\bar{h}_{\nu)\lambda} - \eta_{\mu\nu}\partial^\lambda \partial^\rho \bar{h}_{\lambda\rho} = 0 \,. \qquad (21.34)$$

This equation, and its equivalent for $h^{\mu\nu}$,

$$-\Box h_{\mu\nu} + 2\partial^\lambda \partial_{(\mu}h_{\nu)\lambda} - \partial_\mu \partial_\nu h + \eta_{\mu\nu}\Box h - \eta_{\mu\nu}\partial^\lambda \partial^\rho h_{\lambda\rho} = 0 \,, \qquad (21.35)$$

have identically divergence-free left sides, as required for them to be derivable from a Lagrangian and to be coupled to a conserved source. The field equations for $h^{\mu\nu}$ or $\bar{h}^{\mu\nu}$ may in fact be obtained from the Lagrangian,

$$\mathcal{L}_{\text{FP}} = \tfrac{c^3}{32\pi G}(\tfrac{1}{2}\partial_\mu h_{\lambda\pi}\partial^\mu \bar{h}^{\lambda\pi} - \partial_\mu \bar{h}^{\mu\lambda}\partial^\nu \bar{h}_{\nu\lambda}) \qquad (21.36)$$

in which one factor of $h_{\mu\nu}$ is unbarred and $h^{\mu\nu}$, not $\bar{h}^{\mu\nu}$, is the Lagrangian variable.

Equations 21.34 and 21.35 have a normal gauge invariance of the form $h_{\mu\nu} \to h_{\mu\nu} + \partial_\mu \xi_\nu + \partial_\nu \xi_\mu$ in which $\xi_\mu(x)$ is a completely arbitrary vector function of length dimensions. These equations, fixed to the de Donder gauge, were used earlier to discuss massless spin 2 gravity. In the de Donder gauge, there exists residual gauge invariance which eliminates a further four components leaving two independent.

As soon as one attempts to make use of the identically vanishing divergence of the left side of the Fierz-Pauli equations, 21.34 or 21.35, to couple them to the energy-momentum tensor, one is forced to include in the total energy-momentum tensor, the contribution of the gravitational field itself. This is because the energy-momentum tensor of matter coupled to the field cannot be conserved on its own since that would imply a free equation of motion, $\ddot{z} = 0$ (see Exercise 19.1) contradicting Equation 21.21 obtained from the coupling to $h_{\mu\nu}$.

The inclusion, in the energy-momentum tensor, of a field contribution quadratic in $\partial_\lambda h_{\mu\nu}$, as in Exercise 21.10, gives a non-linear equation for $h_{\mu\nu}$ for which a new Lagrangian cubic in the fields may be constructed. A new contribution to $T^{\mu\nu}$ may then be calculated and added to the source. The Gupta programme [216] for obtaining relativistic gravitation in this way, once started, only leads to a consistent field equation when an effectively infinite number of terms have been included. The resulting theory can be shown to be indistinguishable from general relativity and the necessity of the general theory has been established by Deser [96] in a few steps.

21.5 Einstein strong equivalence principle

Although the systematic development of the general theory of relativity is beyond the scope of this text, a number of its features can be appreciated by comparison with material we have already presented in the context of special relativity. A few of the principles and the key results will therefore be described without proof in this section and the next.

The first hints of a general relativistic theory of gravitation, in which the Principle of relativity is extended beyond frames related by a uniform velocity, were the suggestions added by Einstein, as early as 1907, to a review paper [131] on special relativity and published in preliminary form in another paper [132] in 1911. The foremost principle is the local equivalence between inertial and gravitational acceleration extending the Relativity principle from inertial frames to those with relative acceleration, at that time limited to uniform acceleration. His hope, expressed privately to Habitch [508, p. 76] was the explanation of the perihelion shift of Mercury for which Lorentz' electrodynamic theory of gravity [338, 339] of 1900 was not in agreement. Modern expressions of the essence of Einstein's use of this idea (see Will [586, 588]) refer to,

21.5. Einstein strong equivalence principle

- the **Einstein or Strong equivalence principle**, namely that the Weak equivalence principle holds (see Section 7.11) and that no local non-gravitational measurements depend on the velocity or spacetime location of the freely falling reference frame in which they are performed.

For small spacetime intervals, the second part of the Principle reproduces the Poincaré covariance principle of special relativistic non-gravitational physics with the rôle of a global inertial frame being replaced by a local instantaneous inertial or freely falling frame, namely one not acted upon by non-gravitational interactions.

An isolated particle will appear to have an inertial acceleration if its trajectory is determined in a non-inertial reference frame. A gravitational acceleration, on the other hand is considered to arise as a result of the particle not being isolated from other matter. A consequence of demanding they be equivalent is the exact equality of inertial and gravitational mass. In Newtonian theory, the equality of the two masses is an empirical result based on experiment and has no geometrical origin or consequences.

For Einstein, the equality of the two masses was a consequence of their identity which is postulated to be true as a matter of principle for each and every possible contribution to either one of them. That equivalence of inertial and gravitational mass is embodied in a similar equivalence, in the inertial and gravitational accelerations, in the sense that it is postulated that no local experiment, namely one limited to small distances and time intervals, can ever distinguish between them. This does not mean that gravitation is not a real phenomenon — instead, it implies that genuine gravitational effects are *non-local*, or *global*. (Global is used here with a different connotation to its use in referring to global gauge groups.)

In special relativity, an absolute uniform velocity is not detectable but an absolute acceleration can be determined by applying the relativistic equivalents of Newton's laws. The same is true in the Galilean relativity of Newtonian physics. With the adoption of Einstein's equivalence principle, absolute acceleration involving only two objects is also not measurable. Only relative accelerations involving three or more objects can be determined. Such relative accelerations, which give rise to phenomena such as tides, are the genuine manifestations of gravitational interaction and acceleration.

The observation, from one frame, of at least two other small test particles (in addition to the observer) which are free of electromagnetic and nuclear interactions — we say they are in *free fall* — suffices to determine the existence or not of gravitational effects. This is done by examining whether or not such free particles are accelerated relative to one another. The gravitational attraction between the two test particles is not involved if they are sufficiently small — their motion in the gravitational field of other masses is all that matters.

The Einstein principle of equivalence does more than just embody, as a fundamental identity, the observed equality of inertial and gravitational mass of the Weak equivalence principle. Even without a detailed theory of gravity based on the Principle of equivalence (such as general relativity), the Principle suffices to show, for example, that light must be deflected by a mass which it passes and that the rate of a clock will be affected by a gravitational field. In the absence of matter, light does not follow a linear trajectory in terms of the coordinates of an accelerating frame — it appears deflected. The Equivalence principle states that this is indistinguishable from a measured deflection in an inertial frame owing to the presence of a mass causing gravitational acceleration of the same magnitude in the opposite direction. Thus one expects light deflection due to a nearby mass, and thus curvature of space.

Similar Principle of equivalence arguments with rotating (and therefore accelerating frames) permit one to conclude [203] that clocks should be dilated by gravitational fields. Our analysis of the scalar and tensor theories of gravity shows that such dilation may be interpreted as curvature of time in the sense of requiring a metric tensor whose 00 component depends on position.

These effects imply that spacetime geometry is not a static inert concept but a dynamic entity affected by the presence of matter and therefore also by energy such as electromagnetic radiation. Since the field equations of any long-range relativistic theory of gravitation must, for small perturbations, have similiar space and time dependence to Equations 21.9 and 21.13, gravitational effects in such a theory will propagate as null waves, travelling at the ultimate speed. The energy of those waves will have mass and thus also exert a dynamic effect on the spacetime geometry. Without knowing the details of those field equations, we know at least that gravitation is a non-linear effect closely related to the geometry of spacetime itself and in particular to its curvature.

21.6 General relativity

We have explained the phenomena of the heavens and of our sea by the power of gravity, but have not yet assigned the cause of this power ... .

Isaac Newton   1642–1727

Einstein's *general relativity theory* provides a natural description of the gravitational interaction. It incorporates the Principle of relativity, extended to frames in arbitrary relative motion, as the Principle of equivalence. In it, the effects of gravity are intimately linked to the *geometry of curved spacetime*.

From initial suggestions [131] in 1907, Einstein's ideas on gravitation

Albert Einstein

b. Ulm, Germany
14 March 1879

d. Princeton, New Jersey
18 April 1955

Courtesy of the Mansell Collection, London.

and general relativity progressed to the Equivalence principle [132] in 1911 with the development of the full theory during the next few years [133, 134]. The complete field equations [135, 136, 137, 138] were presented in 1915 and a summary [139] appeared in 1916 which also contained the correct calculation accounting for the anomalous perihelion shifts of planetary orbits. In common with the transition from Galilean physics to special relativity, the Relativity principle of the general theory requires a further revision of what is meant by spacetime measurement, namely in the presence of gravitational fields.

The essence of special relativity is kinematic — other considerations, in particular, the Gauge principle, must be welded to it to give interaction and dynamics. However, general relativity provides equations describing the form of the gravitational interaction and is therefore itself a dynamic theory. The starting point is the 4D equivalent of a smooth surface, namely a *manifold*. The spacetime manifold of general relativity represents the totality of all the events in the universe and, to be consistent with the observed validity of special relativity in any sufficiently small region, it must be locally Minkowskian in the same way that a small part of the surface of a sphere or a saddle is locally Euclidean. In order that parity and time-reversal operations are well-defined, the locally-Minkowskian spacetime must also be spacetime orientable.

Let us suppose that e_a ($a = 0, 1, 2, 3$), previously denoted by e_μ, is a local Lorentz basis ($e_a \cdot e_b = \eta_{ab}$), where Latin indices are used in order to reserve Greek letters for a global coordinate system. Such freely falling Lorentz bases exist at each event. However, without further structure, the physical laws in local Minkowskian frames with bases $e_a(1)$ and $e_a(2)$ (with local coordinates $x^a(1)$ and $x^a(2)$) at two different events (1) and (2) will be unconnected.

The spacetime of general relativity is not in general homogeneous or isotropic. The degree of homogeneity and isotropy of the spacetime is determined by the corresponding properties of the matter distribution. In this respect, it parallels the non-relativistic Newton-Cartan spacetime. In general, it has no global translational or rotational symmetry at all, although some regions may be very highly symmetric as a result of an equivalent symmetry in the matter distribution and thus in the gravitational field and the geometry.

The spacetime of general relativity theory is not flat. On the contrary, general relativistic spacetime has intrinsic curvature as a result of the existence of a global non-Minkowskian metric tensor. The latter is a dynamical field determining the spacetime intervals and governed itself by the matter content. Spacetime is curved by the nearby and distant matter in contrast to the absoluteness of the spacetime of Newton. However, some regions, such as those containing an isolated star, may be so nearly flat at large distances from the star, that the spacetime may be regarded as having large distance or *asymptotic flatness*. The universe as a whole may or may not be asymptotically flat.

If an arbitrary global system of coordinates, $x = \{x^\mu\}$, called *general coordinates*, is introduced to label the events of spacetime, then we let e_μ denote the basis vectors parallel to varying each of the coordinates x^μ, one at a time. The dual basis vectors or 1-forms dx^μ are the normals to the surfaces of constant x^μ. Each of the four general coordinate basis vectors $e_\mu(x)$ of curved spacetime may be expanded in terms of a local Lorentz frame at x according to $e_\mu(x) = e_\mu^a(x) e_a$. The scalar products (with respect to η_{ab}) of e_μ with e_ν provide the components of the metric tensor which are therefore given in terms of the coefficient fields $e_\mu^a(x)$ by

$$g_{\mu\nu} = e_\mu^a(x) e_\nu^b(x) \eta_{ab} . \tag{21.37}$$

The fields $e_\mu^a(x)$ ($a = 0, 1, 2, 3$) transform as Lorentz 4-vectors with respect to the Latin indices and as general 4-vectors (contragredient to dx^μ) with respect to Greek indices. They are referred to as *vierbein fields* ('four legs') and are the general relativistic analogue of the particle einbein $e(\lambda)$ (contragredient to $d\lambda$), the electromagnetic 4-vector potential A_μ and the Yang-Mills gauge fields A_μ^a, with $a = 1, 2, 3$ or $1, 2, \ldots, 8$ for SU(2) or SU(3) symmetries. The einbein is entirely non-propagating but part of the vierbein is dynamic.

21.6. General relativity

The comparison of general relativistic gravitation with the gauge theories of the Standard model is made easier by using a vierbein formulation which also permits the introduction of fermion fields into curved spacetime. The spin 2 gauge boson described by the dynamic part of the vierbein is the graviton analogous to the photon, the weakons and the gluons in the Standard model. Corresponding to the compact SU(2)×U(1)×SU(3) global gauge group of the Standard model, gravitation can be obtained [278, 351, 548] by gauging the Poincaré group ISO(1,3), or more strictly its double cover, the inhomogeneous extension ISL(2,$\mathbb{C}$) of the special group SL(2,$\mathbb{C}$) of 2×2 complex matrices of unit determinant. The vierbeins arise from the translation part of the Poincaré group while the local Lorentz transformations lead to non-propagating fields much as occurs for the entire einbein for a relativistic particle.

In a given coordinate system, the metric tensor will have components $g_{\mu\nu}(x)$ which vary with the coordinates. The non-zero curvature means that no global rectangular Cartesian system of coordinates can be introduced comparable to a global Lorentz frame of special relativity. All such Minkowskian frames are valid only over regions of spacetime which are small compared to the radii of curvature. Using $dx^a = e^a_\mu dx^\mu$ and $ds^2 = \eta_{ab} dx^a dx^b$, the line element from which measurable (proper) space and time intervals are invariantly extracted from coordinate intervals dx^μ using $d\ell = \sqrt{ds^2}$ (for $ds^2 > 0$) and $d\tau = \sqrt{-ds^2}/c$ (for $ds^2 < 0$) can be expressed as,

$$ds^2 = g_{\mu\nu} dx^\mu dx^\nu \ . \tag{21.38}$$

The Standard model of non-gravitational interactions has the purely internal U(1) or Yang-Mills (SU(2) or SU(3)) gauge invariances. The free relativistic particle, of variable $z(\lambda)$, has a re-parameterization gauge invariance, $\lambda \to \bar\lambda = f(\lambda)$, with respect to the evolution parameter λ, for which $d\lambda \to \Lambda d\lambda$ and $\dot z \to \Lambda^{-1}\dot z$. The Fierz-Pauli theory has a mixed coordinate and field variable gauge transformation with respect to arbitrary functions $\xi^\mu(x)$. Correspondingly, the fundamental symmetry of general relativity is *general coordinate covariance* which affects not only the coordinates but also the metric tensor $g_{\mu\nu}$. This symmetry states that the laws of physics are to be form invariant with respect to arbitrary changes of the coordinate labels from one coordinate system $x = x^\mu$ to any other coordinates $x^{\bar\mu}$ given as invertible functions $x^{\bar\mu}(x)$ of the original coordinates, with corresponding gauge transformation of $g_{\mu\nu}$.

Einstein's demand for general coordinate covariance was related [269, vol. 2, p. 328] to his observation that physical measurement is concerned with determining spacetime coincidences matching events of interest with locations of parts of an instrument. Since $dx^\mu \to (\partial x^{\bar\mu}/\partial x^\mu)dx^\mu$, general coordinate transformations leave ds^2 unchanged provided the metric functions $g_{\mu\nu}(x)$ correspondingly transform according to,

$$g_{\bar\mu\bar\nu} = L^\mu{}_{\bar\mu} L^\nu{}_{\bar\nu} g_{\mu\nu} , \qquad (21.39)$$

where the transformation coefficients,

$$L^{\bar\mu}{}_\mu = L^{\bar\mu}{}_\mu(x) = \frac{\partial x^{\bar\mu}}{\partial x^\mu} \quad \text{and} \quad L^\mu{}_{\bar\mu} = L^\mu{}_{\bar\mu}(x) = \frac{\partial x^\mu}{\partial x^{\bar\mu}} , \qquad (21.40)$$

now depend on the coordinates x. The metric is therefore a symmetric covariant *general tensor* $\mathbf{g} = g_{\mu\nu} dx^\mu \otimes dx^\nu$ of second rank, as opposed to a Lorentz tensor in which the transformation coefficients are constant in a Minkowskian frame.

Kretschmann [289] showed in 1917 that any theory may be made generally covariant. Consequently, the physical content of general coordinate covariance is contained in the way in which it is used to couple matter to geometry, in accordance with the Principle of equivalence, with non-local features of the dynamical geometry being interpreted as gravitational phenomena. Other physical quantities are also required to be general invariants, general vectors or general tensors of some rank, whose indices are lowered and raised with $g_{\mu\nu}$ and its inverse $g^{\mu\nu}$. However, since the transformation coefficients depend on x, the partial derivatives of general tensors are not themselves general tensors. Some of the features of general coordinates are illustrated by the curvilinear coordinates widely used in flat 3D Euclidean space in problems where the geometry of an application makes them more suitable than rectangular Cartesian coordinates.

To form field equations with general covariance we must be able to form new general tensors by differentiation. As for the U(1) covariant derivative, $D_\mu \psi = \partial_\mu \psi - iqA_\mu \psi$, one must therefore construct a general coordinate *covariant derivative*, denoted $\nabla = \{\nabla_\mu\}$, related to ∂_μ and $g_{\mu\nu}$, which produces one general tensor from another. Its form depends on the rank of the tensor on which it acts. For the production of mixed second- and third-rank general tensors by covariant differentiation of the contravariant components V^μ of a general vector and a second-rank tensor $T^{\mu\nu}$, its effect [384] is given by,

$$\nabla_\nu V^\mu = \partial_\nu V^\mu + \Gamma^\mu{}_{\nu\kappa} V^\kappa , \qquad (21.41)$$

and,

$$\nabla_\nu T^{\mu\lambda} = \partial_\nu T^{\mu\lambda} + \Gamma^\mu{}_{\nu\kappa} T^{\kappa\lambda} + \Gamma^\lambda{}_{\nu\kappa} T^{\mu\kappa} , \qquad (21.42)$$

respectively. The combination of metric components and their first derivatives in,

$$\Gamma^\mu{}_{\nu\lambda} = \tfrac{1}{2} g^{\mu\pi} (\partial_\lambda g_{\pi\nu} + \partial_\nu g_{\pi\lambda} - \partial_\pi g_{\nu\lambda}) , \qquad (21.43)$$

is called the *affine connection*. Equation 21.43 may be obtained by demanding that the metric itself be covariantly constant, $\nabla \mathbf{g} = 0$ or $\nabla_\lambda g^{\mu\nu} = 0$,

21.6. General relativity

corresponding to constancy of the Minkowskian components $\partial_a \eta^{bc} = 0$ for a Lorentz basis in flat spacetime. The connection coefficients $\Gamma^\mu{}_{\nu\lambda}$, and hence the gravitational potentials $g_{\mu\nu}$, link the special relativistic physics in a basis e_a at one event to similar laws at another event via the global frame x^μ.

As in the U(1) case, one may form the commutator of two covariant derivatives acting on a general vector field V^μ to form the *Ricci identity*,

$$[\nabla_\mu, \nabla_\nu] V^\lambda = R^\lambda{}_{\rho\mu\nu} V^\rho , \qquad (21.44)$$

in which the rank 4 *Riemann tensor* $R^\lambda{}_{\rho\mu\nu}$ can be shown [384] to geometrically characterize, independently of the coordinates, the curvature of the spacetime described by $g_{\mu\nu}$. The components of the Riemann tensor are given in terms of the connection and thus $g_{\mu\nu}$ by a relation [384],

$$R^\rho{}_{\lambda\mu\nu} = \partial_\mu \Gamma^\rho{}_{\lambda\nu} - \partial_\nu \Gamma^\rho{}_{\lambda\mu} + \Gamma^\rho{}_{\kappa\mu}\Gamma^\kappa{}_{\lambda\nu} - \Gamma^\rho{}_{\kappa\nu}\Gamma^\kappa{}_{\lambda\mu} , \qquad (21.45)$$

corresponding to $F_{\mu\nu} = \partial_\mu A_\nu - \partial_\nu A_\mu$ for the electromagnetic interaction. The (symmetric) *Ricci tensor* $R_{\mu\nu}$ is obtained by contraction, $R_{\mu\nu} = R^\lambda{}_{\mu\lambda\nu}$ of the Riemann tensor and its trace, $R = R^\mu{}_\mu$, is the *Ricci scalar*.

An electromagnetically interacting charged scalar or Dirac field satisfies field equations obtainable from the non-interacting equations by replacing ∂_μ by the U(1) gauge-covariant derivatives D_μ with a similar result applying to the fields A^a_μ of Yang-Mills theory. Correspondingly, many special relativistic equations may be converted to a general relativistic equivalent, incorporating gravitational interaction, by replacing $\partial = \{\partial_\mu\}$ by the gravitational covariant derivative $\nabla = \{\nabla_\mu\}$. For example, conservation of energy-momentum of a matter system, which is expressed in flat spacetime by $\partial \cdot T = 0$ or $\partial_\mu T^{\mu\nu} = 0$, will in general relativity take the form $\nabla \cdot T = 0$ or $\nabla_\mu T^{\mu\nu} = 0$, which expands to $\partial_\mu T^{\mu\nu} + \Gamma^\mu{}_{\mu\lambda} T^{\lambda\nu} + \Gamma^\nu{}_{\mu\lambda} T^{\mu\lambda} = 0$. Since this equation introduces the gravitational potentials $g_{\mu\nu}(x)$ via the connection coefficients $\Gamma^\mu{}_{\nu\lambda}(x)$, the conservation of the matter is linked to the gravitational field and the matter is thus interacting gravitationally.

General tensors and manifolds, particularly those which are curved, form a part of the study of *differential geometry* [499], a topic of ever-increasing importance in physics with major applications not only in general relativity, but in theoretical mechanics, thermodynamics and gauge field theory to name only a few. The spacetimes of relativistic gravitation, whether massless spin 2 or general relativistic, where the manifold is endowed with a locally-Minkowskian metric, are examples of pseudo Riemannian curved geometries.

The matter source of gravitation is the symmetric divergence-free energy-momentum tensor $T^{\mu\nu}$. The remarkable property of differential geometry, central to the general theory of relativity, is the existence of a unique, symmetric, identically divergence-free, general tensor [384, 566] obtainable as a function of $g_{\mu\nu}$ and its first and second derivatives and which provides a measure of the amount of intrinsic curvature of the spacetime manifold. That tensor, denoted $G^{\mu\nu}$ and called the *Einstein tensor*, is given by,

$$G_{\mu\nu} = R_{\mu\nu} - \tfrac{1}{2}g_{\mu\nu}R , \qquad (21.46)$$

closely analogous to $\bar{h}_{\mu\nu} = h_{\mu\nu} - \tfrac{1}{2}\eta_{\mu\nu}h$ for the Fierz-Pauli field.

The Einstein tensor is quadratic in the first derivatives $\partial_\lambda g_{\mu\nu}$ and linear in the second derivatives $\partial_\rho \partial_\lambda g_{\mu\nu}$. It thus has dimensions of inverse length squared. The identically vanishing divergence of $G^{\mu\nu}$ is expressed as a generally covariant equation $\nabla \cdot G = 0$ or $\nabla_\mu G^{\mu\nu} = 0$, arising from a differential property of the Riemmann tensor, referred to as the *Bianchi identity*, namely,

$$\nabla_\mu R^\rho{}_{\kappa\nu\lambda} + \nabla_\nu R^\rho{}_{\kappa\lambda\mu} + \nabla_\lambda R^\rho{}_{\kappa\mu\nu} = 0 \quad \text{or} \quad R^\rho{}_{\kappa[\nu\lambda;\mu]} = 0 , \qquad (21.47)$$

where $a;\mu \equiv \nabla_\mu a$. It is directly analogous to the electromagnetic Bianchi identity of Equations 18.16 and 18.18, namely $\partial_{[\mu} F_{\nu\lambda]} = 0$ or $F_{[\nu\lambda,\mu]} = 0$ (where $a,\mu \equiv \partial_\mu a$).

In relativistic gravitation, the constants G and c are available and the dimensions of $GT^{\mu\nu}/c^4$ are the same as those of $G^{\mu\nu}$. The genius of Einstein, in collaboration with Grossmann [133] concerning the mathematics of Riemannian geometry, was to incorporate the universality of gravitation and the Equivalence principle by locking the dynamics of the field $g_{\mu\nu}$ to those of the matter in $T^{\mu\nu}$. This was done using the equations,

$$G^{\mu\nu} = \frac{8\pi G}{c^4} T^{\mu\nu} , \qquad (21.48)$$

now known as *Einstein's field equations* of relativistic gravitation. The factor of 8π is conventional. One thereby demands that energy-momentum and angular momentum conservation, $\nabla_\mu T^{\mu\nu} = 0$ and $T^{\mu\nu} = T^{(\mu\nu)}$, be equivalent to the geometric property, $\nabla_\mu G^{\mu\nu} = 0$, of the vanishing of the divergence of $G^{\mu\nu}$ and its symmetry. The electromagnetic source constraint, $\partial_\mu \partial_\nu F^{\mu\nu} = 0$, is independent of the Bianchi identity. By contrast, in gravitation, the source constraint is intimately connected to the geometrical gauge-related Bianchi identity.

Being linear in the second derivatives of the metric tensor, Einstein's field equations are *quasi-linear* second-order partial differential equations for the relativistic gravitational potentials $g_{\mu\nu}(x)$. They may be solved exactly for certain specific distributions of matter, in particular those which have a high degree of symmetry. One well-known example is the asymptotically flat, static, spherically symmetric Schwarzschild solution [384, 503, 566] for the geometry produced by a mass M, applicable to non-rotating stars and black holes. Another is the everywhere isotropic Robertson-Walker solution [384, 478, 557, 566] for applications in cosmology. Others

21.6. General relativity

are the Reissner-Nordstrøm [401, 471], Kerr [277] and Kerr-Newman [393] solutions for charged, rotating and charged rotating black holes, respectively. All of these geometries are discussed in detail in Misner et al. [384] and an elementary discussion of some of the black hole properties is available in references [113, 114].

The *Schwarzschild solution*,

$$ds^2 = -(1 - \frac{r_M}{r})c^2 dt^2 + (1 - \frac{r_M}{r})^{-1} dr^2 + r^2 dr^2 + r^2 \sin^2\theta \, d\phi^2 , \quad (21.49)$$

where $r_M = 2GM/c^2$ is the Schwarzschild length of the mass M can be applied to the solar system, or to a first approximation for the motion of one star in the field of another in a binary system, to provide elegant demonstrations of the agreement of Einstein's theory with a wide variety of non-Newtonian observational results. It does so without restriction to weak gravity.

The *Robertson-Walker geometry* provides a kinematical basis for the determination of Friedman dynamical models from isotropic cosmological distributions of matter and radiation acting as a source of gravity. They are unrestricted by the asymptotic flatness of a description, such as the Fierz-Pauli theory, based on special relativity.

A typical component of Einstein's equations relates the inverse square of a characteristic length $\ell \approx |G^{\mu\nu}|^{-1/2}$, the radius of curvature of spacetime, to the local density of matter, $\rho \approx |T^{\mu\nu}|/c^2$, according to the approximate relation $\ell \approx c/\sqrt{G\rho}$, as established by dimensional arguments in Exercise 1.3.

Since the covariant constancy of the metric implies it has vanishing divergence, $\nabla_\mu g^{\mu\nu} = 0$, an additional term comprised of a constant Λ times $g^{\mu\nu}$ may be added to Einstein's equations without affecting the automatic conservation of the source. Such a *cosmological term*, with resulting field equations,

$$G^{\mu\nu} + \frac{\Lambda g_{\mu\nu}}{c^2} = \frac{8\pi G}{c^4} T^{\mu\nu} , \quad (21.50)$$

was introduced by Einstein [140] in 1917 with $\Lambda > 0$ (repulsive) in order to obtain a solution corresponding to the apparent staticity of the universe on a large scale. The factor of c^2 is introduced in order that Λ have the same dimensions of inverse time squared as for the Newtonian cosmological constant (see Section 7.9) permitted by Galilean relativity. The subsequent observation, by Hubble [255], of extragalactic red shifts establishing the expansion of the universe, made such a term unnecessary. Nevertheless, it may be important for the unification of gravity with the Standard model of elementary particles and in the quantization of gravity.

One consequence of the linking of the matter content to the gravitational field via Einstein's field equations is that those equations, by a modification of Exercise 19.1 to general coordinates, also determine the form of the equation of motion of a non-quantum test particle. The result is the equation of a *geodesic* in the spacetime of metric $g_{\mu\nu}$, namely,

$$\frac{d}{d\lambda}(g_{\mu\lambda}\dot{z}^\lambda) = \tfrac{1}{2}(\partial_\mu g_{\lambda\pi})\dot{z}^\lambda \dot{z}^\pi , \tag{21.51}$$

where,

$$g_{\mu\nu}\dot{z}^\mu \dot{z}^\nu = 0 \quad \text{or} \quad -c^2 , \tag{21.52}$$

which for a massive (non-null) particle is also the path of extremal proper time. The condition on the 4-velocity $\dot{z} = dz^\mu/d\tau$ of the non-null particle arises from Equation 21.38 for the line element along a timelike path and the definition of proper time in terms of ds^2. Equation 21.43 permits Equation 21.51 to be rewritten in the form,

$$\frac{d^2 z^\lambda}{d\lambda^2} + \Gamma^\lambda{}_{\mu\nu}\dot{z}^\mu \dot{z}^\nu = 0 , \tag{21.53}$$

often used for the geodesic equation in a curved spacetime. Equation 21.51 reveals the close relationship with the equations of the Fierz-Pauli theory arising from special relativistic considerations and massless spin 2 fields. However, there are vital differences, among them the applicability of the present equations to strong gravitational effects, high matter densities and spacetime topologies which are not asymptotically flat.

If one assumes a localized spherically symmetric matter distribution, appropriate to a star, for example, one may introduce coordinates which at large spatial distances $r\to\infty$ are Minkowskian so that the metric tensor in those coordinates satisfies $g_{\mu\nu}\to\eta_{\mu\nu}$. In such an *asymtotically flat spacetime*, one may then form the tensor field,

$$h_{\mu\nu} = g_{\mu\nu} - \eta_{\mu\nu} , \tag{21.54}$$

and re-express Einstein's field equations in terms of $h_{\mu\nu}$. One may consider the case for which the gravitational field is weak, $|h_{\mu\nu}| \ll 1$, which corresponds to considering matter densities ρ satisfying $\rho \ll c^2/G\ell^2$ where ℓ is a characteristic length of the system, such as the radius of a planetary orbit in the perihelion calculation or the solar radius in light deflection by the sun. Under such circumstances, terms which are second-order in the small quantities $h_{\mu\nu}$ in the non-linear expression for $G^{\mu\nu}$ in Einstein's equations may be ignored in first-order approximation.

The resulting *linearized gravity* equations [384] are,

$$-\Box \bar{h}_{\mu\nu} + 2\partial^\lambda \partial_{(\mu} \bar{h}_{\nu)\lambda} - \eta_{\mu\nu}\partial^\lambda \partial^\rho \bar{h}_{\lambda\rho} = \frac{16\pi G}{c^4} T_{\mu\nu} , \tag{21.55}$$

identical to the coupling of a massless spin 2 field in an arbitrary gauge, as described by Equation 21.34, to a conserved source $T^{\mu\nu}$. A corresponding

21.6. General relativity

equation [117, 118, 384] applies to the potential $h_{\mu\nu}$. Selection of the de Donder gauge reproduces Equations 21.9 and 21.13 used earlier to discuss tensor gravity in a special relativistic context. If second-order terms in $h_{\mu\nu}$ are retained, the results are identical, in appropriate coordinates, to the non-linear Fierz-Pauli theory obtained by extending the linear form by including the energy-momentum density of the field $h_{\mu\nu}$ in the source, $T_{\mu\nu}$.

General relativity fits perfectly into a Lagrangian formulation. If one sets out to find a Lagrangian density in the field $g_{\mu\nu}$ and its derivatives $\partial_\lambda g_{\mu\nu}$, then restriction to no higher than second order in the derivatives leads to a unique invariant measure of the curvature, the *Ricci scalar*.

The 4-volume element in curved spacetime is $\sqrt{-g}\, d^4x$, where $g = \det g_{\mu\nu}$, and we may form a natural second-order action for the field $g_{\mu\nu}$ according to,

$$S[g_{\mu\nu}(x)] = \int d^4x \sqrt{-g}\, \mathcal{L} = \frac{c^3}{16\pi G}\int d^4x \sqrt{-g}(R - \frac{2\Lambda}{c^2})\ . \qquad (21.56)$$

With $\Lambda = 0$, $\mathcal{L}$ and S are the *Einstein-Hilbert Lagrangian density* and *action*. The Euler-Lagrange equations corresponding to Equation 21.56 are Einstein's equations with a cosmological term. In the static, weak field limit, these correspond to the Galilean relativistic Equation 7.20.

The extraction of the Hamiltonian formulation of general relativity is a non-trivial process [384] which, as for the Maxwell and Proca fields (see Exercises 18.20, 18.21, 18.22 and 19.12), proceeds by using a first-order formulation of general relativity in which all 50 components $\{g_{\mu\nu}, \Gamma^\mu{}_{\nu\lambda}\}$ of the metric and connection are independent Lagrangian variables. Once satisfactorily constructed, the effective canonical Hamiltonian of general relativity does not lead to a renormalizable theory of *quantum gravity*, for which there does not yet exist a satisfactory theory with experimental confirmation.

The development of general relativity, like the special theory, was not based directly on compelling experimental results unable to be described by existing theories. Much of the testing of the theory has had to await a half century for technological developments capable of measuring the effects with sufficient precision. The genius of Einstein was his ability to build on well-established theory by making generalizations of apparently minor character but which had far-reaching major consequences for our description and understanding of nature.

All physicists base their models of natural processes on the analysis of experimental and observational results. The theories constructed by Newton on interaction and by Einstein on the nature of spacetime are unequalled in the extent to which they generalize from the known observations of their times to provide a basis for predicting new phenomena capable of being tested experimentally.

21.7 Unification and supersymmetry

Matter and energy may be described by fields from which the observed particles arise using the second quantization techniques of quantum field theory.

The fields which conform locally to the Relativity principle may be divided into two theoretical types, tensor fields on the one hand and tensor-spinor fields on the other. Only a few of the simplest of these fields are needed to describe the fundamental particles and interactions.

The observed fields, and the quanta they give rise to, can be divided into two groups in another way. The first set are those Lagrangian potentials which mediate one of the four interactions at the fundamental level, namely the electromagnetic vector potential A_μ, the Yang-Mills fields A_μ^a ($a = 1, 2, 3$ or $1, 2, \ldots, 8$) of the nuclear interactions and the vierbein gauge fields e_μ^a ($a = 1, 2, 3, 4$) of relativistic gravity. Such tensor mediation fields satisfy local Poincaré covariance as a result of being, prior to quantization, null gauge fields of helicity 1 (if non-gravitational) and 2 (for gravity).

Quantization of such fields leads to particles, the *gauge bosons* whose intrinsic angular momentum is 1 or 2 units of the quantum of action, $\hbar$. The Yang-Mills quanta of the weak-nuclear interaction, the intermediate vector bosons or *weakons*, $W^\pm$ and Z^0, acquire mass as a result of spontaneous symmetry breaking. The corresponding Yang-Mills quanta of the strong nuclear interaction, the gluons, cannot exist free due to another quantum field theoretic phenomenon known as *confinement* which also applies to the source particles, the quarks, which interact strongly.

The exchange of such quanta, namely photons, $W^\pm$ and Z^0, gluons and gravitons (the latter as yet unobserved) between matter, provides the quantum description of the four interactions. The integer spin of the exchange quanta means that each is described by a vector or tensor field. To satisfy the requirement of energy positivity, they are quantized using commutators of quantum operators and this in turn means they satisfy Bose-Einstein statistics.

The massless and unconfined gauge bosons — photons and gravitons — give rise to long-range interactions for which a meaningful classical limit exists consisting of large numbers of coherent quantum states with well-defined macroscopic properties. The long range of the electromagnetic interaction of individual charges is nullified by the macroscopic charge neutrality of matter in bulk leaving gravitation dominant on laboratory, terrestrial, astronomical and cosmological scales.

The second group are those fields and quanta which do not mediate any interaction at the fundamental level but act solely as sources which interact via one or more of the four interactions. (The mediating fields and particles themselves can also act as sources to some of the interactions. All act as sources of gravity.) All these non-gauge or matter fields may, at the

21.7. Unification and supersymmetry

fundamental level, be described in terms of spin $\frac{1}{2}$ fields, massive or massless Dirac fields $\psi = \{\psi^\alpha\}$ and their quanta, the six leptons (electrons, muons, tauons and their corresponding neutrinos) and the six quarks. The bulk of the matter of the universe consists of triplets of spin $\frac{1}{2}$ u and d quarks bound into spin $\frac{1}{2}$ nucleons (protons and neutrons) which in turn bind together as atomic nuclei and with electrons to form atoms and molecules.

Quantization of Dirac fields leads to particles with an intrinsic spin projection of $\frac{1}{2}\hbar$ in any specified direction. In order that the matter have stable ground states, positivity of energy must be ensured, and this occurs as a result of quantization with anticommutators of quantum operators. As a consequence, the particles obey Fermi-Dirac statistics which require the particles to satisfy the Pauli exclusion principle limiting the number of quanta in each state to a maximum of one. Thus they do not permit the formation of coherent macroscopic states with large numbers of quanta. No classical limit exists and the matter is therefore intrinsically quantum.

The common spin 1 character of the non-gravitational interaction fields, and their masslessness (prior to spontaneous symmetry breaking) give rise to their gauge properties (electromagnetic and Yang-Mills). These are central to the unification of the electromagnetic and weak-nuclear forces in the Glashow-Salam-Weinberg *electroweak theory* [199, 492, 493, 565]. They are also crucial to schemes for their grand unification [468] with the strong-nuclear interaction via compact global Lie group symmetries large enough to incorporate all the exchange quanta of gauge particles. The massless spin 2, and therefore bosonic, character of the gravitational field also gives rise to gauge properties, although somewhat different to those of the Yang-Mills interactions, and standard techniques of quantization are not successful. These differences are related to the non-compactness of the Poincaré gauge group of gravity and to the fact that the gauge transformations are both internal and external.

On the other hand, the common spin $\frac{1}{2}$ nature of all matter fields means that none of them has local gauge freedom. The matter fields are sufficiently similar that many exact or approximate symmetries exist between them which may be used to construct the large numbers of observed particles in terms of composites and exited states of collections of leptons and quarks. As in the unification of the interaction quanta, enlargement of the symmetries permits the incorporation of greater fractions of the observed particles.

As recently as 1973, the very different bosonic and fermionic character of the mediation and matter quanta seemed to leave little prospect for the unification of the matter and the interaction fields. A number of so-called *no-go theorems* even indicated that no fundamental particle symmetries could incorporate bosons and fermions together as part of one irreducible collection of entities. However, 'no-go' theorems must always be based on certain premises which are invariably vulnerable to new discoveries, whether mathematical or observational.

From 1970 to 1974, the discovery of *supersymmetry*, a kinematic invariance very similar to Poincaré symmetry, but partly external and partly internal, opened up the means to incorporate fermions and bosons in the one interacting theory. Both types of particle appear in a highly symmetrical fashion, each ordinary boson of spin j being accompanied by a fermionic superpartner of spin $j \pm \frac{1}{2}$ and vice versa.

Since such superpartners are not observed at current accelerator energies, a realistic theory must at least partially break the supersymmetry. Spontaneous breaking of the symmetry would leave the partners with rest mass, presumably sufficiently high to be beyond the scope of current accelerator technology.

We would expect the simplest supersymmetric combination of fields to be one comprised of spin 0 and spin $\frac{1}{2}$ parts. A necessary but not sufficient condition for a set of bosonic and fermionic fields to have supersymmetry is equality of the number of independent physical components of each type. This seems to be impossible since the scalar field has one or two components in the real (neutral) and complex (charged) cases compared to four or eight in a real or complex spin $\frac{1}{2}$ Dirac field. The Klein-Gordon scalar field equation determines the spacetime functional dependence of the scalars which satisfy it, constraining them in the free case to be plane waves, but does not affect the number of independent components. Thus a single real scalar field satisfying the Klein-Gordon equation has one physical degree of freedom.

However, the free Dirac equation not only constrains its solutions to be plane waves but relates the four components of a Dirac spinor in such a way that the number which are dynamically independent is half the number of algebraic components. Thus a real Dirac spinor (namely a Majorana spinor) will have two independent physical degrees of freedom.

Even if we choose a real Dirac field, we still seem to have twice as many physical degrees of freedom as a scalar field. However, we must remember that the Dirac and Majorana fields are non-chiral, meaning that the action of the parity operation on an arbitrary Dirac field is meaningful and leaves us with another related part of the same Dirac field. In seeking to match a real Dirac spinor with scalars, we must consider the properties under parity.

Use of the parity operation on scalar fields divides them into two independent types, proper scalars and pseudoscalars. Scalar quantities corresponding to a real Dirac spinor may be the pair comprising a proper scalar and a pseudoscalar. We see therefore that we can match the number of physical degrees of freedom of bosonic and fermionic fields in a natural way and the simplest candidate for a supersymmetric combination is a pair of real spin 0 scalar fields, one proper and the other pseudo, with one real spin $\frac{1}{2}$ Dirac field.

21.7. Unification and supersymmetry

Continuous symmetries lead to conserved Noether quantities and, with localization of their parameters, to gauge theories. We may therefore ask whether a symmetry like rotation can be set up to transform a spinor field into a pair of scalar fields and vice versa. Again, the very different nature of the two types of field seems to rule out this possibility. But this is precisely what was done for the first time in relativistic field theory in 1974 by Wess and Zumino [570].

A small rotation of a 3-vector field $\mathbf{V}$ is defined in terms of four real parameters θ or six real parameters $\omega^{\mu\nu} = -\omega^{\nu\mu}$ in the case of covariant rotations of a 4-vector field V^μ. Similarly, a translation involves four real parameters a^μ. Such parameters, being ordinary real numbers, commute with one another.

The components of the non-quantized bosonic fields also have values at each event which consist of real commuting numbers. By contrast, the non-quantum limit $\hbar \to 0$ of a spinorial quantum operator must consist of components which anticommute, $\psi^\alpha \psi^\beta = -\psi^\beta \psi^\alpha$. Such quantities are referred to as *Grassmann spinors*.

Exercise 21.11 *Show that the product $\psi^\alpha \psi^\beta$ of two anticommuting numbers, ψ^α and ψ^β, commutes with other such pairs, and with single anticommuting numbers.*

The exercise shows that by choosing the parameters ϵ^α in a supersymmetry rotation to be anticommuting spinorial components themselves, one can convert commuting bosonic field components $\phi^{\mu\cdots}$ into anticommuting spinorial fields $\epsilon^\alpha \phi^{\mu\cdots}$ and vice versa.

What Wess and Zumino showed in 1974 was that this can be done in 4D spacetime in such a way that the set of all transformations is a group of operations consistent with relativistic field theory, the *super-Poincaré group*. The Dirac spinor and pair of scalars related by such a supersymmetry transformation is now known as a *Wess and Zumino scalar supermultiplet*. Very soon after the discovery by Wess and Zumino, methods were found to combine almost any of the tensor and tensor-spinor fields of Poincaré-covariant field theory into supersymmetric groupings.

One of these comprises an interacting field theory mediated not just by a spin 2 massless particle, as in general relativity, but by a supersymmetric pair comprised of a spin 2 graviton and a massless spinorial Rarita-Schwinger partner of spin $\frac{3}{2}$ called the *gravitino*, or the 'little graviton'. This theory is now known as *supergravity* [97, 98, 99, 100, 174]. Supergravity played a central role in many unification attempts in the first few years after its discovery but was soon shown to be incapable alone of describing the full spectrum of matter particles of the Standard model and their interactions.

21.8 Relativistic superstrings

> What really interests me is whether God had any choice in the creation of the world.
>
> Albert Einstein   1879–1955

Although we have a strong intuitive notion of the nature of a particle, based largely on macroscopic non-quantum particles, the quantum particles of field theory are not concrete objects which conform to such intuitive ideas. They are, like all of the concepts of theoretical physics, a mathematical model which helps us to bring order into our picture of nature and predict new observations. Many of the outstanding problems of theoretical physics, such as the presence of infinities in the quantum field theory of particles, are now believed to be closely related to the description of phenomena in terms of particles propagating in a continuous spacetime background. A quantum *relativistic string* is an ingredient of an alternative more fundamental theory for describing the quantum physics of nature, especially at very high energies, from which particle properties may be deduced.

The starting point of a relativistic string theory may be sketched by analogy with the motion of a relativistic particle given in Chapter 15. The coordinates $z^\mu(\tau,\sigma)$ of a string depend on two parameters τ and σ. The first is a timelike evolution parameter while the second, spacelike, labels the elements along the length of the string, conventionally given the range $0 \leq \sigma \leq \pi$. The geometric Lagrangian, Equation 15.33, of the relativistic particle gives a parameterization independent action obtained from the invariant length of the timelike path of the particle. If we let $\dot{z}^\mu = \partial z^\mu(\tau,\sigma)/\partial \tau$ and $z'^\mu = \partial z^\mu(\tau,\sigma)/\partial \sigma$, then the *Nambu and Goto action* of the free relativistic string,

$$S = S[z(\tau,\sigma)] = -\frac{1}{2\pi\alpha c}\int_{\tau_i}^{\tau_f} d\tau \int_0^\pi d\sigma \sqrt{(\dot{z}\cdot z')^2 - \dot{z}^2 z'^2}, \qquad (21.57)$$

is correspondingly proportional to the area (which is invariant with respect to Poincaré transformations and reparameterization) of the 2D surface swept out by the string in spacetime during the motion. The constant $T = 1/2\pi\alpha$, required to ensure that the action has the correct dimensions, is interpretable as the tension of the string.

Simple relativistic strings in 4D contain inconsistencies when quantized. Some of these can be eliminated by starting with strings in a higher dimensional spacetime and subsequently compactifying the extra (spatial) dimensions to make predictions about phenomenology in 4D. Without supersymmetry, the most promising string theories in higher dimensions contain acausal propagation corresponding to tachionic particles, thus violating the Principle of causality. The demonstration of anomaly cancellations in quantized supersymmetric 10D gauge theories and *superstring* theory in 1984 by Schwartz and Green [210] raised the possibility of unification via superstrings. Gravity was included in the *heterotic string* in 1985 by Gross et al. [214]. Later that year Candelas et al. [57] provided a plausible scenario for obtaining, from strings, the grand unified theories appropriate to the phenomenology of the Standard model.

The theory of strings is currently being developed. Their possible relevance to the structure and unification of all the fundamental interactions and sources, including the quantization of gravity, has only been discovered since the early years of the present decade. The research that has already been carried out on strings is enormous and has created a great deal of excitement. Despite the lack, to date, of any experimental confirmation and the extreme difficulty at the present time of suggesting experiments to test its predictions, since quantization and unification of gravity necessarily involves energies as high as the Planck energy, 10^{19} GeV, string theory has many very attractive features. The most important are the unification of gravity with the Standard model and the improvements in finiteness, at least at low order, on quantization. It is being pursued intensively by particle physicists in the hope that it may make predictions that can be checked indirectly without the involvement of energies apparently beyond the scope of high-energy acceleration both now and in the foreseeable future.

Give me to learn each secret cause;
Let number's, figure's motion's laws
Revealed before me stand;
These to great Nature's scene apply,
And round the Globe, through the sky,
Disclose her working hand.

Hymn to Science
in Johnson S 1779, vol 55 *Works of the English Poets* (London)

Mark Akenside   1721–1770

Bibliography:

Eguchi T et al., 1980 *Gravitation, gauge theories and differential geometry* [125].
Felsager B 1981 *Geometry, particles and fields* [152].
Green M B, Schwarz J H and Witten E 1987 *Superstring theory* [211].
Griffiths D 1987 *Introduction to elementary particles* [212].
Hawking S W and Israel W 1987 *Three hundred years of gravitation* [228].
Kaku M 1988 *Introduction to superstrings* [271].
Misner C W, Thorne K S and Wheeler J A 1973 *Gravitation* [384].
Ohanian H C 1976 *Gravitation and spacetime* [404].
Parker B 1987 *Search for a supertheory: from atoms to superstrings* [422].

Perkins D H 1987 *Introduction to high energy physics* [435].
Schwartz J 1987 *Superstrings* [502].
Schutz B F 1985 *A first course in general relativity* [500].
Srivistava P P 1986 *Supersymmetry, superfields, ... an introduction* [518].
Wald R M 1984 *General relativity* [556].
Weinberg S 1972 *Gravitation and cosmology* [566].
Wess J and Bagger J 1983 *Supersymmetry and supergravity* [569].
Will C M 1986 *Was Einstein right? — Putting general relativity to the test* [587].

Appendices

A.1 Physical constants

The values of the following constants, in SI units, have been extracted from Cohen and Taylor [74] and Allen [8], which contain details on the uncertainties in the data. The values of c, μ_0, $\epsilon_0 = 1/\mu_0 c^2$ and AU are adopted and therefore exact.

Speed of light	c	$2.997\,924\,580$	$\times 10^8$	$\mathrm{m\,s^{-1}}$
Gravitational constant	G	$6.672\,59$	$\times 10^{-11}$	$\mathrm{N\,m^2\,kg^{-2}}$
Planck constant	h	$6.626\,075\,5$	$\times 10^{-34}$	$\mathrm{J\,s}$
Rationalized Planck constant ($h/2\pi$)	$\hbar$	$1.054\,572\,66$	$\times 10^{-34}$	$\mathrm{J\,s}$
Elementary charge	e	$1.602\,177\,33$	$\times 10^{-19}$	C
Vacuum permeability	μ_0	4π	$\times 10^{-7}$	$\mathrm{N\,A^{-2}}$
Vacuum permittivity	ϵ_0	$8.854\,187\,8\ldots$	$\times 10^{-12}$	$\mathrm{F\,m^{-1}}$
Electron mass	m_e	$9.109\,389\,7$	$\times 10^{-31}$	kg
Proton mass	m_p	$1.672\,623\,1$	$\times 10^{-27}$	kg
Fine structure constant	α	$7.297\,353\,08$	$\times 10^{-3}$	
Boltzmann constant	k	$1.380\,658$	$\times 10^{-23}$	$\mathrm{J\,K^{-1}}$
Electron volt	eV	$1.602\,177\,33$	$\times 10^{-19}$	J
Astronomical unit	AU	1.496	$\times 10^{11}$	m
Light year	ℓy	$9.460\,530$	$\times 10^{15}$	m
Solar mass	$M_\odot$	$1.989\,1$	$\times 10^{30}$	kg
Solar mean radius	$R_\odot$	$6.959\,9$	$\times 10^{8}$	m
Earth mass	$M_\oplus$	5.976	$\times 10^{24}$	kg
Earth mean radius	$R_\oplus$	6.387	$\times 10^{6}$	m
Earth mean orbital radius	$a_\oplus$	$1.495\,979$	$\times 10^{11}$	m
Planck mass, $(\hbar c/G)^{1/2}$	m_P	$2.176\,71$	$\times 10^{-8}$	kg
Planck length, $(G\hbar/c^3)^{1/2}$	ℓ_P	$1.616\,05$	$\times 10^{-35}$	m
Planck time, $(G\hbar/c^5)^{1/2}$	t_P	$5.390\,56$	$\times 10^{-44}$	s
Planck energy, $(\hbar c^5/G)^{1/2}$	E_P	$1.956\,33$	$\times 10^{9}$	J
Planck temperature, $(\hbar c^5/Gk^2)^{1/2}$	T_P	$1.416\,95$	$\times 10^{32}$	K

A.2 Counting components

If there are no symmetries relating the indices, the number of algebraically independent components in an indexed quantity of rank r (number of indices) in dimension N (range of each index) is N^r.

If such an object is completely symmetric in s indices ($2 \leq s \leq r$), the number of independent components will be,

$$N^{r-s}(N+s-1)!/(N-1)!s!\,,$$

being the number of inequivalent ways of choosing s indices from N (including repetitions) times the number N^{r-s} ways of choosing the remaining $r - s$ indices. The number of constraints,

$$N(N+1)\ldots(N+r-1)/r!\,,$$

owing to the vanishing of the completely symmetric part on s indices is given by putting $r = s$ in that result.

The corresponding result if the object is completely antisymmetric on a indices ($2 \leq a \leq N$) is,

$$N^{r-a}N!/(N-a)!a!\,,$$

being the number of ways of choosing a indices from N with no repetitions (since a repeated index from these a would give a zero value of the object by antisymmetry) times N^{r-a} for the remaining indices. Putting $a = r$ the number of constraints due to the vanishing of the completely symmetric part on a indices is:

$$N(N-1)\ldots(N-r+1)/r!\,.$$

These expressions reduce to N^r when $s = 0, 1$ or $a = 0, 1$, as is appropriate for no symmetries. The second result gives 1 for $a = N$ while for $a > N$ all components must be zero as there are no non-trivial indexed objects in dimension N that are completely antisymmetric on $a > N$ indices. In that case at least two indices must coincide causing the object to vanish.

Exercise A.1 Let $\{k, \ell, \ldots\}$ be spatial indices with range $\{1,2,3\}$. Let $\{\mu, \nu, \ldots\}$ be spacetime indices with range $\{0,1,2,3\}$. Let $\{A, B, \ldots\}$ be two-component spinor indices ($A, B, \ldots = 1, 2$). Determine the number of algebraically independent components in the following indexed quantities:
(a) $T_{k\ell} = T_{(k\ell)}$ (b) $J_{k\ell} = J_{[k\ell]}$ (c) $\epsilon_{k\ell m} = \epsilon_{[k\ell m]}$ (d) $T_{\mu\nu} = T_{(\mu\nu)}$
(e) $F_{\mu\nu} = F_{[\mu\nu]}$ (f) $\epsilon_{\mu\nu\lambda\rho} = \epsilon_{[\mu\nu\lambda\rho]}$ (g) $\psi_{AB} = \psi_{(AB)}$ (h) $\epsilon_{AB} = \epsilon_{[AB]}$
(i) $\psi_{A_1 A_2 \ldots A_{2j}} = \psi_{(A_1 A_2 \ldots A_{2j})}$ where j is integral or half-odd-integral.

A.3 Helmholtz decomposition of vector fields

We wish to decompose a 3-vector field $\mathbf{V}(\mathbf{x})$ into the sum $\mathbf{V} = \mathbf{V}^L + \mathbf{V}^T$ of two orthogonal parts, one of which is *longitudinal* satisfying $\nabla \times \mathbf{V}^L = 0$ and the other *transverse* satisfying $\nabla \cdot \mathbf{V}^T = 0$. The projection of $\mathbf{V}$ into these two parts may be achieved using the vector identity,

$$\nabla^2 \mathbf{V} = \nabla(\nabla \cdot \mathbf{V}) - \nabla \times (\nabla \times \mathbf{V}) , \tag{A.1}$$

in which the first part is clearly longitudinal and the second part transverse. Having noted that the second term on the right side is transverse we have no need of its explicit form since given one of the two projectors, for example the one taking $\mathbf{V}$ to $\mathbf{V}^L$ we can obtain the second, taking $\mathbf{V}$ to $\mathbf{V}^T$, by subtracting the first projector from the identity. We may therefore write the decomposition trivially as $\nabla^2 \mathbf{V} = \nabla(\nabla \cdot \mathbf{V}) + (\nabla^2 \mathbf{V} - \nabla(\nabla \cdot \mathbf{V}))$.

We now introduce the inverse ∇^{-2}, of the *Laplacian operator* ∇^2, with the convention that $\phi(\mathbf{x}) = \nabla^{-2} \rho(\mathbf{x})$ means the solution of Poisson's equation, $\nabla^2 \phi = \rho$. This equation always has a solution. Furthermore, for fields which vanish at spatial infinity, the solution is unique since Green's theorem [421] shows that two solutions satisfying the same boundary conditions on a closed surface differ by at most a constant. The inverse is thus well-defined and the Laplacian is referred to as an *invertible* differential operator. For example, since $\nabla^2 \phi = \delta^3(\mathbf{x})$ has the solution $\phi = -1/4\pi|\mathbf{x}|$ we have $\nabla^{-2} \delta^3(\mathbf{x}) = -1/4\pi|\mathbf{x}|$ and hence, more generally,

$$\phi(\mathbf{x}) = \nabla^{-2} \rho(\mathbf{x}) = -\frac{1}{4\pi} \int d^3 x' \frac{\rho(\mathbf{x}')}{|\mathbf{x} - \mathbf{x}'|} . \tag{A.2}$$

The *inverse Laplacian* is a *non-local operator* since it is not expressible as a linear differential operator of finite order. On the contrary, it involves the values of its operand $\rho(\mathbf{x}')/|\mathbf{x} - \mathbf{x}'|$ at locations $\mathbf{x}'$ other than the point $\mathbf{x}$ in question.

Applied to our decomposition equation, we obtain:

$$V_i = \frac{\partial_i \partial_j}{\nabla^2} V_j + \left(\delta_{ij} - \frac{\partial_i \partial_j}{\nabla^2}\right) V_j . \tag{A.3}$$

The indexed notation facilitates application of this result to not just a vector field but also to tensors. The longitudinal and transverse *projection operators* are thus given by:

$$P^L = \{P^L_{ij}\} = \left\{\frac{\partial_i \partial_j}{\nabla^2}\right\} . \tag{A.4}$$

and

$$P^T = \{P^T\} = \{\delta_{ij} - P^L_{ij}\} = \left\{\delta_{ij} - \frac{\partial_i \partial_j}{\nabla^2}\right\} . \tag{A.5}$$

which each operate on arbitrary vectors **V** by contraction, $(PV)_i = P_{ij}V_j$, of one or other index with **V**. It is straightforward matter to verify that they are indeed projection operators; namely that they sum to the identity, are mutually orthogonal ($P^T_{ij}P^L_{jk} = 0$) and *idempotent* ($P^2 = P$).

The orthogonality of P^T and P^L means that the integral over all space of the product of transverse and longitudinal components of 3-vector fields is zero,

$$\int d^3x \; \mathbf{U}^T \cdot \mathbf{V}^L = 0 \; . \tag{A.6}$$

In the special case of a Fourier analysis of the 3-vector field into plane-wave parts:

$$V_j(t,\mathbf{x}) = \int d^3k \; V_j(t,\mathbf{k}) e^{i\mathbf{k}\cdot\mathbf{x}} \tag{A.7}$$

where **k** is the propagation 3-vector, the longitudinal projection operator (which projects parallel to $\hat{\mathbf{k}}$) becomes:

$$P^L = \{P^L_{ij}\} = \left\{\frac{k_i k_j}{k^2}\right\} \tag{A.8}$$

while the transverse projector (projecting into the plane orthogonal to **k**) is:

$$P^T = \{P^T_{ij}\} = \{\delta_{ij} - P^L_{ij}\} = \left\{\delta_{ij} - \frac{k_i k_j}{k^2}\right\} \tag{A.9}$$

A.4 Poincaré Lemma

Consider the well-known result which states that a 3-vector field $\mathbf{V}(\mathbf{x})$ which is curl-free throughout a simply-connected region of $\mathbf{E}^3$ may be re-expressed as the gradient, $\mathbf{V} = \boldsymbol{\nabla}\phi$, of a scalar field $\phi = \phi(\mathbf{x})$.

Since the analogue of the curl in dimensions other than 3 is not a vector (Section 4.7), we recast this result in indexed form. For a 3-vector $\mathbf{V} = \{V^k\}$ satisfying the curl-free condition, $\partial_{[k}V_{\ell]} = 0$, there always exists a single scalar field ϕ for which $V_k = \partial_k \phi$ throughout the simply-connected curl-free region. We say that the 3-vector field is *integrable* and that the curl-free condition is an *integrability condition* for **V**. In fact, the proof of this proposition is based on displaying the above scalar field as a path integral of the vector field from an arbitrary point **x** to the point under consideration and showing that the result is path-independent.

Nothing in the proof of the existence of the scalar field is dependent on the dimension or signature of the space concerned. The key property is that the region over which this result is valid must be simply-connected. Consequently, this result may also be used to conclude that given any 4-vector field $V_\mu(x)$ satisfying $\partial_{[\mu}V_{\nu]} = 0$ throughout a simply-connected region of $\mathbf{E}^{1,3}$, then there exists a single scalar ϕ satisfying $V_\mu = \partial_\mu \phi$ throughout that region.

A.4. Poincaré Lemma

The Poincaré lemma [51, 384] on which the above results are based is not in fact limited to the integration of a vector but may be applied to any completely antisymmetric and curl-free tensor of rank less than the dimension of the space concerned (where for this purpose we regard a vector as a rank 1 antisymmetric tensor). Applied to a second-rank antisymmetric tensor $F_{\mu\nu}=F_{[\mu\nu]}$ satisfying $\partial_{[\lambda}F_{\mu\nu]} = 0$, it states that there exists a 4-vector A_μ for which $F_{\mu\nu} = 2\partial_{[\mu}A_{\nu]} \equiv \partial_\mu A_\nu - \partial_\nu A_\mu$. The curl-free condition $\partial_{[\lambda}F_{\mu\nu]} = 0$ or $\partial_\mu \tilde{F}^{\mu\nu} = 0$ thus guarantees the integrability of $F_{\mu\nu}$. To demonstrate the pattern we describe the next higher result. Given a completely antisymmetric rank 3 tensor $F_{\lambda\mu\nu} = F_{[\lambda\mu\nu]}$ satisfying the curl-free condition, $\partial_{[\pi}F_{\lambda\mu\nu]} = 0$, then there exists an antisymmetric tensor $B_{\mu\nu}$ in terms of which F may be expressed as a covariant curl: $F_{\lambda\mu\nu} = 3!\partial_{[\lambda}B_{\mu\nu]}$.

It should be clear that completely antisymmetric tensors and the operation of taking the curl are of fundamental importance for integrability. In coordinate-free form, these two concepts are powerful differential-geometric tools [164, 384] known, respectively, as a *differential form* of rank $0 \leq p \leq N$, where N is the dimension of the space, and the *exterior derivative* of a form.

References

[1] Adair R K 1988 *A flaw in a universal mirror* Sci. Amer. **258** (Feb.) 30–36. [§ 3.5]

[2] Adams W S 1925 *The relativity displacement of the spectral lines in the companion of Sirius* Proc. Nat. Acad. Sci. **11** 382–387. [§ 7.14]

[3] Adler R, Bazin M and Schiffer M 1975 *Introduction to general relativity* 2nd edn. (McGraw-Hill Kogakusha, Tokyo). [§ 17.6]

[4] Aharoni J 1965 *The special theory of relativity.* 2nd edn. (University Press, Oxford). [§ 5.3]

[5] Aharoni J 1972 *Lectures in mechanics* (University Press, Oxford).

[6] Aharonov Y and Bohm D 1959 *Significance of electromagnetic potentials in the quantum theory.* Phys. Rev. **115** 485–491; 1961 *Further considerations on electromagnetic potentials in the quantum theory.* Phys. Rev. **123** 1511–1524. [§§ 2.4, 20.9]

[7] Aitchison I J R and Hey A J G 1989 *Gauge theories in particle physics.* 2nd edn. (Adam Hilger, Bristol). [§§ 16.3, 16.4, 20.9, 20.13]

[8] Allen C W 1973 *Astrophysical quantities* 3rd edn. (Athlone Press, London). [§§ 6.7, A.1]

[9] Amaldi E 1984 *From the discovery of the neutron to the discovery of nuclear fission* Physics Reports **111** 1–332. [§§ 16, 20.13]

[10] Ampère A-M 1820 *De l'action exercée sur un courant électrique par un autre courant, le globe terrestre ou un aimant* Ann. Chim. Phys. **15** 59–76, 177–208; reprinted pp. 1–74 in Ampère A-M 1921 *Mémoires sur l'électromagnétisme et l'électrodynamique* (Gauthiers-Villars, Paris); part English translation Blunn O M 1965 *The mutual action of two electric currents* and *The interaction between an electrical conductor and a magnet* pp. 140–154 Tricker [547]. [§ 18.1]

[11] Ampère A-M 1822 *Sur la détermination de la formule qui représente l'action mutuelle de deux portions infiniment petites de conducteurs voltaïques (On the determination of the formula giving the interaction of two infinitesimally small portions of voltaic conductors)* L'Acad. Roy. Sci. pp. 75–110. [§ 18.1]

[12] Ampère A-M 1825 (issued with revision in 1827) *Sur la théorie mathématique des phénomènes électrodynamiques, uniquement déduite de*

l'expérience (*On the mathematical theory of electrodynamic phenomena, deduced solely from experiment*) Mém. de l'Acad. des Sci. de Paris **6** 175; reprinted 1885–1887 in *Mémoires sur l'électrodynamique*, 2 vol. (Gauthier-Villars, Paris). Part English translation Blunn O M, pp. 154–200 in Tricker [547]. [§ 18.1]

[13] Ampère A-M 1825 *Memoir on a new electrodynamic experiment* Phil. Mag. **66** 373–387. [§ 18.1]

[14] Anderson C D 1932 *The apparent existence of easily deflectable positives* Science **76** 238–239; 1932 *Energies of cosmic-ray particles* Phys. Rev. **41** 405–421; 1933 *The positive electron* Phys. Rev. **43** 491–494. [§ 3.4]

[15] Anderson J L 1967 *Principles of relativity physics.* (Academic Press, New York). [§§ 9.7, 15.9]

[16] Anderson J L and Bergmann P G 1951 *Constraints in covariant field theories.* Phys. Rev **83** 1018–1025. [§ 15.8]

[17] Angel R B 1980 *Relativity: the theory and its philosophy.* (Pergamon, Oxford).

[18] Archimedes *Sandreckoner*; English translation by Heath T L 1897 *The works of Archimedes* (University Press, Cambridge). [§ 5.1]

[19] Arfken G 1970 *Mathematical methods for physicists*, 2nd edn. (Academic Press, New York). [§ 17.6]

[20] Aristotle 1939 *De caelo* (*On the heavens*), Guthrie W K C (ed./trans.) (Heinemann, London). [§§ 5.1, 5.7]

[21] Arnison G et al. (UA1 collaboration) 1983 *Experimental observation of isolated large transverse energy electrons with associated missing energy at* $\sqrt{s} = 540\,\text{GeV}$. Phys. Lett. **122B** 103–116. [§§ 1.3, 20.13]

[22] Arzélies H 1966 *Relativistic kinematics.* (Pergamon, New York). [§ 5.4]

[23] Augustine of Hippo *On the beginning of time*, in *The city of God*; English translation, Dods M 1948 (Hafner, New York). [§ 2.5]

[24] Bailey J et al., 1979 *Final report on the CERN muon storage ring including the anomolous magnetic moment and the electric dipole moment of the muon, and a direct test of relativistic time dilation.* Nucl Phys. **B150** 1–75. [§ 7.14]

[25] Bailin D and Love A 1986 *Introduction to gauge field theory.* (Adam Hilger, Bristol).

[26] Balazs N L 1971 *Albert Einstein* pp. 312–333 in Gillispie [197], vol. 4. [§ 12.1]

[27] Bargmann V and Wigner E P 1948 *Group theoretical discussion of relativistic wave equations* Proc. Nat. Acad. Sci. (USA) **34** 211–223. [§ 20.3]

[28] Barrow J D and Tipler F J 1986 *The anthropic cosmological principle* (University Press, Oxford). [§ 12.6]

[29] Barut A O 1964 *Electrodynamics and the classical theory of fields* (Macmillan, London).

[30] Barut A O and Rączka R *Theory of group representations and applications* (Polish Scientific, Warsaw, 1980). [§§ 4.6, 4.8, 6.6, 14.2, 20.2, 20.3]

[31] Belinfante F J 1939 *On the spin angular momentum of mesons* Physica **6** 887–897; 1940 *On the current and density of the electric charge, the energy, the linear momentum and the angular momentum of arbitrary fields* Physica **7** 449–474. [§ 19.5]

[32] Bergmann P G 1942 *Introduction to the theory of relativity* (Prentice-Hall, Englewood Cliffs, NJ).

[33] Bernstein J 1962 *A question of parity* New Yorker (12 May) 49–104. [§ 3.3]

[34] Bernstein J 1973 *Einstein* (Penguin, New York).

[35] Bertotti B, Brill D and Krotkov R 1962 *Experiments on gravitation* pp. 1–48 in Witten L (ed.) [590].

[36] Bessel-Hagen E 1921 *Über die Erhaltungssätze der Electrodynamik (On the conservation laws of electrodynamics)* Math. Ann. (Germany) **84** 258–276. [§ 9.6]

[37] Bethe H A and Morrison P 1956 *Elementary nuclear theory* 2nd edn. (Wiley, New York). [§ 20.3]

[38] Biot J-B and Savart F 1820 *Note sur le magnétisme de la pile de Volta (Note on the magnetism of Volta's battery)* Ann. Chim. Phys. **15** 222–223. [§ 18.1]

[39] Biot J-B 1821 *Sur l'aimantation imprimé aux metaux par l'électricité en mouvement (On the magnetization of metals by moving electricity)* Jnl. des savants (1821) pp. 221–235. [§ 18.1]

[40] Biot J-B 1824 *Précis élémentaire de physique expérimentale (Elementary review of experimental physics)*, 2 vol., 3rd edn. (Paris). [§ 18.1]

[41] Blackett P M S and Occhialini G P S 1933 *Some photographs of the tracks of penetrating radiation* Proc. Roy. Soc. (London) **A139** 699–720. [§ 3.4]

[42] Boehm F and Vogel P 1987 *Physics of massive neutrinos* (University Press, Cambridge). [§ 12.5]

[43] Borisenko A I and Tarapov I E 1979 *Vector and tensor analysis with applications* (Dover, New York).

[44] Born M 1956 *Physics and relativity* Helv. Physica Acta Suppl. **4** *Jubilee of relativity theory.* [§ 13]

[45] Bradley J 1728 *A letter from the Reverend Mr James Bradley, Savilian Professor of Astronomy at Oxford and F.R.S. to Dr Edmund Halley, Astron. Reg. &c. giving an account of a new discovered motion of the fix'd stars* Phil. Trans. Roy. Soc. (London) **35** 637–661. [Describes the discovery of stellar aberration.] [§§ 5.1, 7.13, 12.1]

[46] Braginsky V B and Panov V I 1971 *Verification of the equivalence of inertial and gravitational mass* Zh. Eksp. & Teor. Fiz. **61** 873–879; English translation in Sov. Phys. JETP **34** 464–466. [§ 7.11]

[47] Brahe T 1586 *Diarium astrologicum et metheorologium*; 1588 *De mundi aetherei recentioribus phaenomenis* (Own Press, Uraniborg); 1598 *Astronomiae instauratae mechanica* (Wandsbeck); see also Dreyer J L E (ed.) 1913–1929 *Tychonis Brahe Dani opera omnia* 15 vol. (Danske Sprog- og Litteraturselskab, Copenhagen). [§ 7.3]

[48] Brans C and Dicke R H 1961 *Mach's principle and a relativistic theory of gravity* Phys. Rev. **124** 925–935. [§ 17.6]

[49] Brehme R W 1985 *Response to "The conventionality of synchronization"* Am. J. Phys. **53** 56–59. [§ 5.3]

[50] Brewster D 1855 *Memoirs of the life, writings and discoveries of Isaac Newton* vol. 2, chap. 27. (Constable, Edinburgh). [§§ 6, 6.7]

[51] Brittin W E, Smythe W R and Wyss W 1982 *Poincaré gauge in electrodynamics* Am. J. Phys. **50** 693–696. [§§ A.4, 18.7]

[52] Brown L 1978 *The idea of the neutrino* Physics Today **31** (Sep.) 23–28. [§ 20.13]

[53] Brush S 1967 *Note on the history of the FitzGerald-Lorentz contraction* Isis **58** 230–232. [§§ 5.4, 12.1]

[54] Bulmer-Thomas I 1971 *Euclid*, pp. 414–437 in Gillispie [197], vol. 4. [§ 3]

[55] Bunge M 1973 *Philosophy of physics* (Reidel, Dordrecht).

[56] Cabibbo N and Ferrari E 1962 *Quantum electrodynamics with Dirac monopoles* Nuovo Cimento **23** 1147–1154. [§ 18.11]

[57] Candelas P, Horowitz G T, Strominger A and Witten E 1985 *Vacuum configurations for superstrings* Nucl. Phys. **B258** 46–74. [§ 21.8]

[58] Cartan É 1901 *Sur quelques quadratures dont l'élements différentiels contient des fonctions arbitraires (On several quadratures whose differential elements contain arbitrary functions)* Bull. Soc. Math. de Paris **29** 118–130. [§ 7.10]

[59] Cartan É 1923–24 *Sur les variétés à connexion affine et la théorie de la relativité generalisée (Affine manifolds and the theory of general relativity)* (2 parts) Ann. sci. École Norm. Sup. **40** 325–412; **41** 1–25. [§§ 7.10, 7.15]

[60] Cavendish H 1798 *Experiments to determine the density of the earth* Phil. Trans. Roy. Soc. (London) Part II 469–526. [§ 7.4]

[61] Cavendish H 1921 (ed. Thorpe E) *The scientific papers of the honorable Henry Cavendish, F.R.S., vol. 2: Chemical and dynamical* (Cambridge). [§ 7.13]

[62] Chadwick J 1914 *Distribution in intensity in the magnetic spectrum of the β-rays of radium* (B+C) Verh. Dtsch. Phys. Ges. **16** 383–391. [§ 20.13]

[63] Chadwick J 1932 *The existence of a neutron* Proc. Roy. Soc. (London) **A136** 692–708. [§ 20.13]

[64] Chase C 1930 *The scattering of fast electrons by metals. II. Polarization by double scattering at right angles* Phys. Rev. **36** 1060–1065. [§ 3.3]

[65] Chorlton F 1976 *Vector and tensor methods* (Ellis Horwood, Chichester).

[66] Christenson J H, Cronin J W, Fitch V L and Turlay R 1964 *Evidence for the 2π decay of the K_2^0 meson* Phys. Rev. Lett. **13** 138–140. [§ 3.5]

[67] Clark R W 1971 *Einstein: the life and times* (World Publishing, New York); 1979 (Hodder and Stoughton, London).

[68] Clausius R J E 1850 *Über die bewegende Kraft der Wärme und die Gesetze, welche sich daraus für die Wärmelehre selbst ableiten lassen* (*On the moving power of heat and the law which can be derived from the theory of heat itself*) Ann. d. Phys. **79** 386–397; 500–524. [§ 6.4]

[69] Cohen I B 1941 *Benjamin Franklin's experiments* (Harvard University Press, Cambridge, MA). [§ 18.1]

[70] Cohen I B 1971 *Isaac Newton* pp. 42–101 in Gillispie [197], vol. 10. [§ 6.1]

[71] Cohen I B 1978 *Introduction to Newton's Principia* (University Press, Cambridge).

[72] Cohen I B 1985 *Birth of a new physics* (Norton, New York).

[73] Cohen I B 1985 *Revolution in science* (Harvard University Press, Cambridge, MA).

[74] Cohen E R and Taylor B N 1986 *The 1986 adjustment of the fundamental constants*, report of the CODATA Task Group on Fundamental Constants, CODATA Bulletin **63** (Pergamon, New York); 1988 *The fundamental physical constants* Physics Today **41** (Aug.) BG9–13. [§§ 6.7, 7.4, A.1]

[75] Cohen-Tannoudji C, Diu B and Laloë F 1973 *Mécanique quantique* 2 vol. (Hermann, Paris); English translation: 1977 *Quantum mechanics* (Wiley, New York). [§ 10.6]

[76] Collins G P and Doughty N A 1987 *Systematics of arbitrary-helicity Lagrangian wave equations* J. Math. Phys. **28** 448–456. [§ 21.2]

[77] Conway A W and Synge J L (eds.) 1931, 1949, 1967 *The Mathematical papers of sir William Rowan Hamilton* 3 vols: 1931 vol. 1, *Geometrical Optics*; 1949 vol. 2, *Dynamics*; 1967 vol. 3, *Algebra* (University Press, Cambridge).

[78] Copernicus N 1514 *De hypothesibus motuum coelestium a se constitutis commentariolus*; English translation *The commentariolus of Copernicus* in Rosen E 1959 *Three Copernican treatises* (Columbia University Press, New York). [§ 5.1]

[79] Copernicus N 1543 *De revolutionibus orbium coelestium*; English translation *Revolutions of the heavenly spheres* in *Great books of the western world* vol. 16 (Great Books, Chicago). [§§ 5.1, 7.3]

[80] Corson E M 1953 *Introduction to tensors, spinors and relativistic wave equations* (Blackie, Glasgow).

[81] Coulomb C A 1785–1789 (issued 1788–1792) *Sur l'électricité et le magnétisme* (*On electricity and magnetism*) Mém. de l'Acad. Fr. 1785 569–577; 578–611; 612–638; 1786 67–77; 1787 421–467; 1788 617–705; 1789 455–505; reprinted in Potier A (ed.) *Mémoires de Coulomb*, vol. 1 of *Collection de Mémoires relatifs à la physique* (Société Française de physique, Paris). [§ 18.1]

[82] Cox R T, McIlwraith C G and Kurrelmeyer B 1928 *Apparent evidence of polarization in a beam of β-rays* Proc. Nat. Acad. Sci. USA **14** 544–549. [§ 3.3]

[83] Culligan G, Frank S G F and Holt J R 1959 *Longitudinal polarization of the electrons from the decay of unpolarized positive and negative muons* Proc. Phys. Soc. **73** 169–177. [§ 3.4]

[84] Currie D G and Saletan E J 1966 *q-Equivalent particle Hamiltonians I. The classical one-dimension case* J. Math. Phys. **7** 967–974. [§ 8.6]

[85] d'Agostino S 1975 *Hertz's researches on electromagnetic waves* Hist. Stud. Phys. Sci. **6** 261–323. [§ 12.1]

[86] Davies P C W 1974 *Physics of time asymmetry* (Surrey University Press, Leighton Buzzard). [§ 2.5]

[87] Davies P C W 1977 *Space and time in the modern universe* (University Press, Cambridge).

[88] Davies P C W 1980 *Search for gravitational waves* (University Press, Cambridge).

[89] Davies P C W 1982 *Accidental universe* (University Press, Cambridge).

[90] Davies P C W 1983 *God and the new physics* (Penguin, New York).

[91] Davies P C W 1986 *The forces of nature* 2nd edn. (University Press, Cambridge).

[92] Davies P C W 1987 *Superforce, the search for a grand unified theory of Nature* (Unwin, London) (Orig. edn. 1984, Heinemann, London).

[93] Davies P C W and Brown J R (eds.) 1988 *Superstrings: a theory of everything?* (University Press, Cambridge). [§ 1.5]

[94] de Donder T 1921 *La gravifique einsteinienne (Einsteinian gravitation)* (Gauthiers-Villars, Paris). [§ 21.3]

[95] Deser S and Ford K W (eds.) 1965 *Lectures on general relativity: 1964 Brandeis Summer Institute in Theoretical Physics* vol. 1 (Prentice-Hall, Englewood Cliffs, N.J.).

[96] Deser S 1970 *Self-interaction and gauge invariance* Gen. Rel. & Grav. **1** 9–18. [§ 21.4]

[97] Deser S 1978 *Supergravity* pp. 573–595 in Dold and Eckmann [112]. [§§ 16, 20.3, 21.2, 21.7]

[98] Deser S 1979 *The dynamics of supergravity* pp. 461–477 in Lévy and Deser [322]. [§§ 16, 20.3, 21.2, 21.7]

[99] Deser S 1980 *From gravity to supergravity* pp. 357–392 in Held [231]. [§§ 16, 20.3, 21.2, 21.7]

[100] Deser S and Zumino B 1976 *Consistent supergravity* Phys. Lett. **62B** 335–337. [§§ 16, 20.3, 21.2, 21.7]

[101] de Wit B and Smith J 1986 *Field theory in particle physics* (North-Holland, Amsterdam). [§ 20.2]

[102] De Witt C M and Wheeler J A 1967 *Battelle rencontres*, Lectures in mathematics and physics (Benjamin, New York).

[103] Dicke R H 1962 *Mach's principle and invariance under transformation of units* Phys. Rev. **125** 2163-2167. [§ 17.6]

[104] Diels H and Kranz W 1951-52 *Die Fragmente der Vorsokratiker (Fragments from the pre-Socratic philosophers)* 6th edn., 3 vol. (Berlin). [§ 2]

[105] Dieudonné J 1971 *Jules Henri Poincaré* pp. 51-61 in Gillispie [197], vol. 11. [§ 12.1]

[106] Dirac P A M 1928 *The quantum theory of the electrons* Proc. Roy. Soc. (London) **A117** 610-624. [§§ 3.4, 16, 16.7, 20.3]

[107] Dirac P A M 1929 *A theory of electrons and protons* Proc. Roy. Soc. (London) **A126** 360-365. [Hole theory of negative energy states.] [§§ 3.4, 15.8]

[108] Dirac P A M 1930 *Annihilation of electrons and positrons* Proc. Camb. Phil. Soc. **26** 361-375. [§ 3.4]

[109] Dirac P A M 1931 *Quantized singularities in the electromagnetic field* Proc. Roy. Soc. (London) **A133** 60-72; 1948 *The theory of magnetic poles* Phys. Rev. **74** 817-830. [§ 18.11]

[110] Dirac P A M 1950 *Generalised Hamiltonian dynamics* Can. Jnl. Math. **2** 129-148; 1958 *Generalised Hamiltonian dynamics* Proc. Roy. Soc. (London) **A246** 326-332; 1964 *Generalised Hamiltonian dynamics* (Belfer Graduate School of Science, New York). [§§ 15.8, 16.8]

[111] Dirac P A M 1963 *The physicist's picture of nature* Sci. Amer. **208** (May) 45-53. [§ 20]

[112] Dold A and Eckmann B (eds.) 1978 *Differential geometric methods in mathematical physics II* (Springer Verlag, New York).

[113] Doughty N A 1981 *Acceleration of a static observer near the event horizon of a static isolated black hole* Am. J. Phys. **49** 412-416. [§§ 7.2, 21.6]

[114] Doughty N A 1981 *The surface properties of Kerr-Newman black holes* Am. J. Phys. **49** 720-724. [§§ 7.2, 21.6]

[115] Doughty N A and Arnold R A 1989 *Quantization of free Lagrangian gauge fields of arbitrary helicity* Nucl. Phys. **B** (Proc. Suppl.) **6** 390-392. [§ 21.2]

[116] Doughty N A and Arnold R A 1989 *Gupta-Bleuler quantization of free massless Lagrangian gauge fields of arbitrary helicity: the bosonic case* J. Math. Phys. **30** 1545-1553. [§ 21.2]

[117] Doughty N A and Collins G P 1986 *Multispinor symmetries for massless arbitrary spin Fierz-Pauli and Rarita-Schwinger wave equations* J Math. Phys. **27** 1639-1645. [§§ 18.7, 21.2, 21.3, 21.4, 21.6]

[118] Doughty N A and Wiltshire D L 1986 *Weyl field strength symmetries for arbitrary helicity and gauge invariant Fierz-Pauli and Rarita-Schwinger wave equations* J. Phys. A **19** 3727-3739. [§§ 18.7, 21.2, 21.3, 21.4, 21.6]

[119] Drabkin I E and Drake S 1960 *Galileo Galilei on motion and on mechanics* (University of Wisconsin Press, Madison).

[120] Drake S 1971 *Galileo Galilei* pp. 237–250 in Gillispie [197], vol. 5. [§ 5.4]

[121] Dreyer J L E 1890 *Tycho Brahe: a picture of scientific life and work in the sixteenth century* (Adam and Charles Black, Edinburgh).

[122] du Fay C-F de Cistenai 1733 *Sur l'électricité (On electricity)* Mém. de l'Acad. Fr. pp. 23–35, 73–84, 233–254, 457–476. [§ 18.1]

[123] Dyson F W, Eddington A S and Davidson C *A determination of the deflection of light by the sun's gravitational field, from observations made at the total eclipse of May 29, 1919.* Phil. Trans. Roy. Soc. (London) **A220** 291–333. [§ 21.4]

[124] Eddington A S 1928 *The nature of the physical world* (University Press, Cambridge). [§ 2.5]

[125] Eguchi T, Gilkey P B and Hanson A J 1980 *Gravitation, gauge theories and differential geometry* Physics Reports **66**, 213–393. [§ 7.10]

[126] Ehrenberg W and Siday R E 1949 *The refractive index in electron optics and the principles of dynamics* Proc. Phys. Soc. (London) **B62** 8–21. [§§ 2.4, 20.9]

[127] Einstein A 1905 *Über einen die Erzeugung und Verwandlung des Lichtes betreffenden heuristischen Gesichtspunkt (On the production and conversion of light considered from a heuristic point of view)* Ann. Phys. (Germany) **17** 132–148. [Work announcing the equation of the photoelectric effect and which led to the award of the Nobel Prize in 1921.] [§ 15.2]

[128] Einstein A 1905 (September 1905) *Zur Elektrodynamik bewegter Körper* Ann. d. Phys. (Germany) **17** 891–921; English translation: *On the electrodynamics of moving bodies* pp. 35–65 in Lorentz et al. [343]. [This paper, signalling the birth of special relativity theory, was submitted in June 1905, without prior knowledge of the 1904 work of Lorentz [340] and Poincaré [447]. Another of Poincaré's papers [448] was written in June 1905, while a 1906 paper [449] was dated July 1905.] [§§ 5.2, 5.3, 5.4, 12.1, 12.2, 17.5, 21.3]

[129] Einstein A 1905 *Ist die Trägheit eines Körpers von seinem Energieinhalt abhängig?* Ann. d. Phys. (Germany) **18** 639–641; English translation *Does the inertia of a body depend upon its energy content?* pp. 67–71 in Lorentz et al. [343]. [§ 15.2]

[130] Einstein A 1907 *Über die vom Relativitätsprinzip geforderte Trägheit der Energie (On the inertia of energy implied by the principle of relativity)* Ann. d. Phys. (Germany) **23** 371–384. [§ 15.2]

[131] Einstein A 1907 *Über das Relativitätsprinzip und die aus demselben gezogenen Folgerungen (On the Principle of relativity and its consequences.)* Jahrbuch d. Radioakt. und Elektronik **4** 411–462. Erratum: Jahrbuch d. Radioakt. und Elektronik **5** 98–99. [First of Einstein's papers containing comments on the nature of the influence of gravitation on the propagation of light.] [§§ 7.11, 17.5, 21.3, 21.6]

[132] Einstein A 1911 *Über den Einfluss der Schwerkraft auf die Ausbreitung des Lichtes* Ann. d. Phys. (Germany) **35** 898–908; English translation *On the*

influence of gravitation on the propagation of light, pp. 99–108 in Lorentz et al. [343]. [First suggested use of solar eclipse data to test for light deflection by the sun, based on a predicted value half that of general relativity.] [§§ 7.11, 7.13, 17, 17.2, 21.3, 21.4, 21.5, 21.6]

[133] Einstein A 1913 *Entwurf einer verallgemeinerten Relativitätstheorie und einer Theorie der Gravitation. I. Physikalischer Teil* (*A proposed general relativity theory and theory of gravitation. I Physical part*) Z. f. Mathem. u. Phys. **62** 225–244. [First statement of a tensor theory of gravity leading to the general theory of relativity. The mathematical Part II, pp. 245–261, was written by M. Grossmann.] [§ 21.6]

[134] Einstein A 1914 *Die formale Grundlagen der allgemeinen Relativitätstheorie* (*Formal foundations of general relativity theory*) Preuss. Akad. Wiss. Berlin, Sitz.ber., 1030–1085. [§ 21.6]

[135] Einstein A 1915 *Grundgedanken der allgemeinen Relativitätstheorie und Anwendung dieser Theorie in der Astronomie* (*Basic ideas of general relativity theory and its application in astronomy*) Preuss. Akad. Wiss. Berlin, Sitz.ber., p. 315. [Abstract of a paper presented at the meeting of 25 March 1915.] [§§ 7.14, 21.4, 21.6]

[136] Einstein A 1915 *Zur allgemeinen Relativitätstheorie* (*On the general relativity theory*) Preuss. Akad. Wiss. Berlin, Sitz.ber., 778–786 (published 4 November); Nachtrag (Supplement) *Zur allgemeinen Relativitätstheorie* (*On the general relativity theory*) Preuss. Akad. Wiss. Berlin, Sitz.ber., 799–801 (published 11 November); English translation in Lorentz et al. [343]. [§§ 7.14, 21.6]

[137] Einstein A 1915 *Erklärung der Perihelbewegung des Merkur aus der allgemeinem Relativitätstheorie* (*The explanation of the perihelion shift of Mercury using the general relativity theory*) Preuss. Akad. Wiss. Berlin, Sitz.ber., 831–839; English translation in Lorentz et al. [343]. [§§ 7.14, 21.6]

[138] Einstein A 1915 *Die Feldgleichungen der Gravitation* (*The gravitational field equations*) Preuss. Akad. Wiss. Berlin, Sitz.ber., 844–847 (presented 25 November, published 2 December); English translation in Lorentz et al. [343]. [First appearance of the final form of the field equations.] [§§ 7.14, 21.6]

[139] Einstein A 1916 *Die Grundlage der allgemeinen Relativitätstheorie* Ann. d. Phys. (Germany) **49** 769–822; English translation *The foundation of the general theory of relativity*, pp. 109–164 in Lorentz et al. [343]. [This paper, which contains the correct prediction of 45 arc seconds per century for the perihelion advance of Mercury, is also a summary of the work carried out from 1907 culminating in the general relativistic theory of gravitation in 1915.] [§§ 7.14, 21.6]

[140] Einstein A 1917 *Kosmologische Betrachtungen zur allgemeinen Relativitätstheorie* Preuss. Akad. Wiss. Sitz.ber. 142–152; English translation *Cosmological considerations on the general theory of relativity*, pp. 175–188 in Lorentz et al. [343]. [§§ 7.9, 21.6]

[141] Eisenbud L 1958 *On the classical laws of motion* Am. J. Phys. **26** 144–159. [§ 6.1]

[142] Elkana Y 1970 *Helmholtz's 'Kraft': an illustration of concepts in flux* Hist. Stud. Phys. Sci. **2** 263-298. [§ 6.4]

[143] Eötvös R V 1889 *Über die Anziehung der Erde auf verschiedene Substanzen* (*On the earth's attraction for various substances*) Math. Naturw. Ber. aus Ungarn **8** 65-68. [§ 7.11]

[144] Eötvös R V, Pekár D and Fekete E 1922 *Beiträge zum Gesetze der Proportionalität von Trägheit und Gravität* (*Contributions to the law of proportionality of inertia and gravity*) Ann. d. Phys. (Germany) **68** 11-66. [Many earlier and later measurements are summarized and reviewed in Misner et al. [384] and Ohanian [404].] [§ 7.11]

[145] Euler L 1744 *Methodus inveniendi lineas curvas maximi minimive proprietate gaudentes: sive solutio problematis isoperimetrici latissimo sensu accepti* Appendix 2. (Lausanne-Geneva); reprinted pp. 1-308 in *Opera omnia* [148], Series 1, vol. 24. [§§ 8.1, 8.2, 8.5]

[146] Euler L 1764-66 *Elementa calculi variationum* (*Elements of the calculus of variations*) Novi commentarii acad. sci. Petropolitinae **10** 51-93; reprinted pp. 141-176 in *Opera omnia* [148], Series 1, vol. 25. [§ 8.1]

[147] Euler L 1764-66 *Analytica explicatio methodi maximorum et minimorum* (*Analytical description of the method of maxima and minima*) Novi commentarii acad. sci. Petropolitinae **10** 94-134; reprinted pp. 177-207 in *Opera omnia* [148], Series 1, vol. 25. [§ 8.1]

[148] Euler L 1911-82 *Leonhardi Euleri: Opera omnia* 71 vol., (Natural Science Society, Berne).

[149] Everitt C W F 1971 *James Clerk Maxwell* pp. 198-230 in Gillispie [197], vol. 9. [§ 12.1]

[150] Faraday M 1839-55 *Experimental researches in electricity*, 3 vol. (London). [§§ 12.1, 18.1]

[151] Fechner G T 1845 *Über die Verknüpfung der Faraday'schen Inductions-Erscheinungen mit den Ampèreschen elektro-dynamischen Erscheinungen* (*On the connection between the Faraday induction phenomena and Ampère's electrodynamic effect*) Ann. d. Phys. **64** 337-345. [§ 18.1]

[152] Felsager B 1981 *Geometry, particles and fields* (Odense University Press, Gylling Denmark).

[153] Fermat P de 1891 *Oeuvres de Fermat* 1657 *Epist. xlii*, vol. ii, p. 354; 1662 *Epist. xliii*, vol. ii, p. 457; vol. i, pp. 170, 173. [§ 8.1]

[154] Fermi E 1934 *Tentativo di una teoria dei raggi β* (*Attempt at a theory of β-rays*) Nuovo Cim. **11** 1-19. [§ 20.13]

[155] Fermi E 1934 *Versuch einer Theorie der β-Strahlen. I.* (*Attempt at a theory of β-rays. I.*) Z. Phys. **88** 161-171. [§ 20.13]

[156] Fermi E 1932, 1965 *Enrico Fermi Collected papers*, vol. 1 1962 *Italia 1921-1938* and vol. 2 1965 *U.S.A. 1939-1954* (Accademia Nazionale dei Lincei and University of Chicago Press, Rome). [§ 20.13]

[157] Ferretti B 1936 *The absorption of slow mesons by an atomic nucleus*, pp. 55–57, vol. 1, in *Report on the International Conference on Fundamental Particles and Low Temperature*, Cambridge, Jul. 22–27, 1936. [§ 16.8]

[158] Fetter A L and Walecka J D 1980 *Theoretical mechanics of particles and continua* (Wiley, New York). [§ 8.10]

[159] Feynman R P and Gell-Mann M 1958 *Theory of Fermi interaction* Phys. Rev. **109** 193–198. [Introduction of the W intermediate boson.] [§ 20.13]

[160] Feynman R P and Hibbs A R 1965 *Quantum mechanics and path integrals* (McGraw-Hill, New York). [§§ 8.8, 16.1]

[161] Fierz M 1939 *Über die relativistiche Theorie kräftfreier Teilchen mit beliebigem Spin (On the relativistic theory of non-interacting particles of arbitrary spin)* Helv. Phys. Acta **12** 3–37. [Includes the spin-statistics relation.] [§ 20.14]

[162] Fierz M and Pauli W 1939 *On relativistic wave equations for particles of arbitrary spin in an electromagnetic field* Proc. Roy. Soc. (London) **A173** 211–232. [§§ 16, 21.3]

[163] FitzGerald G F 1889 *The ether and the earth's atmosphere* Science **13** 390. [See also Brush [53].] [§§ 5.4, 12.1]

[164] Flanders H 1963 *Differential forms with applications to the physical sciences* (Academic Press, New York). [§ A.4]

[165] Fletcher J G 1960 *Local conservation laws in generally covariant theories* Rev. Mod. Phys. **32** 65–87. [§ 9.7]

[166] Flood R and Lockwood M 1986 *The nature of time* (Basil Blackwell, New York).

[167] Formalont E B and Srameck R A 1976 *Measurements of the solar gravitational deflection of radio waves in agreement with general relativity* Phys. Rev. Lett. **36** 1475–1478; 1975 *A confirmation of Einstein's general theory of relativity by measuring the bending of microwave radiation in the gravitational field of the sun* Astrophys. J. **199** 749–755. [§ 7.13]

[168] Frampton P H 1987 *Gauge field theories* (Benjamin Cummings, Menlo Park, CA). [§ 16.4]

[169] Frank P and Rothe H 1911 *Über die Transformation der Raumzeitkoordinaten von ruhenden auf bewegte Systeme (On the transformation of spacetime coordinates from stationary to moving systems)* Ann. d. Phys. (Germany) **34** 825–855. [§ 5.4]

[170] Franklin B 1750 *New experiments and observations on electricity*, letter ii. [§§ 18.1, 18.3]

[171] Franklin A 1979 *The discovery and nondiscovery of parity nonconservation* Stud. Hist. Phil. Sci. **10** 201–257. [§ 3.3]

[172] Frauenfelder H and Henley E M 1975 *Nuclear and particle physics* (Benjamin, Reading, MA).

[173] Freedman D Z 1979 *Systematics of higher spin gauge fields* pp. 263–268 in van Nieuwenhuizen and Freedman [549]; de Wit B and Freedman D Z 1980 *Systematics of higher spin gauge fields* Phys. Rev. **D21** 358–367. [§ 21.2]

[174] Freedman D Z and van Nieuwenhuizen P 1976 *Progress toward a theory of supergravity* Phys. Rev. **D13** 3214-3218; 1976 *Properties of supergravity theory* Phys. Rev. **D14** 912-916. [§§ 16, 20.3, 21.2, 21.7]

[175] French A P 1968 *Special relativity* (Norton, New York). [§§ 5.4, 12.5, 15.2]

[176] Friedman J I and Telegdi V L 1957 *Nuclear emulsion evidence for parity nonconservation in the decay chain* $\pi^+ - \mu^+ - e^+$ Phys. Rev. **105** 1681-1682. [§ 3.3]

[177] Friedrichs K 1928 *Eine invariante Formulierung des Newtonschen Gravitationsgesetzes und des Grenzüberganges vom Einsteinschen zum Newtonschen Gesetz (An invariant formulation of Newtonian gravitation and the Newtonian limits of Einsteinian laws)* Math. Ann. **98** 566. [§ 7.15]

[178] Fryberger D 1985 *Magnetic monopoles* IEEE Trans. Magn. **21** 84-101; 1989 *On generalized electromagnetism and Dirac algebra* Found. Phys. **19** 125-159. [§ 18.11]

[179] Galilei G 1592 *De Motu*; Unpublished work written at the Universty of Pisa 1589-1592; English translation *On motion* in Drabkin and Drake 1960 [119]. [§ 6.1]

[180] Galilei G 1610 *Siderius nuncius*; English translation, Carlos E S 1964 *The sidereal messenger* (Dawsons of Pall Mall, London).

[181] Galilei G 1632 *Dialogo dei due massini sistemi del mundo* (Landini, Florence); English translation: Drake S 1953 *Dialogue concerning the two chief world systems — Ptolemaic and Copernican* (University of California Press, Berkeley). [§ 5.4]

[182] Galilei G 1638 *Discorsi e dimostriazione matematiche intorno a due nuove scienze* (Elzevier, Leiden); English translation: Drake S 1975 *Dialogues concerning two new sciences* (University of Wisconsin Press, Madison). [§§ 5.4, 7.11]

[183] Galvani L 1791 *De viribus electricitatis in motu musculari commentarius (A commentary on the forces of electricity in muscular movement)* (Bologna); Facsimile edition with English translation 1953 (Burnby Library, Norwalk, CT). [§ 18.1]

[184] Gamow G 1970 *My world line: an informal autobiography* (Viking Press, New York) p. 44. [§ 7.9]

[185] Gardner M 1982 *The ambidextrous universe: mirror asymmetry and time-reversed worlds* (Penguin, Harmondsworth, NY). [§ 3.3]

[186] Gardner M 1967 *Can time go backwards?* Sci. Amer. **216** (Jan.) 98-108.

[187] Garwin R L, Lederman L M and Weinrich M 1957 *Observations of the failure of conservation of parity and charge conjugation in meson decays: the magnetic moment of the free muon* Phys. Rev. **105** 1415-1417. [§ 3.3]

[188] Gauss K F 1827 *Disquisitiones generales circa superficies curvas* in *Werke* [192], vol. 4, pp. 217-258; English translation by Morehead J C and Hiltebeitel A M 1965 *General investigations of curved surfaces of 1827 and 1825* (Raven Press, New York). [§ 2.4]

[189] Gauss K F 1832 *Intensitas vis magneticae terrestris ad mesuram absolutam revocata* (*The intensity of the terrestrial magnetic field referred to an absolute measurement*) Göttingische gelehrte Anzeigen, 24 Dec. 1832, pp. 2041–2058, reprinted in *Werke* [192], vol. 5, pp. 79–118, 293–304. [Gaussian magnetic units.] [§ 18.1]

[190] Gauss K F 1838 *Allgemeine Theorie des Erdmagnetismus — Resultate* (*General theory of the earth's magnetism — results*) reprinted in *Werke* [192], vol. 5, pp. 119–175, on 122–125, 126. [§ 18.1]

[191] Gauss K F 1867 *Zur mathematischen Theorie der elektrodynamischen Wirkungen* (*On the mathematical theory of electrodynamic effects*) pp. 601–626 in Gauss [192]. [§ 18.1]

[192] Gauss K F 1863–1933 *Werke* 12 vol. (Königliche Gesellschaft der Wissenschaft zu Göttingen, Leipzig-Berlin). [§ 18.1]

[193] Georgi H 1988 *Flavour* SU(3) *symmetries in particle physics* Physics Today **14** 29–37. [§ 7.9]

[194] Gibbons G 1979 *The man who invented black holes* New Scientist **82** 1101. [§§ 7.12, 7.13]

[195] Gibson W M and Pollard B R 1976 *Symmetry principles in elementary particle physics* (University Press, Cambridge). [§ 3.3]

[196] Gilbert W 1600 *De magnete, magneticisque corporibus, et de magno magnete tellure* (*On the magnet, magnetic bodies, and the great magnet earth*) (Peter Short, London); English translation Price D J (ed.) 1958 *On the magnet* in *Collector's series in science* (Basic Books, New York). [§ 18.1]

[197] Gillispie C C (ed.) 1971 *Dictionary of scientific biography* (Charle's Scribner's Sons, New York). [§ 12.1]

[198] Gingerich O 1982 *The Galileo affair* Sci. Amer. **247** (Aug.) 132–143. [§ 5.1]

[199] Glashow S L 1961 *Partial-symmetries of weak interactions* Nucl. Phys. **22** 579–588. [§§ 20.13, 21.7]

[200] Glass B 1971 *Pierre Louis Moreau de Maupertuis* pp. 186–189 in Gillispie [197], vol. 9. [§ 8.1]

[201] Goldberg S 1967 *Henri Poincaré and Einstein's theory of relativity* Am. J. Phys. **35** 934–944. [§ 5.4]

[202] Goldberg S 1969 *The Lorentz theory of electrons and Einstein's theory of relativity* Am. J. Phys. **37** 982–994. [§§ 5.4, 12.1]

[203] Goldberg S 1984 *Understanding relativity: Origin and impact of a scientific revolution* (Clarendon Press, Oxford). [§§ 5.4, 21.5]

[204] Goldhaber A S and Nieto M M 1971 *How to catch a photon and measure its mass* Phys. Rev. Lett. **26** 1390–1394. [§§ 18.1, 18.6]

[205] Goldhaber A S and Nieto M M 1974 *Mass of the graviton* Phys. Rev. **D9** 1119–1121. [§ 21.3]

[206] Goldstein H 1950 (1st edn.) *Classical Mechanics*; 1980 (2nd edn.) (Addison-Wesley, Reading Mass.). [§§ 8.5, 8.10, 11.5, 13.6]

[207] Gordon W 1926 *Der Compton Effekt nach der Schrödingerschen Theorie (The Compton effect according to the Schrödinger theory)* Zeits. für Phys. **40** 117–133. [One of the original papers on the Klein-Gordon equation.] [§§ 16, 16.3]

[208] Gray S 1731 *Experiments in electricity* Phil. Trans. Roy. Soc. (London) **37** 18–44, 227–230, 285291, 297–407. [§ 18.1]

[209] Green G 1828 *An essay on the application of mathematical analysis to the theories of electricity and magnetism* (Nottingham); reprinted 1970 in *The mathematical papers of the late George Green*, pp. 3–115 (New York). [§ 18.1]

[210] Green M and Schwartz J 1984 *Anomaly cancellations in supersymmetric $D = 10$ gauge theory and superstring theory* Phys. Lett. **B149** 117–122. [§ 21.8]

[211] Green M B, Schwarz J H and Witten E 1987 *Superstring theory* (University Press, Cambridge).

[212] Griffiths D 1987 *Introduction to elementary particles.* (Wiley, New York).

[213] Grodzins L 1959 *The history of double scattering of electrons and evidence for the polarization of beta rays* Proc. Nat. Acad. Sci. **45** 399–405. [§ 3.3]

[214] Gross D J, Harvey J A, Martinec E and Rohm R 1985 *Heterotic string* Phys. Rev. Lett. **54** 502–505. [§ 21.8]

[215] Gupta S N 1952 *Quantization of Einstein's gravitation field: linear approximation* Proc. Phys. Soc. (London) **A65** 161–169; *Quantization of Einstein's gravitation field: general treatment* Proc. Phys. Soc. (London) **A65** 608–619.

[216] Gupta S N 1957 *Einstein's and other theories of gravitation* Rev. Mod. Phys. **29** 334–336. [§ 21.4]

[217] Hafele J C and Keating R E 1972 *Around-the-world atomic clocks: predicted relativistic time gains* Science **177** 166-167; 1972 *Observed relativistic time gains* Science **177** 168–170; 1972 *Relativistic time for terrestrial circumnavigation* Am. J. Phys. **40** 81–85. [§ 12.5]

[218] Hamel G 1904 *Die Lagrange-Eulerschen Gleichungen der Mechanik (The Euler-Lagrange equations of mechanics)* Z. Math. Phys. **50** 1–57. [§ 9.6]

[219] Hamilton W R 1823 *On caustics.* [Manuscript reprinted in Conway and Synge [77].] [§ 11.3]

[220] Hamilton W R 1828 *Theory of systems of rays* Trans. Roy. Irish Acad. **15** 69–174; **16** 1–61, 93–125; **17** 1–144. [§ 11.3]

[221] Hamilton W R 1833 *On a general method of expressing the paths of light, and of the planets, by the coefficients of a characteristic function* Dublin University Review, pp. 795–826. [§ 11.3]

[222] Hamilton W R 1834 *On a general method in dynamics; by which the study of the motions of all free systems of attracting or repelling points is reduced to the search and differentiation of one central relation, or characteristic function* Phil. Trans. Roy. Soc. (London) **124** 247–308; *Second essay on a*

general method in dynamics Phil. Trans. Roy. Soc. (London) **125** 95–144. [§ 11.2]

[223] Hanson T J, Regge T and Teitelboim C 1976 *Constrained Hamiltonian systems* (Accademia Nazionale dei Lincei, Rome). [§ 15.8]

[224] Haughan M P and Will C M 1987 *Modern tests of special relativity* Physics Today **40** 69–76. [§§ 5.4, 12.3]

[225] Hauser W 1970 *On the fundamental equations of electromagnetism* Am. J. Phys. **38** 80–85; 1971 *Introduction to the principles of electromagnetism* (Addison-Wesley, Reading, Mass.) pp. 547–552.

[226] Hawking S 1988 *A brief history of time from the Big Bang to black holes* (Bantam, London).

[227] Hawking S W and Ellis G F R 1973 *The large-scale structure of space-time* (University Press, Cambridge).

[228] Hawking S W and Israel W 1987 *Three hundred years of gravitation* (University Press, Cambridge).

[229] Heaviside O 1892 *Electrical papers*, 2 vol. (London); reprinted 1970 (Chelsea Publishing, Bronx, NY). [§§ 4.8, 18.1, 18.6]

[230] Heisenberg W 1932 *Über den Bau der Atomkerne. I (On the structure of atomic nuclei. I)* Z. Phys. **77** 1–11. [Isospin.] [§ 20.13]

[231] Held A (ed.) 1980 *General relativity and gravitation* vol. 1 (Plenum, New York).

[232] Helmholtz L F von 1847 *Über die Erhaltung der Kraft, eine physikalische Abhandlung (On the conservation of force [energy], a physical treatise)* 53 page essay read before the Physical Society of Berlin 23 July 1847 (Reimer, Berlin). English translation in Taylor 1854 *Scientific memoirs*, vol. 1, pp. 114–162 (London). [§ 6.4]

[233] Helmholtz L F von 1870 *Über die Bewegungsgleichungen der Elektricität für ruhende leitende Körper (On the equations of electricity for conducting bodies at rest)* Journ. f. d. reine u. angewandte Math. **72** 57–129, reprinted in Helmholtz [234]. [§ 18.1]

[234] Helmholtz L F von 1882-95 *Wissenschaftliche Abhandlungen* 3 vol. (Leipzig).

[235] Hergoltz G 1911 *Über die Mechanik des Deformierbaren vom Standpunkte der Relativitätstheorie (On the mechanics of deformations from the point-of-view of relativity theory)* Ann. d. Phys. (Germany) **36** 493–533. [§ 9.6]

[236] Hero of Alexandria 1899 *Heronis Alexandrini opera omnia* (Teubner, Leipzig). [§ 8.1]

[237] Hertz H 1884 Wiedemann's Ann. **23** 84–103; English translation 1896 *On the relations between Maxwell's fundamental electromagnetic equations and the fundamental equations of the opposing electromagnetics* in *Miscellaneous papers* [246], pp. 273–290. [§ 18.1]

[238] Hertz H 1887 *Über einen Einfluss des ultravioletten Lichtes auf die elektrische Entladung* Sitz.ber. d. Berl. Akad. d. Wiss. reprinted in 1887 Wiedemann's Ann. **31** 983–1000; English translation *On an effect of ultraviolet light on the electric discharge* pp. 63–79 in *Electric waves* [246]. [Photoelectric effect.] [§ 12.1]

[239] Hertz H 1887 *Über sehr schnelle elektrische Schwingungen* Wiedemann's Ann. **31** 421; English translation *On very rapid electric oscillations* pp. 29–53 in *Electric waves* [246]. [§§ 12.1, 18.6]

[240] Hertz H 1888 *Über die Einwirkung einer gradlinigen elektrischen Schwingung auf eine benachbarte Strombahn* Wiedemann's Ann. **34** 155–170, on p. 169; English translation *On the action of a rectilinear electric oscillation upon a neighbouring circuit* pp. 80–94 in *Electric waves* [246]. [§§ 12.1, 18.6]

[241] Hertz H 1888 *Über die Ausbreitungsgeschwindigkeit der electromagnetischen Wirkungen* Sitz.ber. d. Berl. Akad. d. Wiss. 197–210 and Wiedemann's Ann. (Germany) **34** 551–569; English translation *On the finite velocity of propagation of electromagnetic action* pp. 107–123 in *Electric waves* [246]. [§ 12.1]

[242] Hertz H 1888 *Über elektrodynamische Wellen im Luftraume und deren Reflexion* Wiedemann's Ann. **34** 610–623; English translation *On electrodynamic waves in air and their reflection* pp. 124–136 in *Electric waves* [246]. [§ 12.1]

[243] Hertz H 1889 *Über Strahlen elektrische Kraft* Wiedemann's Ann. **36** 769–783; English translation *On electric radiation* pp. 172–185 in *Electric waves* [246]. [§ 12.1]

[244] Hertz H 1890 *Über die Grundgleichungen der Elektrodynamik für ruhende Körper* Gött. Nachr. **19** 106–149, reprinted 1890 in Wiedemann's Ann. **40** 577–624; English translation *On the fundamental equations of electromagnetism for bodies at rest*, pp. 195–240 in *Electric waves* [246]. [A widely used standard form of Maxwell's theory.] [§ 18.1]

[245] Hertz H 1890 *Über die Grundgleichungen der Elektrodynamik für bewegte Körper* Wiedemann's Ann. **41** 369; English translation *On the fundamental equations of electromagnetism for bodies in motion*, pp. 241–268 in *Electric waves* [246]. [§ 18.1]

[246] Hertz H 1895 *Gesammelte Werke*, 3 vol. (Leipzig); vol. 1 English translation by Jones D E 1896 *Miscellaneous papers* (MacMillan, London); 1892 vol. 2 *Untersuchungen über die Ausbreitung der elektrischen Kraft* in English translation, 1893 *Electric waves, being researches on the propagation of electric action with finite velocity through space* (MacMillan, London), reprinted 1962 (Dover, New York).

[247] Hestenes D 1986 *New foundations for classical mechanics* (Reidel, Dordrecht). [§ 8.10]

[248] Higgs P W 1964 *Broken symmetries, massless particles and gauge fields* Phys. Lett. **12** 132–133; 1964 *Broken symmetries and the masses of gauge*

bosons Phys. Rev. Lett. **13** 508–509; 1966 *Spontaneous symmetry breakdown without massless bosons* Phys. Rev. **145** 1156–1163. [§§ 16, 16.3]

[249] Hill E L 1951 *Hamilton's principle and the conservation theorems of mathematical physics* Rev. Mod. Phys. **23** 253–260. [§§ 8.4, 9.7]

[250] Hirosige T 1966 *Electrodynamics before the theory of relativity* Jap. Stud. Hist. Sci. **5** 1–49. [§§ 5.4, 12.1]

[251] Hirosige T 1968 *Theory of relativity and the ether* Jap. Stud. Hist. Sci. **7** 37–53. [§§ 12.1, 12.4]

[252] Hirosige T 1969 *Origins of Lorentz's theory of electrons and the concept of the electromagnetic field* Hist. Stud. Phys. Sci. **1** 151–209. [§§ 5.4, 12.1]

[253] Hoffman B (ed.) 1966 *Perspectives in geometry and relativity* (Indiana University Press, Bloomington).

[254] Holton G 1960 *On the origins of the special theory of relativity* Am. J. Phys. **28** 627–636. [§§ 5.4, 12.1]

[255] Hubble E 1929 *A relation between distance and radial velocity among extragalactic nebulae* Proc. Nat. Acad. Sci. **15** 169–173. [§§ 7.9, 21.6]

[256] Hughes R J, Goldman T and Nieto M M 1989 *Red-shift experiments and non-Newtonian gravitational forces* in Fackler O et al. (eds.) *Proceedings of the VII Moriond Workshop: Tests of fundamental laws in physics; Les Arcs, Savoie* (Editions Française, Paris). [§§ 17.2, 19.4]

[257] Hulse R A and Taylor J H 1974 *Discovery of a pulsar in a binary system* Astrophys. J. **195** L51–53. [§ 17.5]

[258] Huyghens C 1690 *Traité de la lumière (Treatise on light)* (Pierre van der Aa, Leiden) reprinted 1888–1950 in *Oeuvres complètes de Christiaan Huyghens* 22 vol. (Society of Sciences of Holland, The Hague). [§ 7.13]

[259] Ignatowsky W von 1910 *Einige allgemeine Bermerkungen zum Relativitätsprinzip (Some general remarks on the relativity principle)* Phys Z. **11** 972–976; 1911 *Das Relativitätsprinzip (The relativity principle)* Arch. Math. Phys. (Germany) **17** 1–24; **18** 17–40. [§ 5.4]

[260] Isaak G R 1970 *The Mössbauer effect: application to relativity* Phys. Bull. **21** 255–257. [§ 5.4]

[261] Itard J 1971 *Joseph-Louis Lagrange* pp. 559–573 in Gillispie [197], vol. 7. [§§ 8.1, 11.5]

[262] Itzykson C and Zuber J-B 1980 *Quantum field theory* (McGraw-Hill, New York). [§§ 3.4, 3.5, 16.1, 16.4, 16.6, 16.7, 18.13, 20.2, 20.5, 20.10, 20.11, 20.14]

[263] Jackson J D 1975 *Classical electrodynamics* 2nd edn. (Wiley, New York) pp. 251–252 and pp. 547–548. [§§ 18.2, 18.3, 18.11, 19.7, 19.9]

[264] Jagannathan K and Singh L P S 1986 *Attraction/repulsion between like charges and the spin of the classical mediating field* Phys. Rev. **D33** 2475–2477. [§§ 17.1, 17.7]

[265] Jaki S L 1978 *Johann Georg von Soldner and the gravitational bending of light* Found. Phys. **8** 927–950; Jaki S L 1978 *A forgotten bicentenary: Johann Georg Soldner* Sky and Telescope **55** (Jun.) 460–461. [§§ 1.5, 7.13]

[266] Jauch J M and Rohrlich F 1955 *The theory of photons and electrons* (Addison-Wesley, Reading, MA). [§ 14.6]

[267] Joliot F 1933 *Preuve expérimental de l'annihilation des électrons positifs (Experimental confirmation of the annihilation of positive electrons)* Comptes Rendus Acad. Sci. Paris **197** 1622–1625; 1934 **198** 81–85. [§ 3.4]

[268] Jordan P 1959 *Zum gegenwärtigen Stand der Diracschen kosmolgischen Hypothesen (On the present status of the Dirac cosmological hypotheses)* Z. Phys. **157** 112–121. [§ 17.6]

[269] Jungnickel C and McCormach R 1986 *Intellectual mastery of nature: theoretical physics from Ohm to Einstein*, vol. 1: *The torch of mathematics 1800–1870*, vol. 2: *The now mighty theoretical physics 1870–1925* (University of Chicago Press, Chicago, IL). [§§ 6.4, 12.1, 18.1, 21.6]

[270] Kahl R 1971 *Selected writings of H L F von Helmholtz* (Wesleyan University Press, Middletown, CT). [§ 21]

[271] Kaku M 1988 *Introduction to superstrings* (Springer-Verlag, New York).

[272] Kaluza T 1921 Preuss. Akad. Wiss. Berlin, Sitz.ber., 966. [§ 18.1]

[273] Kane G 1987 *Modern elementary particle physics* (Addison-Wesley, Reading, Mass.).

[274] Kaye G N C and Laby T N 1986 *Tables of physical and chemical constants* (Longman, New York). [§§ 5.3, 6.7]

[275] Kemmer N 1938 *Quantum theory of Einstein Bose particles* Proc. Roy. Soc. (London) **166** 127–153. [§ 16.8]

[276] Kepler J 1609 *Astronomia nova aitiologetos seu physica coelestes, tradita commentariis de motibus stellae, ex observationibus G V Tychonis Brahe (The new astronomy: based on causes or celestial physics)* (G Voegelinus, Heidelberg); French translation 1974 Peyroux J (ed.) *Astronomie nouvelle* (Bordeaux).

[277] Kerr R P 1962 *Gravitational field of a spinning mass as an example of algebraically special metrics* Phys. Rev. Lett. **11** 237–238. [§ 21.6]

[278] Kibble T W B 1960 *Lorentz invariance and the gravitational field* J. Math. Phys. **2** 212–221. [§ 21.6]

[279] Kibble T W B 1985 *Classical mechanics* 3rd edn. (Longman, London). [§ 8.10]

[280] Kippenhahn R 1983 *One hundred billion suns* (Weidenfeld and Nicolson). [§ 1.5]

[281] Kirk G S, Raven J E and Schofield M 1983 *The pre-Socratic philosophers* 2nd edn. (University Press, Cambridge). [§ 2]

[282] Kittel C, Knight W D and Ruderman M A 1965 *Mechanics: Berkeley Physics Course* vol. 1 (McGraw-Hill, New York). [§ 17.5]

[283] Klein C F 1918 *Über die Differentialsgesetze für Erhaltung von Impuls und Energie in der Einsteinschen Gravitationstheorie (On the differential conservation laws of momentum and energy in the Einstein gravity theory)* Nachr. kgl. Ges. Wiss. Göttingen pp 171–189. [§ 9.6]

[284] Klein O B 1926 *Quantentheorie und fünfdimensionale Relativitätstheorie (Quantum theory and 5-dimensional relativity theory)* Zeits. für Phys. **37** 895–906. [One of the original papers on the Klein-Gordon equation.] [§§ 16, 16.3, 18.1]

[285] Klein O B 1939 *Sur la théorie des champs associés à des particules chargées (On the field theory of charged particles)* pp. 81–98 in *Les Nouvelles théories de la physique (New theories of physics)*, Proc. Conf. in Warsaw, 1938, of the *Institut International de Cooperation Intellectuelle* (Paris). [§ 20.13]

[286] Klein M J 1966 *Thermodynamics and quanta in Planck's work* Physics Today **19** 23–32. [§§ 1.5, 6.7]

[287] Kobe D H 1984 *Helmholtz theorem for antisymmetric second-rank tensor fields and electromagnetism with magnetic monopoles* Am. J. Phys. **52** 354–358. [§§ 7.6, 18.11]

[288] Kramers H A 1937 *The use of charge-conjugated wave functions in the hole-theory of the electron* Proc. Amst. Acad. **40** 814–823. [§ 3.4]

[289] Kretschmann E 1917 *Über den physicalischen Sinn der Relativitäts postulate, A. Einsteins neue und seine ursprüngliche Relativitätstheorie. (On the physical significance of the relativity postulates, A. Einstein's original and new relativity theories)* Ann. Phys. (Germany) **53** 575–614. [§ 21.6]

[290] Kuhn T S 1957 *The Copernican revolution* (Harvard University Press, Cambridge, MA).

[291] Kuhn T S 1959 *Energy conservation as an example of simultaneous discovery* pp. 321-356 in Clagett M (ed.) *Critical problems in the history of science* (Madison, WI) [§ 6.4]

[292] Kundt W 1966 *Canonical quantization of gauge invariant field theories* Springer Tracts in Mod. Phys. **40** 107–168. [§ 15.8]

[293] Lagrange J-L 1759 *Recherches sur la méthode de maximis et minimis (Research on the method of maxima and minima)* Miscellanea Taurinensia **1**; reprinted in *Oeuvres* [301], vol. 1, pp. 1–20. [§ 8.1]

[294] Lagrange J-L 1760–61 *Essais d'une nouvelle méthode pour déterminer les maxima et les minima des formules intégrales indéfinies (Essay on a new method of determining the maxima and minima of indefinite integral fomulae)* Miscellanea Taurinensia **2**; reprinted in *Oeuvres* [301], vol. 1, pp. 333–362. [§ 8.1]

[295] Lagrange J-L 1760–61 *Applications de la méthode exposée dans le mémoire précédent a la solution de différents problèmes de la dynamique (Application of the method described in the preceding memoir to the solution of various dynamical problems)* Miscellanea Taurinensia **2**; reprinted in *Oeuvres* [301], vol. 1, pp. 363–468. [§ 8.2]

[296] Lagrange J-L 1777 *Remarques générales sur le mouvement de plusieurs corps qui s'attirent mutuellement en raison inverse des carrés des distances* (*General comments on the motion of several bodies with mutual inverse-square attraction*) Nouv. Mém. de l'Acad. de Berlin; reprinted in *Oeuvres* [301], vol. 4, pp. 399–418. [Lagrange's formula. §§ 6.4, 8.2, 18.1]

[297] Lagrange J-L 1780 *Théorie de la libration de la lune* (*Theory of the moon's libration*) Nouv. Mém. de l'Acad. de Berlin; reprinted in *Oeuvres* [301], vol. 5, pp. 3–122. [Constancy of total energy.] [§§ 6.4, 8.2]

[298] Lagrange J-L 1788 *Mécanique analytique* (Chez la Veuve Desaint, Paris); 1965 Édition complète, 2 vol., (Blanchard, Paris). [Introduction of the 'Lagrangian' function.] [§ 8.2]

[299] Lagrange J-L 1809 *Sur la théorie générale de la variation des constantes arbitraire dans tous les problèmes de la mécanique* Mém. de l'Inst. de Fr. reprinted in *Oeuvres* [301], vol. 6, pp. 769–805. [Use of the 'Lagrangian' function.] [§ 8.2]

[300] Lagrange J-L 1809 *Sur la théorie de la variation des constantes arbitraire dans les problèmes de la mécanique dans lequel on simplifie l'application des formules générales a ces problèmes* Mém. de l'Inst. de Fr. reprinted in *Oeuvres* [301], vol. 6, pp. 807–816. [§ 8.2]

[301] Lagrange J-L 1867–82 *Oeuvres* (Serret J-A, ed.) 13 vol. (Gauthier-Villars); reprinted 1910 Serret M S A and Darboux (eds.) (Paris).

[302] Lamoreaux S K et al. 1986 *New limits on spatial anisotropy from optically pumped ^{201}Hg and ^{199}Hg* Phys. Rev. Lett. 57 3125–3128. [§ 18.6]

[303] Lanczos C 1962 *The variational principle of mechanics* (University of Toronto Press, Toronto).

[304] Landau L D 1957 *On the conservation laws for weak interactions* Nucl. Phys. 3 127–131. [§§ 3.4, 16, 20.7]

[305] Landau L D and Lifschitz E M 1976 *Mechanics* 3rd Engl. edn. (Pergamon, Oxford). [§ 8.10]

[306] Landau L D and Lifschitz E M 1980 *The classical theory of fields* 4th Engl. edn. (Pergamon, Oxford).

[307] Landsberg P T (ed.) 1982 *The enigma of time* (Adam Hilger, Bristol).

[308] Laplace P-S de 1785 (for the year 1782) *Théorie des attractions des sphéroïdes et de la figure des planètes* (*Theory of the attraction of spheroids and the figure of the planets*) Mém. de l'Acad. Fr. p. 113; reprinted pp. 341–419 *Oeuvres complètes* [311], vol. 10. [Laplace's equation.] [§§ 6.4, 7.7, 18.1]

[309] Laplace P-S de 1789 (for the year 1787) *Théorie de l'anneau de Saturne* (*Theory of Saturn's rings*) Mém. de l'Acad. Roy. des Sci. de Paris; reprinted pp. 275–292 in *Oeuvres complètes* [311], vol. 11. [§ 7.7]

[310] Laplace P-S de 1796 *Exposition du système du monde* vol. 2, p. 305 (De l'Imprimerie du Circle Social, Paris); 1799 *Proof of the theorem, that the attractive force of a heavenly body could be so large, that light could not flow out of it* in Zach F X von (ed.) Allgemeine geographische Ephemeriden 4 269. [The passage referring to 'black holes' is reproduced in translation in Appendix A, pp. 365–368, of Hawking and Ellis [227].] [§§ 1.5, 7.12]

[311] Laplace P-S de 1895 *Oeuvres complètes: publiées sous les auspices de l'Académie des Sciences* (Gauthier-Villars, Paris).

[312] Laporte O 1924 *Die Structur des Eisenspektrums (The structure of the spectra of iron)* Z. für Phys. **23** 135–175. [§ 3.3]

[313] Laporte O and Uhlenbeck G E 1931 *Application of spinor analysis methods to the Maxwell and Dirac equations* Phys. Rev. **37** 1380–1397. [§ 20.2]

[314] Layzer D 1975 *The arrow of time* Sci. Amer. **233** (Dec.) 56–69.

[315] Lee T D 1971 *History of weak interactions*, an address given at Columbia University, 26 March 1971, unpublished. [§ 3.3]

[316] Lee A R and Kalotas T M 1975 *Lorentz transformations from the first postulate* Am. J. Phys. **43** 434–437. [§§ 5.5, 12.3]

[317] Lee T D and Yang C N 1956 *Question of parity conservation in weak interactions* Phys. Rev. **104** 254–258. [§ 3.3]

[318] Lee T D and Yang C N 1957 *Parity nonconservation and a two-component theory of the neutrino* Phys. Rev. **105** 1671–1675. [§§ 16, 20.7]

[319] Lee T D, Oehme R and Yang C N 1957 *Remarks on possible noninvariance under time reversal and charge conjugation* Phys. Rev. **106** 340–345. [§ 3.4]

[320] Lenz H F E 1834 *Über die Bestimmung der Richtung der durch elektrodynamische Vertheilung erregten galvanischen Ströme (On the determination of the direction of galvanic currents caused by an electrodynamic distribution)* Ann. d. Phys. **31** 485–494. [§ 18.1]

[321] Lévy-Leblond J-M 1976 *One more derivation of the Lorentz transformation* Am. J. Phys. **44** 271–277. [§ 5.5]

[322] Lévy M and Deser S (eds.) 1979 *Recent developments in gravitation (Cargèse Summer School 1978)* (Plenum, New York).

[323] Lewis G N and Tolman R C 1909 *The Principle of relativity and non-Newtonian mechanics* Phil. Mag. **18** 510–523. [§ 5.4]

[324] Lloyd H 1832 *On the phenomena presented by light in its passage along the axes of biaxial crystals* Trans. Roy. Irish Acad. **17** 145; reprinted pp. 1–18 in Lloyd H 1877 *Miscellaneous papers connected with physical science* (Longmans Green, London). [§ 11.3]

[325] Lodge O J 1885 *On the identity of energy: in connection with Mr Poynting's paper on the transfer of energy in an electromagnetic field; and on the two fundamental forms of energy* Phil. Mag. **19** 482–487. [§ 6.6]

[326] Longair M S 1983 *Theoretical concepts in physics: An alternative view of theoretical reasoning in physics for final-year undergraduates* (University Press, Cambridge). [§§ 12.1, 18.1]

[327] Lopez L L and Nanopoulos D V 1989 *Three generations: a prediction fulfilled* Preprint CTP-TAMU-62/89 (10 pp.) Center for Theoretical Physics, Texas A & M University. [§ 20.13]

[328] Lord E 1976 *Tensors, relativity and cosmology* (Tata McGraw-Hill, New Delhi). [§ 9.7]

[329] Lorentz H A 1875 *Over de theorie der terugkaatsing en breking van het licht* (*On the theory of the reflection and refraction of light*), Academisch proefschrift, (Doctoral thesis), University of Leiden (Van der Zande, Arnhem); Z. Math. Phys. **22** 1205; reprinted in *Collected papers* [599], vol. 1, pp. 1–192; French translation: *Sur la théorie de la réflexion et de la réfraction de la lumière* in *Collected papers*, vol. 1, pp. 193–383. [§§ 12.1, 18.6]

[330] Lorentz H A 1878 *De moleculaire theoriën in natuurkunde* Inaugural address at Leiden University, 25 January *Collected papers* [599], vol. 9, pp. 1–25; English translation *Molecular theories in physics* reprinted in *Collected papers* vol. 9, pp. 26–49. [§ 12.1]

[331] Lorentz H A 1887 *De l'influence du mouvement de la terre sur les phénomènes lumineux* (*The influence of the earth's movement on luminous phenomena*) Versl. Kon. Akad. Wetensch. (Amsterdam) **2** 297; 1887 Arch. néerl. **21** 103–176; reprinted in *Collected papers* [599], vol. 4, pp. 153–214. [§ 12.1]

[332] Lorentz H A 1892 *La théorie électromagnétique de Maxwell et son application aux corps mouvants* (*Maxwell's electromagnetic theory and its application to moving bodies*) Arch. néerl. **25** 363–551; reprinted in *Collected papers* [599], vol. 2, pp. 164–343. [The first of a series of papers on the electron theory.] [§§ 5.4, 12.1, 18.1]

[333] Lorentz H A 1892 *De relatieve beweging van de aarde en den aether* Versl. Kon. Akad. Wetensch. (Amsterdam) **1** 74–79; English translation *The relative motion of the earth in the ether* in *Collected papers* [599], vol. 4, pp. 219–223. [This paper contains the first suggestion by Lorentz of length contraction in the direction of motion.] [§ 5.4]

[334] Lorentz H A 1895 *Versuch einer Theorie der electrischen und optischen Erscheinungen in bewegten Körpern* (*Proposed theory for electrical and optical phenomena in moving bodies*) (Brill, Leiden), reprinted in *Collected papers* [599], vol. 5, pp. 1–138. [The introduction is available in English in Schaffner [496], pp. 247–254. The reconciliation of Fresnel's theory of light with the null result of Michelson by calculation of the 'Lorentz-FitzGerald' contraction is given in paragraphs 89–92, reprinted in English as *Michelson's interference experiment* [343], pp. 1–7.] [§§ 5.4, 12.1]

[335] Lorentz H A 1897 *Über den Einfluss magnetischer Kräfte auf die Emission des Lichtes* (*On the influence of magnetic fields on light emission*) Ann. d. Phys. **63** 279–284; French translation 1898 *Influence du champ magnétique sur l'émission lumineuse* Rev. d'électricité **14** 435, reprinted in *Collected papers* [599], vol. 3, pp. 40–46. [§ 5.4]

[336] Lorentz H A 1899 *La théorie simplifiée des phénomènes électriques et optiques dans des corps en mouvement*; French translation from Versl. Kon. Acad. Wetensch. (Amsterdam) **7** 507 in Arch. néerl. **7** 64 reprinted in *Collected papers* [599], vol. 5, pp. 139–155; English translation 1899 *Simplified theory of electrical and optical phenomena in moving systems* Proc. Roy. Acad. Sci. (Amsterdam) **1** 427, reprinted in [496], pp. 255–273. [§§ 5.4, 5.5, 12.1]

[337] Lorentz H A 1900 *Über die scheinbare masse der ionen* (*On the apparent mass of ions*) Physik. Z. **2** 78–80. [§ 5.4]

[338] Lorentz H A 1900 *Elektromagnetische Theorie physikalischer Erscheinungen* (*Electromagnetic theory of physical phenomena*) Phys. Zs. **1** (1899–1900) 498–501, reprinted in *Collected papers* [599], vol. 8, pp. 333–352. [§§ 5.4, 12.1, 21.5]

[339] Lorentz H A 1900 *Considérations sur la pesanteur* (*Concerning weight*) Versl. Kon. Akad. Wet. Amst. **8** 603–620; reprinted in *Collected papers* [599], vol. 5, pp. 198–215. [§§ 12.1, 17.7, 21.5]

[340] Lorentz H A 1904 (27 May) *Electromagnetische verschijnselen in een stelsel dat zich met willekeurige snelheid, kleiner dan die van het licht, beweegt* Versl. Kon. Acad. Wetensch. (Amsterdam) **12** 986–1009; English translation 1904 *Electromagnetic phenomena in a system moving with any velocity smaller than that of light* Proc. Roy. Acad. Sci. (Amsterdam) **6** 809–831; reprinted in Lorentz et al. [343], pp. 11–34 (without section 14) and in *Collected papers* [599], vol. 5, pp. 172–197. [Containing the definitive form of the spacetime transformations now known by the name of Lorentz.] [§§ 5.4, 5.5, 12.1]

[341] Lorentz H A 1904 *Elektronentheorie* (*Electron theory*) in *Encyklopädie der Mathematische Wissenschaften*, vol. 2 (Teubner, Leipzig); 1909 *Theory of electrons and its application to the phenomena of light and radiant heat* (Teubner, Leipzig); revised edn. 1915, reprinted 1952 (Dover, New York). [A course of lectures delivered in Columbia University, New York, March and April 1906.] [§§ 5.4, 18.1]

[342] Lorentz H A, Einstein A and Minkowski H 1913 *Das Relativitätsprinzip: Eine Sammlung von Abhandlungen* (Blumenthal O, ed.) (Teubner, Leipzig) (see [343].)

[343] Lorentz H A, Einstein A, Minkowski H and Weyl H 1920 *Das Relativitätsprinzip* (Sommerfeld A, ed.) (Teubner, Leipzig); English translation: *The Principle of relativity: a collection of original memoirs on the special and general theory of relativity* 1923 (Methuen, London); reprinted 1952 (Dover, New York). [§§ 5.4, 12.4, 14.1]

[344] Lovelock D and Rund H 1975 *Tensors, differential forms and variational principles* (Wiley, New York).

[345] Lubański J K 1942 *Sur la théorie des particules élémentaires de spin quelconque. I.* (*The theory of particles of arbitrary spin. I.*) Physica **9** 310–324. [§ 15.9]

[346] Ludwig W and Falter C 1988 *Symmetry in physics: group theory applied to physical problems* (Springer-Verlag, Berlin). [§§ 6.6, 11.5, 19.6, 19.10]

[347] Luther G G and Towler W R 1982 *Redetermination of the Newtonian gravitational constant G* Phys. Rev. Lett. **48** 121–123. [§ 7.4]

[348] MacCallum M A H (ed.) 1987 *General relativity and gravitation: Proc. of the 11th International Conference on General relativity and gravitation, Stockholm, July 6–12, 1986* (University Press, Cambridge).

[349] MacCurdy E 1938 *The notebooks of Leonardo da Vinci* (Cape, London). [§ 1]

[350] Macdonald A 1980 *Derivation of the Lorentz transformation* Am. J. Phys. **49** 493. [§ 5.4]

[351] MacDowell S W and Mansouri F 1977 *Unified geometric theory of gravity and supergravity* Phys. Rev. Lett. **38** 739–742. [§ 21.6]

[352] Mach E 1883 *Die Mechanik in ihrer Entwickelung historisch-kritisch dargestellt* (Brockhaus, Leipzig); English translation McCormach T J 1893 *The science of mechanics: a critical and historical exposition of its principles* (Chicago); 1960 6th edn. (Open Court, La Salle, IL). [§§ 5.2, 6.3, 17.6, 18.2]

[353] Mackay Alan L 1977 *The harvest of a quiet eye: A selection of scientific quotations* (Institute of Physics, London). [§§ 2, 3, 3.7, 4, 5.8]

[354] Macq P C, Crowe K M and Haddock R P 1958 *Helicity of the electron and positron in muon decay* Phys. Rev. **112** 2061–2071. [§ 3.4]

[355] Maglic B (ed.) 1973 *Adventures in experimental physics*, pp. 93–162, *Gamma volume* (World Science Education, Princeton). [§ 3.3]

[356] Majorana E 1937 *Teoria simmetrica dell'elettrone e del positrone* Nuovo Cimento **5** 171–184; English translation by Maiani L 1981 *Symmetric theory of the electron and the positron* Soryushiron Kenkyu **63** 149–162. [§ 16]

[357] Maricourt P de 1269 *Epistola Petri Peregrini de Maricourt ad Sugerum de Foucancourt, militem, de magnete (Letter on the magnet of Peter Peregrinus of Maricourt to Sygerus of Foucancourt, soldier).* [Introduction of the concept of magnetic poles.] [§ 18.1]

[358] Marion J B 1970 *Classical dynamics of particles and systems* (Academic Press, New York); 3rd edn. Marion J B and Thornton S T 1988 (Harcourt Brace Jovanovich, San Diego). [§ 8.10]

[359] Mariwalla K H 1980 *Uniqueness of classical and relativistic systems* Phys. Lett. **A79** 143–146. [§ 5.5]

[360] Mariwalla K H 1982 *No escaping the second postulate* Am. J. Phys. **50** 583. [§ 5.4]

[361] Marshak R E and Sudarshan E C G 1957 *The nature of the four fermion interaction* pp. V:14–22 in *Proc. Conference on mesons and recently discovered particles*, Padua-Venice 22–27 September 1957; 1958 *Chirality invariance and the universal Fermi interactions* Phys. Rev. **109** 1860–1862. [§ 20.13]

[362] Maupertuis P L M de 1746 *Sur les loix du mouvement (On the laws of motion)* Acad. Roy. Sci. Berlin; reprinted in 1965 *Oeuvres* vol. 2 (Georg Olms, Hildesheim) pp. 270–274. [§§ 8, 8.5]

[363] Maxwell J C 1855, 1856 *On Faraday's lines of force* Trans. Camb. Phil. Soc. **10** Part I 27–83. [Maxwell's first paper on electromagnetism, in two parts.] [§§ 18.6, 18.12]

[364] Maxwell J C 1861-1862 *On physical lines of force* (in 4 parts) Phil. Mag. **21** 161-175; **21** 281-291, 338-348 [containing the phenonological prediction that light consists in electromagnetic vibrations]; **23** 12-24; **23** 85-95; reprinted in Niven [398]. [§§ 12.1, 18.1, 18.12]

[365] Maxwell J C and Jenkin F 1863 *On the elementary relations of electrical quantities* Rep. Brit. Assoc. Adv. Sci. **32** 130-163; reprinted with additions in Jenkin F 1873 Rep. Committee of Electrical Standards (London) pp. 59-96. [This vital paper is unfortunately omitted from the *Scientific papers* [398] of Maxwell and is therefore often overlooked. Apart from providing crucial steps in the development of Maxwell's ideas, it contains the first description of the dual system of electrical units generally referred to to as *Gaussian*.] [§ 18.1]

[366] Maxwell J C 1865 *A dynamical theory of the electromagnetic field* Phil. Trans. Roy. Soc. (London) **155** 459-512. [Providing a theoretical framework for the propagation of electromagnetic waves.] [§§ 12.1, 18, 18.1]

[367] Maxwell J C 1868 *Note on the electromagnetic theory of light* reprinted in [398], pp. 137-143, originally part of a longer article: Phil. Trans. Roy. Soc. (London) **158** 643-657. [Providing a simplified less dynamical version, which became the standard textbook treatment, of the basic electromagnetic equations.] [§ 18.6]

[368] Maxwell J C 1873 *A treatise on electricity and magnetism* 2 vol. (University Press, Oxford); 1891 3rd revised edn. Thomson J J (ed.) 1954 reprint edn. (Dover, New York).

[369] Maxwell J C 1876 *Matter and motion* (Society for promoting Christian knowledge, London); reprinted with notes by Larmor J (ed.) 1952 (Dover, New York). [§ 5.4]

[370] Maxwell J C 1879 *Ether* (article in the *Encyclopaedia Britannica*) reprinted in [398], pp. 763-775. [§ 12.1]

[371] McCormach R 1970 *Einstein, Lorentz, and the electron theory* Hist. Stud. Phys. Sci. **2** 41-87. [§§ 5.4, 12.1]

[372] McCormach R 1970 *H A Lorentz and the electromagnetic view of nature* Isis **61** 459-497. [§§ 5.4, 12.1]

[373] McCormach R 1971 *Hendrik Antoon Lorentz* pp. 487-500 in Gillispie [197], vol. 8. [§ 12.1]

[374] Mermin N D 1984 *Relativity without light* Am. J. Phys. **52** 119-124. [§ 5.4]

[375] Merzbacher E 1970 *Quantum mechanics* 2nd edn. (Wiley, New York). [§ 10.6]

[376] Michell J 1750 *A treatise of artificial magnets; in which is shown an easy and expeditious method of making them superior to the best natural ones.* [§ 18.1]

[377] Michell J 1784 *On the means of discovering the distance, magnitude, etc. of the fixed stars* Phil. Trans. Roy. Soc. (London) **74** 35-57. [§§ 1.5, 7.12]

[378] Michelson A A 1881 *The relative motion of the earth and the luminiferous ether* Am. J. Sci. **22** 120-129. [§§ 5.4, 12.1]

[379] Michelson A A and Morley E W 1887 *On the relative motion of the earth and the lumniferous ether* Am. J. Sci. **34** 333–345; 1887 *On the relative motion of the earth and the lumniferous ether* Phil. Mag. **24** 449–463. [§§ 5.4, 12.1]

[380] Miller A I 1973 *A study of Henri Poincaré's 'Sur la dynamique de l'électron'* Arch. Hist. Ex. Sci. **10** 207–328. [§ 12.1]

[381] Minkowski H 1907 *Das Relativitätsprinzip (The Relativity principle)* Ann. der Phys. **47** 927–938. [§ 12.4]

[382] Minkowski H 1908 *Die Grundgleichungen für die elektromagnetischen Vorgänge in bewegten Körpern (Fundamental equations for electromagnetic processes in a moving body)* Gött. Nachr., p. 53; 1910 *Zwei Abhandlungen über die Grundgleichungen der Elektrodynamik (Two essays on the basic equations of electrodynamics)* (Leipzig). [§ 12.4]

[383] Minkowski H 1909 *Raum und Zeit* Phys. Zs. **10** 104–111; Address delivered at the 80th Assembly of German natural scientists and physicians, Cologne 21 September 1908; reprinted in English *Space and time*; in reference [343], p. 73–91. [§ 12.4]

[384] Misner C W, Thorne K S and Wheeler J A 1973 *Gravitation* (Freeman, San Francisco). [§§ 2.2, 2.4, 4.4, 7.11, 7.14, 7.15, 15.2, 17.1, 17.5, 17.6, 17.7, 18.7, 18.11, 20.11, 21.3, 21.4, 21.6, A.4]

[385] Misner C W and Wheeler J A 1957 *Classical physics as geometry: Gravitation, electromagnetism, unquantized charge and mass as properties of curved empty space* Ann. Phys. (U.S.A.) **2** 525–603. [§ 18.11]

[386] Møller C 1962 *The theory of relativity* (Universtity Press, Oxford).

[387] Mook D E and Vargish T 1987 *Inside relativity* (University Press, Princeton). [§§ 5.1, 12.5]

[388] Moore J H 1928 *Recent spectrographic observations of the companion of Sirius* Publ. Astr. Soc. Pac. **40** 229–233. [§ 7.14]

[389] Morris R 1984 *Time's arrows* (Simon and Schuster, New York).

[390] Morrison P 1957 *The overthrow of parity* Sci. Amer. **196** (Apr.) 45–53.

[391] Muirhead H 1965 *The physics of elementary particles* (Pergamon, Oxford). [§§ 3.3, 3.4, 3.5]

[392] Murdoch D 1987 *Niels Bohr's philosophy of physics* (University Press, Cambridge). [§ 10.6]

[393] Newman E T, Couch E, Chinnapared, Exton A, Prakash A and Torrence R 1965 *Metric of a rotating charged mass* J. Math. Phys. **6** 918–919. [§ 21.6]

[394] Newman J R (ed.) 1956 *The world of mathematics* (Simon and Schuster, New York). [§ 1.6]

[395] Newton I 1687 *Philosophiae naturalis principia mathematica (Mathematical principles of natural philosophy,* referred to as the *Principia)* 1st edn. (Streater, London), 1713 2nd edn. (Cambridge), 1726 3rd edn. (London); final edition (prepared by Newton) in English translation by Motte A 1729

(London); revised translation by Cajori F 1934, 1966 *Sir Isaac Newton's mathematical principles of natural philosophy and his system of the world* (University of California Press, Berkeley). [Our references will be to the Cajori edition. See also Cohen [71] and Brewster [50].] [§§ 3, 5, 5.2, 5.6, 6.1, 6.7, 7.11, 18.1]

[396] Newton I 1704 1st edn.; 1730 *Opticks, or, a treatise of the reflections, refractions, inflections & colours of light* 4th edn. (London); reprint edn. 1952 (Dover, New York). [§§ 6.3, 7.3, 7.13]

[397] Nieto M M, Goldman T and Richards J H 1988 *The principle of equivalence, quantum gravity and new gravitational forces* Austr. Phys. **25** 259–262. [§§ 7.11, 7.14, 17.1, 17.7]

[398] Niven W D (ed.) 1890 *The scientific papers of J. Clerk Maxwell* 2 vol. (University Press, Cambridge); reprinted in 1 vol. 1952 (Dover, New York). [§ 18.1]

[399] Noether E 1918 *Invariante Variationsprobleme (Invariance in variation problems)* Nachr. kgl. Ges. Wiss. Göttingen Math. physik. Kl. pp. 235–257. [§ 9.6]

[400] Nordstrøm G 1913 *Zur Theorie der Gravitation vom Standpunkt des Relativitätsprinzips (On the relativistic theory of gravitation)* Ann. Phys. (Germany) **42** 533–554. [§§ 17.1, 17.6, 21.3]

[401] Nordstrøm G 1918 *On the energy of the gravitational field in Einstein's theory* Proc. Roy. Acad. Sci. (Amsterdam) **20** 1238–1245. [§ 21.6]

[402] Nordtvedt K Jr 1968 *Equivalence principle for massive bodies I. Phenomenology, II. Theory* Phys. Rev. **169** 1014–1016, 1017–1025; 1968 *Testing relativity with laser ranging to the moon* Phys. Rev. **170** 1186–1187. [§ 17.6]

[403] Oersted H C 1820 *Experimenta circa effectum conflictus electrici in acum magneticam (Experiments on the effect of current electricity on the magnetic needle)* (Copenhagen); English translation in Thomson 1820 *Annals of philosophy*, vol. 15, p. 273. [§ 18.1]

[404] Ohanian H C 1976 *Gravitation and spacetime* (Norton, New York).

[405] Ohanian H C 1988 *Classical electrodynamics* (Allyn and Bacon, Boston).

[406] Ohm G S 1825 *Vorläufige Anzeige des Gesetzes, nach welchem Metalle die Contaktelektricität leiten (Provisional indication of the laws governing the contact electricity of metals)* J. f. Chem. u. Phys. **44** 79–88. [§ 18.1]

[407] Ohm G S 1826 *Bestimmung des Gesetzes, nach welchem Metalle die Contaktelektricität leiten, nebst einem Entwurfe zu einer Theorie des Voltaischen Apparates und des Schweiggerschen Multiplicators (Determination of the laws governing the contact electricity of metals together with a scheme for a theory of voltaic apparatuses and Schweigger multipliers)* J. f. Chem. u. Phys. **46** 137–166. [§ 18.1]

[408] Ohm G S 1826 *Versuch einer Theorie der durch galvanische Kräfte hervorgebrachten elektroscopischen Erscheinungen (Toward a theory of galvanic forces from electroscopic phenomena)* Poggendorf's Ann. d. Phys. u. Chem. **6** 459–469; **7** 45–54, 117–118. [§ 18.1]

[409] Ohm G S 1827 *Die galvanische Kette, mathematische bearbeit (The galvanic circuit investigated mathematically)* (Berlin); reprinted 1892 in Lommel E (ed.) *Ohm G S: Gesammelte Abhandlungen*, pp. 61–186 (Leipzig). [§ 18.1]

[410] Ohnuki Y 1988 *Unitary representations of the Poincaré group and relativistic wave equations* (World Scientific, Singapore). [§§ 3.4, 3.5, 14.2, 20.2, 20.3, 20.6, 20.10, 20.14]

[411] Okun L B 1984 *Leptons and quarks* (North-Holland, Amsterdam).

[412] Olver P J *Applications of Lie groups to differential equations* (Springer Verlag, New York).

[413] O'Raifeartaigh L (ed.) 1972 *General relativity* (Clarendon Press, Oxford).

[414] Overseth O E 1969 *Experiments in time reversal* Sci. Amer. **221** (Oct.) 88–101. [§ 3.3]

[415] Page L 1952 *Introduction to theoretical physics* (van Nostrand, Princeton).

[416] Page L and Adams N I, Jr. 1940 *Electrodynamics* (van Nostrand, New York).

[417] Paik H J 1987 *Terrestrial experiments to test theories of gravitation* in MacCallum [348]. [§§ 7.14, 17.1, 17.7]

[418] Pais A 1982 *Subtle is the Lord: the science and the life of Albert Einstein* (University Press, Oxford).

[419] Pais A 1986 *Inward bound: Of matter and forces in the physical world* (University Press, Oxford).

[420] Panofsky W K H, Aamodt R L and Hadley J 1951 *Gamma-ray spectrum resulting from capture of negative π-mesons in hydrogen and deuterium* Phys. Rev. **81** 565–574. [§ 16.8]

[421] Panofsky W K H and Phillips M 1955 *Classical electricity and magnetism* (Addison-Wesley, Cambridge, MA). [§§ 18.1, A.3]

[422] Parker B 1987 *Search for a supertheory: from atoms to superstrings* (Plenum, New York).

[423] Pauli W 1921 *Relativitätstheorie* (Teubner, Leipzig); translation, 1981 *Theory of relativity* (Dover, New York).

[424] Pauli W 1933 *Die allgemeinen Prinzipien der Wellenmechanik (The general principles of wave mechanics)* pp. 83–272 in Smekal [514]. [§§ 20.2, 20.7, 20.11]

[425] Pauli W 1934 *Pauli intervention* pp. 324–325 in *Structure et propriétés des noyaux atomiques* Rapports et discussions du septième Conseil de physique, tenu à Bruxelles du 25 au 29 Octobre 1933 (Gauthier-Villars, Paris). English translation in Brown [52]. [§ 20.13]

[426] Pauli W 1935 *Beiträge zur mathematischen Theorie der Diracshen Matrixes (Contribution to the mathematical theory of Dirac matrices)* pp. 31–35 in *Peter Zeeman 1865-25 Maj-1935* (Martin Nijhoff, 's-Gravenhage). [§ 20.2]

[427] Pauli W 1936 *Contributions mathematiques à la théorie des matrices de Dirac (Mathematical contributions to the theory of Dirac matrices)* Ann. Inst. Henri Poincaré **6** 137–152. [§ 20.2]

[428] Pauli W 1940 *The connection between spin and statistics* Phys. Rev. **58** 716–722. [§ 20.14]

[429] Pauli W 1955 *Niels Bohr and the development of physics* (McGraw-Hill, New York). [CPT theorem.] [§§ 3.5, 16.1]

[430] Pease R and Lindley D 1989 *Is the end of particle proliferation at hand?* Nature **341** 555. [§ 20.13]

[431] Penrose R 1967 *Structure of space-time* pp. 121–235 in De Witt C M and Wheeler J A (eds.) [102]. [§§ 3.1, 7.15]

[432] Penrose R 1972 *Black holes* Sci. Amer. **226** (May) 38–55.

[433] Penrose R 1980 *Physical space-time and non-realizable CR-structures* pp. 401–421 in Browder F E (ed.) *The mathematical heritage of Henri Poincaré*, Part I, Proc. of Symp. in Pure Math., vol. 39 (Amer. Math. Soc., Providence, Rhode Island). [§§ 3.1, 7.15]

[434] Penrose R and Rindler W 1984 *Spinors and spacetime* 2 vol. (University Press, Cambridge). [§ 12.5]

[435] Perkins D H 1987 *Introduction to high energy physics* (Addison-Wesley, Reading, MA).

[436] Planck M 1899 *Über irreversible Strahlungsvorgänge; fünfte Mittheilung (On irreversible radiation phenomena; fifth part)* Sitz. preuss. Akad. Wiss. 440–480; 1900 *Über irreversible Strahlungsvorgänge* Ann. d. Phys. (Germany)**306** 69–122. [§§ 1.5, 6.7]

[437] Planck M 1900 *Über eine Verbesserung der Wien'schen Spektralgleichung* Verhandl. der Deutschen Physikal. Gesellsch. **2** 202–204; English translation *On an improvement of the Wien radiation law* in Kangro H 1972 *Planck's original papers in quantum physics* (Taylor and Francis, London), discussed in Kangro H 1976 *Early history of Planck's radiation law* (Taylor and Francis, London). [§ 12.1]

[438] Poincaré H 1895 *A propos de la théorie de M. Larmor (On Larmor's theory)* L'Éclairage électrique (in 4 parts) **3** 5–13, 289–295; **5** 5–14, 385–392; reprinted in *Oeuvres* [451], vol. 9, pp. 369–382, 383–394, 395–413, 414–426. [§§ 5.4, 15.2]

[439] Poincaré H 1897 *La théorie de Lorentz et les expériences de Zeeman (Lorentz's theory and Zeeman's experiments)* L'Éclairage électrique **11** 481–489, reprinted in *Oeuvres* [451], vol. 9, pp. 427–441.

[440] Poincaré H 1898 *La mesure du temps* Rév. de Métaphys. et de Morales **6** 1–13. [§§ 2.2, 5.2, 5.4]

[441] Poincaré H 1899 *La théorie de Lorentz et le phénomène de Zeeman (Lorentz's theory and the Zeeman effect)* L'Éclairage électrique **19** 5–15, reprinted in *Oeuvres* [451], vol. 9, pp. 442–460.

[442] Poincaré H 1900 *La théorie de Lorentz et le principe de réaction (Lorentz's theory and the principle of reaction)* Arch. néerl. **5** 252–278, reprinted in *Oeuvres* [451], vol. 9, pp. 464–488. [§§ 5.4, 15.2, 19]

[443] Poincaré H 1900 *Relations entre la physique expérimentale et de la physique mathématique* (*The relations between experimental and theoretical physics*). [Report presented to the International Physics Congress in Paris] (Gauthier-Villars, Paris), vol. 1, pp. 1–29, reprinted in 1900 Rév. gén. sci. pures et appl. **11** 1163–1175 and in 1900 Rév. scient. **14** 705–715.

[444] Poincaré H 1901 *Éléctricité et optique*; (Carré et Naud, Paris). [Lectures delivered at the Sorbonne in 1899. Earlier editions date from 1890.]

[445] Poincaré H 1903 *La science y l'hypothèse*; 1904 *La valeur de la science*; 1908 *Science et méthode*; 1914 *Dernières pensées* (Flammarion, Paris); English translations: Halsted G B 1982 *The foundations of science: Science and hypothesis; The value of science; Science and method* (University Press of America, Washington, D.C.) and 1963 *Last thoughts* (Dover, New York). [§§ 2.4, 7, 11]

[446] Poincaré H 1904 *L'état actuel et l'avenir de la physique mathématique* (*The present state and future of mathematical physics*) Bull. des sci. mathém. **28** 302–323, reprinted in *La valeur de la science* [445], pp. 200–211, as *L'avenir de la physique mathématique*. [§§ 5.2, 5.4, 5.5, 15.2]

[447] Poincaré H 1905 *The principles of mathematical physics* Monist **15** 1–24, an address delivered before the International Congress of Arts and Science, St. Louis, September 1904, translated by Halsted G B. [§§ 5.3, 5.4]

[448] Poincaré H 1905 (dated 5 June 1905) *Sur la dynamique de l'électron* (*The dynamics of the electron*) Compt. Rend. **140** 1504–1508, reprinted in *Oeuvres* [451], vol. 9, pp. 489–493. [§§ 5.4, 5.5, 12.3, 13.3, 18.6]

[449] Poincaré H 1906 (dated 23 July 1905) *Sur la dynamique de l'électron* Rend. Circ. Mat. Palermo **21** 129–176, reprinted in [451], vol. 9, pp. 494–550. [§§ 5.5, 12, 12.3]

[450] Poincaré H 1908 *Sur la dynamique de l'électron* Rév. de Sci. **19** 386–402, reprinted in [451], vol. 9, pp. 551–586; 1913 *Sur la dynamique de l'électron: Supplément aux annales de postes, télégraphes et téléphones* (Dumas, Paris).

[451] Poincaré H 1951–56 *Oeuvres: publiées sous les auspices de l'Académie des Sciences* (*Collected papers*) 11 vol. (Gauthier-Villars, Paris). [§§ 5.4, 5.5, 12, 15.2]

[452] Poisson S-D 1809 *Mémoire sur la variation des constantes arbitraires dans les questions de mécanique* (*Memoir on the variation of constants of motion in mechanics*) J. École Polytech. **8** 266–344. [Original article on Poisson Brackets.] [§ 11.5]

[453] Poisson S-D 1812 *Sur la distribution de l'électricité à la surface des corps conducteurs* (*On the surface distribution of electricity on conductors*) Mém. de l'Institut Fr., année 1811 pt. i, pp. 1–92; pt. ii, pp. 163–274. [§§ 6.4, 18.1]

[454] Poisson S-D 1813 *Remarques sur une équation qui se présente dans la théorie des attractions des sphéroïdes* (*Remarks on an equation that appears in the theory of attraction of spheroids*) Bull. de la Soc. Philomathique de Paris **3** 388–392. [§§ 6.4, 18.1]

[455] Polyakov A M 1974 *Particle spectrum in quantum field theory* JETP Lett. **20** 194–195; 1975 *Isomeric states of quantum fields* Sov. Phys. JETP **41** 988–995. [Magnetic monopoles.] [§ 18.11]

[456] Popper D M 1954 *Red shift in the spectrum of 40 Eriadni B* Astrophys. J. **120** 316–321. [§ 7.14]

[457] Pound R V and Rebka G A 1960 *Apparent weight of photons* Phys. Rev. Lett. **4** 337–342. [§ 7.14]

[458] Pound R V and Snider J L 1964 *Effect of gravity on nuclear resonance* Phys. Rev. Lett. **13** 539–540; 1965 *Effect of gravity on gamma radiation* Phys. Rev. **140B** 788–803. [§ 7.14]

[459] Poynting J H 1884 *On the transfer of energy in the electromagnetic field* Phil. Trans. **175** 343–361. [§§ 6.6, 18.12]

[460] Price D J de S 1962 *Science since Babylon* (Yale UP, New Haven, CT). [§ 4]

[461] Priestley J 1767 *The history and present state of electricity with original experiments* (London). [§ 18.1]

[462] Proca A 1936 *Sur les équations fondamentales des particules élémentaires (On the fundamental equations of elementary particles)* Compt. Rend. **202** 1490–1492. [§§ 16, 18.11, 18.13]

[463] Purcell E M and Ramsey N F 1950 *On the possibility of electric dipole moments for elementary particles and nuclei* Phys. Rev. **78** 807. [§ 3.3]

[464] Purcell E M 1963 *Electricity and magnetism* (McGraw-Hill, New York), p. 153. [§ 18.3]

[465] Pyenson L 1985 *The young Einstein* (Adam Hilger, Bristol).

[466] Quinn H R et al., 1989 *Teachers' resource book on fundamental particles and interactions* (University of California, Berkeley). [§ 20.13]

[467] Ramond P 1981 *Field theory: a modern primer* (Benjamin Cummings, Reading, MA). [§§ 9.7, 13.1, 16.5, 20.9]

[468] Ramond P 1983 *Gauge theories and their unification* Ann. Rev. Nucl. Part. Sci. **33** 31–66. [§§ 20.13, 21.7]

[469] Rarita W and Schwinger J 1941 *On a theory of particles with half-integer spin* Phys. Rev. **60** 61. [§§ 16, 21.2]

[470] Reichenbach H 1958 *Philosophy of space and time* (Dover, New York).

[471] Reissner H 1916 *Über die Eigengravitation des elektrischen Feldes nach der Einsteinschen Theorie (On the intrinsic gravitation of electrical fields in Einstein's theory)* Ann. Phys. (Germany) **50** 106–120. [§ 21.6]

[472] Resnick R 1968 *Introduction to special relativity* (Wiley, New York). [§ 5.5]

[473] Riemann B 1858 (unpublished) *Ein Beitrag zur Elektrodynamik (A contribution to electrodynamics)* published posthumously: 1867 Ann. d. Phys. **131** 237–243, reprinted in *Werke* [475], pp. 270–275. [§ 18.1]

[474] Riemann B 1876 *Schwere, Elektricität und Magnetismus, nach den Vorlesungen von Bernard Riemann* (*Weight, electricity and magnetism, from the lectures of Bernard Riemann*), Hattendorf K (ed.) (Hannover). [§ 18.1]

[475] Riemann B 1876 *Gesammelte mathematische Werke und wissenschaftlicher Nachlass* (*Collected mathematical works and scientific papers*), Weber H (ed.) (Leipzig).

[476] Rindler W 1977 *Essential relativity* (Springer-Verlag, New York).

[477] Rindler W 1982 *Introduction to special relativity* (University Press, Oxford).

[478] Robertson H P 1935 *Kinematics and world structure* Astrophys. J. **82** 248–301; 1936 *Kinematics and world structure* **83** 187–201, 257–271. [§ 21.6]

[479] Robertson D S and Carter W E 1984 *Relativistic deflection of radio signals in the solar gravitational field measured with VLBI* Nature **310** 572–574. [§ 21.4]

[480] Robertson H P and Noonan T W 1968 *Relativity and cosmology* (Saunders, Philadelphia). [§ 13.2]

[481] Roll P G, Krotkov R and Dicke R 1964 *The equivalence of inertial and passive gravitational mass* Ann. Phys. (USA) **26** 442–517. [§ 7.11]

[482] Roman P 1969 *Introduction to quantum field theory* (Wiley, New York).

[483] Roos M 1963 *Tables of elementary particle and resonant states* (first in series) Rev. Mod. Phys. **35** 31–323; Roos M et al., (Particle data group) 1976 *Review of particle properties*(last in series) Rev. Mod. Phys. **48** No. 2, Part II pp. S1– S246.

[484] Rosen E 1971 *Nicholas Copernicus* pp. 401–411 in Gillispie [197], vol. 3. [§§ 3.6, 5.1]

[485] Rosenfeld L 1940 *Sur le tenseur d'impulsion-énergie* (*On the energy-momentum tensor*) Acad. Roy. Belgique classe sci. **18** Mém. No. 6, 1–30. [§ 19.5]

[486] Rudenko V N 1978 *Relativistic experiments in gravitational fields* Sov. Phys. Uspecki **21** 893–916. [§§ 7.11, 7.13]

[487] Ruffini R and Wheeler J A 1971 *Introducing the black hole* Physics Today **24** (Jan.) 30–41.

[488] Ryder L H 1985 *Quantum field theory* (University Press, Cambridge). [§§ 8.8, 11.1, 15.3, 16.1, 18.11, 20.11]

[489] Sachs R G 1987 *Physics of time reversal* (University of Chicago Press, Chicago).

[490] Sakurai J J 1964 *Invariance principles and elementary particles* (University Press, Princeton).

[491] Salam A 1957 *On parity conservation and neutrino mass* Nuovo Cimento **5** 299–301. [§§ 16, 20.7]

[492] Salam A 1968 pp. 367–377 in Svartholm N (ed.) *Elementary particle theory; relativistic groups and analyticity*, Proc. Eight Nobel Symposium, 19–25 May, 1968, (Almquist and Wiksells, Stockholm, Wiley New York). [§§ 20.13, 21.7]

[493] Salam A and Ward J C 1964 *Electromagnetic and weak interactions* Phys. Lett. **13** 168–171. [§§ 20.13, 21.7]

[494] Santilli R M 1978 *Foundations of theoretical mechanics* (Springer-Verlag, Heidelberg). [§ 8.6]

[495] Schaffner K F *The Lorentz electron theory of relativity* Am. J. Phys. **37** 498–513. [§§ 5.4, 12.1]

[496] Schaffner K F 1972 *Nineteenth-century aether theories* (University Press, Oxford).

[497] Schilpp P A (ed.) 1949 *Albert Einstein: Philosopher-scientist* (Library of Living Philosophers, Evanston, IL); 2nd edn. 1951 (Tudor, New York). [§ 14]

[498] Schrödinger E 1926 *Quantisierung als Eigenwertproblem* Ann. d Phys. (Germany) (Parts I to IV) **79** 361–376, 489–527; **80** 437–490; **81** 109–139; reprinted in English, *Quantization as a problem of proper values* in Schrödinger E 1928 *Collected papers on wave mechanics* translated by Shearer J F and Deans W M (Blackie and Sons, London). [The first of these papers sets out the non-relativistic Schrödinger equation for the first time. The fourth provides its relativistic generalization, now known as the Klein-Gordon equation.] [§§ 10, 10.6, 16, 16.3]

[499] Schutz B F 1980 *Geometrical methods of mathematical physics* (University Press, Cambridge). [§§ 4.4, 5.8, 7.10, 9.2, 21.6]

[500] Schutz B F 1985 *A first course in general relativity* (University Press, Cambridge).

[501] Schwartz H M 1984 *Deduction of the general Lorentz transformations from a set of necessary assumptions* Am. J. Phys. **52** 346–350. [§ 5.4]

[502] Schwartz J 1987 *Superstrings* Physics Today **40** 33–40.

[503] Schwarzschild K 1916 *Über das Gravitationsfeld eines Massenpunktes nach der Einsteinschen Theorie* (*On the gravitational field of a mass point in Einstein's theory*) Preuss. Akad. Wiss. Berlin, Sitz.ber., 424–434. [§ 21.6]

[504] Schweber S S 1961 *An introduction to relativistic quantum field theory* (Harper and Row, New York).

[505] Schwinger J 1951 *Theory of quantized fields. I.* Phys. Rev. **82** 914–927. [Spin-statistics theorem.] [§ 3.5]

[506] Schwinger J 1953 *Theory of quantized fields I. and II.* Phys. Rev. **91** 713–728; 1953 *Theory of quantized fields VI.* Phys. Rev. **94** 1362–1384. [CPT theorem.] [§ 16.1]

[507] Scott G D and Viner M R 1965 *The geometrical appearance of large objects moving at relativistic speeds* Am. J. Phys. **33** 534–536. [§ 12.5]

[508] Seelig C 1952 *Albert Einstein: Eine dokumentarische Biographie* (Europa-Verlag, Zurich); English translation by Savill M 1956 *Albert Einstein: a documentary biography* (Staples, London). [§§ 17.5, 21.5]

[509] Shankland R S 1964 *Michelson-Morley experiment* Am. J. Phys. **32** 16–35. [§ 15]

[510] Shanmugadhasan S 1973 *Canonical formalism for degenerate Lagrangians* J. Math. Phys. **14** 677–687. [§ 15.8]

[511] Shapiro I I, Counselman C C and King R W 1976 *Verification of the Principle of equivalence for massive bodies* Phys. Rev. Lett. **36** 555–558. [§ 17.6]

[512] Shapiro S L and Teukolsky S A 1983 *Black holes, white dwarfs and neutron stars: the physics of compact objects* (Wiley, New York). [§§ 1.5, 1.5, 17.6]

[513] Skylar L 1974 *Space, time and spacetime* (University of California Press, Berkeley).

[514] Smekal A (ed.) 1933 *Handbuch der Physik*, vol. 24/1 (Springer Verlag, Berlin)

[515] Soldner, Johann Georg von 1801 *Über die Ablenkung eines Lichtstrals von seiner geradlinigen Bewegung, durch die Attraktion eines Weltkörpers, an welchem er nahe vorbei geht* in Bode J E Astron. Jahrbuch (Berl.) für 1804 pp. 161–172; English translation *On the deviation of a light ray from its rectilinear motion, through the attraction of a celestial body which it passes close by* in Jaki S L [265]. [§§ 7.13, 21.4]

[516] Soper D E 1975 *Classical field theory* (Wiley, New York). [§§ 4.8, 12.5]

[517] Srivistava P P 1973 *Conformal symmetry in Lagrangian field theory* Nucl. Phys. **B64** 499–510; 1973 *Conformal symmetry in Lagrangian field theory* Revista Brasileira de Física **3** 577–599. [§§ 9.7, 10.3]

[518] Srivistava P P 1986 *Supersymmetry, superfields, supergravity: an introduction* (Adam Hilger, Bristol).

[519] Stacey F D, Tuck G J, Moore G I, Holding S C, Goodwin B D and Zhou R 1987 *Geophysics and the law of gravity* Rev. Mod. Phys. **59** 157–174. [§§ 7.4, 7.11, 7.14, 17.1, 17.7]

[520] Stedman G E 1989 *Diagram techniques in group theory* (University Press, Cambridge). [§ 4.6]

[521] Stewart I 1987 *The problems of mathematics* (University Press, Oxford). [§§ 2.4, 3, 12.6]

[522] Stratton J A 1941 *Electromagnetic theory* (McGraw-Hill, New York).

[523] Streater R F and Wightman A S *PCT, spin and statistics and all that* (Benjamin, New York). [§ 3.5]

[524] Sudarshan E C G 1968 *The fundamental theorem on the connection between spin and statistics* pp. 379–386 in Svartholm [529]. [Extends the spin statistics relation from local relativistic Lagrangian theory to any situation where spin is well-defined.] [§ 20.14]

[525] Sudarshan E C G and Mukunda N 1974 *Classical dynamics: a modern perspective* (Wiley, New York). [§§ 9.2, 10.1, 10.3, 11.5, 15.8, 18.13, 19.2]

[526] Sundermeyer K 1982 *Constrained dynamics* (Springer-Verlag, Berlin). [§§ 8.4, 9.2, 9.7, 10.3, 15.7, 15.8, 18.6]

[527] Süsskind C 1964 *Observations of electromagnetic wave radiation before Hertz* Isis **55** 32–42. [§ 12.1]

[528] Süssmann G 1969 *Begründung der Lorentz-Gruppe allein mit Symmetrie- und Relativitäts-Annahmen* (*Establishing the Lorentz group solely by symmetry and relativity assumptions*) Z. Naturforsch. **A24** 495–498. [§ 5.4]

[529] Svartholm N (ed.) 1968 *Elementary particle theory: relativistic groups and analyticity*, Proc. Eighth Nobel Symposium, 19–25 May, 1968, Aspengården, Sweden. (Almquist and Wiksells, Stockholm; Wiley, New York).

[530] Taylor J H, Fowler L A and McCulloch P M 1979 *Measurements of general relativistic effects in the binary pulsar PSR 1913+16* Nature **277** 437–440. [§ 17.5]

[531] Terletski Y P 1968 *Paradoxes in the theory of relativity* (Plenum, New York). [§ 5.4]

[532] Thibeau J 1933 *L'annihilation des positrons au contact de la matière et la radiation que en resulte* (*The annihilation of positrons on contact with matter and the radiation produced*) Comptes Rendus Acad. Sci. Paris **197** 1629–1632. [§ 3.4]

[533] Thirring W E 1959 *Lorentz-invariante Gravitationstheorien* (*Lorentz-invariant gravitation theories*) Fortschr. Physik **7** 79–101. [§§ 17.1, 17.2]

[534] Thirring W E 1961 *An alternative approach to the theory of gravitation* Ann. Phys. (USA) **16** 96–117. [§§ 17.1, 17.2, 21.4]

[535] Thomson J J et al., 1931 *James Clerk Maxwell, a commemorative volume, 1831–1931* (University Press, Cambridge). [§ 16]

[536] 't Hooft G 1974 *Magnetic monopoles in unified gauge theories* Nucl. Phys. B **79** 276–284. [§ 18.11]

[537] Thorne K S 1974 *The search for black holes* Sci. Amer. **231** (Dec.) 32–43.

[538] Thorne K S, Price R H and Macdonald D A 1986 *Black holes: the membrane paradigm* (Yale University Press, New Haven).

[539] Thurston W P and Weeks J R 1984 *The mathematics of three-dimensional manifolds* Sci. Amer. **251** (Jul.) 94–106. [§ 2.4]

[540] Trautman A 1962 *Conservation laws in general relativity* pp.169–198 in Witten [590]. [§ 15.8]

[541] Trautman A 1964 *Foundations and current problems of general relativity* pp. 1–248 in Deser and Ford [95]. [§ 15.8]

[542] Trautmann A 1966 *Comparison of Newtonian and relativistic theories of spacetime* in Hoffman [253]. [§ 7.15]

[543] Trautmann A 1972 *Invariance of Lagrangian systems* pp. 85–99 in O'Raifeartaigh [413].

[544] Trautmann A 1980 *Fiber bundles, gauge fields, and gravitation* in Held [231], vol. 1. [§ 7.15]

[545] Trefil J S 1980 *From atoms to quarks: an introduction to the strange world of particle physics* (Athlone Press, London).

[546] Trefil J S 1983 *The moment of creation* (Charles Scribner's Sons, New York).

[547] Tricker R A R (ed.) 1965 *Early electrodynamics: the first law of circulation* (Pergamon, Oxford).

[548] Utiyama R 1959 *Theory of invariant variation and generalized canonical dynamics* Progr. Theor. Phys. Suppl. **9** 19–44. [§§ 9.7, 15.8, 21.3, 21.6]

[549] van Nieuwenhuizen P and Freedman D Z (eds.) 1979 *Supergravity* (North-Holland, Amsterdam).

[550] van der Waerden B L 1929 *Spinor Analyse* Göttinger Nachrichten pp. 100–109. [§§ 13.3, 20.2]

[551] van der Waerden B L 1932 *Die Gruppentheoretische Methode in der Quantenmechanik (The group-theoretical method in quantum mechanics)* (Springer, Berlin). [§§ 13.3, 20.2]

[552] Voigt W 1887 *Über das Doppler'sche Prinzip (On the Doppler effect)* Nachr. Ges. Wiss. Göttingen pp. 41–51, reprinted in 1915 Phys. Zs. **16** 381–386. [§ 5.5]

[553] Volta E G A A 1800 *Description du nouvel appareil galvanique (Description of the new galvanic apparatus)* Journal de Phys. **51** 344–354. [§ 18.1]

[554] Volta E G A A 1801 *De l'électricité dite galvanique (On the so-called galvanic electricity)* Annales de Chim. **40** 225–256. [§ 18.1]

[555] Vessot R F C and Levine M W 1979 *A test of the Equivalence principle using a space-borne clock* Gen. Rel. Grav. **10** 181–204; Vessot R F C et al., 1980 *Test of relativistic gravitation with a space-borne hydrogen maser* Phys. Rev. Lett. **45** 2081–2084. [§§ 7.14, 17.2]

[556] Wald R M 1984 *General relativity* (University of Chicago Press, Chicago).

[557] Walker A G 1936 *On Milne's theory of world structure* Proc. London Math. Soc. **42** 90–127. [§ 21.6]

[558] Watson W 1746–47 *Experiments and observations tending to illustrate the nature and properties of electricity* Phil. Trans. Roy. Soc. (London) **43** 481–501, **44** 41–50, 388–395, 695–749 on p. 718. [§ 18.1]

[559] Watson A 1988 *Mathematics of a fake world* New Scientist **118** No. 1615 41–45. [§ 12.6]

[560] Watanabe S 1951 *Reversibility of quantum electrodynamics* Phys. Rev. **84** 1008–1025. [§ 14.6]

[561] Weber W E 1846 *Elektrodynamische Massbestimmungen. Über ein allgemeines Grundgesetze der Elektrischen Wirkung (Electrodynamic mass determination. On a general law of electrical action)* Abh. sächs. Ges. Wiss. pp. 211–378, reprinted in Weber [563], vol. 3, p. 25–214 *Galvanismus und Elektrodynamik, erster Theil (Galvanism and electrodynamics, Part I)*. [§ 18.1]

[562] Weber W E and Kohlrausch R 1856 *Über die Elektrizitätsmenge, welche bei galvanische Strömen durch den Querschnitt der Kette fliesst* (*On the quantity of electricity which flows by galvanic current through the cross-section of a circuit*) Ann. d. Phys. **99** 10–25, reprinted in Weber [563], vol. 3, pp. 597–608. [§ 12.1]

[563] Weber W E 1893 *Werke* (Heinrich Weber, Berlin).

[564] Weichert E 1911 *Relativitätsprinzip und Äther I* (*The relativity principle and the ether*) Phys. Z. **17** 689–707. [§ 5.4]

[565] Weinberg S 1967 *A model of leptons* Phys. Rev. Lett. **19** 1264–1266. [§§ 20.13, 21.7]

[566] Weinberg S 1972 *Gravitation and cosmology* (Wiley, New York). [§§ 2.4, 15.7, 15.9, 21.6]

[567] Weinstock R 1961 *Laws of classical motion: what's F? what's m? what's a?* Am. J. Phys. 698–702. [§ 6.1]

[568] Wentzel G 1949 *Quantum theory of fields* (Interscience, New York).

[569] Wess J and Bagger J 1983 *Supersymmetry and supergravity* (Princeton University Press, Princeton).

[570] Wess J and Zumino B 1974 *Supergauge transformations in four dimensions* Nucl. Phys. **B70** 39–50. [§§ 21.2, 21.7]

[571] Weyl H 1918 *Raum Zeit Materie* (Berlin). English translation: Brose H L 1922 *Space Time Matter* (Methuen, London); reprint edn. 1952 (Dover, New York). [§ 1.2]

[572] Weyl H 1918 *Gravitation und Elektrizität* Preuss. Akad. Wiss. Sitz.ber. 465–480; reprinted in English as *Gravitation and electricity* in [343], pp. 199–216; 1919 *Eine neue Erweiterung der Relativitätstheorie* (*A new extension of the theory of relativity*) Ann. d. Phys. (Germany) **59** 101–133; 1920 *Elektrizität und Gravitation* Phys. Z. **21** 649–651. [Extension of general relativity by gauge invariance to include electrodynamics.] [§§ 7.10, 18.9]

[573] Weyl H 1929 *Electron und Gravitation. I.* (*The electron and gravitation*) Z. Physik **56** 330–352. [Containing the 'Weyl equation' now used to describe neutrino propagation.] [§§ 16, 20.7]

[574] Weyl H 1931 *Theory of groups and quantum mechanics* (Methuen, London). [§ 3.4]

[575] Wheeler J A 1968 *Our universe: the known and the unknown* Am. Sci. **56** 1-20 and Am. Scholar **37** 248–274. [§ 1.5]

[576] Whittaker E 1910 (and rev. edn. 1951) *A history of the theories of the aether and electricity* (Thomas Nelson & Sons, London). [§§ 5.4, 12.1, 18.1]

[577] Wick G C 1934 *Sugli elementi radioattivi di F. Joliot e I. Curie* (*On the radioactive elements of F. Joliot and I Curie*) Rend. Lincei **19** 319–324. [Artificial radioactivity from nuclear reactions.] [§ 20.13]

[578] Wick G C, Wightman A S and Wigner E P 1952 *The intrinsic parity of elementary particles* Phys. Rev. **88** 101–105. [§§ 3.3, 13.3, 20.3]

[579] Wick G C *Invariance principles of nuclear physics* Annual Rev. Nucl.Sci. **8** 1–48. [§ 3.5]

[580] Wien W 1900 *Über die Möglichkeit einer elektromagnetischen Begründung der Mechanik* (*On the possibility of an electromagnetic foundation of mechanics*) Arch. néerl. **5** 96–104; reprinted 1901 Ann. d. Phys. **5** 501–513. [§ 12.1]

[581] Wigner E P 1927 *Einige Folgerungen aus der Schrödingerschen Theorie für die Termstructuren* (*Some deductions from the Schrödinger theory of term structures*) Z. Phys. **43** 624–652. [Laporte' rule and the parity symmetry of the transition dipole moment.] [§ 3.3]

[582] Wigner E P 1932 *Über die Operation der Zeitumkehr in der Quantenmechanik* (*On the operation of time-reversal in quantum mechanics*) Nach. Ges. Wiss. Gött. **32** 546–559. [§ 3.5]

[583] Wigner E P 1939 *On unitary representations of the inhomogeneous Lorentz group* Annals of Math. **40** 149–204. [§§ 20.2, 20.3]

[584] Wigner E P 1965 *Violations of symmetries in physics* Sci. Amer. **213** (Dec.) 28–36. [§ 3.3]

[585] Will C M 1981 *Theory and experiment in gravitational physics* (University Press, Cambridge). [§ 5.4, 7.11, 17.6]

[586] Will C M 1984 *The confrontation between general relativity and experiment: an update* Physics Reports **113** 345–422. [§§ 7.11, 7.13, 17.6, 21.5]

[587] Will C M 1986 *Was Einstein right?* — *Putting general relativity to the test* (Basic Books, New York). [§ 5.4]

[588] Will C M 1987 *Experimental gravitation from Newton's Principia to Einstein's general relativity* pp. 80–127 in Hawking and Israel [228]. [§§ 5.4, 12.3, 17.6, 18.6, 21.5]

[589] Williams J G and Dicke R et al., 1976 *New test of the eqivalence principle from lunar laser ranging* Phys. Rev. Lett. **36** 551–554. [§ 17.6]

[590] Witten L (ed.) 1962 *Gravitation: an introduction to current research* (Wiley, New York).

[591] Witten E 1981 *A new proof of the positive energy theorem* Commun. Math. **80** 381–402. [§§ 2.1, 17.1]

[592] Wu C S, Ambler E, Hayward R W, Hoppes D D and Hudson R P 1957 *Experimental test of parity conservation in beta decay* Phys. Rev. **105** 1413–1415. [§ 3.3]

[593] Wu C S and Moszkowski S A 1966 *Beta decay. Monographs and texts in physics and astronomy* (Interscience, London). [§ 20.13]

[594] Yang C N 1964 *Nobel prize address* in *Nobel lectures in physics 1942-62* (Elsevier, Amsterdam). [§ 3.3]

[595] Yang C N and Mills R L 1954 *Conservation of isotopic spin and isotopic gauge invariance* Phys. Rev. **96** 191–195. [§§ 16, 20.13]

[596] Youschkevitch A P 1971 *Leonhard Euler* pp. 467–484 in Gillispie [197], vol. 4. [§ 8.1]

[597] Yukawa H 1935 *On the interaction of elementary particles. I.* Proc. Phys-Math. Soc. Japan **17** 48–57. [Interaction via particle exchange.] [§§ 1.4, 7.14]

[598] Yukawa H and Kikuchi C 1950 *The birth of the meson theory* Am. J. Phys. **18** 154–156. [§ 1.4]

[599] Zeeman P and Fokker A D (eds.) 1934–39 *H A Lorentz, Collected papers* 9 vol., (Martinus Nijhoff, The Hague). [§ 6.3]

> Get a thorough insight into the Index, by which the whole book is governed and turned, like fishes by the tail. For to enter the palace of Learning at the great gate, requires an expense of time and forms; therefore men of much haste and little ceremony are content to get in by the back-door.
>
> Jonathan Swift 1667–1745

Index

3-momentum, relativistic, 299, 302
3-volume, 62, 325, 331, 384
4-acceleration, 298
4-momentum, 299, 302, 311, 331, 358
4-vector potential, electromagnetic, 393
4-velocity, 298
4-volume, 278

Abelian symmetry, 8, 222, 350
aberration, 82, 154, 303
absolute
    acceleration, 119, 481
    simultaneity, 84, 102, 111, 249
    spacetime, 80, 84, 94
acceleration, 64, 87, 108, 119, 298
    gravitational, 480
    inertial, 480
action, 62, 71, 103, 165, 180
    and reaction, 116, 124, 150
    at a distance, 18, 135, 209, 241
    functional, 175
        covariant, 181
        dimension, 129, 180
        einbein, 312
        Einsteinian mechanics, 307
        electrodynamic, 416
        free particle, 309
        hermiticity, 181
        mechanical, 174
        Nambu and Goto string, 496
        Newtonian field, 206, 233
        Newtonian gravity, 211
        phase space, 230
        relativistic field, 337
        tensor gravity, 473
    quantum of, 415, 492
active
    gravitational mass, 150
    transformation
        parity, 46
        rotation, 63
        time reversal, 54
Adams W S, 158
additivity, 111
    angular momentum, 114

centre of mass motion, 113
colour charge, 17
dynamic variables, 109
electric charge, 9
energy, 114
Lie group parameter, 272
mass, 113
momentum, 113
adjoint spinor, Dirac, 438
Adler R, Bazin M and Schiffer M, 373
affine connection, 486
affine spacetime, 105
Aharonov Y and Bohm D, 36, 452
A'h-mosé the Scribe, 26
Akenside, Mark, 497
algebra, Clifford, Dirac, Pauli, 432
algebra, Lie, of angular momentum, 233
Ampère's law, 390, 414
Ampère, André-Marie, 378
ampere, SI base unit, 128
Anderson J L, 200
Anderson J L and Bergmann P G, 316
angle, 33
angular
    momentum, 108
        additivity, 114
        algebra, 233
        conservation, 27, 34, 57, 126,
            169, 293, 332, 341, 419, 470
        covariant, 311
        density, 221, 332, 421
        operator, 113
        orbital, 72, 110, 116, 320
        scalar field, 355
        spin, 72, 116, 320, 341
        tensor, 117, 304, 319, 421
anomaly, 341
anticommutation
    Dirac matrices, 432
    quantization rules, 444
anticommuting components, 444
antineutrino, 51, 458
antiparticle, 9, 52, 257, 444
antiselfduality, 50, 292
antisymmetric

metric tensor, 283
  part of a tensor, 67, 285
  tensor, 67, 285
Archimedes, 81
Aristarchus of Samos, 81, 133
Aristotelean spacetime, 41, 84, 94
Aristotle, 80, 104, 150
arrow of time, 37
associativity, group axiom, 59
asymptotic flatness, 484, 490
atom, 27, 493
atomistic philosophy, 27, 123
attractiveness of gravity, 132, 213, 361
Augustine of Hippo, 38
axial vector, 70, 71, 287

Bailey J et al., 159
baryon, 12, 200
base manifold, fibre bundle, 105
basis spinor, 441
basis, Lorentz, 263
Belinfante, F J, 420
bending of light
  general relativistic, 482
  Newtonian, 154
  scalar gravity, 367, 372
  tensor gravity, 476
Bentley, Richard, 135
Bergmann P G, 316
Bessel-Hagen E, 199
β-decay, 13, 458
Bianchi identity
  electromagnetic, 391, 488
  general relativistic, 488
Big Bang, 22
Biot, Jean-Baptiste, 378
Biot-Savart law, 414
bipolarity of charge, 9, 19, 385
black hole, 20, 21, 23, 371
  Kerr, 489
  Kerr-Newman, 489
  Newtonian, 23, 152
  Reissner-Nordstrøm, 489
  Schwarzschild, 488
  stellar and galactic, 132
blackbody radiation, 245
Bohr, Niels, 217
Boltzmann's constant k, 127
Bólyai, Farkas, 60
Bólyai, János, 32, 35
boost, 62, 93
  covariance, 109, 201
  electromagnetic field strength, 398
  Galilean, 98, 102, 125, 169
  Lorentz, 99, 247, 270, 272, 285

  matrix, 270, 274
Born, Max, 217
Bose, Satyendranath, 281
Bose-Einstein
  statistics, 10, 20, 22, 281, 465, 492
boson, 22, 265, 281, 344, 439
bottom (flavour), 17
boundary condition, 118
bracket, Poisson, 232
Bradley, James, 82, 154, 240
Brahe, Tycho, 40, 133
Brans-Dicke-Jordan gravity, 374
Brewster D, 107
Bulmer-Thomas, Ivor, 40
bundle of Galilean frames, 105
Byron, Lord, 161

calculus of variations, 166, 175, 207
canonical
  dimension
    bosonic, 344
    fermionic, 439
    first-order field, 217
  energy-momentum tensor, 341, 419
    Dirac, 440
    electromagnetic, 407
  Hamiltonian, 308
    free particle, singular, 316
    general relativity, singular, 316
  invariance
    Euler-Lagrange equations, 191
    Lagrangian, 423
  momentum, 308, 338, 342
  quantization, 464
  spin density, 420
  transformation, 190
Carnot, Nicolas Leonard Sadi, 123
Cartan, Élie, 148
cartesian coordinate, 42
causal
  interval, 88
  signal, 100, 254
  structure
    conical, $\mathbf{E}^{1,3}$, 253
    planar, $\mathbf{G} = \{\mathbf{E}^1, \mathbf{E}^3\}$, 252
causality
  and interaction, 118
  Newtonian, 115
  principle of, 83, 87, 93, 98, 100, 248
Cavendish, Henry, 135, 156
censorship hypothesis, cosmic, 23
central extension, group, 126
central interaction, 120, 121
centre of mass, 113

Index 545

frame, 112
motion, 110
    additivity, 113
    conservation, 125, 169, 201
characteristic function of Hamilton, 227
charge
    baryonic, 200
    colour, 17, 200
    conjugation, 52, 348, 350, 411, 454
    conservation, electric, 9, 255, 385
    density, 385
    electric, 9, 200, 383
    leptonic, 200
    magnetic, 72, 288, 380, 390, 401
    Noether, 199, 221, 231, 339
    parity, 53
    quantization, 11, 385
    space, 350
charm (flavour), 17
chiral object, 48
chirality, 38, 46
    electromagnetic field, 391
    invariance, 459
    matrix, 445
    of fields and quanta, 50, 444
    of nature, 49, 257
chromodynamics, quantum, 18
circle S^1, 30
Clarke, Samuel, 84
classical
    electrodynamics, 10, 412
    limit, 465, 492
    particle, 28, 29
    symmetry, 340
    vacuum, 27
Clausius, Rudolph Julius Emmanuel, 123
Clifford
    algebra, 432
    William Kingdon, 432
clock
    paradox, 256
    synchronization, 85, 246
closed system, 118
closure, group axiom, 59
colour charge, 17, 200
commutation, quantization rules, 444
commutator algebra, 226
commutator, covariant derivatives, 455
commuting observables, 442
compactness
    symmetry, 8, 13, 17, 222, 350, 448
        Abelian or non-Abelian, 459
completeness of a basis, 426
complex
    conjugation, 260

Minkowskian spacetime, 260
    scalar field, 350
complexion, electromagnetic, 403
Compton wavelength, 30, 343, 407
configuration
    mechanical, 171
    space
        Hamiltonian, 172, 209, 308
        Lagrangian, 227
confinement, 16, 492
conformal transformation, 254, 265
conical
    causal structure, 253
    refraction, 229
connected group manifold, 57
connection, affine, 486
connectivity
    R and S^1, 30
    E^3, 121
    restricted Lorentz group, 269
conservation
    4-momentum, 331
    angular momentum, 27, 34, 57, 126,
        169, 293, 332, 341, 419, 470
    centre of mass, 125, 169, 201
    colour charge, 17
    electric charge, 9, 385
    energy, 34, 57, 124, 169, 225, 364
    energy-momentum, 19, 341, 470
    in Einsteinian dynamics, 299, 304
    in Newtonian dynamics, 109, 305
    local and global, 328
    momentum, 27, 57, 126, 169, 201
    Newtonian helicity, 116
    Noether charge, 221
    on-shell, 308
    parity, 50, 73, 288
    weak, 200
conservative interaction, 120
conservative vector, 120
constant
    cosmological, 146, 211
    fine structure, 137, 415
    of motion, 200
    Table of numerical values, 499
    universal, 127, 415
constrained Hamiltonian, 316
constraint, 7, 149, 170, 184, 422, 467
    configuration space, 312
    dynamic, 317
    Einsteinian free particle, 310
    holonomic, 185
    Lagrange, 187, 312, 317
    mass shell, 353
    Maxwell field, 422

null particle, 313
phase space, 317
Proca field, 407
proper time, 298, 305
source, 389
contact interaction, 13, 458
Continental model, electric fluid, 244
continuity equation, 220, 329, 339, 385
continuous symmetry, 195
Einsteinian mechanics, 307
infinitesimal, 196, 220
contraction, 281
Lorentz-FitzGerald, 90, 243, 257
of indices, 67, 263
contragredient matrices, 267
contravariant components, 260, 277, 280
coordinate
external and internal, 143
general, 484
generalized, 171, 190
ignorable, 368
interval, 32
isotropic, 369, 478
length, 256
local and global, 31
Minkowski, 263
rectangular Cartesian, 34
rectangular cartesian, 42
singularity, 23
spacetime, 30, 259
time interval, 255
time, integrability, 297
Copenhagen interpretation, 217
Copernican revolution, 81
Copernicus, Nicolas, 56, 81, 133
corpuscle of light, 154
cosmic
censorship hypothesis, 23
microwave radiation, 22
cosmological
constant, 146, 147, 211, 489
interaction, Newtonian, 146, 213
red shift, 147
cosmology, 20, 134, 488
Coulomb
Charles Augustin de, 377
gauge
Einsteinian particle, 310
electromagnetic, 427
law, 7, 380
counting tensor components, 57, 69, 500
coupling, 179
constant, 11, 183, 346, 415, 456
minimal, 415, 455, 456
covariance

boost, 109
Euclidean, 78, 140
Galilean, 103, 109, 121, 124, 134, 168, 171, 212, 215
Galilean boost, 201
general coordinate, 485
Lagrangian density, 218
manifest, 78, 283
parity, 50, 73, 120, 140, 288, 382
Poincaré, 92, 247, 249, 284, 382, 391
rotational, 78, 273
time-reversal, 55
translational, 78, 284
covariant
3-volume, 325, 331, 384
action, 181
angular momentum, 304, 311
canonical momentum, 342
components, 260, 277, 280
curl, 291, 394
derivative, 456, 486, 487
gauge, 424
Lagrangian, 181, 192, 308, 339
Maxwell's equations, 389, 393
momentum, 299, 302
notation, 282
rotation, 311
CPT, 454
CPT theorem, 339, 455
cross product, 72, 76
curl, 76, 291, 394
current
density, 385
electromagnetic, 383, 413
Klein-Gordon, 351
Noether, 220, 339
Currie D G and Saletan E J, 178
curvature
constant, 35, 36
extrinsic and intrinsic, 34
intrinsic, 484, 488
Newton-Cartan spacetime, 160
spacetime, 477, 482
cyclic group of 2nd order, 59

da Vinci, Leonardo, 3
d'Alembert, Jean le Rond, 78
d'Alembertian operator, 277, 387
de Donder gauge, spin 2, 472
decoupled gauge, 424
degeneracy pressure, 132
degree of freedom, 149
Dirac field, 435
dynamical, 184, 305, 467

Index   547

field, 204
   physical, 205, 423, 424, 494
Democritus of Abdera, 27
density
   4-momentum, 330
   angular momentum, 332
   Noether, 220
   transformation, 190
derivative
   exterior, 503
   gauge-covariant, 456
   general covariant, 486
   of a tensor, 281
determinant of a matrix, 73, 289
diagonal interaction, 10
diagram
   Feynman, 453
   Minkowski, 253
differential
   equation, 130
      ordinary, 78, 119
      partial, 78, 284, 432
      quasi-linear, 372, 488
   form, 503
   geometry, 190, 487
   identity, 468
dilatation, 265
dimension
   action, 180
   canonical bosonic, 344
   canonical fermionic, 439
   natural units, 128
   spacetime, 257
   topological invariant, 30
dimensional analysis, 15, 23, 127, 136, 155, 202, 359
Dirac
   adjoint spinor, 438
   algebra, 432
   delta function, 212, 330
   electric current, 443
   energy relation, 303
   equation, 354, 436, 468, 494
   field, 139, 336, 384, 421, 431, 493
   Hamiltonian density, 441
   matrix, 432
   Paul Adrien Maurice, 316, 336, 402, 430, 434
   plane wave expansion, 443
   probability current, 443
   spin matrix, 433, 442
   spinor, 283, 337, 433
direct product
   group, 60, 350
   matrix, 65, 279
   of vectors, 66, 279
discrete
   symmetry, 348
   transformation, 46, 53
distance, 33, 42
divergence
   of a 3-vector, 63
   of a 4-vector, 277, 387
   of quantum fields, 11, 30, 346
   term in a Lagrangian, 178, 192
double valuedness, spinorial, 281
down (flavour), 16
du Fay, Charles-François, 376
dual
   covariant components, 260
   of an antisymmetric tensor, 291
   vector and tensor in 3D, 76
duality
   parts of a tensor, 292
   rotation, 403
   symmetry, 403
dummy index, 45, 263
dynamic
   constraint, 317
   equation, 186, 227, 422
   Lagrangian, 187, 199, 216
   theory, 19, 24, 108, 247, 483

Eddington, Arthur Stanley, 37
effective Hamiltonian, 318
einbein, 312, 313, 318, 416, 473
   Lagrangian, 313, 359
Einstein
   Albert, 41, 62, 83, 85, 86, 88, 90, 91, 99, 147, 156, 159, 245, 246, 256, 258, 259, 276, 296, 300, 335, 357, 363, 364, 381, 465, 470, 482, 483, 491, 496
   field equation, 23, 372, 488
   Hilbert Lagrangian, 491
   summation convention, 43, 261
   tensor, gravitation, 488
Einsteinian
   dynamics, 304
   kinematics, 100, 245, 256, 296
   relativity, 111
electric
   charge, 9, 62, 71, 103, 200, 288, 348, 383
      bipolarity, 9
      conservation, 9, 385
      fractional, 15
      invariance, 384
      natural dimension, 129
   current, 244

field strength, 65, 390
electricity and the ether, 245
electrodynamics, 6, 285, 410
    classical, 10, 412
    diagonal, 10
    gauge invariance, 7, 306, 449
    linearity, 10
    long range, 7, 257
    non-linearity, 452
    parity covariance, 6, 382, 406
    quantization, 11
    renormalizable, 11
    scalar, 457
    source, 382
    spinor, 451
    weakness, 415
electromagnetic
    complexion, 403
    coupling constant, 415
    energy density, 404
    energy-momentum tensor, 403
    field, 291, 306
        generic, 399
        pure electric, 399
        pure magnetic, 399
        radiation, 399
        tensor, 389
    fine structure constant, 137
    interaction, 6, 415
    potentials, 393, 394
    radiation, 393
    stress tensor, 405
electron, 12, 127, 281, 336, 493
    non-relativistic, 283
    relativistic, 283
    theory of Lorentz, 90, 243
electrostatic field equations, 149
electroweak interaction, 6, 13, 344, 461
elements, ancient, 80, 130
embedding of a subspace, 34
end-point contribution, action, 176
energy, 62, 108, 124
    additivity, 114
    conservation, 34, 57, 124, 169, 364
    density, 330, 404
    flux, 329
    kinetic, 111
        Einsteinian, 299
        Newtonian, 300
    mechanical, 122
    Newtonian, 110
    positivity, 19, 132, 213, 344–346, 348, 353, 359, 439, 444, 465, 473, 492, 493
    relativistic, 302
    rest, 299
    self-gravitational, 370
energy-momentum
    conservation, 19, 341, 470
    scalar field, 355
    tensor, 358, 470, 487
        Belinfante, 293, 420
        canonical, 341, 419
        Dirac, 440
        electromagnetic, 403
        Fierz-Pauli gravity, 479
        particle, 330
        scalar field, 345
        symmetric, 332, 420
        trace, 405
        vector, relativistic field, 421
Eötvös experiment, 151
equation of motion, 78, 118
equivalence
    mass and energy, 301, 371, 463
    of matrices, 267
    principle
        Einstein, 362
        strong, 152, 370, 371, 481
        weak, 104, 151, 360, 370, 481
    relation, 60, 96
Erlangen programme, 41, 195
ether
    and electricity, 245
    celestial, 80, 130
    drag, 240
    drift or wind, 240
    electric, magnetic, gravitational, 240
    luminiferous, 90, 240
Euclid, 32, 40, 41, 80
    fifth postulate, 32, 40
Euclidean
    covariance, 78, 140
    geometry, 62
    group, 393
    isometry group, IO(3), 57
    restricted group, ISO(3), 75
    scalar, 62, 102
    space
        E^1, 42
        E^3, 44
        E^4, 258
        E^n, 33
    spacetime, $E^1 \times E^3$, 41, 77, 82, 111
    tensor field, 66
    vector, 63
euclidicity of 3D space, 33, 42
Euler
    derivative, 173, 179, 191, 199, 208, 218, 468

Lagrange equations, 167, 177, 179
    canonical invariance, 191
    Dirac field, 440
    Einsteinian mechanics, 307
    electromagnetic, 406
    Fierz-Pauli gravity, 473
    form invariance, 191
    general relativity, 491
    mechanical, 172
    Newtonian field, 208
    relativistic field, 339
    scalar field, 345
    scalar gravity, 360
    spinor electrodynamics, 451
  Leonhard, 166, 167, 174
  theorem, 172
event, 29
  horizon, 23
  spacetime, 259
evolution parameter, 87, 307, 315
  Newtonian, 112
exchange of particles, 10, 138, 447, 462, 492
exclusion principle, 132, 444
exterior derivative, 503
external
  coordinate, 143
  interaction, 117
  potential energy, 124
  source, 211, 345, 447
  symmetry, 265, 494
extrinsic curvature, 34

factoring a Lorentz transformation, 274
family, lepton, 462
family, lepton and quark, 12
Faraday
  law, 390, 414
  Michael, 138, 204, 241, 379
  rotation, 241
Fermat's principle, 165
Fermat, Pierre de, 165
Fermi
  Dirac statistics, 22, 281, 444, 465, 493
  Enrico, 13, 281, 458
  theory of weak interaction, 13, 458
fermion, 22, 265, 281, 439
Fetter A L and Walecka J D, 185
Feynman
  graph, 453
  slash, 436
fibre, 105
  bundle, 105, 148, 160, 252
field, 28, 138, 139, 202

  chiral, 444
  Dirac, 139, 336, 431
  equations, 78, 284
    electrostatic, 149, 214
    Galilean, 203, 208
    magnetostatic, 149, 214
    Newtonian gravity, 141
    Newtonian potential, 142
    non-dynamic, 205
    non-linear, 18, 211
  Fierz-Pauli, massless, 336, 472
  Galilean, 139, 202
  gauge, 148
  Klein-Gordon, 277, 335
  Lagrangian, 340
  local, 138
  mediating, 10, 138, 447
  non-chiral, 431
  non-gauge, 492
  quantum, 10
  Rarita-Schwinger, 466
  redefinition, 471
  relativistic Lagrangian, 335
  scalar, 277
  source, 10, 139
  strength, 149, 390
    electromagnetic, 389
    Newtonian gravity, 142
    Proca, 408
  tensor, 280
  vector, 277
  Yang-Mills, 460
Fierz M and Pauli W, 336, 473
Fierz-Pauli field, 336, 391, 421, 472, 491
fine structure
  weak, 459
fine structure constant, 11
  electromagnetic, 137, 415
  gravitational, 137
  strong, 462
first
  law of Newton, 112
  order Lagrangian, 408
  quantization, 353, 463
FitzGerald, George Francis, 90, 93, 100, 243, 300
flatness
  asymptotic, 466, 484
  spacetime, 484
flatness, extrinsic and intrinsic, 33, 34
flavour, 12, 16
flux, 126
  energy, 329
  momentum, 329
force, 108, 115, 117, 120, 126

form invariance
    Euclidean, 78
    Euler-Lagrange equations, 191
    Galilean, 124
    Lorentz, 283
    Poincaré, 284
    rotational, 78
    spacetime reflection, 50, 55
    translational, 78
form, differential, 503
four-momentum, 278
fractional electric charge, 15
frame
    absolute, 91
    centre of mass, 112
    inertial, 90, 108, 110, 112, 470
    Lorentz, 263
    of reference, 28
    privileged, 470
    proper, spacelike interval, 256
    proper, timelike interval, 255
    rest, 255
    spacetime, 30
Frank P and Rothe H, 89
Franklin, Benjamin, 377
free
    field, 24, 210
    index, 44, 261
    part of Lagrangian, 180
    particle, 160
        einbein Lagrangian, 318
        Einsteinian, 305
        generalized momentum, 315
        geometric Lagrangian, 309
        Hamiltonian, 315
        phase space Lagrangian, 318
free fall, 475, 481
French A P, 256
Fresnel, Augustin Jean, 154, 240
functional
    action, 175
    derivative, 178, 206, 234
    Lagrangian field, 206
    variation, 176, 192, 338
fundamental
    interaction, 4
    Poisson bracket, 234, 319
future, 29
    of an event, Galilean, 252
    of an event, Lorentz, 253
    pointing, 325

Galilean
    boost, 98, 102
    boost invariance, 125
    covariance, 103, 109, 121, 124, 134, 168, 171, 203, 212, 215
    field, 139, 202
        action functional, 206
        Euler-Lagrange equations, 208
        free, 210
        Hamiltonian density, 209, 234
        Lagrangian density, 206
        local, 209
        non-dynamic, 209
        self interaction, 210
        stress tensor, 220
        symmetry, 217
    hypothesis, 102
    invariant, 102
    kinematics, 102, 393
    principle of inertia, 112
    proper length, 102
    relativity, 83, 91, 93, 107, 110, 247
    spacetime, 102, 160
    symmetry, 189
    vector, 103
Galilei
    Galileo, ii, 90, 92, 94, 101, 104, 107, 150, 153
    group, 101, 105, 126, 232
Galle, Johann, 157
Galois, Evariste, 60
Galvani, Luigi, 377
gamma 5 condition, 445
gamma trace, 467
Gardiner, Martin, 51
gauge, 148, 398
    condition, 306, 366, 367, 424, 471
    Coulomb, 427
    coupling, principle of, 6, 449
    covariant, 424
    covariant derivative, 456, 457, 461, 487
    decoupled, 424
    electrodynamic, 423
    field, 148, 258, 460, 493
    fixing, 306, 310, 398, 423
    freedom, 148, 149, 187, 227, 305, 317, 345, 393, 397, 408
    function, 148
    function, spinorial, 468
    group, 453, 461
    Hamilton, electromagnetic, 423, 428
    invariance, 6, 7, 148, 187, 310, 316, 342, 396, 466
        electrodynamic, 7
        first and second kind, 448, 452
        free particle, 305
        gravitational, 21

Index   551

    local, 200, 391, 449
    residual, 425, 427, 472
    strong-nuclear interaction, 17
    trivial, 397
    weak-nuclear interaction, 13
    Yang-Mills, 13, 15, 17
  laboratory, Einsteinian particle, 310
  Lorentz, 424, 435, 471
  non-covariant, 427
  particle, 493
  potential, 148
  principle, 6, 8, 24, 349, 448
  proper time, 310
  radiation, 428
  transverse, electromagnetic, 427
  variance, 148, 393
gauging a global symmetry, 9, 449
Gauss
  Carl Friedrich, 32, 35, 37, 378
  law, electric, 379, 390, 414, 422
  law, magnetic, 376, 390, 414
  theorem, 208
    Euclidean, 327
    Minkowskian, 327
general
  coordinate, 484
    transformation, 190, 265, 309
  covariance, principle of, 485
  relativity, 19, 131, 247, 372, 391,
    467, 477, 482
    1-dimensional, 314
    canonical Hamiltonian, 316
    classic tests, 159
    energy positivity, 361
    square root, 437
  tensor, 486
generalized
  coordinate, 171, 190, 205
  Kronecker delta, 74, 290
  momentum, 172, 226, 308
    charged particle, 417
    free particle, 315
    Newtonian field, 206
  Newtonian mechanics, 119, 184
  velocity, 171
    covariant, 337
    Newtonian field, 205
    relativistic field, 337
generation, lepton, 12
generator
  Dirac spinor representation, 434
  Lorentz transformation, 293
  space translation, 227
  spacetime translation, 299
  time translation, 227, 229

  vector representation, 294
  Yang-Mills fields, 461
genus, of a space, 30
geodesic, 33
  curved spacetime, 490
  Minkowskian spacetime, 310
geometric
  Lagrangian, particle, 309, 313, 315
geometric space, 42
geometry
  3D Euclidean, 62
  differential, 487
  non-Euclidean, 32, 41
  of gauge invariance, 41
  Riemannian, 487
  Robertson-Walker, 488
  Schwarzschild, 489
  spacetime, 32, 250, 482
Gilbert, William, 376
Glashow-Salam-Weinberg theory, 6, 461
global
  coordinate, 31
  gauge group, 453
  phenomena, 481
  symmetry, 8, 222, 349, 448
gluon, 16, 18, 281, 336, 462, 492
Goldberg S, 89
Goldstein H, 174, 185, 273
Gordon, George (Lord Byron), 161
Gordon, Walter, 335, 343
gradient
  Euclidean space, 65, 67
  Minkowskian spacetime, 278, 282
grand unification, 6, 9, 402, 462
Grassmann
  Hermann Günther, 465
  spinor, 465, 495
  variable, 444, 495
graviphoton, 375
graviscalar, 375
gravitation, 18, 130, 131
  Aristotelean, 80
  attractiveness, 132, 144, 213, 361
  Brans-Dicke-Jordan, 374
  classic tests, 475
  constant G, 127, 135
  Copernican, 82
  covariant source, 470
  dynamic theory, 247
  equivalence principle, 481
  fine structure constant, 137
  gauge invariance, 21, 449
  general relativistic, 28, 145, 147,
    314, 482
  global character, 481

inverse square law, 213
light deflection, 476
linearized, 285, 490
long range, 19, 131, 213, 257
many-body, 18
Newton's law, 133, 144
Newtonian, 18, 82, 131, 473
   field strength, 142
   potential, 142
non-inverse square factor, 160
non-linear Fierz-Pauli, 479
non-linearity, 20, 370, 478
non-renormalizability, 137, 491
parity-covariance, 18
quantization, 22, 24, 30, 491
radius of curvature, 23, 32
red shift, 19, 158, 363
relativistic, 131, 470
scalar light deflection, 367
scalar-tensor, 374
short-range term, 160, 358
source, relativistic, 358
speed of propagation, 135, 255, 482
spin 2 theory, 145, 466
tensor character, 20, 466
time dilation, 86, 363, 476, 482
unification, 466
uniqueness of free fall, 104, 151, 481
universality, 18, 131, 135, 358, 370, 471, 478, 488
weakness, 137, 336
Yukawa contribution, 160
gravitational
   length of mass, 153
   mass, 150, 370, 481
   test body, 151
gravitino, 467, 495
graviton, 20, 281, 467, 492, 495
Gray, Stephen, 376
Greek indices, 259
Green's theorem, 378, 501
Green, George, 378
Grossmann M, 488
ground state
   classical, 28
   quantum, 14
group, 59, 93, 96
   1D translation of fields, 349
   Abelian, 222
   central extension, 126
   compact, 222
   cyclic of 2nd order, 59
   direct product, 60, 350
   Euclidean, 393
   Galilei, 101, 105, 126, 232
   inhomogeneous Lorentz, 265
   irreducible representation, 350
   isometry, 265
   isometry, Euclidean, 57
   isomorphism, 222
   Klein, $Z_2 \otimes Z_2$, 60
   Lie, 198, 269, 349
   Lorentz, 100, 265, 435
   of transformations, 7
   orthogonal, $O(N)$, $SO(N)$, 283
   Poincaré, 21, 249, 319, 383, 431, 435
   representation, 281, 431, 439
   restricted
      Euclidean, $ISO(3)$, 75
      Galilei, 101
      Lorentz, $SO(1,3)$, 249, 269
      Poincaré, $ISO(1,3)$, 249
   restriction, 278
   rotation, $SO(2)$, 222, 350
   rotation, $SO(3)$, 75
   Schrödinger, 126
   spacetime translation, 265, 435
   special linear, $SL(2,\mathbb{C})$, 460
   unitary, $SU(2)$, 460
   unitary, $U(1)$, 222
   unitary, $U(N)$, $SU(N)$, 283
   Yang-Mills, 21

hadron, 16
Hafele J C and Keating R E, 256
half-odd-integer spin, 50, 185, 217, 281, 439, 444, 465
Hamel G, 199
Hamilton
   characteristic function, 227
   equations, 228, 231
      Newtonian field, 234
      Newtonian mechanics, 228
      relativistic field, 342
      relativistic mechanics, 318
   gauge, electromagnetic, 423, 428
   Jacobi equation, 231
   principle, 195
      modified, 195, 225
      Newtonian field, 207
      Newtonian mechanics, 177
      of stationary action, 177, 179, 194, 307, 339
      phase space, 229, 231
      relativistic field, 339
      relativistic mechanics, 307
   William Rowan, 37, 226
Hamiltonian, 5, 62, 71, 168

canonical, 308
configuration space, 172, 209, 225
constrained, 316
density
   Dirac field, 441
   Klein-Gordon, 354
   Newtonian, 209
   Proca, 408
   relativistic, 342
effective, particle, 318
Einsteinian mechanics, 308
formulation, 227, 233
general relativity, 491
Maxwell field, 423
mechanical, 227
non-canonical, 318
phase space, 233, 315
proper time, 308
quantum operator, 113
relativistic field, 342
relativistic particle, 315
handedness, 38, 446
Hanson T J et al., 317
Heaviside, Oliver, 79, 93, 243, 380
Heisenberg
   uncertainty principle, 463
   Werner, 217
helicity, 447, 472, 492
   conservation, 116
   Einsteinian particle, 285
   massive Dirac particle, 442
   massive field or particle, 322
   massless field or particle, 324
   Newtonian particle, 116
   states, photon, 429
helicity $\frac{1}{2}$:, 447
helicity 1:, 395
helicity 2:, 472, 492
heliocentric cosmology, 81, 134
Helmholtz
   Hermann L F, 120, 123, 379, 380, 466
   theorem, 120, 501
Heraclides Ponticus, 81, 133
Hergoltz G, 199
hermiticity
   Dirac Lagrangian, 439
   Dirac matrix, 433
   Lagrangian and action, 181
   scalar Lagrangian, 351
   Schrödinger Lagrangian, 216
Hero of Alexandria, 165
Hertz, Heinrich Rudolf, 243, 380, 393
Hessian matrix, 186
Hestenes D, 185

hidden symmetry, 14
Higgs field, 277, 335, 344
Hill E L, 200
history
   of a field system, 338
   of a particle, 53
hole theory, 52
holonomic constraint, 185
homogeneity
   of tensor transformation, 283
   spacetime, 299, 484
   spatial, 33, 66, 93, 126, 169
   temporal, 34, 124, 169
homogeneous wave equation, 424
Hooke, Robert, 134, 154
Hubble, Edwin, 147
Hulse R A and Taylor J H, 367
Huyghens, Christiaan, 154
hyperon, 52
hypersurface
   3D in spacetime, $\mathbb{E}^{1,3}$, 325
   oriented, 325
   spacelike, 325
hypothesis of Galilean relativity, 102

idempotent operator, 48, 502
identity, group axiom, 59
Ignatowsky W von, 89
ignorable coordinate, 368
imaginary angle, Lorentz boost, 273
imaginary time coordinate, 260
improper tensor, 71
indefinite metric, 251, 257, 259, 311
index
   dummy, 45, 263
   free, 44
   Greek, spacetime, 259
   Latin, spatial, 43
   lowering, 43, 262
   notation
      3D Euclidean, 43
      4D Minkowski, 259
   raising, 44, 262
induction, electromagnetic, 379
inertia, 108, 109, 299
inertial
   frame, 90, 108, 110, 112, 470
   mass, 150, 247, 370, 481
inhomogeneous Lorentz group, 265
inhomogeneous wave equation, 424
initial data, 7, 87, 149, 184–186
   constraints, 306, 422, 467
   Newtonian field, 205
   relativistic field, 337
integer spin, 50, 281, 344, 439, 465, 492

integrability
    condition, 394, 401, 502
    coordinate time, 297
interaction, 87, 108, 284, 300, 491, 492
    and causality, 118
    central, 120, 121
    conservative, 120
    contact, 13, 458
    cosmological, 146
    diagonal, 10
    electromagnetic, 6, 306, 415
    electroweak, 6, 13, 461
    external, 28, 117
    field and particle, 212
    fundamental, 4
    gauge principle of, 8, 448
    global, 481
    gravitation, 18, 247, 487
    internal, 34, 121
    inverse-square, 131
    locality of, 118
    long-range, 7, 131
    mutual, 114, 348
    'non-conservative', 173
    non-diagonal, 15, 16
    parity conserving, 50
    parity covariance, 73
    part of Lagrangian, 180
    point, 13
    potential energy of, 121
    prescribed, 28
    self, 13, 115, 346, 348
    short-range, 11
    strong-nuclear, 15, 462
    superposition law, 115
    time-reversal invariance, 56
    unification, 24
    universality, 384
    weak-nuclear, 11, 446
intermediate vector boson, 15, 462, 492
internal
    coordinate, 143
    interaction, 121
    potential energy, 124
    structure, 109
    symmetry, 143, 221, 265, 349, 494
intrinsic
    curvature, 34, 484, 488
    parity, Dirac spinor, 438
    spin, 72, 492
invariance, 7, 143
    canonical, 191
    colour charge, 17
    electric charge, 384
    Galilean boost, 169
    gauge, 7, 148, 187
    Lagrangian density, 218
    local gauge, 200
    of Lagrangian, 191
    parity, 47
    reparameterization, einbein, 312
    rotation, 169
    time reversal, 54
    translation, 169
invariant
    Lorentz, 276
    of electromagnetic field, 395
    part of a tensor, 286
    relativistic, 276
    rotational, 395
inverse
    square, gravity, 131, 146, 358
inverse, group axiom, 59
invertible differential operator, 501
irreducible
    group representation, 350, 453, 472
    parts of a tensor, 68
irreversibility, 38
Isaak G R, 92
isolated system, 34, 108, 118, 121, 179
    Einsteinian, 299
    Newtonian, 110
isometry group
    3D Euclidean, 46, 57
    Minkowskian spacetime, 265
isospin, 14, 460
isotropic
    coordinates, 369, 478
    tensor, 66
isotropy, 126, 488
    3D Euclidean, 33, 66, 93
    spacetime, 484
Itzykson C and Zuber J-B, 408

Jacobi
    Carl Gustav Jacob, 231
    identity, 232
Jauch J M and Rohrlich F, 288
Johnson, Samuel, 497

Kaluza-Klein reduction, 381
Kant, Immanuel, 77, 123
kaon, 17, 51, 55, 281, 335, 344, 347
    time reversal violation, 38
Kaye G N C and Laby T N, 127
Kelvin, Lord (William Thomson), 38, 244
kelvin, SI base unit, 128
Kepler
    Johannes, 40, 133, 183

laws of planetary motion, 130, 133
problem, 367, 368
Kerr black hole, 489
Kerr-Newman black hole, 489
Kibble T W B, 185
kilogram, SI base unit, 128
kinematic
  theory, 24, 108, 247, 483
  time dilation, 256
kinematics
  Einsteinian, 100, 245, 256, 296
  Galilean, 102, 393
kinetic energy
  Einsteinian, 300
  Newtonian, 119
Klein
  bottle, 37
  Christian Felix, 41, 199
  Gordon current, 351
  Gordon equation, 343, 358, 396, 435, 472, 494
    square root, 437
  Gordon field, 277, 335, 383
  Gordon Hamiltonian, 354
  Gordon Lagrangian, 456
  group, $Z_2 \otimes Z_2$, 60
  Oskar, 335
Klein, Oskar, 343
Kronecker delta, 43, 45, 73, 262
  generalized, 74, 290
Kundt W, 316

laboratory gauge, 310
  relativistic particle, 315
Lagrange
  constraint, 187, 312, 317
  equations, 167
  Joseph-Louis, 40, 166, 170, 231, 378
  multiplier, 312
Lagrangian, 5, 62, 71, 103, 181
  canonical, 423
  configuration space, 167
  covariant, 181, 192, 218, 308, 339
  density, 181, 190
  derivative, 173
  Dirac field, 440
  divergence term in, 178
  dynamic, 187, 199, 216
  einbein, 313
  Einstein-Hilbert, 491
  Einsteinian field, 337
  Einsteinian mechanics, 307
  field, 340
  first-order, 408
  formulation, 119, 172, 202, 225
  free Newtonian particle, 182
  free part, 180
  functional, 206
  Galilean dimension, 206
  Galilean field, 206
  geometric, particle, 309, 315
  hermiticity, 181, 216, 351, 440
  higher derivative, 207
  interaction, 179
  interaction part, 180
  invariance of, 191
  invariant, 218
  Laplace's equation, 210
  Maxwell field, 406
  mechanical, 172
  natural dimension, 129
  Newtonian gravity, 184, 211
  Newtonian mechanics, 167
  non-dynamic, 187, 201, 210
  non-singular, 186, 206, 208, 216, 227, 233, 308, 337, 408, 423, 440
  non-uniqueness, 178, 340
  phase space, 318
  philosophy, 177, 180, 415
  Poisson's equation, 211
  Proca field, 408
  quasi-invariant, 192
  scalar field, 344
  Schrödinger, 216
  singular, 186, 209, 227, 309, 408
  unconstrained, 227
Landau L D and Lifschitz E M, 185
Laplace's equation, 121, 143, 210
Laplace, Pierre-Simon de, 23, 143, 152, 378
Laplacian, 63, 65, 141, 143, 501
Laporte's rule, 49
Latin indices, spatial, 43
Le Verrier, Urbain Jean Joseph, 157
Lee A R and Kalotas T M, 93
Lee T D and Yang C N, 51
Legendre
  Adrien-Marie, 228
  transformation, 228, 315
Leibniz, Gottfried Wilhelm, 84
lemma, Poincaré, 394
length
  coordinate, 256
  proper
    Galilean, 102
    Lorentzian, 255
Lenz's law, 379
Lenz, Heinrich Friedrich Emil, 379

Leonardo da Vinci, 3
lepton, 12, 200, 493
Leucippus, 27
Lévy-Leblond J M, 93, 96
Lewis G N and Tolman R C, 89
Lie
    algebra
        Euclidean, 233
        Galilei, 232, 235
        Lorentz, 319
        Poincaré, 232, 319, 355
        rotation, 75, 233
    canonical parameter, 272
    group, 269, 349, 449
    symmetry, 198, 308, 339
light
    cone, 253, 271
    corpuscle, 154
    deflection, 19
        general relativistic, 482
        Newtonian, 154
        post-Newtonian, 478
        scalar gravity, linear, 367
        scalar gravity, non-linear, 372
        tensor gravity, 476
    reflection, 165, 393
    refraction, 165, 393
    speed of propagation, 255, 392
    wave or corpuscular, 229, 239
lightlike interval, 255
line element
    curved spacetime, 485
    Euclidean, 42, 44
    indefinite, 257
    Minkowski, 250, 261
linear
    connection, 148, 160
    superposition, 10, 380, 412, 453
linearity
    of tensor transformation, 78, 283
linearized gravity, 285, 490
Lobatchevsky, Nicolai, 32, 35
local
    coordinate, 31
    field, 138, 206, 241
        Newtonian, 209, 234
        relativistic, 337
    gauge group, 453
    gauge invariance, 200, 391, 449
    phenomena, 36, 249, 481
    symmetry, 7
    theory, Newtonian, 118
locality
    of interaction, 118
    spatial, 204, 206

    temporal, 203, 206
localization of a global group, 449, 456
locally
    Euclidean space, 36
    Minkowskian spacetime, 483
Lodge, Oliver Joseph Lodge, 126
long-range interaction, 7, 131, 257, 358
Longair M S, 242
longitudinal vector, 120, 501
Lord E, 200
Lorentz
    basis, 263
    boost, 99, 247
        imaginary angle, 273
        matrix, 270
        non-Abelian, 275
        rapidity, 272
    covariance, manifest, 283
    electron theory, 90, 243, 245
    FitzGerald contraction, 90, 243, 257, 278
    gauge, 424, 435, 471
    generator
        Dirac spinor, 434
        matrix, 293, 419
        tensor representation, 295
        vector representation, 294
    group, 100, 265, 270
        connectivity, 269
        inhomogeneous, 265
        restricted, 269, 435
    Hendrik Antoon, 90, 99, 100, 120, 242, 243, 248, 300
    index, 259
    invariant, 276
    invariant momentum space, 352
    invariants, quadratic, 292
    law, charged particle, 6, 72, 382, 411, 417
    orthogonality, 251, 284
        condition, 265, 434
    Poincaré relativity, 247
    scalar product, 282
    transformation, 99, 249
    transformation, factoring, 274
lowering of indices, 43, 64, 262
Ludwig W and Falter C, 126, 232
luminiferous ether, 240
Luther G G and Towler W R, 135

Mach's principle, 374
Mach, Ernst, 85, 115, 374, 383
Machado, Antonio, ii
Mackay, Alan L, 40, 62, 105
macroscopic charge neutrality, 377, 492

Index

magnetic
    charge, 72, 288, 380, 389, 390, 401
    charge quantization, 402
    induction, 65, 390
    monopole, 72, 288, 402
magnitude of a 4-vector, 278
main sequence star, 371
Majorana, Ettore, 336
Majorana, spinor, 336, 431, 444
manifest covariance
    Euclidean, 78
    Lorentz, 283
    Maxwell's equations, 389
    Poincaré, 284, 297, 299
    rotational, 78
manifold, 30, 105, 483
many particle theory, 464
Maricourt, Pierre de, 376
Mariwalla K H, 94
Maskelyne, Nevil, 135
mass, 62, 71, 103
    active gravitational, 150
    additivity, 113
    energy equivalence, 300, 371
    gravitational, 370, 481
    inertial, 108, 150, 247, 370, 481
    measurement, 115
    Newtonian inertial, 109, 110
    passive gravitational, 150
    relativistic, 301
    rest, 299
    shell, 303
    shell condition, 317, 353
    velocity dependence, 245
massive
    field, 343, 407, 436
    helicity, 322, 442
    particle, 8, 299, 431
massless
    field, 358
    helicity, 324
    particle, 7, 14, 18, 21, 311
matrix
    covariant indexing, 262
    Dirac, 432
    Dirac spin, 433
    direct product, 279
    equivalence, 267
    Lorentz boost, 270
    Lorentz generator, 293
    Minkowski, 261
    multiplication, 45, 262
    orthogonal, 46, 71
    partitioning, 274
    Pauli, 433
    transpose, 45
matter
    Einsteinian, 131
    field, 460, 492
    Newtonian, 131
    particle, 493
Maughan, William Somerset, 105
Maupertuis, Pierre L M de, 165, 174
Maxwell
    displacement current, 390
    equations, 6, 9, 11, 91, 381, 406
        1 + 2 dimensions, 395
        covariant, first-order, 389
        covariant, second-order, 393, 418
        extended, 401
        Poincaré covariance, 100, 391
    field, 336
        degrees of freedom, 429
    James Clerk, 90, 136, 138, 241, 242, 335, 336, 376, 380, 388
    Lagrangian, first-order, 409
    potential, constraint structure, 421
Mayer, Julius Robert, 123
measurement, 83, 282, 346
    mass, 115
    spacetime, 85, 246, 300, 381, 483, 491
mechanical energy, 122
mechanics, 138
mediating
    field, 10, 138, 204, 447, 452, 492
    particle, 493
membrane, relativistic, 346
metre
    SI base unit, 128
    SI definition, 85
metric
    space, 33, 42
    tensor, 366, 484, 485
        antisymmetric, 283
        Minkowskian, 265
        scalar gravity, 367
        symmetric, 44
Michell, John, 23, 135, 152, 156, 377
Michelson, Albert A, 91, 242
microwave radiation, cosmic, 22
minimal coupling, 415, 455, 456, 461
Minkowski
    coordinate, 263
    diagram, 253
    geodesic, 310
    Hermann, 41, 252, 276, 279
    matrix, 261
    metric tensor, 265

spacetime, $E^{1,3}$, 84, 250, 252, 258, 276, 301
spacetime, complex, 260
tensor, 280
mixed components, tensor, 280
Möbius, Augustus, 37
momentum, 64, 108, 110
    additivity, 113
    canonical, 308
    conservation, 27, 57, 126, 169, 201
    density, particle, 329
    flux, particle, 329
    generalized, 172, 308
    operator, 113
momentum-space interval
    Lorentz-invariant, 352
    non-null fermionic, 441
monad, 84
monopole, magnetic, 72, 288, 402
Mook D E and Vargish T, 257
Moore J H, 158
Morgan, Augustin de, 40
Morley E W, 243
Morrison P, 51
motion, 29
multiparticle mechanics, 413
muon, 12, 493
muon decay, 159
mutual interaction, 114, 348

Näbauer, Martin, 157
naked singularity, 23
Nambu and Goto action, 496
natural
    dimension, 128
        action, 129
        electric charge, 129
        gravitational constant, 138
        Lagrangian, 129
    equation system, 127
    topology, 30
    units, 127, 343
nature, chirality of, 49, 257
negative
    curvature, 36
    energy, 463
    frequency, 353
    probability, 463
neutralino, 462
neutrino, 12, 49, 51, 281, 446, 458, 493
    massless, 283, 285, 336
    speed of propagation, 255, 446
neutron, 16, 493
    star, 13, 20, 23, 132, 367, 371
Newton
    Isaac, 40, 80, 90, 100, 107, 108, 112, 114, 117, 129, 130, 132, 135, 154, 155, 161, 183, 239, 381, 482, 491
    laws, 90, 107, 126
        first, 112
        of gravity, 133, 144
        of mechanics, 130, 304
        second, 115, 122
        third, 116, 117, 124
Newton-Cartan spacetime, 160
Newtonian
    black hole, 23, 152
    causality, 115
    conservation laws, 110
    energy, 300
    gravity, 18, 131, 473
        action, 211
        Lagrangian, 184
        Lagrangian density, 211
    inertial mass, 109
    light deflection, 154
    mechanics
        Noether's theorem, 200
        symmetry relations, 199
    potential energy, 122
    rest frame, 112
    spacetime, $\{E^1, E^3\}$, 80
Noether
    Amelia Emmy, 198, 199
    charge, 221
        Einsteinian mechanics, 307
        Newtonian field, 234
        Newtonian mechanics, 199
        on-shell, 200, 308
        phase space, 231
        relativistic field, 339, 421
        spacetime rotation, 311
        spacetime translation, 311, 316
    current
        Dirac, 443
        Newtonian, 220
        relativistic, 339, 421
    density
        complex field, 223
        Newtonian field, 220
    relations, 199
    theorem, 9, 125, 200, 219, 338, 348, 384, 419, 421
nogo theorem, supersymmetry, 493
non-Abelian symmetry, 13, 17, 21
non-canonical Hamiltonian, 318
non-chiral
    field, 71, 494
    object, 47

particle, 431, 467
non-compactness
    group parameter, 273
    interval, 99
    symmetry, 21
'non-conservative' interaction, 173
non-covariant gauge, 427
non-diagonal interaction, 15, 16
non-dynamic
    field equation, 205
    Lagrangian, 187, 201
    Newtonian field, 209
    variable, 187
    variable, einbein, 313
non-Euclidean geometry, 32
non-finiteness, of quantum fields, 11, 22
non-gauge field, 492
non-integrability, proper time, 297
non-linearity
    field equations, 13, 18, 20, 211, 346, 371
    gravity, 370, 478, 482
    spinor electrodynamics, 452
non-local operator, 427, 501
non-local phenomena, 481
non-propagating field, 209
non-renormalizability, 22
    gravitation, 137, 491
    scalar fields, 347
non-simple connectivity, 57, 269
non-singular Lagrangian, 186, 227, 233, 308, 337, 408, 423, 440
    density, 206, 208, 216
Nordtvedt K, 370
nuclear binding, 16
nucleon, 15, 493
nucleus, 493
null
    4-momentum, 311, 358
    4-vector, 278
    cone, 253
    field, 285
    field, electromagnetic, 399
    field, tensor gravity, 472
    interval, 255
    orthogonal vectors, 285, 323
    particle, 312, 330
        constraint, 313
        locus, 296
    propagation, 311
    ray, 312
    wave front, 312
numerical tensor, 66, 280

observable, quantum, 226

Occam
    razor, 214, 386
    William of, 214
Oersted, Hans Christian, 377, 378
off-shell configuration path, 182, 197
Ohm, Georg Simon, 378
on-shell
    configuration path, 182, 197
    conservation, 308
    Noether charge, 200
operator
    angular momentum, 113
    d'Alembertian, 387
    Hamiltonian, 113
    hermitian, 214
    idempotent, 502
    Laplacian, 63, 65, 143
    momentum, 113
    non-local, 501
    projection, 48, 68
orbital angular momentum, 72, 116, 221
orientability, 30
    spacetime, 289, 483
    spatial, 36, 46, 74
    temporal, 36, 53
orientation
    of spatial direction, 28
    of time, 29
oriented hypersurface, 325
orthochronous transformation, 268
orthogonal
    condition, Lorentz, 265
    group, SO(2), 222
    groups, O(N), SO(N), 283
    matrix, 71
    null vectors, 285, 323
    transformation, 46, 56
orthogonality, Lorentz, 284
orthonormal basis, 34
orthonormality
    Euclidean basis, 42
    Lorentz basis, 263
    polarization basis, 426
Overseth, O E, 51

Paik H J, 160
paradox, clock or twin, 256
parallax, stellar, 82
parallel
    axiom of Euclid, 40
    transport, 35
parameter space
    3D rotations, 57
    connected, 57
    non-simply connected, 57

parity, 59, 70, 269, 289, 433, 494
  active, 46
  conjugate, 47
  conservation, 50, 73, 288, 446
  covariance, 50, 71, 73, 120, 140, 286, 288, 382, 446
    Dirac equation, 437
    electrodynamics, 389, 406
    gravity, 131
  Dirac equation, 445
  even, 47
  intrinsic, 52
  intrinsic, Dirac spinor, 438
  invariance
    of interaction, 47
    of spacetime, 95
  matrix, 56, 268
  mixed, 48
  non-covariance, 11, 286
  odd, 48
  passive, 49
  reversal, 46
  self-inverse, 46
  transformation, 56
  violation, 11, 51
  well-defined, 47
Parmenides of Elea, 27
particle
  classical, 28, 29
  creation, 463
  dynamics, Einsteinian, 304, 305
  dynamics, Newtonian, 109
  Einsteinian, 309
  exchange, 10, 138, 447, 462, 492
  gauge, 493
  kinematics, Einsteinian, 296
  kinematics, Galilean, 109
  massive, 8, 431
  massless, 7, 14, 18, 21, 311
  matter, 493
  mediating, 10, 138, 493
  non-chiral, 431
  null, 296, 330
  self-conjugate, 21, 400, 431
  spin 2, 467
  swarm, 331
  timelike, 296, 299
  trajectory, 53
  virtual, 29, 463
  $W^{\pm}$, 15, 336
  Z^0, 15
particular integral, 373
passive
  gravitational mass, 150
  transformation, 54
  rotation, 63
  spatial reflection, 49
past, 29
  of an event, Lorentz, 253
past-pointing
  hypersurface normal, 325
Pasteur, Louis, 383
path
  integral quantization, 181, 225, 340
  length, proper, 309
  of a field system, 338
  of a particle, 53
  on-shell and off-shell, 182, 197
  virtual, 175
Pauli
  algebra, 433
  coupling, 455
  exclusion principle, 132, 444
  Lubański tensor, 322
  Lubański vector, 288, 322, 472
  matrix, 433
  neutrino hypothesis, 458
  Schrödinger equation, 215
  spinor, 283
  Wolfgang, 132
periastron shift, 369
perihelion shift, 19, 157, 369
  general relativistic, 483
  linear scalar gravity, 367
  non-linear scalar gravity, 373
  non-linear tensor gravity, 478
  post-Newtonian, 478
permeability, vacuum, 127
permittivity, vacuum, 127, 390
permutation pseudotensor, 74, 289, 290
phase
  of a Hamiltonian system, 228
  space, 227
    constraint, 317
    Hamilton's principle, 231
    Hamiltonian, 315
    Hamiltonian density, 233
    Lagrangian, free particle, 318
    transformation, U(1), 8, 443
ϕ^4 theory, 346, 351
photon, 7, 9, 281, 336, 492
  helicity states, 429
  self conjugacy, 386
  virtual, 453
physical singularity, 23
pion, 16, 17, 72, 277, 281, 288, 335, 344, 347
planar causal structure, 252
Planck
  constant $\hbar = h/2\pi$, 127, 415

Index   561

energy, 24, 26, 497
length, 24, 31
Max Karl Ernst Ludwig, 24, 245
time, 147
plane wave
   electromagnetic, 399, 425
   free Dirac field, 441
   relativistic, 435
Poincaré
   covariance, 249, 283, 310, 371, 382
   group, 21, 249, 251, 265, 269, 350, 383, 431, 435, 472
   Henri, ii, 29, 32, 83, 86, 89, 91, 92, 99, 100, 130, 188, 225, 239, 245, 266, 410
   lemma, 394, 401, 503
   Lie algebra, 232, 319
   scalar, 277, 299, 383
   tensor, 280, 304
   transformation, 249, 419
   vector, 278, 299, 302, 383
point
   interaction, 13
   transformation, 190, 217
Poisson
   bracket, 11, 231, 316, 318
      algebra, 232
      fundamental, 232, 319
      Newtonian field, 234
   equation, 142, 143, 213, 427, 501
   Lagrangian density, 211
   Siméon-Denis, 142, 231, 378
polar vector, 69
polarization
   basis
      4-vector, 425
      Dirac spinor, 441
      tensor, 426
   Dirac wave, 442
   electromagnetic wave, 429
   gravitational wave, 19
Popper D M, 158
positive
   curvature, 35
   definiteness, 42
   energy, 19, 132, 144, 213, 344–346, 348, 359, 361, 439, 444, 465, 473, 492, 493
   frequency, 353
positron, 52
post-Newtonian
   approximation, 478
   light deflection, 478
   perihelion shift, 478
postulates

Euclid's fifth, 32
first of relativity, 90, 246
of relativity, 88, 246
second of relativity, 83, 86, 246
potential
   electromagnetic, 3-vector, 394
   electromagnetic, 4-vector, 393
   electromagnetic, scalar, 63
   energy
      gravitational, 145
      internal and external, 124
      Newtonian, 122
      of interaction, 121
   gauge, 148
   gravitational, 142
   scalar electromagnetic, 394
   velocity dependent, 173
Pound R V and Rebka G A, 159
Pound R V and Snider J L, 159
Poynting vector, 405
Poynting, John Henry, 126
PPN parameters, gravity, 478
practical units, 127
precession
   general, 157
   planetary perihelia, 157
   planetary perihelia, 19, 369
   stellar periastra, 19, 369
prescribed source, 447
pressure, electromagnetic, 405
Priestley, Joseph, 377
principle of
   causality, 19, 21, 83, 87, 98, 248
   covariance, 77, 92
   equivalence
      strong, 152, 370, 371, 481
      weak, 104, 151, 481
   exclusion, 444
   Galileo, 89, 107, 112
   gauge coupling, 8, 449
   general covariance, 485
   inertia, 107, 112
   linear superposition, 10, 412
   Poincaré covariance, 249
   relativity, 4, 7, 24, 57, 78, 83, 90, 93, 105, 245, 269, 283, 383, 470
   stationary action, 5, 24, 165, 177, 225, 231, 307, 339
   uniqueness of free fall, 151
privileged frame, 470
probability density
   Dirac field, 443
   Klein-Gordon, 354
   positive-definite, 443

Schrödinger, 223
Proca
    A, 336, 400
    constraint, 407
    equations, 407
    field, 400
    Hamiltonian density, 408
    Lagrangian, 408
projection operator, 48, 68, 445, 502
propagation vector, 352
proper
    frame, spacelike interval, 256
    frame, timelike interval, 255
    length
        curved spacetime, 485
        Galilean, 102
        Lorentzian, 255
    path length, particle, 309
    scalar, 70, 287
    tensor, 70, 287
    time, 297
        constraint, 305
        gauge, 310
        Hamiltonian, 308
        interval, 255, 485
        non-integrability, 297
    vector, 69, 70, 287
proton, 16, 127, 493
pseudo
    scalar, 71, 287
    tensor, 71, 287
    vector, 70, 287
Ptolemy (Claudius Ptolemaeus), 80
pulsar, 367

quantization, 319
    ansatz, 214
    anticommutation rules, 444, 493
    canonical, 225, 464
    commutation rules, 444, 463, 492
    constrained system, 316
    electric charge, 11, 385
    first, 353, 463
    magnetic charge, 402
    path-integral, 181, 225, 340
    principles of, 4
    second, 354, 464
    single particle, 463
quantum
    chromodynamics, 18
    cosmological constant, 147
    electrodynamics, 11, 451
    field
        divergence, 11
        operator, 464
    theory, 10, 354, 444, 464, 492
    gravity, 22, 24, 491
    ground state, 14
    mechanics, 463
    observable, 226
    of action, 415, 492
    symmetry, 340
    vacuum, 14, 147
quark, 10, 12, 15, 16, 281, 283, 336, 385, 431, 462, 492, 493
    confinement, 16
quasar, 21
quasi-invariant Lagrangian, 192
quasi-linear equation, 372, 488

radiation
    black body, 245
    cosmic microwave, 22
    electromagnetic, 393
    field, electromagnetic, 399
    gauge, 428
radioactive decay
    α particle, 16
    β particle, 13, 458
    γ particle, 6
radius of curvature, 32
    gravitational, 23
raising of indices, 44, 64, 262
Ramond P, 200
range of interaction
    electromagnetic, 7
    gravitation, 19, 131, 213
    strong-nuclear, 16
    weak-nuclear, 15
rank of a tensor, 66, 281
Rankine, William John Macquorn, 123
rapidity of Lorentz boost, 272
Rarita W and Schwinger J, 336
Rarita-Schwinger
    equation
        massive, 468
        massless, 468
    field, 336, 421, 466
    potential, 468
    quanta, 495
ray, 312
Rayleigh, Lord (John Strutt) , 136
reaction, Newtonian, 116
real line R, 30
realization
    group, 60
    physical variable, 46
rectangular Cartesian coordinates, 34
red shift
    cosmological, 147

Index   563

    gravitational, 19, 86, 158, 363
reference frame, 28, 30
reflection
    spacetime, 92, 101
    spatial, 46
    temporal, 54
reflection of light, 165, 242, 393
reflexivity of equivalence, 60
refraction of light, 165, 242, 393
    conical, 229
Reissner-Nordstrøm black hole, 489
relativistic
    3-momentum, 302
    energy, 302
    gravitation, 470
    mass, 301
    notation
        1 + 3, 259, 282, 291, 385, 399
        covariant, 259, 291
    quantum mechanics, 353, 463
    string, 496
relativity
    Einsteinian, 29, 57, 93, 111
    Galilean, 29, 57, 83, 91, 93, 107, 110, 247
    Galilean hypothesis, 102
    general, 19, 131, 247, 470, 482
    of simultaneity, 84, 250
    of uniform motion, 119
    postulates, 88, 246, 249
    principle of, 4, 7, 83, 88, 90, 93, 105, 245, 269, 383, 418, 439, 448, 472
    special, 19, 93, 247, 470, 483
renormalizability, 351, 448
    electrodynamics, 11, 137, 307
    gravitation, 22
    quantum fields, 346
    strong-nuclear, 18
    weak-nuclear, 15
    Yang-Mills, 307
reparameterization invariance, 312
representation, 46
    group, 281, 431, 439
residual gauge invariance, 425, 427, 472
Resnick R, 94
rest
    energy, 299
    frame
        Einsteinian, 255, 302
        Newtonian, 112
    mass, 288, 299
restricted group
    Euclidean, 75
    Galilei, 101

    Lorentz, 249, 269
    Poincaré, 249
restriction of a group, 278
reversal of time, 53
Ricci identity, 456
Ricci scalar, 487, 491
Ricci tensor, 487
Riemann tensor, 487
Riemann, Bernhard, 32, 241, 379, 380
Riemannian geometry, 487
rigid body, 249
Robertson H P and Noonan T W, 264
Robertson-Walker geometry, 488
Römer, Ole Christensen, 154, 240
Rosen, Edward, 56, 82
Rosenfeld L, 420
rotation
    3D, 56, 62, 66, 69
    active, 63
    composition of, 69
    covariance, 78, 273, 399
    covariant, 257, 311
    duality, 403
    electrodynamic field, 391
    Faraday, 241
    group, SO(3), 75
    hyperbolic, 273
    invariance, 169
    Lorentz, 257
    matrix, Lorentz form, 272
    parameters, 285
    passive, 63
Russell, Bertrand, 245
Rutherford, Ernest, 136

Saint Augustine of Hippo, 38
Santilli R M, 177
Savart, Félix, 378
scalar, 62, 276, 278, 280
    electrodynamics, 457
    Euclidean, 62
    field, 63, 277
        complex, 350
        Lagrangian density, 344
    gravity, relativistic, 357
    Poincaré, 277, 299, 383
    product
        Euclidean, 42, 44, 62, 104
        Minkowskian, 250, 262, 282
    proper, 70, 287
    pseudo, 71, 287
scalar-tensor gravity, 374
scale invariance, 148, 254
Schrödinger
    equation, 214, 223, 343, 396

Erwin, 202, 215, 335, 343, 430, 463
group, 126
wave function, 139
Schwarzschild, geometry, 489
Schwarzschild, Karl, 153
Schwarzschild, length, 153, 476, 489
Scott G D and Viner M R, 257
sea-gull interaction term, 457
second
law of Newton, 115, 122
law of thermodynamics, 38
quantization, 354, 464
relativity postulate
Einsteinian, 246
Galilean, 102
SI base unit, 128
SI definition, 85
self interaction, 13, 115, 346, 348
Newtonian field, 210
quartic, 351
scalar field, 345
self-conjugate particle, 21, 386, 400, 444
Majorana, 431
self-gravitational energy, 370
self-inverse operation
parity, 46
time reversal, 54
selfduality
antisymmetric tensor, 4D, 292
of fields and particles, 50
Shanmugadhasan S, 317
short-range interaction
gravity, 358
strong-nuclear, 16
weak-nuclear, 11
SI units, viii, 85, 127, 389
signal coherence, 257
signature, spacetime metric, 251
simple connectivity, 30, 121, 327
simultaneity, 29, 83, 84, 102, 111, 160, 244, 246, 249, 250
single-valuedness, tensorial, 281
singular Lagrangian, 186, 209, 227, 309, 408
singularity
Big Bang, 22
black hole, 22
coordinate, 23
naked, 23
physical, 23
SL(2,C) symmetry, 460
Snell, law of refraction, 165
Snell, Willebrord van Royen, 165
SO(2) symmetry, 222, 350

SO(3) symmetry, 233
SO(N) symmetry, 460
Soldner, Johann Georg von, 156, 476
solenoidal vector, 120
Sophocles, viii
source
condition, 468
constraint, 389, 424
electrodynamic field, 244, 382, 384
external, 211, 447
field, 10, 139, 492
Newtonian gravity, 131
prescribed, 447
relativistic gravity, 358, 470, 487
space, 27, 279
3D Euclidean, E^3, 27
4D Euclidean, E^4, 258
configuration, 171
dimension, 30
locally-Euclidean, 36
metric, 33, 42
spacelike
4-vector, 278
hypersurface, 325
interval, 255
signature, 251
spacetime, 29
affine, 105
Aristotelean, 41, 84, 94
boost invariance, 257
coordinate, 30, 259
curvature, 477, 482
diagram, 253
dimension, 257
direction, 251
dynamic, 19, 482
Euclidean, $E^1 \times E^3$, 41, 77, 82, 111
event, 259
frame, 87
Galilean, 102, 160
geodesic, 310
geometry, 32, 83, 250, 482
homogeneity, 484
isotropy, 484
measurement, 85, 246, 381, 483, 491
Minkowskian, $E^{1,3}$, 84, 252, 258
Minkowskian, $E^{1,3}$, 279
Newton-Cartan, 160
Newtonian, 80
parity invariance, 95, 257
radius of curvature, 23
reference frame, 30
reflection, 92
rotation invariance, 257

Index 565

symmetry, 484
time reversal invariance, 257
topology, 24, 490
translation invariance, 257, 419
spatial
   curvature, 478
   index, 43
special relativity, 93, 247, 470, 483
speed
   of light, 127, 255, 392
      SI definition, 85
   of propagation
      electromagnetic, 255, 392
      gravitation, 135, 255
      neutrino, 255
   ultimate, 7, 100, 245, 255, 296, 301, 392, 415, 446
spherical symmetry, 488
spin, 323, 343, 472
   angular momentum, 72, 341
      spacelike, 321
      tensor, 117, 320
      vector, 72, 116, 321
   density, canonical, 420
   density, Galilean, 293
   density, total, 420
   even integer, 361
   half-odd-integer, 50, 185, 217, 444, 465
   integer, 50, 344, 465
   intrinsic, 72
   massless field, 324
   multiplicity, 68
   Pauli-Lubański vector, 322
spin 0:, 15, 68, 204, 277, 295, 335, 344, 357, 494
spin $\frac{1}{2}$:, 8, 15, 51, 204, 215, 336, 344, 353, 431, 435, 446, 467, 493, 494
spin 1:, 7, 14, 18, 68, 295, 336, 353, 395, 400, 407, 425, 432, 467, 492
spin $\frac{3}{2}$:, 336, 467, 495
spin 2:, 20, 21, 68, 336, 361, 391, 426, 466, 471, 490, 492, 495
spin s, 68
spinor, 281, 439
   basis, Dirac, 441
   Dirac, 283, 337, 433, 434
   electrodynamics, 451
   Grassmann, 465, 495
   Majorana, 336, 431, 444
   Pauli, 204, 283
   Weyl, 283, 337, 446
spontaneous symmetry breaking, 14, 346, 407, 492

square root
   general relativity, 437
   Klein-Gordon equation, 437
Srivistava P P, 200
standard model, 462, 485
star, 20
   main sequence, 20, 371
   neutron, 13, 20, 132, 367, 371
   white dwarf, 20, 132, 158, 371
stationary action
   principle of, 5, 165, 231, 307, 339
statistics
   Bose-Einstein, 10, 20, 22, 281, 492
   Fermi-Dirac, 22, 281, 444, 493
steady-state field, 414
stellar
   aberration, 82, 154, 240
   energy generation, 13
   parallax, 82
Stewart I, 258
Stokes G, 242
strange (flavour), 17
strangeness, 344, 348
stress tensor, 71, 126
   Einsteinian, 329
   electromagnetic, 405
   Newtonian field, 220
stress-energy tensor, 330
string
   relativistic, 346, 496
   supersymmetric, 22
   tension, 496
strong equivalence principle, 152, 370, 371, 481
strong-nuclear interaction, 15, 288, 462
   confinement, 16
   gauge invariance, 17
   non-diagonal, 16
   non-linearity, 18
   quantization, 18
   renormalizable, 18
   short range, 16
Strutt, John W (Lord Rayleigh), 136
SU(2) symmetry, 14, 148, 460
SU(3) symmetry, 18, 149, 462
SU(N) symmetry, 460
subspace, embedded, 34
Sudarshan E C G and Mukunda N, 190, 203, 232, 317, 408, 413
summation convention, Einstein, 43, 261
Sundermeyer K, 200, 317, 318
super
   gravity, 336, 437, 466, 495
   interaction, 5
   renormalizable, 347

string, 22, 381
   symmetry, 22, 265, 336, 467, 494
   unification, 9, 22, 466
supernova, 1987a, 13
superposition
   of interaction, 115
   principle of, 10, 412, 453
surface term
   action, 177, 178, 194
      Einsteinian field, 340
      Newtonian field, 208
swarm of particles, 331
Swift, Jonathan, 542
Switzer J E, 62
symmetric
   part of a tensor, 67, 285
   tensor, 67, 285
symmetry, 7, 143, 189
   Abelian, 8, 350
   breaking
      dynamical, 189
      spontaneous, 14, 189, 346
   charge conjugation, 348
   classical, 340
   compact, 8, 13, 17, 350
   continuous, 195, 495
   cyclic, Z_2, 59, 348
   discrete, 348
   duality, 403
   dynamic, 195
   energy-momentum tensor, 332
   external, 265, 494
   Galilean, 189
   global, 8, 222, 349
   hidden, 14
   internal, 143, 221, 265, 349, 494
   Lie, 198, 308, 339
   local, 7
   Newtonian field, 217
   non-Abelian, 13, 17, 21
   non-compact, 21
   of equation system, 195
   of equivalence, 60
   quantum, 340
   spacetime, 484
   spatial translation, 197
   spherical, 488
   static system, 195
   transformation, 196
   unitary, 8, 14, 18
   Yang-Mills, 13
synchronization of clocks, 85, 246
Système International units, viii, 127

tachionic signals, 254

τ^+ particle, 51
tauon, 12, 493
temporal homogeneity, 124
tension, electromagnetic, 405
tensor, 280
   4D dual, 291
   algebra, 280
   angular momentum, 117
   antisymmetric, 67, 285
   antisymmetric part, 67, 285, 286
   arbitrary rank, 66
   calculus, 280
   derivative, 281
   dual, 75
   duality parts, 292
   Euclidean, 66
   field, 66, 280, 439
   gravity, 466
   improper, 71
   invariant, 66
   invariant part, 68, 286
   irreducible part, 68
   isotropic, 66
   Lorentz, 281
   Minkowskian metric, 265
   mixed components, 280
   numerical, 66, 280
   Pauli-Lubański, 322
   Poincaré, 280, 304
   product, 67
   proper, 70, 287
   rank, 66, 281
   spin, 117, 320
   stress, 71, 126, 329
   symmetric, 67, 285
   symmetric part, 67, 68, 285, 286
   trace, 67
   trace part, 68, 286
   traceless part, 68, 286
   transformation, homogeneity, 283
   transformation, linearity, 283
tensor-spinor, 281, 439
tetrad, 314
thermodynamics, first law, 123
thermodynamics, second law, 38, 123
θ^+ particle, 51
third law of Newton, 116, 117
Thomson, Joseph John, 136
Thomson, William (Lord Kelvin), 38, 244
time, 27, 29, 279
   arrow of, 37
   dilation
      gravitational, 86, 363, 476, 482
      kinematic, 256, 278, 297

Index   567

evolution parameter, 112, 229, 297
interval
   coordinate, 255, 297
   proper, 255
   reversal, 46, 53, 59, 269, 288, 433
      4D matrix, 268
      active, 54
      covariance, 55
      Dirac spinor, 454
      invariance, 54
      invariance of spacetime, 95
      passive, 54
      self-inverse, 54
      violation, 55
   river of, 29
timelike
   4-vector, 278
   interval, 255
   particle locus, 296
   signature, 251
Todd D P, 242
top (flavour), 17
topology, 30
   1D connectivity, 30
   black hole, 32
   cosmological, 31
   invariant, 30
   natural, 30, 33
   of a torus in nD, 31
   of spacetime, 31
   simply-connected, 30
   spacetime, 490
torus, 31
trace
   energy-momentum tensor, 405
   of a tensor, 67, 285
   part of a tensor, 68, 286
trajectory of a particle, 53
transformation
   active, 46, 54
   canonical, 190
   conformal, 265
   discrete, 46, 53
   general coordinate, 190, 265
   isometry, 46
   Lagrangian density, 190
   Legendre, 228
   Lorentz, 99
   orthochronous, 268
   orthogonal, 46
   passive, 49, 54
   Poincaré, 419
   point, 190, 217
   restricted Lorentz, 249
   restricted Poincaré, 249

supersymmetry, 265
symmetry, 196
transitivity of equivalence, 60
translation
   covariance, 78
   group, 435
   invariance, 221
      spacetime, 419
      spatial, 169, 197, 200
      temporal, 169
   spacelike, 276, 284
   spacetime, 299
   spatial, 46, 62, 66
   temporal, 46, 62, 124
transpose of a matrix, 45
transverse
   gauge, electromagnetic, 427
   vector, 120, 427, 501
   vector potential, 428
   vector-spinor, 468
Trautman A, 317
true dynamical variable, 184, 186, 306, 408
twin paradox, 256

$U(1)$ phase invariance, 8, 222, 350, 384, 443
ultimate speed, 7, 100, 245, 247, 255, 296, 301, 392, 415, 482
uncertainty principle, 463
unconstrained
   dynamics, 227
   Lagrangian, 227
unification, 147, 454
   electroweak, 6
   grand, 6
   of interaction, 5, 24
   optics and electromagnetism, 241
   principles, 134
   super, 5, 466
   terrestrial and celestial, 130
uniform motion, 95
uniqueness of free fall, 104, 151, 481
unitary
   group, $U(1)$, 222
   groups, $U(N)$, $SU(N)$, 283
   symmetry, 8
units, natural and practical, 127
units, SI, 85, 127
universal constant, 127, 415
universality of gravitation, 18, 19, 131, 135, 370, 471, 478, 488
up (flavour), 16
Utiyama R, 200, 317

vacuum
    classical, 27, 28
    instability, 353
    permeability, 127, 389
    permittivity, 127, 390
    quantum, 14, 29, 147
    quantum field theory, 346
    stability, 144, 346
variational calculus, 166, 175
vector, 280
    axial, 70, 71, 287
    boson, intermediate, 15, 281, 462
    conservative part in 3D, 120
    Euclidean, 43, 63, 65, 278
    Galilean, 103
    invariant, 64
    longitudinal part in 3D, 120, 422
    Lorentz, 277
    Pauli-Lubański, 288, 322
    Poincaré, 278, 299, 302, 383
    polar, 69, 287
    potential
        3D electromagnetic, 65
        4D electromagnetic, 393
    proper, 69, 70, 287
    pseudo, 70
    solenoidal part in 3D, 120
    spacetime, 260
    timelike, spacelike or null, 278
    transverse part in 3D, 120, 422
vector-spinor, 281, 467
    gamma-traceless, 467
    transverse, 468
velocity, 64, 87
    absolute, 481
    addition, 98, 99
    dependent
        mass, 301
        potential, 173
    generalized, 171
    reversal, 54
Vessot R F C et al., 159
vierbein field, 314, 492
virtual
    particle, 29, 463
    path, 175
    photon, 453
vis viva, 123
Volta, Alessandro Guiseppe, 377
volume, 103
    3D Euclidean, 62
    3D in 4D spacetime, 325
    4D Minkowski, 278

$W^{\pm}$ particle, 15, 281, 336, 402, 460,
462, 492
Watanabe S, 288
Watson A, 258
Watson, William, 377
wave
    equation
        electromagnetic potential, 394
        homogeneous, 343, 424
        inhomogeneous, 424
    front, null, 312
    function
        Klein-Gordon, 353
        Schrödinger, 139, 384
    mechanics, non-relativistic, 463
    mechanics, relativistic, 353
    operator, 277, 387
wavelength, Compton, 30, 343, 407
weak
    conservation law, 200
    equivalence principle, 151, 360, 481
    gravitation, 336, 490
weak-nuclear interaction, 11
    Fermi theory, 13
    gauge invariance, 13
    parity violation, 11, 446
    renormalizability, 15
weakon, 15, 492
Weber, Wilhelm Eduard, 241, 379
Weichert E, 89
Weizsäcker, Carl-Friedrich von, 105
Wess and Zumino multiplet, 495
Wess J and Zumino B, 495
Weyl
    equation, 446
    Hermann, 8, 148, 189, 336, 396, 422
    spinor, 283, 337, 446
Wheeler, John Archibald, 23, 29
white dwarf star, 20, 132, 158, 371
Whittaker E, 241
Wigner, Eugene, 51
Wolfe, Thomas Clayton, 38
world view
    Aristotelean, 80, 104
    Cartesian, 240
    Copernican, 81
    Einsteinian, 246, 247
    Galilean, 102
    Kantian, 123
    Laplacian, 123
    Leibnizian, 84
    Lorentzian, 245
    Machian, 85, 115
    Minkowskian, 252
    Newtonian, 85, 245

Ptolemaic, 81
worldline of a particle, 53

Yang C S and Mills R L, 336, 460
Yang-Mills
    field, 336, 391, 407, 460, 487, 492
    gauge invariance, 15, 17, 306
    group, 21
    non-linearity, 13
    symmetry, 13, 148, 461
Young, Thomas, 154, 240
Yukawa
    Hideki, 15
    potential
        Newtonian gravity, 160, 375
        relativistic gravity, 358
    theory, 15, 288

Z^0 particle, 15, 281, 336, 462, 492
Z_2 cyclic group, 59
Z_2 symmetry, 348
$Z_2 \otimes Z_2$ Klein group, 60
Zeeman effect, 244

For Product Safety Concerns and Information please contact our EU
representative GPSR@taylorandfrancis.com
Taylor & Francis Verlag GmbH, Kaufingerstraße 24, 80331 München, Germany

www.ingramcontent.com/pod-product-compliance
Ingram Content Group UK Ltd.
Pitfield, Milton Keynes, MK11 3LW, UK
UKHW021426080625
459435UK00011B/179